Algorithms and Combinatorics 14

Springer
Berlin
Heidelberg
New York
Barcelona
Budapest
Hong Kong
London
Milan
Paris
Santa Clara
Singapore
Tokyo

Paul Erdös Multi(media)
by Jiří Načeradský and Jarik Nešetřil

Ronald L. Graham
Jaroslav Nešetřil (Eds.)

The Mathematics of Paul Erdös II

Springer

Ronald L.Graham
AT&T Bell Laboratories
600 Mountain Avenue
Murray Hill, NJ 07974
USA
e-mail: rlg@research.att.com

Jaroslav Nešetřil
Department of Applied Mathematics
Charles University
Malostranske nam. 25
11800 Praha
Czech Republic
e-mail: nesetril@kam.ms.mff.cuni.cz

Library of Congress Cataloging-in-Publication Data
The mathematics of Paul Erdös / Ronald L. Graham, Jaroslav Nešetřil (eds.)
p. cm. – (Algorithms and combinatorics, ISSN 0937-5511 ; 13–14)
Includes bibliographical references
ISBN 3-540-61032-4 (Berlin : v. 1 : hardcover : acid-free paper). –
ISBN 3-540-61031-6 (Berlin : v. 2 : hardcover : acid-free paper).
1. Mathematics. 2. Erdös, Paul, 1913– . I. Erdös, Paul, 1913– . II. Graham, Ronald L., 1935– . III. Nešetřil, Jaroslav. IV. Series.
QA3.M423 1996 510–dc20 96-31947 CIP

The Mathematics of Paul Erdös I has been published
as Volume 13 of this series with the ISBN 3-540-61032-4

Mathematics Subject Classification (1991):
00A05, 05-XX, 05Cxx, 60-XX, 03Exx, 04-XX, 51-XX, 11-XX

ISSN 0937-5511
ISBN 3-540-61031-6 Springer-Verlag Berlin Heidelberg New York

Printed in Germany

Typesetting: By the editors using a Springer TEX macro package.

SPIN: 10534899 41/3143 - 5 4 3 2 1 0 - Printed on acid-free paper

IN MEMORIAM

PAUL ERDŐS

26. 3. 1913 – 20. 9. 1996

The week before these volumes were scheduled to go to press, we learned that Paul Erdős died on September 20, 1996. He was 83. Paul died while attending a conference in Warsaw, on his way to another meeting. In this respect, this is the way he wanted to "leave". In fact, the list of his last month's activities alone inspires envy in much younger people.

Paul was present when the completion of this project was celebrated by an elegant dinner in Budapest for some of the authors, editors and Springer representatives attending the European Mathematical Congress. He was especially pleased to see the first copies of these volumes and was perhaps surprised (as were the editors) by the actual size and impact of the collection. We hope that these volumes will provide a source of inspiration as well as a last tribute to one of the great mathematicians of our time. And because of the unique lifestyle of Paul Erdős, a style which did not distinguish between life and mathematics, this is perhaps a unique document of our times as well.

R. L. G.
J. N.

Preface

In 1992, when Paul Erdős was awarded a Doctor Honoris Causa by Charles University in Prague, a small conference was held, bringing together a distinguished group of researchers with interests spanning a variety of fields related to Erdős' own work. At that gathering, the idea occurred to several of us that it might be quite appropriate at this point in Erdős' career to solicit a collection of articles illustrating various aspects of Erdős' mathematical life and work. The response to our solicitation was immediate and overwhelming, and these volumes are the result.

Regarding the organization, we found it convenient to arrange the papers into six chapters, each mirroring Erdős' holistic approach to mathematics. Our goal was not merely a (random) collection of papers but rather a thoroughly edited volume composed in large part by articles explicitly solicited to illustrate interesting aspects of Erdős and his life and work. Each chapter includes an introduction which often presents a sample of related Erdős' problems "in his own words". All these (sometimes lengthy) introductions were written jointly by editors.

We wish to thank the nearly 70 contributors for their outstanding efforts (and their patience). In particular, we are grateful to Béla Bollobás for his extensive documentation of Paul Erdős' early years and mathematical high points (in the first part of this volume); our other authors are acknowledged in their respective chapters. We also want to thank A. Bondy, G. Hahn, I. Ouhel, K. Marx, J. Načeradský and Ché Graham for their help and for the use of their works. At various stages of the project, the book was supported by AT&T Bell Laboratories, GAČR 2167 and GAUK 351. We also are indebted to Dr. Joachim Heinze and Springer Verlag for their encouragement and support. Finally, we would like to record our extreme debt to Susan Pope (at AT&T Bell Laboratories) who somehow (miraculously) managed to convert more than 50 manuscripts of all types into the attractive form they now have.

Here then is a unique portrait of a man who has devoted his whole being to "proving and conjecturing" and to the pursuit of mathematical knowledge and understanding. We hope that this will form a lasting tribute to one of the great mathematicians of our time.

R. L. Graham
J. Nešetřil

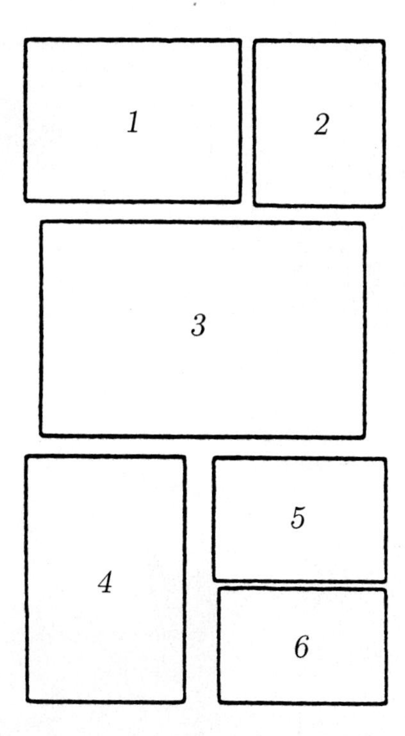

1 P. Erdős relaxing with mother in Mátraháza
2 P. Erdős with an epsilon
3 P. Erdős in an elementary school in Hungary (Dunapentele, later Stalinváros, now Dunaujváros; 1955)
4 1947
5 50's
6 Yosemite late 40's

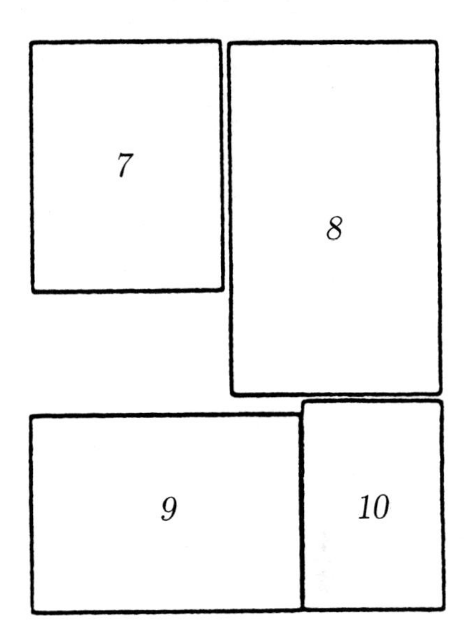

7 Portrait by I. Ouhel (oil, 40 × 60 cm, 1992)
8 Dr. h. c. Prague 1992 (photo G. Hahn)
9 1992 (photo G. Hahn)
10 Portrait by K. Marx (oil, 15 × 24 cm, 1993)

Contents of
The Mathematics of Paul Erdős II

Contents of
The Mathematics of Paul Erdős I

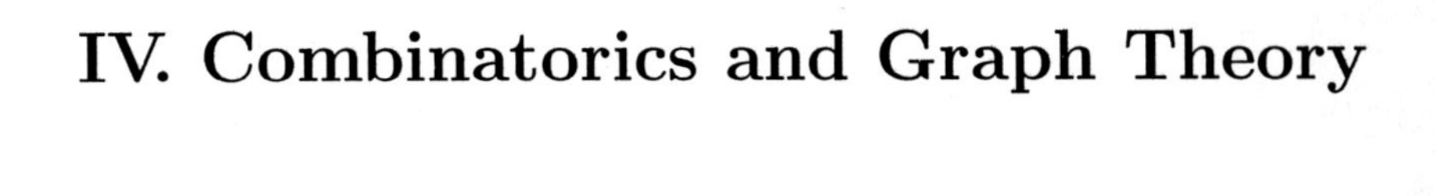

IV. Combinatorics and Graph Theory

Introduction

Erdős' work in graph theory started early and arose in connection with D. König, his teacher in prewar Budapest. The classic paper of Erdős' and Szekeres from 1935 also contains a proof in "graphotheoretic terms." The investigation of the Ramsey function led Erdős to probabilistic methods and seminal papers in 1947, 1958 and 1960. It is perhaps interesting to note that three other very early contributions of Erdős' to graph theory (before 1947) were related to infinite graphs: infinite Eulerian graphs (with Gallai and Vászoni) and a paper with Kakutani On nondenumerable graphs (1943). Although the contributions of Erdős to graph theory are manifold, and he proved (and always liked) beautiful structural results such as the Friendship Theorem (jointly with V. T. Sós and Kövári), and compactness results (jointly with N. G. de Bruijn), his main contributions were in asymptotic analysis, probabilistic methods, bounds and estimates. Erdős was the first who brought to graph theory the experience and rigor of number theory (perhaps being preceded by two papers by V. Jarnik, one of his early coauthors). Thus he contributed in an essential way to lifting graph theory up from the "slums of topology." This chapter contains a "special" problem paper not by Erdős but by his frequent coauthors from Memphis: R. Faudree, C. C. Rousseau and R. Schelp (well, there is actually an Erdős supplement there as well). We encouraged the authors to write this paper and we are happy to include it in this volume. This chapter also includes two papers coauthored by Béla Bollobás, who is one of Erdős' principal disciples. Bollobás contributed to much of Erdős' combinatorial activities and wrote important books about them. (*Extremal Graph Theory*, *Introduction to Graph Theory*, *Random Graphs*). His contributions to this chapter (coauthored with his two former students G. Brightwell and A. Thomason) deal with graphs (and thus are in this chapter) but they by and large employ random graph methods (and thus they could be contained in Chapter 3). The main questions there may be considered as extremal graph theory questions (and thus they could fit in Chapter 5). Other contributions to this chapter, which are related to some aspect to Erdős' work or simply pay tribute to him are by N. Alon, Z. Füredi, M. Aigner and E. Triesch, S. Bezrukov and K. Engel, A. Gyárfás, S. Brandt, N. Sauer and H. Wang, H. Fleischner and M. Stiebitz, and D. Beaver, S. Haber and P. Winkler.

Perhaps the main love of Erdős in graph theory is the chromatic number. Let us close this introduction with a few of Erdős' recent problems related mostly to this area in his own words:

Many years ago I proved by the probability method that for every k and r there is a graph of girth $\geq r$ and chromatic number $\geq k$. Lovász when he was still in high school found a fairly difficult constructive proof. My proof still had the advantage that not only was the chromatic number of $G(n)$ large but the largest independent set was of size $< \epsilon n$ for every $\epsilon > 0$ if $n > n_0(\epsilon, r, k)$. Nešetřil and V. Rödl later found a simpler constructive proof.

There is a very great difference between a graph of chromatic number $\aleph_0$ and a graph of chromatic number $\geq \aleph_1$. Hajnal and I in fact proved that if G has chromatic number $\aleph_1$ then G must contain a C_4 and more generally G contains the complete bipartite graph $K(n, \aleph_1)$ for every $n < \aleph_0$. Hajnal, Shelah and I proved that every graph G of chromatic number $\aleph_1$ must contain for some k_0 every odd cycle of size $\geq k_0$ (for even cycles this was of course contained in our result with Hajnal), but we observed that for every k and every m there is a graph of chromatic number m which contains no odd cycle of length $< k$. Walter Taylor has the following very beautiful problem: Let G be any graph of chromatic number $\aleph_1$. Is it true that for every $m > \aleph_1$ there is a graph G_m of chromatic number m all finite subgraphs of which are contained in G? Hajnal and Komjáth have some results in this direction but the general conjecture is still open. If it would have been my problem, I certainly would offer 1000 dollars for a proof or a disproof. (To avoid financial ruin I have to restrict my offers to my problems.)

Let k be fixed and $n \to \infty$. Is it true that there is an $f(k)$ so that if $G(n)$ has the property that for every m every subgraph of m vertices contains an independent set of size $m/2 - k$ then $G(n)$ is the union of a bipartite graph and a graph of $\leq f(k)$ vertices, i.e., the vertex set of $G(n)$ is the union of three disjoint sets S_1, S_2 and S_3 where S_1 and S_2 are independent and $|S_3| \leq f(k)$. Gyárfás pointed out that even the following special case is perhaps difficult. Let m be even and assume that every m vertices of our $G(n)$ induces an independent set of size at least $m/2$. Is it true then that $G(n)$ is the union of a bipartite graph and a bounded set? Perhaps this will be cleared up before this paper appears, or am I too optimistic?

Hajnal, Szemerédi and I proved that for every $\epsilon > 0$ there is a graph of infinite chromatic number for which every subgraph of m vertices contains an independent set of size $(1-\epsilon)m/2$ and in fact perhaps $(1-\epsilon)m/2$ can be replaced by $m/2 - f(m)$ where $f(m)$ tends to infinity arbitrarily slowly. A result of Folkman implies that if G is such that every subgraph of m vertices contains an independent set of size $m/2 - k$ then the chromatic number of G is at most $2k + 2$.

Many years ago Hajnal and I conjectured that if G is an infinite graph whose chromatic number is infinite, then if $a_1 < a_2 < \ldots$ are the lengths of the odd cycles of G we have

$$\sum_i \frac{1}{a_i} = \infty$$

and perhaps $a_1 < a_2 < \ldots$ has positive upper density. (The lower density can be 0 since there are graphs of arbitrarily large chromatic number and girth.)

We never could get anywhere with this conjecture. About 10 years ago Mihók and I conjectured that G must contain for infinitely many n cycles of length 2^n. More generally it would be of interest to characterize the infinite sequences $A = \{a_1 < a_2 < \ldots\}$ for which every graph of infinite chromatic number must contain infinitely many cycles whose length is in A. In particular, assume that the a_i are all odd.

All these problems we unattackable (at least for us). About three years ago Gyárfás and I thought that perhaps every graph whose minimum degree is ≥ 3 must contain a cycle of length 2^k for some $k \geq 2$. We became convinced that the answer almost surely will be negative but we could not find a counterexample. We in fact thought that for every r there must be a G_r every vertex of which has degree $\geq r$ and which contains no cycle of length 2^k for any $k \geq 2$. The problem is wide open.

Gyárfás, Komlós and Szemerédi proved that if k is large and $a_1 < a_2 < \ldots$ are the lengths of the cycles of a $G(n, kn)$, that is, an n-vertex graph with kn edges, then

$$\sum \frac{1}{a_i} > c \log n \ .$$

The sum is probably minimal for the complete bipartite graphs.

(Erdős-Hajnal) If G has large chromatic number does it contain two (or k if the chromatic number is large) edge-disjoint cycles having the same vertex set? It surely holds if $G(n)$ has chromatic number $> n^\epsilon$ but nothing seems to be known.

Fajtlowicz, Staton and I considered the following problem (the main idea was due to Fajtlowicz). Let $F(n)$ be the largest integer for which every graph of n vertices contains a regular induced subgraph of $\geq F(n)$ vertices. Ramsey's theorem states that $G(n)$ contains a trivial subgraph, i.e., a complete or empty subgraph of $c \log n$ vertices. (The exact value of c is not known but we know $1/2 \leq c \leq 2$.) We conjectured $F(n)/\log n \to \infty$. This is still open. We observed $F(5) = 3$ (since if $G(5)$ contains no trivial subgraph of 3 vertices then it must be a pentagon). Kohayakawa and I worked out the $F(7) = 4$ but the proof is by an uninteresting case analysis. (We found that this was done earlier by Fajtlowicz, McColgan, Reid and Staton, see Ars Combonatoria vol 39.) It would be very interesting to find the smallest integer n for which $F(n) = 5$, i.e., the smallest n for which every $G(n)$ contains a regular induced subgraph of ≥ 5 vertices. Probably this will be much more difficult than the proof of $F(7) = 4$ since in the latter we could use properties of perfect graphs. Bollobás observed that $F(n) < c\sqrt{n}$ for some $c > 0$.

Let $G(10n)$ be a graph on $10n$ vertices. Is it true that if every index subgraph of $5n$ vertices of our $G(10n)$ has $\geq 2n^2 + 1$ edges then our $G(10n)$

contains a triangle? It is easy to see that $2n^2$ edges do not suffice. A weaker result has been proved by Faudree, Schelp and myself at the Hakone conference (1992, I believe) see also a paper by Fan Chung and Ron Graham (one of the papers in a volume published by Bollobás dedicated to me).

A related forgotten conjecture of mine states that if our $G(10n)$ has more than $20n^2$ edges and every subgraph of $5n$ vertices has $\geqq 2n^2$ edges then our graph must have a triangle. Simonovits noticed that if you replace each vertex of the Petersen graph by n vertices you get a graph of $10n$ vertices, $15\ n^2$ edges, no triangle and every subgraph of $5n$ vertices contains $\geq 2n^2$ edges.

Problems in Graph Theory from Memphis

R. J. Faudree*, C. C. Rousseau and R. H. Schelp**

University of Memphis, Memphis, Tennessee 38112

Summary. This is a summary of problems and results coming out of the 20 year collaboration between Paul Erdős and the authors.

1. Introduction (by Paul Erdős)

Over a 20-year period, my friends at the University of Memphis and I have done a great deal of work in various branches of combinatorics and graph theory. We have published more than 40 papers and posed many problems. Here I only mention two examples. We wrote the first paper on the size Ramsey number [34] and stated many problems and conjectures in this subject. Some of our questions have been answered by Beck [3] but many of them are still open, and the subject is very much alive. We also proved the conjecture of Hajnal and myself concerning monochromatic domination in edge colored complete graphs.

During my present visit, many new problems were raised. I state only one. Let $f(n,k)$ be the smallest integer for which there is a graph with k vertices and $f(n,k)$ edges in which every set of $n+2$ vertices induces a subgraph with maximum degree at least n. This problem was raised in trying to settle an old conjecture of ours: is it true that for every $m \geq 2n$ every graph with m vertices and $\binom{2n+1}{2} - \binom{n}{2} - 1$ edges is the union of a bipartite graph and a graph every vertex of which has degree less than n. Faudree has a very nice proof for $m = 2n+1$. We tried to prove it for $m = 2n+2$ and this led us to our problem.

I hope that if I live we will have many more new problems and results.

2. Generalized Ramsey Theory

2.1 Terminology and Notation

Let $G_1, G_2, \ldots, G_k$ be a sequence of graphs, each having no isolated vertices. The *Ramsey number* $r(G_1, G_2, \ldots, G_k)$ is the smallest integer r so that in every k-coloring of the edges of K_r, there is a monochromatic copy of G_i in the ith color for at least one i. Most of the known results concern the two-color case. In case $k = 2$ and $G_1 = G_2 = G$, we denote the Ramsey number by $r(G)$. Throughout this paper, the notation will generally follow that of the well-known text by Chartrand and Lesniak [20]. In particular, the graph theoretic functions $\delta(G)$, $\Delta(G)$, $\beta(G)$ and $\chi(G)$ are used to represent the minimum degree, maximum degree, independence number, and chromatic number, respectively, of the graph G.

---

* Partial support received under ONR grant N00014-91-J-1085.
** Partial support received under NSF grant DMS-9400530.

2.2 Cycle vs. Complete Graph

The cycle-complete graph Ramsey number $r(C_m, K_n)$ was first studied by Bondy and Erdős in [5]. They obtained the exact value in case m is large in comparison with n.

Theorem 2.1 ([5]). *For $m \geq n^2 - 2$,*

$$r(C_m, K_n) = (m-1)(n-1) + 1.$$

Question 2.21 ([33]). What is the smallest value of m such that

$$r(C_m, K_n) = (m-1)(n-1) + 1?$$

It is possible that $r(C_m, K_n) = (m-1)(n-1) + 1$ for all $m \geq n$.

Additional results concerning cycle-complete graph Ramsey numbers were obtained in [33], among them the following upper bound.

Theorem 2.2 ([33]). *For all $m \geq 3$ and $n \geq 2$,*

$$r(C_m, K_n) \leq \{(m-2)(n^{1/k} + 2) + 1\}(n-1),$$

where $k = \lfloor (m-1)/2 \rfloor$.

For m fixed and $n \to \infty$, the probabilistic method [54] yields

$$r(C_m, K_n) > c\left(\frac{n}{\log n}\right)^{(m-1)/(m-2)} \qquad (n \to \infty). \tag{2.1}$$

The case of $m = 3$ is well studied. By a recent result of Kim [50] $r(C_3, K_n) > cn^2/\log n$, and in [53] Shearer improves the technique of Ajtai, Komlós and Szemerédi [1] to obtain

$$r(C_3, K_n) < \left\lceil \frac{n^2}{\log(n/e)} \right\rceil.$$

The case of $m = 4$ is comparatively open. As above, the probability method yields

$$r(C_4, K_n) > c\left(\frac{n}{\log n}\right)^{3/2} \qquad (n \to \infty).$$

In [33] we proved that $r(C_4, K_n) = o(n^2)$, and an unpublished observation of Szemerédi yields

$$r(C_4, K_n) < c\left(\frac{n}{\log n}\right)^2. \tag{2.2}$$

Conjecture 2.1. There exist an $\epsilon > 0$ such that $r(C_4, K_n) < n^{2-\epsilon}$ for all sufficiently large n.

From Theorem 2.2 and (2.1) it follows that if m is fixed,

$$r(C_m, K_n) > r(C_{2m-1}, K_n) \qquad \text{and} \qquad r(C_m, K_n) > r(C_{2m}, K_n),$$

for all sufficiently large n. In view of this, it may be that for n appropriately large, $r(C_m, K_n)$ first decreases as m increases. In view of Theorem 2.1, we know that $r(C_m, K_n)$ ultimately increases with m. It may be that there is a unique value of m at which the minimum is attained.

Question 2.22. [33] For n fixed, where is the minimum value of $r(C_m, K_n)$ attained?

2.3 "Goodness" in Generalized Ramsey Theory

Let F be a graph with chromatic number $\chi(F)$. The *chromatic surplus* of F, denoted $s(F)$, is the least number of vertices in a color class under any proper $\chi(F)$-coloring of the vertices of F. If G is any connected graph of order $n \geq s(F)$, then

$$r(F,G) \geq (\chi(F)-1)(n-1)+s(F).$$

The example which establishes this bound is simple.

Example 2.1. With $p = (\chi(F)-1)(n-1)+s(F)-1$, two-color the edges of K_p so that the blue graph consists of $\chi(F)$ disjoint complete graphs, $\chi(F)-1$ of order $n-1$ and one of order $s(F)-1$. In this two-coloring, there is no connected blue subgraph of order n and there is no red copy of F.

In case $r(F,G) = (\chi(F)-1)(n-1)+s(F)$, we shall say that G is F-*good*. Chvátal observed that for each m and each tree T of order n,

$$r(K_m,T) = (m-1)(n-1)+1.$$

In other words, every tree is K_m-good. In large part, the study of "goodness" in Ramsey theory is an attempt to see how far this simple result of Chvátal can be generalized. Known cases where G is F-good involve some restriction on the number of edges and/or maximum degree of G [14], [15], [36].

Theorem 2.3 ([37]). *Given k, Δ and $s = p_1 \leq \cdots \leq p_m$, there exists a corresponding number n_0 such that every connected graph G with $n \geq n_0$ vertices, $q \leq n+k$ edges and maximum degree $\leq \Delta$ is $K(p_1, p_2, \ldots, p_m)$-good.*

Corollary 2.1 ([37]). *Let F be a fixed graph with chromatic number χ and chromatic surplus s. Set $\alpha = 1/(2|V(F)|-1)$. Then there are constants C_1 and C_2 such that for all sufficiently large n, every connected graph G of order n satisfying $|E(G)| \leq n + C_1 n^\alpha$ and $\Delta(G) \leq C_2 n^\alpha$ is F-good.*

The *edge density* of a graph G is defined to be $\max\{q(H)/p(H)\}$, where $p(H)$ and $q(H)$ denote the number of vertices and edges, respectively, of H, and the maximum is taken over all subgraphs $H \subset G$. As usual, $\Delta(G)$ denotes the maximum degree in G.

Conjecture 2.2 ([37]). Let F be an arbitrary graph and let (G_n) be a sequence of connected graphs such that (i) G_n is of order n, (ii) $\Delta(G_n) = o(n)$ and (iii) each graph in (G_n) has edge density at most ρ (a constant). Then G_n is F-good for all sufficiently large n.

Conjecture 2.3 ([37]). If F is an arbitrary graph and (G_n) is a sequence of connected graphs such that $\Delta(G_n)$ is bounded, then G_n is F-good for all sufficiently large n.

If all trees are F good, it is natural to ask for the largest integer $q = f(F,n)$ so that every connected graph with n vertices and at most q edges is F-good. For $F = K_3$ we have the following result.

Theorem 2.4 ([9]). *Let $f(K_3,n)$ be the largest integer q so that every connected graph with n vertices and at most q edges is K_3-good. Then $f(K_3,n) \geq (17n+1)/15$ for all $n \geq 4$ and $f(K_3,n) < (27/4+\epsilon)n(\log n)^2$ for all sufficiently large n.*

Question 2.31 ([9]). Does $f(K_3,n)/n$ tend to infinity with n?

Also, it is appropriate to ask for the largest integer $q = g(F,n)$ so that there exists a connected graph with n vertices and q edges which is F-good. The following result is known concerning $f(K_m,n)$ and $g(K_m,n)$ where $m \geq 3$.

Theorem 2.5 ([9]). *For each $m \geq 3$, there exist positive constants A,B,C,D such that*

$$n + A n^{\frac{2}{m-1}} < f(K_m,n) < n + B n^{\frac{4}{m+1}} (\log n)^2,$$

and

$$C n^{\frac{m}{m-1}} < g(K_m,n) < D n^{\frac{m+2}{m}} (\log n)^{\frac{(m+1)(m-2)}{m(m-1)}},$$

for all sufficiently large n.

The example which provides the lower bound for $g(K_3,n)$ is easily described. (A similar example works in general to provide a lower bound for $g(K_m,n)$.)

Example 2.2 ([9]). Choose t and s so that $r(K_3,tK_s) \leq n-1$. Then set $G = K_1 + H$ where H is the graph consisting of t disjoint copies of K_s together with $n-1-ts$ isolated vertices. Then G is a connected graph with n vertices and $t\binom{s}{2} + n - 1$ edges which is K_3-good.

Are there cases where this example is best possible?

Question 2.32. For what values of n (if any) is $g(K_3,n)$ the number of edges in the above example where s and t have been chosen so that $t\binom{s}{2}$ is maximized (or possibly a generalized version in which there are t complete graphs $K_{s_1}, K_{s_2}, \ldots, K_{s_t}$, not all of the same order)?

2.4 Graph vs. Tree

If T is any tree, $r(C_4,T)$ is determined by $r(C_4,K_{1,\Delta(T)})$, where $\Delta(T)$ denotes the maximum degree in T.

Theorem 2.6 ([16]). *Let T be any tree of order n. Then*

$$r(C_4,T) = \max\{4,\, n+1,\, r(C_4,K_{1,\Delta(T)})\}.$$

In view of the importance of $r(C_4,K_{1,n})$, it would be helpful to have more precise information concerning this Ramsey number. It is easy to prove that

$$r(C_4,K_{1,n}) \leq n + \lceil\sqrt{n}\rceil + 1,$$

and in [16] we obtain

$$r(C_4,K_{1,n}) > n + \sqrt{n} - 6n^{11/40}, \tag{2.3}$$

for all sufficiently large n. The proof of the later result uses the fact that for any prime p there is a C_4-free graph $\mathcal{G}_p$ of order p^2+p+1 in which each vertex has degree p or $p+1$. By letting p be the smallest prime greater than $\sqrt{n}$ and randomly deleting a suitable number of vertices from $\mathcal{G}_p$, we obtain a graph which yields (2.3). The bound is is based on information about the distribution of primes. Let p_k denote the k-th prime. It has been conjectured that $p_{k+1} - p_k < (\log p_k)^\alpha$ for some constant α. If this conjecture is true, the lower bound in (2.3) can be improved to $n + \sqrt{n} - 6\left(\frac{1}{2}\log n\right)^\alpha$.

Question 2.41 ([16]). Is it true that $r(C_4, K_{1,n}) < n + \sqrt{n} - c$ holds infinitely often, where c is an arbitrary constant?

Paul Erdős offers \$100 for a proof (one way or the other).

Question 2.42 ([16]). Is it true that $r(C_4, K_{1,n+1}) \leq r(C_4, K_{1,n}) + 2$ for all n?

It has been communicated to the authors that Guantao Chen has verified that the answer to the previous question is yes.

In [40] it is proved that if $n \geq 3m - 3$ then $r(K(1,1,m), T) = 2n - 1$ for every tree T of order n. This was shown to be the case when T is a star $(K_{1,n-1})$ by Rousseau and Sheehan [52] who also proved that the condition $n \geq 3m - 3$ cannot be weakened. This is one of many examples in Ramsey problems involving trees where the star turns out to be the worst, i.e. largest Ramsey number, case. How general is this phenomenon?

Conjecture 2.4. If G is fixed and n is sufficiently large, $r(G,T) \leq r(G, K_{1,n-1})$ for every tree T of order n.

What are the graphs G relative to which all sufficiently large trees are good? An early result showed that this class of graphs includes all those in which some $\chi(G)$-coloring has two color classes consisting of a single vertex.

Theorem 2.7 ([6]). *Let G be a graph with $\chi(G) \geq 2$ which has a vertex $\chi(G)$-coloring with at least two color classes consisting of a single vertex. Then*

$$r(G,T) = (\chi(G) - 1)(n - 1) + 1$$

for every tree T of order $n \geq n_0(G)$.

A complete characterization for the case of $s(G) = 1$ has been obtained by Burr and Faudree [17].

Theorem 2.8 ([17]). *Let G be a graph with chromatic number k. Then for all sufficiently large n, every tree T of order n satisfies $r(G,T) = (k-1)(n-1)+1$ iff there is an integer m, such that G is a subgraph of both the chromatic number k graphs $K_{1,m,m,\ldots,m}$ and $mK_2 + K_{m,m,\ldots,m}$.*

If $G \cong K_{m_1,m_2,\ldots,m_k}$ with $m_1 = 1$, the fact that G has chromatic number k and chromatic surplus 1 gives the lower bound $(k-1)(n-1)+1$ and there is an upper bound which depends only on $r(K_{1,m_2}, T)$.

Theorem 2.9 ([41]). *For all sufficiently large n, every tree T of order n satisfies*

$$(k-1)(n-1) + 1 \leq r(K_{1,m_2,\ldots,m_k}, T) \leq (k-1)(r(K_{1,m_2}, T) - 1) + 1.$$

For $m_1 \neq 1$ it is natural to suspect the following generalization.

Conjecture 2.5 ([41]). If $m_1 \leq m_2 \leq \cdots \leq m_k$ and T is any sufficiently large tree,

$$r(K_{m_1,m_2,\ldots,m_k}, T) \leq (k-1)(r(K_{m_1,m_2}, T) - 1) + m_1.$$

The *decomposition class* of a graph G, denoted $\mathcal{B}(G)$, is defined as follows: a given bipartite graph B belongs to $\mathcal{B}(G)$ whenever there exists a $\chi(G)$ vertex coloring such that B is the bipartite graph induced by some pair of color classes. Let $r(\mathcal{B}(G), T)$ denote the smallest integer r so that in every two-coloring of $E(K_r)$ there is at least one member of $\mathcal{B}(G)$ in the first color or a copy of T in the second color. In [39] it is shown that if B is a fixed bipartite graph and $T_1, T_2, T_3, \ldots$ is any sequence of trees where T_n is of order n, then $r(B, T_n) = n + o(n)$ as $n \to \infty$. Thus $r(\mathcal{B}(G), T_n) = n + o(n)$. The result of Faudree and Burr can be written as follows: if $\mathcal{B}(G)$ contains K_2, then for all sufficiently large n, every tree T of order n satisfies $r(G,T) = (\chi(G) - 2)(n-1) + r(\mathcal{B}(G), T)$. It would be interesting to know how generally this formula holds.

Question 2.43. What graphs G satisfy

$$r(G,T) = (\chi(G) - 2)(n-1) + r(\mathcal{B}(G), T)$$

for every tree T of order $n \geq n_0(G)$?

2.5 Trees

Some very interesting open problems concern the diagonal Ramsey number $r(T)$ where T is a tree. It has been conjectured by Erdős and Sós that each graph with m vertices and at least $(n-2)m/2$ edges contains every tree T of order n; if true, this yields $r(T) \leq 2n - 2$. A general lower bound for $r(T)$ is given by the following example.

Example 2.3. Assume $a \leq b$. Construction (1): Two-color the edges of K_{2a+b-2} so that the red graph is $K_{a-1} \cup K_{a+b-1}$. Construction (2): Two-color the edges of K_{2b-2} so that the red graph is $2K_{b-1}$. In each of these two constructions, there is no monochromatic connected bipartite graph with a vertices in one class and b vertices in the other.

In view of this example, we see that any tree T with a vertices in one color class and b in the other satisfies

$$r(T) \geq \max\{2a + b - 1,\ 2b - 1\}.$$

The *broom* $B_{k,l}$ is the tree on $k+l$ vertices obtained by identifying an endvertex of a path P_l on l vertices with the central vertex of a star $K_{1,k}$ on k edges.

Theorem 2.10 ([35]). *(1)* $r(B_{k,l}) = k + \lceil 3l/2 \rceil - 1$ *for* $k \geq 1$ *and* $l \geq 2k$, *(2)* $r(B_{k,l}) \leq 2k + l$ *for* $5 \leq l \leq 2k$.

From the lower bound given by Example 2.3,

$$r(B_{k,l}) \geq \begin{cases} 2k + 2\lfloor l/2 \rfloor - 1 & \text{when} \quad l < 2k - 1, \\ 2k + 2\lfloor l/2 \rfloor & \text{when} \quad l = 2k - 1. \end{cases}$$

Thus in case (2) of Theorem 2.10, $r(B_{k,l})$ differs from the upper bound by at most 2.

The smallest value of $\max\{2a + b - 1,\ 2b - 1\}$ occurs when $2a = b$, so Example 2.3 shows that $r(T) \geq \lceil 4n/3 \rceil$ for every tree of order n. Moreover, this bound cannot be improved since equality holds for certain booms.

Question 2.51 ([35]). Is $r(T) = 4a$ for every tree with a vertices in one color class and $2a$ in the other?

A special case of the previous question is when T is the tree obtained from subdividing $a - 1$ edges of the star $K_{1,2a}$ [23].

2.6 Multicolor Results

When the list of graphs $(G_1, G_2, \ldots, G_c)$ is restricted to complete bipartite graphs, odd cycles, and exactly one large cycle (even or odd), there is an exact formula for $r(G_1, G_2, \ldots, G_c)$.

Theorem 2.11 ([32]). *Let $[B]$ and $[C]$ denote the following fixed sequences of graphs:*

$$\begin{aligned} [B] &= (K_{b_1,c_1}, \cdots, K_{b_s,c_s}) \qquad b_i \le c_i \ \ (i = 1, 2, \ldots, s), \\ [C] &= (C_{2d_1+1}, \ldots, C_{2d_t+1}). \end{aligned}$$

Further, let $\ell = \sum_{i=1}^{s}(b_i - 1)$, and require that $d_i \ge 2^{t-1}$ for $i = 1, \cdots, t$. Then, if n is sufficiently large, $r(C_n, [B], [C]) = 2^t(n + \ell - 1)$.

All of the results in [32] are for the case where exactly one cycle length is large in comparison with the orders of the remaining graphs $([B], [C])$. It would be quite helpful to have corresponding results when there are two or more large cycles.

Question 2.61 ([32]). What is $r(C_n, C_m, [B], [C])$ when both n and m are large in comparison with the orders of the remaining graphs?

Let $[C_{odd}] = (C_{2d_1+1}, \ldots, C_{2d_t+1})$. Then $r(\le [C_{odd}])$ denotes the smallest number r such that in every t-coloring of the edges of K_r, there is an odd cycle of length at most $2d_i + 1$ in the ith color for some i. Using the proof technique of Theorem 2.11, we have the following multicolor Ramsey result for cycles.

Corollary 2.2 ([32]). *Let $[C_{even}] = (C_{2b_1}, \ldots, C_{2b_s})$, $[C_{odd}] = (C_{2d_1+1}, \ldots, C_{2d_t+1})$, and let $\ell = \sum_{i=1}^{s}(b_i - 1)$. If n is sufficiently large,*

$$r(C_n, [C_{even}], [C_{odd}]) = (r(\le [C_{odd}]) - 1)(n + \ell - 1) + 1.$$

For the case of three colors, the following results were found for $r(\le [C_{odd}])$.

Theorem 2.12 ([32]). *Let $\ell \ge k \ge m$. Then*

$$r(\le (C_{2k+1}, C_{2l+1}, C_{2m+1})) = \begin{cases} 9 & k \ge 4,\ l \ge 2,\ m \ge 1, \\ 11 & k \ge 5,\ l = m = 1. \end{cases}$$

In the remaining cases,

$$r(\le (C_{2k+1}, C_{2l+1}, C_{2m+1})) = \begin{cases} 9 \\ 10 \\ 12 \\ 17 \end{cases} \quad \textit{for } (k, l, m) = \begin{cases} (3,3,3),(3,3,2),(3,3,1),(3,2,2),(2,2,2), \\ (3,2,1),(2,2,1), \\ (4,1,1),(3,1,1),(2,1,1), \\ (1,1,1). \end{cases}$$

By considering a bipartite decomposition of a complete graph, it is clear that, in general,

$$r(\le (C_{2m_1+1}, \cdots, C_{2m_t+1})) \ge 2^t + 1.$$

Note that $r(\le (C_{2k+1}, C_{2l+1}, C_{2m+1})) = 2^3 + 1 = 9$ except for $k \ge 5$, $l = m = 1$ and six special cases.

Question 2.62. When does $r(\le (C_{2m_1+1}, \cdots, C_{2m_t+1})) = 2^t + 1$ hold? In particular, what is true for $t = 4$?

2.7 Ramsey Size Linear Graphs

A graph G is *Ramsey size linear* if there is a constant C such that

$$r(G, H) \leq Cn$$

for every graph H with n edges. The following result is easily proved using the Erdős-Lovász local lemma and the techniques of Spencer [54].

Theorem 2.13 ([38], [43]). *Given a graph G with p vertices and q edges, there exists a constant C such that for n sufficiently large,*

$$r(G, K_n) > \left(\frac{n}{\log n}\right)^{\frac{q-1}{p-2}}.$$

Corollary 2.3. *If G is a graph with p vertices and $2p - 2$ edges, then G is not Ramsey size linear.*

In the other direction, we have the following result.

Theorem 2.14 ([43]). *If G is any graph with p vertices and at most $p+1$ edges, then G is Ramsey size linear.*

There are various small graphs for which the question of Ramsey size linearity is open.

Question 2.71. Are the following graphs Ramsey size linear: $K_{3,3}$, H_5 (the graph obtained from C_5 by adding two vertex disjoint chords), Q_3 (the 3-dimensional cube)?

Question 2.72 ([43]). If G is a graph such that each subgraph on p vertices has at most $2p - 3$ edges, then is G Ramsey size linear?

The graph K_4 is not Ramsey size linear, but the deletion of any edge gives a Ramsey size linear graph. It would be of interest to find a specific infinite family of graphs with this property.

2.8 Repeated Degrees and Degree Spread

If $n \geq r(G)$, and the edges of K_n are two-colored, there must be at least one monochromatic copy of G. If there must be such a monochromatic copy of G in which two vertices have the same degree in the two-colored K_n, at least for all sufficiently large n, then we say that G has the *Ramsey repeated degree property.* In [2] it was shown that K_m where $m \geq 4$ *does not* have this property.

Theorem 2.15 ([19]). *For each $m \geq 1$, the graphs $K_{m,m}$ and C_{2m+1} have the Ramsey repeated degree property.*

Question 2.81. Are there graphs other than K_m for $m \geq 4$ which fail to have the Ramsey repeated degree property?

The *degree spread* of $X \subseteq V(\mathcal{G})$ is the difference between the largest and smallest degree (in $\mathcal{G}$) over all vertices in X. Given graphs G and H and a two-coloring of the edges of K_n with $n \geq r(G, H)$, the degree spread of the coloring is the minimum degree spread over all vertex sets of copies of G and/or H which are appropriately monochromatic. Then $\Phi_n(G, H)$ is the maximum degree spread over all two-colorings of $E(K_n)$.

Theorem 2.16 ([19]). *If* $n \geq 4(r-1)(r-2)$ *where* $r = r(G,H)$, *then* $\Phi_n(G,H) = r(G,H) - 2$.

Question 2.82 ([19]). What is the smallest n for which $\Phi_n(G,H) = r(G,H)-2$? Does this relation hold for all $n \geq r(G,H)$?

3. Size Ramsey Numbers

3.1 Definitions and Notation

Write $F \to (G,H)$ to mean that in every two-coloring of the edges of F, there is a copy of G in the first color or a copy of H in the second. The *size Ramsey number* of the pair (G,H), denoted $\hat{r}(G,H)$, is the smallest number $q = \hat{r}(G)$ for which there exists a graph F with q edges satisfying $F \to (G,H)$. In case $G = H$ we write $F \to G$ for the "arrow" relation and $\hat{r}(G)$ for the size Ramsey number. The size Ramsey number was introduced in [34]. Interestingly enough, the size Ramsey number of K_n is simply expressed in terms of its ordinary Ramsey number:

$$\hat{r}(K_n) = \binom{r(K_n)}{2}.$$

(See [34].) Questions concerning $\hat{r}(P_n)$ were raised in [34]. In [3] Beck proved the rather surprising fact that $\hat{r}(P_n) \leq Cn$ for some constant C

3.2 Complete Bipartite Graphs

The asymptotic behavior of the size Ramsey number of $K_{n,n}$ is studied in [47].

Theorem 3.1 ([47]). *For all* $n \geq 6$,

$$\frac{1}{60}\, n^2\, 2^n < \hat{r}(K_{n,n}) < \frac{3}{2}\, n^3\, 2^n.$$

The upper bound comes from the fact that $K_{a,b}$ with $a = \lfloor n^2/2 \rfloor$ and $b = 3n2^n$ arrows $K_{n,n}$ for $n \geq 6$. The lower bound comes from a simple application of the probabilistic method using a bound for the number of copies of $K_{n,n}$ in any graph with q edges. Of course, the immediate goal is to bring the bounds for $\hat{r}(K_{n,n})$ to within a constant factor.

Question 3.21. Is there a choice of $a = a(n)$ and $b = b(n)$ so that $ab = O(n^2 2^n)$ and $K_{a,b} \to K_{n,n}$?

3.3 Stars and Star-Forests

Theorem 3.2 ([7]). *For positive integers* $k, l, m,$ *and* n,

$$\hat{r}(mK_{1,k}, nK_{1,\ell}) = (m+n-1)(k+l-1).$$

Moreover if $F \to (mK_{1,k}, nK_{1,\ell})$ *and* F *has* $(n+m-1)(k+l-1)$ *edges, then* $F = (m+n-1)K_{1,k+\ell-1}$ *or* $k = l = 2$ *and* $F = tK_3 \cup (m+n+t-1)K_{1,3}$ *for some* $1 \leq t \leq m+n-1$.

This leaves open the size Ramsey number for a pair of star forests. Assume that

$$F_1 = \bigcup_{i=1}^{s} K_{1,n_i} \quad (n_1 \geq n_2 \geq \cdots \geq n_s) \qquad \text{and} \qquad F_2 = \bigcup_{i=1}^{t} K_{1,m_i}, \quad (m_1 \geq m_2 \geq \cdots \geq m_t).$$

For $2 \leq k \leq s+t$ set $l_k = \max\{n_i + m_j - 1 : i + j = k\}$. It is not difficult to prove that $\cup_{k=2}^{s+t} K_{1,\ell_k} \to (F_1, F_2)$ so $\hat{r}(F_1, F_2) \leq \sum_{k=2}^{s+t} \ell_k$.

Conjecture 3.1 ([7]). $\hat{r}(F_1, F_2) = \displaystyle\sum_{k=2}^{s+t} l_k$.

Note that if $n_i = n$ for all i and $m_j = m$ for all j, then the conjectured value agrees with the number $\hat{r}(sK_{1,n}, tK_{1,m}) = (m+n-1)(s+t-1)$ given in Theorem 3.2.

3.4 Star vs. Complete Graph

The following result is easily established.

Theorem 3.3. *The arrow relation* $K_{2n+1} - K_n \to (K_{1,n}, K_3)$ *holds for each* $n \geq 2$.

Possibly $K_{2n+1} - K_n$ is an arrowing graph with the minimum possible number of edges, and thus $\hat{r}(K_{1,n}, K_3) = \binom{2n+1}{2} - \binom{n}{2} = 3\binom{n+1}{2}$. To verify this, it would be sufficient to prove the following conjecture.

Conjecture 3.2 ([22]). For $n \geq 3$, any graph with $\binom{2n+1}{2} - \binom{n}{2} - 1$ edges is the union of a bipartite graph and a graph with maximum degree less than n.

It is not difficult to verify this conjecture if the graph has at most $2n+1$ vertices.

3.5 Graph vs. Matching

No general result is known for $\hat{r}(G, tK_2)$, but some information about this number can be obtained by considering

$$\hat{r}_\infty(G) = \lim_{t \to \infty} \frac{\hat{r}(G, tK_2)}{t\,\hat{r}(G, K_2)} = \lim_{t \to \infty} \frac{\hat{r}(G, tK_2)}{t\,|E(G)|}.$$

Clearly, $0 < \hat{r}_\infty(G) \leq 1$. The following results were obtained in [24].

Theorem 3.4 ([24]). *The set* $\{\hat{r}_\infty(G) : G$ *is a connected graph*$\}$ *is dense in* $[0,1]$.

Theorem 3.5 ([24]). *For each* $n \geq 2$, *there is a corresponding positive constant* a *such that*

$$\frac{1}{n} \leq \hat{r}_\infty(P_n), \hat{r}_\infty(C_n) \leq \frac{a}{n}.$$

Also

$$\frac{1}{n} \leq \hat{r}_\infty(K_{n,n}) \leq \frac{2+\sqrt{2}}{n} \qquad \textit{and} \qquad \frac{2}{n} \leq \hat{r}_\infty(K_n) \leq \frac{8}{n}.$$

Thus $\lim_{n\to\infty} \hat{r}_\infty(H_n) = 0$ for $H_n = P_n$, C_n, $K_{n,n}$ or K_n. On the other hand, we can show that $\lim_{n\to\infty} \hat{r}_\infty(K_{1,n}) = 1$. This suggests the following question.

Question 3.51 ([24]). Does $\lim_{n\to\infty} \hat{r}_\infty(G_n) = 0$ hold for every sequence of graphs (G_n) such that $|V(G_n)| \to \infty$ and $\Delta(G_n)$ is bounded as $n \to \infty$? What sequences (G_n) yield $\lim_{n\to\infty} \hat{r}_\infty(G_n) = 1$?

4. Other Ramsey Problems

4.1 Multiple Copies

Assume $F \to (mG, H)$ where mG denotes m vertex disjoint copies of G. How many copies of G must F contain?

Theorem 4.1 ([7]). *If $F \to (mG, H)$, then $tG \subseteq F$ where*

$$t = \left\lfloor \frac{m|V(G)| + |V(H)| - \beta(H) - 1}{|V(G)|} \right\rfloor .$$

Question 4.11 ([7]). If $F \to (nG)$, then must F contain at least $\left\lfloor \frac{r(nG)}{|V(G)|} \right\rfloor$ copies of G?

We have shown that if G and $\epsilon > 0$ are fixed then for all sufficiently large n, every graph F satisfying $F \to (nG)$ contains at least $r(nG)(1-\epsilon)/|V(G)|$ copies of G.

4.2 Ramsey Minimal Problems

Let F, G, H be graphs without isolates. The graph F is (G,H)-*minimal* if $F \to (G,H)$ but $F - e \not\to (G,H)$ for any edge e of F. The pair (G,H) is called *Ramsey-finite* or *Ramsey-infinite* according to whether the class of all (G,H)-minimal graphs is a finite or infinite set. It has been shown by Nešetřil and Rödl [51] that (G,H) is Ramsey-infinite when at least one of the following hold:

(i) G and H are both 3-connected.
(ii) $\chi(G)$ and $\chi(H)$ are both ≥ 3.
(iii) G and H are both forests, neither of which is a union of stars.

Statement (iii) of the above result has been strengthened in the following way.

Theorem 4.2. *(i) [13] The pair (G,H) is Ramsey-infinite when both G and H are forests, at least one having a non-star component.*
(ii) [10] If G and H are star forests with no single edge star (K_2), then (G,H) is Ramsey-finite if and only if both G and H are single stars, each with an odd number of edges.
(iii) [11] The pair $(K_{1,m} \cup kK_2, K_{1,n} \cup \ell K_2)$ is Ramsey-finite when both m and n are odd.

Subsequently, Faudree characterized Ramsey-finite pairs of forests.

Theorem 4.3 ([48]). *If G and H are forests, then (G,H) is Ramsey-finite if and only if*

$$G=\left(\bigcup_{i=1}^{s}K_{1,m_i}\right)\bigcup mK_2, \qquad m_1\geq m_2\geq\ldots\geq m_s\geq 2,\ m\geq 0,$$

$$H=\left(\bigcup_{i=1}^{t}K_{1,n_i}\right)\bigcup nK_2, \qquad n_1\geq n_2\geq\ldots\geq n_t\geq 2,\ s\geq t\geq 0,$$

and one of the following hold:
(1) $t=0$ and $n>0$,
(2) $s=t=1$ and m_1,n_1 are odd,
(3) $s\geq 2$, $t=1$, m_1,n_1 are odd, $m_1\geq n_1+m_2-1$ and $n\geq n_0=n_0(G,H)$.

It should be noted that the precise value of $n_0(G,H)$ in (3) is not known. This is the only place where the Ramsey-finiteness question for a pair of star forests is not explicitly settled.

The above results leave a considerable gap when either G or H has connectivity two or less. It is likely that (G,H) is Ramsey-infinite for all pairs not identified above as *Ramsey-finite.*

Conjecture 4.1 ([12]). The pair (G,H) is Ramsey-infinite unless both G and H are stars with an odd number of edges or at least one of G and H contains a single edge component.

An interesting case of the above conjecture is when G is a cycle and H is two-connected. No technique is presently known for showing such a pair is Ramsey-infinite.

Other results concerning Ramsey finiteness were obtained in [8] and [10].

Theorem 4.4. *(i) [8] If G is a matching and H is an arbitrary graph, then (G,H) is Ramsey-finite.*
(ii) [10] The pair $(K_{1,k},G)$ is Ramsey-infinite if $k\geq 2$ and G is a two-connected graph. Also $(K_{1,2},H)$ is Ramsey-infinite if H is a bridgeless connected graph.

Conjecture 4.2 ([9]). The pair $(K_{1,2},G)$ is Ramsey-finite only if G is a matching.

Conjecture 4.3 ([9]). If (G,H) is Ramsey-finite for each graph H, then G must be a matching.

There is no known case where adding independent edges to G (or H) changes a Ramsey-finite pair (G,H) to a Ramsey-infinite pair.

Conjecture 4.4 ([12]). If (G,H) is Ramsey-finite then $(G\cup K_2,H)$ is Ramsey-finite.

5. Extremal Problems

5.1 Edge Density and Triangles

The classical result of Turán implies that any graph with n vertices and at least $\lfloor n^2/4\rfloor+1$ edges contains a triangle. A natural generalization is obtained by fixing α between 0 and 1 and asking for the minimum number $f_\alpha(n)$ so that a graph of order n must contain a triangle if each set of $\lfloor \alpha n\rfloor$ vertices induces a subgraph with at least $f_\alpha(n)$ edges.

Example 5.1 ([44]). Let $M_1 = K_2$, $M_2 = C_5$, and let M_3 be the graph obtained from C_8 by adding all of the long chords. From M_i obtain a graph H_i of order $n = k|V(M_i)|$ by replacing each vertex of M_i by an independent set of k vertices, and making each pair of vertices of H_i in different expanded sets adjacent if and only if the replaced vertices are adjacent in M_i. Each graph H_i is extremal triangle-free in that the addition of any edge creates a triangle.

Note that the minimum number of edges induced by a set of $n/2$ vertices in H_2 is $n^2/50$. This leads to the following conjecture of Erdős.

Conjecture 5.1 ([44]). If each set of $\lfloor n/2 \rfloor$ vertices in a graph G of order n spans more than $n^2/50$ edges, then G contains a triangle.

An analysis of the graphs H_1, H_2, and H_3 leads to the following more general conjecture.

Conjecture 5.2 ([44]). Suppose $17/30 \le \alpha \le 1$ and $\beta > (2\alpha - 1)/4$ or $43/120 \le \alpha < 17/30$ and $\beta > (5\alpha - 2)/25$. For all sufficiently large n, a graph of order n in which each set of $\lfloor \alpha n \rfloor$ vertices spans at least βn^2 edges must contain a triangle.

In [44] this conjecture is proved for $.648\ldots \le \alpha \le 1$; otherwise it is open. The case of $\alpha = 1/2$ is of particular interest.

One direction would be to extend these results from triangles to arbitrary complete graphs. For $\alpha = 1/2$, the following conjecture has been made by Chung and Graham.

Conjecture 5.3 ([21]). Let $b_t(n)$ denote the minimum number of edges induced by any set of $n/2$ vertices in the Turán graph on n vertices for K_t. If each set of $\lfloor n/2 \rfloor$ vertices in a graph G of order n spans more than $b_t(n)$ edges, then G contains a K_t.

5.2 Minimal Degree

The wheel $W_n = K_1 + C_{n-1}$ has n vertices, $2n - 2$ edges, and minimum degree 3, but any proper subgraph of W_n has a vertex of degree less than 3. On the other hand, it is easy to prove that any graph with n vertices and $2n - 1$ edges has a proper subgraph of minimum degree at least 3 [26].

Theorem 5.1 ([42]). *If G is a graph with n vertices and $2n - 1$ edges, then G contains a subgraph with less than $n - \sqrt{n/48}$ vertices that has minimum degree at least 3.*

A stronger result has been conjectured.

Conjecture 5.4 ([42]). There is an $\epsilon > 0$ such that every graph with n vertices and $2n - 1$ edges has a subgraph with $(1 - \epsilon)n$ or fewer vertices that has minimum degree at least 3.

A more general version of Theorem 5.1 was proved in [42] and independently by other authors.

Theorem 5.2 ([42]). *For $k \ge 2$, any graph with n vertices and $(k-1)(n-k+2) + \binom{k-2}{2} + 1$ edges contains a subgraph with at most $n - \lfloor \sqrt{n/(6k^3)} \rfloor$ vertices that has minimum degree at least k.*

The generalized wheel $W(k-2,n) = K_{k-2} + C_{n-k+2}$ shows that the number of edges in the last result cannot be reduced. Conjecture 5.4 has the following generalization for subgraphs of minimum degree at least $k \geq 2$.

Conjecture 5.5 ([42]). For each $k \geq 2$, there is a corresponding $\epsilon > 0$ such that every graph with n vertices and $(k-1)(n-k+2)+\binom{k-2}{2}+1$ edges has a subgraph with $(1-\epsilon)n$ or fewer vertices that has minimum degree at least k.

5.3 Odd Cycles in Graphs of Given Minimum Degree

Example 5.2 ([28]). Let $k \geq 3$ be an odd integer and let H be the two-connected nonbipartite graph obtained from C_{k+2} by replacing each of its $k+2$ vertices by an independent set of size s. Then H is a graph of order $n = (k+2)s$ that contains no odd cycles of length $\leq k$ and contains all possible odd cycles of larger length. Also H is regular of degree $2s$.

Theorem 5.3 ([28]). *Let $k \geq 3$ be an odd positive integer. There exists $f(k)$ such that if $n \geq f(k)$ and G is a two-connected nonbipartite graph of order n with minimum degree $\geq 2n/(k+2)$, then either G contains a C_k or $G \cong H$.*

If G is a two-connected nonbipartite graph of appropriately large order n and $\delta(G) \geq 2n/(k+2)$, what range of odd cycles must G contain?

Example 5.3 ([28]). Set $a = \lceil 2n/(k+2) \rceil$ and $b = n - a$. The graph G obtained by adding an edge to the smaller part of $K_{a,b}$ contains C_{2t+1} for $1 \leq t \leq a-1$, but no odd cycle of length greater than $2a-1$.

Thus the best possible result would show that, aside from H, any two-connected nonbipartite graph with large order n and minimum degree at least $2n/(k+2)$ contain odd cycles of all lengths in a range up to about $4n/(k+2)$. The next result shows that such a result is true (asymptotically) if the factor 4 in the last statement is replaced by $8/3 - o(n)$.

Theorem 5.4 ([28]). *Given $0 < c < \frac{1}{3}$ and $0 < \epsilon < 1$, there exist $h_1(c,\epsilon)$ and $h_2(c,\epsilon)$ so that for all $n \geq h_2(c,\epsilon)$ every two-connected nonbipartite graph G of order n and minimum degree $\delta(G) \geq cn$ contains C_{2t+1} for $h_1(c,\epsilon) \leq 2t+1 \leq 4(1-\epsilon)cn/3$.*

Is Example 5.3 really the truth?

Question 5.31 ([28]). If n is appropriately large and $G \ncong H$ is a two-connected nonbipartite graph of order n satisfying $\delta(G) \geq 2n/(k+2)$, must G contain C_{2t+1} for $k \leq 2t+1 \leq 2\lceil 2n/(k+2) \rceil - 1$?

5.4 Vertices and Edges on Odd Cycles

Every graph with $n \geq 3$ vertices and at least $\lfloor n^2/4 \rfloor + 1$ edges contains a triangle, and in fact a cycle of length $2k+1$ for every $k \leq \lfloor (n+1)/4 \rfloor$. For such a graph, how many vertices (edges) must lie on triangles and, more generally, on cycles of length $2k+1$?

Theorem 5.5 ([31]). *Let G be a graph with n vertices and at least $\lfloor n^2/4 \rfloor + 1$ edges. (a) At least $\lfloor n/2 \rfloor + 2$ vertices of G are on triangles, and this result is sharp. (b) At least $2\lfloor n/2 \rfloor + 1$ edges of G are on triangles, and this result is sharp. (c) If $k \geq 2$ and $n \geq \max\{3k(3k+1), 216(3k-2)\}$, then at least $2(n-k)/3$ vertices of G are on cycles of length $2k+1$, and this result is asymptotically best possible. (d) If $k \geq 2$ is fixed, then at least $11n^2/144 - O(n)$ edges of G are on cycles of length $2k+1$.*

The example used to show that the result in (c) is asymptotically best possible suggests what the sharp result should be in place of (d).

Example 5.4. Consider the graph $K_s \cup T_2(n-s)$ where $s = \lceil 2n/3 \rceil + 1$ and $T_2(m)$ denotes the complete bipartite graph with parts $\lfloor m/2 \rfloor$ and $\lceil m/2 \rceil$ (the Turán graph). This graph has $\lceil 2n/3 \rceil + 1$ vertices on cycles of length $2k+1$ and $\sim 2n^2/9$ edges on cycles of length $2k+1$.

Conjecture 5.6. If $k \geq 2$ is fixed, then as $n \to \infty$ every graph with n vertices and $\lfloor n^2/4 \rfloor + 1$ or more edges has at least $2n^2/9 - O(n)$ edges on cycles of length $2k+1$.

5.5 Extremal Paths

Let m, n and k be fixed positive integers with $m > n \geq k$. We wish to find the minimum value of l such that each m vertex graph with at least l vertices of degree $\geq n$ contains a P_{k+1}. A plausible minimum value for l is suggested by the following construction.

Example 5.5. Let $m = t(n+1) + r$ where $0 \leq r < n+1$ and assume $k < 2n+1$. Let $s = \lfloor (k-1)/2 \rfloor$. The graph consisting of t vertex disjoint copies of $H = \bar{K}_{n+1-s} + K_s$ and r isolated vertices contains ts vertices of degree n and no P_{k+1}. When k is even and $r + s \geq n$, the number of vertices of degree $\geq n$ in this graph can be increased by 1 to $ts+1$ without forcing the graph to contain a P_{k+1}. Simply take one of the vertices of degree s and make it adjacent to each of the r isolated vertices.

Theorem 5.6 ([45]). *Let m and n be positive integers such that $n+1 \leq m \leq 2n+1$. If G is of order m and contains at least k vertices of degree $\geq n$, then G contains (i) a P_{2k-5} if $k \leq n/2+3$, (ii) a P_{n+1} if $(n+1)/2+3 \leq k \leq n$, and (iii) a P_k if $n+1 \leq k$.*

Conjecture 5.7 ([45]). Let m, n and k be positive integers with $m > n \geq k$ and set $\delta = 2$ if k is even and $\delta = 1$ if k is odd. If G is a graph of order m and at least $l = \lfloor (k-1)/2 \rfloor \lfloor m/(m+1) \rfloor + \delta$ vertices of degree $\geq n$, then G contains a P_{k+1}.

5.6 Degree Sequence and Independence in K_4-Free Graphs

For a graph G with $\delta(G) \geq 1$ let $f(G)$ denote the largest number of times any entry in the degree sequence of G is repeated. If G is K_r-free graph and $f(G) \leq k$, what is the smallest possible value of the independence number $\beta(G)$?

Theorem 5.7 ([30]). *(i) If G is a K_3-free graph of order n and $f(G) \leq k$, then $\beta(G) \geq n/k$.*

(ii) If $r \geq 5$ and $k \geq 2$, then there exists a sequence (G_n) of K_r-free graphs such that $|V(G_n)| = n$ and $f(G_n) \leq k$ for each n and $\beta(G_n) = o(n)$ as $n \to \infty$.
(iii) If G is a K_4-free graph of order n and $f(G) = 2$, then $\beta(G) \geq n/12$.
(iv) There exists no sequence (G_n) of K_4-free graphs such that $|V(G_n)| = n$ and $f(G_n) = 3$ for each n and $\beta(G_n) = o(n)$ as $n \to \infty$.

Bollobás has shown that there exists a sequence (G_n) of K_4-free graphs such that $|V(G_n)| = n$ and $f(G_n) \geq 5$ for each n and $\beta(G_n) = o(n)$ as $n \to \infty$ [4]. The following questions are open.

Question 5.61 ([30]). Does there exist a sequence (G_n) of K_4-free graphs such that $|G_n| = n$ and $f(G_n) = 4$ for each n and $\beta(G_n) = o(n)$ as $n \to \infty$?

Question 5.62 ([30]). If G is a K_3-free graph of order n with $f(G) \leq k$, is the lower bound $\beta(G) \geq n/k$ best possible?

6. Other Problems

6.1 Monochromatic Coverings

If A and B are sets of vertices in a given graph G, we say that A *dominates* or *covers* B if for every $y \in B \setminus A$ there is an $x \in A$ such that $xy \in E(G)$. The following result was conjectured by Erdős and Hajnal.

Theorem 6.1 ([27]). *For any fixed t, in every two-coloring of the edges of K_n there exists a set of t or fewer vertices which monochromatically dominates at least $n(1 - 2^{-t})$ vertices. This result is essentially sharp.*

In [25] we studied monochromatic domination when more than two colors are involved. For three colors, we have the following result.

Theorem 6.2 ([25]). *Three-color the edges of K_n. Then in at least one color, there is a set consisting of 22 or fewer vertices which dominates a set of at least $2n/3$ vertices.*

The fact that no small set will, in general, monochromatically dominate more than two-thirds of the vertices comes from the following example due to Kierstead.

Example 6.1 (Kierstead). Three-color the edges of K_{3m} as follows. Form the partition $V(K_{3m}) = (A_1, A_2, A_3)$ and for $i \equiv 1, 2, 3 \pmod 3$ color the internal edges of A_i and all the edges between A_i and A_{i+1} with color i. Then each monochromatic subgraph is isomorphic to $K_m + \overline{K}_m$ and there is no set which monochromatically dominates more than $2m$ vertices.

It is possible that Theorem 6.2 holds with "22" replaced by "3".

Question 6.11. Is it true that in every three-coloring of the edges of K_n there is a set three vertices which monochromatically dominates at least two-thirds of all the vertices?

6.2 Spectra

A nonnegative integer q belongs to the *k-spectrum* of a graph G if there is some set of k vertices in G spanning q edges. Denote the k-spectrum of G by $s_k(G)$. Then $s_k(G) \subseteq \{0, 1, \cdots \binom{k}{2}\}$, but not every such subset is an example of a k-spectrum. Let n_k denote the number of distinct subsets of $\{0, 1, \cdots, \binom{k}{2}\}$ which are realizable as k-spectra. The k-spectra of all large trees were characterized in [46].

Theorem 6.3 ([46]). *If $n \geq \max\{2k-1, 3k-5\}$, then for each tree T of order n there exist corresponding integers r and s with $0 \leq \lceil r/2 \rceil \leq s \leq r \leq k-1$ such that $s_k(T_n) = [0, r] \cup [s, k-1]$.*

The following bounds have been obtained for number of realizable k-spectra.

Theorem 6.4 ([46]). *For any integer $k \geq 2$, the number of distinct realizable k-spectra satisfies $\frac{1}{16}\left(\frac{5}{2}\right)^{k-1} < n_k < 2^{\binom{k}{2}+1}$.*

For $k = 2, 3$ and 4, the realizable k-spectra have been determined [49].

Question 6.21. For $k \geq 5$, what is n_k, the number of distinct realizable k-spectra?

A similar problem was asked by Erdős and Faudree concerning the *cycle spectrum* of a graph, namely the set of distinct cycle lengths of a graph. Let c_n be the number of distinct cycle spectra for graphs of order n. The following family of graphs gives a lower bound for the c_n.

Example 6.2. Let v be a vertex in C_n and let $S \subseteq \{2, 3, \ldots, \lfloor n/2 \rfloor\}$. Add to C_n all chords extending from v whose lengths belong to S. Distinct sets S will give distinct cycle spectra, since the cycles of length at least $\lceil n/2+1 \rceil$ will be different.

Using this example, we see that $2^{n/2} - 1 \leq c_n < 2^n - 2$.

Question 6.22. What is the number of distinct cycle spectra of a graph of order n?

6.3 Clique Coverings and Partitions

A *clique covering* of a graph G is a set of cliques that together contain all of the edges of G; if each edge is contained in precisely one of these cliques, we have a *clique partition*. The *clique covering number* $cc(G)$ of G is the smallest cardinality of any clique covering, and the *clique partition number* $cp(G)$ is the smallest cardinality of any clique partition. The relationship between $cc(G)$ and $cp(G)$ was investigated in [29], where the following example was described.

Example 6.3. For any integer n divisible by 8, let $G_n = K_{n/2} + 4K_{n/8}$. Then, clearly $cc(G_n) = 4$, and it was shown in [29] that $cp(G_n) = n^2/16 + 3n/4$. Thus,

$$\frac{cp(G_n)}{cc(G_n)} > n^2/64.$$

Question 6.31. What is the largest C for which there is a sequence of graphs (G_n) such that $|V(G_n)| = n$ and

$$\frac{cp(G_n)}{cc(G_n)} > Cn^2?$$

In [18] the authors exhibit a sequence of graphs (G_n) such that $|V(G_n)| = n$ and

$$cp(G_n) - cc(G_n) = \frac{n^2}{4} - \frac{n^{3/2}}{2} + \frac{n}{4} + O(1), \qquad (n \to \infty).$$

Paul Erdős has asked whether the $n^{3/2}$ term is really necessary.

Question 6.32. Is there a sequence of graphs (G_n) such that $|V(G_n)| = n$ and

$$cp(G_n) - cc(G_n) = n^2/4 + O(n), \qquad (n \to \infty)?$$

References

1. M. Ajtai, J. Komlos and E. Szemerédi, *A note on Ramsey numbers*, **J. Combin. Theory Ser. A** 29, (1980), 354-360.
2. M. O. Albertson and D. M. Berman, *Ramsey graphs without repeated degrees*, **Cong. Numer.** 83 (1991), 91-96.
3. J. Beck, *On size Ramsey numbers of paths, trees and circuits, I*, **J. Graph Theory** 7, (1983), 115-129.
4. B. Bollobás, *Degree multiplicities and independent sets in K_4-free graphs*, preprint.
5. J. A. Bondy and P. Erdős, *Ramsey numbers for cycles in graphs*, **J. Combin. Theory Ser. B** 14, (1973), 46-54.
6. S. A. Burr, P. Erdős, R. J. Faudree, R. J. Gould, M. S. Jacobson, C. C. Rousseau, and R. H. Schelp, *Goodness of trees for generalized books*, **Graphs Combin.** 3 (1987), 1-6.
7. S. A. Burr, P. Erdős, R. J. Faudree, C. C. Rousseau, and R. H. Schelp, *Ramsey minimal graphs for multiple copies*, **Proc. Koninklijke, Nederlandse Akad. Van Wetenschappen, Amsterdam, Series A.** 81(2) (1978), 187-195.
8. S. A. Burr, P. Erdős, R. J. Faudree, and R. H. Schelp, *A class of Ramsey-finite graphs*, **Proc. 9th S. E. Conf. on Combinatorics, Graph Theory, and Computing** (1978), 171- 178.
9. S. A. Burr, P. Erdős, R. J. Faudree, C. C. Rousseau, and R. H. Schelp, *An extremal problem in generalized Ramsey theory*, **Ars Combin.** 10 (1980), 193-203.
10. S. A. Burr, P. Erdős, R. J. Faudree, C. C. Rousseau, and R. H. Schelp, *Ramsey minimal graphs for the pair star - connected graph*, **Studia Scient. Math. Hungar.** 15 (1980), 265-273.
11. S. A. Burr, P. Erdős, R. J. Faudree, C. C. Rousseau, and R. H. Schelp, *Ramsey minimal graphs for star forests*, **Discrete Math.** 33 (1981), 227-237.
12. S. A. Burr, P. Erdős, R. J. Faudree, C. C. Rousseau, and R. H. Schelp, *Ramsey minimal graphs for matchings*, **The Theory and Applications of Graphs**, G. Chartrand, editor, John Wiley (1981) 159-168.
13. S. A. Burr, P. Erdős, R. J. Faudree, C. C. Rousseau, and R. H. Schelp, *Ramsey minimal graphs for forests*, **Discrete Math.** 38 (1982), 23-32.
14. S. A. Burr, P. Erdős, R. J. Faudree, C. C. Rousseau, and R. H. Schelp, *Ramsey numbers for the pair sparse graph-path or cycle*, **Trans. Amer. Math. Soc.** 2 (269) (1982), 501-512.
15. S. A. Burr, P. Erdős, R. J. Faudree, C. C. Rousseau, and R. H. Schelp, *The Ramsey number for the pair complete bipartite graph - graph with limited degree*, **Graph Theory with Applications to Algorithms and Computer Sciences** G. Chartrand, ed. Wiley-Interscience, New York, (1985), 163-174.

16. S. A. Burr, P. Erdős, R. J. Faudree, C. C. Rousseau, and R. H. Schelp, *Some complete bipartite graph - tree Ramsey numbers*, **Ann. Discrete Math.** 41 (1989), 79-90.
17. S. A. Burr and R. J. Faudree, *On graphs G for which all large trees are G-good*, **Graphs Combin.** (to appear).
18. L. Caccetta, P. Erdős, E. T. Ordman and N. J. Pullman, *The difference between the clique numbers of a graph*, **Ars Combin.** 19A (1985), 97-106.
19. G. Chen, P. Erdős, C. C. Rousseau and R. H. Schelp, *Ramsey problems involving degrees in edge-colored complete graphs of vertices belonging to monochromatic subgraphs*, **European J. Combin.** 14 (1993), 183-189.
20. G. Chartrand and L. Lesniak, **Graphs and Digraphs**, Wadsworth and Brooks/Cole, Pacific Grove, California, 1986.
21. F. R. K. Chung and R. L. Graham, *On graphs not containing prescribed induced subgraphs*, **A tribute to Paul Erdős**, (eds. A. Baker, B. Bollobás, and A. Hajnal), Cambridge University Press, Cambridge, (1990), 111-120.
22. P. Erdős, *Problems and results in graph theory*, **The Theory and Applications of Graphs**, G. Chartrand, editor, John Wiley (1981) 331-341.
23. P. Erdős, *Some recent problems and results in graph theory, combinatorics, and number theory*, **Proceedings 7th S-E Conf. Comb. Graph Theory, and Computing**, (1976) 3-14.
24. P. Erdős and R. J. Faudree, *Size Ramsey numbers involving matchings*, **Colloquia Mathematica Societatis Janos Bolyai** 37 (1981), 247-264.
25. P. Erdős, R. J. Faudree, R. J. Gould, A. Gyárfás, and R. H. Schelp, *Monochromatic coverings in colored complete graphs*, **Congressus Numerantium** 71 (1990), 29-38.
26. P. Erdős, R. J. Faudree, A. Gyárfás, and R. H. Schelp, *Cycles in graphs without proper subgraphs of minimal degree 3*, **(Proceedings of the Eleventh British Combinatorial Conference), Ars Combin.** 25B (1988), 195-202.
27. P. Erdős, R. J. Faudree, A. Gyárfás, and R. H. Schelp, *Domination in colored complete graphs*, **J. Graph Theory** 13 (1989), 713-718.
28. P. Erdős, R. J. Faudree, A. Gyárfás, and R. H. Schelp, *Odd cycles in graphs of given minimal degree*, **Graph Theory, Combinatorics, and Applications**, Wiley and Sons, New York, *Proceedings of the Sixth International Conference on Graph Theory and Applications*, (1991), 407-418.
29. P. Erdős. R. J. Faudree, and E. Ordman, *Clique partitions and clique coverings*, **Discrete Math.** 72 (1988), 93-101.
30. P. Erdős, R. J. Faudree, T. J. Reid, R. H. Schelp, and W. Staton, *Degree sequence and independence in K_4-free graphs*, to appear in **Discrete Math.**
31. P. Erdős, R. J. Faudree, and C. C. Rousseau, *Extremal problems involving vertices and edges on odd cycles in graphs*, **Discrete Math.** 101, (1992), 23-31.
32. P. Erdős, R. J. Faudree, C. C. Rousseau, and R. H. Schelp, *Generalized Ramsey theory for multiple colors*, **J. of Comb. Theory B** 20 (1976), 250-264.
33. P. Erdős, R. J. Faudree, C. C. Rousseau, and R. H. Schelp, *Cycle-complete graph Ramsey numbers*, **J. of Graph Theory** 2 (1978), 53-64.
34. P. Erdős, R. J. Faudree, C. C. Rousseau, and R. H. Schelp, *The Size Ramsey number, a new concept in generalized Ramsey theory*, **Periodica Mathematica Hungarica** 9 (1978), 145-161.
35. P. Erdős, R. J. Faudree, C. C. Rousseau, and R. H. Schelp, *Ramsey numbers for brooms*, **Proc. 13th S.E. Conf. on Comb., Graph Theory and Computing** 283-294, (1982).
36. P. Erdős, R. J. Faudree, C. C. Rousseau, and R. H. Schelp, *Tree - multipartite graph Ramsey numbers*, **Graph Theory and Combinatorics - A Volume in Honor of Paul Erdős**, Bela Bollobás, editor, Academic Press, (1984), 155-160.

37. P. Erdős, R. J. Faudree, C. C. Rousseau, and R. H. Schelp, *Multipartite graph - sparse graph Ramsey numbers*, **Combinatorica** 5, (1985), 311-318. (with P. Erdős, C. C. Rousseau, and R. H. Schelp.
38. P. Erdős, R. J. Faudree, C. C. Rousseau, and R. H. Schelp, *A Ramsey problem of Harary on graphs with prescribed size*, **Discrete Math.** 67 (1987), 227-234.
39. P. Erdős, R. J. Faudree, C. C. Rousseau, and R. H. Schelp, *Extremal theory and bipartite graph - tree Ramsey numbers*, **Discrete Math.** 72 (1988), 103-112.
40. P. Erdős, R. J. Faudree, C. C. Rousseau, and R. H. Schelp, *Book - tree Ramsey numbers*, **Scientia, A: Mathematics** 1 (1988), 111-117.
41. P. Erdős, R. J. Faudree, C. C. Rousseau, and R. H. Schelp, *Multipartite graph - tree Ramsey numbers*, **Annals of the New York Academy of Sciences**, 576 (1989), 146-154, Proceedings of the First China - USA International Graph Theory Conf.
42. P. Erdős, R. J. Faudree, C. C. Rousseau, and R. H. Schelp, *Subgraphs of minimal degree k*, **Discrete Math.** 85, (1990), 53-58.
43. P. Erdős, R. J. Faudree, C. C. Rousseau, and R. H. Schelp, *Ramsey size linear graphs*, to appear in Proceedings of Cambridge Combinatorics Colloquium
44. P. Erdős, R. J. Faudree, C. C. Rousseau and R. H. Schelp, *A local density condition for triangles*, to appear in **Discrete Math.**
45. P. Erdős, R. J. Faudree, R. H. Schelp, and M. Simonvits *An extremal result for paths*, **Annals of The New York Academy of Sciences**, 576 (1989), 155-162. Proceedings of the China - USA Graph Theory Conf.
46. P. Erdős, R. J. Faudree, and V. T. Sós, *k-spectrum of a graph*, to appear in **Proceedings of the Seventh International Conference on Graph Theory, Combinatorics, Algorithms, and Applications**, Kalamazoo, Michigan
47. P. Erdős and C. C. Rousseau, *The size Ramsey number of a complete bipartite graph*, **Discrete Math.** 113, (1993), 259-262.
48. R. J. Faudree, *Ramsey minimal graphs for forests*, **Ars Combin.**, 31, (1991), 117-124.
49. R. J. Faudree, R. J. Gould, M. S. Jacobson, J. Lehel, and L. M. Lesniak *Graph spectra*, manuscript.
50. J. K. Kim, *The Ramsey number* $R(3.t)$ *has order of magnitude* $t^2/\log t$, preprint.
51. J. Nešetřil and V. Rödl, *The structure of critical graphs*, **Acta. Math. Acad. Sci. Hungar.**, 32, (1978), 295-300.
52. C. C. Rousseau and J. Sheehan, *A class of Ramsey problems involving trees*, **J. London Math. Soc.** (2) 18, (1978), 392-396.
53. J. B. Shearer, *A note on the independence number of a triangle-free graph*, **Discrete Math.** 46, (1983), 83-87.
54. J. Spencer, *Asymptotic lower bounds for Ramsey functions*, **Discrete Math.** 20, (1977), 69-76.

Neighborly Families of Boxes and Bipartite Coverings

Noga Alon*

[1] Institute for Advanced Study, Princeton, NJ 08540, USA
[2] Department of Mathematics, Tel Aviv University, Tel Aviv, Israel

Summary. A *bipartite covering of order k* of the complete graph K_n on n vertices is a collection of complete bipartite graphs so that every edge of K_n lies in at least 1 and at most k of them. It is shown that the minimum possible number of subgraphs in such a collection is $\Theta(kn^{1/k})$. This extends a result of Graham and Pollak, answers a question of Felzenbaum and Perles, and has some geometric consequences. The proofs combine combinatorial techniques with some simple linear algebraic tools.

1. Introduction

Paul Erdős taught us that various extremal problems in Combinatorial Geometry are best studied by formulating them as problems in Graph Theory. The celebrated Erdős de Bruijn theorem [3] that asserts that n non-collinear points in the plane determine at least n distinct lines is one of the early examples of this phenomenon. An even earlier example appears in [4] and many additional ones can be found in the surveys [5] and [12]. In the present note we consider another example of an extremal geometric problem which is closely related to a graph theoretic one. Following the Erdős tradition we study the graph theoretic problem in order to deduce the geometric consequences.

A finite family $\mathcal{C}$ of d-dimensional convex polytopes is called k-*neighborly* if $d-k \leq dim(C \cap C') \leq d-1$ for every two distinct members C and C' of the family. In particular, a 1-neighborly family is simply called *neighborly*. In this case the dimension of the intersection of each two distinct members of the family is precisely $d-1$. Neighborly families have been studied by various researchers, see, e.g., [10], [14], [15], [16], [17]. In particular it is known that the maximum possible cardinality of a neighborly family of d-simplices is at least 2^d ([16]) and at most 2^{d+1} ([14]). The maximum possible cardinality of a neighborly family of *standard boxes* in R^d, that is, a neighborly family of d-dimensional boxes with edges parallel to the coordinate axes, is precisely $d+1$. This has been proved by Zaks [17], by reducing the problem to a theorem of Graham and Pollak [8] about bipartite decompositions of complete graphs. In the present note we consider the more general problem of k-neighborly families of standard boxes. The following result determines the asymptotic behaviour of the maximum possible cardinality of such a family.

Theorem 1.1. *For $1 \leq k \leq d$, let $n(k,d)$ denote the maximum possible cardinality of a k-neighborly family of standard boxes in R^d. Then*

(*i*) $\quad d+1 = n(1,d) \leq n(2,d) \leq \cdots \leq n(d-1,d) \leq n(d,d) = 2^d.$

(*ii*) $$\left(\frac{d}{k}\right)^k \leq \prod_{i=0}^{k-1}\left(\left\lfloor\frac{d+i}{k}\right\rfloor+1\right) \leq n(k,d) \leq \sum_{i=0}^{k} 2^i \binom{d}{i} < 2\left(\frac{2ed}{k}\right)^k.$$

* Research supported in part by the Sloan Foundation, Grant No. 93-6-6.

This answers a question of Felzenbaum and Perles [6], who asked if for fixed k, $n(k, d)$ is a nonlinear function of d.

As in the special case $k = 1$, the function $n(k, d)$ can be formulated in terms of bipartite coverings of complete graphs. A *bipartite covering* of a graph G is a family of complete bipartite subgraphs of G so that every edge of G belongs to at least one such subgraph. The covering is of *order* k if every edge lies in at most k such subgraphs. The *size* of the covering is the number of bipartite subgraphs in it. The following simple statement provides an equivalent formulation of the function $n(k, d)$.

Proposition 1.1. *For $1 \leq k \leq d$, $n(k, d)$ is precisely the maximum number of vertices of a complete graph that admits a bipartite covering of order k and size d.*

The rest of this note is organized as follows. In Section 2 we present the simple proof of Proposition 1.1. The main result, Theorem 1.1, is proved in Section 3. Section 4 contains some possible extensions and open problems.

2. Neighborly families and bipartite coverings

There is a simple one to one correspondence between k-neighborly families of n standard boxes in R^d and bipartite coverings of order k and size d of the complete graph K_n. To see this correspondence, consider a k-neighborly family $\mathcal{C} = \{C_1, \ldots, C_n\}$ of n standard boxes in R^d. Since any two boxes have a nonempty intersection, there is a point in the intersection of all the boxes (by the trivial, one dimensional case of Helly's Theorem, say). By shifting the boxes we may assume that this point is the origin O. If $n \geq 2$, O must lie in the boundary of each box, since it belongs to all boxes, and the dimension of the intersection of each pair of boxes is strictly smaller than d. Put $V = \{1, 2, \ldots, n\}$. For each coordinate x_i, $1 \leq i \leq d$, let H_i be the complete bipartite graph on V whose sets of vertices are $V_i^+ = \{j : C_j$ is contained in the half space $x_i \geq 0\}$, and $V_i^- = \{j : C_j$ is contained in the half space $x_i \leq 0\}$. It is not difficult to see that if the dimension of $C_p \cap C_q$ is $d - r$, then the edge pq of the complete graph on V lies in exactly r of the subgraphs H_i. Therefore, the graphs H_i form a bipartite covering of order k and size d.

Moreover, the above correspondence is invertible; given a bipartite covering of the complete graph on $V = \{1, 2, \ldots, n\}$ by complete bipartite subgraphs $H_1, \ldots, H_d$ one can define a family of n standard boxes as follows. Let V_i^+ and V_i^- denote the two color classes of H_i. For each j, $1 \leq j \leq n$, let C_j be the box defined by the intersection of the unit cube $[-1, 1]^d$ with the half spaces $x_i \geq 0$ for all i for which $j \in V_i^+$ and the half spaces $x_i \leq 0$ for all i for which $j \in V_i^-$. If the given covering is of order k, the family of standard boxes obtained is k-neighborly.

The correspondence above clearly implies the assertion of Proposition 1.1, and enables us to study, in the next section, bipartite coverings, in order to prove Theorem 1.1.

3. Economical bipartite coverings

In this section we prove Theorem 1.1. In view of Proposition 1.1 we prove it for the function $n(k, d)$ that denotes the maximum number of vertices of a complete graph that admits a bipartite covering of order k and size d.

Part (i) of the theorem is essentially known. The fact that $n(1,d) = d+1$ is a Theorem of Graham and Pollak [8], [9]. See also [7], [11], [18], [13], [1] and [2] for various simple proofs and extensions. The statement that for every fixed d, $n(k,d)$ is a non-decreasing function of k is obvious and the claim that $n(d,d) = 2^d$ is very simple. Indeed, the chromatic number of any graph that can be covered by d bipartite subgraphs is at most 2^d, implying that $n(d,d) \leq 2^d$. To see the lower bound, let V be a set of 2^d vertices denoted by all the binary vectors $\epsilon = (\epsilon_1, \ldots, \epsilon_d)$, and let H_i be the complete bipartite graph whose classes of vertices are all the vertices labelled by vectors with $\epsilon_i = 0$ and all the vertices labelled by vectors with $\epsilon_i = 0$. Trivially $H_1, \ldots, H_d$ form a bipartite covering (of order d and size d) of the complete graph on V, showing that $n(d,d) = 2^d$, as claimed.

The lower bound in part (ii) of the theorem is proved by a construction, as follows. For each i, $0 \leq i \leq k-1$, define $d_i = \lfloor (d+i)/k \rfloor$ and $D_i = \{1, 2, \ldots, d_i, d_i + 1\}$. Observe that $\sum_{r=0}^{k-1} d_r = d$. Let V denote the set of vectors of length k defined as follows

$$V = \{ \ (\epsilon_0, \epsilon_1, \ldots, \epsilon_{k-1}) : \epsilon_i \in D_i \ \}.$$

For each r, $0 \leq r \leq k-1$, and each j, $1 \leq j \leq d_r$, let $H_{r,j}$ denote the complete bipartite graph on the classes of vertices

$$A_{r,j} = \{ \ (\epsilon_0, \epsilon_1, \ldots, \epsilon_{k-1}) : \epsilon_r = j \ \}$$

and

$$B_{r,j} = \{ \ (\epsilon_0, \epsilon_1, \ldots, \epsilon_{k-1}) : \epsilon_r \geq j+1 \ \}.$$

Altogether there are $\sum_{r=0}^{k-1} d_r = d$ bipartite subgraphs $H_{r,j}$. It is not too difficult to see that they form a bipartite covering of the complete graph on V. In fact, if $(\epsilon_0, \ldots, \epsilon_{k-1})$ and $(\epsilon'_0, \ldots, \epsilon'_{k-1})$ are two distinct members of V, and they differ in s coordinates, then the edge joining them lies in precisely s of the bipartite graphs. Since $1 \leq s \leq k$ for each such two members, the above covering is of order k, implying the lower bound in part (ii) of the theorem.

The upper bound in part (ii) is proved by a simple algebraic argument. Let $H_1, \ldots, H_d$ be a bipartite covering of order k and size d of the complete graph on the set of vertices $N = \{1, 2, \ldots, n\}$. Let A_i and B_i denote the two vertex classes of H_i. For each $i \in N$, define a polynomial $P_i = P_i(x_1, \ldots, x_d, y_1, \ldots y_d)$ as follows:

$$P_i = \prod_{j=1}^{k} \Big(\sum_{p:\ i \in A_p} x_p + \sum_{q:\ i \in B_q} y_q - j \Big).$$

For each $i \in N$ let $e_i = (b_{i1}, \ldots, b_{id}, a_{i1}, \ldots, a_{id})$ be the zero-one vector in which $a_{ip} = 1$ if $i \in A_p$ (and $a_{ip} = 0$ otherwise), and, similarly, $b_{iq} = 1$ if $i \in B_q$ (and $b_{iq} = 0$ otherwise). The crucial point is the fact that

$$P_i(e_j) = 0 \ \ for \ \ all \ \ i \neq j \ \ and \ \ P_i(e_i) \neq 0. \tag{3.1}$$

This holds as the value of the sum

$$\sum_{p:\ i \in A_p} x_p + \sum_{q:\ i \in B_q} y_q$$

for $x_p = b_{jp}$ and $y_q = a_{jq}$ is precisely the number of bipartite subgraphs in our collection in which i and j lie in distinct color classes. This number is 0 for $i = j$ and is between 1 and k for all $i \neq j$, implying the validity of (3.1).

Let $\overline{P}_i = \overline{P}_i(x_1, \ldots, x_d, y_1, \ldots, y_d)$ be the multilinear polynomial obtained from the standard representation of P_i as a sum of monomials by replacing each monomial of the form $c\prod_{s\in S} x_s^{\delta_s} \prod_{t\in T} y_t^{\gamma_t}$, where all the δ_s and γ_t are positive, by the monomial $c\prod_{s\in S} x_s \prod_{t\in T} y_t$. Observe that when all the variables x_p, y_q attain 0, 1-values, $P_i(x_1, \ldots, y_d) = \overline{P}_i(x_1, \ldots, y_d)$, since for any positive δ, $0^\delta = 0$ and $1^\delta = 1$. Therefore, by (3.1),

$$\overline{P}_i(e_j) = 0 \ \ for \ \ all \ \ i \neq j \ \ and \ \ \overline{P}_i(e_i) \neq 0. \tag{3.2}$$

By the above equation, the polynomials $\overline{P}_i$ $(i \in N)$ are linearly independent. To see this, suppose this is false, and let

$$\sum_{i\in N} c_i \overline{P}_i(x_1, \ldots, y_d) = 0,$$

be a nontrivial linear dependence between them. Then there is an $i' \in N$ so that $c_{i'} \neq 0$. By substituting $(x_1, \ldots, y_d) = e_{i'}$ we conclude, by (3.2), that $c_{i'} = 0$, contradiction. Thus these polynomials are indeed linearly independent. Each polynomial $\overline{P}_i$ is a multilinear polynomial of degree at most k. Moreover, by their definition they do not contain any monomials that contain both x_i and y_i for the same i. It thus follows that all the polynomials $\overline{P}_i$ are in the space generated by all the monomials $\prod_{s\in S} x_s \prod_{t\in T} y_t$, where S and T range over all subsets of N satisfying $|S| + |T| \leq k$ and $S \cap T = \emptyset$. Since there are $m = \sum_{i=0}^{k} 2^i \binom{d}{i}$ such pairs S, T, this is the dimension of the space considered, and as the polynomials $\overline{P}_i$ are n linearly independent members of this space it follows that $n \leq m$. This completes the proof of part (ii) and hence the proof of Theorem 1.1. □

4. Concluding remarks and open problems

The proof of the upper bound for the function $n(k, d)$ described above can be easily extended to the following more general problem. Let K be an arbitrary subset of cardinality k of the set $\{1, 2, \ldots, d\}$. A bipartite covering $H_1, \ldots, H_d$ of size d of the complete graph K_n on n vertices is called a covering of *type* K if for every edge e of K_n, the number of subgraphs H_i that contain e is a member of K. The proof described above can be easily modified to show that the maximum n for which K_n admits a bipartite covering of type K and size d, where $|K| = k$, is at most $\sum_{i=0}^{k} 2^i \binom{d}{i}$. There are several examples of sets K for which one can give a bigger lower bound than the one given in Theorem 1.1 for the special case of $K = \{1, \ldots, k\}$. For example, for $K = \{2, 4\}$, there is a bipartite covering $H_1, \ldots, H_d$ of type K of a complete graph on $n = 1 + \binom{d}{2}$ vertices. To see this, denote the vertices by all subsets of cardinality 0 or 2 of a fixed set D of d elements and define, for each $i \in D$, a complete bipartite graph whose classes of vertices are all subsets that contain i and all subsets that do not contain i. Similar examples exist for types K of bigger cardinality.

One can consider bipartite coverings of prescribed type of other graphs besides the complete graph, and the algebraic approach described above can be used to supply lower bounds for the minimum possible number of bipartite subgraphs in such a cover, as a function of the rank of the adjacency matrix of the graph (and the type).

The main problem that remains open is, of course, that of determining precisely the function $n(k, d)$ for all k and d. Even the precise determination of $n(2, d)$ seems difficult.

References

1. N. Alon, *Decomposition of the complete r-graph into complete r-partite r-graphs*, Graphs and Combinatorics 2 (1986), 95-100.
2. N. Alon, R. A. Brualdi and B. L. Shader, *Multicolored forests in bipartite decompositions of graphs*, J. Combinatorial Theory, Ser. B (1991), 143-148.
3. N. G. de Bruijn and P. Erdős, *On a combinatorial problem*, Indagationes Math. 20 (1948), 421-423.
4. P. Erdős, *On sequences of integers none of which divides the product of two others, and related problems*, Mitteilungen des Forschungsinstituts für Mat. und Mech., Tomsk, 2 (1938), 74-82.
5. P. Erdős and G. Purdy, *Some extremal problems in combinatorial geometry*, in: Handbook of Combinatorics (R. L. Graham, M. Grötschel and L. Lovász eds.), North Holland, to appear.
6. A. Felzenbaum and M. A. Perles, Private communication.
7. R. L. Graham and L. Lovász, *Distance matrix polynomials of trees*, Advances in Math. 29 (1978), 60-88.
8. R. L. Graham and H. O. Pollak, *On the addressing problem for loop switching*, Bell Syst. Tech. J. 50 (1971), 2495-2519.
9. R. L. Graham and H. O. Pollak, *On embedding graphs in squashed cubes*, In: Lecture Notes in Mathematics 303, pp 99-110, Springer Verlag, New York-Berlin-Heidelberg, 1973.
10. J. Kasem, *Neighborly families of boxes*, Ph. D. Thesis, Hebrew University, Jerusalem, 1985.
11. L. Lovász, *Combinatorial Problems and Exercises* , Problem 11.22, North Holland, Amsterdam 1979.
12. J. Pach and P. Agarwal, **Combinatorial Geometry**, DIMACS Tech. Report 41–51, 1991 (to be published by J. Wiley).
13. G. W. Peck, *A new proof of a theorem of Graham and Pollak*, Discrete Math. 49 (1984), 327-328.
14. M. A. Perles, *At most* 2^{d+1} *neighborly simplices in* E^d, Annals of Discrete Math. 20 (1984), 253-254.
15. J. Zaks, *Bounds on neighborly families of convex polytopes*, Geometriae Dedicata 8 (1979), 279-296.
16. J. Zaks, *Neighborly families of* 2^d *d-simplices in* E^d, Geometriae Dedicata 11 (1981), 505-507.
17. J. Zaks, Amer. Math. Monthly 92 (1985), 568-571.
18. H. Tverberg, *On the decomposition of* K_n *into complete bipartite graphs*, J. Graph Theory 6 (1982), 493-494.

Cycles and Paths in Triangle-Free Graphs

Stephan Brandt*

Fachbereich Mathematik, Freie Universität Berlin, Arnimallee 2–6,
D-14195 Berlin, Germany

Summary. Let G be a triangle-free graph of order n and minimum degree $\delta > n/3$. We will determine all lengths of cycles occurring in G. In particular, the length of a longest cycle or path in G is exactly the value admitted by the independence number of G. This value can be computed in time $O(n^{2.5})$ using the matching algorithm of Micali and Vazirani. An easy consequence is the observation that triangle-free non-bipartite graphs with $\delta \geq \frac{3}{8}n$ are hamiltonian.

1. Introduction

In recent years a lot of research was performed on sufficient conditions for cycles in graphs which do not contain certain graphs as induced subgraphs. Mostly one of the forbidden subgraphs is the claw $K_{1,3}$.

A completely different class of graphs are triangle-free graphs, i.e. graphs containing no K_3 (which is always an induced subgraph). Note that no graph with maximum degree more than 2 is claw-free and triangle-free at the same time. Most attention found a very special class of triangle-free graphs, namely bipartite graphs. Our main objective are cycle lengths in triangle-free non-bipartite graphs.

Häggkvist [14] proved that computing the circumference (i.e. the length of a longest cycle) in general graphs of order n with minimum degree $\delta > (\frac{1}{2} - \varepsilon)n$ is $\mathcal{NP}$-hard for every $\varepsilon > 0$. Note that for $\varepsilon = 0$ the circumference is n by Dirac's celebrated result [12]. It seems even difficult to determine precisely the circumference or the length of a longest path in terms of other invariants of the graph (whose determination might be $\mathcal{NP}$-hard themselves).

The situation changes if we consider triangle-free graphs. The main observation in this paper is that in triangle-free graphs with minimum degree $\delta > n/3$ the circumference is exactly $\min\{n, 2(n-\alpha)\}$ and the length of a longest path is $\min\{n-1, 2(n-\alpha)\}$ where α denotes the independence number of the graph, and both values can be computed in polynomial time. A simple derivation of the circumference result is that triangle-free non-bipartite graphs with minimum degree $\delta \geq 3n/8$ are hamiltonian.

Moreover, we obtain that triangle-free non-bipartite graphs with $\delta > n/3$ are **weakly pancyclic**, i.e. they contain all cycles between their girth (length of a shortest cycle) and circumference. This settles the triangle-free part of a result of Brandt, Faudree and Goddard (see [7]) who prove that every graph with minimum degree $\delta \geq (n+2)/3$ is weakly pancyclic or bipartite.

For 2-connected graphs with an odd cycle of length at most 5, Brandt, Faudree and Goddard [8] proved that even the degree condition $\delta > n/4 + c$ for a moderately large constant c suffices to ensure that the graph is weakly pancyclic.

Very recently, substantial progress was obtained concerning cycles in triangle-free non-bipartite graphs. Dingjun Lou [15] proved that triangle-free non-bipartite graphs $\neq C_5$ satisfying the famous Chvátal-Erdős condition for hamiltonian graphs

* Supported by Deutsche Forschungsgemeinschaft, grant We 1265.

[11] are weakly pancyclic with girth 4 and circumference n, thereby answering a conjecture of Amar et. al. [3] in the affirmative. Bauer, van den Heuvel and Schmeichel [5] proved that there are triangle-free graphs with arbitrary large toughness which contradicts a conjecture of Chvátal [10] saying that there is a constant t_0 such that t_0-tough graphs are pancyclic. Subsequently Alon [1] determined graphs with arbitrary large girth and toughness.

Since there are still extensive variations in the notation of standard invariants of a graph $G = (V(G), E(G))$ in graph-theoretical literature we give a brief collection of the notation used in this paper:

$\alpha(G)$ (vertex-)independence number of G,
$\kappa(G)$ (vertex-)connectivity of G,
$\nu(G)$ edge-independence number (or matching number) of G,
$\omega(G)$ number of components of G,
$\delta(G)$ minimum degree of G,
$\Delta(G)$ maximum degree of G,
$c(G)$ circumference of G, i.e. the length of a longest cycle,
$p(G)$ length (i.e. number of edges) of a longest path of G,

If there are no ambiguities we frequently omit the explicit reference to the graph, by simply writing α, κ, ν, etc. For the meaning of these parameters we refer the reader to introductory graph theory literature, e.g. [9]. As usual $|G|$ denotes the order of G. Moreover, for a subset $S \subseteq V(G)$ we denote the induced subgraph of $V(G) \setminus S$ by $G - S$, and for $v \in V(G)$ we denote the number of neighbors of v in S by $d(v, S)$. If H is a fixed subgraph of G we briefly write $G - H$ and $d(v, H)$ instead of $G - V(H)$ and $d(v, V(H))$, respectively.

The **toughness** of $G \neq K_n$ is

$$\tau(G) := \min_S |S|/\omega(G - S)$$

where the minimum is taken over all separating vertex sets, i. e. sets $S \subseteq V(G)$ with $\omega(G - S) > 1$. A graph is called t-tough if $\tau \geq t$. Cycles and paths on k vertices are denoted by C_k (P_k, resp.).

Finally, for a graph G and a fixed subgraph H we say that H is **matching compatible**, if $\nu(G - H) \geq \nu(G) - \lfloor \frac{1}{2}|H| \rfloor$. Note that for graphs G with a perfect matching there are only even order subgraphs which are matching compatible. Matching compatible paths and cycles in bipartite graphs and digraphs were considered by Amar and Manoussakis in [4].

2. Main results

Theorem 2.1. *Let G be a triangle-free non-bipartite graph of order n and independence number α. If the minimum degree $\delta > n/3$ then G is weakly pancyclic with $c(G) = \min\{n, 2(n - \alpha)\}$ and girth 4 unless $G = C_5$.*

Theorem 2.2. *Let G be a triangle-free graph of order n and independence number α. If $\delta \geq n/3$ then $p(G) = \min\{n - 1, 2(n - \alpha)\}$.*

Consider the graphs $G_1(r)$ and $G_2(r)$ obtained from C_5 by replacing four vertices by a set of r independent vertices and the fifth vertex by $2r$ ($2r + 1$, resp.) independent vertices and by joining two sets by a complete bipartite graph whenever the original vertices in C_5 were adjacent (see figure 2.1).

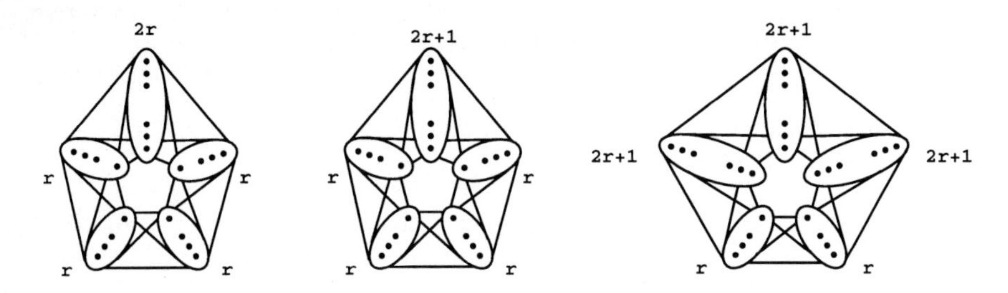

Fig. 2.1. The graphs $G_1(r)$, $G_2(r)$ and $G_3(r)$.

The graph $G_1(r)$ has minimum degree $\delta = n/3$ and independence number $\alpha = n/2$ but no hamiltonian cycle, thus Theorem 2.1 is best possible. Moreover the Petersen graph has $\delta = (n-1)/3$ and $\alpha = 4$ but it contains only cycles of lengths $5, 6, 8, 9$, so it is not weakly pancyclic. The graph $G_2(r)$ shows that Theorem 2.2 is best possible, since it has $\delta = (n-1)/3$ and $\alpha = (n+1)/2$ but no hamiltonian path.

As an easy consequence of Theorem 2.1 we obtain a degree bound for triangle-free non-bipartite graphs to be hamiltonian.

Theorem 2.3. *Let G be a triangle-free non-bipartite graph of order n. If $\delta \geq 3n/8$ then G is hamiltonian.*

Consider the graph $G_3(r)$ obtained from C_5 by replacing two consecutive vertices of C_5 by r independent vertices and the remaining three vertices by $2r+1$ vertices each and by joining sets in the same way as for $G_1(r)$ and $G_2(r)$ (see figure 2.1). This graph has $\delta = (3n-1)/8$ but is non-hamiltonian, thus Theorem 2.3 is best possible.

There is a related result for bipartite graphs involving smaller degree constraints, which is easily obtained by combining results from [2] and [4]. We will denote the graph consisting of two complete balanced bipartite graphs $K_{r,r}$ intersecting in one vertex by $H(r)$.

Theorem 2.4. *Let G be a bipartite graph of order $n \geq 3$. If $\delta > n/4$ then $c(G) = 2(n-\alpha)$ and G contains all even length cycles between 4 and $c(G)$ unless $G = C_6, H(r)$.*

Now, by combining Theorem 2.1 with Theorem 2.4 we immediately obtain the following Corollary.

Corollary 2.1. *Let G be a triangle-free graph of order $n \geq 3$. If $\delta > n/3$ then $c(G) = \min\{n, 2(n-\alpha)\}$ and G contains all even length cycles between 4 and $c(G)$ unless $G = C_5$.* □

As a consequence of this observation we will show that the circumference of such graphs can be computed in polynomial time.

Theorem 2.5. *The circumference of every triangle-free graph of order n with $\delta > n/3$ can be computed in time $\mathcal{O}(n^{2.5})$.*

This result is based on the matching algorithm of Micali and Vazirani [16]. H. J. Veldman [19] proved that computing the circumference in the class of bipartite graphs with $\delta > (\frac{1}{4} - \varepsilon)n$ is $\mathcal{NP}$-hard for every $\varepsilon > 0$ by a variation of Häggkvist's simple construction for general graphs [14]. It would be interesting to know whether there is a constant $c < 1/3$ such that the determination of the circumference in triangle-free graphs with $\delta > cn$ is still polynomial.

3. Matchings and independence number

The number of edges in a maximum matching $\nu(G)$ of any graph G of order n is bounded by

$$\nu(G) \leq \min\{\lfloor \tfrac{1}{2}n \rfloor, n - \alpha(G)\}. \tag{3.1}$$

Our first step will be to show that for triangle-free graphs with $\delta(G) \geq n/3$ equality holds in (3.1).

For the further results in this paragraph it is helpful to observe the following sequence of inequalities for a triangle-free graph G:

$$\alpha(G) \geq \Delta(G) \geq \delta(G) \geq \kappa(G).$$

The second and third inequality hold in arbitrary graphs. The following result ensures that for triangle-free graphs with $\delta \geq n/3$ the third inequality is an equality.

Lemma 3.1. *Let G be a triangle-free graph. Then $\kappa(G) \geq \min\{\delta(G), 4\delta(G) - n\}$. In particular, if $\delta(G) \geq n/3$ then $\kappa(G) = \delta(G)$.*

Proof. If $\kappa(G) = \delta(G)$ there is nothing to prove. So assume $\kappa(G) \leq \delta(G) - 1$ and let S be a disconnecting set of cardinality $\kappa(G)$ and G_1 and G_2 components of $G - S$. Let $v_i \in V(G_i)$, $i = 1, 2$. Since $\kappa(G) < \delta(G)$, v_i has a neighbor $w_i \in V(G_i)$. Since G is triangle-free v_i and w_i have no common neighbor. So from

$$4\delta(G) \leq \sum_{i=1,2} (d(v_i) + d(w_i)) \leq |G_1| + |G_2| + 2|S| \leq n + \kappa(G)$$

we derive $\kappa(G) \geq 4\delta(G) - n$. □

In the next result it will be shown that for triangle-free graphs with $\delta > n/3$ which are not 1-tough the complements of maximum independent sets determine the toughness.

Theorem 3.1. *Let G be a triangle-free graph with $\delta \geq n/3$ and toughness $\tau < \delta/(n - 2\delta + 1)$. Then $\tau = (n - \alpha)/\alpha$.*

Proof. Obviously $\tau \leq (n - \alpha)/\alpha$ so it remains to prove "$\geq$". Suppose S is a separating set with $\tau = |S|/\omega(G - S) < (n - \alpha)/\alpha$. Then $G - S$ is not the empty graph, so there is an edge vw in $G - S$. By Lemma 3.1 we have

$$|S| \geq \kappa(G) = \delta(G),$$

and since all neighbors of v and w are distinct, and lie in the same component of $G - S$ or in S we get

$$\omega(G - S) \leq n - 2\delta + 1$$

thus $\tau = |S|/\omega(G - S) \geq \delta/(n - 2\delta + 1)$, a contradiction. □

Note that $G_1(r)$ is a graph with $\delta = n/3$ and $\alpha = n/2$ which is not 1-tough. So its toughness is not determined by the complements of maximum independent sets. While 1-toughness is a necessary condition for being hamiltonian it is a sufficient condition for even order graphs to contain a perfect matching. This follows from a famous result of Tutte [18] saying that a graph contains a perfect matching if and only if for every subset S of the vertex set $o(G-S) \leq |S|$ holds, where $o(G-S)$ denotes the number of odd order components of $G-S$. We will need the defect version of this result, which was first discovered by Berge [6].

Theorem 3.2. *The maximum number of independent edges of a graph G is*

$$\nu(G) = \min_{S \subseteq V(G)} \frac{1}{2}(n - o(G-S) + |S|).$$

Now the main result of this paragraph is an easy consequence of the two previous results:

Theorem 3.3. *If G is triangle-free with $\delta \geq n/3$ then*

$$\nu(G) = \min\{\lfloor \tfrac{1}{2}n \rfloor, n - \alpha\}.$$

Proof. By (3.1) it suffices to show $\nu(G) \geq \min\{\lfloor \frac{1}{2}n \rfloor, n-\alpha\}$. If the toughness $\tau \geq \delta/(n-2\delta+1)$ then for every S we have

$$\omega(G-S) \leq |S|(\frac{1}{3}n+1)/\frac{1}{3}n = |S| + 3|S|/n < |S| + 2$$

since $|S| < 2n/3$ (otherwise $\omega(G-S) \leq n/3 < |S|$). So $o(G-S) \leq \omega(G-S) \leq |S|+1$ and using Theorem 3.2 we get $\nu(G) \geq (n-1)/2$, thus $\nu(G) = \lfloor \frac{1}{2}n \rfloor$.

Otherwise using Theorem 3.1 we have for every S

$$o(G-S) - |S| \leq \omega(G-S) - |S| \leq \alpha - (n-\alpha) = 2\alpha - n$$

implying $\nu(G) \geq n - \alpha$ again using Theorem 3.2. □

Paths

We are now going to prove Theorem 2.2.

Proof (Theorem 2.2). Clearly a longest path has at most $s = 1 + \min\{n-1, 2(n-\alpha)\}$ vertices. Moreover, by Theorem 3.3, $\nu(G) = \min\{\lfloor \frac{1}{2}n \rfloor, (n-\alpha)\}$. First observe that G contains a path which is matching compatible, having odd order if $\nu(G) < n/2$. Let $P = P_\ell$ be a longest such path and suppose $\ell < s$. Let v and w be the endvertices of P. Note that the subgraph induced by P cannot be hamiltonian, since by Lemma 3.1 the graph G is connected, yielding a longer matching compatible path. Thus, by Ore's Lemma [17], $\delta \leq \min\{d(v,P), d(w,P)\} \leq (\ell-1)/2$, implying $\ell \geq \frac{2}{3}n+1$. Since $\ell < s$ we have $\nu(G-P) \geq 1$ so there is an edge xy of a maximum matching in $G-P$. Again we calculate $d(x, G-P) + d(y, G-P) \leq n - \ell$ which implies

$$d(x,P) + d(y,P) \geq 2\delta - (n-\ell) \geq \ell - n/3 > (\ell-1)/2$$

since $\ell > \frac{2}{3}n - 1$. By the maximality of ℓ neither v nor w can have a neighbor in $\{x,y\}$. So there must be two consecutive vertices in $P - v - w$ both having a neighbor in $\{x,y\}$. Since G is triangle-free this contradicts the maximality of ℓ. Thus $\ell = s$ so, indeed, $p(G) = s - 1$. □

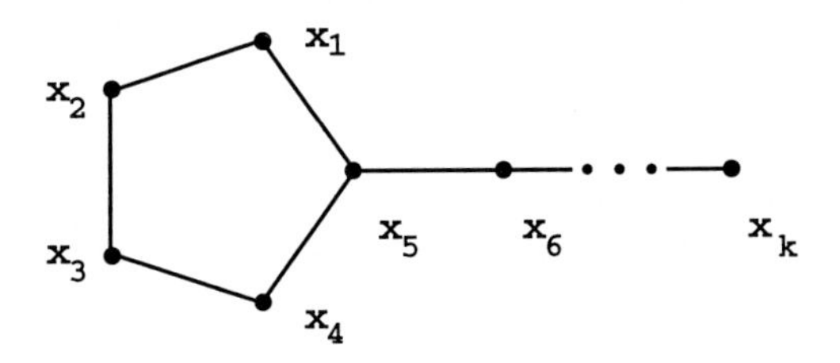

Fig. 4.1. The 5-lasso.

4. Lassos

An **ℓ-lasso** $L_{k,\ell}$, $k \geq \ell \geq 3$, consists of a path $x_1x_2\ldots x_{k-1}x_k$ with the additional edge x_1x_ℓ and we refer to k as the **length** of $L_{k,\ell}$. A lasso is **even** or **odd** according to ℓ being even or odd. In our proofs 5-lassos $L_{k,5}$ are very important, and we will always assume that the vertices of a 5-lasso are labeled as in figure 4.1. Note that in graph theoretical literature the lassos are also called cups or lollipops.

Lassos and cycles

The only aim of this part is to discover a surprising property of triangle-free graphs. The existence of a subgraph $L_{k,5}$ in a triangle-free graph with $\delta \geq n/3$ implies the existence of cycles of all lengths between 5 and $k-1$. Here $n/3$ is best possible since the Petersen graph has $\delta = (n-1)/3$ and contains $L_{10,5}$ but does not contain C_7. For the proof it is helpful to define 5-lassos with the additional edge x_3x_6 which we will denote by $L^+_{k,5}$ (see figure 4.2).

Theorem 4.1. *Let G be a triangle-free graph containing a lasso $L = L_{k,5}$ for $k \geq 5$. If $\delta \geq n/3$ then G contains C_ℓ for all ℓ, $5 \leq \ell \leq k-1$. Moreover, if L is a longest 5-lasso then G contains $C = C_{k-2}$ such that $G - C$ contains an edge.*

Proof. We will show that the following two claims are true whenever G contains $L_{k,5}$:

(1) G contains C_{k-1},
(2) G contains $L_{k+1,5}$ or G contains a C_{k-2} where $G - C_{k-2}$ contains an edge.

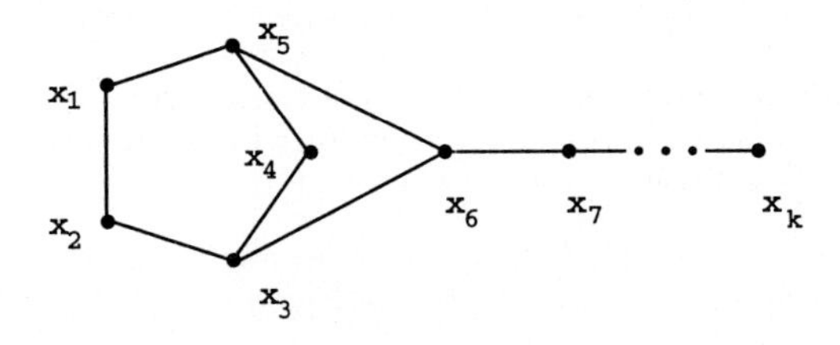

Fig. 4.2. The lasso $L^+_{k,5}$.

Since with $L_{k,5}$ all shorter 5-lassos are contained in G the claims (1) and (2) prove our statement.

For a lasso $L = L_{k,5}$ define $X_k = \{i|x_k x_{i-1} \in E(G)\}$, and $X_j = \{i|x_j x_i \in E(G)\}$ for $1 \leq j < k$.

First suppose that G contains a lasso $L = L_{k,5}$ but no subgraph isomorphic to $L^+_{k,5}$. So x_k has at most one neighbor in $\{x_1, x_2, \ldots, x_5\}$ and by symmetry we may assume $x_k x_1 \notin E(G)$. Moreover, x_6 is adjacent to neither x_2 nor x_3. If a vertex $v \in V(G - L)$ is adjacent to two vertices of x_2, x_3 and x_k then it is easily seen that G contains a longer 5-lasso and a C_{k-1} because G is triangle-free. So $d(x_2, G - L) + d(x_3, G - L) + d(x_k, G - L) \leq n - k$.

Consider $X = X_2 \cup X_3 \cup X_k$. By the above reasoning on the neighbors of x_6 and x_k and by the observation that x_5 is adjacent to neither x_2 nor x_3 we get that either $5 \notin X$ or $6 \notin X$ thus $|X| \leq k - 1$. On the other hand we have

$$d(x_2, L) + d(x_3, L) + d(x_k, L) \geq 3\delta - (n - k) \geq k > |X|.$$

Thus there is an index i such that $i \in X_k$ and $i \in X_j$ for $j \in \{2, 3\}$ since G is triangle-free. So, again, it is easily verified that G contains C_{k-1} and a C_{k-2} with an edge in $G - C_{k-2}$.

Now assume $L^+ = L^+_{k,5} \subseteq G$. For $k = 6$ the lasso $L^+_{6,5}$ contains C_5 and a C_4 with an edge in $G - C_4$ so assume $k \geq 7$. Consider this time x_1, x_2 and x_k. By the same reasoning as above we are done if there is a vertex in $G - L^+$ adjacent to two of them. Since G is triangle-free $X_1 \cap X_2 = \emptyset$ and $X_1 \cup X_2 \subseteq \{1, 2, 3, 5, 7, 8, \ldots, k\}$. Moreover, we are done if x_k is adjacent to one vertex of x_1, x_2, x_3 or x_5, so we may assume $X_k \subseteq \{5, 7, 8, \ldots, k\}$. Thus we have $X = X_1 \cup X_2 \cup X_k \subseteq \{1, 2, 3, 5, 7, 8, \ldots, k\}$, in particular $|X| \leq k - 2$. Again we calculate

$$d(x_1, L^+) + d(x_2, L^+) + d(x_k, L^+) \geq 3\delta - (n - k) \geq |X| + 2$$

so X_k and $X_1 \cup X_2$ must intersect in an index $i \neq 5$, yielding a C_{k-1} and a C_{k-2} with an edge in $G - C_{k-2}$. □

Lassos and matchings

The previous observation suggests looking for long 5-lassos in triangle-free graphs in order to obtain many cycles at once. We will now examine how to determine the length of a longest 5-lasso in such a graph. The starting point will be a powerful result relating the length of a longest path to the length of a longest cycle.

Theorem 4.2 (Enomoto, van den Heuvel, Kaneko, Saito [13]). *Let G be a graph of order n. If every set of three independent vertices u, v, w satisfies $d(u) + d(v) + d(w) \geq n$ then $c(G) \geq p(G)$ or G is a spanning subgraph of one of six exceptional families of graphs each of which has connectivity $\kappa < n/3$.*

For our purposes it suffices to replace the degree sum condition by $\delta \geq n/3$. Note that the condition $c(G) \geq p(G)$ for connected graphs implies equality unless G is hamiltonian where $c(G) = p(G) + 1$.

Lemma 4.1. *If G is a triangle-free non-bipartite graph of order $n \geq 6$ with $\delta \geq n/3$ then G contains a lasso $L = L_{k,5}$ which is matching compatible and where k is odd if n is odd and $\alpha < n/2$.*

Proof. First assume $\alpha \geq n/2$. Take an independent set A, $|A| = \alpha$, and a maximum matching M, $|M| = n - \alpha$, which exists by Theorem 3.3. Note that M saturates every vertex of $G - A$. Since G is not bipartite $G - A$ contains an edge vw. Moreover, because $2\delta > n - \alpha$ the matching neighbors v' and w' of v and w, respectively, must have a common neighbor $u \in G - A$. Now the graph spanned by the set $\{u, u', v, v', w, w'\}$ where u' is the matching neighbor of u contains a lasso $L = L_{6,5}$ and $G - L$ contains a matching on $\nu - 3$ edges.

For $\alpha < n/2$ we will first show that G contains an odd lasso $L_{k,\ell}$ where $G - L_{k,\ell}$ has a perfect matching. From Theorem 2.2 we obtain that $p(G) \geq n - 1$. Since G satisfies the hypothesis of Theorem 4.2 and using Lemma 3.1 we infer that G cannot be an exceptional graph, so $c(G) \geq n - 1$. If $c(G) = n$ then every hamiltonian cycle has an even length chord since G is non–bipartite. This yields a spanning odd lasso $L_{n,\ell}$. If $c(G) = n - 1$ then we have $L_{n,n-1}$ and we are done if $n - 1$ is odd. Otherwise recall the standard labeling of the lasso vertices. If x_n is adjacent to an odd labeled vertex x_ℓ then we obtain $L_{n,\ell+2}$. If x_n is adjacent only to even labeled vertices then since $\alpha < n/2$ two even labeled vertices must be adjacent. So we get an even length chord splitting the cycle into an odd cycle where x_n has a neighbor and an odd length path. So we get a perfect matching from the path and an odd lasso.

Take a smallest order odd lasso $L = L_{k,\ell}$ where $G - L$ has a perfect matching. Such a lasso exists by the above reasoning. Clearly we may assume $k \leq \ell + 1$ since otherwise we add the edge $x_k x_{k-1}$ to the matching. Note that the cycle of the lasso is an induced cycle in G since every chord yields a shorter odd lasso and a perfect matching. If $\ell = 5$ we are done so suppose $\ell \geq 7$. Now neither x_3 nor x_4 can be adjacent to $x_{\ell+1}$ so for $X = \{x_3, x_4, x_\ell\}$ we calculate

$$\sum_{x \in X} d(x, G - L) \geq 3\delta - 6 - (k - \ell) > n - k.$$

So one of the $(n - k)/2$ matching edges in $G - L$ has at least 3 neighbors in X. It is now easily checked that in all cases we obtain a shorter odd lasso L' where $G - L'$ has a perfect matching. □

It should be mentioned that there is a direct way of proving Lemma 4.1 without using the powerful Theorem 4.2, at the expense of some tedious case analysis. However, this might be of interest for giving an algorithm which actually finds cycles of given length in triangle-free graphs with $\delta > n/3$.

Proposition 4.1. *Every triangle-free non-bipartite graph G with $\delta > n/3$ contains a lasso $L_{s,5}$ where $s = \min\{n, 2(n - \alpha)\}$.*

Proof. Clearly every 5-lasso in G can have at most s vertices since at most half of its vertices form an independent set. Let r be the length of a longest lasso $L = L_{r,5}$ contained in G which is matching compatible and which has odd order if $\alpha < n/2$ and n is odd. By Lemma 4.1 such a lasso exists and it is easily observed that L has even order if $\alpha \geq n/2$ and order of the same parity as the order of the graph otherwise. Consider the endvertex x_r of the lasso. If x_r has a neighbor in $G - L$ then the maximality requirement for r is obviously violated. So from $(n + 1)/3 \leq d(x_r, L) \leq (r - 1)/2$ we obtain $r \geq (2n + 5)/3$.

Now consider a matching edge vw in $G - L$. First observe that v and w have together at most 4 neighbors in $\{x_1, \ldots, x_6\}$, since if one vertex, say v, is adjacent to x_6 the other vertex w cannot be adjacent to a vertex with a smaller index without enlarging the length of the lasso by two. Since $d(v, G - L) + d(w, G - L) \leq n - r$ we obtain

$$d(v, L) + d(w, L) \geq (2n + 2)/3 - n + r = r - (n - 2)/3 \geq (r + 3)/2.$$

Thus

$$d(v, \{x_7, \ldots, x_r\}) + d(w, \{x_7, \ldots, x_r\}) \geq (r-5)/2.$$

Since G is triangle-free, and v and w are both non-adjacent to x_r they must be adjacent to two consecutive vertices on the path $x_6x_7\ldots x_{r-1}$, contradicting the maximality of the lasso. So we conclude that the lasso must have at least $2\nu(G)$ vertices and one vertex more if $\alpha < n/2$ and n is odd. Now Theorem 3.3 completes the proof. □

5. Proofs of the main results

Non-bipartite graphs

In order to carry out the proofs of Theorems 2.1 and 2.3 we need a better bound for the independence number than the obvious bound $\alpha \leq n - \delta$ holding for arbitrary graphs.

Lemma 5.1. *For any triangle-free non-bipartite graph the inequality* $\alpha \leq \min\{2(n-2\delta), n-\delta-1\}$ *holds.*

Proof. Let A be an independent set on α vertices. Since G is non-bipartite $G - A$ contains an edge xy. Since no vertex from A is adjacent to both x and y we get $\alpha+\delta < n$ implying $\alpha \leq n-\delta-1$. Now assume $d(x, G-A) \geq d(y, G-A)$. Since every neighbor of x in A has all its neighbors in $G - A$ we have $d(x, G-A) \leq (n-\alpha)-\delta$. Furthermore x and y have no common neighbor in A. Thus

$$2\delta \leq d(x) + d(y) \leq 2(n-\alpha-\delta) + |A|,$$

implying $\alpha \leq 2(n-2\delta)$. □

We are now prepared to prove the main result.

Proof (Theorem 2.1). First note that there is no graph of order $n < 8$ satisfying the hypothesis. By Proposition 4.1 the graph G contains a lasso $L = L_{r,5}$ where $r = \min\{n, 2(n-\alpha)\}$. In particular G contains C_5. A simple counting argument proves that there must be a vertex in $G - C_5$ having two neighbors in C_5, hence G contains C_4.

Now consider L. Theorem 4.1 implies that G contains all cycles of lengths $5 \leq \ell \leq r-1$ and a cycle $C = C_{r-2}$ where $G-C$ contains an edge vw. Using Lemma 5.1 we get $\alpha \leq (2n-4)/3$, thus $r \geq (2n+8)/3$. Since $d(v, G-C)+d(w, G-C) \leq n-r+2$ we get

$$d(v, C) + d(w, C) \geq 2\delta - (n-r+2) > |C|/2.$$

Thus there must be two consecutive vertices on the cycle each having a neighbor in vw. Since G is triangle-free this provides the missing C_r. □

Proof (Theorem 2.3). This is immediate from Theorem 2.1 since by Lemma 5.1 we obtain $\alpha(G) \leq n/2$. □

Bipartite graphs

The following result of Amar and Manoussakis [4, Corollary 5] settles the circumference part of Theorem 2.4.

Theorem 5.1. *Let G be bipartite with bipartition $A \cup B$, $|A| \leq |B|$. If $\delta > |A|/2$ then $c(G) = 2|A|$ or $G = H(r)$.*

Now the following local condition of Amar [2] provides all the even length cycles but the 4-cycle (actually Amar's result gives more information than we mention here).

Theorem 5.2. *Suppose C is a hamiltonian cycle in a bipartite graph G and x and x^{++} are vertices at distance 2 in C such that the neighborhoods of x and x^{++} are not identical. If $d(x) + d(x^{++}) \geq (n+1)/2$ then G contains all even length cycles between 6 and n.*

Proof (Theorem 2.4). For a bipartition $A \cup B$, $|A| \leq |B|$, fix a cycle C of length $2|A|$ which is guaranteed by Theorem 5.1 and consider the subgraph of G induced by C. This graph contains every vertex of A. If $|A| = 2$ the conclusion is certainly true so assume $|A| \geq 3$. Thus $|B \cap C| \geq 3$ and it is easily calculated that among any three vertices of $B \cap C$ there is a pair whose neighborhoods intersect in at least two vertices (unless $G = C_6$), so G contains C_4. If the neighborhoods of all vertices in $B \cap C$ are identical then the subgraph induced by C is complete balanced bipartite otherwise there are two vertices in $B \cap C$ at distance 2 in C satisfying the hypothesis of Theorem 5.2. So in any case G contains cycles from 4 through $2|A|$. Since B is an independent set $2|A| \geq 2(n - \alpha)$. □

Computing the circumference

Proof (Theorem 2.5). Determine a maximum matching M in $\mathcal{O}(n^{2.5})$ time using the algorithm of Micali and Vazirani [16]. By Corollary 2.1 and Theorem 3.3 we have $c(G) = 2\nu(G)$ unless $\nu(G) = (n-1)/2$ and $\alpha(G) \neq (n+1)/2$, in which case $c(G) = n$. So if $\nu = (n-1)/2$ select a set S' of vertices as follows: take the vertex outside the matching and the matching neighbors of its neighbors. Define S as the set of vertices having no neighbor in S'. Since $|S'| \geq \delta + 1$ and $|N(v) \cup N(w)| \geq 2\delta$ for every edge vw, at most one vertex of every matching edge is in S. It is easily checked that $\alpha = (n+1)/2$ iff S is an independent set of cardinality $(n+1)/2$. Finally observe that computing S and checking its independence can be done in a straightforward way in $\mathcal{O}(n^2)$ time, once the matching is given. □

Acknowledgment: Part of this research was performed while the author was visiting the Charles University in Prague. The author would like to thank Jarík Nešetřil and his group for their kind hospitality.

References

1. N. Alon, Tough Ramsey graphs without short cycles, Manuscript (1993).
2. D. Amar, A condition for a hamiltonian graph to be bipancyclic, *Discr. Math.* **102** (1991), 221–227.

3. D. Amar, I. Fournier and A. Germa, Pancyclism in Chvátal–Erdős graphs, *Graphs and Combinatorics* **7** (1991), 101–112.
4. D. Amar and Y. Manoussakis, Cycles and paths of many lengths in bipartite digraphs, *J. Combin. Theory Ser. B* **50** (1990), 254–264.
5. D. Bauer, J. van den Heuvel and E. Schmeichel, Toughness and triangle-free graphs, Preprint (1993).
6. C. Berge, Sur le couplage maximum d'un graphe, *C. R. Acad. Sci. Paris (A)* **247** (1958), 258–259.
7. S. Brandt, Sufficient conditions for graphs to contain all subgraphs of a given type, Ph.D. thesis, Freie Universität Berlin, 1994.
8. S. Brandt, R. J. Faudree and W. Goddard, Weakly pancyclic graphs, in preparation.
9. G. Chartrand and L. Lesniak, Graphs & Digraphs (2nd edition), Wadsworth & Brooks/Cole, Pacific Grove, 1986.
10. V. Chvátal, Tough graphs and hamiltonian cycles, *Discr. Math.* **5** (1973), 215–228.
11. V. Chvátal and P. Erdős, A note on hamiltonian circuits, *Discr. Math.* **2** (1972), 111–113.
12. G. A. Dirac, Some theorems on abstract graphs, *Proc. London Math. Soc. (3)* **2** (1952), 69–81.
13. H. Enomoto, J. van den Heuvel, A. Kaneko, and A. Saito, Relative length of long paths and cycles in graphs with large degree sums, Preprint (1993).
14. R. Häggkvist, On the structure of non-hamiltonian graphs, *Combin. Prob. and Comp.* **1** (1992), 27–34.
15. Dingjun Lou, The Chvátal–Erdős condition for cycles in triangle-free graphs, to appear.
16. S. Micali and V. V. Vazirani, An $\mathcal{O}(V^{1/2}E)$ algorithm for finding maximum matching in general graphs, *Proc. 21st Ann. Symp. on Foundations of Computer Sc.* IEEE, New York (1980), 17–27.
17. O. Ore, Note on Hamiltonian circuits, Amer. Math. Monthly **67** (1960), 55.
18. W. T. Tutte, The factorization of linear graphs, *J. London Math. Soc.* **22** (1947), 107–111.
19. H. J. Veldman, Personal communication, 1994.

Reconstruction Problems for Digraphs

M. Aigner[1] and E. Triesch[2]

[1] Math. Institut, Freie Universität, Berlin, Arnimallee 3, D–14195 Berlin
[2] Forschungsinstitut für Diskrete Mathematik, Nassestraße 2, D–53113 Bonn

Summary. Associate to a finite directed graph $\mathbf{G}(V, E)$ its out-degree resp. in-degree sequences d^+, d^- and the corresponding neighborhood lists N^+, N^- (when $\mathbf{G}$ is a labeled graph). We discuss various problems when sequences resp. lists of sets can be realized as degree sequences resp. neighborhood lists of a directed graph.

1. Introduction

Consider a finite graph $G(V, E)$. Let us associate with G a finite list $P(G)$ of parameters, e.g. the degrees, the list of cliques, the chromatic polynomial, or whatever we like. For any set P of invariants there arise two natural problems:

(R) **Realizability**. Given P, when is $P = P(G)$ for some graph G? We then call P *graphic*, and say that G *realizes* P.

(U) **Uniqueness**. Suppose $P(G) = P(H)$. When does this imply $G \cong H$? In other words, when is P a complete set of invariants?

The best studied questions in this context are probably the reconstruction conjecture for (U), and the degree realization problem for (R). This latter problem was solved in a famous theorem of Erdős-Gallai [4] characterizing graphic sequences. Their theorem reads as follows: Let $d_1 \geq \ldots \geq d_n \geq 0$ be a sequence of integers. Then $(d_1 \geq \ldots \geq d_n)$ can be realized as the degree sequence of a graph if and only if

$$\sum_{i=1}^{k} d_i \leq k(k-1) + \sum_{j=k+1}^{n} \min(d_j, k) \quad (k = 1, \ldots, n).$$

A variant of this problem concerning neighborhoods was first raised by Sós [13] and studied by Aigner–Triesch [2]. Consider a finite labeled graph $G(V, E)$ and denote by $N(u)$ the neighborhood of $u \in V$. $\mathcal{N}(G) = \{N(u) : u \in V\}$ is called the *neighborhood list* of G. Given a list (multiset) $\mathcal{N} = (N_1, \ldots, N_n)$ of sets. When is $\mathcal{N} = \mathcal{N}(G)$ for some graph G? In contrast to the polynomial verification of (1), it was shown in [2] that the neighborhood list problem NL is NP-complete for arbitrary graphs. For bipartite graphs, NL turns out to be polynomially equivalent to the GRAPH ISOMORPHISM problem. A general survey of these questions appears in [3].

In the present paper we consider directed graphs $\mathbf{G}(V, E)$ on n vertices with or without loops and discuss the corresponding realizability problems for the degree resp. neighborhood sequences. We assume throughout that there is at most one directed edge (u, v) for any $u, v \in V$. To every $u \in V$ we associate its *out-neighborhood* $N^+(u) = \{v \in V : (u, v) \in E\}$ and its *in-neighborhood* $N^-(u) = \{v \in V : (v, u) \in E\}$ with $d^+(u) = |N^+(u)|$ and $d^-(u) = |N^-(u)|$ being the *out-degree* resp. *in-degree* of u.

For both the degree realization problem and the neighborhood problem we have three versions in the directed case:

($\mathcal{D}^+$) Given a sequence $d^+ = (d_1^+, \ldots, d_n^+)$ of non-negative integers. When is d^+ realizable as the out-degree sequence of a directed graph?

Obviously, ($\mathcal{D}^-$) ist the same problem.

($\mathcal{D}_-^+$) Given a sequence of pairs $d_-^+ = ((d_1^+, d_1^-), \ldots, (d_n^+, d_n^-))$. When is there a graph $\mathbf{G}$ with $d^+(u_i) = d_i^+$, $d^-(u_i) = d_i^-$ for all i?

($\mathcal{D}^+, \mathcal{D}^-$) Given two sequences $d^+ = (d_1^+, \ldots, d_n^+)$, $d^- = (d_1^-, \ldots, d_n^-)$. When is there a directed graph such that d^+ is the out-degree sequence (in some order) and d^- the in-degree sequence?

In an analogous way, we may consider the realization problems ($\mathcal{N}^+$), ($\mathcal{N}_-^+$), ($\mathcal{N}^+, \mathcal{N}^-$) for neighborhood lists.

In section 2 we consider the degree problems and in section 3 the neighborhood problems. Section 4 is devoted to simple directed graphs when there is at most one edge between any two vertices and no loops.

2. Degree Sequences

Depending on whether we allow loops or not there are six different reconstruction problems whose solutions are summarized in the following diagram:

degrees	($\mathcal{D}^+$)	($\mathcal{D}_-^+$)	($\mathcal{D}^+, \mathcal{D}^-$)
with loops	trivial	Gale-Ryser	Gale-Ryser
without loops	trivial	Fulkerson	Fulkerson

The problems ($\mathcal{D}^+$) have the following trivial solutions: (d^+) is realizable with loops if and only if $d_i^+ \leq n$ for all i, and without loops if and only if $d_i^+ \leq n-1$ for all i.

Consider ($\mathcal{D}_-^+$) with loops. We represent $\mathbf{G}(V, E)$ as usual by its adjacency matrix M where the rows and columns are indexed by the vertices $u_1, \ldots, u_n$ with $m_{ij} = 1$ if $(u_i, u_j) \in E$ and 0 otherwise. To realize a given sequence (d_-^+) is therefore equivalent to constructing a $0,1$–matrix with given row-sums d_i^+ and column-sums d_i^- which is precisely the content of the Gale-Ryser Theorem [7, 11]. In fact, the Gale-Ryser theorem applies to the situation ($\mathcal{D}^+, \mathcal{D}^-$) as well by permuting the columns. If we do not allow loops, then the realization problem ($\mathcal{D}_-^+$) is settled by an analogous theorem of Fulkerson [5, 6]. He reduces the problem of constructing a $0,1$-matrix with zero trace and given row and column sums to a network flow problem, an approach which can also be used in the case of the Gale-Ryser theorem thus showing that both problems are polynomially decidable. Finally, we remark that the case ($\mathcal{D}^+, \mathcal{D}^-$) can be reduced to the case ($\mathcal{D}_-^+$) in view of the following proposition.

Proposition 2.1. *Suppose two sequences $d^+ = (d_1^+, \ldots, d_n^+)$, $d^- = (d_1^-, \ldots, d_n^-)$ are given. Denote by $\overline{d}^+$ (resp. $\overline{d}^-$) a non-increasing (resp. non-decreasing) rearrangement of d^+ (resp. d^-). If there exists a $0,1$-matrix $M = (m_{ij})$ with $\sum_{j=1}^n m_{jj} = 0$, $\sum_{j=1}^n m_{ij} = d_i^+$, $\sum_{j=1}^n m_{ji} = d_i^-$, $1 \leq i \leq n$, then there exists a $0,1$-matrix $\overline{M} = (\overline{m}_{ij})$ satisfying $\sum_j \overline{m}_{jj} = 0$, $\sum_j \overline{m}_{ij} = \overline{d}_i^+$, $\sum_j \overline{m}_{ji} = \overline{d}_i^-$, $1 \leq i \leq n$.*

Proof. Suppose M is given as above. By permuting the rows and columns of M by the same permutation (which does not change the trace) we may assume that $d^+ = \overline{d}^+$.

Now suppose that for some indices $i < j$, $d_i^- > d_j^-$. We will show that M can be transformed into a matrix $\hat{M}$ with zero trace, row sum vector d^+ and column sum vector $\hat{d}^-$, where $\hat{d}^-$ arises from d^- by exchanging d_i^- and d_j^-. Since each permutation is generated by transpositions, this will obviously complete the proof. To keep notation simple, we give the argument only for the case $i = n-1$, $j = n$ but it is immediately clear how the general case works. Suppose

$$M = \left(\begin{array}{c|c|c} & a & b \\ \hline & 0 & y \\ \hline & x & 0 \end{array}\right)$$

(i) If $x \leq y$, then we exchange (a_i, b_i) when $a_i = 1$ and $b_i = 0$ for $d_{n-1}^- - d_n^-$ indices $i \leq n-2$.

(ii) If $x = 1$, $y = 0$, then, since $d_{n-1}^+ \geq d_n^+$ there exists some $\ell < n-1$ such that

$$\begin{pmatrix} m_{n-1,\ell} \\ m_{n,\ell} \end{pmatrix} = \begin{pmatrix} 1 \\ 0 \end{pmatrix}.$$

Now let

$$\hat{M} = \left(\begin{array}{c|c|c|c|c} & & & b & a \\ \hline & 0 & & 0 & 1 \\ \hline & 1 & & 0 & 0 \end{array}\right)$$

$$\ell$$

3. Neighborhood Lists

As in the undirected case the neighborhood problems are more involved. Again, we summarize our findings in a diagram and discuss then the proofs.

neighbors	$(\mathcal{N}^+)$	$(\mathcal{N}_-^+)$	$(\mathcal{N}^+, \mathcal{N}^-)$
with loops	trivial	$\geq$ GRAPH ISOM.	$\approx$ GRAPH ISOM.
without loops	bipartite matching	NP-complete	NP-complete

Let us consider $(\mathcal{N}^+)$ first. Allowing loops, any list (N_i^+) can be realized. In the absence of loops, (N_i^+) can be realized if and only if $(N_1^{+^c}, \ldots, N_n^{+^c})$ has a transversal, where N^c is the complement of N. So, this problem is equivalent to the bipartite matching problem and, in particular, polynomially decidable.

Let us treat next the problem $(\mathcal{N}_-^+)$ without loops. The special case $N_i^+ = N_i^-$ for all i clearly reduces to the (undirected) neighborhood list problem NL which as mentioned is NP-complete. Accordingly, $(\mathcal{N}_-^+)$ is NP-complete as well.

We show next that the decision problem $(\mathcal{N}_-^+)$ with loops is polynomially equivalent to the matrix symmetry problem MS defined as follows:

The input is an $n \times n$–matrix A with $0,1$-entries, with the question: Does there exist a permutation matrix P such that $(PA)^T = PA$ holds?

Let us represent $\mathbf{G}(V,E)$ again by its adjacency matrix. Then, clearly, MS is the special case of $(\mathcal{N}_-^+)$ where $N_i^+ = N_i^-$ for all i. To see the converse denote by x^i (resp. y^i) the incidence vectors of N_i^+ (resp. N_i^-) as row vectors, and set $X = \begin{pmatrix} x^1 \\ \vdots \\ x^n \end{pmatrix}$, $Y = \begin{pmatrix} y^1 \\ \vdots \\ y^n \end{pmatrix}$. The problem $(\mathcal{N}_-^+)$ is thus equivalent to the following decision problem: Does there exist a permutation matrix P such that

$$PX = (PY)^T?$$

Note that $PX = (PY)^T \iff (PX)^T = PY$. Now consider the $4n \times 4n$–matrix

$$\Gamma = \begin{array}{|c|c|c|c|} \hline O & I & X & U \\ \hline I & O & I & W \\ \hline Y & I & O & Z \\ \hline U^T & W^T & Z^T & O \\ \hline \end{array}$$

where O, I are the zero-matrix and identity matrix, respectively, and U, W, Z are matrices with identical rows each which ensure that a permutation matrix R satisfying $(R\Gamma)^T = R\Gamma$ must be of the form

$$R = \begin{array}{|c|c|c|c|} \hline R_1 & O & O & O \\ \hline O & R_2 & O & O \\ \hline O & O & R_3 & O \\ \hline O & O & O & R_4 \\ \hline \end{array}$$

Clearly, such matrices U, W, Z exist. Now $(R\Gamma)^T = R\Gamma$ if and only if

$$(R_1X)^T = R_3Y, \quad R_2^T = R_1, \quad R_3 = R_2^T = R_1,$$

i.e. if and only if $(R_1X)^T = R_1Y$, and the result follows.

It was mentioned in [2] that MS is at least as hard as GRAPH ISOMORPHISM, but we do not know whether they are polynomially equivalent.

Let us, finally, turn to the problems $(\mathcal{N}^+, \mathcal{N}^-)$. We treat $(\mathcal{N}^+, \mathcal{N}^-)$ with loops, the other version is settled by a matrix argument as above. Using again the notation x^i, y^i, X, Y, our problem is equivalent to the following matrix problem: Do there exist permutation matrices P and Q such that

$$PX = Y^TQ, \quad \text{i.e.} \quad PXQ^T = Y^T?$$

This latter problem is obviously polynomially equivalent to HYPERGRAPH ISOMORPHISM which is known to be equivalent to GRAPH ISOMORPHISM (see [2]).

4. Simple Directed Graphs and Tournaments

Let us now consider simple directed graphs and, in particular, tournaments. In contrast to the non-simple case, where $(\mathcal{D}^+)$ and $(\mathcal{N}^+)$ are trivial resp. polynomially solvable (bipartite matching), the problems now become more involved.

As for $(\mathcal{D}^+)$, a necessary and sufficient condition for (d_i^+) to be realizable as out-degree sequence of a *tournament* was given by Landau [9]. Assume $d_1^+ \geq \ldots \geq d_n^+$, then (d_i^+) is realizable if and only if

$$\sum_{i=1}^{k} d_i^+ \leq (n-1) + \ldots + (n-k) \quad (k = 1, \ldots, n)$$

with equality for $k = n$.

Deleting the last condition yields the corresponding result for arbitrary simple digraphs.

Theorem 4.1. *A sequence $(d_1^+ \geq \ldots \geq d_n^+)$ is realizable as out-degree sequence of a simple directed graph if and only if*

$$\sum_{i=1}^{k} d_i^+ \leq \sum_{i=1}^{k} (n-i) \quad (k = 1, \ldots, n). \tag{4.1}$$

Proof. The condition is obviously necessary. For the converse we make use of the dominance order of sequences $p = (p_1 \geq \ldots \geq p_n)$, $q = (q_1 \geq \ldots \geq q_n)$:

$$p \leq q \Longleftrightarrow \sum_{i=1}^{k} p_i \leq \sum_{i=1}^{k} q_i \quad (k = 1, \ldots, n). \tag{4.2}$$

Suppose $m = \sum_{i=1}^{n} d_i^+$, and denote by $L(m)$ the lattice of all sequences $p = (p_1 \geq \ldots \geq p_n)$, $\sum_{i=1}^{n} p_i = m$, ordered by (2), see [1] for a survey on the uses of $L(m)$. Let $S(m) \subseteq L(m)$ be the set of sequences which are realizable as out-degree sequences of a simple digraph with m edges.

Claim 1. $S(m)$ is a down-set, i.e. $p \in S(m)$, $q \leq p \Longrightarrow q \in S(m)$.

It is well-known that the order $\leq$ in $L(m)$ is transitively generated by successive "pushing down boxes", i.e. it suffices to prove the claim for $q \leq p$ with $q_r = p_r - 1$, $q_s = p_s + 1$ for some $p_r \geq p_s + 2$, and $q_i = p_i$ for $i \neq r, s$. Now suppose $\mathbf{G}$ realizes the sequence p, with $d^+(u_i) = p_i$. Since $d^+(u_r) \geq d^+(u_s) + 2$, there must be a vertex u_t with $(u_r, u_t) \in E$, $(u_s, u_t) \notin E$. If $(u_t, u_s) \in E$, replace $(u_r, u_t), (u_t, u_s)$ by $(u_t, u_r), (u_s, u_t)$, and if $(u_t, u_s) \notin E$, replace (u_r, u_t) by (u_s, u_t). In either case, we obtain a simple directed graph $\mathbf{G}'$ with q as out-degree sequence, and the claim is proved.

Claim 2. Suppose $m = (n-1) + \ldots + (n-\ell+1) + r$ with $r \leq n - \ell$, then $\overline{p}_m = (n-1, \ldots, n-\ell+1, r)$ is the only maximal element of $S(m)$ in $L(m)$.

By the definition (2), the sequence $\overline{p}_m$ clearly dominates any sequence in $S(m)$, and since $\overline{p}_m$ can obviously be realized as out-degree sequence of a simple digraph, it is the unique maximum of $S(m)$.

Taking Claims 1 and 2 together yields the characterization

$$d \in S(m) \Longleftrightarrow d \leq \overline{p}_m,$$

and this latter condition is plainly equivalent to (1). □

As is to be expected the neighborhood list problem ($\mathcal{N}^+$) is considerably more involved. We consider the problem NLSD (neighborhood list of a simple digraph):

Instance: A list $N^+ = (N_1^+, \dots, N_n^+)$.

Question: Does there exist a simple digraph $\mathbf{G}(V, E)$ such that N^+ is the list of its out-neighborhoods?

In terms of the adjacency matrix, the problem reads as follows: We are given an $n \times n$–matrix Γ (whose rows are the incidence vectors of the sets (N_i^+). Does there exist a permutation matrix R such that $R\Gamma$ is the adjacency matrix of a simple digraph, i.e. such that

$$R\Gamma + (R\Gamma)^T \leq J - I$$

where J is the all-ones matrix, and $\leq$ is to be understood coordinatewise.

Theorem 4.2. *The decision problem NLSD is NP-complete.*

For the proof we need an auxiliary result. Consider the decision problem k-Aut:

Instance: A simple undirected graph G.

Question: Does G admit an automorphism all of whose orbits have length at least k?

The problem 2-Aut is the FIXED-POINT-FREE AUTOMORPHISM problem which was shown to be NP-complete by Lubiw [10].

Lemma 4.1. *3-Aut is NP-complete.*

Proof. We give a transformation from 2-Aut. Note first that the construction given by Lubiw shows that we may assume an instance G of 2-Aut which satisfies the following conditions:

(a) For $u \neq v \in V$, the sets $N(u) = \{w \in V : uw \in E\}$ and $N(v)$ are distinct.

(b) For $u \neq v \in V$, $N(u) \cup \{u\} \neq N(v) \cup \{v\}$.

For such graphs G, Sabidussi has shown in [12] that the *composition* $G[G]$ has the automorphism group Aut$G \circ$ Aut$G$ where $\circ$ denotes the *wreath product* and $V(G[G]) = V \times V$,

$$E(G[G]) = \{\{u_1, v_1\}, \{u_2, v_2\} : u_1u_2 \in E \quad \text{or} \quad u_1 = u_2, v_1v_2 \in E\}.$$

That is, the composition arises from G by replacing each vertex by an isomorphic copy of G and joining either all pairs in different copies or none according to whether the original vertices are adjacent or not. The wreath product Aut$G\circ$Aut$G$ is defined as follows:

$$\text{Aut}G \circ \text{Aut}G = \{(\varphi; \psi_1, \dots, \psi_n) : \varphi, \psi_i \in \text{Aut}G\}$$

where the action on $V \times V$ is given by

$$(\varphi; \psi_1, \dots, \psi_n)(u, v) = (\varphi(u), \psi_u(v)) \quad \text{with} \quad V = \{1, \dots, n\}.$$

We claim that G has a fixed-point free automorphism if and only if $G[G]$ admits an automorphism with all orbit lengths at least 3. This will then prove our Lemma.

Suppose first that there exists $\varphi \in$ Aut$G$, $\varphi(u) \neq u$ for all $u \in V$, and let $O_1, \dots, O_k$ be the orbits of φ on V. Suppose $|O_1| = \ldots = |O_h| = 2$ and $|O_i| \geq 3$ for $h+1 \leq i \leq k$. Define $\rho = (\varphi; \psi_1, \dots, \psi_n) \in$ Aut$G \circ$ Aut$G$ such that $\psi_u \in \{\text{id}, \varphi\}$ for all u and $\psi_u = \varphi$ for exactly one u in each orbit O_i, $1 \leq i \leq h$. Let $(u, v) \in V \times V$ and denote by $\overline{O}$ its ρ-orbit. If the orbit O_i of u under φ has at least three elements, then the same obviously holds for $\overline{O}$. In case $O_i = \{u, w\}$, suppose $\psi_u = \varphi$. Then $\{(u, v), (w, \varphi(v)), (u, \varphi(v))\} \subseteq \overline{O}$ and thus $|\overline{O}| \geq 3$. The case $\psi_w = \varphi$ is analogous.

Now assume, conversely, that each $\varphi \in \text{Aut}G$ has a fixed point $u = u_\varphi$. Then each $\rho = (\varphi; \psi_1, \ldots, \psi_n)$ has a fixed point as well, namely the pair $(u_\varphi, u_{\psi_{u_\varphi}})$. Since the transformation $G \longrightarrow G[G]$ is clearly polynomial, the proof of the Lemma is complete.

Remark 4.1. By iterating the construction of the Lemma, it can be shown that k-Aut is NP-complete for every fixed $k \geq 2$.

Proof (of Theorem 4.2). We provide a transformation from 3-Aut. Suppose G is a simple undirected graph with incidence matrix $A \in \{0,1\}^{n \times m}$, where m is the number of edges. We may assume $n \geq 4$, $m \geq 3$. We denote by $t_{r \times s}$ the $r \times s$–matrix with all entries equal to t, and set $\overline{A} = 1_{n \times m} - A$.

Now consider the $(n+m+2) \times (n+m+2)$–matrix

$$\Gamma = \begin{array}{|c|c|c|c|} \hline I_n & A & 1_{n\times 1} & 1_{n\times 1} \\ \hline \overline{A}^T & 0_{m\times m} & 1_{m\times 1} & 0_{m\times 1} \\ \hline 0_{1\times n} & 0_{1\times m} & 0 & 1 \\ \hline 0_{1\times n} & 1_{1\times m} & 0 & 0 \\ \hline \end{array}$$

Our theorem will be proved by the following claim: There exists a permutation matrix R with $R\Gamma + (R\Gamma)^T \leq J - I$ if and only if G admits an automorphism with all orbit lengths at least 3.

Note first that row $n+m+1$ is the only row with at least $n+m$ zeros. Since the $(n+m+1)$-st column contains $n+m$ ones, R must fix row $n+m+1$. But then row $n+m+2$ must also be fixed, since the entry in position $(n+m+2, n+m+1)$ in $R\Gamma$ must be 0. Furthermore, the rows with numbers $\{1, \ldots, n\}$ and $\{n+1, \ldots, n+m\}$ are permuted among themselves. Hence, R is of the form

$$R = \left(\begin{array}{cc|cc} P & 0 & 0 & 0 \\ 0 & Q & 0 & 0 \\ \hline 0 & 0 & 1 & 0 \\ 0 & 0 & 0 & 1 \end{array}\right)$$

where P and Q are permutation matrices of order n and m, respectively. Now we have

$$R\Gamma + (R\Gamma)^T \leq J - I \iff \begin{array}{l} P + P^T \leq J - I \text{ and} \\ PA + (Q\overline{A}^T)^T \leq 1_{n\times m}. \end{array}$$

It is easily seen, that $P + P^T \leq J - I$ holds if and only if the permutation π corresponding to P has no cycle of length two or less. For the second condition, we note

$$PA + \overline{A}Q^T = PA + (1_{n\times m} - A)Q^T \leq 1_{n\times m}$$

if and only if $PA \leq AQ^T$ if and only if $PA = AQ^T$.

Denoting by Perm_n the set of permutation matrices of order n, we have

$$\text{Aut}G = \{S \in \text{Perm}_n : \exists U \in \text{Perm}_m \text{ with } SA = AU^T\},$$

and the result follows. □

The corresponding decision problem ($\mathcal{N}^+$) for *tournaments* is open, but we surmise that it is also NP-complete.

References

1. M. Aigner: Uses of the diagram lattice. Mitteilungen Math. Sem. Giessen 163 (1984), 61–77.
2. M. Aigner, E. Triesch: Reconstructing a graph from its neighborhood list. Combin. Probab. Comput. 2 (1993), 103–113.
3. M. Aigner, E. Triesch: Realizability and uniqueness in graphs. Discrete Math. 136 (1994), 3–20.
4. P. Erdős, T. Gallai: Graphen mit Punkten vorgeschriebenen Grades. Math. Lapok 11 (1960), 264–274.
5. D. R. Fulkerson: Zero-one matrices with zero trace. Pacific J. Math. 10 (1960), 831–836.
6. D. R. Fulkerson, H. J. Ryser: Multiplicities and minimal widths in $(0, 1)$–matrices. Canad. J. Math. 11 (1962), 498–508.
7. D. Gale: A theorem on flows in networks. Pacific J. Math. 7 (1957), 1073–1082.
8. M. R. Garey, D. S. Johnson: Computers and Intractability: A guide to the theory of NP-completeness. Freeman, San Francisco, 1979.
9. H. G. Landau: On dominance relations and the structure of animal societies, III: the condition for a score structure. Bull. Math. Biophys. 15 (1955), 143–148.
10. A. Lubiw: Some NP-complete problems similar to graph isomorphism. SIAM J. Computing 10 (1981), 11–21.
11. H. J. Ryser: Combinatorial properties of matrices of zeros and ones. Canad. J. Math. 9 (1957), 371–377.
12. G. Sabidussi: The composition of graphs. Duke Math. J. 26 (1959), 693–696.
13. V. T. Sós: Problem. In: Combinatorics (A. Hajnal, V.T. Sós, eds.), Coll. Math. Soc. J. Bolyai 18, North Holland, Amsterdam (1978), 1214.

The Dimension of Random Graph Orders

Béla Bollobás[1] and Graham Brightwell[2]

[1] Department of Pure Mathematics and Mathematical Statistics, University of Cambridge, 16, Mill Lane, Cambridge CB2 1SB, U.K.
[2] Department of Mathematics, London School of Economics and Political Science, Houghton Street, London WC2A 2AE, U.K.

Summary. The random graph order $P_{n,p}$ is obtained from a random graph $G_{n,p}$ on $[n]$ by treating an edge between vertices i and j, with $i \prec j$ in $[n]$, as a relation $i < j$, and taking the transitive closure. This paper forms part of a project to investigate the structure of the random graph order $P_{n,p}$ throughout the range of $p = p(n)$. We give bounds on the dimension of $P_{n,p}$ for various ranges. We prove that, if $p \log\log n \to \infty$ and $\epsilon > 0$ then, almost surely,

$$(1-\epsilon)\sqrt{\frac{\log n}{\log(1/q)}} \le \dim P_{n,p} \le (1+\epsilon)\sqrt{\frac{4\log n}{3\log(1/q)}}.$$

We also prove that there are constants c_1, c_2 such that, if $p \log n \to 0$ and $p \ge \log n/n$, then

$$c_1 p^{-1} \le \dim P_{n,p} \le c_2 p^{-1}.$$

We give some bounds for various other ranges of $p(n)$, but several questions are left open.

1. Introduction

The *random graph order* $P_{n,p}$ is defined as follows. The vertex set is $[n] \equiv \{1, \dots, n\}$: we use the symbol $\prec$ to denote the standard linear order on this set. We take a random graph $G_{n,p}$ on $[n]$, and interpret an edge between vertices i and j with $i \prec j$ as a relation $i < j$ of the order. (For the theory of random graphs, founded by Erdős and Rényi [10, 11], see Bollobás [4].) The full order $<$ is then defined by taking the transitive closure. Put another way, if $i \prec j$, then we have $i < j$ in $P_{n,p}$ iff there is a $\prec$-increasing sequence $i = i_1, i_2, \dots, i_m = j$ of vertices such that each of $i_1i_2, i_2i_3, \dots, i_{m-1}i_m$ is an edge of the underlying random graph.

In general, the probability p will be a function of n. We say that $P_{n,p}$ has a property *almost surely* if, as $n \to \infty$, the probability that $P_{n,p(n)}$ has the property tends to 1. Throughout the paper, any inequalities we state are only claimed to hold for n sufficiently large. Also throughout the paper, we set $q = 1 - p$.

Random graph orders have been studied by Barak and Erdős [3], Albert and Frieze [1], Alon, Bollobás, Brightwell and Janson [2], Newman and Cohen [17], Newman [16] and Simon, Crippa and Collenberg [18]. The first three of these papers deal exclusively with the case where the probability p is constant; the next two are concerned with the height of the random order, and the final one is essentially concerned with the number of incomparable pairs of elements. See also the survey by Brightwell [8]. This paper is part of a project to investigate the structure of the random graph order more fully throughout the range of $p = p(n)$. The authors have produced a paper [5] dealing with the width of $P_{n,p}$ throughout the range, and a further paper [6] on the general structure of random graph orders.

This paper is concerned with the dimension of $P_{n,p}$. Recall that a *realiser* of a partial order $(X,<)$ is a set of linear orders on X whose intersection is exactly $(X,<)$. The *dimension* of $(X,<)$ is the minimum cardinality of a realiser. Alternatively, the dimension is the minimum d such that $(X,<)$ can be embedded in $\mathbb{R}^d$ with the co-ordinate order. Trotter's book [21] is a thorough treatment of dimension theory for partial orders. The dimension has already been studied to some effect for another model of random orders by Erdős, Kierstead and Trotter [9]. We discuss their result here, partly because we shall use it later, and also because it motivates our approach to dimension.

Define a *random bipartite order* $B_{n,p}$ by taking a random bipartite graph on the union of two disjoint sets V and W of vertices, with $|V| = |W| = n$, and joining each pair of vertices $(v, w) \in V \times W$ with probability $p = p(n)$, independently. Then the relations of $B_{n,p}$ are exactly the relations $v < w$, for a vertex $v \in V$ adjacent to a vertex $w \in W$.

The dimension of $B_{n,p}$ is certainly at most n, since the dimension of any partial order is bounded above by its width (see, for instance, Trotter [21]). The result of Erdős, Kierstead and Trotter [9] implies that, even if p is fairly small, the dimension of $B_{n,p}$ is almost surely almost as large as n.

Theorem 1.1. *For every $\epsilon > 0$, there is a $\delta > 0$ such that, if $\log^{1+\epsilon} n/n < p < 1 - n^{-1+\epsilon}$, then, almost surely,*

$$\dim B_{n,p} > \frac{\delta pn \log pn}{1 + \delta p \log pn}.$$

In particular, there is a constant c such that, if $1/\log n < p < 1 - n^{-1/2}$, then, almost surely,

$$\dim B_{n,p} \geq n(1 - c/p \log n). \qquad \square$$

The proof of Theorem 1.1 can be found in the original paper [9], or in Trotter's book [21, Chapter 7].

Theorem 1.1 suggests a method of proving that a partial order has large dimension; namely that one should try to find a large "random" bipartite order within it. We shall see this idea applied in due course.

We also note here a powerful upper bound for the dimension, due to Füredi and Kahn [12] (see also Trotter [21, Chapter 7]). The *maximum degree* of a partial order $(X,<)$ is the maximum, over all vertices $x \in X$, of the number of vertices of X comparable with x.

Theorem 1.2. *Let $(X,<)$ be a partial order with maximum degree Δ. Then*

$$\dim(X,<) \leq 50\Delta \log^2 \Delta. \qquad \square$$

A random bipartite order with $p = 2c/\log n$, where c is the constant of Theorem 1.1, almost surely has dimension at least $\epsilon\Delta \log \Delta$, for some fixed $\epsilon > 0$, where Δ is the maximum degree.

Returning now to random graph orders, the only previous result concerning the dimension is due to Albert and Frieze [1], who showed the following,

Theorem 1.3. *Let p be a fixed constant with $0 < p < 1$, and set $q = 1 - p$. Then, for any $\epsilon > 0$, almost surely,*

$$(1-\epsilon)\sqrt{\frac{\log n}{\log(1/pq)}} \leq \dim P_{n,p} \leq \sqrt{\frac{2\log n}{\log(1/q)}} + \frac{1}{2} + \epsilon. \qquad \square$$

Thus if $p = q = 1/2$, the upper bound is essentially twice the lower bound; and the ratio of upper to lower bounds increases as p becomes smaller. The upper bound in Theorem 1.3 is an immediate consequence of the same upper bound for the width of $P_{n,p}$, as proved by Barak and Erdős [3] (and subsequently sharpened by Bollobás and Brightwell [5]). The lower bound is obtained by showing that a specific partial order of the given dimension almost surely exists as a suborder of $P_{n,p}$. Albert and Frieze conjectured that this lower bound was correct. One of our principal aims here is to improve both the upper and lower bounds in Theorem 1.3, thus disproving that conjecture. We prove the following result.

Theorem 1.4. *Suppose $p(n)$ is a function of n such that $p \log\log n \to \infty$ and $p \leq 1 - 1/\sqrt{\log n}$. For any $\epsilon > 0$ we have, almost surely,*

$$(1-\epsilon)\sqrt{\frac{\log n}{\log(1/q)}} \leq \dim P_{n,p} \leq (1+\epsilon)\sqrt{\frac{4\log n}{3\log(1/q)}}.$$

The gap between the upper and lower bounds for the dimension of $P_{n,p}$ is thus reduced to a multiplicative factor of $(1+\epsilon)2/\sqrt{3}$, for each constant value of p, and indeed whenever $p(n)$ tends to 0 more slowly than $1/\log\log n$. We strongly suspect that the lower bound here is essentially correct, at least for constant p, but have been unable to prove it. Section 2 is devoted to a proof of Theorem 1.4: in fact, we prove slightly more general results.

Our other main aim is to estimate the dimension for rather lower ranges of $p = p(n)$. In [5], we showed that there is a type of "phase transition" as $p(n)$ passes through the range $p = c/\log n$; this is investigated further, and to some extent explained, in [6]. There is a second, equally radical, change of behaviour near $p = \log n/n$, the nature of which will hopefully become clear in the course of our proofs. A third transition phase, as p decreases through c/n, is readily understood, since the underlying random graph changes from being almost all in one component, to being a collection of small trees. The following result of Bollobás and Brightwell [5] gets close to pinning down the behaviour of the width of $P_{n,p}$ between the first and last of these three transitions.

Theorem 1.5. *If $p\log n \to 0$ and $pn \to \infty$, then the width of $P_{n,p}$ almost surely lies between $1.455p^{-1}$ and $2.428p^{-1}$.*

We thus have an upper bound for $\dim P_{n,p}$ of $2.428p^{-1}$ in this range. It is natural to ask whether this is roughly the right order of magnitude: we shall prove that it is, provided we are above the second phase transition point. The main result of Section 3 is as follows.

Theorem 1.6. *There is an $\epsilon > 0$ such that, if $p \geq \log n/n$ and $p\log n \to 0$, then, almost surely,*

$$\epsilon p^{-1} \leq \dim P_{n,p} \leq 2.428p^{-1}.$$

Below $p = \log n/n$, the maximum degree of the random partial order is smaller than p^{-1}, so the Füredi-Kahn bound, Theorem 1.2, gives us a better upper bound on the dimension. We show that, at least in some range below $p = \log n/n$, this upper bound is not too far off being correct.

Theorem 1.7. *There are positive constants c_1, c_2 such that, if $\frac{4}{5}\log n/n \leq p \leq (\log n - \log\log n)/n$, then, almost surely,*

$$c_1 e^{pn} \leq \dim P_{n,p} \leq c_2 e^{pn} \log^3 n.$$

In particular, there is a genuine change of behaviour near $p = \log n/n$, which is associated with a sudden drop in the value of the maximum degree from ϵn to $o(n)$. A later result, Lemma 3.1, gives some more information on the distribution of the number of vertices above a given vertex in $P_{n,p}$.

The proofs of Theorem 1.6 and the lower bound in Theorem 1.7 are practically identical, and will be dealt with together in Section 3. The constant 4/5 appearing in Theorem 1.7 can be improved, and indeed it appears that, with some difficulty, it can be replaced by any $\epsilon > 0$. The upper bound in Theorem 1.7 remains valid for much smaller values of p, and it would be of some interest to determine if a lower bound of the form given in Theorem 1.7 holds further down. The following easy result deals with very small values of p.

Theorem 1.8. *(i) If $pn \to 0$, but $pn^{7/6} \to \infty$, then, almost surely,* $\dim P_{n,p} = 3$. *(ii) If $pn^{7/6} \to 0$, then, almost surely,* $\dim P_{n,p} = 2$.

Proof. This result is an immediate consequence of well-known structural results for random graphs $G_{n,p}$ with $pn \to 0$ (see, for instance, Bollobás [8]).

If $pn \to 0$, then all the components of the underlying random graph are trees, so the dimension of the random order is at most 3 (see, for instance, Trotter [21, Chapter 5, (2.6)]). If $pn^{7/6} \to \infty$, then some of the components are seven-vertex trees of dimension 3, whereas if $pn^{7/6} \to 0$, all components are trees of at most six vertices, so $\dim P_{n,p} = 2$. □

If $p = c/n$ with $c < 1$, then all components are either trees or unicyclic graphs. The following conjecture would imply that Theorem 1.8(i) can be extended to this range of $p(n)$.

Conjecture 1.1. If P is a partial order whose covering graph is unicyclic, then $\dim P \leq 3$.

As pointed out to us by Tom Trotter, Theorem (4.4) of Chapter 4 of [21] implies that, if P is as above, then $\dim P \leq 5$. What happens if $p = c/n$ with $c > 1$?

The reader will perhaps have noticed that there is one intermediate range of $p(n)$ that we have not yet discussed, namely the range where $p \log n \to \infty$ but $p \log \log n \to 0$. In this range, the width of $P_{n,p}$ is still almost surely about $\sqrt{2p^{-1} \log n}$, and indeed the upper bound for $\dim P_{n,p}$ in Theorem 1.4 still holds. Our methods can be adapted to give apparently rather weak lower bounds for the dimension, but we have been unable to answer the following question.

Question 1.1. Is there an $\epsilon > 0$ such that, if $p \geq \log n/n$, then, almost surely,

$$\dim P_{n,p} \geq \epsilon \text{ Width}(P_{n,p})?$$

Note that we have an affirmative answer if either $p \log n \to 0$ or $p \log \log n \to \infty$. However, we suspect that the answer is "no".

2. Large $p(n)$

This section is devoted to the proof of Theorem 1.4. In fact, we shall prove two slightly sharper and more general results, which together imply Theorem 1.4. We first consider the lower bound.

Theorem 2.1. *There is a constant k such that, if $2/\log\log n \le p \le 1 - 1/\sqrt{\log n}$, then, almost surely,*

$$\dim P_{n,p} \ge \sqrt{\frac{\log n}{\log(1/q)}} \left(1 - \frac{k}{p \log\log n}\right).$$

Proof. We shall use Theorem 1.1, due to Erdős, Kierstead and Trotter [9]. Set

$$m = \left\lfloor \sqrt{\frac{\log n}{\log(1/q)}} \left(1 - \frac{1}{\log\log n}\right) \right\rfloor,$$

and note that $m \le \sqrt{\log n \log\log n}$. Our aim is to find a "random $B_{m,p}$" inside our random graph order $P_{n,p}$, and to apply Theorem 1.1.

We break our vertex set $[n]$ into $l = \lfloor n/2m \rfloor$ consecutive pieces $A_1, \ldots, A_l$, so that A_{j+1} consists of the vertices between $2mj + 1$ and $2m(j+1)$ inclusive. Each set A_i thus consists of $2m$ vertices; we subdivide it further into two consecutive sets B_i and C_i of m vertices each, so that B_{j+1} consists of the vertices between $2mj+1$ and $2mj + m$ inclusive, and C_{j+1} contains the remainder.

Note that, if there are no edges of the underlying random graph inside a set B_i or C_i, then that set forms an antichain of size m in $P_{n,p}$. We call a set A_i *good* if both B_i and C_i form antichains. Note that, conditional on A_i being good, the random order $P_{n,p}$ restricted to A_i is distributed as a random bipartite order $B_{m,p}$ (with vertex classes B_i and C_i).

The probability that A_i is good is equal to $q^{2\binom{m}{2}} \ge q^{m^2}$. Also, the events that the various A_i are good are independent. Thus the probability P that there is no good A_i is at most

$$\left(1 - q^{m^2}\right)^l.$$

Substituting our chosen value for m, and using the inequality $l \ge n/3m$, we obtain:

$$\begin{aligned} P &\le \exp\left(-\frac{n}{3m} q^{\frac{\log n}{\log(1/q)}(1-1/\log\log n)^2}\right) \\ &\le \exp\left(-\frac{1}{3\sqrt{\log n \log\log n}} n^{1/\log\log n}\right) \\ &= o(1). \end{aligned}$$

Therefore there is almost surely some good A_i. Thus we have, almost surely, $\dim P_{n,p} \ge \dim B_{m,p}$.

Note that $1/\log m < p < 1 - m^{-1/2}$, so, by Theorem 1.1, we almost surely have

$$\begin{aligned} \dim P_{n,p} &\ge \dim B_{m,p} \ge m\left(1 - \frac{c}{p \log m}\right) \\ &\ge \sqrt{\frac{\log n}{\log(1/q)}} \left(1 - \frac{2}{\log\log n}\right)\left(1 - \frac{3c}{p\log\log n}\right), \end{aligned}$$

where c is the constant of Theorem 1.1. The result now follows on choosing an appropriate value of k. □

We now turn our attention to the upper bound. The main point here is that the dimension is bounded away from the width. Informally, very large antichains are rare, and one needs more than one large antichain to have large dimension.

Furthermore, for p above our first transition phase around $1/\log n$, it is shown in [6] that the random graph order has the structure of many small orders arranged one above another. Thus we shall make use of the following two results.

Theorem 2.2. *Let A be a maximal antichain in $(X,<)$, with $A \neq X$. Set $U = \{x \in X : x > a$ for some $a \in A\}$ and $D = \{x \in X : x < a$ for some $a \in A\}$, so that X is the disjoint union of A, U and D. Let W_U and W_D denote the widths of the partial order $(X,<)$ restricted to U and D respectively. Then*

$$\dim(X,<) \leq 1 + W_U + W_D. \qquad \square$$

Trotter [19] proves Theorem 2.2 in the cases where (i) $W_D = 0$ or (ii) $W_D = W_U$, and in fact the proof of the latter result generalises immediately to give Theorem 2.2. This proof can also be found in Trotter's book [21, Chapter 1 (11.3)].

A *post* in a partial order is an element comparable with all others. Posts in $P_{n,p}$ are considered in Alon, Bollobás, Brightwell and Janson [2], and in more detail in Bollobás and Brightwell [6]. Assume for the moment that $P_{n,p}$ has at least one post. If i and j are posts in $P_{n,p}$ such that there are no posts k with $i \prec k \prec j$, then the partial order restricted to the interval $[i,j)$ of $[n]$ is called a *factor* of $P_{n,p}$. If i is the first and j the last post, then the partial order restricted to either of $[1,i)$ or $[j,n]$ is also regarded as a factor. If the factors of $P_{n,p}$ are $F_1,\ldots,F_m$, with $F_1 \prec \ldots \prec F_m$ in $[n]$, then $P_{n,p}$ is the *linear sum* of $F_1,\ldots,F_m$, i.e., the partial order defined by taking the union of the F_i, and putting in each relation of the form $x < y$, where x is an a strictly lower factor than y. Note that the dimension of a linear sum is equal to the maximum of the dimensions of its constituent factors.

Set $\eta(p) = \prod_{i=1}^{\infty}(1-q^i)$. We note that there is a constant c such that, for $p \leq 1/2$, $\eta(p) \geq c\exp(-\pi^2/6\log(1/q))$. This can be read out of Hall [13, equation 4.2.11]. We quote the following result from [6].

Theorem 2.3. *Suppose $p\log n \to \infty$. Then there is almost surely no set of $\eta(p)^{-1}\log^3 n$ consecutive vertices of $[n]$ that does not contain a post in $P_{n,p}$.* $\square$

We shall apply this when $p\log n \to \infty$, so that $\eta(p)^{-1}\log^3 n = o(n^{\epsilon})$ for every $\epsilon > 0$. The significance of Theorem 2.3 is then that the partial order $P_{n,p}$ is almost surely the linear sum of factors, all of which have at most $\eta(p)^{-1}\log^3 n$ elements. In view of Theorem 2.2, this implies that, if $P_{n,p}$ does have dimension at least d, then one of the factors must contain both an antichain A of size d, and a pair of disjoint antichains, also disjoint from A, of sizes summing to at least $d-1$.

Our plan is to estimate the expected number of occurrences of such structures.

Theorem 2.4. *Let $\epsilon > 0$ be any fixed constant. Suppose $p\log n \to \infty$. Then, almost surely,*

$$\dim P_{n,p} \leq (1+\epsilon)\sqrt{\frac{4\log n}{3\log(1/q)}}.$$

Proof. In [5], it is proved that the expected number of antichains of size w whose $\prec$-first vertex is x is at most $\eta(p)^{-2}q^{\binom{w}{2}}$. One can readily adapt the argument there to show that, for any triple (x,y,z) of vertices, the expected number of triples of antichains of sizes w_1, w_2, w_3 with $\prec$-first vertices x,y,z respectively is at most

$$\eta(p)^{-6}q^{\binom{w_1}{2}+\binom{w_2}{2}+\binom{w_3}{2}}.$$

The number of triples of vertices (x,y,z) in the same factor of $P_{n,p}$ is almost surely at most $n(\eta(p)^{-1}\log^3 n)^2$, from Theorem 2.3.

Hence, by Theorem 2.2, the expected number of factors of dimension at least d, conditional on the conclusion of Theorem 2.3 holding, is at most

$$n\eta(p)^{-8}(\log n)^6 q^{\binom{d}{2}} \sum_{i=0}^{d-1} q^{\binom{i}{2}+\binom{d-i-1}{2}}.$$

The largest term in the sum is the term with $i = \lfloor (d-1)/2 \rfloor$, so this expectation is at most

$$n\eta(p)^{-8}(\log n)^6 dq^{d^2/2+2(d/2)^2/2-2d} = n\eta(p)^{-8}(\log n)^6 dq^{3d^2/4-2d}.$$

Setting $d = (1+\epsilon)\sqrt{4\log n/3\log(1/q)}$, for any $\epsilon > 0$, we see that this is at most

$$n^{1-(1+\epsilon)^2}\eta(p)^{-8}(\log n)^7 e^{-3\sqrt{p^{-1}\log n}} = o(1).$$

Since the dimension of $P_{n,p}$ is at most the maximum dimension of its factors, this implies that, almost surely, $\dim P_{n,p} \leq d$, as required. □

As mentioned earlier, we do not expect that the constant 4/3 in Theorem 2.4 is correct. One possible way to decrease it would be to try to improve on Theorem 2.2, under suitable extra hypotheses. Although Trotter [20, 21] has shown that Theorem 2.2 is sharp, even when $W_D = W_U$, the example he constructs looks completely unlike anything that is likely to arise as a suborder of $P_{n,p}$: in particular the size of the antichain A is exponential in W_D. On the other hand, one can almost surely find in $P_{n,p}$ three consecutive sets of vertices of sizes $w/2$, w and $w/2$ successively, where $w = (1-\epsilon)\sqrt{4\log n/3\log(1/q)}$; however the (random) order induced by these sets almost surely has dimension at most $w/2$ – see Erdős, Kierstead and Trotter [9].

We suspect that the lower bound in Theorem 2.1 gives the correct constant, at least for constant values of p, so we make the following conjecture.

Conjecture 2.1. Let p be a fixed constant with $0 < p < 1$. For every $\epsilon > 0$, we have, almost surely,

$$\dim P_{n,p} \leq (1+\epsilon)\sqrt{\frac{\log n}{\log(1/q)}}.$$

3. Small $p(n)$

Our principal aim in this section is to prove Theorem 1.6, giving a lower bound of ϵp^{-1} on the dimension for the major range of $p(n)$. As mentioned earlier, this bound is best possible up to the value of ϵ. We shall work in slightly greater generality so as to prove Theorem 1.7 as well.

We start with an overview of the proof, assuming for simplicity that $p \geq \log n/n + \omega(n)\log\log n/n$, where $\omega(n) \to \infty$. Here and later, we treat integer and real quantities interchangeably, for the sake of clarity: the proof is not materially affected.

Throughout this section, we set $m = p^{-1}$. We restrict attention to the set $[N]$ consisting of the $\prec$-first $N = m\log m + 6m$ vertices of $[n]$, so we are dealing with a random graph order $P_{N,p}$. We partition $[N]$ into sets $W \prec U \prec U' \prec W'$, with $|U| = |U'| = \frac{1}{2}m\log m$, and $|W| = |W'| = 3m$. Let A be the set of maximal elements of the partial order restricted to W, and let A' be the set of minimal elements in the partial order restricted to W'. It was noted in our earlier paper [5] that, almost surely, $|A|, |A'| \geq 9m/10$. Our aim is to prove that the random order $P_{N,p}$ restricted to $A \cup A'$ almost surely has dimension at least ϵm.

Intuitively, the size of $U \cup U'$, $m \log m$, has been fixed so that, for each pair $(a, a') \in A \times A'$, the probability that $a < a'$ in $P_{N,p}$ is a constant. If these relations were independent, then the partial order we consider would be a random bipartite order, and, by the result of Erdős, Kierstead and Trotter [9], Theorem 1.1, would almost surely have dimension at least $(1-\epsilon)9m/10$. However of course the relations are far from independent. We overcome this, at the expense of finishing with a rather small constant, in the following manner. We prove that, almost surely, there are at least γm^2 pairs $(a, a') \in A \times A'$ which are not related in $P_{N,p}$. If we have a realiser $L_1, \ldots, L_d$ of $P_{N,p}$, then, for each such unrelated pair, there must be an $i \in \{1, \ldots, d\}$ such that a is above a' in L_i – we say L_i *reverses* (a, a'). On the other hand, we prove that, for any subset T of A of (small) size t, the number of elements of A' incomparable with all elements of T is at most $(1 - \delta)^t m$. This will imply that the total number of pairs (a, a') reversed by a linear extension L is at most $2m/\delta$, and so we must have $\dim P_{N,P} \geq d \geq \gamma\delta m/2$.

For X a subset of $[N]$ and B a subset of A, let $\Gamma_X(B)$ be the set of vertices $x \in X$ such that x is above some vertex of B in $P_{N,p}$, i.e., there is a $\prec$-increasing path in the underlying random graph from some vertex of B to x. Our strategy will be to prove that, for every (small) subset B of A, $\Gamma_U(B)$ is about the "right" size, namely $n^{1/2}|B|$. Unfortunately, this will fail as it stands because a few vertices a in A are likely to be below far too few or too many vertices in U. So our first step will be to exclude such vertices, and certain others, from consideration.

Formally, we shall begin by looking just at the set $A \cup U$. We shall prove that the partial order restricted to this set almost surely has certain properties, and by symmetry the partial order restricted to $A' \cup U'$ has analogous properties. These properties will of course be independent of edges between U and U'. Then we prove that, conditional on these properties, the partial order $P_{N,p}$ almost surely has large dimension.

For the sake of convenience, we begin by working in a slightly altered model of random orders. Define the model $P'(s,p)$ by taking two disjoint vertex sets R and $S = [s]$, with $|R| = 9p^{-1}/10$. Then we put in edges independently, with probability p, between each pair of vertices, not both in R. Thus the model restricted to S is just a copy of $P_{s,p}$. All vertices in R are to be thought of as "below" S, so a vertex $x \in R$ is below $y \in S$ if there is a vertex $z \in S$ such that xz is an edge of the graph, and $z < y$ in $P_{s,p}$. We shall apply this with $R = A$ and $S = U$. For the greater generality required to prove Theorem 1.7, we need to consider cases where s is a little smaller than $\frac{1}{2}m \log m$, where again $m = p^{-1}$.

For the next few steps, we work in the model $P'(s,p)$, assuming that $\frac{2}{5}m \log m \leq s \leq \frac{1}{2}m \log m$. As mentioned above, we wish to show that there is almost surely a large subset R_G of R such that, for every (small) subset T of R_G, $\Gamma_S(T)$ is not too far from its expected size. There are essentially two steps here, first to ensure that this holds for every single-element subset of R_G, and then to extend the argument to larger subsets.

We start with a lemma which is perhaps of independent interest. Let the random variable $X(s,p)$ be the number of vertices of S above a given vertex $x \in R$ in $P'(s,p)$. Of course, this is distributed as the number of vertices above vertex 1 in $P_{s+1,p}$. The lemma below gives the exact distribution of $X(s,p)$, and also some convenient estimates. Note that our estimates are only good in the case where $pq^{-s} = o(1)$, which is when s is rather less than $m \log m$. The identity given in part (i) has been obtained independently by Simon, Crippa and Collenberg [18], who study in more detail the mean and variance of $X(s,p)$ and the behaviour in the limit as $s \to \infty$ with p fixed.

In the lemma below, $\mathrm{Geom}(a^{-1})$ denotes a geometric random variable with mean a^{-1}, i.e., with $\Pr(\mathrm{Geom}(a^{-1}) = t) = a(1-a)^t$ for t a non-negative integer. The *total variation distance* $d_{\mathrm{TV}}(X,Y)$ between two real-valued random variables X and Y is the supremum over $x \in \mathbb{R}$ of $|\Pr(X \le x) - \Pr(Y \le x)|$.

Lemma 3.1. *(i) For* $0 \le t \le s$,

$$\Pr(X(s,p) = t) = q^{s-t} \prod_{i=s-t+1}^{s} (1-q^i).$$

(ii) Suppose $p \le 1/2$. *Then*

$$d_{\mathrm{TV}}\left(X(s,p), \mathrm{Geom}(q^{-s})\right) \le 3pq^{-s}.$$

(iii) For every t,

$$\Pr(X(s,p) > t) \le (1-q^s)^t.$$

(iv) If $p \le 1/2$, *then, for every* t,

$$\Pr(X(s,p) \le t) \le \exp(2pq^{-s})\left(1-(1-q^s)^t\right).$$

Proof. Of course, the formula in (i) can be proved by induction on s, using the recurrence

$$r(t,s) = r(t,s-1)q^{t+1} + r(t-1,s-1)(1-q^t),$$

where $r(t,s) = \Pr(X(s,p) = t)$: see Simon, Crippa and Collenberg [18] for details. We prefer to give a combinatorial proof, which is hopefully slightly more informative.

The probability that the vertices of $S = [s]$ above x are exactly $k_1 + 1$, $k_1 + k_2 + 2, \ldots, k_1 + \ldots + k_t + t$ is

$$\left(\prod_{i=1}^{t}(1-q^i)\right)\left(\prod_{i=1}^{t+1} q^{ik_i}\right),$$

where $k_{t+1} = s - \sum_{i=1}^{t} k_i - t$. As the possible vectors $(k_1, \ldots, k_{t+1})$ are exactly the ordered partitions of $s-t$ into $t+1$ parts, we have

$$\Pr(X(s,p) = t) = \left(\prod_{i=1}^{t}(1-q^i)\right)\left[X^{s-t}\right]\prod_{i=1}^{t+1}\frac{1}{1-q^iX}.$$

By Theorem 349 of Hardy and Wright [14], the coefficient of X^{s-t} in the product is

$$q^{s-t}\prod_{i=1}^{s-t}\frac{1-q^{t+i}}{1-q^i},$$

which implies our formula for $\Pr(X(s,p) = t)$.

For the estimates (ii)-(iv), we have that

$$\begin{aligned} a_t &\equiv \Pr(X(s,p) = t) = q^{s-t}\prod_{j=0}^{t-1}(1-q^{s-j}) \quad (0 \le t \le s) \\ b_t &\equiv \Pr(\mathrm{Geom}(q^{-s}) = t) = q^s(1-q^s)^t \quad (0 \le t). \end{aligned}$$

We claim that:

(1) $a_t \geq b_t$ for all t at most some t_0, and $a_t < b_t$ thereafter;
(2) if $p \leq 1/2$, then $a_t/b_t \leq \exp(2pq^{-s})$ for all t.

Claim (1) will imply (iii) immediately, since the right hand side in (iii) is the probability that $\mathrm{Geom}(q^{-s})$ is greater than t. Similarly Claim (2) implies (iv) immediately. For (ii), the two claims together give that

$$\begin{aligned} d_{\mathrm{TV}}\left(X(s,p), \mathrm{Geom}(q^{-s})\right) &= \sum_{t=0}^{t_0}(a_t - b_t) \\ &\leq \sum_{t=0}^{t_0} b_t\left(\exp(2pq^{-s}) - 1\right) \\ &\leq \exp(2pq^{-s}) - 1. \end{aligned}$$

This final expression is at most $3pq^{-s}$ whenever $3pq^{-s} < 1$. Since the total variation distance is certainly at most 1, we thus have the result as stated.

For $j = 1, \ldots, s$, set $\alpha_j = q^{-1}(1 - q^{s-j})/(1 - q^s)$. Note that $a_t/b_t = \prod_{j=0}^{t-1} \alpha_j$. Observe that the sequence α_j starts greater than 1 and decreases to 0 by $j = s$. This establishes Claim (1). (Note that the claim is certainly true if $t > s$, when $a_t = 0 < b_t$.)

For Claim (2), we have

$$\begin{aligned} \log \alpha_j &= -\log(1-p) + \log\left(1 - \frac{q^s}{1-q^s}(q^{-j} - 1)\right) \\ &\leq p + p^2 - q^s(q^{-j} - 1) \\ &\leq p + p^2 - jpq^s, \end{aligned}$$

and hence

$$\log(a_t/b_t) \leq p\left(t(1+p) - q^s \sum_{j=0}^{t-1} j\right) = \frac{tp}{2}\left(2 + 2p - q^s(t-1)\right). \qquad (*)$$

The expression on the right hand side of $(*)$ is maximised for $2t - 1 = 2(1+p)q^{-s}$, when it is equal to

$$\frac{pq^{-s}}{2}\left(1 + 2p + p^2 + q^s + pq^s + q^{2s}/4\right) \leq 2pq^{-s},$$

as required to establish Claim (2) and complete the proof of the lemma. (Here, we have used the weak bounds $p \leq 1/2$ and $q^s \leq 1$.) □

Returning to our main thread, we wish to prove that "many" vertices of R have about the expected number, namely about q^{-s}, of vertices of S above them. Lemma 3.1 will suffice for that, and we shall simply ignore vertices of R below too few or too many vertices. We also wish to exclude certain other vertices $a \in R$, namely those which only have sufficiently many vertices of S above them because of the large number of directed paths through one particular neighbour a^*: the problem with such vertices is that if another vertex b of R is also adjacent to a^*, the combined neighbour-set $\Gamma_S(\{a,b\})$ may be too small.

It turns out to be sufficient for our purposes to ensure that vertices of R are below about the right number of vertices from the bottom portion of S. Define then

C to be the set of the bottom m vertices of S, and W to be the set consisting of the next $w \equiv 6m \log\log m$, with $Y = C \cup W$.

We begin by choosing a large matching among those edges of the random graph between R and C. A result in our earlier paper [5, Theorem 14], implies that there is almost surely such a matching of size at least $m/3$. Given such a matching M, let R_M be the set of vertices of R incident with an edge of M. For $a \in R_M$, let a^* be the vertex of C with $aa^* \in M$. For $a \in R_M$, let $N(a)$ be the set of vertices of W sending a path down to a avoiding a^*, and $N(a^*)$ be the set of vertices of W sending a path down to a^*; i.e, $N(a^*) = \Gamma_W(a^*)$. The idea is that, if both $N(a)$ and $N(a^*)$ are large, then any subset of R including a will receive its due "contribution" from a.

We shall call a vertex $a \in R_M$ *good* if it satisfies
(i) $5\log^6 m \geq |\Gamma_Y(a)|$,
(ii) both $|N(a)|$ and $|N(a^*)|$ are at least $\frac{1}{8}\log^6 m$.

Lemma 3.2. *There are almost surely at least $m/12$ good vertices in R_M.*

Proof. Let D be the set of vertices incident with M. Note that the choice of D depends only on the edges of the random graph between R and C. For $x \in D$, let $Z(x)$ be the number of vertices of W having a directed path to x whose penultimate vertex is in W. Thus $Z(x) \subseteq N(x)$ for $x \in D$. The random variable $Z(x)$ is distributed as $X(w,p)$, so by Lemma 3.1 the probability that $Z(x)$ is at most $\frac{1}{8}\log^6 m$ is at most

$$e^{2pq^{-w}}\left(1-(1-q^{6m\log\log m})^{\log^6 m/8}\right) = 1 - e^{-1/8} + o(1).$$

For $x, y \in D$, the probabilities that $Z(x)$ and $Z(y)$ are too small are not independent, so we cannot immediately deduce that, almost surely, many vertices $a \in R$ have $Z(a)$ and $Z(a^*)$ large enough. However, the problem is easily overcome. For any subset V of W, and any $x \in D$, the event that $\{v \in W : v > x\} = V$ depends only on the set of edges of the random graph upwards from $V \cup \{x\}$, so is independent of the random graph restricted to $(W \setminus V) \cup \{y\}$, for any other element $y \in D$. Thus the random variable $Z(y)$, conditioned on the event that $Z(x) < \frac{1}{8}\log^6 m$, dominates a random variable distributed as $X(w',p)$, where $w' = w - \frac{1}{8}\log^6 m$. The probability that such a random variable is at most $\frac{1}{8}\log^6 m$ is again at most $1 - e^{-1/8} + o(1)$. Thus the variance of the number of vertices of D below too few vertices in W is $o(m^2)$, so the number of such vertices is almost surely at most $\frac{2m}{3}(1 - e^{-1/8} + \epsilon) < m/12$. Therefore, almost surely, the number of vertices $a \in R_M$ such that a and a^* both have sufficient neighbours in W is at least $m/3 - 2m/12 > m/6$.

Finally, the expected number of vertices $a \in R_M$ below more than $5\log^6 m$ vertices of $C \cup W$ can also be estimated from Lemma 3.1 to be at most

$$\frac{1}{3}m\exp(-5\log^6 m q^{m+6m\log\log m}) = \frac{1}{3}m\exp(-5e^{-1} + o(1)) < m/13.$$

As before, the variance of this number is $o(m^2)$, so there are almost surely at most $m/12$ vertices in R_M with too many neighbours. Hence the number of good vertices in R is almost surely at least $m/6 - m/12$, as desired. □

Let R_G be a set of $m/12$ good vertices in R_M. We have that each vertex of R_G is below about $\log^6 m$ vertices of W. We need a little more, namely that each smallish set T of good vertices has about $|T|\log^6 m$ neighbours in W. This follows from the principle that random graphs do not possess small dense subgraphs.

Lemma 3.3. *Almost surely, for every subset T of R_G with $|T| = t \leq m^{1/4}$, $|T \cup \Gamma_Y(T)| \geq t \log^6 m/16$.*

Proof. For any subset T of the whole set R, define $G(T) = T \cup \Gamma_Y(T)$ and $g(T) = |G(T)|$. Define a spanning forest $F(T)$ in the graph restricted to $G(T)$ by choosing, for each element of $\Gamma_Y(T)$, one edge down to an element of $G(T)$.

We first prove that, almost surely, for each subset T of R, either $g(T) > g(t) \equiv t \log^6 m/16$, or the number of edges induced by the random graph on $G(T)$ is at most $|F(T)|+3t/2$. Indeed, the probability that a given set T of size t fails is at most the probability that (a) $g(T) \leq g(t)$ and (b) of a given set of at most $\binom{g(t)}{2} - |F(T)|$ pairs of vertices, at least $3t/2$ of them are in the random graph. The probability of (b) is at most

$$\binom{g(t)^2}{3t/2} p^{3t/2} \leq \left(\frac{2eg(t)^2}{3tm}\right)^{3t/2} \leq \left(\frac{t \log^{12} m}{m}\right)^{3t/2} \leq \left(\frac{\log^{12} m}{m^{3/4}}\right)^{3t/2},$$

for $t \leq m^{1/4}$. The expected number of failing t-sets is thus at most

$$m^t \left(\frac{\log^{12} m}{m^{3/4}}\right)^{3t/2} = \left((\log m)^{18} m^{-1/8}\right)^t \leq (\log m)^{-t}.$$

Hence, almost surely, there are no over-dense sets $G(T)$. We assume from now on that this is the case.

Now let T be a subset of R_G of size t, with $g(T) \leq t \log^6 m/10$. Set $T^* = \{a^* : a \in T\}$, and $D(T) = T \cup T^*$. By the above, $G(T)$ spans at most $3t/2$ edges other than those in $F(T)$. For each such edge, take the lower endpoint, and then follow down the edges of $F(T)$ until a vertex of $D(T)$ is reached. Thus there are at least $t/2$ vertices of $D(T)$ that are not hit. Now consider the collection of sets $N(a)$, for a a vertex of $D(T)$ that is not hit. Each such set has size at least $\log^6 m/8$, by our choice of R_G. We claim that these sets are mutually disjoint. If not, there is a minimal vertex w in some pair of them, so w sends edges down to both sets, and one of these edges is not in $F(T)$: the path thus generated hits the vertex a defining one of the two sets, leading to a contradiction. Thus the union of these sets has size at least $t \log^6 m/16$, and this union is certainly contained in $G(T)$, which is a contradiction. □

Once we have all our small sets "spreading" into the bottom $6m \log\log m + m$ elements of S, it is not hard to show that they continue to spread at a very steady rate through S.

Lemma 3.4. *Almost surely, for every subset T of R_G with $|T| = t \leq m^{1/4}$,*

$$2te^{s/m} \geq |T \cup \Gamma_S(T)| \geq te^{s/m}/80.$$

Proof. We break $S \backslash Y$ into consecutive sets of size $m/\log m$, say $A_{k+1} < A_{k+2} < \ldots < A_l$, where $l = s \log m/m \leq \frac{1}{2} \log^2 m$ and $k = 6 \log m \log\log m + \log m$. For $j = k, \ldots, l$, set $C_j = Y \cup \bigcup_{i=k+1}^{j} A_i$. For a set T of size t, and $k \leq j < l$, we estimate the probability that

$$\left| |\Gamma_{C_{j+1}}(T)| - |\Gamma_{C_j}(T)|(1 + 1/\log m) \right| \geq |\Gamma_{C_j}(T)| \log^{-5/2} m,$$

given that $m^{3/4} \geq \gamma \equiv |\Gamma_{C_j}(T)| \geq t \log^6 m/16$.

The number $N_j(T)$ of vertices of A_{j+1} sending an edge to $\Gamma_{C_j}(T)$ is a binomial random variable with parameters $m/\log m$ and

$$1-(1-p)^\gamma = \frac{\gamma}{m}(1+O(m^{-1/4})).$$

Hence we have

$$\Pr\left(\left|N_j(T) - \frac{\gamma}{\log m}\right| \geq \frac{\gamma}{\log^{5/2} m}\right) \leq 2e^{-\gamma/3\log^4 m} \leq 2e^{-t\log^2 m/48},$$

where we used the Chernoff bounds on the tail of the Binomial distribution, the lower bound on γ, and implicitly the fact that the variation we tolerate is much greater than the error in our estimate of $\gamma/\log m$ for the mean of $N_j(T)$.

Hence, for each fixed t, the probability that, for some T of size t, and some j, the set $\Gamma_{C_j}(T)$ fails to spread within the prescribed bounds is at most

$$2m^t le^{-t\log^2 m/48} = o(m^{-1/4}).$$

Thus, almost surely, every set $\Gamma_{C_j}(T)$ spreads at the required rate.

Hence, almost surely, for every set T of size at most $m^{1/4}$,

$$\begin{aligned}
|\Gamma_S(T)| &\geq \tfrac{1}{16} t\log^6 m\left(1+\tfrac{1}{\log m} - \tfrac{1}{\log^{5/2} m}\right)^{l-k} \\
&\geq \tfrac{t}{16}\exp\left(6\log\log m + \left(\tfrac{s\log m}{m} - 6\log m\log\log m - \log m\right)\left(\tfrac{1}{\log m} - \tfrac{1}{\log^2 m}\right)\right) \\
&\geq \tfrac{t}{16}\exp\left(\tfrac{s}{m} - 1 - \tfrac{s}{m\log m}\right) \\
&\geq \tfrac{t}{16}e^{s/m-3/2} \geq \tfrac{t}{80}e^{s/m}.
\end{aligned}$$

Similarly, almost surely, for every set T,

$$\begin{aligned}
&|\Gamma_S(T)| \leq 5t\log^6 m(1+1/\log m + 10^5/\log^{5/2} m)^{l-k} \\
&\leq 5t\exp\left(6\log\log m + \left(\tfrac{s\log m}{m} - 6\log m\log\log m - \log m\right)\tfrac{1}{\log m}\right) \\
&= 5te^{s/m-1} \leq 2te^{s/m}.
\end{aligned}$$

This completes the proof. □

There is one more property we want from our model $P'(s,p)$, namely that no single element of S is above too many members of R. A reasonably sharp bound follows readily from Lemma 3.1(iii): the number $J(x)$ of elements of $R\cup S$ below any member x of S is dominated by a random variable distributed as $X(s+r,p)$, so the probability that $J(x)$ is greater than $3e^{s/m}\log m$ is at most

$$\begin{aligned}
(1-q^{s+r})^{3e^{s/m}\log m} &\leq \left(1-(1-1/m)^{s+m}\right)^{3e^{s/m}\log m} \\
&\leq \exp\left(\frac{-3e^{s/m}\log m}{e^{s/m+1}(1+o(1))}\right) = o(m^{-11/10}) = o(s^{-1}),
\end{aligned}$$

so there is almost surely no vertex $x\in S$ above more than $3e^{s/m}\log m$ elements of R.

Collecting together our results so far, we have the following.

Lemma 3.5. *Suppose $p \to 0$, and $s \leq \frac{1}{2} m \log m$. Then, almost surely, the random order $P'(s,p)$ has the following properties.*

(1) There is a subset R_G of R of size at least $m/12$ such that, for every subset T of R_G of size $t \leq m^{1/4}$,

$$2te^{s/m} \geq |T \cup \Gamma_S(T)| \geq te^{s/m}/80.$$

(2) No element of S is above more than $3e^{s/m} \log m$ elements of R. □

Our strategy for the next few steps is as follows. We take four sets A, A', S, S' with $|A| = |A'| = m/12$ and $|S| = |S'| = s$. (One should think of $R \prec S \prec S' \prec R'$, with a $P'(s,p)$ random partial order on $R \cup S$ and another independent copy on the dual of $R' \cup S'$, then $A = R_G$ and $A' = R'_G$.) We then fix *any* partial orders $<$ and $<'$ on $A \cup S$ and $A' \cup S'$ respectively, satisfying the conclusions of Lemma 3.5. To be precise, we state the following three properties.

(P1) Every element of A is minimal in $<$, and every element of A' is maximal in $<'$.

(P2) For every subset T of A of size $t \leq m^{1/4}$,

$$2te^{s/m} \geq |T \cup \Gamma_S(T)| \geq te^{s/m}/80;$$

and for every subset T of A' of size $t \leq m^{1/4}$,

$$2te^{s/m} \geq |T \cup \Gamma'_{S'}(T)| \geq te^{s/m}/80,$$

where $\Gamma'_S(T)$ is the set of elements of S' below some element of T in $<'$.

(P3) No element of S is above more than $3e^{s/m} \log m$ elements of A in $<$, and no element of S' is below more than $3e^{s/m} \log m$ elements of A' in $<'$.

Given orders $<, <', <_B$ with: (i) $<$ and $<'$ orders on $S \cup A$ and $S' \cup A'$ respectively, satisfying (P1)-(P3), (ii) $<_B$ a bipartite order, in which $x <_B y$ implies that $x \in A \cup S$ and $y \in A' \cup S'$, we define the order $Q(<, <', <_B)$ to be the transitive closure of the union of the three orders. Thus, for $a \in A, a' \in A'$, a is below a' in $Q(<, <', <_B)$ iff $x <_B y$ for some $x \in \Gamma_S(a)$ and $y \in \Gamma'_{S'}(a')$.

Given $<$ and $<'$ as above, we define a random partial order $P'' = P''(<, <', p)$ on $A \cup S \cup S' \cup A'$ by taking a random bipartite order $<_B = <_B (s,p)$, with edge-probability p, and vertex sets $A \cup S, S' \cup A'$, and forming $Q(<, <', <_B)$.

(Note that, if $<$ and the dual of $<'$ were chosen as random $P'(s,p)$ orders, with S immediately below S' in $\prec$, and then A and A' selected as sets of good vertices, then the random partial order just defined is distributed as the restriction of the random graph order to the chosen vertex set, conditional on $<$ and $<'$.)

Our aim is to prove that, almost surely, the random order P'' has large dimension.

We first show that, almost surely, many pairs $(a, a') \in A \times A'$ are unrelated in P''. For this, we may as well assume that $s = \frac{1}{2} m \log m$.

A pair $(a, a') \in A \times A'$ is related exactly when there is some relation of $<_*$ between an element of $\Gamma_S(a)$ and an element of $\Gamma'_{S'}(a')$. These sets both have sizes at most $2e^{s/m} = 2m^{1/2}$, by property (P2), so the probability that the pair is unrelated is at least $q^{4m} = e^{-4}(1 + o(1))$. Therefore the expected number of unrelated pairs is at least $\left(\frac{m}{12}\right)^2 e^{-4}(1 + o(1)) \geq m^2/8000$.

Again, we have to confront the problem that the relations between these pairs are far from independent. We deal with this by using the following "isoperimetric inequality" of Bollobás and Leader [7] (see also Leader [15]). Let $\mathcal{Q}_k(p)$ denote the "weighted" k-dimensional cube, i.e., the set of subsets of $\{1,\dots,k\}$ with a set V given weight $p^{|A|}q^{k-|A|}$. This weight is a probability measure, so $\mathcal{Q}_k(p)$ is thus a probability space. Our intention here is to identify $\{1,\dots,k\}$ with the elements of $(S\cup A)\times(S'\cup A')$, when the probability measure $\Pr_p$ on $\mathcal{Q}_k(p)$ coincides with our probability measure on bipartite orders $<_B$. For $\mathcal{A}$ a subset of $\mathcal{Q}_k(p)$, and $l\in\mathbb{N}$, we define

$$\mathcal{A}_{(l)}=\{B\in\mathcal{Q}_k(p):|B\triangle A|\le l \text{ for some } A\in\mathcal{A}\}.$$

Finally, for $r\in\mathbb{N}$, we define $\mathcal{B}^{(r)}$ to be the set of subsets of $\{1,\dots,k\}$ with at most r elements. The inequality of Bollobás and Leader [7] is as follows.

Lemma 3.6. *(i) Let $\mathcal{A}\subseteq\mathcal{Q}_k(p)$ be a down-set with $\Pr_p(\mathcal{A})\ge\Pr_p(\mathcal{B}^{(r)})$. Then, for every $l\in\mathbb{N}$, we have $\Pr_p(\mathcal{A}_{(l)})\ge\Pr_p(\mathcal{B}^{(r+l)})$.*
(ii) For $z\ge 0$, let $\mathcal{A}\subseteq\mathcal{Q}_k(p)$ be either a down-set or an up-set, such that $\Pr_p(\mathcal{A})\ge e^{-z}$. Then $\Pr_p(\mathcal{A}_{(l)})\ge 1-e^{-z}$, where $l=\sqrt{12zpk}$. □

Part (ii) of Lemma 3.6 follows immediately from part (i) on applying the Chernoff bounds for the binomial distribution.

As indicated, we apply Lemma 3.6 with $z=\log m$, $k=(s+m/12)^2\le\frac{1}{3}m^2\log^2 m$, and the set $\{1,\dots,k\}$ identified with the pairs $(x,y)\in(S\cup A)\times(S'\cup A')$. Let $\mathcal{A}$ be the set of those bipartite orders $<_B$ giving rise to a partial order $Q(<,<',<_B)$ in which at least $m^2/10000$ pairs $(a,a')\in A\times A'$ are unrelated. Note that $\mathcal{A}$ is a down-set. Since the number of unrelated pairs has expectation at least $m^2/8000$ and is bounded above by m^2, the set $\mathcal{A}$ has probability at least $1/4000\ge 1/m=e^{-z}$. Therefore, by Lemma 3.6, $\Pr_p(\mathcal{A}_{(l)})$ is at least $1-o(1)$, i.e., a random order $<_B$ is almost surely within $l\le 2m^{1/2}\log^{3/2}m$ edges of a bipartite order in $\mathcal{A}$.

By property (P3) of the orders $<$ and $<'$, the addition or removal of any edge (x,y) between $S\cup A$ and $S'\cup A'$ can only change the number of unrelated $A-A'$ pairs by at most $(3e^{s/m}\log m)^2\le 9m\log^2 m$. Thus the number of unrelated pairs in P'' is almost surely at least $m^2/10000-(2m^{1/2}\log^{3/2}m)(9m\log^2 m)\ge m^2/20000$. We say that a partial order P'' with at least $m^2/20000$ unrelated $A-A'$ pairs satisfies property (Q1).

To complete our project, we now have to show that, almost surely, every small subset of A has fairly many elements of A' above it, and vice versa. The proof of this is very similar to the previous part.

Consider any subset T of A of size $t\le m^{1/4}$, and any element x of A'. By property (P2), $|T\cup\Gamma_S(T)|\ge te^{s/m}/80$, and $|\Gamma'_{S'}(x)|\ge e^{s/m}/80$. Thus the probability that x is above some element of T in P'' is at least $1-(1-1/m)^{te^{2s/m}/6400}\ge 1-\exp\left(-te^{2s/m}/6500m\right)$. Hence the number $N(T)$ of elements of A' above some element of T has mean at least $(m/12)(1-x^t)$, where $x=\exp(-e^{2s/m}/6500m)$. Note that $1-x\ge e^{2s/m}/7000m$.

We again use the result of Bollobás and Leader [7], Lemma 3.6, to show that $N(T)$ is almost surely never far from its mean. This time, we identify $\{1,\dots,k\}$ with the set of pairs $(x,y)\in(T\cup\Gamma_S(T))\times(A'\cup S')$, so $k\le 3ste^{s/m}$: clearly $N(T)$ depends only on the relations of $<_B$ in this set.

Let $\mathcal{A}=\mathcal{A}(T)$ be the set of bipartite orders $<_B$ such that $N(T)\ge(m/12)(1-x^t)-1$. Note that $\mathcal{A}$ is an up-set. Also, $\mathbb{E}N(T)<(m/12)(1-x^t)-1+m\Pr_p(\mathcal{A})$, so $\Pr_p(\mathcal{A})\ge 1/m$.

We apply Lemma 3.6 with $z=t\log m$, and deduce that

$$\Pr_p(\mathcal{A}_{(l)}) \geq 1 - e^{-t\log m} = 1 - o\left(\binom{m}{t}^{-1} m^{1/4}\right).$$

Therefore, with at least this probability, the addition of at most $l = \sqrt{12zpk}$ relations to $<_B$ will ensure that $N(T) \geq (m/12)(1-x^t) - 1$. Note that

$$l = \sqrt{12zpk} \leq \sqrt{36t \log m\, m^{-1} st e^{s/m}} \leq 6te^{s/2m} \log m.$$

By property (P3) of $<$, the addition of any one relation only increases $N(T)$ by at most $3e^{s/m} \log m$.

Therefore, almost surely, every set T of size $t \leq m^{1/4}$ has

$$N(T) \geq \frac{m}{12}(1-x^t) - 1 - 18te^{3s/2m} \log^2 m.$$

Hence, for each T, the number of elements of A' *not* above some element of T is almost surely at most

$$\frac{m}{12}x^t + 20te^{3s/2m} \log^2 m.$$

We say that a partial order P'' satisfying this condition has property (Q2). Of course, the analogous property (Q2') also holds almost surely for all small subsets T' of A'.

Let (Q3) be the property that, for every pair of sets $T \subset A$, $T' \subset A'$, of sizes

$$t_0 \equiv 10^6 m \log m e^{-2s/m},$$

we have x below y in P'' for some $x \in T$, $y \in T'$. We prove that P'' almost surely has property (Q3). Recall that $s \geq \frac{2}{5} m \log m$, so that $t_0 \leq m^{1/4}$ and therefore, by property (P2), $|T \cup \Gamma_S(T)|$ and $|T' \cup \Gamma_{S'}(T')|$ are both at least $t_0 e^{s/m}/80$. The probability that a particular pair (T, T') fails is at most

$$(1 - 1/m)^{t_0^2 e^{2s/m}/6400} \leq e^{-10t_0 \log m} = o\left(\binom{m}{t_0}^{-2}\right).$$

Thus the expected number of failing pairs is $o(1)$.

Now we assume that P'' does have properties (Q1), (Q2), (Q2') and (Q3). Thus there is a relation between every pair of subsets of size $t_0 = 10^6 m \log m e^{-2s/m}$, every subset T of A of size $t \leq m^{1/4}$ has $N(T) \geq (m/12)(1-x^t) - 20te^{3s/2m} \log^2 m$, and the analogous inequality holds for subsets of A', and there are at least $m^2/20000$ unrelated pairs in $A \times A'$. Recall here that $x \leq 1 - e^{2s/m}/7000m$. We now show that these assumptions imply that the dimension of P'' is large.

We bound the number of pairs $(a, a') \in A \times A'$ reversed by a linear extension L of P''. Consider the top t_0 elements of A in L, say $a_1 >_L a_2 >_L \cdots >_L a_{t_0}$, and similarly the bottom t_0 elements of A', say $a'_1 <_L \cdots <_L a'_{t_0}$. By property (Q3), there is some pair i, j $(1 \leq i, j \leq t_0)$ such that $a_i < a'_j$ in P'', so every pair reversed by L involves one of these $2t_0$ elements.

For $1 \leq j \leq t_0$, the number of reversed pairs involving a_j is at most the number $N(\{a_1, \ldots, a_j\})$ of elements of A' incomparable with all of $a_1, \ldots, a_j$, and so by (Q2) is at most $\frac{m}{12}x^j + 20je^{3s/2m} \log^2 m$. Hence the total number of reversed pairs is at most

$$2\sum_{j=1}^{t_0}\left(\tfrac{m}{12}x^j+20je^{3s/2m}\log^2 m\right)$$
$$\le \tfrac{m}{6}\sum_{j=0}^{\infty}\left(1-\tfrac{e^{2s/m}}{7000m}\right)^t + 20t_0^2e^{3s/2m}\log^2 m$$
$$= \tfrac{7000m^2}{6e^{2s/m}} + 2\times 10^{13}m^2e^{-5s/2m}\log^4 m$$
$$\le 1200m^2e^{-2s/m}.$$

Therefore, under our assumptions, the dimension of P'' is at least

$$\frac{m^2}{20000}\Big/\frac{1200m^2}{e^{2s/m}} \ge 10^{-8}e^{2s/m}.$$

As we have been indicating throughout, this suffices to prove Theorem 1.6, and indeed also Theorem 1.7. We fill in the details below.

Proof. [of Theorem 1.6] The upper bound follows immediately from the upper bound on the width given in Theorem 1.5.

For the lower bound, we set $m = p^{-1}$ as usual, and take $s = \frac{1}{2}m\log m$. We are given $p \ge \log n/n$, so $n \ge 2s + 6m$. We define four sets as follows: B consists of the $\prec$-first $3m$ elements of $[n]$, U consists of the $\prec$-next s elements, U' of the $\prec$-next s elements, and B' of the $\prec$-next $3m$. There are almost surely at least $9m/10$ maximal elements in the random order $P_{n,p}$ restricted to B, and $9m/10$ minimals in the order restricted to B', as noted in [5]. If this is the case, then, to be definite, take A to be the set of the $9m/10$ $\prec$-largest maximals from B, and A' to consist of the $9m/10$ $\prec$-smallest minimals from B'. Note that the choice of A and A' depends only on the order restricted to B and to B'.

We next consider the random order restricted to the sets $A\cup U$ and $A'\cup U'$. These are distributed as a random order $P'(s,p)$ and its dual, so the conclusions of Lemma 3.5 hold almost surely for $R = A$ and $S = U$, and dually for $R = A'$ and $S = U'$. In other words, the restrictions $<$ and $<'$ of the random order $P_{n,p}$ to $A\cap U$ and to $A'\cap U'$ almost surely have properties (P1)-(P3).

Conditioned on the restrictions of $P_{n,p}$ being equal to any particular pair $<, <'$ satisfying (P1)-(P3), we have seen that the dimension of the random order restricted to $A\cup A'$, and hence dim$P_{n,p}$, is almost surely at least $10^{-8}e^{2s/m} = 10^{-8}m$.

Combining all the above, we see that the dimension of $P_{n,p}$ is almost surely at least $10^{-8}p^{-1}$, as claimed. □

Proof. [of Theorem 1.7] Set $m = p^{-1}$ and $s = \frac{1}{2}(n-3m)$. Note that $s \ge \frac{2}{5}m\log m$. The proof of Theorem 1.6 goes through without alteration, and we deduce that

$$\dim P_{n,p} \ge 10^{-8}e^{2s/m} \ge 10^{-10}e^{n/m},$$

almost surely, as required.

For the upper bound, we apply Theorem 1.2, the upper bound of Füredi and Kahn [12]. Thus we need an estimate on the maximum degree Δ of an element of $P_{n,p}$. A bound on Δ follows from Lemma 3.1(iii), just as in the proof of Lemma 3.5(2), namely, almost surely, $\Delta \le 3e^{pn}\log n$. Theorem 1.2 now tells us that

$$\dim P_{n,p} \le 50\Delta\log^2\Delta \le 150e^{pn}\log n(pn+\log\log n+2)^2 \le 40e^{pn}\log^3 n,$$

as required. □

We have proved Theorem 1.6 with 10^{-8} for the value of ϵ; clearly this can be improved substantially, but tinkering with the method is unlikely to produce a reasonable constant. We suspect the result is true with $\epsilon = 1$, at least, and maybe it is even true that the dimension is almost surely $(1 - o(1))$ times the width.

On a slightly more abstract note, it is likely that there is a constant c such that, if $pn/\log n \to \infty$, $p \log n \to 0$ and $\epsilon > 0$, then, almost surely,

$$(c - \epsilon)p^{-1} \leq \dim P_{n,p} \leq (c + \epsilon)p^{-1},$$

but this may be rather hard to prove.

References

1. M.Albert and A.Frieze, Random graph orders, *Order* **6** (1989) 19–30.
2. N.Alon, B.Bollobás, G.Brightwell and S.Janson, Linear extensions of a random partial order, *Annals of Applied Prob.* **4** (1994) 108–123.
3. A.Barak and P.Erdős, On the maximal number of strongly independent vertices in a random acyclic directed graph, *SIAM J. Algebraic and Disc. Methods* **5** (1984) 508–514.
4. B.Bollobás, *Random Graphs*, Academic Press, London, 1985, xv+447pp.
5. B.Bollobás and G.Brightwell, The width of random graph orders, submitted.
6. B.Bollobás and G.Brightwell, The structure of random graph orders, submitted.
7. B.Bollobás and I.Leader, Isoperimetric inequalities and fractional set systems, *J. Combinatorial Theory (A)* **56** (1991) 63–74.
8. G.Brightwell, Models of random partial orders, in *Surveys in Combinatorics 1993, Invited papers at the 14th British Combinatorial Conference*, K.Walker Ed., Cambridge University Press (1993).
9. P.Erdős, H.Kierstead and W.T.Trotter, The dimension of random ordered sets, *Random Structures and Algorithms* **2** (1991) 253–275.
10. P.Erdős and A.Rényi, On random graphs I, *Publ. Math. Debrecen* **6** (1959) 290–297.
11. P.Erdős and A.Rényi, On the evolution of random graphs, *Publ. Math. Inst. Hungar. Acad. Sci.* **5** (1960) 17–61.
12. Z.Füredi and J.Kahn, On the dimensions of ordered sets of bounded degree, *Order* **3** (1986) 17–20.
13. M.Hall, *Combinatorial Theory*, 2nd Edn., Wiley-Interscience Series in Discrete Mathematics (1986) xv+440pp.
14. G.H.Hardy and E.M.Wright, *An Introduction to the Theory of Numbers*, 5th Edn., Oxford University Press (1979) xvi+426pp.
15. I.Leader, Discrete isoperimetric inequalities, in *Probabilistic Combinatorics and its Applications*, Proceedings of Symposia in Applied Mathematics 44, American Mathematical Society, Providence (1991).
16. C.M.Newman, Chain lengths in certain directed graphs, *Random Structures and Algorithms* **3** (1992) 243–253.
17. C.M.Newman and J.E.Cohen, A stochastic theory of community food webs: IV. Theory of food chain lengths in large webs, *Proc. R. Soc. London Ser. B* **228** (1986) 355–377.
18. K.Simon, D.Crippa and F.Collenberg, On the distribution of the transitive closure in random acyclic digraphs, *Lecture Notes in Computer Science* **726** (1993) 345–356.
19. W.T.Trotter, Inequalities in dimension theory for posets, *Proc. Amer. Math. Soc.* **47** (1975) 311–316.

20. W.T.Trotter, Embedding finite posets in cubes, *Discrete Math.* **12** (1975) 165–172.
21. W.T.Trotter, *Combinatorics and Partially Ordered Sets: Dimension Theory*, The Johns Hopkins University Press, Baltimore (1992) xiv+307pp.

Hereditary and Monotone Properties of Graphs

Béla Bollobás and Andrew Thomason

Department of Pure Mathematics and Mathematical Statistics, 16 Mill Lane, Cambridge CB2 1SB, England

Summary. Given a hereditary graph property $\mathcal{P}$ let $\mathcal{P}^n$ be the set of those graphs in $\mathcal{P}$ on the vertex set $\{1,\ldots,n\}$. Define the constant c_n by $|\mathcal{P}^n| = 2^{c_n\binom{n}{2}}$. We show that the limit $\lim_{n\to\infty} c_n$ always exists and equals $1 - 1/r$, where r is a positive integer which can be described explicitly in terms of $\mathcal{P}$. This result, obtained independently by Alekseev, extends considerably one of Erdős, Frankl and Rödl concerning principal monotone properties and one of Prömel and Steger concerning principal hereditary properties.

AMS Subject Classification: Primary 05C35, Secondary 05C30.

1. Introduction

A *property* $\mathcal{P}$ of graphs is an infinite class of graphs which is closed under isomorphism. A property $\mathcal{P}$ is *hereditary* if every *induced subgraph* of every member of $\mathcal{P}$ is also in $\mathcal{P}$, and is *monotone* if every *subgraph* of every member of $\mathcal{P}$ is also in $\mathcal{P}$; a monotone property is therefore also hereditary. Let $\mathcal{P}^n$ be the set of graphs in $\mathcal{P}$ with vertex set $[n] = \{1,\ldots,n\}$. In this paper we are interested in the rate of growth of $\mathcal{P}^n$ with n, so it is convenient to define the constant $c_n = c_n(\mathcal{P})$ by $|\mathcal{P}^n| = 2^{c_n\binom{n}{2}}$. Note that eventually $c_n < 1$ unless $\mathcal{P}$ is the *trivial* property consisting of all graphs.

Scheinermann and Zito [22] asked if, for a hereditary property, $\lim_{n\to\infty} c_n$ always exists, and if so what the possible values are. The limit had been evaluated earlier for the monotone property of K_n-free graphs by Erdős, Kleitman and Rothschild [10], by using the method of Kleitman and Rothschild [12]. Their result was generalized by Erdős, Frankl and Rödl [9], who considered the monotone property of graphs not containing a given graph F as a subgraph. The structure of K_n-free graphs was investigated by Kolaitis, Prömel and Rothschild [13].

It is considerably more difficult to determine the asymptotic size of the hereditary, but non-monotone, property $\mathcal{P}$ of not containing a given graph F as an induced subgraph; this problem has been studied by Prömel and Steger in a series of papers [14, 15, 16, 17, 18]. In particular in [15] they gave sharp estimates for the property of not containing an induced quadrilateral, and in [16] they evaluated $\lim_{n\to\infty} c_n$ for every given graph F.

Our purpose in this paper is to evaluate $\lim_{n\to\infty} c_n$ for *every* hereditary property $\mathcal{P}$, not only for *principal* properties, namely those defined by a single forbidden subgraph (induced or otherwise). Our main result (Theorem 4.1) claims that this limit equals $1-1/r$, where $r = r(\mathcal{P})$ is an integer which we call the *colouring number* of $\mathcal{P}$, to be defined below. Since writing this paper we have discovered that we were anticipated in the result by Alekseev [2], but the present proof appears to be simpler and more natural. In particular, it is reasonable to suspect a relationship between the main theorem and the Erdős-Stone theorem. In our proof this relationship is established and is made transparent.

The proof of Theorem 4.1 contains an implicit demonstration of the existence of $\lim_{n\to\infty} c_n$. In fact, Alekseev [1] already showed that the limit exists. Moreover it was shown in [6] that the sequence (c_n) is monotone decreasing (and so the limit must therefore exist). The analogous sequence monotonicity property was shown to pertain to uniform hypergraph properties, and so the corresponding limits again exist. However, for hypergraphs we are unable to say anything about the value of the limits, or even whether every real number between zero and one is the limit for some property.

The colouring number $r(\mathcal{P})$ of a property $\mathcal{P}$ is defined as follows. Let $0 \le s \le r$ be integers. An (r,s)-*colouring* of a graph H is a map $\psi : V(H) \to [r]$ such that $H[\psi^{-1}(i)]$ is complete for $1 \le i \le s$ and is empty otherwise. Note that H is $(r,0)$-colourable if and only if $\chi(H) \le r$, since an $(r,0)$-colouring is just an r-colouring in the usual sense. Note too that a graph is (r,r)-colourable if and only if $\chi(\overline{H}) \le r$ (all notation used but not defined in this paper is described in [3]). Now let

$$\mathcal{C}_k(r,s) = \{H \ : \ |H| = k \text{ and } H \text{ is } (r,s)\text{-colourable}\}.$$

Then the *colouring number* $r(\mathcal{P})$ of a hereditary property $\mathcal{P}$ is defined by

$$r(\mathcal{P}) = \max\{r \ : \ \text{there exists } 0 \le s \le r \text{ such that } \mathcal{P} \supset \bigcup_{k\ge 1} \mathcal{C}_k(r,s)\},$$

that is, $r(\mathcal{P})$ is the largest integer r such that, for some s, $\mathcal{P}$ contains every (r,s)-colourable graph. Since a graph of order r is (r,s)-colourable for every s, $0 \le s \le r$, it follows that $r(\mathcal{P})$ is finite if $\mathcal{P}$ is non-trivial. Note also that the only $(1,0)$-colourable graphs are empty, and the only $(1,1)$-colourable graphs are complete, so by Ramsey's theorem $r(\mathcal{P}) \ge 1$.

Consideration of the property complementary to $\mathcal{P}$ gives us another way to view the colouring number of $\mathcal{P}$. A property $\mathcal{P}$ is hereditary if, and only if, for some sequence $F_1, F_2, \dots$ of graphs, $\mathcal{P}$ is the collection of graphs having no induced subgraph isomorphic to an F_i. Then

$$r(\mathcal{P}) = \max\{r : \text{ for some } 0 \le s \le r \text{ no } F_i \text{ is } (r,s)\text{-colourable}\}.$$

Although monotone properties $\mathcal{P}$ are also hereditary, and so the above serves to define $r(\mathcal{P})$, it is worth giving the definition in a third and simpler form for these properties. A property $\mathcal{P}$ is monotone if, and only if, for some sequence $F_1, F_2, \dots$ of graphs, $\mathcal{P}$ is the collection of graphs having no subgraph isomorphic to an F_i. Since any (r,s)-colourable graph contains an $(r,0)$-colourable subgraph, we see that for a *monotone* property $\mathcal{P}$ the colouring number is

$$r(\mathcal{P}) = \max\{r : \text{ no } F_i \text{ is } r\text{-colourable}\}.$$

Therefore for monotone properties our result is that $\lim_{n\to\infty} c_n = 1 - 1/r$, where $r = \min\{\chi(F_i)\} - 1$. This is in contrast to the case of hereditary properties in general, for which the colouring number is not merely the minimum of those colouring numbers of the properties defined by excluding a single F_i.

It is immediate that $\liminf_{n\to\infty} c_n \ge 1 - 1/r(\mathcal{P})$ for any hereditary property $\mathcal{P}$. For let s be an integer for which $\mathcal{P}$ contains every (r,s)-colourable graph. Partition the set $[n]$ into r disjoint classes $V_1, \dots, V_r$, where $\lfloor n/r \rfloor \le |V_i| \le \lceil n/r \rceil$. Every graph with vertex set $[n]$ in which the subgraph spanned by V_i is complete for $1 \le i \le s$, and in which the subgraph spanned by V_i is empty for $s < i \le r$, is (r,s)-colourable and is therefore in $\mathcal{P}^n$. Hence $|\mathcal{P}^n| \ge 2^{(1-1/r+O(1/n))\binom{n}{2}}$, which shows that $\liminf_{n\to\infty} c_n \ge 1 - 1/r$.

Let us denote by $\mathrm{ex}_{\mathrm{ind}}(n, \mathcal{P})$ the maximal number of edges in a graph G_0 of order n, for which there is an edge-disjoint graph G_1 on the same vertex set, such that every graph G with $G_1 \subseteq G \subseteq G_0 \cup G_1$ is in the hereditary property $\mathcal{P}$. This invariant was introduced (for principal properties) by Prömel and Steger [16]. Trivially $|\mathcal{P}^n| \geq 2^{\mathrm{ex}_{\mathrm{ind}}(n,\mathcal{P})}$, and it is clear from the construction in the paragraph above that $\mathrm{ex}_{\mathrm{ind}}(n, \mathcal{P}) \geq (1 - \frac{1}{r(\mathcal{P})} + o(1))\binom{n}{2}$; it is, therefore, a consequence of our result that this last inequality is, in fact, an equality.

The proof of Theorem 4.1 is based on three well-known results, namely those of Ramsey [19], of Szemerédi [23] and of Erdős and Stone [11]. Ramsey's theorem states that for each positive integer k there exists an integer $R(k)$, such that if $n \geq R(k)$ and the edges of the complete graph K_n are coloured with two colours then there will be a monochromatic complete subgraph of order k. Szemerédi's lemma will be stated and discussed in section 3. The Erdős-Stone theorem will be discussed in section 2; in fact, for the present purpose we have to prove a slight extension of that theorem. This extension will be examined more fully elsewhere [7].

Each of the three cited fundamental results asserts the existence of certain constants. For the achievement of our aim, which is to describe $\lim_{n\to\infty} c_n$, it is sufficient that these constants exist. To investigate the rate of convergence of the sequence (c_n) we would need effective versions of these theorems, but we make no attempt whatsoever to carry out this investigation.

2. An extension of the Erdős-Stone theorem

Let $r, t \geq 1$ and $\epsilon > 0$. The well-known theorem of Erdős and Stone [11] states that every graph of order n and size at least $(1 - 1/r + \epsilon)\binom{n}{2}$ contains $K_{r+1}(t)$, the complete $(r+1)$-partite graph with t vertices in each class, provided n is large enough. Exactly how large n needs to be, as a function of r, t and ϵ, was investigated by Bollobás and Erdős [4], Bollobás, Erdős and Simonovits [5], and Chvátal and Szemerédi [8].

We shall prove here an extension of the Erdős-Stone theorem, in which a certain number of 'forbidden' edges are added to the graph, and it is required that the $K_{r+1}(t)$ span no forbidden edge. To be precise, let G and F be two graphs on the same vertex set. We say that a subgraph H of G is *F-avoiding* if $V(H)$ spans no edge of F. Our theorem shows that if G satisfies the conditions of the Erdős-Stone theorem, and if $e(F)$ is sufficiently small, then G will contain an F-avoiding $K_{r+1}(t)$.

An alternative proof of the theorem has been pointed out by Rödl [20]; his proof depends on a 'supersaturated' version of the Erdős-Stone theorem, stating that every graph of order n (sufficiently large) and size at least $(1-1/r+\epsilon)\binom{n}{2}$ contains not just one but $cn^{(r+1)t}$ copies of $K_{r+1}(t)$. Given this theorem, our theorem follows at once since if $e(F)$ is small it cannot meet all of the copies of $K_{r+1}(t)$. However, our purpose here is to give a short self-contained proof. We will not need to know the rate of growth of n with r, t and ϵ, so the examination of this rate of growth will be made elsewhere [7]. For the proof it will be convenient to have the following weak form of Turán's theorem [24], wherein $\overline{G}$ denotes the complement of G.

Lemma 2.1. *If G is a graph of order $m \geq t^2$ and $e(\overline{G}) \leq m^2/2t$ then $G \supset K_t$.*

Proof. By Turán's theorem, if $G \not\supset K_t$ then

$$e(\overline{G}) \geq \sum_{i=0}^{t-2} \binom{\lfloor (m+i)/(t-1) \rfloor}{2} \geq (t-1)\binom{m/(t-1)}{2} = \frac{m(m-t+1)}{2(t-1)} > \frac{m^2}{2t},$$

as claimed. □

Theorem 2.1. *Given $r \geq 0$, $t \geq 1$ and $\epsilon > 0$, there exist $\delta = \delta(r,t,\epsilon)$ and $n_0 = n_0(r,t,\epsilon)$ such that the following holds. Let F and G be graphs on the same vertex set of order $n \geq n_0$, with $e(F) \leq \delta n^2$ and, if $r \geq 1$, with*

$$e(G) \geq (1 - \frac{1}{r} + \epsilon)\binom{n}{2}.$$

Then G contains an F-avoiding $K_{r+1}(t)$ subgraph.

Proof. Note first that we may assume that F and G share no edges. We shall apply induction on r. If $r = 0$ then Lemma 2.1 applied to $\overline{G}$ shows that $\delta = 1/(2t)$ and $n_0 = t^2$ will do.

Suppose then that $r \geq 1$ and that the assertion holds for smaller values of r. Let us first make the customary observation that for some m, $(\epsilon/2)^{1/2}n \leq m \leq n$, G has an induced subgraph G_m of order m and minimal degree at least $(1 - 1/r + \epsilon/2)m$. Indeed, if this were not the case then, with $s = \lfloor (\epsilon/2)^{1/2} n \rfloor$, we could find subgraphs $G_n = G \supset G_{n-1} \supset \ldots \supset G_s$ such that, for $n > i \geq s$, G_i is a subgraph induced by i vertices and the only vertex of G_{i+1} not in G_i has degree less than $(1 - 1/r + \epsilon/2)(i+1)$ in G_{i+1}. But then

$$e(G_s) > (1 - \frac{1}{r} + \epsilon)\binom{n}{2} - (1 - \frac{1}{r} + \frac{\epsilon}{2})\sum_{s}^{n-1}(i+1) \geq \epsilon n^2/4 \geq s^2/2$$

if n is large, which is a contradiction.

Let F_m be the subgraph of F induced by $V(G_m)$. Note that $e(F_m) \leq \delta n^2 \leq (2\delta/\epsilon)m^2$. Let $T = \lceil 4t/\epsilon r \rceil$. By the induction hypothesis, if n is large enough and δ is small enough, G_m contains an F-avoiding $K_r(T)$, say K. The bound on the minimal degree of G_m implies that each vertex of K sends at least $(1 - 1/r + \epsilon/2)m - rT$ edges to $G_m - K$. We claim that the set U of vertices of $G_m - K$ sending at least $(r - 1 + \epsilon r/4)T$ edges to K has at least $\epsilon rm/5$ members. Indeed, if this were not the case then the number of edges f between K and $G_m - K$ would satisfy

$$\frac{\epsilon rm}{5} rT + \left(1 - \frac{\epsilon r}{5}\right) m \left(r - 1 + \frac{\epsilon r}{4}\right) T > f \geq \left\{\left(1 - \frac{1}{r} + \frac{\epsilon}{2}\right) m - rT\right\} rT,$$

which is false if n is large.

Each vertex of U is joined to a $K_r(t)$ subgraph of K. There are only $\binom{T}{t}^r$ such subgraphs, so for some $K_r(t)$ subgraph K' the set W of vertices of U joined to every vertex of K' has at least $\frac{\epsilon rm}{5}\binom{T}{t}^{-r}$ members. Now W spans at most $(2\delta/\epsilon)m^2$ edges of F, so again by Lemma 2.1 if n, and hence m, is large enough and δ is small enough, the set W contains t vertices spanning an independent set in F, and hence forming with K' an F-avoiding $K_{r+1}(t)$ subgraph of G. □

3. Universal Graphs

Given a finite class $\mathcal{G}$ of graphs, we say that a graph G is $\mathcal{G}$-*universal* if every member of $\mathcal{G}$ is an induced subgraph of G. The classes of graphs of interest to us are the classes $\mathcal{C}_k(r,s)$ of (r,s)-colourable graphs defined earlier. Our aim in this section is to show that every suitably large collection of graphs has a $\mathcal{C}_k(r,s)$-universal member, for some s.

For the proof of our theorem we will need a lemma of Szemerédi [23]. In order to state this lemma the notion of uniformity must be defined: a pair of subsets U and W of the vertex set of a graph G is said to be *η-uniform* if $|d(U,W) - d(U',W')| < \eta$ whenever $U' \subseteq U$, $|U'| > \eta|U|$ and $W' \subseteq W$, $|W'| > \eta|W|$, where $d(U,W) = e(U,W)/|U||W|$. Szemerédi's Uniformity Lemma is (equivalent to) the statement that, given $\eta > 0$ and an integer l, there is an integer $L = L(l,\eta)$ such that the vertices of every graph of order n can be partitioned into m classes $V_1,\dots,V_m$, for some $l \le m < L$, so that $\lfloor n/m \rfloor \le |V_i| \le \lceil n/m \rceil$ and all but at most $\eta\binom{m}{2}$ of the pairs (V_i,V_j) are η-uniform, $1 \le i < j \le m$.

The following lemma is a standard application of the notion of uniformity (see for example [21]); once again we shall give a self-contained proof in a form convenient for our needs.

Lemma 3.1. *Let H be a graph with vertex set $\{x_1,\dots,x_k\}$. Let $0 < \lambda, \eta < 1$ satisfy $k\eta \le \lambda^{k-1}$. Let G be a graph with vertex set $\bigcup_{i=1}^k V_i$ where the V_i are disjoint sets each of order $u \ge 1$. Suppose that each pair (V_i,V_j), $1 \le i < j \le k$, is η-uniform, that $d(V_i,V_j) \le 1-\lambda$ if $x_ix_j \notin E(H)$ and that $d(V_i,V_j) \ge \lambda$ if $x_ix_j \in E(H)$. Then there exist vertices $v_i \in V_i$, $1 \le i \le k$, such that the map $x_i \mapsto v_i$ gives an isomorphism between H and the subgraph of G spanned by $\{v_1,\dots,v_k\}$.*

Proof. Observe that by replacing, if necessary, the set of V_i–V_j edges of G by the complementary set of V_i–V_j edges, we may assume that H is the complete graph and that $d(V_i,V_j) \ge \lambda$ for all $1 \le i < j \le k$. We shall select, one by one, the vertices $v_1,\dots,v_k$ so that after $v_1,\dots,v_l$ have been chosen there are sets $U_j^l \subset V_j$, $l < j \le k$, such that each of $v_1,\dots,v_l$ is joined to every vertex of U_j^l and $|U_j^l| \ge (\lambda-\eta)^l u$. Clearly we can begin (with $l = 0$) by taking $U_j^0 = V_j$.

In order to find v_{l+1}, having found $v_1,\dots,v_l$, let

$$W_j = \{v \in U_{l+1}^l : |\Gamma(v) \cap U_j^l| < (\lambda-\eta)|U_j^l|\}$$

for $l+1 < j \le k$. Since $d(W_j,U_j^l) < \lambda - \eta$, the definition of η-uniformity applied to the pairs (W_j,U_j^l) implies that either $|W_j| < \eta u$ or $|U_j^l| < \eta u$. However the latter is ruled out because $l \le k-1$ and hence $|U_j^l| \ge (\lambda-\eta)^{k-1}u > (\lambda^{k-1} - (k-1)\eta)u \ge \eta u$. Hence

$$|U_{l+1}^l \setminus \bigcup_{j>l+1} W_j| > (\lambda-\eta)^l u - (k-l-1)\eta u \ge (\lambda^l - l\eta - (k-l-1)\eta)u > 0.$$

We may therefore choose a vertex $v \in U_{l+1}^l$ so that if we let $U_j^{l+1} = \Gamma(v) \cap U_j^l$ then $|U_j^{l+1}| \ge (\lambda-\eta)^{l+1}u$ for $l+1 < j \le k$. In this way we may proceed to find each of the vertices $v_1,\dots,v_k$. □

We define a *coloured partition* π to be an edge colouring of the complete graph on the vertex set $[m]$ which uses four colours, namely grey, red, green and blue. The number m is called the *order* of π and is denoted by $|\pi|$. We associate with π two graphs, F_π and G_π, both on the vertex set $[m]$; the edge set of F_π is $\{ij : ij \text{ is grey}\}$ and the edge set of G_π is $\{ij : ij \text{ is green}\}$.

Given a graph G and constants $0 < \lambda, \eta < 1$, we say that G *satisfies* the coloured partition π with respect to λ and η, if there is a partition of the vertices of G into $|\pi|$ classes $V_1,\dots,V_{|\pi|}$, with $|V_1| \le |V_2| \le \dots \le |V_{|\pi|}| \le |V_1| + 1$, such that the pair (V_i, V_j) is not η-uniform only if ij is grey, and otherwise $0 \leq d(V_i, V_j) \leq \lambda$, $\lambda < d(V_i, V_j) < 1 - \lambda$ or $1 - \lambda \leq d(V_i, V_j) \leq 1$ according as ij is red, green or blue. Szemerédi's Uniformity Lemma asserts that, given λ, η and some integer l, there exists an integer $L = L(l, \eta)$ such that any graph G satisfies some coloured partitition π with respect to λ and η, with $l \leq |\pi| < L$ and $e(F_\pi) \leq \eta\binom{|\pi|}{2}$.

The following theorem shows that a graph satisfying π will be $\mathcal{C}_k(r, s)$-universal if the size of G_π is large. We shall use this result only to prove Theorem 3.2, but we state it because it may be useful in further investigations of hereditary graph properties. The dependencies of some of the parameters on other ones gives the theorem a technical appearance, but these dependencies are likely to be crucial in applications. In this theorem and the next one, we will use $r - 1$ where in Theorem 2.1 we used r. Despite the potential confusion, we adopt this usage because it seems more natural here.

Theorem 3.1. *Let $r, k \in \mathbb{N}$, $r \geq 2$, $\epsilon > 0$ and $0 < \lambda < 1$ be given. Then there exist positive constants $l_1 = l_1(r, k, \epsilon)$ and $\eta_1 = \eta_1(r, k, \epsilon, \lambda)$ with the following property. Let $0 < \eta \leq \eta_1$ and let π be a coloured partition with $|\pi| \geq l_1$, $e(F_\pi) \leq \eta\binom{|\pi|}{2}$ and*

$$e(G_\pi) \geq \left(1 - \frac{1}{r-1} + \epsilon\right)\binom{|\pi|}{2}.$$

Then there is an integer $s = s(\pi)$, $0 \leq s \leq r$, such that every graph of order $n \geq |\pi|$ that satisfies π with respect to λ and η is $\mathcal{C}_k(r, s)$-universal.

Proof. Let t be the Ramsey number $R(k)$, let $l_1 = n_0(r - 1, t, \epsilon)$ and let $\eta_1 = \min\{\delta(r - 1, t, \epsilon), k^{-1}\lambda^{k-1}\}$, where n_0 and δ are the functions appearing in Theorem 2.1. We shall show that these functions have the required properties.

By Theorem 2.1 there is an F_π-avoiding copy of $K_r(t)$ in G_π. Consider one of the r vertex classes of this $K_r(t)$, say T where $|T| = t$. The edges ij, where $\{i, j\} \subset T$, are coloured either red, green or blue. Let us recolour *orange* the edges coloured red or green. By the definition of t, there is a subset $T' \subset T$, $|T'| = k$, such that the edges ij where $\{i, j\} \subset T'$, are either all orange or all blue. This argument can be applied to each class of the $K_r(t)$. Therefore our $K_r(t)$ in G_π contains a subgraph $K = K_r(k)$ such that the edge ij is green if i and j are in different classes of K, the edge ij is blue if i and j are both in one of the first s classes of K, and the edge ij is orange if i and j are both in one of the remaining $r - s$ classes of K, where $s = s(\pi)$ lies in the range $0 \leq s \leq r$.

Let $H \in \mathcal{C}_k(r, s)$, so that $|H| = k$ and H is (r, s)-colourable. Choose an (r, s)-colouring of H and label the vertex set of H by $x_1, \ldots, x_k$ so that the first k_1 vertices have colour 1, the next k_2 vertices have colour 2 and so on. Now construct a subgraph K' of K by choosing k_1 vertices from the first class of K, k_2 vertices from the second class, and so on. Thus $|K'| = k$. Let G be a graph of order $n \geq |\pi|$ that satisfies π with respect to λ and η, and let G' be the subgraph of G induced by $\bigcup\{V_i : i \in V(K')\}$. The definition of the colouring and the choice of K' mean that Lemma 3.1 can be applied to H and G'. This shows that H is an induced subgraph of G' and so also of G. □

We are now able to prove the main theorem of this section, which shows that any suitably large collection of graphs will have a $\mathcal{C}_k(r, s)$-universal member.

Theorem 3.2. *Let $r, k \in \mathbb{N}$, $r \geq 2$ and $\epsilon > 0$ be given. Then there exists $n_1 = n_1(r, k, \epsilon)$ such that if $n > n_1$ and $\mathcal{P}^n$ is a collection of at least $2^{(1-1/(r-1)+\epsilon)\binom{n}{2}}$ labelled graphs with vertex set $[n] = \{1, \ldots, n\}$, then $\mathcal{P}^n$ contains a $\mathcal{C}_k(r, s)$-universal graph for some s, $0 \leq s \leq r$.*

Proof. We begin by choosing constants λ, l and η as follows. Choose $0 < \lambda < 1/4$ small enough so that $(e/\lambda)^\lambda < 2^{\epsilon/10}$. Now choose $0 < \eta < \min\{\epsilon/8, \eta_1(r,k,\epsilon/8,\lambda)\}$ and an integer l larger than $\max\{20/\epsilon, l_1(r,k,\epsilon/8)\}$, where the functions η_1 and l_1 are those appearing in Theorem 3.1.

Apply Szemerédi's Uniformity Lemma, with the usual parameters l and η, to each graph G in $\mathcal{P}^n$, thus obtaining, for each such graph, a coloured partition $\sigma(G)$ which G satisfies with respect to λ and η, with $l \le |\sigma(G)| < L = L(l,\eta)$ and $e(F_{\sigma(G)}) \le \eta\binom{|\sigma(G)|}{2}$. Since there are at most $4^{\binom{L}{2}}$ coloured partitions of order less than L and at most n^L ways to split the set $[n]$ into fewer than L parts, there is some coloured partition π satisfied by at least $n^{-L}4^{-\binom{L}{2}}2^{(1-1/(r-1)+\epsilon)\binom{n}{2}}$ graphs of $\mathcal{P}^n$, which number being at least $2^{(1-1/(r-1)+\epsilon/2)\binom{n}{2}}$ if n is large. From now on we consider only this particular coloured partition π which is satisfied by many graphs. We shall show that any graph in $\mathcal{P}^n$ satisfying π is $\mathcal{C}_k(r,s)$-universal, where $s = s(\pi)$, provided n is both larger than L and is large enough to ensure the validity of the estimates below. This assumption on the size of n will be made without further mention throughout the remainder of the proof.

Our aim is now to show that the coloured partition π must have many green edges. This can be done by estimating the number of graphs which can satisfy the coloured partition π. Within such a graph there are at most $2^{\binom{N}{2}}$ possible distributions of edges inside a vertex class V_i, where $N = \lceil n/|\pi| \rceil$. Between two classes V_i and V_j the possible distributions of edges of the graph number at most 2^{N^2} for grey and green edges ij of π, and at most $\sum_{i=0}^{\lambda N^2}\binom{N^2}{i}$ for red and blue edges. Let f be the number of grey and green edges in the coloured partition π. Then the number of graphs which may satisfy π with respect to λ and η is at most

$$2^{|\pi|\binom{N}{2}} \times \left[\sum_{i=0}^{\lambda N^2}\binom{N^2}{i}\right]^{\binom{|\pi|}{2}} \times 2^{fN^2}.$$

Each of the first two factors here is bounded above by $2^{\frac{\epsilon}{10}\binom{n}{2}}$; the first because

$$|\pi|\binom{N}{2} < \frac{2}{|\pi|}\binom{n}{2} < \frac{\epsilon}{10}\binom{n}{2}$$

and the second because

$$\left[\sum_{i=0}^{\lambda N^2}\binom{N^2}{i}\right]^{\binom{|\pi|}{2}} < \left[2\binom{N^2}{\lambda N^2}\right]^{\binom{|\pi|}{2}} < \left(\frac{e}{\lambda}\right)^{\lambda N^2\binom{|\pi|}{2}} \le \left(\frac{e}{\lambda}\right)^{\lambda n^2/2} < 2^{\frac{\epsilon}{10}\binom{n}{2}}.$$

It follows that the number of graphs satisfying the coloured partition π is at most $2^{\frac{\epsilon}{5}\binom{n}{2}+fN^2}$, and since we have at least $2^{(1-1/(r-1)+\epsilon/2)\binom{n}{2}}$ such graphs we see that

$$f \ge \left(1 - \frac{1}{r-1} + \frac{3\epsilon}{10}\right)\binom{n}{2}N^{-2} > \left(1 - \frac{1}{r-1} + \frac{\epsilon}{4}\right)\binom{|\pi|}{2}.$$

Now $f = e(F_\pi) + e(G_\pi)$, and Szemerédi's lemma asserts that $e(F_\pi) \le \eta\binom{|\pi|}{2}$. Since $\eta < \epsilon/8$, we see that $e(G_\pi) \ge (1 - 1/(r-1) + \epsilon/8)\binom{|\pi|}{2}$. Theorem 3.1 now implies that any graph of order n which satisfies π is $\mathcal{C}_k(r, s(\pi))$-universal, as claimed. □

4. Hereditary properties

Recall that for a non-trivial hereditary property $\mathcal{P}$ of graphs the colouring number $r(\mathcal{P})$ is defined by

$$r(\mathcal{P}) = \max\{r \ : \text{ there exists } 0 \le s \le r \text{ such that } \mathcal{P} \supset \bigcup_{k \ge 1} \mathcal{C}_k(r,s)\}.$$

We can now state and prove the main result of this paper.

Theorem 4.1. *Let $\mathcal{P}$ be a non-trivial hereditary property of graphs and let $\mathcal{P}^n$ be the set of graphs in $\mathcal{P}$ with vertex set $[n] = \{1, \ldots, n\}$. Set $|\mathcal{P}^n| = 2^{c_n \binom{n}{2}}$. Then*

$$\lim_{n\to\infty} c_n = 1 - 1/r(\mathcal{P}),$$

where $r(\mathcal{P})$ is the colouring number of $\mathcal{P}$.

Proof. Let $r = r(\mathcal{P})$. We saw in the introduction that $\liminf_{n\to\infty} c_n \ge 1 - 1/r$. Suppose that the assertion of the theorem is false, so that $\limsup_{n\to\infty} c_n > 1-1/r$, which is to say there exists $\epsilon > 0$ such that $|\mathcal{P}^n| > 2^{(1-1/r+\epsilon)\binom{n}{2}}$ for infinitely many values of n. It follows from Theorem 3.2 that for each integer k there is an integer s, $0 \le s \le r+1$, such that for some n the set $\mathcal{C}_k(r+1,s)$ is contained in $\mathcal{P}^n$ and therefore in $\mathcal{P}$. Consequently, for some value of s, we have $\mathcal{C}_k(r+1,s) \subset \mathcal{P}$ for infinitely many k, and hence for all k. But this contradicts the definition of $r(\mathcal{P})$, so our theorem is proved. □

As an example, let $\mathcal{P}$ be the property consisting of planar graphs and let $\mathcal{P}'$ be those graphs containing no induced subgraph isomorphic to either K_5 or $K_{3,3}$. Clearly $\mathcal{P} \subset \mathcal{P}'$. Since K_5 is $(2,s)$-colourable for $s = 1,2$ and $K_{3,3}$ is $(2,0)$-colourable, it follows that $r(\mathcal{P}) = r(\mathcal{P}') = 1$, so $|\mathcal{P}'| = 2^{o(n^2)}$.

For a slightly less simple example, let $\mathcal{P}_1$ be the (monotone) property of containing no complete graph K_4, and let $\mathcal{P}_2$ be the property of containing no induced subgraph isomorphic to the 7-cycle C_7. Now let $\mathcal{P}$ be $\mathcal{P}_1 \cap \mathcal{P}_2$; that is, let $\mathcal{P}$ be the property of having no induced subgraph isomorphic to either K_4 or C_7. Each of K_4 and C_7 is $(4,s)$-colourable for every s, $0 \le s \le 4$. Now K_4 is $(3,s)$-colourable for $1 \le s \le 3$ but is not $(3,0)$-colourable, so $r(\mathcal{P}_1) = 3$. On the other hand, C_7 is $(3,0)$-colourable but has no $(3,s)$-colouring if $s = 2,3$, so $r(\mathcal{P}_2) = 3$. Finally, neither K_4 nor C_7 has a $(2,0)$-colouring so $r(\mathcal{P}) = 2$. Consequently $|\mathcal{P}_1^n| \approx 2^{n^2/3}$ and $|\mathcal{P}_2^n| \approx 2^{n^2/3}$, whereas $|\mathcal{P}^n| \approx 2^{n^2/4}$.

References

1. V.E. Alekseev, Hereditary classes and coding of graphs, *Probl. Cybern.* **39** (1982) 151–164 (in Russian).
2. V.E. Alekseev, On the entropy values of hereditary classes of graphs, *Discrete Math. Appl.* **3** (1993) 191–199.
3. B. Bollobás, *Extremal Graph Theory*, Academic Press (1978), xx+488pp.
4. B. Bollobás and P. Erdős, On the structure of edge graphs, *Bull. London Math. Soc.* **5** (1973) 317–321.

5. B. Bollobás, P. Erdős and M. Simonovits, On the structure of edge graphs II, *J. London Math. Soc.* **12**(2) (1976) 219–224.
6. B. Bollobás and A. Thomason, Projections of bodies and hereditary properties of hypergraphs (submitted).
7. B. Bollobás and A. Thomason, An extension of the Erdős-Stone theorem, (in preparation).
8. V. Chvátal and E. Szemerédi, On the Erdős-Stone theorem, *J. London Math. Soc.* **23** (1981) 207–214.
9. P. Erdős, P. Frankl and V. Rödl, The asymptotic enumeration of graphs not containing a fixed subgraph and a problem for hypergraphs having no exponent, *Graphs and Combinatorics* **2** (1986), 113–121.
10. P. Erdős, D.J. Kleitman and B.L. Rothschild, Asymptotic enumeration of K_n-free graphs, in *International Coll. Comb.*, Atti dei Convegni Lincei (Rome) **17** (1976) 3–17.
11. P. Erdős and A.H. Stone, On the structure of linear graphs, *Bull. Amer. Math. Soc.* **52** (1946) 1087–1091.
12. D.J. Kleitman and B.L. Rothschild, Asymptotic enumeration of partial orders on a finite set, *Trans. Amer. Math. Soc.* **205** (1975) (205–220).
13. Ph.G. Kolaitis, H.J. Prömel and B.L. Rothschild, K_{l+1}-free graphs: asymptotic structure and a 0-1-law, *Trans. Amer. Math. Soc.* **303** (1987) 637–671.
14. H.J. Prömel and A. Steger, Excluding induced subgraphs: quadrilaterals, *Random Structures and Algorithms* **2** (1991) 55–71.
15. H.J. Prömel and A. Steger, Excluding induced subgraphs II: Extremal graphs, *Discrete Applied Mathematics*, (to appear).
16. H.J. Prömel and A. Steger, Excluding induced subgraphs III: a general asymptotic, *Random Structures and Algorithms* **3** (1992) 19–31.
17. H.J. Prömel and A. Steger, The asymptotic structure of H-free graphs, in *Graph Structure Theory* (N. Robertson and P. Seymour, eds), Contemporary Mathematics **147**, Amer. Math. Soc., Providence, 1993, pp. 167-178.
18. H.J. Prömel and A. Steger, Almost all Berge graphs are perfect, *Combinatorics, Probability and Computing* **1** (1992) 53-79.
19. F.P. Ramsey, On a problem of formal logic, *Proc. London Math. Soc.* **30**(2) (1929) 264–286.
20. V. Rödl (personal communication).
21. V. Rödl, Universality of graphs with uniformly distributed edges, *Discrete Mathematics* **59** (1986) 125–134.
22. E.R. Scheinerman and J. Zito, On the size of hereditary classes of graphs, *J. Combinatorial Theory (B)*, (to appear).
23. E. Szemerédi, Regular partitions of graphs, in *Proc. Colloque Inter. CNRS* (J.-C. Bermond, J.-C. Fournier, M. las Vergnas, D. Sotteau, eds), (1978).
24. P. Turán, On an extremal problem in graph theory (in Hungarian), *Mat. Fiz. Lapok* **48** (1941) 436–452.

Properties of Graded Posets Preserved by Some Operations

Sergej Bezrukov[1] and Konrad Engel[2]

[1] Department of Math. & CS, University of Paderborn, D–33198 Paderborn, Germany
[2] Department of Math., University of Rostock, D–18051 Rostock, Germany

Summary. We answer the following question: Let P and Q be graded posets having some property and let $\circ$ be some poset operation. Is it true that $P \circ Q$ has also this property? The considered properties are: being Sperner, a symmetric chain order, Peck, LYM, and rank compressed. The studied operations are: direct product, direct sum, ordinal sum, ordinal product, rankwise direct product, and exponentiation.

1. Introduction and overview

Throughout we will consider finite *graded* partially ordered sets, i.e. finite posets in which every maximal chain has the same length. For such posets P there exists a unique function $r : P \mapsto \mathbb{N}$ (called *rank function*) and a number m (called *rank of* P), such that $r(x) = 0$ ($r(x) = m$) if x is a minimal (resp., maximal) element of P, and $r(y) = r(x) + 1$ if y covers x in P (denoted $x \lessdot y$). The set $P_{(i)} := \{x \in P : r(x) = i\}$ is called *i-th level* and its cardinality $|P_{(i)}|$ the *i-th Whitney number*. If S is a subset of P, let $r(S) := \sum_{x \in S} r(x)$, in particular $r(P) := \sum_{x \in P} r(x)$. Let us emphasize, that $r(P)$ is here not the rank of P.

A symmetric chain is a chain of the form $C = (x_0 \lessdot x_1 \lessdot \ldots \lessdot x_s)$, where $r(x_0)+r(x_s) = m$. A subset A of P is called a *k-family*, if there are no $k+1$ elements of A lying on one chain in P. Further, $F \subseteq P$ is called a *filter*, if $y \geq x \in F$ implies $y \in F$, and $I \subseteq P$ is said to be an *ideal*, if $y \leq x \in I$ implies $y \in I$. Let $d_k(P) := \max\{|A| : A \text{ is a } k\text{-family}\}$ and $w_k(P)$ denotes the largest sum of k Whitney numbers. Obviously, $w_k(P) \leq d_k(P)$, for $k \geq 1$. The poset P is said to be:

i) *Sperner* (S), if $d_1(P) = w_1(P)$,
ii) *symmetric chain order* (SC), if P has a partition into symmetric chains,
iii) *Peck*, if $d_k(P) = w_k(P)$, for $k \geq 1$, and $|P_{(0)}| = |P_{(m)}| \leq |P_{(1)}| = |P_{(m-1)}| \leq \cdots \leq |P_{(\lfloor m/2 \rfloor)}| = |P_{(\lceil m/2 \rceil)}|$,
iv) *LYM*, if $\sum_{x \in A} \frac{1}{|P_{(r(x))}|} \leq 1$ for every antichain A of P,
v) *rank compressed* (RC), if $\mu_r := \frac{r(F)}{|F|} \geq \frac{r(P)}{|P|} =: \mu_P$ for every filter $F \neq \emptyset$ of P.

Since F is a filter iff $P \setminus F$ is an ideal, one can define equivalently

v)' P is rank compressed, if $\mu_I := \frac{r(I)}{|I|} \leq \mu_P$ for every ideal $I \neq \emptyset$ of P.

Let us mention that P.Erdös [6] proved already in 1945 that finite Boolean lattices are Peck.

In the following we will study which of these properties are preserved by usual poset operations, i.e. the question is: if P and Q have some property, is it true that $P \circ Q$ has this property either (here $\circ$ is some operation)?

Throughout let m (resp., n) be the rank of P (resp., Q). If it is not clear from the context whether r is the rank function of P, Q, or $P \circ Q$ we will write r_P, r_Q, $r_{P \circ Q}$, respectively.

A widely studied operation is the *direct product* $P \times Q$, i.e. the poset on the set $\{(x, y) : x \in P \text{ and } y \in Q\}$, such that $(x, y) \leq (x', y')$ in $P \times Q$ if $x \leq_P x'$ and $y \leq_Q y'$. It is well-known, that the direct product preserves the properties SC (de Bruijn et al. [2] and Katona [10]), Peck (Canfield [3]), and RC (Engel [4]), and it does not preserve the properties S and LYM (but with an additional condition it does (Harper [8] and Hsieh, Kleitman [9])), see Figure 1.

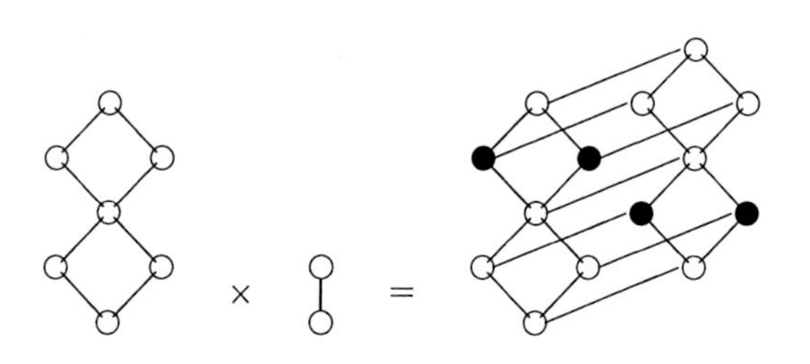

Fig. 1.1.

A simple operation is the *direct sum* $P + Q$, i.e. the poset on the union $P \cup Q$, such that $x \leq y$ in $P + Q$ if either $x, y \in P$ and $x \leq_P y$, or $x, y \in Q$ and $x \leq_Q y$. In order to obtain again a graded poset we will suppose here $m = n$. Then it is easy to see, that SC and Peck properties are preserved, but S, LYM, and RC not, see Figure 2.

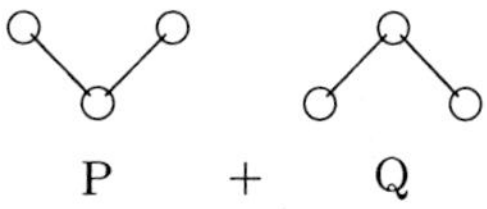

P + Q

Fig. 1.2.

Another easy operation is the *ordinal sum* $P \oplus Q$, i.e. the poset on the union $P \cup Q$, such that $x \leq y$ in $P \oplus Q$, if $x, y \in P$ and $x \leq_P y$, or $x, y \in Q$ and $x \leq_Q y$, or $x \in P$ and $y \in Q$. To draw the Hasse diagram of $P \oplus Q$, put Q above P and connect each maximal element of P with each minimal element of Q. Then it is obvious, that properties S and LYM are preserved (note that any antichain in $P \oplus Q$ is either contained completely in P or completely in Q), and also property RC is preserved (see Theorem 1), but properties SC and Peck are not preserved, see Figure 3.

An interesting operation is the *ordinal product* $P \otimes Q$, i.e. the poset on the set $\{(x, y) : x \in P \text{ and } y \in Q\}$, such that $(x, y) \leq (x', y')$ in $P \otimes Q$, if $x = x'$ and $y \leq_Q y'$, or $x <_P x'$. To draw the Hasse diagram of $P \otimes Q$, replace each element x of P by a copy Q_x of Q, and then connect every maximal element of Q_x with every

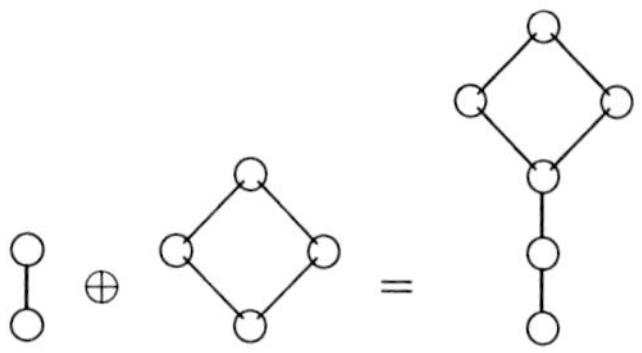

Fig. 1.3.

minimal element of Q_y whenever y covers x in P. In Theorem 2 we will prove, that properties S, LYM, and RC are preserved. Figure 4 shows that properties SC and Peck are not preserved.

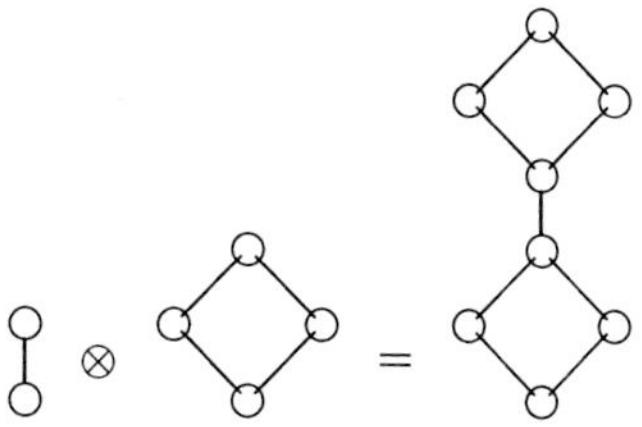

Fig. 1.4.

Studying posets like square submatrices of a square matrix, Sali [11] introduced the *rankwise direct product* $P \times_r Q$. We will suppose here again $m = n$. Then $P \times_r Q$ is the subposet of $P \times Q$, induced by $\bigcup_{i=0}^{m} P_{(i)} \times Q_{(i)}$. Sali [11] showed, that properties SC, Peck, and LYM are preserved and gave an example that property S is not preserved. Here we present an example, which shows that also RC property is not preserved. Look at the poset P of Figure 5, which is easily seen to be rank compressed.

The indicated elements form a filter F. Now it is easy to see that the filter $F \times_r F$ in $P \times_r P$ does not verify the filter inequality of v).

Finally, we will consider also the *exponentiation* Q^P, i.e. the poset on the set of all order-preserving maps $f : P \mapsto Q$ (that is, $x \leq_P y$ implies $f(x) \leq_Q f(y)$), such that $f \leq g$ if $f(x) \leq_Q g(x)$ for all $x \in P$. In Theorem 3 we will prove that none of the 5 properties is preserved.

2. Main results

Theorem 2.1. *If P and Q are rank compressed, then $P \oplus Q$ is rank compressed either.*

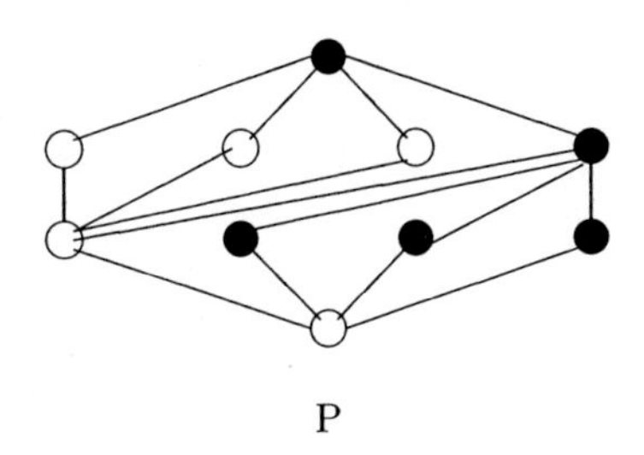

Fig. 1.5.

Proof. Obviously, if y has rank i in Q, then it has rank $i+m+1$ in $P \oplus Q$. Hence $\mu_{P\oplus Q} = \frac{r_P(P)+r_Q(Q)+(m+1)|Q|}{|P|+|Q|}$. Let I be an ideal in $P \oplus Q$ and $I \neq \emptyset$.
Case 1. Assume $I \cap Q = \emptyset$. Since $\mu_P \leq \mu_{P\oplus Q}$, and $\mu_I \leq \mu_P$ as p is rank compressed, it follows $\mu_I \leq \mu_{P\oplus Q}$.
Case 2. Let no $I \cap Q \neq \emptyset$. Then $P \subseteq I$ and $\tilde{I} := Q \cap I$ is an ideal in Q. One has $|I| = |P| + |\tilde{I}|$, $r(I) = r_P(P) + r_Q(\tilde{I}) + (m+1)|\tilde{I}|$. Therefore, $\mu_I \leq \mu_{P\oplus Q}$ is equivalent to

$$\Big(|\tilde{I}|r_Q(Q) - |Q|r_Q(\tilde{I})\Big) + |Q \setminus \tilde{I}|\Big((m+1)|P| - r(P)\Big) + |P|r(Q \setminus \tilde{I}) \geq 0.$$

This inequality is true, since Q is rank compressed and any element of P has rank at most m. ■

Theorem 2.2. *If P and Q are Sperner or LYM or rank compressed, then $P \otimes Q$ is resp., Sperner or LYM or rank compressed either.*

Proof. Let A be an antichain in $P \otimes Q$. Denote $A_x = \{y \in Q : (x,y) \in A\}$ and $\tilde{A} = \{x \in P : A_x \neq \emptyset\}$. Obviously, $\tilde{A}$ and A_x are antichains in P and Q_x, respectively.

If P and Q are Sperner, then

$$|A| = \sum_{x\in\tilde{A}} |A_x| \leq \sum_{x\in\tilde{A}} w_1(Q) = |\tilde{A}|w_1(Q) \leq w_1(P)w_1(Q) = w_1(P\otimes Q),$$

hence $P \otimes Q$ is Sperner.

Now let P and Q be LYM. Obviously, the level containing (x,y) has $|P_{(r(x))}||Q_{(r(x))}|$ elements. We have

$$\sum_{(x,y)\in A} \frac{1}{|(P\otimes Q)_{(r(x,y))}|} = \sum_{(x,y)\in A} \frac{1}{|P_{(r(x))}||Q_{(r(y))}|} = \sum_{x\in\tilde{A}} \frac{1}{|P_{(r(x))}|} \sum_{y\in A_x} \frac{1}{|Q_{(r(y))}|} \leq 1.$$

Finally, let P and Q be rank compressed. Let I be an ideal in $P \otimes Q$ and A be the set of maximal elements of I (note that A is an antichain). We use the notation $\tilde{A}$ from above and define further $I_x := I \cap Q_x$, $F_x := Q_x \setminus I_x$, $\tilde{I} := \{x \in P : I_x \neq \emptyset\}$. Then I_x (F_x) is an ideal (resp., a filter) in Q_x and $\tilde{I}$ is an ideal in P. It is easy to see that:

$$
\begin{aligned}
|I| &= |\tilde{I}||Q| - \sum_{x\in\tilde{A}} |F_x|, \\
r(I) &= |\tilde{I}|r_Q(Q) + (n+1)|Q|r_P(\tilde{I}) - \sum_{x\in\tilde{A}} \Big(r_Q(F_x) + (n+1)|F_x|r_P(x)\Big) \\
|P\otimes Q| &= |P||Q| \\
r(P\otimes Q) &= |P|r_Q(Q) + (n+1)|Q|r_P(P).
\end{aligned}
$$

Now $\frac{r(I)}{|I|} \le \frac{r(P\otimes Q)}{|P\otimes Q|}$ iff

$$
|P|\sum_{x\in\tilde{A}}\Big(|F_x|r_Q(Q) - |Q|r_Q(F_x)\Big) \le (n+1)|Q|\times
$$

$$
\left[|Q|\Big(|\tilde{I}|r_P(P) - |P|r_P(\tilde{I})\Big) + \sum_{x\in\tilde{A}}\Big(|P|r_P(x) - r_P(P)\Big)|F_x|\right].
$$

The LHS is not greater than 0 since Q is rank compressed. So it is sufficient to verify, that the formula in brackets is not smaller than 0. Denote $\tilde{A}' = \{x \in \tilde{A} : |P|r_P(x) - r_P(P) \le 0\}$. Since one can omit the positive summands in the formula and in view of $|F_x| \le |Q_x|$ it is enough to show that

$$
|\tilde{I}|r_P(P) - |P|r_P(\tilde{I}) + \Big(|P|r_P(\tilde{A}') - |\tilde{A}'|r_P(P)\Big) \ge 0,
$$

which is equivalent to

$$
|\tilde{I}\setminus\tilde{A}'|r_P(P) - |P|r_P(\tilde{I}\setminus\tilde{A}') \ge 0.
$$

This inequality is true, since P is rank compressed and $\tilde{I}\setminus\tilde{A}'$ is an ideal in P. ∎

Let P^l be the direct product of l copies of P. The investigation of rank compressed posets was initiated by the following result of Alekseev [1]:

$$
P \text{ is rank compressed iff } d_1(P^l) \sim w_1(P^l) \text{ as } l \mapsto \infty. \tag{2.1}
$$

Moreover, from the Local Limit Theorem of Gnedenko one can easily derive

$$
w_1(P^l) \sim \frac{|P|^l}{\sqrt{2\pi l}\sigma_P} \quad \text{if } P \text{ is not an antichain,}
$$

where $\sigma_P^2 = \frac{1}{|P|}\sum_{x\in P} r^2(x) - \mu_p^2$ (see Engel, Gronau [5]).

Remark 2.1. Straight-forward computations give us the following results:

$$
\begin{aligned}
\sigma^2_{P\times Q} &= \sigma_P^2 + \sigma_Q^2, \\
\sigma^2_{P\otimes Q} &= \sigma_Q^2 + (n+1)^2\sigma_P^2, \\
\sigma^2_{P\oplus Q} &= \frac{|P||Q|}{(|P|+|Q|)^2}\Big(\mu_Q + (m+1-\mu_P)\Big)^2 + \frac{1}{|P|+|Q|}\Big(|P|\sigma_p^2 + |Q|\sigma_Q^2\Big).
\end{aligned}
$$

∎

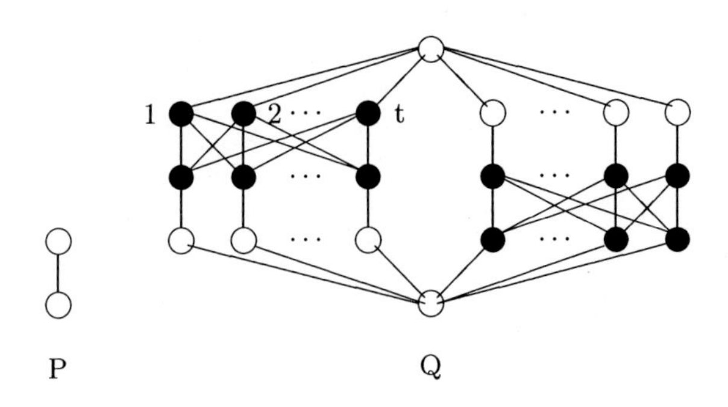

Fig. 2.1.

Theorem 2.3. *The exponentiation does not preserve any of the properties S, SC, Peck, LYM, or RC.*

Proof. First take P and Q from Figure 6.
Here we mean that the complete bipartite graphs $K_{t,t}$ are forms on the indicated vertices.

Obviously, P and Q have properties S, SC, Peck, LYM, and RC. Now Q^P is isomorphic to the subposet of $Q \times Q$ induced by the set $\{(x,y) : x \leq_Q y\}$. Consider the ideal

$$I := \{(x,y) \in Q^P : x \leq i \text{ and } y \leq i \text{ for some } i = 1, \dots, t\}.$$

Easy calculations give us

$$\mu_I = \frac{9t^2 + 21t}{2t^2 + 7t + 1} \quad \text{and} \quad \mu_{Q^P} = 4.$$

But $\mu_I > \mu_{Q^P}$ iff $t \geq 8$. Hence, Q^P is not rank compressed if $t \geq 8$, and consequently has not properties SC, Peck, LYM, since these properties imply property RC (see [5]).

Finally let $t \geq 8$ and denote $P' = P + \cdots + P$ (l times). Again, P' has all of the properties above. It is known (see Stanley [12]), that $Q^{P+\ldots+P} \cong Q^P \times \ldots \times Q^P$, hence $Q^{P'} \cong (Q^P)^l$. Since Q^P is not rank compressed, by (1)

$$d_1\left(Q^{P'}\right) = d_1\left((Q^P)^l\right) > w_1\left((Q^P)^l\right) = w_1\left(Q^{P'}\right)$$

if l is sufficiently large. Thus, $Q^{P'}$ is not Sperner. ■

Concerning the exponentiation, let us mention that if Q is a distributive lattice, then so is Q^P. Since distributivity implies rank compression (see [5]), in a lot of cases the exponentiation provides a rank compressed poset. In particular, if Q is a two-element chain, Q^P is isomorphic to the lattice of ideals of P, which is consequently rank compressed for any poset P.

3. Summary

In the following table we have summarized which of the considered properties are preserved and which not:

	$P+Q$ $m=n$	$P \oplus Q$	$P \times Q$	$P \otimes Q$	$P \times_r Q$ $m=n$	Q^P
Sperner	no	yes	no	yes	no	no
Symm. chain	yes	no	yes	no	yes	no
Peck	yes	no	yes	no	yes	no
LYM	no	yes	no	yes	yes	no
Rank compr.	no	yes	yes	yes	no	no

References

1. V.B. Alekseev. The number of monotone k-valued functions. *Problemy Kibernet.*, 28:5–24, 1974.
2. N.G.de Bruijn, C.A.v.E. Tengbergen, and D. Kruyswijk. On the set of divisors of a number. *Nieuw Arch. Wiskunde*, 23:191–193, 1951.
3. E.R. Canfield. A Sperner property preserved by product. *Linear and Multilinear Algebra*, 9:151–157, 1980.
4. K. Engel. Optimal representations of partially ordered sets and a limit Sperner theorem. *European J. Combin.*, 7:287–302, 1986.
5. K. Engel and H.-D.O.F. Gronau. Sperner theory in partially ordered sets. *BSB B.G. Teubner Verlagsgesellschaft*, Leipzig, 1985.
6. P.Erdös. On a lemma of Littlewood and Offord. *Bull. Amer. Math. Sos.*, 51:898–902, 1945.
7. J.R. Griggs. Matchings, cutsets, and chain partitions in ranked posets. *preprint*, 1991.
8. L.H. Harper. The morphology of partially ordered sets. *J. Combin. Theory, Ser. A*, 17:44–58, 1974.
9. W.N. Hsieh and D.J. Kleitman. Normalized matching in direct products of partial orders. *Studies in Appl. Math.*, 52:285–289, 1973.
10. G.O.H. Katona. A generalization of some generalizations of Sperner's theorem. *J. Combin. Theory, Ser. B*, 12:72–81, 1972.
11. A. Sali. Constructions of ranked posets. *Discrete Math.*, 70:77–83, 1988.
12. R.P. Stanley. *Enumerative combinatorics*, volume 1. Wadsworth & Brooks, Monterey, California, 1986.

Intersection Representations of the Complete Bipartite Graph

Zoltán Füredi

[1] Department of Mathematics, University of Illinois, Urbana, IL 61801–2917, USA
[2] Mathematical Institute of the Hungarian Academy of Sciences, P.O.B. 127, Budapest 1364 Hungary

Summary. A p-representation of the complete graph $\mathcal{K}_{n,n}$ is a collection of sets $\{S_1, S_2, \ldots, S_{2n}\}$ such that $|S_i \cap S_j| \geq p$ if and only if $i \leq n < j$. Let $\vartheta_p(\mathcal{K}_{n,n})$ be the smallest cardinality of $\cup S_i$. Using the Frankl-Rödl theorem about almost perfect matchings in uncrowded hypergraphs we prove the following conjecture of Chung and West. For fixed p while $n \to \infty$ we have $\vartheta_p(\mathcal{K}_{n,n}) = (1 + o(1))n^2/p$. Several problems remain open.

1. The p-intersection number of $\mathcal{K}(n, n)$

One of the important topics of graph theory is to represent graphs, or an interesting class of graphs, using other simple structures. One approach is to represent the vertices by sets so that vertices are adjacent if and only if the corresponding sets intersect (line graphs). More generally, the p-intersection number of a graph is the minimum t such that each vertex can be assigned a subset of $\{1, \ldots, t\}$ in such a way that vertices are adjacent if and only if the corresponding sets have at least p common elements. Such a system is called a *p-representation* (or p-intersection representation) of the graph $\mathcal{G}$, and the minimum t is denoted by $\vartheta_p(\mathcal{G})$.

For any graph of v vertices Erdős, Goodman, and Pósa [8] showed that $\vartheta_1 \leq \lfloor v^2/4 \rfloor$, and here equality holds for $\mathcal{K}_{\lfloor v/2 \rfloor, \lceil v/2 \rceil}$. Myung S. Chung and D. B. West [5] conjectured that the complete bipartite graph also maximizes ϑ_p. Their lower bound for $p > 1$ is

$$\vartheta_p(\mathcal{K}_{n,n}) \geq (n^2 - n)/p + 2n. \tag{1.1}$$

In this note we determine ϑ_p for these and a few more graphs. The *complete k-partite graph*, $\mathcal{K}^{(k)}_{n,\ldots,n}$, has kn vertices, k disjoint independent sets of sizes n and all the $\binom{k}{2}n^2$ edges between different classes. $\mathcal{K}(n \times k)$ denotes a graph with vertex set $V^1 \cup \ldots \cup V^k$, $V^\ell = \{v^\ell_1, \ldots, v^\ell_n\}$ and v^ℓ_i is joined to v^m_j if and only if $i \neq j$ and $\ell \neq m$. So $\mathcal{K}(n \times 2)$ is obtained from $\mathcal{K}_{n,n}$ by deleting a one factor.

Theorem 1.1. *For fixed p and k, the p-intersection number of the complete k-partite graph $\mathcal{K}^{(k)}_{n,\ldots,n}$ is $(1 + o(1))\, n^2/p$.*

Note that the asymptotic is independent from the fixed value of k. In Section 4 we will give a partial proof for Theorem 1.1 using classical design theory and obtain a better error term. An *Hadamard matrix* of order n is a square matrix M with ± 1 entries such that $MM^t = nI_n$. It is conjectured that it exists for all $n \equiv 0 (\text{mod } 4)$. The smallest undecided case is larger than 184. An $S_\lambda(v, l, t)$ *block design* is a l-uniform (multi)hypergraph on v vertices such that each t-subset is contained in exactly λ hyperedges (blocks). Block designs with $\lambda = 1$ are called *Steiner systems.* All notions we use about designs can be found, e.g., in Hall's book [12]. Wilson [19] proved that for any l there exists a bound $v_0(l)$ such that for all $v \geq v_0$ there exists a Steiner system $S(v, l, 2)$ if $\binom{v}{2}/\binom{l}{2}$ and $(v - 1)/(l - 1)$ are integers.

Theorem 1.2. (a) *If there exists an Hadamard matrix of size* $4p$*, and a Steiner system* $S(n, 2p, 2)$*, then* $\vartheta_p(\mathcal{K}(n \times 2)) = (n^2 - n)/p$.

(b) *If* $p = q^d$ *where* q *is a prime power,* d *a positive integer,* $k \leq q$*, and there exists a Steiner system* $S(n, q^{d+1}, 2)$*, then* $\vartheta_p(\mathcal{K}(n \times k)) = (n^2 - n)/p$.

Corollary 1.1. *For all* p *in Theorem 1.2, and* $n > n_0(p)$

$$n^2/p < \vartheta_p(\mathcal{K}_{n,n}) \leq \vartheta_p(\mathcal{K}(n \times 2)) + pn \leq (n^2/p) + 4pn$$

This covers all cases $p \leq 46$. A construction from the finite projective space is given in Section 3. In Section 5 we list a few open problems.

2. A random construction

A hypergraph $\mathcal{H}$ with edge set $\mathcal{E}(\mathcal{H})$ and vertex set $V(\mathcal{H})$ is called *r-uniform* (or an r-graph) if $|E| = r$ holds for every edge $E \in \mathcal{E}(\mathcal{H})$. The *degree*, $\deg_{\mathcal{H}}(x)$, of the vertex $x \in V$ is the number of edges containing it. The degree of a pair, $\deg_{\mathcal{H}}(x, y)$, is the number of edges containing both vertices x and y. The *dual*, $\mathcal{H}^*$, of $\mathcal{H}$ is the hypergraph obtained by reversing the roles of vertices and edges keeping the incidencies, i.e., $V(\mathcal{H}^*) = \mathcal{E}(\mathcal{H})$. A *matching* $\mathcal{M} \subset \mathcal{E}(\mathcal{H})$ is a set of mutually disjoint edges, $\nu(\mathcal{H})$ denotes the largest cardinality of a matching in $\mathcal{H}$.

We are going to use a theorem of Frankl and Rödl [9]. The following slightly stronger form is due to Pippenger and Spencer [17]: For all integer $r \geq 2$ and real $\varepsilon > 0$ there exists a $\delta > 0$ so that: If the r-uniform hypergraph $\mathcal{H}$ on z vertices has the following two properties (i) $(1-\delta)d < \deg_{\mathcal{H}}(x) < (1+\delta)d$ holds for all vertices, (ii) $\deg_{\mathcal{H}}(x, y) < \delta d$ for all distinct x and y, then there is a matching in $\mathcal{H}$ almost as large as possible, more precisely

$$\nu(\mathcal{H}) \geq (1 - \varepsilon)(z/r). \tag{2.1}$$

Far reaching generalizations of (2.1) has been recently proved by Kahn [15].

Suppose $\mathcal{G}$ is a graph and $\mathcal{F} = \{F_1, ...F_t\}$ is a family of subsets of the vertex set $V(\mathcal{G})$, repetition allowed. Such a system $\mathcal{F}$ is called a *p-edge clique cover* if every edge of $\mathcal{G}$ is contained in at least p members of $\mathcal{F}$ and the non-edge pairs are covered by at most $p-1$ F_i's. A p-edge clique cover is the dual of a p-representation (and vice versa), so the smallest t for which there is a p-edge clique cover is $\vartheta_p(\mathcal{G})$. This was the way Kim, McKee, McMorris and Roberts [16] first defined and investigated $\vartheta_p(\mathcal{G})$.

Proof (of Theorem 1.1). To construct a p-edge clique cover of the complete k-partite graph with n-element classes $V^1, \ldots, V^k$ consider the following multigraph $\mathcal{M}$. Every edge contained in a class V^i has multiplicity $p-1$, and all edges joining distinct classes (crossing edges) have multiplicity p. The total number of edges is

$$|\mathcal{E}(\mathcal{M})| = (p-1)k\binom{n}{2} + p\binom{k}{2}n^2 = (1+o(1))\frac{n^2}{p}\left(k\binom{p}{2} + \binom{k}{2}p^2\right) \tag{2.2}$$

Let $r = |\mathcal{E}(\mathcal{K}^{(k)}_{p,\ldots,p})| = k\binom{p}{2} + \binom{k}{2}p^2$. Define the r-uniform hypergraph $\mathcal{H}$ with vertex set $\mathcal{E}(\mathcal{M})$ as follows. The hyperedges of $\mathcal{H}$ are those r-subsets of $\mathcal{E}(\mathcal{M})$ which form a complete k-partite subgraph with p vertices in each V^i. The number of such subgraphs is

$$|\mathcal{E}(\mathcal{H})| = \binom{n}{p}^k (p-1)^{k\binom{p}{2}} p^{\binom{k}{2}p^2}.$$

Let $e \in \mathcal{E}(\mathcal{M})$ be an edge contained in a class V^i. The number of $\mathcal{K}^{(k)}_{p,\dots,p}$'s, i.e., the number of hyperedges of $\mathcal{H}$ containing e is exactly

$$\deg_{\mathcal{H}}(e) = \binom{n-2}{p-2}\binom{n}{p}^{k-1} (p-1)^{k\binom{p}{2}-1} p^{\binom{k}{2}p^2}. \tag{2.3}$$

For any crossing edge $f \in \mathcal{E}(\mathcal{M})$ connecting two distinct classes we have

$$\deg_{\mathcal{H}}(f) = \binom{n-1}{p-1}^2\binom{n}{p}^{k-2} (p-1)^{k\binom{p}{2}} p^{\binom{k}{2}p^2-1}. \tag{2.4}$$

The ratio of the right hand sides of (2.3) and (2.4) is $n/(n-1)$, so the hypergraph $\mathcal{H}$ is nearly regular, it satisfies the first condition in the Frankl-Rödl theorem for any $\delta > 0$ if n is sufficiently large. For two distinct edges, $e_1, e_2 \in \mathcal{E}(\mathcal{M})$, obviously $\deg_{\mathcal{H}}(e_1, e_2) = O(n^{kp-3})$, so condition (ii) is fulfilled, too. Apply (2.1) to $\mathcal{H}$. We get a system $\mathcal{F} = \{F_1, \dots, F_\nu\}$ of kp-element subsets of $\cup V^i$ such that every pair e contained in a class V^i is covered at most $p-1$ times, every pair f joining two distinct classes is covered at most p times. Moreover, $\nu = (1-o(1))n^2/p$, by (2.2). It follows that almost all edges of $\mathcal{K}^{(k)}_{n,\dots,n}$ are covered exactly p times, so the system $\mathcal{F}$ can be extended to a p-edge clique cover by adding sufficiently many (but only $o(n^2)$) edges.

3. Exact results from the finite projective space

Proposition 3.1. *For all $p \geq 1$, $\vartheta_p(\mathcal{K}(n \times 2)) \geq (n^2 - n)/p$.*

Proof. Let V^1 and V^2 be the two parts of the vertex set of the graph, $|V^1| = |V^2| = n$, let $\{A_i \cup B_i : 1 \leq i \leq t\}$ be a p-edge clique cover, their average size on one side is $\ell := \sum_i (|A_i| + |B_i|)/(2t)$. Using the inequalities (3.1a) $\sum_i \binom{|A_i|}{2} \leq (p-1)\binom{n}{2}$, and (3.1b) $\sum_i \binom{|B_i|}{2} \leq (p-1)\binom{n}{2}$, and the fact that (3.1c) all the $n^2 - n$ crossing edges are covered at least p times we have

$$p(n^2-n) \leq \sum_i |A_i||B_i| \leq \sum_i \left(\binom{|A_i|}{2} + \binom{|B_i|}{2}\right) + \ell t \leq (p-1)(n^2-n) + \ell t.$$

This gives $\ell t \geq (n^2 - n)$. On the other hand, (3.1a-b) give $2t\binom{\ell}{2} \leq 2(p-1)\binom{n}{2}$, hence $\ell \leq p$ and $t \geq (n^2-n)/p$ follows.

Replacing (3.1c) by $pn^2 \leq \sum_i |A_i||B_i|$ the above proof gives

$$\vartheta_p(\mathcal{K}_{n,n}) \geq (n+p-1)^2/p, \tag{3.1}$$

which is better than (1.1) for $n < (p-1)^2$.

Consider a $\mathcal{K}(n \times k)$ with classes $V^1, \dots, V^k$, $V^\ell = \{v^\ell_1, \dots, v^\ell_n\}$. We call the p-edge clique cover $\mathcal{F} = \{F_1, \dots, F_t\}$ *perfect* if the sets $\{F_i \cap V^\ell : 1 \leq i \leq n\}$ form an $S_{p-1}(n, p, 2)$ design for all ℓ. It follows, that $|F_i| = kp$ for all F_i, every edge of $\mathcal{K}(n \times k)$ is contained in exactly p sets, every pair from V^ℓ is covered $p-1$ times, every pair of the form $\{v^\ell_\alpha, v^m_\alpha\}$ is uncovered, and $t = (n^2 - n)/p$.

Proposition 3.2. *If $p = q^d$, where q is a prime power, d a positive integer and $k \le q$, then there exists a perfect p-edge clique cover of $\mathcal{K}(q^{d+1} \times k)$. Hence, in this case, $\vartheta_p(\mathcal{K}(q^{d+1} \times k)) = q^{d+2} - q$.*

Proposition 3.3. *If $p = q^d + q^{d-1} + \ldots + 1$, where q is a prime power, d a positive integer and $k \le q+1$, then $\vartheta_p(\mathcal{K}^{(k)}_{q^{d+1},\ldots,q^{d+1}}) = q^2(q^d + q^{d-1} + \ldots + 1)$.*

Proof. The lower bounds for ϑ_p are implied by Proposition 3.1 and (3.2), respectively. The upper bounds are given by the following construction. Let X be the point set of a $(d+2)$-dimensional projective space of order q, $PG(d+2,q)$, let $Z \subset X$ be a subspace of dimension d, and let $Y^1, \ldots, Y^{q+1}$ be the hyperplanes containing Z, $V^\ell = Y^\ell \setminus Z$. The sets V^ℓ partition $X \setminus Z$ into q^{d+1}-element classes. Choose a point $c \in V^{q+1}$ and label the vertices of $V^\ell = \{v_i^\ell : 1 \le i \le q^{d+1}\}$ in such a way that $\{v_i^\ell : 1 \le \ell \le q\} \cup \{c\}$ form a line for all i.

The hyperplanes not containing Z and avoiding c induce a perfect p-edge clique cover of $\mathcal{K}(p^{d+1} \times k)$ with classes $V^1, \ldots, V^k$ $(k \le q)$. Indeed, $PG(d+2,q)$ contains $q^{d+2} + q^{d+1} + \ldots + q + 1$ hyperplanes and they cover each pair of points exactly p times. The point c is contained by exactly $q^{d+1} + \ldots + q + 1$ of the hyperplanes, Z is contained in $q+1$ of them, and $Z \cup \{c\}$ is contained in a unique one. So the above defined cover consists of $q^{d+2} - q$ sets. These sets still cover each pair of the form $\{v_i^\ell, v_j^m\}$, $i \ne j$ exactly p times. However, the pairs of the form $\{v_\alpha^\ell, v_\alpha^m\}$ are uncovered, because any subspace containing these two points must contain the line through them, so it must contain the element c.

Similarly, considering all the hyperplanes not containing Z, we get a p-edge clique cover of $\mathcal{K}^{(k)}_{n,\ldots,n}$ with classes $V^1, \ldots, V^k$ where $n = q^{d+1}$ and $k \le q+1$.

The dual (perfect) p-representation of $\mathcal{K}(q^{d+1} \times k)$ can be obtained by considering a line, L, in the affine space of dimension $d+2$, and assigning all sets of the form $Y \setminus \{v^\ell\}$ to the vertices of the ℓ'th color class, where $v^\ell \in L$, and Y is a hyperplane with $Y \cap L = \{v^\ell\}$. Similarly, the dual p-representation of $\mathcal{K}^{(k)}_{n,\ldots,n}$ on the underlying set $X \setminus L$ can be obtained by assigning the sets $Y \setminus \{v^\ell\}$ to the ℓ'th color class. There might be more optimal constructions using higher dimensional spaces.

4. Constructions from Steiner systems

Proposition 4.1. *If there exists an Hadamard matrix of size $4p$, then there exists a perfect p-edge clique cover of $\mathcal{K}(2p \times 2)$, so its $\vartheta_p = 4p - 2$.*

Proof. We are going to give a perfect p-intersection representation with underlying set $\{1, \ldots, 4p-2\}$. Its dual is a perfect p-edge clique cover. Let M be an Hadamard matrix of order $4p$. We may suppose that the last row contains only $+1$'s. The ± 1's in any other row define a partition of $\{1, \ldots, 4p\}$ into two $2p$-element sets $P_i^+ \cup P_i^-$. We may also suppose that the last two entries are $M_{i,4p-1} = 1$, $M_{i,4p} = -1$ for $1 \le i \le 2p$. Finally, assign the set $P_i^+ \setminus \{4p-1\}$ to the vertex v_i^1, $P_i^- \setminus \{4p\}$ to v_i^2.

Note that both in Proposition 3.2 and here we have got the perfect p-edge clique cover from a resolvable $S_\lambda(sp, p, 2)$ design, where s is an integer, $\lambda = (p-1)/(s-1)$.

Proof (of Theorem 1.2(b)). First, we consider a perfect p-edge clique cover, $\mathcal{F}$, in the case $n = q^{d+1}$ given by Proposition 3.2. Consider k identical copies of a Steiner system $S(n, q^{d+1}, 2)$ over the n-element sets V^ℓ. Replace each block and its corresponding pairs by a copy of $\mathcal{F}$. Then we obtain a system is a perfect p-edge clique cover.

The proof of the case (a) is similar, we put together a perfect p-edge clique cover using a building block of size $2p$ supplied Proposition 4.1 and a Steiner system $S(n, 2p, 2)$. Taking the sets $\{v_i^1, v_i^2, \ldots, v_i^k\}$ p times, we get that $\vartheta_p(K_{n,\ldots,n}^{(k)}) \leq \vartheta_p(\mathcal{K}(n \times k)) + pn$. As $\vartheta_p(K_{n,\ldots,n}^{(k)})$ is a monotone function of n we got Corollary 1.3.

Conjecture 4.1. If $(n^2 - n)/p$ is an integer and $n > n_0(p, k)$, then there exists a perfect p-edge clique cover of $\mathcal{K}(n \times k)$, hence its p-intersection number $\vartheta_p = (n^2 - n)/p$.

The case $p = 1$ corresponds to the fact that there are transversal designs $T(n, k)$ (i.e., mutually orthogonal Latin squares of sizes n) for $n > n_0(k)$ (Chowla, Erdős, Straus [4], also see Wilson [20]).

Chung and West [5] proved the case $k = p = 2$. They showed $\vartheta_2(\mathcal{K}(n \times 2)) = (n^2 - n)/2$ by constructing a perfect 2-edge cover (they call it a perfect 2-generator) for the cases $n \equiv 1, 2, 5, 7, 10$, or $11 \pmod{12}$. This and (1.1) imply that

$$\vartheta_2(\mathcal{K}_{n,n}) = (n^2 + 3n)/2 \tag{4.1}$$

holds for these cases. Their conjecture about the so-called orthogonal double covers (a conjecture equivalent to the existence of a perfect 2-edge cover of $\mathcal{K}(n \times 2)$) which conjecture had appeared in [6], too, is true for all $n > 8$. This was proved by Ganter and Gronau [11] and independently by Bennett and Wu [1]. So $n_0(2, 2) = 8$ and (4.1) holds for all $n > 8$.

There are two more values proved in [5], namely the special cases $q = 2$ and $q = 3$ of the following conjecture. $\mathcal{K}(q^2 + q + 1 \times 2)$ has a perfect q-edge cover whenever a projective plane of order q exists. This would imply that equality holds in (1.1) for $(p, n) = (q + 1, q^2 + q + 1)$.

5. Further problems, conjectures

The first nontrivial lower bound for $\vartheta_p(\mathcal{K}_{n,n})$ was proved by Jacobson [13]. He and Kézdy and West [14] also investigated $\vartheta_2(\mathcal{G})$ for other classes of graphs, like paths and trees.

How large $\vartheta_p(\mathcal{K}(n \times k))$ and $\vartheta_p(\mathcal{K}_{n,\ldots,n}^{(k)})$ if n is fixed and $k \to \infty$?

Estimate ϑ_p for complete bipartite graph with parts of sizes a and b when $a \to \infty$, p is fixed and a/b goes to a finite limit.

Another interesting graph where one can expect exact results is a cartesian product, its vertex set is $I_1 \times \ldots \times I_\ell$, and $(i_1, \ldots, i_\ell)$ is joined to $(i'_1, \ldots, i'_\ell)$ if and only if $i_\alpha \neq i'_\alpha$ for all $1 \leq \alpha \leq \ell$.

For a matching, $\mathcal{M}$, of size n it easily follows that $\vartheta_p(\mathcal{M}) = \min\{t : \binom{t}{p} \geq n\}$.

One can ask the typical value of $\vartheta_p(\mathcal{G})$, i.e., the expected value of ϑ_p for the random graph of n vertices. The case of $p = 1$ was proposed in [10], and the best bounds are due to Bollobás, Erdős, Spencer, and West [3]: For almost all graphs its edge set can be covered by $O(n^2 \log\log n / \log n)$ cliques. The conjecture is, that here the term $\log\log n$ can be deleted. A counting argument gives the lower bound $E(\vartheta_1(\mathcal{G})) \geq (2 - o(1))n^2/(2\log_2 n)^2$. Obviously,

$$\vartheta_p(\mathcal{G}) \leq \vartheta_{p-1}(\mathcal{G}) + 1 \leq \vartheta_1(\mathcal{G}) + p - 1,$$

so the order of magnitude of $E(\vartheta_p)$ is at most that of $E(\vartheta_1)$. Until there is such a large gap between the lower and upper bounds of $E(\vartheta_1)$, one cannot expect better bounds for $E(\vartheta_p)$.

The notion of ϑ_p was generalized from the study of the *p-competition* graphs. Another generalization, also having several unsolved questions, is the clique coverings by *p rounds.* Let $\mathcal{G}$ be a simple graph and let $\varphi_p(\mathcal{G})$ be the minimum of $\sum_{1\leq i\leq p} n_i$ such that there are families $\mathcal{A}_1, \ldots, \mathcal{A}_p$, $|\mathcal{A}_i| = n_i$, such that each edge $e \in \mathcal{E}(\mathcal{G})$ is covered by each family (i.e., there exists an $A \in \mathcal{A}_i$ with $e \subset A$), but this does not hold for the non-edges. It is known [10], that for all graphs on n vertices $\varphi_2 \leq 3n^{5/3}$ and for almost all graphs $\varphi_2 > 0.1n^{4/3}/(\log n)^{4/3}$. For further problems and questions, see [10].

Bollobás [2] generalized the Erdős-Goodman-Pósa result as follows. The edge set of every graph on n vertices can be *decomposed* into $t(k-1, n)$ parts using only $\mathcal{K}_k$'s and edges, where $t(k-1, n)$ is the maximum number of edges in a $(k-1)$-colored graph on n vertices, e.g., $t(2, n) = \lfloor n^2/4 \rfloor$. There are many beautiful results and problems of this type, the interested reader can see the excellent survey by Pyber [18]. Most of the problems can be posed to multigraphs, obtaining new, interesting, non-trivial problems.

Let $\vartheta_p^*(\mathcal{G})$ the minimum t such that each vertex can be assigned a subset of $\{1, \ldots, t\}$ in such a way that the intersection of any two of these sets is at most p, and vertices of $\mathcal{G}$ are adjacent if and only if the corresponding sets have *exactly* p common elements. Note that in all of the results in this paper were proved an upper bound for ϑ_p^*. If $\mathcal{G}^n$ is a graph on n vertices with with $2n-3$ edges such that two vertices are connected to all others, then one can show that $\lim_{n\to\infty} \vartheta_p^*(\mathcal{G}^n) - \vartheta_p(\mathcal{G}^n) = \infty$ for any fixed p. What is $\max(\vartheta_p^*(\mathcal{G}) - \vartheta_p(\mathcal{G}))$ and $\max(\vartheta_p^*/\vartheta_p)$ for different classes of graphs? Is it true that $\vartheta_p^*(\mathcal{G}) \leq (1 + o(1))n^2/(4p)$ for every n-vertex graph?

6. Acknowledgments

The author is indebted to D. West for several fruitful conversations.

This research was supported in part by the Hungarian National Science Foundation, Grant No. OTKA 1909.

This paper was presented in the 886th AMS Regional Meeting (DeKalb, IL, USA, May 1993) in honour to Paul Erdős' 80'th birthday. (See *Abstracts of AMS* **14** (1993), p. 412, # 882-05-182.)

One of our main constructions (Theorem 1.1) along with the lower bound (3.2) was independently discovered by Eaton, Gould and Rödl [6]. They also considered 2-representations of bounded degree trees.

References

1. F. E. Bennett and Lisheng Wu, On the minimum matrix representation of closure operations, *Discrete Appl. Math.* **26** (1990), 25–40.
2. B. Bollobás, On complete subgraphs of different orders, *Math. Proc. Cambridge Philos. Soc.* **79** (1976), 19–24.

3. B. Bollobás, P. Erdős, J. Spencer, and D. B. West, Clique coverings of the edges of a random graph, *Combinatorica* **13** (1993), 1–5.
4. S. Chowla, P. Erdős, E. Straus, On the maximal number of pairwise orthogonal Latin squares of a given order, *Canad. J. Math.* **12** (1960), 204–208.
5. M. S. Chung and D. B. West, The p-intersection number of a complete bipartite graph and orthogonal double coverings of a clique, *Combinatorica* **14** (1994), 453–461.
6. J. Demetrovics, Z. Füredi, and G. O. H. Katona, Minimum matrix representation of closure operations, *Discrete Appl. Math.* **11** (1985), 115–128.
7. N. Eaton, R. J. Gould, and V. Rödl, On p-intersection representations, manuscript, 1993.
8. P. Erdős, A. Goodman, and L. Pósa, The representation of a graph by set intersections, *Canad. J. Math.* **18** (1966), 106–112.
9. P. Frankl and V. Rödl, Near-perfect coverings in graphs and hypergraphs, *European J. Combin.* **6** (1985), 317–326.
10. Z. Füredi, Competition graphs and clique dimensions, *Random Structures and Algorithms* **1** (1990), 183–189.
11. B. Ganter and H.-D. O. F. Gronau, Two conjectures of Demetrovics, Füredi, and Katona concerning partitions, *Discrete Mathematics* **88** (1991), 149–155.
12. M. Hall, Jr., *Combinatorial Theory*, Wiley–Interscience, New York, 1986.
13. M. S. Jacobson, On the p-edge clique cover number of complete bipartite graphs, *SIAM J. Disc. Math.* **5** (1992), 539–544.
14. M. S. Jacobson, A. E. Kézdy, and D. B. West, The 2-intersection number of paths and bounded-degree trees, manuscript
15. J. Kahn, On a theorem of Frankl and Rödl, in preparation.
16. Suh-ryung Kim, T. A. McKee, F. R. McMorris, and F. S. Roberts, p-competition graphs, *DIMACS Technical Report* 89-19, to appear in *SIAM J. Discrete Math.*
17. N. Pippenger and J. Spencer, Asymptotic behavior of the chromatic index for hypergraphs, *J. Combin. Th., Ser. A* **51** (1989), 24–42.
18. L. Pyber, Covering the edges of a graph by, *Colloq. Math. Soc. J. Bolyai* **60** pp. 583–610. *Graphs, Sets and Numbers*, G. Halász et al. Eds., Budapest, Hungary, January 1991.
19. R. M. Wilson, An existence theory for pairwise balanced designs I–III, *J. Combinatorial Th., Ser. A* **13** (1972), 220–273 and **18** (1975), 71–79.
20. R. M. Wilson, Concerning the number of mutually orthogonal Latin squares, *Discrete Math.* **9** (1974), 181–198.

Reflections on a Problem of Erdős and Hajnal

András Gyárfás*

Computer and Automation Institute, Hungarian Academy of Sciences, Budapest, Hungary

Summary. We consider some problems suggested by special cases of a conjecture of Erdős and Hajnal.

1. Epsilons

The problem I am going to comment reached me in 1987 at Memphis in a letter of Uncle Paul. He wrote: 'We have the following problem with Hajnal. If $G(n)$ has n points and does not contain induced C_4, is it true that it has either a clique or an independent set with n^ϵ points? Kind regards to your boss + colleagues, kisses to the ϵ -s. E.P.' After noting that ϵ have been used in different contexts I realized soon that $\frac{1}{3}$ is a good ϵ (in both senses since I have three daughters). About a month later Paul arrived and said he meant C_5 for C_4. And this minor change of subscript gave a problem still unsolved. And this is just a special case of the general problem formulated in the next paragraph.

2. The Erdős-Hajnal problem (from [7])

Call a graph H-free if it does not contain induced subgraphs isomorphic to H. Complete graphs and their complements are called *homogeneous sets*. As usual, $\omega(G)$ and $\alpha(G)$ denotes the order of a maximum clique and the order of a maximum independent set of G. It will be convenient to define $hom(G)$ as the size of the largest homogeneous set of G, i.e. $hom(G) = max\{\alpha(G), \omega(G)\}$ and

$$hom(n, H_1, H_2, ...) = min\{hom(G) : |V(G)| = n, G \text{ is } H_i\text{-free}\}$$

A well-known result of Paul Erdős ([5]) says that there are graphs of n vertices with $hom(G) \leq 2\log n$ (log is of base 2 here). The following problem of Erdős and Hajnal suggests that in case of forbidden subgraphs $hom(G)$ is much larger: Is it true, that for every graph H there exists a positive ϵ and n_0 such that every H-free graph on $n \geq n_0$ vertices contains a homogeneous set of n^ϵ vertices? If such ϵ exists for a particular H, one can define the 'best' exponent, $\epsilon(H)$ for H as

$$\epsilon(H) = sup\{\epsilon > 0 : hom(n, H) \geq n^\epsilon \text{ for } n \geq n_0\}$$

The existence of $\epsilon(H)$ is proved in [7] for P_4-free graphs (usually called *cographs* but in [7] the term *very simple graphs* have been used). In fact, a stronger statement is proved in [7]: if $\epsilon(H_i)$ exists for $i = 1, 2$ and H is a graph formed by putting all or no edges between vertex disjoint copies of H_1 and H_2 then $\epsilon(H)$ is also exists. Combining this with the well known fact that P_4-free graphs are perfect ([14]), it follows that $\epsilon(H)$ exists for those graphs H which can be generated from the one-vertex graph and from P_4, using the above operations. In the spirit of [7], call this

* Supported by OTKA grant 2570

class SVS (still very simple). In terms of graph replacements (see 4. below), SVS is generated by replacements into two-vertex graphs starting from K_1 and P_4. As far as I know, the existence of $\epsilon(H)$ is not known for any graph outside SVS.

3. Large perfect subgraphs

A possible approach to find a large homogeneous set in a graph is to find a large perfect subgraph. It was shown in [7] that any H-free graph of n vertices has an induced cograph of at least $e^{c(H)\sqrt{\log n}}$ vertices for sufficiently large n. This shows that the size of the largest homogeneous set makes a huge jump in case of any forbidden subgraph. In a certain sense it is not so far from $n^{c(H)}$... What happens if cographs are replaced by other perfect graphs? A deep result of Prömel and Steger ([13]) says that almost all C_5-free graphs are perfect (generalized split graphs). This suggest the possibility to find a large (n^ϵ) generalized split graph in a C_5-free graph of n vertices and prove the existence of $\epsilon(C_5)$ this way.

4. Replacements

A well-known important concept in the theory of perfect graphs is the replacement of a vertex by a graph. The replacement of vertex x of a graph G by a graph H is the graph obtained from G by replacing x with a copy of H and joining all vertices of this copy to all neighbors of x in G. According to a key lemma (Replacement Lemma) of Lovász (see for example in [12]), perfectness is preserved by replacements. The property of being H-free is obviously preserved by replacements if (and in some sense only if) H can not be obtained from a smaller graph by a nontrivial (at least two-vertex) replacement. For such an H, replacements can be applied to get an upper bound on $\epsilon(H)$. Analogues of the Replacement Lemma can be also useful to find large homogeneous sets (an example is Lemma 7.1 below).

5. Partitions into homogeneous sets

For certain graphs H, the existence of $\epsilon(H)$ follows from stronger properties. An H-free graph G may satisfy $\chi(G) \leq p(\omega(G))$ or $\theta(G) \leq p(\alpha(G))$ or more generally $cc(G) \leq p(\alpha(G), \omega(G))$ where p is a polynomial of constant degree and $\chi, \theta, \alpha, \omega, cc$ denote the chromatic number, clique cover number, independence number, clique number and cochromatic number of graphs. Using terminology from [10], p is a *polynomial binding function* (for χ, θ, cc, respectively). It is clear that if H-free graphs have a polynomial binding function of degree k then $\epsilon(H) \geq \frac{1}{k+1}$. Binding functions for χ (for θ) in H-free graphs may exist only if H ($\overline{H}$) is acyclic. However, the existence of a polynomial binding function for cc in H-free graphs is equivalent with the existence of $\epsilon(H)$.

6. Small forbidden subgraphs

The existence of $\epsilon(H)$ follows if H has at most four vertices since these graphs are all in SVS. But, as will be shown below, to find $\epsilon(H)$ for these small graphs is not always that simple...

Since $hom(n, H) = hom(n, \overline{H})$ from the definition, it is enough to consider one graph from each complementary pair. For $H = K_m$, finding $hom(n, H)$ is the classical Ramsey problem. In case of $m = 2, 3, 4$, $\epsilon(K_2) = 1$ (trivial), $\epsilon(K_3) = \frac{1}{2}$ (from Uncle Paul's lower bound on $R(3, m)$ in [4]), $\frac{1}{3} \leq \epsilon(K_4) \leq 0.4$ (the upper bound is due to Spencer [15]). If $H = P_3$ or $H = P_4$ then an H-free graph is perfect and thus $\epsilon(H) = \frac{1}{2}$.

There are four more graphs with four vertices to look at. Let H_1 be $K_{1,3}$, the *claw*, and let H_2 be K_3 with a pendant edge. It is not difficult to see that $\epsilon(H_i) = \frac{1}{3}$ in this case. The construction is simple: let G be a graph on m vertices with no independent set of three vertices and with no complete subgraph of much more than $\sqrt{m}$ vertices ([4]). Take about $\frac{\sqrt{m}}{2}$ disjoint copies of G. This graph is H_1-free, has $\frac{m^{\frac{3}{2}}}{2}$ vertices and has no homogeneous subset with much more than $m^{\frac{1}{2}}$ vertices. The complement of this graph is good for H_2. On the the other hand, let G be an H_i-free graph with n vertices (i is 1 or 2). If the degree of a vertex v is at least $n^{\frac{2}{3}}$ then $\Gamma(v)$ (the set of vertices adjacent to v) contains a homogeneous set of at least $n^{\frac{1}{3}}$ vertices (in case of H_1 by Ramsey's theorem, in case of H_2 by perfectness). Otherwise G has an independent set of at least $n^{\frac{1}{3}}$ vertices. This gives

Proposition 6.1. *If H is the claw or K_3 with a pendant edge then $\epsilon(H) = \frac{1}{3}$.*

The remaining two H-s are the C_4 and K_4 minus an edge (the *diamond*). The following argument is clearly discovered by many of us, could be heard from Uncle Paul too. It was used for example in [9], [16]. Let $S = \{v_1, \ldots, v_\alpha\}$ be a maximum independent set of a C_4-free or diamond-free graph G. Then $V(G)$ is covered by the following $\binom{\alpha+1}{2}$ sets: $A(i)$, $B(i,j)$, $1 \leq i < j \leq \alpha$ where

$$A(i) = \{v \in V(G) - S : \Gamma(v) \cap S = \{v_i\}\} \cup \{v_i\}$$

and

$$B(i,j) = \{v \in V(G) - S : \Gamma(v) \cap S \supseteq \{v_i, v_j\}\}$$

The sets $A(i)$ induce complete subgraphs by the maximality of S and the sets $B(i,j)$ induce homogeneous sets (complete if G is C_4-free, independent if G is diamond-free). This gives

Proposition 6.2. *If G is C_4-free or diamond-free then $cc \leq \binom{\alpha+1}{2}$.*

Corollary 6.1. *If H is either C_4 or the diamond then $hom(n, H) \geq (2n)^{\frac{1}{3}}$. Therefore $\epsilon(H) \geq \frac{1}{3}$.*

Vertex disjoint union of complete graphs shows that $\epsilon(H) \leq \frac{1}{2}$ for any connected graph H. In case of $H = C_4$ this upper bound can be improved as follows. Let $R(C_4, m)$ be the smallest integer k such that any graph on k vertices either contains a C_4 (not necessarily induced C_4!) or contains an independent set on m vertices. F.K.Chung gives a graph G_m in [3] which shows that $R(C_4, m) \geq m^{\frac{4}{3}}$ for infinitely many m. Replacing each vertex of G_m by a clique of size $\frac{m}{3}$ we have a graph with no induced C_4 and with no homogeneous subset larger than m. This gives

Proposition 6.3. $\epsilon(C_4) \leq \frac{3}{7}$

Notice that if $R(C_4, m) \geq m^{2-\epsilon}$ with every $\epsilon > 0$ as asked by Uncle Paul then the replacement described above would show that $\epsilon(C_4) = \frac{1}{3}$.

Perhaps the next construction has a chance to improve the upper bound on $\epsilon(H)$ if H is the diamond. The vertices of G_q are the points of a *linear complex*

([11]) of a 3-dimensional projective space of order q. Two points are adjacent if and only if they are on a line of the linear complex. The graph G_q has $q^3 + q^2 + q + 1$ vertices, $\omega(G_q) = q+1$ and G_q is diamond-free. But is it true that $\alpha(G_q) < q^{\frac{3}{2}-\epsilon}$ for some positive ϵ and for infinitely many q? Thanks for the conversations to T.Szőnyi who thinks this is not known.

Problem 6.1. Improve the exponents in the above estimates of $hom(n, H)$ if H is either C_4 or the diamond.

What happens if G is C_4-free *and* diamond-free? In these graphs each four cycle induces a K_4. The sets $B(i, j)$ collapse implying

Corollary 6.2. *$hom(n, C_4, Diamond) \geq \sqrt{\frac{2}{3}n} - 1$*

Problem 6.2. Is it true that $hom(n, C_4, Diamond) = \sqrt{n} + o(\sqrt{n})$?

During the years between the submission and publication of this paper, Problem 6.2 had been answered affirmatively, in fact $hom(n, C_4, Diamond) = \lceil \sqrt{n} \rceil$ ([6]).

There are eight five-vertex graphs outside SVS. Keeping one from each complementary pair reduces the eight to five: $K_{1,3}$ with a subdivided edge, the *bull* (the self-complementary graph different from C_5), C_4 with a pendant edge, P_5, C_5. The existence of $\epsilon(H)$ is open for all of them, perhaps the list is about in the order of increasing difficulty. The construction in Proposition 1 shows that $\epsilon(H)$ is at most $\frac{1}{3}$ for all but C_5. In case of C_5 repeated replacements of $\overline{C_7} \cup K_3$ into itself shows $\epsilon(C_5) \leq \frac{\log 3}{\log 10}$. (Any C_5-free graph G with $hom(G) = 3$ and with at least 11 vertices would improve this.)

7. Forbidden complementary pairs

Perhaps an interesting subproblem is to find bounds on $hom(n, H, \overline{H})$. In case of four-vertex H, the structure of graphs which are both H-free and $\overline{H}$-free is well understood and values of $hom(n, H, \overline{H})$ can be determined as follows: $n^{\frac{1}{2}}$ if $H = P_4$ (from perfectness); $n^{\frac{1}{2}} - 1$ if $H = P_3 + K_1$ (from structure, [10]); $\frac{n-1}{2}$ if $H = C_4$ (from structure, [2]); $n - 4$ if H is a diamond (from structure, [10]); $\frac{2n}{5}$ if H is a claw (from structure, [10]).

The rest of this section is devoted to the case $H = P_5$. The upper bound $hom(n, P_5, \overline{P_5}) \leq n^{\frac{1}{\log 5}}$ is shown by replacing repeatedly C_5 into itself. The lower bound $n^{\frac{1}{3}}$ will follow from Corollary 7.1 which is the consequence of the following result.

Theorem 7.1. *If G is P_5-free and $\overline{P_5}$ -free then G satisfies the following property SP^*: there is an induced perfect subgraph of G whose vertices intersect all maximal cliques of G.*

Notice that property SP^* is a generalization of *strong perfectness* introduced by Berge and Duchet in [1]. (Maximal clique is a clique which is not properly contained in any other clique.) By Theorem 7.1, if G is both P_5-free and $\overline{P_5}$-free then G can be partitioned into at most $\omega(G)$ vertex disjoint perfect subgraphs. Each of these perfect graphs has clique number at most $\omega(G)$ thus each has chromatic number at most $\omega(G)$. This gives the next corollary.

Corollary 7.1. *If a graph is both P_5-free and $\overline{P_5}$-free then $\chi \leq \omega^2$.*

The proof of Theorem 7.1 is combining a result of Fouquet [8] with the following analogue of Lovász replacement lemma.

Lemma 7.1. *Property SP^* is preserved by replacements.*

Proof. The proof of the lemma is along the same line as the replacement lemma of Lovász. Assume that G and H have property SP^* and R is the graph obtained by replacing $v \in V(G)$ by H. Let G_1 and H_1 be perfect subgraphs of G and H such that $V(G_1)$ intersects all maximal cliques of G and $V(H_1)$ intersects all maximal cliques of H.

Case 1. $v \notin V(G_1)$. We claim that $V(G_1)$ intersects all maximal cliques of R. Let K be a maximal clique of R. If $V(K) \cap V(H)$ is empty then the claim follows from the definition of G_1. Otherwise $\{v\} \cup (K \cap V(G))$ is a clique of G which can be extended in G to a maximal clique K' intersecting $V(G_1)$. Since K is obtained by replacing $v \in K'$ by $K \cap V(H)$, K intersects $V(G_1)$.

Case 2. $v \in V(G_1)$. By the Lovász replacement lemma, the subgraph Z of R induced by $(V(G_1) \cup V(H_1)) - \{v\}$ is perfect. If a maximal clique K of R intersects $V(H)$, it intersects it in a maximal clique of H which (by the definition of H_1) intersects $V(H_1)$. If K does not intersect H then it does not contain v so it intersects $V(G_1) - \{v\}$ by the definition of G_1. Therefore K intersects Z.

Theorem 7.2 ((Fouquet, [8])). *Each graph from the family of P_5-free and $\overline{P_5}$-free graphs is either perfect or isomorphic to C_5 or can be obtained by a nontrivial replacement from the family.*

Now Theorem 7.1 follows by induction from Lemma 7.1 and Theorem 7.2.

8. Berge graphs

These are graphs which do not contain induced subgraphs isomorphic to C_{2k+1} or to $\overline{C_{2k+1}}$ for $k \geq 2$. According to the Strong Perfect Graph Conjecture (of Berge), Berge graphs are perfect. The following weaker form of this conjecture is attributed to Lovász in [7] (illustrating the difficulty of proving the existence of $\epsilon(C_5)$).

Problem 8.1. There exists a positive constant ϵ such that Berge graphs with n vertices contain homogeneous subsets of n^ϵ vertices.

Similar problems can be asked for subfamilies of Berge graphs for which the validity of SPGC is not known. One of them is the following.

Problem 8.2. Show that C_4-free Berge graphs with n vertices contain homogeneous subsets of $n^{\frac{1}{2}}$ (or at least $cn^{\frac{1}{2}}$) vertices.

Thanks to an unknown referee for valuable remarks.

References

1. C.Berge, D.Duchet, Strongly perfect graphs, in Topics on Perfect graphs, Annals of Discrete Math. Vol 21 (1984) 57-61
2. Z.Blázsik, M.Hujter, A.Pluhár, Zs.Tuza, Graphs with no induced C_4 and $2K_2$, Discrete Math. 115 (1993) 51-55.
3. F.R.K.Chung, On the covering of graphs, Discrete Math. 30 (1980) 89-93.
4. P.Erdős, Graph Theory and Probability II., Canadian J.Math. 13, (1961) 346-352.
5. P.Erdős, Some Remarks on the Theory of Graphs, Bulletin of the American Mathematics Society, 53 (1947), 292-294.
6. P.Erdős, A.Gyárfás, T.Luczak, Graphs in which each C_4 spans K_4, submitted.
7. P.Erdős, A.Hajnal, Ramsey Type Theorems, Discrete Applied Math. 25 (1989) 37-52.
8. J.L.Fouquet, A decomposition for a class of $(P_5, \overline{P_5})$-free graphs, preprint.
9. A.Gyárfás, A Ramsey type theorem and its applications to relatives of Helly's theorem, Periodica Math. Hung. 3 (1973) 299-304.
10. A.Gyárfás, Problems from the world surrounding perfect graphs, Zastowania Matematyki, Applicationes Mathematicae, XIX 3-4 (1987) 413-441.
11. Hirschfeld, Finite Projective Spaces of three dimensions, Clarendon Press, Oxford, 1985.
12. L.Lovász, Perfect Graphs, in Selected Topics in Graph Theory 2. Academic Press (1983) 55-87.
13. H.J.Prömel, A.Steger, Almost all Berge graphs are perfect, Report 91715, Forschungsinstitut fur Diskrete Mathematik, Bonn.
14. D.Seinsche, On a property of the class of n-colorable graphs, Journal of Combinatorial Theory B. 16 (1974) 191-193.
15. J.Spencer, Ten lectures on the Probabilistic method, CBMS-NSF Conference Series, 52.
16. S.Wagon, A bound on the chromatic number of graphs without certain induced subgraphs, J.Combinatorial Theory B. 29 (1980) 345-346.

The Chromatic Number of the Two-packing of a Forest

Hong Wang and Norbert Sauer*

Department of Mathematics and Statistics, The University of Calgary, Calgary, 2500 University Drive N.W, T2N 1N4, Alberta, Canada

Summary. A two-packing of a graph G is a bijection $\sigma : V(G) \to V(G)$ such that for every two adjacent vertices $a, b \in V(G)$ the vertices $\sigma(a)$ and $\sigma(b)$ are not adjacent. It is known [2], [6] that every forest G which is not a star has a two-packing σ. If F_σ is the graph whose vertices are the vertices of G and in which two vertices a, b are adjacent if and only if a, b or $\sigma^{-1}(a), \sigma^{-1}(b)$ are adjacent in G then it is easy to see that the chromatic number of F_σ is either 1, 2, 3 or 4. We characterize, for each number n between one and four, all forests F which have a two-packing σ such that F_σ has chromatic number n.

1. Introduction

We only discuss finite simple graphs and use standard terminology and notation from [1] except as indicated. For any graph G, we use $V(G)$ and $E(G)$ to denote the set of vertices and the set of edges of G, respectively. A forest is a graph without cycles. A tree is a connected forest. The *length* of a path or cycle is the number of its edges. A path of length n is denoted by P_n and a cycle of length n is denoted by C_n. The complete graph on n vertices is denoted by K_n. A vertex of a graph G is an *isolated point* if its degree is zero and it is an *endpoint of* G if its degree is one. A vertex of a forest F adjacent to an endpoint of F is called a *node of* F. The set of vertices of a forest F which are adjacent to at least one of the endpoints of a set A of endpoints is called the set of *nodes belonging* to the set A of endpoints.

The *distance* between two vertices x and y of a graph G is the length of the shortest path in G from x to y. The *diameter* of a graph G is the largest distance between any two vertices of G. Observe that if T is a tree and the distance between the vertices x and y of T is the diameter of T, then the vertices x and y are endpoints of T. A vertex x of a tree T is in the *center* of T, or is a *central point* of T, if the maximal distance from x to any other vertex of T is minimal for the vertex x. Observe that if the diameter of T is odd then T has exactly two central points and if the diameter of T is even, then T has exactly one central point.

A tree of order at least two has at least two endpoints and if it has exactly two endpoints it is a path. A tree with exactly two nodes is called a *dragon.* Note that every tree of diameter three is a dragon. A tree of diameter two or one with $n \geq 1$ edges is called a *star* S_n. Hence an isolated point is not a star. If A is a set of vertices of the graph G then $G - A$ is the graph obtained from G by removing the vertices in A together with all of the incident edges from G. We will write $G - a$ for $G - \{a\}$. If F is a forest and A a set of endpoints of F then the forest $F - A$ is called a *derived* forest of F. If A is the set of all endpoints of F then $F - A$ is the *completely derived* forest of F. A tree T is a *crested dragon* if T has a path

* Supported by NSERC of Canada Grant #691325.
AMS–Classification : 05C70
Keywords: Packing, placement, factorization, tree, forest, chromatic number

$P = a_1, a_2, a_3, a_4$ of length three as a derived tree, the vertices a_1, a_4 and possibly a_3 are nodes of T, every node of T is in the set $\{a_1, a_3, a_4\}$ and the number of endpoints adjacent to a_1 is strictly larger than the number of endpoints adjacent to a_3. Observe that the crested dragon T is a dragon of diameter five if and only if a_3 is not a node of T. See Figure 1.

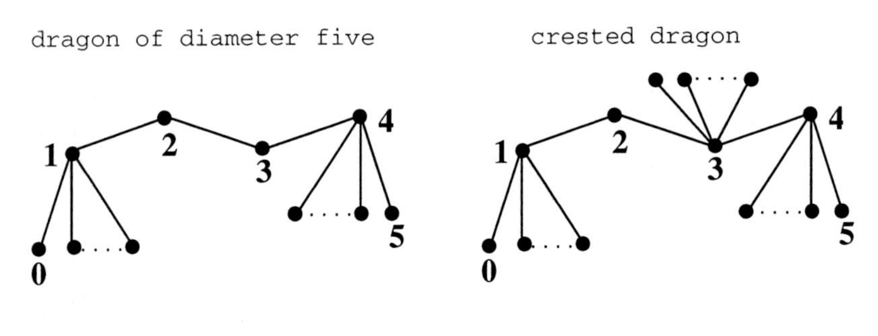

Figure 1

Let G be a graph and σ a bijection from $V(G)$ to $V(G)$. If σ has the property that for every pair a and b of adjacent vertices of G the vertices $\sigma(a)$ and $\sigma(b)$ are not adjacent then σ is called a *two-packing of the graph G*. If σ is a two-packing of G then the graph $\sigma(G)$ has the same vertices as the graph G and two vertices a and b of $\sigma(G)$ are adjacent in $\sigma(G)$ if and only if the vertices $\sigma^{-1}(a)$ and $\sigma^{-1}(b)$ are adjacent in the graph G. If E is the set of edges of G then $\sigma(E)$ denotes the set of edges of $\sigma(G)$. Note that $E \cap \sigma(E) = \emptyset$. The graph G_σ has the same vertices as the graph G and the edges of G_σ are given by $E(G_\sigma) = E(G) \cup E(\sigma(G))$. The graph G_σ can be *factorized* into the graphs G and $\sigma(G)$. Hence results about two-packings of graphs can also be understood as results about factorizations of graphs.

Burns and Schuster proved that if a graph G of order n does not contain a vertex of degree $n-1$ and contains no cycles of length 3 or 4 and $|E(G)| = n-1$ then there is a two-packing of G, [2]. Faudre, Rousseau, Schelp and Schuster proved that if a graph G of order n does not contain a vertex of degree $n-1$ and contains no cycles of length 3 or 4 and $|E(G)| \leq 6n/5 - 2$, then there is a two packing of G, [3]. St. Brandt proved that if G is a non-star graph of girth larger than or equal to seven then there is a packing of two copies of G, [4]. It follows from [2] that every tree which is not a star has a two-packing and we recently [6] characterised those trees which have a "three-packing".

The results mentioned above establish the existence of two-packings of a graph G but are not concerned with any properties of the graph G_σ. From the point of factorizations of graphs this would be a very interesting question. Given the difficulties in establishing the existence of two-packings we looked for a simple case in which we might be able to determine the possible chromatic numbers of the two-packings of a graph. Note first that for any two-packing σ of G the chromatic number of G_σ satisfies the inequality $\chi(G) \leq \chi(G_\sigma) \leq \chi(G)^2$. This means that if F is a forest which has at least one edge and σ is a two-packing of F then

$2 \leq \chi(F_\sigma) \leq 4$. If F_σ has chromatic number less than or equal to two we will call the two-packing σ a bipartite two-packing.

If G is a connected bipartite graph then the partition of G into the two color classes is unique. We will call the color class of a vertex x the *parity* of x and will often call the vertices in one class the vertices of even parity or even vertices and the vertices in the other class the vertices of odd parity or the odd vertices. We will then take care to label the even vertices with even and the odd vertices with odd numbers. A consequence of the uniqueness of the color-classes of a connected bipartite graph G is, that if σ is a bipartite two-packing of G then either for every vertex x of G, $\sigma(x)$ has the same parity as x or for every vertex x of G, $\sigma(x)$ has a different parity than x. In the first case we will say that the bipartite two-packing σ is an equal parity packing and in the second case we will say that the bipartite two-packing σ is an unequal parity packing. If F is a forest and σ a two-packing of F then σ is an *equal parity two-packing* of F if for every vertex a of F, whenever $\sigma(a)$ and a are in the same connected component of F then, in this connected component of F, the parity of $\sigma(a)$ is the same as the parity of a.

It will be often necessary to exhibit a two-packing σ for some small forest F. We usually will do this in a figure in which the edges of F are solid lines and the edges of $\sigma(F)$ are interrupted lines. In order to check the claim that the exhibited map σ is indeed a two-packing one has to check that the graph with the interrupted edges is indeed isomorphic to the graph having solid edges and that the graph with the solid edges is the intended graph. It is then visually clear that the two edge-sets have no edge in common. In the case of a bipartite two-packing σ, if in the text a vertex is described by a labeled letter then in the figure the label alone will be used to name the vertex. The labels of the vertices of F will be in bold times italic typset and the vertices of $\sigma(F)$ in gray courier oblique. All vertices of one parity will have even and all the other vertices will have odd labels. In order to check that σ is an equal parity two-packing it suffices to check that the parity of the two labels of each vertex is the same.

We will, for each number n between one and four, completely characterize all forests F which have a two-packing σ such that F_σ has chromatic number n. More exactly we will prove the following theorems:

Theorem 1.1. *A forest has a bipartite two-packing if and only if it is either a singleton vertex or disconnected or a tree of diameter at least five which is not a crested dragon. Every forest which has a bipartite two-packing has an equal parity two-packing.*

Theorem 1.2. *A forest F has a two-packing σ such that F_σ has chromatic number three if and only if either F is a tree with diameter at least four or F is disconnected and contains a path of length two.*

Theorem 1.3. *A forest F has a two-packing σ such that F_σ has chromatic number four if and only if F contains a path of length three and F does not consist of exactly two connected components where one of the two components is a path of length three and the other is a star.*

2. Preliminary results

If F is a forest, B a set of vertices of F and σ a two-packing with the property that for all $b \in B$, $\sigma(b) \neq b$ holds, then we say that σ *splits* the vertices of B. If B consists of a single vertex b we say that the two-packing σ splits the vertex b.

If σ is an equal parity two-packing which splits all vertices of $V(F)$ then we say that σ is a *perfect packing*. Let A be a set of endpoints of a forest G and B the set of nodes which belong to A. Assume that σ is a two-packing of $G - A$ which splits the vertices of B and λ the extension of σ from $G - A$ to F such that for all $a \in A$, $\lambda(a) = a$ holds. Obviously λ is a two-packing of G. We will call λ a *trivial extension* of σ to the forest G. The following *First Preservation Lemma* holds for trivial extensions of two-packings.

Lemma 2.1 (First Preservation Lemma). *Let σ be a two-packing of a forest F and λ a trivial extension of σ to the forest G. If F_σ has chromatic number four then G_λ has chromatic number four. If F_σ has chromatic number three then G_λ has chromatic number three. If σ is an equal parity two-packing then λ is an equal parity two-packing.*

Proof. Assume that A is the set of endpoints of G such that $G - A = F$ and that B is the set of nodes of G which belong to the set A of endpoints. Then by the definition of trivial extension, the two-packing σ splits the nodes in B. The chromatic number of G_λ is larger than or equal to the chromatic number of F_σ, and because the chromatic number of any two-packing of a forest is at most four, the first of the assertions follows. The endpoints in A will have degree two in G_λ, which implies the second assertion. Let us now assume that σ is an equal parity two-packing. Then, for every node $b \in B$ and every two coloring of F_σ, the vertices b and $\sigma(b)$ will be colored with the same color. This of course implies the third assertion. ∎

Corollary 2.1. *If some derived forest of a forest F has a perfect packing then F itself has an equal parity two-packing.*

Let F be a forest, a and b two endpoints of F, a_1 the node adjacent to a and b_1 the node adjacent to b. If a_1 and b_1 are different vertices then the pair of endpoints a and b is called a *reducing pair of endpoints* of the forest F. If the nodes a_1 and b_1 are in the same connected component of F and have the same parity then clearly also the endpoints a and b have the same parity and then a and b is an *equal parity reducing pair* of endpoints of the forest F. Let a, b be a reducing pair of endpoints of the forest F, a_1 and a_2 the nodes adjacent to a and b respectively and σ a two-packing of the forest $F - \{a,b\}$. Let λ be the extension of σ from $F - \{a,b\}$ to F such that if $\sigma(a_1) = a_1$ or $\sigma(b_1) = b_1$ then $\lambda(a) = b$ and $\lambda(b) = a$. In all other cases we put $\lambda(a) = a$ and $\lambda(b) = b$. Clearly λ is a two-packing of the forest F. We will call λ the *extension* of σ to the reducing pair a, b of endpoints. The following *Second Preservation Lemma* holds for extensions of two-packings to reducing pairs of endpoints.

Lemma 2.2 (Second Preservation Lemma). *Let F be a forest, a, b a reducing pair of endpoints of F, σ a two-packing of the forest $F - \{a,b\}$ and λ an extension of σ to the reducing pair a, b of endpoints of F. If F_σ has chromatic number four then F_λ has chromatic number four. If F_σ has chromatic number three then F_λ has chromatic number three. If σ is an equal parity two-packing and a_1 and b_1 have the same parity then λ is an equal parity two-packing. If σ splits a set B of vertices then λ splits the set B of vertices.*

Proof. The chromatic number of F_λ is larger than or equal to the chromatic number of F_σ, and because the chromatic number of any two-packing of a forest is at most four, the first of the assertions follows. The endpoints a and b will have degree two in F_λ, which implies the second assertion. Let us now assume that σ is an equal parity two-packing and that the nodes a_1 and b_1 have the same parity. Then, for every proper two coloring of F_σ, the vertices a_1, b_1, $\sigma(a_1)$ and $\sigma(b_1)$ will be colored with the same color. This of course implies the third assertion. Because σ and λ agree on $F - \{a, b\}$ the two-packing λ splits all vertices which are split by σ. ∎

Observe the following fact:

Lemma 2.3. *If T is a tree with diameter one or two then T does not have a two-packing.* ∎

Lemma 2.4. *For every two-packing σ of a tree T with diameter three the graph T_σ has chromatic number four.*

Proof. Denote the two central points of T by a and b. If $\sigma(a) = a$ then $\sigma(b)$ is an endpoint not adjacent to a and hence adjacent to b. We will show that $\sigma(b)$ and b are also ajacent in $\sigma(T)$, a contradiction. Because σ is an onto map and σ maps all vertices adjacent to a to endpoints adjacent to b there is some endpoint x adjacent to b such that $\sigma(x) = b$. But then $\sigma(x) = b$ is adjacent to $\sigma(b)$ in $\sigma(T)$. If $\sigma(a) = b$ then $\sigma(b)$ is an endpoint not adjacent to b and hence adjacent to a. We will show that $\sigma(b)$ and a are also ajacent in $\sigma(T)$, a contradiction. Because σ is an onto map and σ maps all vertices adjacent to a to endpoints adjacent to a there is some endpoint x adjacent to b such that $\sigma(x) = a$. But then $\sigma(x) = a$ is adjacent to $\sigma(b)$ in $\sigma(T)$. We conclude that $\{\sigma(a), \sigma(b)\} \cap \{a, b\} = \emptyset$.

We will now prove that $T \cup \sigma(T)$ induces the complete graph on the set $\{\sigma(a), \sigma(b), a, b\}$ of four vertices. It follows from the result of the previous paragraph that there are two endpoints x and y of T such that $\sigma(a) = x$ and $\sigma(b) = y$. Again because σ is onto there are two endpoints u and v of T such that $\sigma(u) = a$ and $\sigma(v) = b$. The graph $T \cup \sigma(T)$ induces then six different edges in the set $\{\sigma(a), \sigma(b), a, b\}$ of vertices. Those are first the two edges from a to b and $\sigma(a)$ to $\sigma(b)$. Then each of the four different points x, y which are endpoints of T and $\sigma(u), \sigma(v)$ which are endpoints of $\sigma(T)$ is in the graph $T \cup \sigma(T)$ adjacent to at least one of the vertices $\{\sigma(a), \sigma(b), a, b\}$. ∎

Lemma 2.5. *If T is a tree with diameter at least two and at most four then T does not have a bipartite two-packing.*

Proof. Observe that every tree T of diameter at most four contains a vertex a which is adjacent to every vertex of T whose parity is different than the parity of the vertex a. If T has a bipartite two-packing σ then because σ is a bijection, there is some vertex b of T such that $\sigma(b) = a$. This means that the parity of a is different from the parity of b, otherwise b would have to be an isolated point (and if the diameter of a tree is larger than one it does not contain an isolated point). Because the vertex a is adjacent to every vertex whose parity is different from the parity of a, the vertices a and b are adjacent in T and therefore $\sigma(a)$ and $\sigma(b)$ are adjacent in $\sigma(T)$. Remember that every bipartite two-packing of T either changes the parity of all vertices or maps each parity class into itself. Hence $\sigma(a)$ is a vertex c adjacent to a. This now leads to a contradiction. The vertices a and c are adjacent in T, and because $a = \sigma(b)$ and $c = \sigma(a)$ they are adjacent in $\sigma(T)$. ∎

3. Bipartite two-packings

Lemma 3.1. *Every path of length at least seven has a perfect packing.*

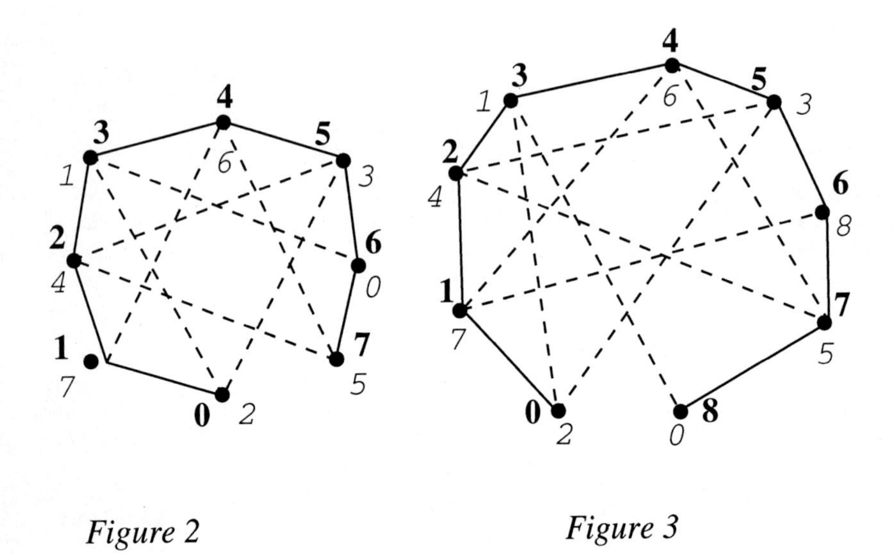

Figure 2 *Figure 3*

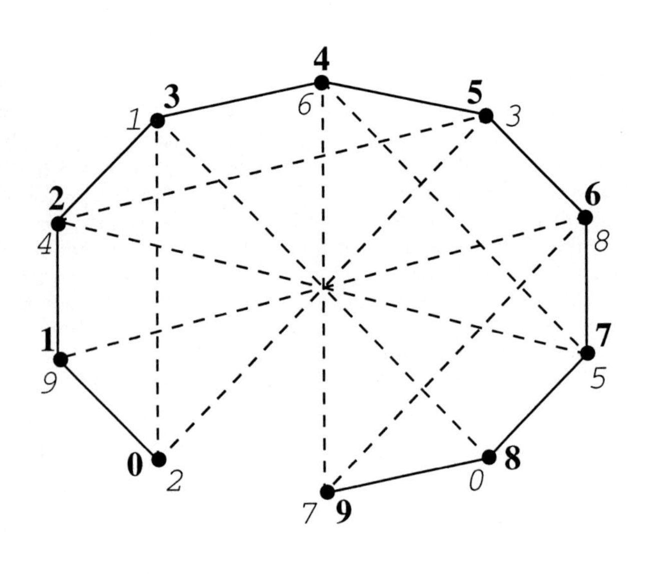

Figure 4

Proof. If the length r of the path is seven, eight or nine Lemma 3.1 follows from Figure 2, Figure 3 and Figure 4. So assume that $P_{r+1} = a_0, a_1, \ldots, a_{r+1}$ is a path with $r \geq 9$ and σ is a perfect packing of the path $P_r = a_0, a_1, \ldots, a_r$. Assume that for x in P_r, $\sigma(x) = a_r$. There are at least five vertices in the path P_r which have the same parity as the vertex a_{r+1}. This means that at least one of them, say a_i with $\sigma(y) = a_i$ has the following properties : $a_{i+1} \neq \sigma(a_r) \neq a_{i-1}$ and y is not adjacent to x. The following bijection λ from (P_{r+1}) to (P_{r+1}) is then a perfect packing of P_{r+1}. First of all λ agrees with σ on $(P_r - y)$ and then $\lambda(y) = a_{r+1}$ and $\lambda(a_{r+1}) = a_i$. It is easy to check that λ is a perfect packing of P_{r+1}. ∎

Lemma 3.2. *If a forest G has an equal parity two-packing which splits the vertices of some set B, then the forest F which consists of G together with an additional isolated point e has an equal parity two-packing which splits every vertex in B and which also splits the isolated point e.*

Proof. Let σ be an equal parity two-packing of G which splits every vertex in B. Let a be any vertex of G and λ a function from $V(G) \cup \{e\}$ to $V(G) \cup \{e\}$ which agrees with σ on all vertices of $V(G) - a$ and for which $\lambda(e) = \sigma(a)$ and $\lambda(a) = e$. Clearly λ is an equal parity two-packing of G which splits the vertices of $B \cup \{e\}$. ∎

Lemma 3.3. *Every forest F with two components where one of the components is an isolated point e and the other component T is a dragon, a star, or an isolated point, has an equal parity two-packing which splits the islolated point e.*

Proof. If T is an isolated point then F clearly has an equal parity two-packing which splits e. If T is a star then a derived forest of F consists of two isolated vertices and hence there is a derived forest of F which has a perfect packing. Lemma 3.3 follows from the Corollary to Lemma 2.1. Let the path $P_r = a_1, a_2, \ldots, a_r$ be the completely derived tree of T and denote the nonempty set of endpoints adjacent to a_1 by A and the nonempty set of endpoints adjacent to a_r by B. Let $a \in A$ and $b \in B$ be two endpoints of T. The path P_r together with a and b is then a path P of length $r+1$. If the forest whose connected components are P_r and e has an equal parity two-packing in which the vertices a_1, a_r and e are split, then according to the First Preservation Lemma, the forest F has an equal parity two-packing which splits e. It follows from Lemma 3.1 and Lemma 3.2 that the forest whose connected components are P and e has an equal parity two-packing in which the vertices a_1, a_r and e are split if $r \geq 6$. If $r = 3$ or $r = 5$ then P_r is a star S_2 or has an equal parity reducing pair of endpoints a and b such that $P_r - \{a, b\}$ is a star S_2. Using the Second Preservation Lemma we see that Lemma 3.3 holds in this case. This leaves the cases $r = 2$ and $r = 4$. If $r = 2$ Figure 5 exhibits an equal parity two-packing of P together with an isolated point which splits the two nodes of P and the isolated point.

Consider $r = 4$. Let $a_0 \in A$, $a_5 \in B$. See Figure 6 in which $1, 4$ and e are split. Then by the First Preservation Lemma, Lemma 3.3 holds for this case as well. ∎

Lemma 3.4. *Every forest F with two components, where one of the components is an isolated vertex e and the other a tree T, has an equal parity two-packing which splits e.*

Proof. By the Second Preservation Lemma we may assume that T does not contain two different nodes which have the same parity. This means that T can not have three different nodes, hence T is a dragon, a star or an isolated point. Lemma 3.4 follows then from Lemma 8. ∎

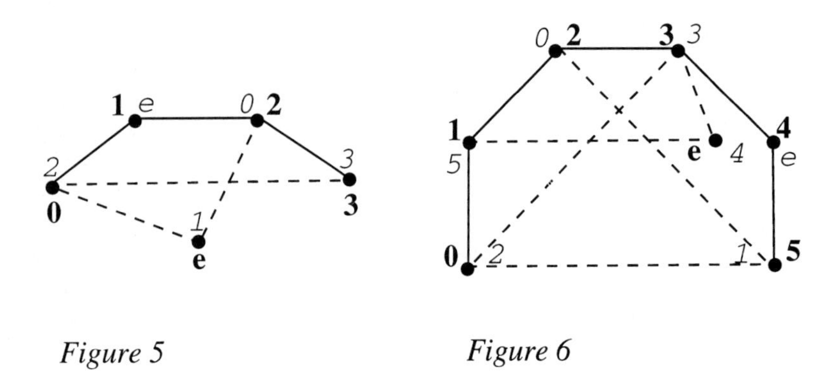

Figure 5

Figure 6

Lemma 3.5. *Every forest F with at least two connected components has an equal parity two-packing.*

Proof. Observe first that if we obtain the forest G from F by adding an edge which is adjacent to vertices in different connected components of F, and if G has an equal parity two-packing, then F has an equal parity two-packing. We may therefore assume without loss of generality that the forest F has exactly two components G and H. Let G_1 be the forest which consists of G together with an additional vertex e_1 and H_1 the forest which consists of H together with an additional vertex e_2. Let γ be an equal parity two-packing of G_1 which splits e_1 and λ be an equal parity two-packing of H_1 which splits e_2. Identify the vertex $\gamma(e_1)$ with the vertex e_2 and the vertex $\lambda(e_2)$ with the vertex e_1. It is not difficult to check that $\gamma \cup \lambda = \sigma$ is an equal parity two-packing of the forest F. ∎

Lemma 3.6. *A crested dragon T does not have a bipartite two-packing.*

Proof. Let $P = a_1, a_2, a_3, a_4$ be the completely derived tree of the crested dragon T. Assume that the nodes of T are a_1, a_4 and possibly a_3. Let A be the set of endpoints adjacent to a_1, B the set of endpoints adjacent to a_4 and C the set of endpoints adjacent to a_3. Then $|A| > |C|$ and C might be empty if T is a dragon of diameter five. Assume for a contradiction that σ is a bipartite two packing of T.

Case 3.1. The two-packing σ is an equal parity two-packing.

Every even vertex of T is adjacent to at least one of the odd vertices a_1 and a_3. If the distance in T between $\sigma(a_1)$ and $\sigma(a_3)$ is two, then for some even vertex x of T, $\sigma(x)$ would be in T adjacent to both vertices $\sigma(a_1)$ and $\sigma(a_3)$ and in $\sigma(T)$ to at least one of the vertices $\sigma(a_1)$ and $\sigma(a_3)$. This is not possible because T and $\sigma(T)$ would then have an edge in common. Hence the distance between $\sigma(a_1)$ and $\sigma(a_3)$ is larger than two and because this distance must be an even number it is at least four. Also both vertices $\sigma(a_1)$ and $\sigma(a_3)$ are odd vertices of T. Hence one of the vertices $\sigma(a_1)$ or $\sigma(a_3)$ must be equal to a_1. If $\sigma(a_1) = a_1$ then the $|A|+1$ even vertices adjacent to a_1 must be mapped by σ to even vertices not adjacent to a_1. But there are only $|C|+1 < |A|+1$ even vertices not adjacent to a_1. If $\sigma(a_3) = a_1$ then the $|C|+2$ even vertices adjacent to a_3 must be mapped by σ to even vertices not adjacent to a_1. But there are only $|C|+1 < |C|+2$ even vertices not adjacent to a_1.

Case 3.2. The two packing σ is an unequal parity two-packing.

The vertex a_4 is adjacent to all odd vertices except a_1. Hence $\sigma(a_4)$ must be an odd endpoint of T, that is $\sigma(a_4)$ is some endpoint adjacent to a_4 and $\sigma(a_1) = a_4$. But this implies that $\sigma(a_1)$ can only be adjacent to a_1 in $\sigma(T)$. Hence a_1 must be an endpoint of T, a contradiction. ∎

Lemma 3.7. *If T is a tree of diameter five which is not a crested dragon but every equal parity reduction of T is a crested dragon or a tree of diameter less than five then T has an equal parity two-packing.*

Proof. We assume first that T contains two equal parity reducing endpoints x and y such that $T - \{x, y\}$ is a crested dragon D. Let m be the node adjacent to x and n the node adjacent to y. Choose a path $P = a_0, a_1, a_2, a_3, a_4, a_5$ in the crested dragon D such that the vertices a_1, a_4 and possibly a_3 are nodes of D but no other vertices are nodes of D. Note that P is a derived tree of T. Denote by A the set of endpoints adjacent to a_1, by B the set of endpoints adjacent to a_4 and by C the set of endpoints adjacent to a_3. Then $|A| > |C|$ and C might be empty. Observe that the nodes m and n are vertices of D. We are going to discuss serveral cases depending on which vertices of D are the nodes m and n of T. Clearly $\{m, n\} \cap (A \cup B) = \emptyset$ otherwise the diameter of T would be larger than five.

We assume first that both nodes m and n are elements of C. Let then R be the subtree of T which is spanned by the vertices of P together with the vertices m, n, x and y. Clearly R is a derived tree of T. Figure 7 exhibits a perfect packing of R. It follows then from the First Preservation Lemma that T has an equal parity two-packing. (The two-packing σ maps a bold labelled vertex i to the gray labeled vertex i. The reason for using numbers and not letters in Figure 7 is to make it obvious that the graph F_σ is bipartite. Indexed letters take too much space in the figure).

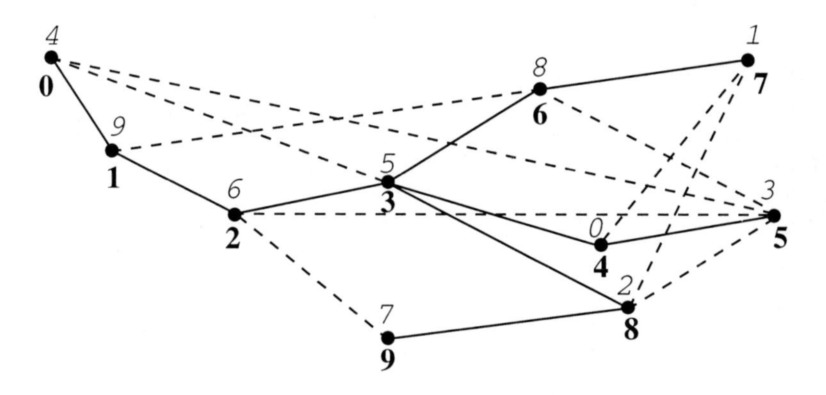

Figure 7

If only one of the nodes m and n, say m, is an element of C, then the other node n must be equal to some vertex of even parity in D. Hence n is equal to a_2 or a_4. If $n = a_2$ let R then be the subtree of T which is spanned by the vertices of

P together with the vertices m and y. Clearly R is a derived tree of T. Figure 8 exhibits a perfect packing of R. It follows from the First Preservation Lemma that T has an equal parity two-packing. If m is an element of C and $n = a_4$ we consider

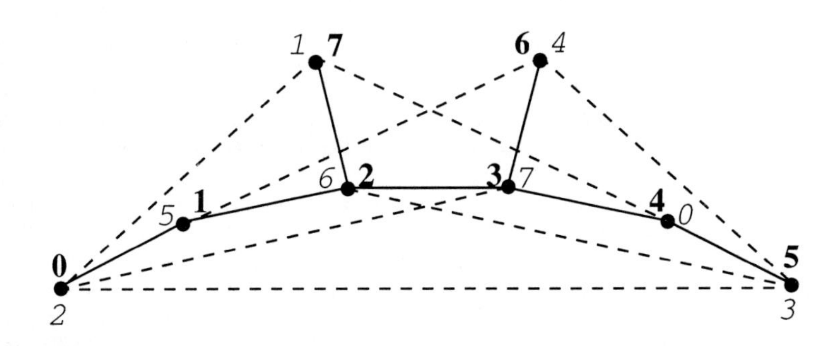

Figure 8

R to be the subtree of T which is spanned by the vertices of P and the vertices m and x. Note that a_2 is now not a node of T. Clearly R is a derived tree of T. Figure 9 exhibits a bipartite two-packing of R in which every vertex splits except the vertex a_2, which corresponds to 2 in the figure. It follows then from the First Preservation Lemma that T has an equal parity two-packing. We arrived now at

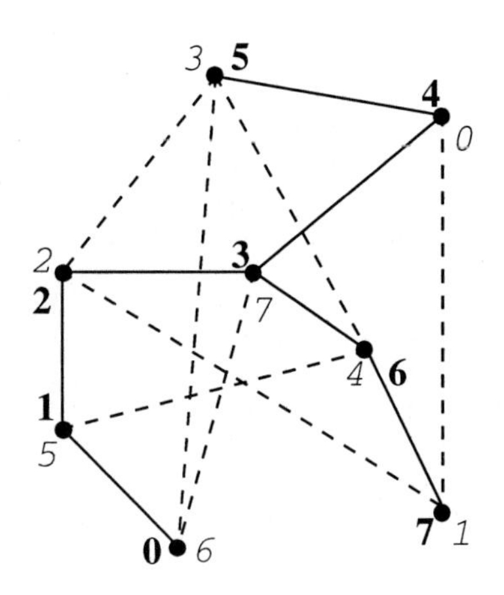

Figure 9

the situation that neither m nor n is an element of C. Because m and n have the same parity it follows that either $\{m, n\} = \{a_1, a_3\}$ or that $\{m, n\} = \{a_2, a_4\}$ holds. If $\{m, n\} = \{a_1, a_3\}$ then T is a crested dragon, contrary to our assumptions on T. (Because $|A| > |C|$ clearly $|A| + 1 > |C| + 1$). If $\{m, n\} = \{a_2, a_4\}$ and C is not empty then the same graph R as used for Figure 8 is a derived tree of T. If C is empty then T is a crested dragon, (label the path P beginning at a_5 with a_0),contrary to the assumption on T.

We can now assume that whenever x and y is an equal parity reducing pair of endpoints of T, then the diameter of the tree $K = T - \{x, y\}$ is smaller than five. Because T has diameter five it contains a path $P = a_0, a_1, a_2, a_3, a_4, a_5$ of length five such that the vertices a_0 and a_5 are endpoints of T and a_1 and a_4 are nodes of T. Since T is not a crested dragon, the tree T has at least one node different from a_1 and a_4. The tree T has at most two more nodes besides a_1 and a_4, otherwise two of those nodes would have the same parity and T could be further equal parity reduced to a tree which has diameter five. If T has two nodes besides a_1 and a_4 we will denote them by n and m and if T has only one such a node we will denote it by m. Note that the two nodes m and n have different parity otherwise T could be equal parity reduced to a tree which has diameter five. For a vertex a in T the distance of a from P is the length of the shortest path from a to any vertex of P. The vertex of P which has shortest distance from a is called the point of attachment of a. Because T has diameter five there is no vertex in T which has distance greater than or equal to three from P. This means that the nodes m and n have distance at most one from P.

If both nodes m and n have distance one from P the points of attachment of m and n are in the set $\{a_2, a_3\}$. Otherwise T would have diameter at least six. If the points of attachment of m and n are different we may assume without loss of generality that the point of attachment of m is a_2 and the point of attachment of n is a_3. If we remove all endpoints from T except the endpoints a_0 and a_5 we arrive at the subtree of T spanned by the vertices $a_0, a_1, a_2, a_3, a_4, a_5, m, n$. We see from Figure 8 that this tree has a perfect packing. Hence by the First Preservation Lemma, T has an equal parity two-packing. If m and n have the same vertex as point of attachment then m and n would have the same parity.

Hence at most one node, say m, has distance one from P while the other node is a vertex in P. As above the point of attachment of m is one of the points a_2 or a_3 and we may assume without loss of generality that a_3 is the point of attachment of the node m. The node n can not be one of the endpoints of T adjacent to a_1 or a_4 because the diameter of T is five. By assumption n is not equal to a_1 or a_4. If n is equal to a_2 then m and n would have the same parity. We conclude that $n = a_3$ holds. The nodes $a_3 = n$ and a_1 of T have the same parity. If a_1 would be adjacent to more than the one endpoint a_0 then T could be further equal parity reduced while still containing a path of length five. Hence a_1 is only adjacent to the one endpoint a_0. If we remove all endpoints from T except the endpoints a_0 and a_5 we arrive at the subtree of T spanned by the vertices $a_0, a_1, a_2, a_3 = n, a_4, a_5, m$. We see from Figure 10 that this tree has an equal parity two-packing in which every vertex, except the vertex a_1, splits. Hence by the First Preservation Lemma the tree T has an equal parity two-packing.

The next case to investigate is the one in which both of the nodes m and n of T are vertices of the path P. Because the diamter of T is five the nodes m and n can not be endpoints of P and by assumption m and n are different from a_1 and a_4. Also, m and n are different from each other. Hence without loss, $m = a_2$ and $n = a_3$. There is then at least one endpoint of T adjacent to a_2 and at least one endpoint of T adjacent to a_3. We can now use Figure 8 and the First Preservation Lemma to conclude that T has a bipartite packing.

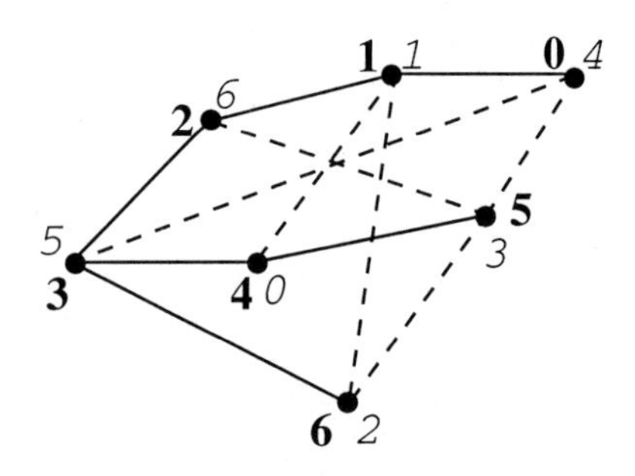

Figure 10

We assume from now on that there is only the one node m in T which is not equal to a_1 or a_4. If m has distance one from P we may assume without loss that the point of attachment of m is a_3. Because m is a node there exists at least one endpoint a_7 of T adjacent to m. If we remove all endpoints from T except the endpoints a_0, a_5 and a_7 we arrive at the subtree of T spanned by the vertices $a_0, a_1, a_2, a_3, a_4, a_5, a_7, m$. We see from Figure 9 that this tree has an equal parity bibartite packing in which every vertex splits, except for the vertex a_2 which is not a node of T. Hence by the First Preservation Lemma, T has an equal parity two-packing.

The last case is when m is a vertex in P. As before we may assume without loss that $m = a_3$ holds. The node $m = a_3$ as adjacent to at least one endpoint, say a_6. The parity of $m = a_3$ is the same as the parity of the node a_1. Hence the node a_1 is adjacent to at most one endpoint of T, otherwise T could be further equal parity reduced while still containing a path of length five. If we remove all endpoints from T except the endpoints a_0, a_5 and a_6 we arrive at the subtree of T spanned by the vertices $a_0, a_1, a_2, a_3 = m, a_4, a_5, a_6$. We see from Figure 10 that this tree has an equal parity bipartite packing in which every vertex, except the vertex a_2, splits. Hence by the First Preservation Lemma, T has an equal parity two-packing.

Lemma 3.8. *A tree T of diameter five has an equal parity two-packing if and only if T is not a crested dragon.*

Proof. If T is a crested dragon it follows from Lemma 3.6 that T does not have a bipartite two-packing. If T has an equal parity reducing pair a and b of endpoints such that $T-\{a,b\}$ is not a crested dragon and has diameter at least five then, using the Second Preservation Lemma, we have to consider the tree $T - \{a,b\}$ instead. Because T is finite we can assume that for every equal parity reducing pair a and b of endpoints of T, $T - \{a,b\}$ is either a crested dragon or has diameter smaller than five. By Lemma 3.7, the tree T has then an equal parity two-packing. ∎

Lemma 3.9. *Every tree T of diameter larger than or equal to five which is not a crested dragon has an equal parity two-packing.*

Proof. By the Second Preservation Lemma we may assume that any equal parity reduction of T either yields a tree of diameter less than five or a crested dragon. Because of Lemma 3.8 we may also assume that the diameter of T is larger than five. If T does not contain an equal parity reducing pair of endpoints then because the diameter of T is larger than five, T would have to be a dragon of diameter at least seven. Because every path of length at least seven has a perfect packing (Lemma 3.1), every dragon of length at least seven has an equal parity two-packing. We may therefore further assume that the tree T contains a pair of endpoints a, b of equal parity which are adjacent to two different nodes.

Case 3.3. The tree T contains endpoints a, b of equal parity which are adjacent to two different nodes such that the tree $T - \{a, b\}$ is a crested dragon D.

We assume that the path a_1, a_2, a_3, a_4 is a derived tree of the crested dragon D such that a_1, a_4 and possibly a_3 are the nodes of D. Let A be the set of endpoints of D adjacent to a_1, B the set of endpoints of D adjacent to a_4, C the set of endpoints of D adjacent to a_3 and assume that a_0 is an endpoint in A and that a_5 is an endpoint in B. Because the diameter of T is larger than five it follows that at least one of the endpoints a or b of the equal parity reducing pair of endpoints of T is adjacent to one of the endpoints in the set $A \cup B$.

If both vertices a and b are adjacent to vertices in the set $A \cup B$ then, because a and b have the same parity, they are either both adjacent to vertices in A or they are both adjacent to vertices in B. In either case the tree R, for which Figure 10 exhibits an equal parity two-packing σ, is a derived tree of T. The two-packing σ splits all of the vertices of R except the vertex labelled 2. Using the First Preservation Lemma we conclude that T has an equal parity two-packing except in the case where both endpoints a and b are adjacent to vertices in B and the set C is not empty. In this case we use the tree for which Figure 11 exhibits a perfect packing and which is then a derived tree of T. We are therefore now in the situation that exactly one of

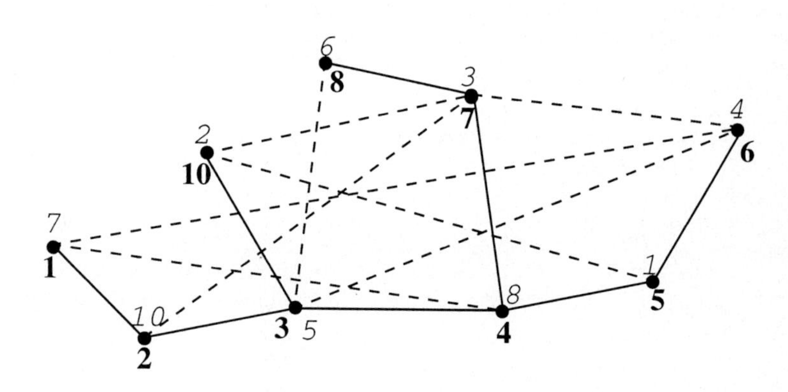

Figure 11

the endpoints a and b, say a, is in the set $A \cup B$. Assume first that $a \in B$. Because a and b have the same parity the endpoint b is not adjacent to a vertex in C. The tree T then contains the path $P = a_0, a_1, a_2, a_3, a_4, a_5, a_6 = a$ of length six and every endpoint of T is adjacent to some vertex in this path. If the center a_3 of P

is not a node of T we are done by the First Preservation Lemma and the following Figure 12 which exhibits a bipartite two-packing of P_6 in which every vertex except the center splits. If a_3 is a node of T we may assume that a_3 is adjacent to a_8. The

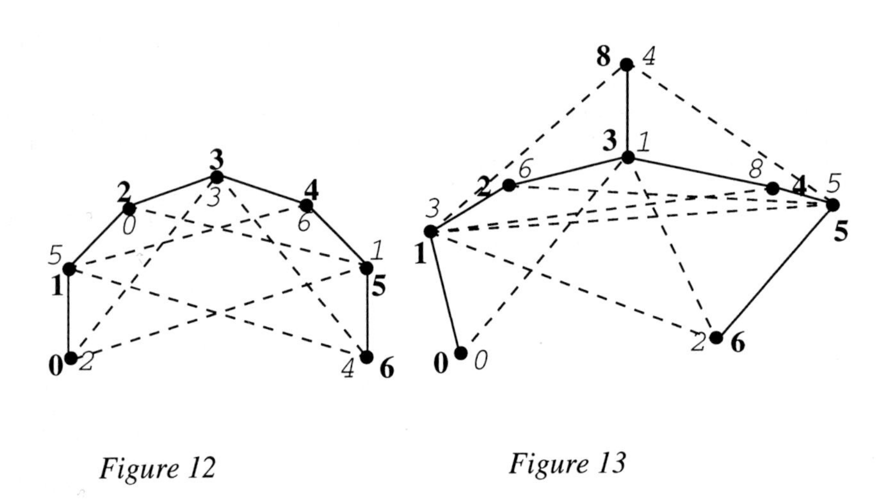

Figure 12 *Figure 13*

vertices of the path P together with a_8 span a subtree R of T which is a derived tree of T. If the node a_5 of T is adjacent to more than one endpoint then T would have an equal parity reducing pair of endpoints x and y such that $T - \{x, y\}$ has diameter six. Hence we assume wihout loss of generality that a_5 is adjacent to only one endpoint. Figure 13 exhibits an equal parity two-packing of R in which all of the nodes which belong to endpoints of T which are not in R are split. Hence T has an equal parity two-packing.

Assume next that a is adjacent to some endpoint, say a_0 in A. The tree T contains then the path $P = a, a_0, a_1, a_2, a_3, a_4, a_5$ of length six. If b is adjacent to an element of C then a_2 is not a node of T. Hence we can use the tree exhibited in Figure 10 together with the First Preservation Lemma to construct an equal parity two-packing of T. If b is not adjacent to a vertex in C but to the center a_2 of P we use Figure 14 and if b is neither adjacent to a_2 nor to a vertex in C we use Figure 13 together with the First Preservation Lemma to construct an equal parity two-packing of T.

Case 3.4. For any two endpoints a, b of equal parity which are adjacent to two different nodes, the tree $T - \{a, b\}$ has diameter four.

Let P be the path $a = a_0, a_1, a_2, a_3, a_4, a_5, a_6 = b$ of length six which joins the endpoints a and b in T. Observe that T does not contain a node, say m, different from the nodes a_1 and a_5 which has parity 1. Otherwise some endpoint x adjacent to m would form an equal parity reducing pair with a_0 and $T - \{x, a_0\}$ would have diameter five. If T would contain two nodes, say m and n, besides the nodes a_1 and a_5 they would both have to have even parity. But then T could be equal parity reduced by some two endpoints adjacent to m and n respectively and still have diameter six. If T contains no other node besides the nodes a_1 and a_5 then T is a dragon of diameter six which has an equal parity two-packing by Figure 12 and the First Preservation Lemma.

It follows that the tree T has exactly three nodes, namely a_1, a_5 and a third one, say m which has even parity. The distance from m to either one of a_1 and a_5 is smaller than four. Otherwise we could reduce T by the pair a_0, a_6 of endpoints and obtain a tree of diameter at least five. Hence m is either a vertex in P or has distance one from a_3. If m is a vertex in P different from the center a_3 we are done by Figure 12 and if m is adjacent to a_3 or equal to a_3 we are done by Figure 13 and the First Preservation Lemma. ∎

Theorem 3.1. *A forest has a bipartite packing if and only if it is either a singleton vertex or disconnected or a tree of diameter at least five which is not a crested dragon. Every forest which has a bipartite packing has an equal parity packing.*

Proof. Follows immediately from Lemma 2.5, Lemma 3.5, Lemma 3.8 and Lemma 3.9. ∎

4. Two-packings which yield a graph of chromatic number three

Lemma 4.1. *Every tree with diameter at least four has a three chromatic two-packing.*

Proof. Using the Second Preservation Lemma we may assume that if a, b is any reducing pair of endpoints of T then $T - \{a, b\}$ does not have diameter at least four. This implies that T has at most three different nodes.

Case 4.1. The tree T has exactly three nodes.

Assume that the nodes are the vertices a_1, b_1 and c_1 and that a is an endpoint adjacent to a_1, b is an endpoint adjacent to b_1 and that c is an endpoint adjacent to c_1. Assume also that the distance from a to b is the diameter of T. The diameter of T can not be larger than or equal to five because then the tree $T - \{b, c\}$ would still have diameter at least four. Hence there is a path $P = a, a_1, d, b_1, b$ of length four from a to b. Let the point of attachment of c_1 be the vertex in the path P which has shortest distance to c_1. The point of attachment of c_1 can not be a or b because a and b are endpoints of T. If the point of attachment of c_1 is a_1 then c_1 is not a vertex in P, otherwise $c_1 = a_1$ contrary to the assumption that the three nodes a_1, b_1 and c_1 are pairwise different. But c_1 has then at least distance one from a_1 and the distance from c to b would be at least five. Similarly the point of attachment of c_1 is not equal to b_1. Hence the point of attachment of c_1 must be the central point d of the path P. The distance from c_1 to d is at most one otherwise the distance from c to a would be at least five. Hence the distance from c_1 to d is either one or zero. In either case the tree R which consists of the path P of length four together with one endpoint x adjacent to the central point d will be a derived tree of the tree T. Figure 14 shows a two-packing of the tree R of chromatic number three in which every vertex splits. Using the First Preservation Lemma we conclude that T has a three chromatic two-packing.

Case 4.2. The tree T has exactly two nodes.

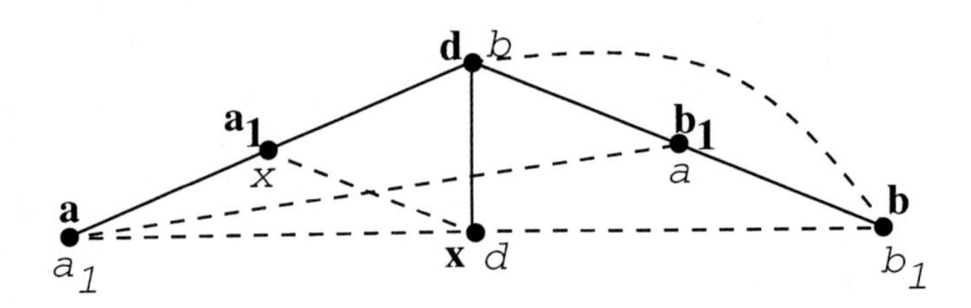

Figure 14

This means that T is a dragon. Assume that the nodes are the vertices a_1 and b_1 and that a is an endpoint adjacent to a_1 and b is an endpoint adjacent to b_1. Observe that either a_1 or b_1 is adjacent to exactly one endpoint. Otherwise T can be further reduced to a tree which has diameter at least four. For the same reason the dragon T has diameter four or five. In one case a path of length four and in the other case a path of length five is a derived tree of T. Because at least one of the two nodes is adjacent to only one endoint it is sufficient to find a two-packing of a path of length four and a two-packing of a path of length five in which at least one of the nodes splits and which has chromatic number three. Such two-packings are exhibited in Figure 15. ∎

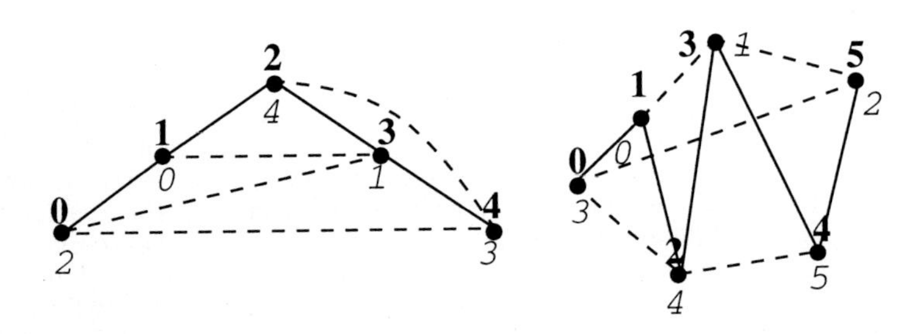

Figure 15

Lemma 4.2. *A forest F which is not connected has a three chromatic two-packing if and only if F contains a path of length two.*

Proof. If F does not contain a path of length two then the degree of every vertex of F is at most one and F consists of a partial matching (that is a set of pairwise not adjacent edges), together with some isolated vertices. Hence for any two-packing σ of F the maximal degree of F_σ is at most two and the edges of F_σ can be factorized into two matchings. This means that every connected component of F_σ is either a path or a circuit. Since an odd circuit can not be factorized into two matchings each of the circuits has even length which implies that the chromatic number of F_σ is at most two.

Let us now assume that F contains a path of length two. Using the Second Preservation Lemma we may assume that F can not be reduced without yielding a forest which has no path of length two. That implies that F has at most two connected components which contain a path of length one or two and all other components of F are isolated points. Otherwise F could be further reduced. Observe that every isolated point has a bipartite two-packing. This means that we can assume that F has at most two connected components and that if F contains an isolated point then F has exactly two components, one the isolated point and the other a tree of diameter two or three. There are three more cases to be discussed. In each case, a two-packing σ of F is exhibited such that every vertex of F is split and F_σ has chromatic number three.

Case 4.3. The forest F has two connected components A and B which contain a path of length two. Since the forest F can not be further reduced, each of the two connected components A and B is a path of length two. Figure 16 exhibits a two-packing σ such that F_σ has chromatic number three. ∎

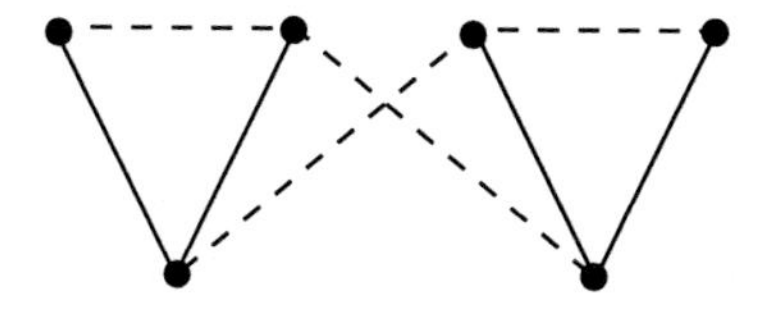

Figure 16

Case 4.4. The forest F has only one component A which contains a path of length two and an isolated point e.

The forest F consists then of two components, A and the isolated point e. The diameter of A is at most three, otherwise F could be further reduced. We conclude that A is either a star or a dragon of diameter three. If A is a dragon of diameter three we can assume that at most one of the two nodes has degree larger than or equal to three, otherwise F could be further reduced. Figure 17 exhibits for both cases a two-packing σ such that F_σ has chromatic number three.

Case 4.5. The forest F has only one component A which contains a path of length two and no isolated point. Then the other component of F contains exactly one edge. Since F can not be reduced further, F must consist of a path of length two and a second component of diameter one. Figure 18 exhibits a two-packing σ such that F_σ has chromatic number three.

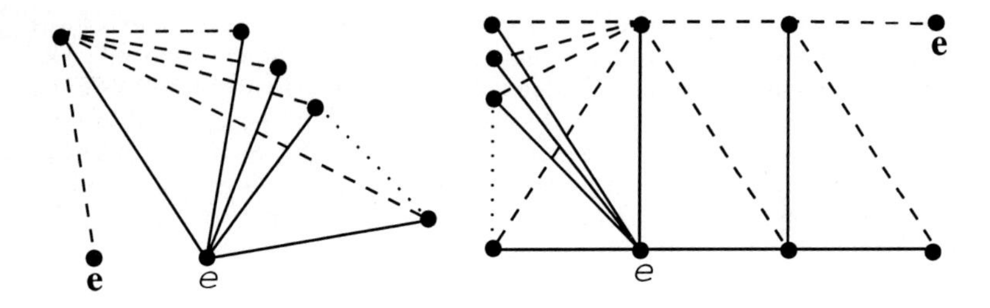

Figure 17

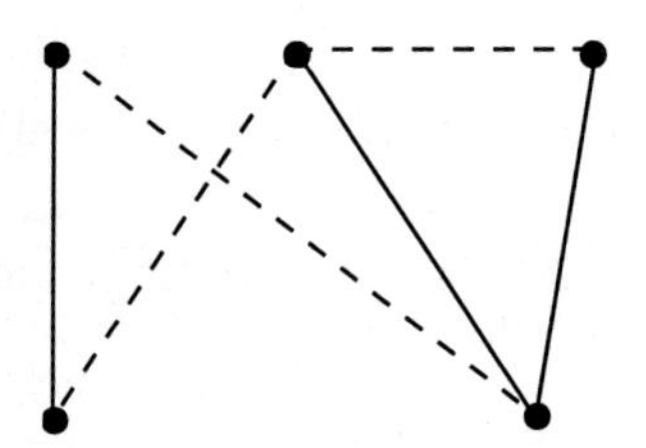

Figure 18

Theorem 4.1. *A forest F has a two-packing σ such that F_σ has chromatic number three if and only if F is a tree with diameter at least four or F is disconnected and contains a path of length two.*

Proof. If F is disconnected then the theorem follows from Lemma 4.2. If F is connected and contains a path of length four the theorem follows from Lemma 4.1. If F is connected and has diameter zero, that is F is an isolated point, then for every two-packing σ of F, the graph F_σ has chromatic number one. If F is connected and has diameter one or diameter two, then F does not have a two-packing, by Lemma 2.3. If F is connected and has diameter three then it follows from Lemma 2.4 that for every two packing σ of F, the graph F_σ has chromatic number four. ∎

5. Two-packings which yield a graph of chromatic number four

For $n \geq 3$, the *wheel* W_n is a graph which consists of a cycle with n vertices and an additional vertex which is adjacent to all of the vertices in the cycle. The complete graph K_4 for example is the wheel W_3.

Lemma 5.1. *A graph G with at most six vertices and chromatic number four contains the complete graph K_4 or the wheel W_5 as a subgraph.*

Proof. We only need consider the cases that G has five vertices and that G has six vertices. If G has six vertices and does not contain a triangle, then because G has chromatic number four it must contain a circuit C of length five. Because G does not contain a triangle C has no chords. If the sixth vertex is not adjacent to all of the vertices of C, G would have chromatic number three and if the sixth vertex is adjacent to all vertices of C then G is the wheel W_5. If G contains a triangle with vertices a, b, c then color the vertex a with color 0 the vertex b with color 1 and the vertex c with color 2. If G does not contain the complete graph K_4, then each of the other three vertices x, y, z is adjacent to at most two of the vertices in the triangle spanned by a, b, c. If two of the vertices x, y and z are adjacent then they are not adjacent to the same two vertices of the triangle spanned by a, b, c. We color the vertices x, y and z with the colors $0, 1$ and 2 in such a way that none of the vertices x, y, z is adjacent to a vertex in the set $\{a, b, c\}$ of vertices which has the same color. This can be done because each of the vertices x, y, z is adjacent to at most two of the vertices in the triangle spanned by a, b, c. We also assume that the number of adjacent pairs of vertices which received the same color is as small as possible under the coloring condition given above. Observe that if all three colors are used to color the vertices x, y and z then G would have a good three coloring. Hence we can assume that at least two of the three vertices x, y and z have the same color.

Assume first that all three vertices x, y and z have the same color, say 2. If no two of them are adjacent then we arrived at a three coloring of the graph G. Hence assume that the vertices x and y are adjacent. At least one of the two vertices x and y, say y is not adjacent to one of the two vertices, say b. We can then color y with the color 1 instead of with the color 2 and obtain a coloring in which there are fewer adjacent pairs of vertices which have the same color.

Assume next that the vertices x and y have the same color, say 2, and that the vertex z has a different color, say 1. The vertices x and y are adjacent, otherwise the given coloring would be a good three coloring of the graph G. If one of the two vertices x and y, say y, is not adjacent to the vertex a, then y could be colored with the color 0 to obtain a good three coloring of G. Hence both vertices x and y are adjacent to the vertex a. This means that one of the two vertices x and y, say y is not adjacent to the vertex b. If the vertices z and y would not be adjacent then coloring y with the color 1 instead of 2 would give a good three coloring of G. Hence y is adjacent to z. If the vertex z would not be adjacent to the vertex a we would obtain a good three coloring of G by changing the color or z from 1 to 0 and the color of y from 2 to 1. Hence z is adjacent to a. The vertices z and x are not adjacent otherwise the vertices x, y, z and a would form a K_4. This implies that x is adjacent to b otherwise x could be colored with color 1 to give a good three coloring of G. Also the vertices z and c are adjacent, otherwise we could change the color of z from 1 to 2 and the color of y from 2 to 1 to obtain a good three coloring of G. This now finally implies that G contains the wheel W_5 because x, y, z, c, b, x form a cycle of length five and the vertex a is adjacent to the other five vertices.

If G has five vertices let G_1 be the graph obtained from G by adding an additional isolated point. We can then apply the result for six vertices to G_1 and obtain the lemma in the case of five vertices as well. ∎

Lemma 5.2. *If A and B is a factorization of the complete graph K_4 into two forests, then both of the factors A and B are isomorphic to the path P_3 of length three.*

Proof. If one of the factors A or B contains less than three edges the other factor must contain at least four edges and hence a circuit. This means that both of the factors A and B contain exactly three edges. Hence each of the factors A and B is either the path P_3 or the star S_3. The complement of S_3 in K_4 is a triangle. Hence both factors A and B are isomorphic to the path P_3. ∎

Lemma 5.3. *A forest F with exactly two connected components P and S, where P is a path of length three and S is a star S_n with $n \geq 1$, does not have a two-packing σ such that F_σ has chromatic number four.*

Proof. Assume for a contradiction that F_σ has chromatic number four. Let G be the graph obtained from F_σ by removing all vertices from F_σ which have degree at most two. Clearly, F_σ has chromatic number four if and only if G has chromatic number four. Because F contains at most three vertices which have degree larger than one, the graph G contains at most six vertices. Hence by Lemma 5.1, G and therefore F_σ contains the complete graph K_4 or the wheel W_5. According to Lemma 5.2, if A, B is a two-factorization of K_4 into two forests then both factors A and B are isomorphic to P_3. Because F contains only one path of length three, the two factors A and B are a two-packing of P_3. But this is not possible because the star S does not have a two-packing.

Let us now assume that G contains the wheel W_5. The wheel W_5 contains six vertices each of which has degree at least three. This means that the vertices of this wheel consist of the three vertices of F which are not endpoints of F together with their σ images. Notice that if y is a vertex in W_5 which is not an endpoint of F then both $\sigma^{-1}(y)$ and $\sigma(y)$ is an endpoint of F. Let a be the center of the star S. The vertex x of W_5 which is adjacent to all other vertices of W_5 has degree five. We may assume without loss of generality that three of the edges adjacent in G to x are edges of F and hence that $x = a$. The two central points c, d of P are then in W_5 adjacent to $x = a$. The two edges of W_5 from c, d to $x = a$ must be edges of $\sigma(F)$. But then $\sigma^{-1}(a)$ could not be an endpoint of F. ∎

Lemma 5.4. *A forest F which does not contain a path of length three does not have a two-packing σ such that F_σ has chromatic number four.*

Proof. Let us direct the edges of F in such a way that every directed edge points towards an endpoint of F. Then we direct the edges of $\sigma(F)$ such that σ preserves the directions. This makes F_σ into a directed graph. Observe that a vertex in F_σ has degree larger than or equal to three if and only if its indegree is at most one. If F_σ has chromatic number four then the graphs which we obtain from F_σ by successively removing vertices of degree at most two will also have chromatic number four. Hence this process of removing vertices of degree at most two ends in a graph G of chromatic number four in which every vertex has degree at least three. This is a contradiction because in G every vertex has indegree at most one and hence outdegree at least two. ∎

Lemma 5.5. *Every tree T of diameter at least three has a two packing σ such that F_σ has chromatic number four.*

Proof. By the Second Preservation Lemma we may assume that T is reduced, that means that whenever a, b is a reducing pair of endpoints then $T-\{a,b\}$ has diameter smaller than three. This implies that T has diameter either three or four and if T has diameter four then T has exactly two endpoints, that is T is a path of length four, and if T has diameter three then T is a dragon of diameter three because every tree of diameter three is a dragon. Observe that in both cases the path of length three is a derived tree. Any factorization of the complete graph K_4 into two paths of length three results in a two packing σ of the path of length three, in which every vertex splits. ∎

Lemma 5.6. *If F is a forest which contains at least one path of length three and if F does not have exactly two components one of which is a star and the other a path of length three then F has a two-packing σ such that F_σ has chromatic number four.*

Proof. By the Second Preservation Lemma we may assume that F is reduced, that is if a and b is a reducing pair of vertices then $F-\{a,b\}$ is a forest which either does not contain a path of length three or $F-\{a,b\}$ has exactly two components one of which is a star and the other a path of length three. If F has more than two components A, $B_1, B_2, \ldots, B_n$ with $n \geq 2$, where A contains a path of length three then let σ be a two-packing of A such that F_σ has chromatic number four, (Lemma 5.5), and λ a two-packing of the forest which consists of the other components. Clearly $\sigma \cup \lambda$ has then the required properties. We may therefore assume that F has exactly two components A and B where A has diameter at least three. If B is not a star then again the union of a four chromatic two-packing of A and a two-packing of B will be a four chromatic two-packing of F. We assume therefore that B is a star. If A has diameter larger than or equal to five then F could be reduced further by a pair of endpoints, one from B and the other from A. If A has diameter four then A has exactly two endpoints or else we could reduce F by a pair of endpoints one from A and the other from B. We arrived then at the case where A is a path of length four and B a star with at least two vertices. Figure 19 exhibits a two packing σ for this case such that F_σ has chromatic number four. If A has diameter three then A is a dragon of diameter three. If A would have more than three endpoints then F could be further reduced. Hence A has exactly three endpoints, otherwise A would be a path of length three. Figure 20 exhibits a two packing λ such that F_λ has chromatic number four. ∎

Theorem 5.1. *A forest F has a two-packing σ such that F_σ has chromatic number four if and only if F contains a path of length three and F does not consist of exactly two connected components where one of the two components is a path of length three and the other is a star.*

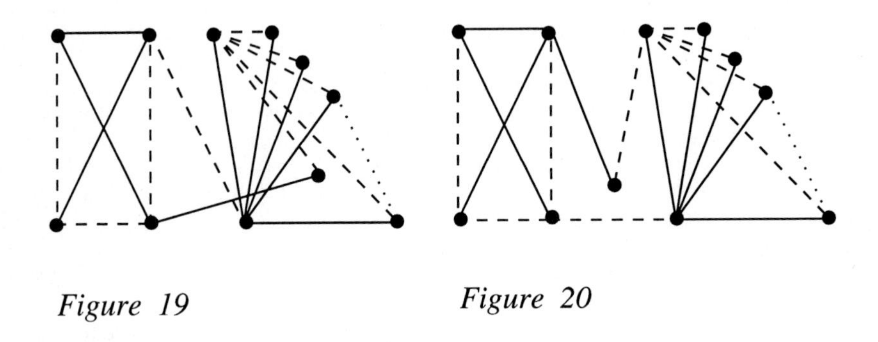

Figure 19 *Figure 20*

Proof. Theorem 5.1 follows immediately from Lemma 5.3, Lemma 5.4, Lemma 5.5 and Lemma 5.6. ∎

References

1. B. Bollobás, *Extremal Graph Theory*, Academic Press, London (1978)
2. D. Burns, S. Schuster, *Embedding $(n, n-1)$-graphs in their complements*, Israel J. Math., **30** (1978), 313 - 320.
3. R.J. Faudree, C.C. Rousseau, R.H. Schelp, S. Schuster, *Embedding graphs in their complements*, Czecho-slovac J. Math., **31** 106 (1981), 53 - 62.
4. Stephan Brandt, *Embedding graphs without short cycles in their complements*, to appear in: Proceedings of the seventh International Conference. (Kalmazoo, 1992)
5. N. Sauer, J. Spencer, *Edge disjoint placement of graphs* J. Combinatorial Theory B, **25**, (1978), 295 - 302.
6. H. Wang, N. Sauer, *Packing three copies of a tree into a complete graph*, Europ. J. Combinatorics (1993) **14**.

On the Isolation of a Common Secret

Don Beaver[1], Stuart Haber[2], and Peter Winkler[3]

[1] Pennsylvania State University, University Park, PA, USA
[2] Surety Technologies Inc. and Bellcore, Morristown NJ, USA
[3] AT&T Bell Laboratories, Murray Hill, NJ, USA

Summary. Two parties are said to "share a secret" if there is a question to which only they know the answer. Since possession of a shared secret allows them to communicate a bit between them over an open channel without revealing the value of the bit, shared secrets are fundamental in cryptology.

We consider below the problem of when two parties with shared knowledge can use that knowledge to establish, over an open channel, a shared secret. There are no issues of complexity or probability; the parties are not assumed to be limited in computing power, and secrecy is judged only relative to certainty, not probability. In this context the issues become purely combinatorial and in fact lead to some curious results in graph theory.

Applications are indicated in the game of bridge, and for a problem involving two sheriffs, eight suspects and a lynch mob.

1. Introduction

Suppose two parties—let us call them "Alice" and "Bob"—share a secret, that is, they have common knowledge possessed by no one else; then Alice may use her secret to transmit a bit to Bob in such a way that no eavesdropper can deduce the value of the bit. For example, if Alice and Bob are the only people in the world who know whether President Bush wears a wig, Alice may send Bob the following message (or the same message, with 0 and 1 interchanged):

"My bit is 0 if the President wears a wig, 1 otherwise."

While Eve (an eavesdropper) may believe that the President probably does not wear a wig, and therefore that Alice's bit is more likely to be 1 than 0, her inability to determine the value of the bit with certainty is all that concerns us here.

This method of encryption is called a "one-time pad"; Alice and Bob share a bit of information, and they can use it (once) to pass a bit in secret.

There are, however, situations where even though Alice and Bob appear to possess shared information not available to the public, this information does not take the form of a shared secret. Nonetheless, Alice and Bob may be able to *isolate* a shared secret by communicating with each other, even though their messages are public. Since this is precisely the situation where cryptologic methods are needed (communication lines are available, but not private), Alice and Bob are almost as well off here as if they had begun with a shared secret; they must merely spend a few preliminary rounds of communication in establishing the secret.

Let us give two examples of such situations, before proceeding further.

(1) **The game of bridge:** Here two partners wish to communicate in private, but the rules of the game require that all communication be done by legal bids and plays, about which there may be no prior private understandings. Thus, there are initially no shared secrets. But there is private information: each player knows, by virtue of looking at his own hand, 13 cards that do *not* belong to his partner. Can they make use of this information to communicate in private?

(2) **The 'two sheriffs' problem:** Two sheriffs in neighboring towns are on the track of a killer, in a case involving eight suspects. By virtue of independent, reliable detective work, each has narrowed the list to only two. Now they are engaged in a telephone call; their object is to compare information, and if their pairs overlap in just one suspect, to identify him (the killer) and put out a.p.b.'s so as to catch him in either town.

The difficulty is that their telephone line has been tapped by the local lynch mob, who know the original list of suspects but not which pairs the sheriffs have arrived at. If they are able to identify the killer with certainty as a result of the phone call, he will be lynched before he can be arrested.

Can the sheriffs accomplish their objective without tipping off the mob?

2. The Mathematical Model

One natural model for common knowledge is obtained by imagining that in any situation there is an underlying finite space S of *possibilities* of which any one element may be "the truth." Alice's knowledge concerning S at any point consists of some subset X of S, meaning that X is precisely the set of truths consistent with what Alice knows. As Alice communicates with Bob she obtains more information, and her knowledge set X shrinks accordingly.

At any time the "true point" must lie both in X and in Bob's knowledge set Y, but if there are two or more points in $X \cap Y$ Alice and Bob will never be able to choose among them by communicating with each other. Hence for our purposes if X is a *possible* knowledge set for Alice and Y for Bob, our only concern is whether they intersect.

Consequently we choose to model common knowledge by using a mere vertex to represent a possible knowledge set of Alice's, and similarly for Bob; we connect vertex x of Alice's with vertex y of Bob's when the corresponding knowledge sets intersect, that is, when the two vertices are simultaneously possible. The "truth" is thus represented by some adjacent pair of vertices, i.e. an edge.

Alice and Bob's knowledge at any time thus constitutes a graph, which, in accordance with cryptographic tradition, is assumed to be known to everyone in the world. The interpretation of this graph will, we hope, become clear to the reader after some examples.

It is convenient to formalize our model as follows.

Definition 2.1. *A bigraph is a finite, non-empty collection H of ordered pairs such that if (x, y) is in H then (y, z) is not.*

Elements of the set $A(H) := \{x : (x, y) \in H \text{ for some } y\}$ will be termed "Alice's vertices" and are perforce distinct from the symmetrically defined "Bob's vertices" in $B(H)$. Thus the elements of H are edges of a bipartite graph, but note that the vertices come equipped with a labelled left-right (Alice-Bob) partition and that isolated vertices cannot arise.

Our model now consists of a bigraph H, known to all, the edges of which represent possible truths. Alice knows the endpoint in $A(H)$ of the true edge, Bob its endpoint in $B(H)$; in other words, if the true edge is (x, y) then Alice knows x and Bob knows y.

We say that Alice and Bob *share a secret* if there is a question to which they know the answer and Eve does not. In the wig example, the question "Does the President wear a wig?" can be answered only by Bob and Alice, so they indeed share a secret in this case. Here, the bigraph H consists of a pair of disjoint edges, one corresponding to "the President wears a wig" and the other to "the President does not wear a wig." The disconnectivity of H is its crucial property:

Theorem 2.1. *Two parties share a secret if and only if their bigraph is disconnected.*

Proof. It is immediate that Alice and Bob share a secret whenever their bigraph is disconnected, since if C is one of its connected components, only they can answer the question "Is the true edge in C ?". To see the converse, let Q be the given question and let $a_1, a_2, \ldots$ be its possible answers (from Eve's point of view). Write "$(u,v)\#a_i$" if it is simultaneously possible for a_i to be the answer to Q, and (u,v) to be the true edge of H.

We now note that if $(x,y)\#a_i$ and $(u,v)\#a_j$ for $i \neq j$, then (x,y) and (u,v) can be neither identical nor adjacent; if, for example, $x = u$ then when Alice's end of the true edge is x she will be unable to decide between answers a_i and a_j.

It follows that the edges consistent with the various answers a_i determine a partition of H, each part of which is a non-empty union of connected components; since the number of possible answers must be at least two, H is disconnected. □

If H is connected, are Alice and Bob doomed never to share a secret? Of course, if they can arrange a private (secure) conversation, they can agree on some string of bits and thus share as many secrets as they wish; this indeed is often done in traditional cryptography, in the name of agreeing on or distributing *key*. Unfortunately this phase is often dangerous and sometimes impossible; else, cryptography would be unnecessary. However, Alice and Bob may be able to use their common knowledge (reflected in the structure of H) to isolate a common secret by means of a *public* conversation; and it is just this process which we wish to investigate.

Consider, for example, the bigraph $H = \{(a_1,b_1), (a_1,b_2), (a_2,b_2), (a_2,b_3), (a_3,b_3), (a_3,b_4), (a_4,b_4), (a_4,b_1)\}$. H is an 8-cycle, thus connected, but Alice and Bob can disconnect it as follows: if Alice's end of the true edge is a_1 or a_3 she says so: "My end of the true edge is either a_1 or a_3." Bob can tell by looking at his end which of the two possibilities is the case, hence they now share a secret; this is reflected in the fact that after Alice's announcement, their bigraph is disconnected (see Fig. 1). Of course, had Alice's end been a_2 or a_4, an announcement to that effect would also have done the job. (We shall see later that two-sided conversations may be necessary to disconnect some bigraphs.)

The next example is inspired by the work of Winkler [4, 5] on cryptologic techniques for the game of bridge, and the recent works of Fischer, Paterson and Rackoff [2] and Fischer and Wright [3] using random deals of cards for cryptographic key. A card is dealt at random, face down, to each of Alice and Bob, from a three-card deck consisting (say) of an Ace, a King, and a Queen. The remaining card is discarded unseen. Here H is a 6-cycle (see Fig. 2). The six edges correspond to the six possible deals; the set $A(H)$ consists of Alice's three possible holdings ("A", "K" or "Q") and similarly for $B(H)$. The fact that "K" in $A(H)$ and "K" in $B(H)$ are not adjacent corresponds to the fact that Alice and Bob cannot both hold the King. Clearly Alice and Bob share *some* knowledge in this situation, but as we shall see it is not enough for them to be assured of being able to isolate a shared secret.

For a third example, suppose H is a 10-cycle on vertices $v_0, v_1, \ldots, v_9$ with v_i adjacent to v_j just when $|i-j| = 1 \bmod 10$. (See Fig. 3.)

Let $A(H)$ consist of the vertices of even index. Then the following protocol allows Alice to disconnect the bigraph: if she holds v_i, she chooses j to be *either* $i+4$ or $i+6 \bmod 10$, then tells Bob:

"I hold either v_i or v_j."

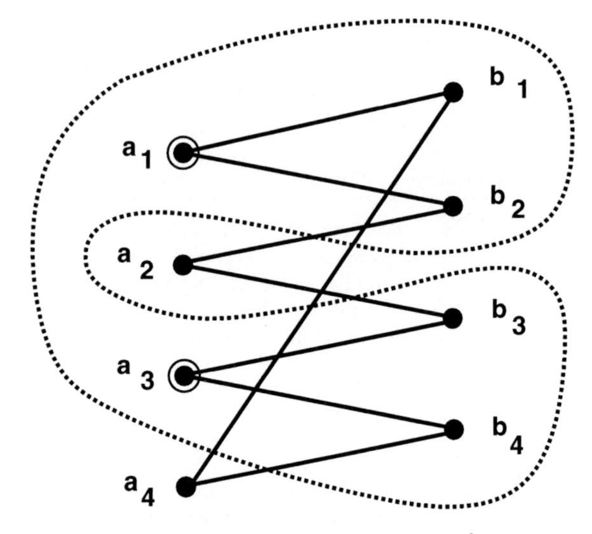

Fig. 2.1. Alice separates an 8-cycle

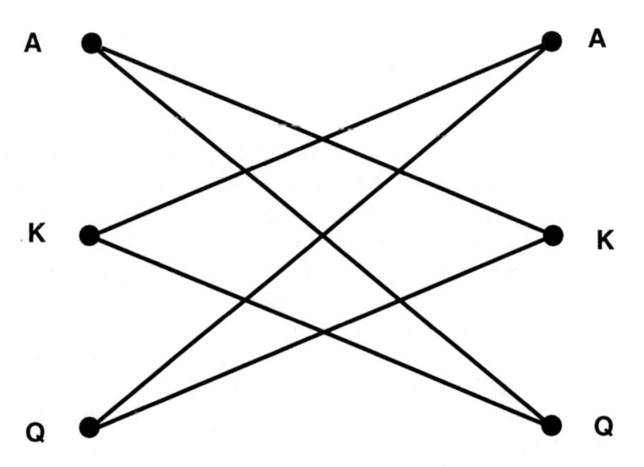

Fig. 2.2. Bigraph for dealing 2 cards from a 3-card deck

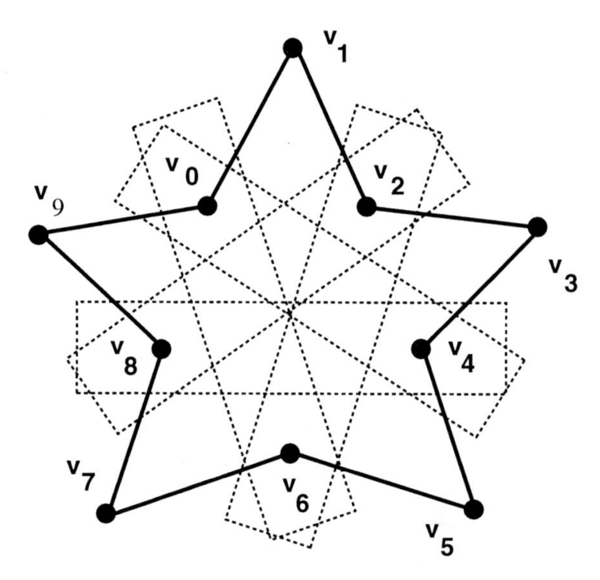

Fig. 2.3. Non-deterministically separating a 10-cycle

This protocol is said to be *non-deterministic* because Alice has more than one choice of message for a given holding. Non-determinism as used here is thus quite different from its use in complexity theory, and in fact is more closely related in some respects to randomization (despite the absence of probability in our model). In particular non-deterministic communication protocols are quite practical.

Incidentally, we assume nothing is given away by the order in which objects are named in a semantically symmetric expression such as "v_i or v_j." In a deterministic protocol we can insure this by a naming convention, such as enforced alphabetical order, but our mathematical model for communication will obviate the problem.

In order to define "communication protocol" rigorously we need to define "conversation", even though the latter definition needs the former for interpretation. Accordingly, a *conversation* will be a finite string $m_1, \ldots, m_t$ of positive integers, communicated alternately by Alice and Bob beginning with Alice.

Although this seems perhaps a limited vocabulary for communication, it is in fact completely general because meanings can be assigned to the numbers via an agreed-upon protocol. The communication protocol specifies *under precisely what circumstances a given number may be uttered.* Thus, we operate in effect in the "political" theory of communication: when someone says something we ask not "what does that mean" but "under what circumstances would he/she have said that." This model is both stronger and simpler than one in which messages are sentences. To see its effect, consider solving the C_{10} bigraph above by using always vertices i and $i+4$. This looks reasonable at first glance but is actually a sham. For example, Alice would not then say "My end is 4 or 8" if she held 8. Hence an eavesdropper can eliminate 8 if she hears this, preventing disconnection of the bigraph and ruining the protocol.

3. Deterministic Separation

Definition 3.1. *A* deterministic communication protocol *for the bigraph H is a sequence of functions $f_1, \ldots$ into the positive integers, such that f_1 is a function of x (Alice's end of the true edge), and each subsequent f_i depends on the values of $f_1, \ldots, f_{i-1}$ and either on x, if i is odd, or on y (Bob's end of the true edge) if i is even.*

Thus, a deterministic communication protocol, combined with a true edge, produces a unique conversation given by $m_i = f_i(x; m_1, \ldots, m_{i-1})$ for i odd and $m_i = f_i(y; m_1, \ldots, m_{i-1})$ for i even. After step i of the conversation, the situation can be again described by a bigraph H^i consisting of those edges (u, v) such that if (u, v) had been the true edge the conversation would have been as seen. H^i is a sub-bigraph of H which contains the actual true edge; in fact H^i is obtained from H^{i-1} by deleting all edges incident to certain left-hand vertices (for i odd) or certain right-hand vertices (for i even). The vertices which survive on the left (when i is odd) are just $\{u : f_i(u; m_1, \ldots, m_{i-1}) = m_i\}$.

For our purposes a conversation $(m_1, \ldots, m_t)$ will be deemed "successful" if H^t is disconnected, and a communication protocol for H will be said to "separate H" or to be a "separation protocol for H" if it always produces a successful conversation. Finally, H itself will be termed *deterministicallyseparable* if there is a deterministic communication protocol which separates H.

We can also give a recursive characterization of the deterministically separable bigraphs. It will be useful to introduce notation for the *sub-bigraph* $H|A'$ of a bigraph H induced by a subset A' of $A(H)$; namely,

$$H|A' := H \cap (A' \times B(H)) \ .$$

Thus from the standpoint of ordinary graph theory, $H|A'$ is obtained by discarding isolated vertices from the subgraph of H induced by $A' \cup B(H)$. The definition of $H|B'$ for $B' \subset B(H)$ is similar.

Theorem 3.1. *Let* **DS** *be the smallest symmetric class of bigraphs which contains the disconnected bigraphs and has the following property: for any bigraph H, if there is a partition $A(H) = A_1 \cup A_2$ of Alice's vertices such that $H|A_1$ and $H|A_2$ are both in* **DS**, *then H is also in* **DS**. *Then* **DS** *is the class of deterministically separable bigraphs.*

Proof. Let us check first that the class of deterministically separable bigraphs is indeed closed under the operation defined in the statement of the theorem. It is certainly symmetrical, since a protocol which separates H can be modified to one which separates the dual of H (i.e. the result of reversing the ordered pairs in H) by switching the roles of Alice and Bob, and adding a meaningless message from Alice to the beginning. If a partition is given along with separation protocols P_1 and P_2 for the two sub-bigraphs of H, we design a separation protocol P for H as follows: first Alice sends "i" iff her end of the true edge lies in A_i, then Bob sends back a meaningless "1", then protocol P_i is followed.

It remains to show that any symmetric class **C** containing the disconnected bigraphs and closed under the stated operation contains all deterministically separable bigraphs. This is done by induction on the number of edges.

Let H be connected but deterministically separable via a communication protocol P; then sooner or later P must call for a first meaningful message, say from Alice. We may assume that her options at that point are to send "1", "2", etc. up to "k" for some $k \geq 2$, according to whether her end of the true edge lies in E_1, E_2, etc; there is no dependence on previous conversation here because by assumption said conversation has up to now been predictable. Let us modify P slightly by having Alice send only a 1 or 2 at this point, the former just when her end is in E_1; if she sends a "2" Bob sends a meaningless "1" back, then Alice sends the "2", "3",... or "k" that would have been sent before and the protocol P resumes. We have thus found protocols which separate each of $H|E_1$ and $H|((A(H) - E_1)$. Each of these is thus in $\mathbf{C}$ by the induction assumption, hence $H \in \mathbf{C}$ by the closure condition. □

4. Non-Deterministic Separation

We now introduce two ways of weakening the definition of a communication protocol. First, if the functions f_i are permitted to be *multi*-valued, so that at each point Alice or Bob has one *or more* possible messages to send, we say that the protocol is *non-deterministic.* Note that during a non-deterministic communication protocol, the bigraph shrinks as before but this time each party, at his or her turn to speak, is provided with a (message-labelled) *cover* of his or her vertices instead of a *partition*. It is still the case, however, that the knowledge of Alice and Bob is expressed at each point by the state of their bigraph.

Second, if the functions are permitted to depend for odd i on a random number m known only to Alice, and for even i on a random number n known only to Bob, then the protocol is *randomized.* Here the bigraph does *not* any longer describe the situation completely, as Bob and Alice may learn things about each other's random number.

Whether a communication protocol is non-deterministic, randomized or both, however, we continue to insist that the protocol *always* produce a successful conversation in order to qualify for separating H.

Fortunately, the three new categories of separation protocol which arise result in only one new category of bigraph.

Theorem 4.1. *The following are equivalent, for any bigraph H:*

(a) H is separable by a non-deterministic communication protocol;
(b) H is separable by a randomized communication protocol;
(c) H is separable by a randomized, non-deterministic communication protocol.

Proof. We need to show (c)→(b) and (c)→(a), the reverse implications being trivial. Of these the former is easy: by extending the range of the random numbers, Alice and Bob can use them to decide which message to send when there is more than one choice.

Turning random numbers into non-deterministic choices looks awkward because a random number may be used many times in the protocol, whereas there is no "consistency" built in to nondeterminism. However, this problem is illusory. Suppose, at Alice's first turn to speak, that she is supplied with a randomized separation protocol but no random number; then she chooses a random number and acts accordingly. She cannot "remember" that number at her next turn and use it again, but she can compute which random numbers are consistent with her previous action and choose one of those upon which to base her next message. Bob behaves similarly; at each turn he determines which values of his non-existent random number are consistent with his own previous actions (and his end of the true edge), then picks one such value and acts accordingly. □

A bigraph will be called, simply, *separable* if one (thus all) of the conditions of Theorem 4.1 obtains. The following recursive characterization is analogous to Theorem 3.1. although a small additional subtlety arises in the proof.

Theorem 4.2. *Let* $\mathbf{S}$ *be the smallest symmetric class of bigraphs which contains the disconnected bigraphs and has the following property: for any bigraph* H*, if there is a covering* $A(H) = A_1 \cup A_2$ *of Alice's vertices such that* $H|A_1$ *and* $H|A_2$ *are both in* $\mathbf{S}$*, and both* A_1 *and* A_2 *are strictly contained in* $A(H)$*, then* H *is also in* $\mathbf{S}$*. Then* $\mathbf{S}$ *is the class of separable bigraphs.*

Proof. The proof that the class of separable bigraphs is symmetrical and closed under the operation defined in the statement of the theorem is as in Theorem 3.1, except that if Alice's end of the true edge lies in $A_1 \cap A_2$ she may send *either* message "1" or message "2".

It remains to show that any symmetric class $\mathbf{C}$ containing the disconnected bigraphs and closed under the stated operation contains all separable bigraphs; this is again done by induction on the number of edges.

Let H be connected but separable via a non-deterministic communication protocol P, and suppose that H is the smallest separable bigraph not in $\mathbf{C}$. Let us call a vertex u in $A(H)$ (or, dually, in $B(H)$) *weak* if there is no proper subset A' of $A(H)$ containing u for which $H|A'$ is separable.

We claim that there is some weak vertex in $A(H)$. For, if not, define for each $x \in A(H)$ a proper subset A_x which does yield a separable sub-bigraph of H. Since the A_x's cover $A(H)$, we can find $x_1, x_2, \ldots, x_k$ such that the A_{x_i}'s cover $A(H)$ with k minimal (but necessarily greater than 1). Set $A_1 := A_{x_1}$, $A_2 := A_{x_2} \cup \cdots \cup A_{x_k}$. Then A_1 and A_2 are a proper cover of $A(H)$, and $H|A_1$ is separable by assumption. However, $H|A_2$ is also separable, since Alice can reduce it to a separable bigraph by sending some i for which her end of the true edge lies in A_{x_i}, $2 \leq i \leq k$. These bigraphs are thus both in $\mathbf{C}$ by the induction assumption, contradicting the fact that H is not in $\mathbf{C}$.

Since the class of separable bigraphs is symmetric, the dual of H is also separable but not in $\mathbf{C}$; hence the same argument produces a weak vertex in $B(H)$. It may seem to the reader that weak vertices cause trouble only if found on the true edge, and thus that Alice and Bob can't *both* be stymied as long as no two weak vertices are adjacent. However, it turns out that the mere *presence* of weak vertices on both sides is enough to render H inseparable.

To see this, let u be a weak vertex in $A(H)$; there must be some message (say m) which Alice is permitted to send when her end of the true edge is u. Let A' be the set of vertices in $A(H)$ which, like u, allow the message m; then $H|A'$ must be separable, since this is the bigraph which results when m is sent. Thus A' must not be a proper subset of $A(H)$, that is, $A' = A(H)$ and the message m is meaningless.

If m is indeed sent, the protocol turns to Bob who still has all of H before him. By the same reasoning as above, he must also have a meaningless message (i.e. a message he can send regardless of which vertex is his end of the true edge) available to him.

Now we're back to Alice with H still intact. We thus see that Alice and Bob must be allowed by the protocol to pass meaningless messages back and forth ad infinitum, irrespective of which edge of H is the true edge; but then we have a contradiction, since H is required to have a communication protocol which always separates. $\square$

Theorem 4.2 is often useful in determining separability via case analysis. For example, it is easy to check that no path with fewer than 5 edges is separable, nor is the 6-cycle (Fig. 2) separable because for any proper subset S of Alice's or Bob's vertices, $H|S$ would be a path of length 2 or 4.

There is one class of bigraphs which is easily seen to be disjoint from the class of separable bigraphs, a fact which helps in obtaining negative results.

Theorem 4.3. *Suppose that there is an edge of H which is adjacent to all other edges of H. Then H is not separable.*

Proof. Such an edge cannot be contained in any disconnected sub-bigraph of H; thus, if it happens to be the true edge, no protocol can separate H. □

Note that, in particular, no separable bigraph can have a vertex which is adjacent to all the vertices on the other side.

We are now in a position to prove that that the class of separable bigraphs is strictly larger than the class of deterministically separable bigraphs.

Let $A(H) = \{a_1, \ldots, a_7\}$ and $B(H) = \{b_1, \ldots, b_7\}$, and put $(a_i, b_j) \in H$ iff $j - i = 1$, 2 or 4, where the indices are interpreted always modulo 7. Then H is the incidence graph of a Fano plane (see Fig. 4) and we have:

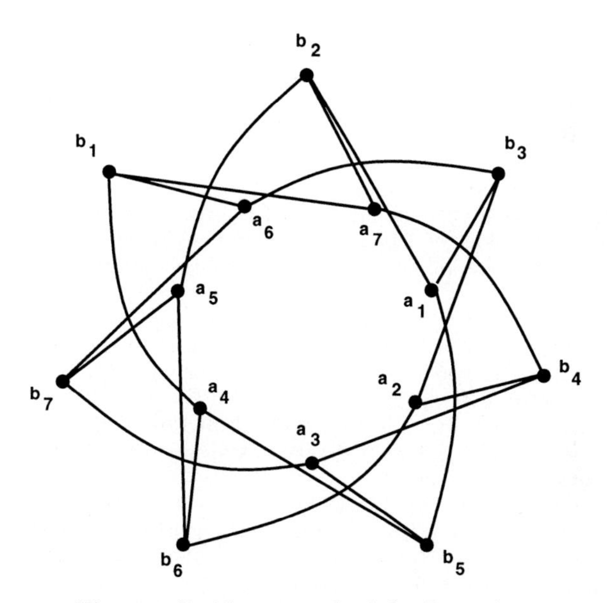

Fig. 4.1. Incidence graph of the Fano plane

Theorem 4.4. *The incidence graph of the Fano plane is separable but not deterministically separable.*

Proof. We first provide a non-deterministic separation protocol. Alice begins by sending a number k such that her end of the true edge is in the set $\{a_k, a_{k+1}, a_{k+2}\}$. Bob now (deterministically) sends back "1" if his end is b_{k+2} or b_{k+6}; "2" if it is b_{k+4} or b_{k+5}; "3" if it is b_{k+3} or b_{k+1}. (It cannot be b_k). This separates H into a 2-edge component and a 1-edge component.

If, on the other hand, there were a deterministic separation protocol for H, then one of the parties would eventually have to send a meaningful message, thus effecting a partition of his or her vertices. This may as well be Alice since H is symmetrical. If one of the parts has fewer than 3 of Alice's vertices in it, or has 3 vertices whose neighborhoods intersect, then one of Bob's remaining points will be of full degree, contradicting Theorem 4.3. Otherwise the partition must be isomorphic to $\{a_1, a_2, a_3\}$ versus $\{a_4, a_5, a_6, a_7\}$. The former part induces a deterministically separable sub-bigraph as we have seen from the above protocol, but the sub-bigraph induced by the latter part contains a vertex (b_1) whose neighborhood intersects the neighborhoods of all other vertices of $B(H)$. Thus, if Bob's end of the true edge is b_1, he cannot separate the bigraph at this time. However, Alice is also stymied because she has only one vertex available (a_5) not adjacent to b_1, thus her vertices can never be partitioned so as to induce disconnected sub-bigraphs. □

The situation changes if we consider bigraphs which are separable *in one round*, that is, by just one message from Alice. (This is not, by the way, a symmetric class; the path on 7 vertices, for example, can be separated only by a message from the party with 4 vertices.) Before proceeding, we need a curious graph-theoretic result.

Theorem 4.5. *Let G be any graph with no vertex adjacent to all others. Then there is a partition $V_1, \ldots, V_k$ of the vertices of G such that for each $i = 1, \ldots, k$ the subgraph $\langle V_i \rangle$ induced by V_i is disconnected.*

Proof. If not let G be a counterexample with smallest possible number of vertices. For any subset U of the set of vertices V, let $\omega(U)$ be the number of vertices in $V - U$ which are adjacent to all other vertices in $V - U$.

Note first that if $\langle U \rangle$ is disconnected, then $\omega(U)$ must be non-zero; else we may apply the induction hypothesis to get a suitable partition of $V - U$, and appending U itself to this partition yields a partition of V suitable for G.

Choices of U for which $\langle U \rangle$ is disconnected do exist, of course, since U can be taken to be a pair of non-adjacent vertices. Hence we may choose a U for which $\langle U \rangle$ is disconnected and $\omega(U)$ is minimal.

Now let x be any full vertex in $V - U$, that is, any vertex in $V - U$ which is adjacent to all other vertices in $V - U$. By assumption x has *some* non-neighbor, say y, in V; let C be the set of vertices of the component of $\langle U \rangle$ into which y falls.

Suppose first that y is not the only vertex in C, and let $W = U \setminus \{u\}$. Then $\langle W \rangle$ is still disconnected, but x is no longer full in $V - W$ since a non-neighbor y has "moved in." Of course y is not full either, and any other vertex which is full in $V - W$ must already have been full in $V - U$. Hence $\omega(W) < \omega(U)$, a contradiction.

We are reduced to the case where y is an isolated point of $\langle U \rangle$; now we let $W = U \cup \{x\}$. Since x and y are not adjacent y is still isolated in $\langle W \rangle$. Any full vertex of $V - W$ was adjacent to x in $V - U$ and therefore already full in $V - U$; but x itself is now gone from $V - W$ and so we again have $\omega(W) < \omega(U)$, and this contradiction proves the theorem. □

Note that the induction hypothesis, and thus the theorem itself, can be strengthened to read "each $\langle V_i \rangle$ has at least two vertices and contains an isolated point" without changing the proof. However, we will not need the stronger statement.

Theorem 4.6. *The following are equivalent for a bigraph H.*

(a) H is separable in one round;
(b) H is deterministically separable in one round;
(c) for every vertex u of $A(H)$ there is a vertex v of $A(H)$ such that the neighborhoods of u and v (in $B(H)$) are disjoint.

Proof. It is enough to show (a)→(c)→(b). Suppose that H is separable in one round, let u be any vertex of $A(H)$, and let i be a message that can be sent by Alice when her end of the true edge is u. Since the bigraph that results from sending "i" is disconnected, there is a vertex v on Alice's side of it which is in a different component from u; then u and v must originally have had disjoint neighborhoods.

Now suppose that (c) is satisfied and form a graph G on the vertices in $A(H)$ by defining $\{u, v\}$ to be an edge whenever u and v have *intersecting* neighborhoods. Condition (c) says precisely that G has no vertex of full degree, hence we may apply Theorem 4.5 to obtain a partition $V_1, \ldots, V_k$ of the vertices of $A(H)$ each part of which induces a disconnected subgraph of G, hence also of H. Sending "i" when Alice's end of the true edge lies in V_i thus yields a deterministic separation protocol. □

We have said that a disconnected bigraph is sufficient to enable Alice and Bob to communicate a bit in secret; we are now in a position to show that separability is in fact *necessary* for such a communication, thus completing the reduction of the original cryptologic problem to a graph-theoretical one.

Let us fix a bigraph H and suppose that Alice (say) has been supplied with a bit ε which she must communicate in secret to Bob, over our usual public channel. The effect of the bit is to double the vertices on Alice's side of H; that is, each vertex $a \in A(H)$ now becomes a pair $a(0), a(1)$ each with the same neighborhood that a had in $B(H)$. The edge $(a(0), b)$ corresponds to the original (a, b) together with the statement "$\varepsilon = 0$".

At the conclusion of a successful, non-randomized communication protocol the question "What is the value of ε?" must be answerable by Bob but not by Eve, hence the bigraph must now be disconnected—and moreover (although we shall not need this fact) the vertices from $A(H)$ in each component must either all correspond to $\varepsilon = 0$ or all to $\varepsilon = 1$.

Theorem 4.7. *Alice can communicate a bit to Bob in secret, via a randomized and/or non-deterministic communication protocol, iff their bigraph is separable.*

Proof. Sufficiency has already in effect been demonstrated; Bob and Alice can co-operatively disconnect the bigraph, ignoring the bit value, then a message of the form "My bit is 0 if the true edge lies in component C, 1 otherwise" does the trick.

For the converse, we double the bigraph as above so that the protocol may be regarded as a special communication protocol, which we denote by P. If P is randomized, then we may replace the random inputs by non-determinism as in Theorem 4.1; thus we may assume P is merely non-deterministic.

Now we construct from P a randomized (!) communication protocol P' which operates on the original, undoubled bigraph plus a single random bit α for Alice. P' operates by the rule $f'_i(x; \alpha; m_1, \ldots, m_{i-1}) = f_i(x(\alpha); m_1, \ldots, m_{i-1})$ for i odd, and $f'_i = f_i$ for i even.

At each stage, the bigraph associated with P' will be precisely the image of the bigraph associated with P under the collapsing map Φ which sends $x(0)$ and $x(1)$ to x. But the image of a disconnected bigraph under this mapping is again disconnected, so P' is a separation protocol for the original bigraph, completing the proof of the theorem. □

Theorem 4.7 says, in effect, that if Alice and Bob know that they will need to communicate a bit in secret, then they can disconnect their bigraph *in advance*; when the bit comes in, it can then be communicated (in either direction) by a single message.

5. The 'Two Sheriffs' Problem

Let us now look now at the two sheriffs problem, but generalized as follows: one sheriff (whom we shall call "Lew") has narrowed his list of suspects to p, the other

("Ralph") to q, and the total number of suspects is n. The edges of the bigraph H here represent all possible pairs (L, R) of subsets of the set $N = \{1, 2, \ldots, n\}$ of suspects, with $|L| = p$, $|R| = q$ and $|L \cap R| \neq \varnothing$. Lew's side $A(H)$ of the bigraph will thus contain $\binom{n}{p}$ vertices and Ralph's side $\binom{n}{q}$ vertices, adjacency arising when the corresponding subsets intersect.

If Lew and Ralph succeed in determining the identity of the killer without tipping off the mob, they will share a secret and thus must have disconnected H. Conversely, suppose they manage to disconnect the bigraph; then Lew and Ralph can reduce further to two non-adjacent edges, one of which is the true edge. If that edge represents an overlap of one, the sheriffs will have found the killer.

Theorem 5.1. *If $n = 2pq$ then there is a deterministic separation protocol for solving the two sheriffs problem.*

Proof. We make use of Baranyai's Theorem [1], which says the following:

If k divides n then there is an array $\{K_{i,j}\}$, $1 \leq i \leq rn/k$, $1 \leq j \leq \binom{n}{k}/(n/k)$ of subsets of $N = \{1, 2, \ldots, n\}$ such that each $|K_{i,j}| = k$, each column $K_{1,j}, \ldots, K_{n/k,j}$ is a partition of N, and each subset of N of size k appears exactly once in the array.

Such an array (known as a 1-factorization of the complete k-uniform hypergraph on n vertices) is fixed by Lew and Ralph (publicly) for $k = p$, and Lew proceeds to tell Ralph on which column his end L of the true edge can be found.

Ralph is thus presented with a partition $L_1, \ldots, L_{n/p}$ of N, one of whose parts is Lew's narrowed-down suspect set. His job will be to split the index set $I = \{1, 2, \ldots, n/p = 2q\}$ into two parts, say I_1 and I_2, so that his suspect set R is contained in $\bigcup_{i \in I_j} L_i$. This will disconnect the bigraph.

To do this the sheriffs employ a fixed (but arbitrary) map Φ from the set of subsets of I of size at most q to the set of subsets of I of size *exactly* q, such that $\Phi(S) \cap S = \varnothing$ for every set S in the domain of Φ. Ralph forms the set $F := \{i : R \cap L_i \neq \varnothing\}$, then puts $F' := \Phi(F)$ and $F'' := \Phi(F')$. F' and F'' are thus complementary subsets of I of cardinality q; let I_1 be the one containing the element "1" of I, and let I_2 be the other. Ralph now identifies I_1 and I_2 and announces that his set F, defined as above, is contained either in I_1 or I_2.

The resulting bigraph will contain all vertices of $B(H)$ for which the resulting F would have been contained in I_1 or I_2 *and* would have had cardinality q, since in those cases F' and F'' are not dependent on the choice of Φ. Those vertices for which F is contained in I_e will form a connected component, for $e = 1, 2$.

Let k be the index of Lew's end of the true edge, that is, let L_k be Lew's suspect set; suppose $k \in I_e$. Let i be such that k is the ith smallest member of I_e, and let j be the ith smallest element of I_{3-e}. Lew now announces that his set of suspects is in fact either L_j or L_k.

Ralph (but not the mob) will know which of the two is Lew's suspect set; say it is L_j. If $|R \cap L_j| > 1$ then Ralph announces that the killer cannot be identified; otherwise, however, he now knows the killer (say, x). Choosing (again by order of numbers) the corresponding element y of L_k, he announces that the killer is one of x and y. This completes the protocol. □

Let us see how this works in the original case $p = q = 2$, $n = 8$. The following Baranyai array can be used:

$$\begin{array}{ccccccc} \{1,2\} & \{1,3\} & \{1,4\} & \{1,5\} & \{1,6\} & \{1,7\} & \{1,8\} \\ \{3,4\} & \{2,4\} & \{2,3\} & \{2,6\} & \{2,5\} & \{2,8\} & \{2,7\} \\ \{5,6\} & \{5,7\} & \{5,8\} & \{3,7\} & \{3,8\} & \{3,5\} & \{3,6\} \\ \{7,8\} & \{6,8\} & \{6,7\} & \{4,8\} & \{4,7\} & \{4,6\} & \{4,5\} \end{array}$$

Suppose that the true edge is either ({1,2},{1,3}) or ({5,6},{5,7}). Then Lew will announce that his suspect set belongs to the first column, that is, is one of {1,2}, {3,4}, {5,6} or {7,8}. If Ralph himself had one of these sets he would simply announce at this point that the killer cannot be identified; as it is, he splits the index set, telling Lew that his suspect set is either contained in $\{1,2\} \cup \{3,4\}$ or in $\{5,6\} \cup \{7,8\}$. Lew now says "My set is either {1,2} or {5,6}" and Ralph comes back with "The killer is either 1 or 5".

Non-deterministic versions of the above protocol are more easily described; Lew merely picks some partition of which his suspect set is a part, and Ralph can reduce to two possible suspect sets whose intersections with the partition indices are complementary. Here just one more message, from Lew to Ralph, completes the protocol. Moreover, this can be made to work for any $n > 2pq$ as well.

However, we can do even better when non-determinism is permitted; for example, here is a non-deterministic separation protocol for solving the case where $n = k^2$, $p = (k-1)^2 + 1$ and $q = 1$, for any $k \geq 2$.

Lew begins by choosing a $k \times k$ array $\{s_{i,j}\}$ of all the suspects, such that for some j', Lew's suspect set consists precisely of $s_{i',1}$ and all $s_{i,j}$ such that $i \neq i'$ and $j \neq 1$. Ralph, who began knowing the identity of the killer, replies as follows: if the killer is $s_{i,j}$ for $j \neq 1$ and any i, he says "The killer is either $s_{1,j}$ or $s_{i,j}$." If the killer is some $s_{i,1}$ then he picks any $j \neq 1$ and makes the same statement.

By first partitioning the suspects into possible vertices (as in the p, q, $2pq$ case) and then making a $k \times k$ array of the sort described above, but where the array elements are members of a partition instead of single suspects, we may combine the techniques for the following result:

Theorem 5.2. *The two sheriffs problem is solvable non-deterministically whenever* $n \geq q(1 + \sqrt{p-1})^2$.

It is perhaps interesting to note that we have separated a very dense bigraph here, regular on each side. In fact, related to these are the following dense bigraphs, which are *deterministically* separable: fix a large k and set H equal to

$$\{((a,b),(c,d)) : 1 \leq a,b,c,d \leq k \text{ and either } a = c \text{ and } b = d, \text{ or } a \neq c \text{ and } b \neq d\}\,.$$

To separate H, Alice's announces the first coordinate of her pair and Bob the second coordinate of his. Then each will know whether their edge is based on the equalities or the inequalities in the definition above.

6. Multi-Party Generalization

It is evident that many of our definitions and results can be extended to the case where there are more than two conversants. In this case conversation protocols, in order to remain general, allow the identity of the next speaker to depend, at each turn, on the previous conversation.

In [3] Fischer and Wright suggest using random deals to facilitate secret key exchange within a group of persons wishing to communicate privately in yet-to-be-specified subgroups. Among the negative results in [3] is a theorem (Theorem 9,

p. 11) which states that no communication protocol for 3 players, each dealt one card of a 3-card deck, can enable them to isolate a secret bit. Fischer and Wright indicate that our methods can be used to generalize the result to $k > 3$ players; we show here how that can be done.

The definition of "bigraph"' extends easily as follows: a *k-graph* is a finite collection H of k-tuples $x = (x_1, \ldots, x_k)$ (which we call "blocks" to avoid confusion) such that $x, y \in H$ implies that x_i and y_j are distinct for $i \neq j$. A k-graph is thus a particular special case of k-uniform hypergraph in which the sets $\{x_i : x \in H\}$ partition the vertex set of H.

The proof of Theorem 2.1 goes through, as does an appropriate version of Theorem 4.2; thus we are once more reduced to showing that the players, say X_1 through X_k, cannot cooperatively disconnect their k-graph H_k which in this case consists of a block for each permutation of the cards.

Theorem 6.1. *The "permutation k-graph" H_k is inseparable for $k \geq 3$.*

Proof. H_2 is of course separable, indeed disconnected to begin with. Let us assume that P is a (non-deterministic) separation protocol for H_k, for some $k > 2$, and let $H^0, H^1, \ldots, H^t$ be the state of the k-graph for $X_1, \ldots, X_k$ at each stage of some (successful) conversation using P. Then $H^0 = H_k$ and $H^0, \ldots, H^{t-1}$ are connected k-graphs. We claim first that H^t must consist only of two blocks, which up to permutation of the players and cards, may as well be $1, 2, 3, \ldots, k$ and $2, 3, \ldots, k, 1$. To see this let x and y be two blocks of H^t which lie in different components; then in particular x and y are not adjacent so $x_i \neq y_i$ for $i = 1, 2, \ldots, k$. Let G be the graph on vertices 1,2 , ..., k with j adjacent to j' when $\{j, j'\} = \{x_i, y_i\}$ for some i; then since x and y are each permutations, G is regular of degree 2. If ϕ is an automorphism of G such that $\phi(j)$ is adjacent to j for all j then $(\phi(x_1), \ldots, \phi(x_k))$ is a block of H^k which is adjacent to x if ϕ has any fixed points and to y if ϕ is not the identity. Since no block of H^t can be adjacent to both x and y, every such ϕ must fix all vertices or none; hence G consists of a single cycle. By relabelling we may assume $x = (1, 2, 3, \ldots, k)$ and $y = (2, 3, \ldots, k, 1)$.

Now if H^t contains any other block some player, say X_k, must have another vertex, say $j \neq k, 1$. But then the block $(1, 2, \ldots, j-2, j-1, j+1, j+2, \ldots, k-1, k, j)$ lies in H^t; and it is adjacent to both x (at player X_1) and y (at player X_k), a contradiction. This proves the claim.

We may now assume that H^t consists exactly of the above blocks x and y, and that the last player to speak was X_1; then the vertices of H^{t-1} are exactly those appearing in x and y, plus some additional vertices held by X_1 which she eliminated in her last message. The fact that H^{t-1} contains those additional vertices means that if one of them had been X_1's "true" vertex, the conversation might have gone exactly as it did until the last message. But H^{t-1} contains no pair of non-adjacent blocks other than x and y, since in every block not equal to x or y, player X_2 holds a 2 and player X_k holds a 1. Hence, the protocol P has failed in this case and this contradiction proves the theorem. □

7. Final Comments

In this work we have only begun to study the combinatorial cryptology of isolating a common secret; unsettled questions abound. For example: Are there separable bigraphs requiring arbitrarily many rounds in order to separate them? Is there a good way to characterize those bigraphs that are separable in n rounds? What about the bigraphs that are deterministically separable in n rounds? Is it possible to prove that the two sheriffs problem cannot be solved deterministically when $n < 2pq$?

We hope that our "bigraphs" may prove to be a useful way of representing common knowledge, even for applications unrelated to the problem of isolating a common secret. Although they carry the same information as do other representations, they may help attract graph-theorists to knowledge problems and thus bring some powerful theorems and sharp combinatorial minds to bear.

References

1. Zs. Baranyai, On the factorization of the complete uniform hypergraph, *Infinite and finite sets (Keszthely, 1973; dedicated to P. Erdős on his 60th birthday)*, Vol. I (1975) 91–108.
2. M. Fischer, M. Paterson, and C. Rackoff, Secret bit transmission using a random deal of cards, *Distributed Computing and Cryptography*, American Mathematical Society (1991) 173–181.
3. M. Fischer and R. Wright, Multiparty secret key exchange using a random deal of cards, *Proceedings of CRYPTO '91*, Springer-Verlag Lecture Notes in Computer Science, Vol 576 (1992) 141–155.
4. P. Winkler, Cryptologic techniques in bidding and defense (Parts I, II, III, IV), *Bridge Magazine* (April-July 1981) 148–149, 186–187, 226–227, 12–13.
5. P. Winkler, The advent of cryptology in the game of bridge, *Cryptologia*, vol. 7 (1983) 327–332.

Some Remarks on the Cycle Plus Triangles Problem

H. Fleischner[1] and M. Stiebitz[2]

[1] Institut für Infomationsverarbeitung, Österr. Akad. d. Wiss., Domus Universitatis, Sonnenfelsgasse 19, A-1010 Wien, Austria
[2] Institut für Mathematik, Technische Universität Ilmenau, PSF 327, D-98684 Ilmenau, Germany

1. Introduction and Main Results

All (undirected) graphs and digraphs considered are assumed to be finite (if not otherwise stated) and loopless. Multiple edges (arcs) are permitted. For a graph G, let $V(G)$, $E(G)$, and $\chi(G)$ denote the vertex set, the edge set, and the chromatic number of G, respectively. If $X \subseteq V(G)$ and $F \subseteq E(G)$, then $G - X - F$ denotes the subgraph H of G satisfying $V(H) = V(G) - X$ and $E(H) = \{xy \mid xy \in E(G) - F \text{ and } x, y \notin X\}$.

A system of non-empty subgraphs $\{G_1, \ldots, G_m\}$, $m \geq 1$, in a graph G is called a *decomposition* of G iff $E(G) = \bigcup_{i=1}^{m} E(G_i)$ and $E(G_i) \cap E(G_j) = \emptyset$, $1 \leq i < j \leq m$. For an integer $k \geq 2$, let $\mathcal{G}_k$ denote the family of all graphs G which have a decomposition into a Hamiltonian cycle and $m \geq 0$ pairwise vertex disjoint complete subgraphs each on k vertices. Note that every vertex of a graph $G \in \mathcal{G}_k$, $k \geq 2$, has degree either $k+1$ or 2. In particular, $\mathcal{G}_k$ contains every cycle.

The authors proved the following result providing an affirmative solution to a colouring problem of P. Erdős, which became known as the "cycle–plus–triangles problem" (see e.g. [5] or [7]).

Theorem 1.1 ([9]). *If $G \in \mathcal{G}_3$ then $\chi(G) \leq 3$.* □

Actually, Theorem 1.1 was proved only for 4–regular graphs. But every non-regular graph $G' \in \mathcal{G}_3$ can be obtained from a 4–regular graph $G \in \mathcal{G}_3$ by subdividing some edges of the Hamiltonian cycle of G, and hence every legal 3–colouring of G can be extended to a legal 3–colouring of G'. Recently, H. Sachs [10] found a purely elementary proof of Theorem 1.1.

Theorem 1.2. *Let $k \geq 4$ be an integer and let G be a graph. Assume that G has a decomposition $\mathcal{D}$ into a Hamiltonian cycle and $m \geq 0$ pairwise vertex disjoint complete subgraphs each on at most k vertices. Then $\chi(G) \leq k$.*

Proof (by induction on k). Let $X = \bigcup V(K)$ where the union is taken over all complete subgraphs $K \in \mathcal{D}$ of size $\leq k-2$. Then every vertex of X has degree $\leq k-1$ in G implying that every legal k–colouring of $G - X$ can be extended to a legal k–colouring of G. Therefore, we need only to show that

$\chi(G - X) \leq k$. Clearly, $G - X$ is a subgraph of some graph G' which has a decomposition $\mathcal{D}'$ into a Hamiltonian cycle and a (possibly empty) set of pairwise vertex disjoint complete subgraphs on $k-1$ or k vertices each. We claim that $\chi(G') \leq k$. Let $\mathcal{D}_1$ be the set of all complete subgraphs from $\mathcal{D}'$ with exactly k vertices. If $\mathcal{D}_1 = \emptyset$ then from the induction hypothesis or, in case $k = 4$, from Theorem 1.1 we conclude that $\chi(G') \leq k - 1 < k$. If $\mathcal{D}_1 \neq \emptyset$ then we argue as follows. First, choose from every complete subgraph $K \in \mathcal{D}_1$ an arbitrary vertex $y = y(K)$ and let Y be the set of all these vertices. Then $G' - Y$ is a subgraph of some graph in $\mathcal{G}_{k-1}$ and therefore, the induction hypothesis or, in case $k = 4$, Theorem 1.1 implies that $G' - Y$ has a legal colouring with $k - 1$ colours, say $1, 2, \ldots, k - 1$. Obviously, each complete subgraph $K - y(K)$ of $G' - Y$, $K \in \mathcal{D}_1$, contains precisely one vertex coloured 1 (note that $K - y(K)$ has $k - 1$ vertices). Now, let Z be the set of all such vertices. Then Z is an independent set in $G' - Y$, and hence also in G'. By the same argument as before we conclude that $G' - Z$ has a legal $(k-1)$–colouring implying that G' has a legal k–colouring, i.e. $\chi(G') \leq k$. This proves our claim, and hence the desired result. □

Theorem 1.2 is not true for $k = 3$. A counterexample is the complete graph on 4 vertices, a further one due to H. Sachs is shown in Fig. 1.1.

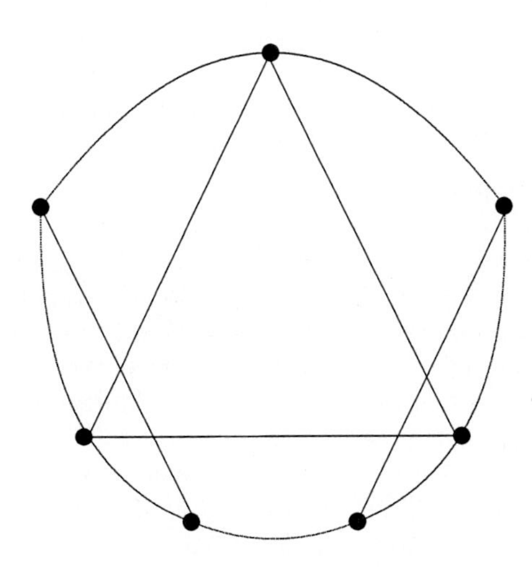

Fig. 1.1.

As a consequence of Theorems 1.1 and 1.2 we obtain the following proposition.

Corollary 1.1. *If $G \in \mathcal{G}_k$ for some $k \geq 3$, then $\chi(G) \leq k$.* □

The proof of Theorem 1.1 given in [9] uses a recent result of N. Alon and M. Tarsi [2] which provides a colouring criterion for a graph G based on orientations of G. On the one hand, this yields a somewhat stronger result than Theorem 1.1, namely that every graph $G \in \mathcal{G}_3$ has list chromatic number at most 3. On the other hand, this implies that the proof of Theorem 1.1 is not constructive in the sense that it does not yield any algorithm which, given $G \in \mathcal{G}_3$, allows a legal 3–colouring of G to be constructed. For a graph $G \in \mathcal{G}_k$, $k \geq 4$, a legal k–colouring of G can be easily obtained using the induction step of Theorem 1.2 and the following colouring procedure for regular graphs from $\mathcal{G}_4$.

Let $G \in \mathcal{G}_4$ be any 5–regular graph, and let $\mathcal{D}$ be a decomposition of G into a Hamiltonian cycle C and $m \geq 1$ pairwise vertex disjoint complete subgraphs $K^1, \ldots, K^m$ each on 4 vertices. For $E' \subseteq E(G)$, let $G(E')$ denote the spanning subgraph of G with vertex set $V(G)$ and edge set E'. By a linear factor of G we mean a set $L \subseteq E(G)$ such that $G(L)$ is 1–regular.

To obtain a legal 4–colouring of G, we first choose an arbitrary linear factor L of C (note that the cycle C has length 0 mod 4) and, for each $i \in \{1, 2, \ldots, m\}$, an arbitrary linear factor L_1^i of K^i. Obviously, the subgraph $G(F)$, where $F = L \cup L_1^1 \cup \ldots \cup L_1^m$, is the disjoint union of even cycles, and hence we easily find a legal 2–colouring c_1 of $G(F)$. Now it is easy to check that we can choose a linear factor L_2^i of $K^i - L_1^i$ $(i = 1, \ldots, m)$ such that c_1 remains a legal 2–colouring of $G(F_1)$, where $F_1 = F \cup L_2^1 \cup \ldots \cup L_2^m$. Then $F_2 = E(G) - F_1$ consists of the linear factor $L' = E(C) - L$ and the linear factors $L_3^i = E(K^i) - L_1^i - L_2^i$ $(i = 1, \ldots, m)$. Therefore, $G(F_2)$ is the disjoint union of even cycles and has a legal 2–colouring c_2. Eventually, because of $G = G(F_1 \cup F_2)$, the mapping c defined by $c(x) = (c_1(x), c_2(x))$, $x \in V(G)$, is a legal 4–colouring of G.

Noga Alon [1] proposed a more general situation. Let H be a graph on n vertices. If k divides n, then H is said to be *strongly k–colourable* if, for any partition of $V(H)$ into pairwise disjoint sets V_i each having cardinality k, there is a legal k–colouring of H in which each colour class intersects each V_i by exactly one vertex. Notice that H is strongly k–colourable if and only if the chromatic number of any graph G obtained from H by adding to it a union of vertex disjoint complete subgraphs (on the set V(H)) each having k vertices is k. If k does not divide n, then H is said to be strongly k–colourable if the graph obtained from H by adding to it $k\lceil n/k \rceil - n$ isolated vertices is strongly k–colourable. Of course, the *strong chromatic number* of a graph H, denoted by $s\chi(H)$, is the minimum k such that H is strongly k–colourable. In this terminology, Theorem 1.1 says that, for every cycle C_{3m} on $3m$ vertices, $s\chi(C_{3m}) \leq 3$. Moreover, Theorem 1.2 implies that, for every cycle C, $s\chi(C) \leq 4$. For cycles on $4m$ vertices this statement was established by several researchers including F. de la Vega, M. Fellows and N. Alon (see [7]). The 4–chromatic graph depicted in Fig. 1.1 shows that $s\chi(C_7) = 4$.

Noga Alon [1] investigated the function $s\chi(d) = \max(s\chi(H))$, where H ranges over all graphs with maximum degree at most d. On the one hand, using similar arguments as above he gives a short proof that $s\chi(H) \leq 2^{\chi'(H)}$, where $\chi'(H)$ denotes the chromatic index of H. Because of Vizing's Theorem, this yields $s\chi(d) \leq 2^{d+1}$. On the other hand, using probabilistic arguments, he proved that there is a (very large) constant c such that $s\chi(d) \leq cd$ for every d. This leaves the following problem. What is the minimum value k such that every graph with maximum degree d is strongly k–colourable? Even for $d = 2$, the answer is not known. Clearly, for $d = 1$ we have $s\chi(1) = 2$. To obtain a lower bound for $s\chi(d)$, we construct a graph H_d for every $d \geq 1$ in the following way. Let X_1, X_2, X_3, X_4 be disjoint sets on r vertices each, $r \geq 1$. Join X_i to X_{i+1} $(i = 1, 2, 3)$ and X_4 to X_1 by all possible edges. The resulting graph is H_{2r}. The graph H_{2r-1} is obtained from H_{2r} by removing two adjacent vertices. Then the graph H_d, $d \geq 1$, is regular of degree d and it is easy to check that H_d is not strongly k–colourable for $k \leq 2d - 1$, i.e. $s\chi(H_d) \geq 2d$. This yields $s\chi(d) \geq 2d$ for every $d \geq 1$.

In [5] P. Erdős asked whether the statement of Theorem 1.1 remains true if the Hamiltonian cycle of the decomposition is replaced by a 2–regular graph not containing a cycle of length 4. The graph G depicted in Fig. 1.2 provides a negative answer. This graph belongs to an infinite family of 4-regular 4-chromatic (in fact, 4-critical) graphs due to Gallai [6] all of which (except one) are counterexamples with respect to Erdős's question.

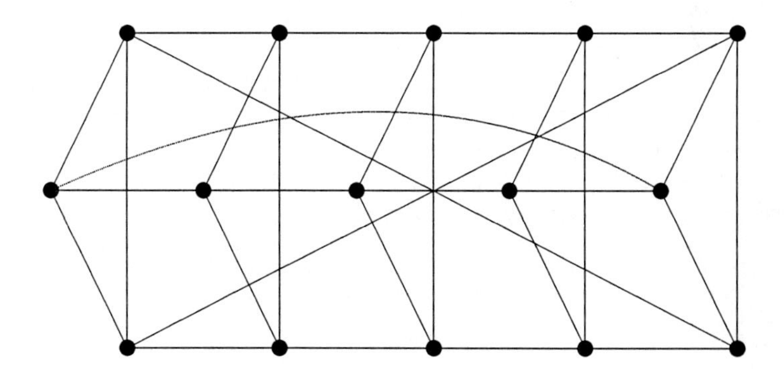

Fig. 1.2.

We define a (not necessarily connected) graph to be *Eulerian* iff each vertex has even degree. Analogously, a digraph is said to be *Eulerian* iff each vertex has equal out-degree and in-degree. Let G be an Eulerian graph. Clearly, G has an orientation D (this means, that D is a digraph whose underlying graph is G) which is Eulerian. Such an orientation is briefly called an *Eulerian orientation* of G. We denote the number of all Eulerian orientations of G by $e(G)$. If D is an arbitrary Eulerian orientation of the Eulerian graph G, then (see [8]) $e(G)$ is equal to the number of all spanning Eulerian subdigraphs of D. Therefore, Theorem 2.1 in [9] implies the following.

Theorem 1.3 **([9]).** *If $G \in \mathcal{G}_3$, then $e(G) \equiv 2 \bmod 4$.* □

Theorem 1.3 may be considered as the main result of [9]. To deduce Theorem 1.1 from Theorem 1.3, the colouring criterion of Alon & Tarsi (see [2] or [9]) is needed. As an immediate consequence of Theorem 1.3 we obtain the following result. This was first noted by Alon, Gutner and Tarsi.

Corollary 1.2. *Let G be a graph, which has a decomposition into a Hamiltonian cycle C and $m \geq 1$ pairwise vertex disjoint triangles $T_1, \ldots, T_m$, i.e. $G \in \mathcal{G}_3$. Then G has an Eulerian orientation D such that no triangle T_i ($i = 1, \ldots, m$) is cyclically oriented in D. Moreover, if V_1 (V_2) is the set of all those vertices from G which are sinks (sources) of some triangle T_i with respect to D, then both V_1 and V_2 are independent sets in G each of size m.*

Proof. Let $\mathcal{O}$ denote the set of all Eulerian orientations of G, and let $\mathcal{O}_i$ ($i = 1, \ldots, m$) denote the set of all Eulerian orientations of G in which the triangle T_i is cyclically oriented. To prove that a desired orientation D of G exists, suppose the contrary. Then $\mathcal{O} = \bigcup_{i=1}^{m} \mathcal{O}_i$, and hence

$$e(G) = |\mathcal{O}| = \sum_{I \subseteq \{1,\ldots,m\}} (-1)^{|I|+1} o(I),$$

where $o(I) = |\bigcap_{i \in I} \mathcal{O}_i|$. It is easy to check that, for $I \neq \emptyset$, we have $o(I) = 2^{|I|} e(G')$, where $G' = G - \bigcup_{i \in I} E(T_i)$. Since $G' \in \mathcal{G}_3$, from Theorem 1.3 it then follows that $o(I) \equiv 0 \bmod 4$, and hence $e(G) \equiv 0 \bmod 4$, a contradiction. Therefore, the desired orientation D of G exists. Obviously, the sets V_1 and V_2 satisfying the hypothesis of Corollary 1.2 are both independent sets of size m in G. This proves Corollary 1.2. □

In particular, from Theorem 1.3 we obtain immediately that every 4-regular graph $G \in \mathcal{G}_3$ with $3m$ vertices has independence number m, as conjectured by D. Z. Du and D. F. Hsu in 1986 (see [7]).

In connection with the Cycle Double Cover Conjecture R. Goddyn (private communication) suspects that there is no snark H with a chordless dominating cycle C (this means, that $V(H) - V(C)$ is an independent set in H). The first author and M. Tarsi observed that the following result can be deduced from Corollary 1.2.

Corollary 1.3. *Assume that H is a 3-regular graph which has a dominating cycle C. Then there is a matching M of $H - E(C)$ such that $H - M$ is 2-regular or isomorphic to a subdivision of a 3-regular bipartite graph.*

Proof. Put $X = V(H) - V(C)$, and let M_1 be the set of all chords of C in H. From the cycle C we construct a new graph in the following way. For every vertex $x \in X$, join the three neighbours of x on C by three new edges forming a triangle T_x. Clearly, the resulting graph G belongs to $\mathcal{G}_3$. From Corollary 1.2 it then follows that there is an Eulerian orientation D of G such that no triangle T_x is cyclically oriented. Now, let V_1 (V_2) be the set of all those vertices from G which are sinks (sources) of some triangle T_x with respect to D. Then, for every vertex $x \in X$, there is exactly one edge in H joining x to some vertex of $V(C) - V_1 - V_2$. Denote the set of all such edges by M_2, and put $M = M_1 \cup M_2$. Clearly, M is a matching of $H - E(C)$. Since both V_1 and V_2 are independent sets in G (see Corollary 1.2), it is easy to check that $H - M$ is 2-regular (in case of $X = \emptyset$) or isomorphic to a subdivision of a 3-regular bipartite graph (in case of $X \neq \emptyset$). This proves Corollary 1.3. □

One can easily show that if a 3-regular graph H with a dominating cycle contains two disjoint matchings as described in Corollary 1.3, then H has a nowhere-zero 5-flow.

Finally, let us mention the following infinite version of Corollary 1.1. A set of cardinality k is briefly called a k–set.

Theorem 1.4. *Let $k \geq 3$ be an integer. For any partition $\mathcal{D}$ of the integers Z into k–sets, there is another partition $\{X_1, \ldots, X_k\}$ of Z such that X_i ($i = 1, \ldots, k$) contains a member of each k–set from $\mathcal{D}$ but no consecutive pair of integers.*

Proof. Define G to be the infinite graph with vertex set Z, where $xy \in E(G)$ iff $\mid x - y \mid = 1$ or $x, y \in K \in \mathcal{D}$. Consider an arbitrary finite subgraph H of G. Then it is easy to check that H is a subgraph of some graph from $\mathcal{G}_k$ and hence, using Corollary 1.1, we conclude that H is k–colourable. From a well–known result of Erdős & de Bruijn [4] we then obtain that G is itself k–colourable. This immediately implies that $\chi(G) = k$. Obviously, any legal k–colouring of G provides a partition $\{X_1, X_2, \ldots, X_k\}$ of Z having the desired properties. □

References

1. N. Alon, The strong chromatic number of a graph, Random Structures and Algorithms 3(1) (1992), 1-7.
2. N. Alon and M. Tarsi, Colourings and orientations of graphs, Combinatorica 12(2) (1992), 125-134.
3. D. Z. Du, D. F. Hsu and F. K. Hwang, The Hamiltonian property of consecutive-d digraphs, Math. Comput. Modelling 17 (1993), no. 11, 61-63.
4. P. Erdős and N. G. de Bruijn, A colour problem for infinite graphs and a problem in the theory of relations, Nederl. Akad. Wetensch. Proc. Ser. A 54 (=Indag. Math. 13) (1951), 371-373.

5. P. Erdős, On some of my favourite problems in graph theory and block designs, Le Matematiche, Vol. XLV (1990) Fasc.I, 61-74.
6. T. Gallai, Kritische Graphen I, Publ. Math. Inst. Hung. Acad. Sci. 8 (1963), 373-395.
7. M. R. Fellows, Transversals of vertex partitions of graphs, SIAM J. Discrete Math 3 (1990), 206-215.
8. H. Fleischner, Eulerian Graphs and Related Topics, Part 1, Vol.1, Ann. Discrete Math. 45; Vol. 2, Ann. Discrete Math. 50 (North-Holland, Amsterdam 1990/91).
9. H. Fleischner and M. Stiebitz, A solution to a colouring problem of P. Erdős, Discrete Math. 101 (1992), 39-48.
10. H. Sachs, An elementary proof of the cycle-plus-triangles theorem, manuscript.

V. Ramsey and Extremal Theory

Introduction

It is perhaps evident from several places of this volume that Ramsey theorem played a decisive role in Erdős' combinatorial activity. And perhaps no other part of combinatorial mathematics is so dear to him as Ramsey theory and extremal problems. He was not creating or even aiming for a theory. However, a complex web of results and conjectures did, in fact, give rise to several theories. They all started with modest short papers by Erdős, Szekeres and Turán in the thirties. How striking it is to compare these initial papers with the richness of the later development, described, e.g., by the survey articles of Miki Simonovits (extremal graph theory) and Jeff Kahn (extremal set theory). In addition, the editors of these volumes tried to cover in greater detail the development of Ramsey theory mirrored and motivated by Erdős' papers. In a way (and this certainly is one of the leitmotivs of Erdős' work), there is little difference between, say, density Ramsey type results and extremal problems.

One can only speculate on the origins of density questions. It is clear that in the late 30's and 40's, the time was ripe in ideas which later developed into extremal and density questions: we have not only the Erdős-Turán 1941 paper but also, Erdős and Tomsk 1938 paper on number theory which anticipated extremal theory by determining $n^{3/2}$ as upper bound for C_4-free graphs, the Sperner paper and also the Erdős-Ko-Rado work (which took several decades to get into print). All these ideas, together with Turán's extremal results provided a fruitful cross-interaction of ideas from various fields which, some 30 years later, developed into density Ramsey theorems and extremal theory. We are happy to include in this chapter papers by Gyula Katona (which gives a rare account of Erdős method of encouraging and educating young talented students), and by Vojtěch Rödl and Robin Thomas related to an aspect of Erdős' work. And we include an important paper by Alexander Kostochka which gives a major breakthrough to the Erdős-Rado Δ-system problem (for triples; meanwhile Kostochka succeeded in generalizing the result to k-tuples). Finally, we have the paper of Saharon Shelah, solving a model theoretic Ramsey question of Väänänen, illustrating (once again) that Ramsey theory is alive and well. Let us complement these lines by Erdős' own words:

Hajnal, Rado and I proved

$$2^{c\,k^2} < r_3(k,k) < 2^{2^k} ,$$

we believe that the upper bound is correct or at least is closer to the truth, but Hajnal and I have a curious result: If one colors the triples of a set of n elements by two colors, there always is a set of size $(\log n)^{1/2}$ where the distribution is not just, i.e., one of the colors has more than $0.6\binom{t}{3}$ of the triples for $t = (\log n)^{1/2}$. Nevertheless we believe that the upper bound is correct. It would perhaps change our minds if we could replace 0.6 by $1-\epsilon$ for some $t, t > (\log n)^\epsilon$. We never had any method of doing this.

Let H be a fixed graph and let n be large. Hajnal and I conjectured that if $G(n)$ does not contain an induced copy of H then $G(n)$ must contain a trivial subgraph of size $> n^\epsilon$, $\epsilon = \epsilon(H)$. We proved this for many special cases but many problems remain. We could only prove $\exp(c\sqrt{\log n})$.

Extremal graph theory

Denote by $T(n;G)$ the smallest integer for which every graph of n vertices and $T(n;G)$ edges contains G as a subgraph. Turán determined $T(n;G)$ if G is $K(r)$ for all r, $T(n,C_4) = (\frac{1}{2}+o(1))n^{3/2}$ was proved by V. T. Sós, Rényi and myself. The exact formula for $T(n,c_4)$ is known if $n = p^2+p+1$ (Füredi).

I have many asymptotic and exact results for $T(n,C_4)$ and many results and conjectures with Simonovits. I have to refer to the excellent book of Bollobás and the excellent survey of Simonovits and a very good recent survey article of Füredi. Here I only state two results of Simonovits and myself: $T(n,G) < cn^{8/5}$ where G is the edge graph of the three-dimensional cube. Also we have fairly exact results for $T(n,K(2,2,2))$.

Ramsey-Turán theorems

The first papers are joint with V. T. Sós and then there are comprehensive papers with Hajnal, Simonovits, V. T. Sós and Szemerédi. Here I only state a result with Bollobás, Szemerédi and myself: For every $\epsilon > 0$ and $n > n_o(\epsilon)$ there is a graph of n vertices $\frac{n^2}{8}(1-\epsilon)$ edges with no $K(4)$ and having largest independent set is $o(n)$, but such a graph does not exist if the number of edges is $\frac{n^2}{8}(1+\epsilon)$. We have no idea what happens if the number of edges is $\frac{n^2}{8}$.

One last Ramsey type problem: Let n_k be the smallest integer (if it exists) for which if we color the proper divisors of n_k by k colors then n_k will be a monochromatic sum of distinct divisors, namely a sum of distinct divisors in a color class. I am sure that n_k exists for every k but I think it is not even known if n_2 exists. It would be of some interest to determine at least n_2. An old problem of R. L. Graham and myself states: Is it true that if m_k is sufficiently large and we color the integers $2 \le t \le m_k$ by k colors then

$$1 = \sum \frac{1}{t_i}$$

is always solvable monochromatically? I would like to see a proof that m_2 exists. (Clearly $m_k \geq n_k$.) Perhaps this is really a Turán type problem and not a Ramsey problem. In other words, if m is sufficiently large and $1 < a_1 < a_2 < \cdots < a_\ell \leq m$ is a sequence of integers for which $\sum_\ell 1/a_\ell > \delta \log m$ then

$$1 = \sum \frac{\epsilon_i}{a_i} \quad (\epsilon_i = 0 \text{ or } 1)$$

is always solvable. I offer 100 dollars for a proof or disproof. Perhaps it suffices to assume that

$$\sum_{a_i < m} \frac{1}{a_i} > C(\log \log m)^2$$

for some large enough C. For further problems of this kind as well as for related results see my book with R. L. Graham. I hope before the year 2000 a second edition will appear.

Paul Erdős' Influence on Extremal Graph Theory

Miklós Simonovits*

Mathematical Institute of the Hungarian Academy of Sciences, Budapest, Hungary

Dedicated to Paul Erdős on the occasion of his 80th birthday

Summary. Paul Erdős is 80[1] and the mathematical community is celebrating him in various ways. Jarik Nešetřil also organized a small conference in Prague in his honour, where we, combinatorists and number theorists attempted to describe in a limited time the enourmous influence Paul Erdős made on the mathematics of our surrounding (including our mathematics as well). Based on my lecture given there, I shall to survey those parts of Extremal Graph Theory that are connected most directly with Paul Erdős's work.

In *Turán type extremal problems* we usually have some sample graphs $L_1, \ldots, L_r$, and consider a graph G_n on n vertices not containing any L_i. We ask for the maximum number of edges such a G_n can have. We may ask similar questions for hypergraphs, multigraphs and digraphs.

We may also ask, how many copies of forbidden subgraphs L_i must a graph G_n contain with a given number of edges superseding the maximum in the corresponding extremal graph problems. These are the problems on *Supersaturated Graphs.*

We can mix these questions with Ramsey type problems, (Ramsey-Turán Theory). This topic is the subject of a survey by V. T. Sós [162].

These topics are definitely among the favourite areas in Paul Erdős's graph theory.

Keywords: graphs, extremal graphs, graph theory.

1. Introduction

Extremal graph theory is a wide and fast developing area of graph theory. Having many ramifications, this area can be defined in a broader and in a more restricted sense. In this survey we shall restrict our considerations primarily to **"Turán Type Extremal Graph Problems"** and some closely related areas.

Extremal graph theory is one of the wider theories of graph theory and – in some sense – one of those where Paul Erdős's profound influence can really be seen and appreciated.

What is a Turán Type Extremal Problem?

We shall call the **Theory of Turán type extremal problems** the area which – though being much wider – still is originated from problems of the following type:

* Supported by GRANT "OTKA 1909".

[1] This refers to the time of the conference, not to when this volume appears.

Given a family $\mathcal{L}$ of sample graphs, what is the maximum number of edges a graph G_n can have without containing subgraphs from $\mathcal{L}$.

Here "subgraph" means "not necessarily induced". In Section 11 we shall also deal with the case of "excluded induced subgraphs", as described by Prömel and Steger.

Below K_t, C_t and P_t will denote the complete graph, the cycle and the path of t vertices and $e(G)$ will be the number of edges of a graph G. G_n will be a graph of n vertices, $G(X,Y)$ a bipartite graph with colour classes X and Y.

The first result in our field may be that of Mantel [130] back in 1907, asserting that if a graph G_n contains no K_3, then

$$e(G_n) \le \left[\frac{n^2}{4}\right].$$

Mantel's result soon became forgotten. The next extremal problem was the problem of C_4.

The C_4–Theorem and Number Theory

In 1938 Erdős published a paper [43]

> P. Erdős: On sequences of integers no one of which divides the product of two others and related problems, Mitt. Forsch. Institut Mat. und Mech. Tomsk 2 (1938) 74-82.

In this paper Erdős investigated two problems:

(A) Assume that $n_1 < \ldots < n_k$ are positive integers such that n_i does not divide $n_h n_\ell$, except if either $i = h$ or $i = \ell$. What is the maximum number of such integers in $[1,n]$? Denote this maximum by $A(n)$.

Let $\pi(x)$ denote the number of primes in $[2,x]$. Clearly, the primes in $[2,n]$ satisfy our condition, therefore $A(n) \ge \pi(n)$. One could think that one can find much larger sets of numbers satisfying this condition. Surprisingly enough, the contrary is true: Erdős has proved that $A(n) \approx \pi(n)$. More precisely,

$$\pi(n) + \frac{n^{2/3}}{80\log^2 n} \le A(n) \le \pi(n) + O\left(\frac{n^{2/3}}{\log^2 n}\right).$$

For us the other problem of [43] is more important:

(B) Assume that $n_1 < \ldots < n_k$ are positive integers such that $n_i n_j \ne n_h n_\ell$ unless $i = h$ and $j = \ell$ or $i = \ell$ and $j = h$. What is the maximum number of such integers in $[1,n]$? Denote this maximum by $B(n)$.

Here Erdős proved that

$$\pi(n) + \frac{cn^{3/4}}{(\log n)^{3/2}} \le B(n) \le \pi(n) + O\left(n^{3/4}\right).$$

Later Erdős improved the upper bound to

$$B(n) \le \pi(n) + O\left(\frac{n^{3/4}}{\log^{3/2} n}\right),$$

see [55]. It is still open if, for some $c \neq 0$,

$$B(n) = \pi(n) + (1 + o(1)) \frac{cn^{3/4}}{(\log n)^{3/2}}$$

or not.

Solving this unusual type of number theoretical problem, Erdős (probably first) applied Graph Theory to Number Theory. He did the following: Let $\mathcal{D}$ be the set of integers in $[1, n^{2/3}]$, $I\!P$ be the set of primes in $(n^{2/3}, n]$ and $\mathcal{B} = \mathcal{D} \cup I\!P$.

Lemma (Erdős). *Every integer $a \in [1, n]$ can be written as*

$$a = bd \ : \ b \in \mathcal{B}, \ d \in \mathcal{D}.$$

Let $\mathcal{A}$ be a set satisfying the condition in (B). Let us represent each $a \in \mathcal{A}$ as described in the Lemma: $a_i = b_j(i)d_j(i)$. We may assume that $b_i > d_i$. Build a bipartite graph $G(\mathcal{B}, \mathcal{D})$ by joining b_i to d_j if $a = b_i d_j \in \mathcal{A}$. Thus we represent each $a \in \mathcal{A}$ by an edge of a bipartite graph $G(\mathcal{B}, \mathcal{D})$. Erdős observed that

the number theoretic condition in (B) implies that $C_4 \not\subseteq G(\mathcal{B}, \mathcal{D})$.

Indeed, if we had a 4–cycle $(b_1 d_1 b_2 d_2)$ in our graph, then $a_1 = b_1 d_1$, $a_2 = d_1 b_2$, $a_3 = b_2 d_2$ and $a_4 = d_2 b_1$ all would belong to $\mathcal{A}$ and $a_1 a_3 = a_2 a_4$ would hold, contradicting our assumption. So the graph problem Erdős formulated was the following:

> Given a bipartite graph $G(X, Y)$ with m and n vertices in its colour classes. What is the maximum number of edges such a graph can have without containing a C_4?

Erdős proved the following theorem:

Theorem 1.1. *If $C_4 \not\subseteq G(X, Y)$, $|X| = |Y| = k$, then*

$$e(G(X, Y)) \leq 3k^{3/2}.$$

Here the constant 3 is not sharp (see Section 4). Basically this theorem implied the upper bound on $B(n)$. To get the lower bound Erdős used finite geometries. Erdős writes:

> "... Now we prove that the error term cannot be better than $O\left(\frac{cn^{3/4}}{(\log n)^{3/2}}\right)$.
> First I prove the following lemma communicated to me by Miss E. Klein.[2]

Lemma. *Given $p(p+1)+1$ elements, (for some prime p) we can construct $p(p+1)+1$ combinations, taken $(p+1)$ at a time[3] having no two elements in common."*

Clearly, this is a finite geometry, and this seems to be the first application of Finite Geometrical Constructions in proving lower bounds in Extremal Graph Theory. Yet, Erdős does not speak here of finite geometries, neither of lower bounds for the maximum of $e(G(X, Y))$ in Theorem 1.1.

---

[2] Eszter Klein, later Mrs. Szekeres.

[3] Meaning: $p+1$-tuples.... The text here does not tell us if the role of Mrs. Szekeres was important here or not, but somewhere else Erdős writes: "Mrs. Szekeres and I proved ..."

In the last years Erdős, András Sárközy and V. T. Sós started applying similar methods in similar number theoretical problems, which, again, led to new extremal graph problems, [80]. I mention just one of them:

Let $F_k(N)$ be the maximum number of integers $a_1 < a_2 < \ldots < a_t$ in $[1, N]$ with the property that the product of k different ones is never a square.

Theorem 1.2 (Erdős–A. Sárközy–T. Sós, [80]). *There exists a positive absolute constant $c > 0$ and for every $\varepsilon > 0$ an $N_0(\varepsilon)$ such that for $N > N_0(\varepsilon)$ we have*

$$\frac{(\sqrt{2}-\varepsilon)N^{2/3}}{\log^{4/3} N} < F_6(N) - \pi(N) - \pi(N/2) < cN^{7/9}\log N. \tag{1.1}$$

Taking all the primes of $[2, N]$ and all the numbers $2p$ where p is a prime in $[1, N/2]$ we get $\pi(N) + \pi(N/2)$ such numbers and the above theorem suggests that this construction is almost the best.

The solution of this problem depends on extremal graph theorems connected to excluding C_6. Analogous theorems hold for the even values of k, and somewhat different ones for odd values of k. The question which was asked is:

> What is the maximum number of edges a bipartite graph $G(U, V)$ with u RED and v BLUE vertices can have if $G(U, V)$ contains no C_6 and $uv \le N$?

In [80] the following conjecture was formulated:

Conjecture 1.3. *If $G(U, V)$ is a bipartite graph with $u = |U|$ RED vertices and $v = |V|$ BLUE ones, and $G(U, V)$ contains no C_6, and $v \le u \le v^2$, then $e(G(U, V)) \le c(uv)^{2/3}$.*

The above upper bound of [80] has been improved first by Gábor Sárközy [146]. Then E. Győri proved the above conjecture, which in turn brought down the upper bound of (1.1) to the lower bound, apart from some log-powers.

As related references, [80], [32], [146], [122], [123] should be mentioned.

Theorem 1.4 (Győri). *If $C_{2k} \not\subseteq G(m, n)$, then $e(G(m, n)) \le (k-1)n + O(m^2)$.*

Conjecture 1.5 (Győri). *There exists a $c > 0$ for which, if $C_{2k} \not\subseteq G(m, n)$, then $e(G(m, n)) \le (k-1)n + O(m^{2-c})$.*

Perhaps even $e(G(m, n)) \le (k-1)n + O(m^{3/2})$ could be proved for C_6.

"How Did Crookes Miss to Invent the X–Ray?"

Erdős feels that he "should have invented" Extremal Graph Theory, back in 1938. He has failed to notice that his theorem was the root of an important and beautiful theory. 2-3 years later Turán proved his famous theorem and right after that he posed a few relevant questions, thus initiating a whole new branch of graph theory. Erdős often explains his "blunder" by telling the following story.

> Crookes observed that leaving a photosensitive film near the cathod-ray-tube causes damage to the film: it becomes exposed. He concluded that

"Nobody should leave films near the cathod-ray-tube." Röntgen observed the same phenomenon a few years later and concluded that this can be used for filming the inside of various objects. His conclusion changed the whole Physics.[4] "It is **not enough** to be in the right place at the right time. You should also have an open mind at the right time," Paul concludes his story.

Erdős's influence on the field is so thorough that we do not even attempt to describe it in its full depth and width. We shall neither try to give a very balanced description of the whole, extremely wide area. Instead, we pick a few topics to illustrate Erdős's role in developing this subject, and his vast influence on others.

Also, I shall concentrate more on the new results, since the book of Bollobás [10], or the surveys of myself, [154], [156], [157], or the surveys of Füredi [99] and Sidorenko [149] provide a lot of information on the topic and some problem-papers of Erdős, e.g., [51] [59], [61], [62] are also highly recommended for the reader wishing to learn about the topic. Also, wherever it was possible, I selected newer results, or older but less known theorems (partly to avoid *unnecessary* repetition compared to the earlier surveys).

I had to leave out quite a few very interesting topics. Practically, I skipped all the hypergraph theorems, [99], [149] the covering problems connected to the Erdős–Goodman-Pósa theorem, applications of finite geometrical methods in extremal graph theory, see, e.g. [161], [156], ... application of Lazebnik–Ustimenko type constructions, [122] and many more... Among others, I had to leave out that part of Ramsey Theory, which is extremely near to Extremal Graph Theory, (see [94]) ... and for many other things see Bollobás [10], [13], [14] ...

"The Complete List of Theorems"

If one watches Erdős in work, beside of his great proving power and elegance, one surprising feature is, how he poses his conjectures. This itself would deserve a separate note. Sometimes one does not immediately understand the importance of his questions. Slightly mockingly, once his friend, Hajnal told to him: "You would like to have a Complete List of Theorems". I think there is some truth in this remark, still one modification should be made.

Erdős does not like to state his conjectures immediately in their most general forms. Instead, he picks very special cases and attacks first these ones. Mostly he picks his examples "very fortunately". Therefore, having solved these special cases he very often discovers whole new areas, and it is difficult for the surrounding to understand how can he be so "fortunate". So, the reader of Erdős and the reader of this survey should keep in mind that Erdős's method is to attack always **important special cases**.

---

[4] When I asked Paul, why did he think that Röntgen's discovery changed the whole Physics, he answered that Röntgen's findings had led to certain results of Curie and from that point it was only a short step to the A-bomb.

Notation

We shall primarily consider simple graphs: graphs without loops and multiple edges but later there will be paragraphs where we shall consider digraph- and hypergraph extremal problems.

Given a family $\mathcal{L}$ of – so called – **excluded** or **forbidden** subgraphs, $\text{ex}(n,\mathcal{L})$ will denote the maximum number of edges a graph G_n can have without containing forbidden subgraphs. (Containment does not assume "induced subgraph" of the given type.) The family of graphs attaining the maximum will be denoted by $\text{EX}(n,\mathcal{L})$. If $\mathcal{L}$ consists of a single L, we shall use the notation $\text{ex}(n,L)$ and $\text{EX}(n,L)$ instead of $\text{ex}(n,\{L\})$ and $\text{EX}(n,\{L\})$.

For a set Q, $|Q|$ will denote its cardinality. Given a graph G, $e(G)$ will denote the number of its edges, $v(G)$ the number of its vertices, $\chi(G)$ and $\alpha(G)$ its chromatic and independence numbers, respectively. For graphs the (first) subscript will mostly denote the number of vertices: G_n, S_n, $T_{n,p}$, ... denote graphs on n vertices. There will be one exception: speaking of excluded graphs $L_1,\dots,L_r$ we use superscripts just to enumerate these graphs. Given two disjoint vertex sets, X and Y, in a graph G_n, $e(X,Y)$ denotes the number of edges joining X and Y. Given a graph G and a set X of vertices of G, the number of edges in a subgraph spanned by a set X of vertices will be denoted by $e(X)$, the subgraph of G spanned by X is $G(X)$.

Special graphs. K_p will denote the complete graph on p vertices, $T_{n,p}$ is the so called Turán graph on n vertices and p classes: n vertices are partitioned into p classes as uniformly as possible and two vertices are joined iff they belong to different classes. This graph is the (unique) p–chromatic graph on n vertices with the maximum number of edges among such graphs. $K_p(n_1,\dots,n_p)$ (often abbreviated to $K(n_1,\dots,n_p)$) denotes the complete p–partite graph with n_i vertices in its ith class, $i=1,2,\dots,p$.

We shall say that X is **completely joined** to Y if every vertex of X is joined to every vertex of Y. Given two vertex-disjoint graphs, G and H, their **product** $G\otimes H$ is the graph obtained by joining each vertex of G to each one of H.

Quoting. Below sometimes I quote some paragraphs from other papers, but the references and occasionally the notations too are changed to comply with mines.

2. Turán's Theorem

Perhaps Turán was the third to arrive at this field. In 1940 he proved the following theorem, [173] (see also [174], [172]):

Theorem 2.1 (Turán). *(a) If G_n contains no K_p, then $e(G_n)\le e(T_{n,p-1})$. In case of equality $G_n=T_{n,p-1}$.*

Turán's original paper contains much more than just this theorem. Still, the main impact coming from Turán was that he asked the general question:

> *What happens if we replace K_p with some other forbidden graphs, e.g., with the graphs coming from the Platonic polyhedra, or with a path of length ℓ, etc.*

Turán's theorem also could have sunk into oblivion. However, this time Erdős was more open-minded. He started proving theorems, talked to people about this topic and people started realizing the importance of the field.

Turán died in 1976. The first issue of Journal of Graph Theory came out around that time. Both Paul [60] and I were asked to write about Turán's graph theory [154]. (In the introductory issue of the journal Turán himself wrote a Note of Welcome, also mentioning some historical facts about his getting involved in graph theory [176].) Let me quote here some parts of Paul Erdős's paper [60].

> "In this short note I will restrict myself to Turán's work in graph theory, even though his main work was in analytic number theory and various other branches of real and complex analysis. Turán had the remarkable ability to write perhaps only one paper in various fields distant from his own; later others would pursue his idea and new subjects would be born.
>
> In this way Turán initiated the field of extremal graph theory. He started this subject in [1941], (see [173], [174]) He posed and completely solved the following problem ..."

Here Erdős describes Turán Theorem and Turán's hypergraph conjecture, and a result of his own to which we shall return later. Then he continues:

> "Turán also formulated several other problems on graphs, some of which were solved by Gallai and myself [66]. *I began a systematic study of extremal graph theory in 1958 on the boat from Athens to Haifa and have worked on it since then.* The subject has a very large literature; Bollobás has written a comprehensive book on extremal problems in graph theory which will appear soon." (Paul meant [10].)

One final remark should be made here. As I stated in other places, Paul Turán's role was crucial in the development of Extremal Graph Theory. Still, even here there is a point, where Erdős's influence should be mentioned again. More precisely, the influence of an Erdős–Szekeres paper. As today it is already well known, Erdős and Szekeres tried to solve a problem in convex geometry and rediscovered Ramsey Theorem [91]. They informed Turán about their theorem, according to which either the graph or the complementary graph contains a large complete graph. Turán regarded this result as a theorem where one ensures the existence of a large complete subgraph in G_n by assuming something about the complementary graph. So Turán wanted to change the condition and still arrive at the same conclusion. This is why he supposed that a lower bound is given on the number of edges and deduced the existence of a large complete subgraph of G_n. Turán writes in [173]:

> "Theorem I gives a condition to guarantee the existence of a complete subgraph on k vertices in a graph on a finite number of vertices. The only related theorem – as far as I know – can be found in a joint paper of Pál Erdős and György Szekeres [91] and essentially states that if a graph A on n vertices is such that its complement $\overline{A}$ contains only complete subgraphs having "few" vertices, then the graph A contains a complete subgraph on "many" vertices. Their theorem contains only bounds in the place of the expressions "few" and "many", in fact it gives almost only the existence; the exact solution seems to be very interesting but difficult ...

Some (Further) Historical Remarks

(a) Turán's paper contains an infinite Ramsey theorem. I quote:

Theorem II. *Let us suppose that for the infinite graph A containing countably many vertices $P_1, P_2, \ldots$ there is an integer $d > 1$ such that if we choose arbitrary d*

different vertices of A, there will be at least two among these vertices joined by an edge in A. Then A has at least one complete subgraph of infinitely many vertices.

This theorem is weaker than the one we usually teach in our courses, nevertheless, historically it is interesting to see this theorem in Turán's paper.

(b) Turán's theorem is connected to the Second World War in two ways. On the one hand, Turán, sent to forced labour service and deprived of paper and pencil, started working on problems that were possible to follow without writing them down. Also he made his famous hypergraph conjecture, thinking that would he have paper and pencil, he could have easily proved it.

On the other hand it is worth mentioning that Turán's Theorem was later rediscovered by A. A. Zykov [181, 1949] who (because of the war) learned too late that it had already been published.

(c) As to Mantel's result, I quote the last 4 lines of [173]:

"**Added in proof.** ... Further on, I learned from the kind communication of Mr. József Krausz that the value of $d_k(n)$ $(= \mathrm{ex}(n, K_3))$ is given on p438, for k=3 was found already in 1907 by W. Mantel, (Wiskundige Opgaven, vol. **10**, p60-61). I know his paper only from the reference of Fortschritte d. Math., vol **38** p.270."

(d) During the war Turán was trying to prove that either a graph G_n or its complementary graph contains a complete graph of order $[\sqrt{n}]$. He writes in [176]:

"I still have the copybook in which I wrote down various approaches by induction, all they started promisingly, but broke down at various points. I had no other support for the truth of this conjecture, than the symmetry and some dim feeling of beauty: ... In one of my first letters to Erdős after the war I wrote of this conjecture to him. In his answer he proved that my conjecture was utterly false ..."

Of course, all we know today, what Erdős wrote to Turán: the truth is around $c \log n$. This was perhaps the first application of probability to Graph Theory, though many would deny that Erdős's elegant answer uses more than crude counting. Probably this is where the Theory of Random Graphs started. (To be quite precise, one should mention, that T. Szele had a similar proof for Rédei's theorem on directed Hamiltonian cycles in tournaments, [166], already in 1943, however, Erdős's proof was perhaps of more impact and it was the first where no other approach could replace the counting argument. Another early breakthrough of the Random Graph Method was when Erdős easily answered the following problem of Schütte [49]: Is there a tournament where for every k players there is a player which beats all of them?)

I would suggest the reader to read also the beautiful paper of Turán [176], providing a lot of information on what I have described above shortly.

(e) For a longer account on the birth of the Erdős–Szekeres version of Ramsey theorem see the account of Gy. Szekeres in the introduction of the Art of Counting, [57].

3. Erdős–Stone Theorem

Setting out from a problem in topology, Erdős and A. H. Stone proved the following theorem in 1946 [90]:

Theorem 3.1 (Erdős–Stone). *For every fixed p and m*

$$\mathrm{ex}(n, K_{p+1}(m, \dots, m)) = \left(1 - \frac{1}{p}\right)\binom{n}{2} + o(n^2). \tag{3.1}$$

Moreover, if p is fixed and $m := \sqrt{\ell_p(n)}$ where $\ell_p(x)$ denotes the p times iterated logarithm of x, (3.1) still holds.

Here $m = \sqrt{\ell_p(n)}$ is far from being the best possible. The sharp order of magnitude is $c \log n$. Let $m = m(n, \varepsilon)$ be the largest integer such that, if $e(G_n) > e(T_{n,p}) + \varepsilon n^2$, then G_n contains the regular $(p+1)$–partite graph $K_{p+1}(m, m, \dots, m)$. One can ask how large is $m = m(n, \epsilon)$, defined above? This was determined by Bollobás, Erdős, Simonovits, [15], [17] Chvátal and Szemerédi [37].

The first breakthrough was that the p–times iterated log was replaced by $K \log n$, where K is a constant [15]. In the next two steps the dependence of this constant on p and ε were determined.

Theorem 3.2 (Bollobás–Erdős–Simonovits [17]). *There exists an absolute constant $c > 0$ such that every G_n with*

$$e(G_n) \geq \left(1 - \frac{1}{p} + \varepsilon\right)\binom{n}{2}$$

contains a $K_{p+1}(m, m, \dots, m)$ with

$$m > \frac{c \log n}{p \log(1/\varepsilon)}.$$

The next improvement, essentially settling the problem completely is the result of Chvátal and Szemerédi, providing the exact dependence on all the parameters, up to an absolute constant.

Theorem 3.3 (Chvátal–Szemerédi [37]).

$$\frac{\log n}{500 \log(1/\varepsilon)} < m(n, \varepsilon) < \frac{c \log n}{\log(1/\varepsilon)}.$$

One could have thought that the problem is settled but here is a nice result of Bollobás and Kohayakawa, improving Theorem 3.3.

Conjecture 3.4 (Bollobás–Kohayakawa [18]). *There exists an absolute constant $\alpha > 0$ such that for all $r \geq 1$ and $0 \leq \varepsilon \leq 1/r$ every G_n of sufficiently large order satisfying*

$$e(G_n) \geq \left(1 - \frac{1}{r} + \varepsilon\right)\binom{n}{2}$$

contains a $K_{r+1}(s_0, m_0, \dots, m_0)$, where

$$s_0 = s_0(n) = \left\lfloor \frac{\alpha \log n}{\log(1/\varepsilon)} \right\rfloor \qquad \text{and} \qquad m_0 = m_0(n) = \left\lfloor \frac{\alpha \log n}{\log r} \right\rfloor.$$

Bollobás and Kohayakawa [18] succeded in proving that under the above conditions, if $0 < \gamma < 1$, then G_n contains a $K_{r+1}(s_1, m_1, \dots, m_1, \ell)$, where

$$s_1 = \left\lfloor \frac{\alpha(1-\gamma)\log n}{r \log(1/\varepsilon)} \right\rfloor, \qquad m_1 = \left\lfloor \frac{\alpha(1-\gamma)\log n}{\log r} \right\rfloor, \qquad \text{and} \qquad \ell = \lfloor \alpha \varepsilon^{1+\gamma/2} n^{\gamma} \rfloor.$$

Observe that this result is fairly near to proving Conjecture 3.4: the first class is slightly smaller and the last class much larger than in the conjecture.

The Kővári–V. T. Sós–Turán Theorem

The Kővári–T. Sós–Turán theorem [121] solves the extremal graph problem of $K_2(p,q)$, at least, provides an upper bound, which in some cases proved to be sharp.[5] This theorem is on the one hand a generalization of the C_4–problem, since $C_4 = K(2,2)$, and, on the other hand, is a special case of the Erdős–Stone theorem, apart from the fact that we get sharper estimates.[6,7]

Theorem 3.5 (Kővári–T. Sós–Turán). *Let $2 \le p \le q$ be fixed integers. Then*

$$\mathrm{ex}(n, K(p,q)) \le \frac{1}{2}\sqrt[p]{q-1}\,n^{2-1/p} + O(n).$$

The exponent $2-(1/p)$ is conjectured to be sharp but this is known only for $p=2$ and $p=3$, (see Erdős, Rényi, V. T. Sós, [78] and independently W. G. Brown [25]). Random graph methods [89] show that

$$\mathrm{ex}(n, K(p,p)) > c_p n^{2-\frac{2}{p+1}}.$$

Recently Füredi [101] improved the constant in the upper bound, showing that

$$\mathrm{ex}(n, K(2,t+1)) = \frac{1}{2}\sqrt{t}n^{3/2} + O(n^{4/3}),$$

and that the constant provided by Brown's construction is sharp. While one conjectures that $\mathrm{ex}(n,K(4,4))/n^{7/4}$ converges to a positive limit, we know only, by the Brown construction, that $\mathrm{ex}(n,K(4,4)) > \mathrm{ex}(n,K(3,3)) > cn^{5/3}$. It is unknown if

$$\frac{\mathrm{ex}(n,K(4,4))}{n^{5/3}} \to \infty.$$

The Matrix Form

The problem of Zarankiewicz is to determine the maximum integer $k_p(n)$ such that if A_n is a matrix with n rows and n columns consisting exclusively 0's and 1's, and the number of 1's is at least $k_p(n)$, then A_n contains a minor B_p of p rows and columns so that all the entries of B_p are 1's.

One can easily see that this problem is equivalent with determining the maximum number of edges a bipartite graph $G(n,n)$ can have without containing $K(p,p)$.

In [121] the authors remark that the problem can be generalized to the case of general matrices: when A has m rows and n columns and B has p rows and q columns. Denote the maximum by $k(m,n,p,q)$. There are many results on estimating this function but we shall not go here into details. Rather, we explain the notion of symmetric and asymmetric bipartite graph problems.

---

[5] A footnote of [121] tells us that the authors have received a letter from Erdős in which Erdős informed them that he also had proved most of the results of [121].

[6] Estimates, sharper than in the original Erdős–Stone.

[7] Very recently J. Kollár, L. Rónyai and T. Szabó showed that if $q > p!$, then $\mathrm{ex}(n, K_2(p,q)) > c_p n^{2-(1/p)}$. The paper: "Norm-graphs and Bipartite Turán numbers" will appear in Combinatorica, 1996.

As Erdős pointed out,

Theorem 3.6. *Every graph* G_n *has a bipartite subgraph* $H(U,V)$ *in which each vertex has at least half of its original degree:* $d_H(x) \geq \frac{1}{2}d_G(x)$, *and therefore* $e(H(U,V)) \geq \frac{1}{2}e(G_n)$.

One important consequence of this (almost trivial) fact is that (as to the order of magnitude), it does not matter if we optimize $e(G_n)$ over *all* graphs or only over the *bipartite* graphs. Another important consequence is that some matrix extremal problems are equivalent to graph extremal problems. Conversely, many extremal graph problems with bipartite excluded subgraphs have equivalent matrix forms as well:

As usually, having a bipartite graph, $G(U,V)$ we shall associate with it a matrix A, where the rows correspond to U, the columns to V and $a_{ij} = 1$ if the ith element of U is joined to the jth element of V, otherwise $a_{ij} = 0$.

Given a bipartite graph $L = L(X,Y)$ and another bipartite graph $G(U,V)$, $|U| = m$ and $|V| = n$, take the $m \times n$ adjacency matrix A of G and the adjacency-matrix B of L. Assume for a second that the colour-classes of L are symmetric (in the sense that there is an automorphism of L exchanging the two colour-classes). Then the condition that $L \not\subseteq G(U,V)$ can be formulated by saying that the matrix A has no submatrix equivalent to B, where equivalency means that they are the same apart from some but same permutation of the rows and columns. So Turán type problems lead to problems of the following forms:

> *Given an* $m \times n$ *0-1 matrix, how many 1's ensure a submatrix equivalent to* B*?*

If, on the other hand, the forbidden graph $L = L(X,Y)$ has no automorphism exchanging X and Y, then the matrix-problem and the graph-problem may slightly differ. Excluding the submatrices equivalent to B means that we exclude that $G(U,V)$ contains an L with $X \subseteq U$ and $Y \subseteq V$, but we do not exclude $L \subseteq G(U,V)$ in the opposite position. Denote by $\text{ex}^*(n,L)$ the maximum in this asymmetric case. Clearly, $\text{ex}^*(n,L) \geq \text{ex}(n,L)$, and they are equal if L has a colour-swapping automorphism.

Conjecture 3.7 (Simonovits). *If* L *is bipartite, then* $\text{ex}^*(n,L) = O(\text{ex}(n,L))$.

We do not know this even for $K(4,5)$. The difficulty in disproving such a conjecture is partly that in all the proofs of upper bounds on degenerate extremal graph problems, we use only "one-sided" exclusion. Therefore the upper bounds we know are always upper bounds on $\text{ex}^*(n,L)$.

Conjecture 3.8 (Erdős–Simonovits). *For every* $\mathcal{L}$ *with a bipartite* $L \in \mathcal{L}$ *there is a bipartite* $L^* \in \mathcal{L}$ *for which* $\text{ex}(n,\mathcal{L}) = O(\text{ex}(n,L^*))$.

We close this part with a beautiful but probably difficult problem of Erdős.

Conjecture 3.9. $\text{ex}(n,\{C_3,C_4\}) = \frac{1}{2\sqrt{2}}n^{3/2} + o(n^{3/2})$.

The meaning of this conjecture is that excluding C_3 beside C_4 has the same effect as if we excluded all the odd cycles. If we replace C_3 by C_5, then this is true, see [84]. Erdős risks the even sharper conjecture that the exact equality may hold:

$$\text{ex}(n,\{C_3,C_4\}) = \text{ex}(n,\{C_3,C_4,C_5,C_7,C_9,C_{11}\ldots\}).$$

For some further information, see a survey paper of Richard Guy, [103] and also a paper of Guy and Znam [104] on $K(p,q)$ and the results of Lazebnik, Ustimenko and Woldar on cycles [122], [123].

Applications of Kővári–T. Sós–Turán Theorem

It is interesting to observe that the C_4–theorem and its immediate generalizations (e.g. the Kővári–T. Sós–Turán theorem) have quite a few applications. Some of them are in geometry. For example, as Erdős observed, if we have n points in the plane, and join two of them if their distance is exactly 1, then the resulting graph contains no $K(2,3)$. So the number of unit distances among n points of the plane is $O(n^{3/2})$. Similarly, the unit-distance-graph of the 3-dimensional space contains no $K(3,3)$, therefore the number of unit distances in the 3-space is $O(n^{5/3})$. There are deeper and sharper estimates on this subject, see Spencer, Szemerédi and Trotter [163] or Clarkson, Edelsbrunner, Guibas, Sharir and Welzl [38].

Conjecture 3.10 (Erdős). *For every $\varepsilon > 0$, the number of unit distances among n points of the plain is $O(n^{1+\varepsilon})$.*

We mention one further application: the **chromatic number of the product** hyper-graph. Claude Berge was interested in calculating the chromatic number of the product of two graphs, $\mathcal{H}$ and $\mathcal{H}'$. Generally there are various ways to define the product of r–uniform hypergraphs. **This** product is defined as the r^2–uniform hypergraph whose vertex-set is the Cartesian product $V(\mathcal{H}) \times V(\mathcal{H}')$ and the edge-set is

$$\left\{ H \times H' \ : \ H \in E(\mathcal{H}) \qquad \text{and} \qquad H' \in E(\mathcal{H}') \right\}.$$

The chromatic number of the graph is the least k such that the vertices can be coloured in k colours without having monochromatic r^2–tuples. Berge and I [8] estimated the chromatic number of products of graphs (hypergraphs) using Kővári–T. Sós–Turán theorem. The same time a student of Berge, F. Sterboul [164],[165] have proved the same theorem and an *earlier* paper of V. Chvátal [36] used the same technique to prove some assertions roughly equivalent with this part of our paper [8].

4. Graph Theory and Probability

Erdős wrote two papers with the above title, one in 1959, [45], and the other in 1961, [46]. These papers were of great importance. In the first one Erdős proved the following theorem.

Theorem 4.1. *For fixed k and sufficiently large ℓ, if $n > \ell^{1+1/(2k)}$, then there exist (many) graphs G_n of girth k and independence number $\alpha(G_n) < \ell$.*

Clearly, the chromatic number of such a graph is at least $\frac{v(G_n)}{\alpha(G_n)}$. So, as Erdős points out, a corollary of his theorem is that

Corollary 4.2. *For every integer k for $n > n_0(k)$ there exist graphs G_n of girth $\geq k$ and chromatic number $\geq n^{\frac{1}{2k+1}}$.*

This theorem seems to be a purely Ramsey theoretical result, fairly surprising in those days, but, it has many important consequences in Extremal Graph Theory as well. The same holds for the next theorem, too:

Theorem 4.3 ([46]). *Assume that $n > n_0$. Then there exist graphs G_n with $K_3 \not\subseteq G_n$ and $\alpha(G_n) = O(\sqrt{n}\log n)$.*

One important corollary of Theorem 4.1, more precisely, of its proof is that

Theorem 4.4. *If $\mathcal{L}$ contains no trees, then $\mathrm{ex}(n, \mathcal{L}) > c_{\mathcal{L}}^* n^{1+c_{\mathcal{L}}}$, for some constants $c_{\mathcal{L}}^*, c_{\mathcal{L}} > 0$.*

On the other hand, it is easy to see that if $L \in \mathcal{L}$ is a tree (or a forest), then $\mathrm{ex}(n, \mathcal{L}) = O(n)$.

These theorems use random graph methods. They and some of their generalizations play also important role, in obtaining lower bounds in Turán–Ramsey Theorems. (See also Füredi-Seress, [102].) For general applications of the probabilistic methods in graph theory see, e.g. Erdős–Spencer [89], Bollobás [13], Alon–Spencer [6].

Of course, speaking of Graph Theory and Probability, one should also mention the papers of Erdős and Rényi, perhaps above all, [77].

5. The General Theory

In this section, we present the asymptotic solution to the general extremal problem.

> **General Extremal Problem.** Given a family $\mathcal{L}$ of forbidden subgraphs, find those graphs G_n that contain no subgraph from $\mathcal{L}$ and have the maximum number of edges.

The problem is considered to be "completely solved" if all the extremal graphs have been found, at least for $n > n_0(\mathcal{L})$. Quite often this is too difficult, and we must be content with finding $\mathrm{ex}(n, \mathcal{L})$, or at least good bounds for it.

It turns out that a parameter related to the chromatic number plays a decisive role in many extremal graph theorems. The **subchromatic number** $p(\mathcal{L})$ of $\mathcal{L}$ is defined by

$$p(\mathcal{L}) = \min\{\chi(L) : L \in \mathcal{L}\} - 1.$$

The following result is an easy consequence of the Erdős–Stone theorem [90]:

Theorem 5.1 (The Erdős–Simonovits Theorem [81]). *If $\mathcal{L}$ is a family of graphs with subchromatic number p, then*

$$\mathrm{ex}(n, \mathcal{L}) = \left(1 - \frac{1}{p}\right)\binom{n}{2} + o(n^2).$$

The meaning of this theorem is that $\mathrm{ex}(n, \mathcal{L})$ depends only very loosely on $\mathcal{L}$; up to an error term of order $o(n^2)$, it is already determined by the minimum chromatic number of the graphs in $\mathcal{L}$.

Classification of Extremal Graph Problems

By the above theorem,

$$\mathrm{ex}(n, \mathcal{L}) = o(n^2)$$

if and only if $p(\mathcal{L}) = 1$, i.e. there exist bipartite graphs in $\mathcal{L}$. From the Kővári-T. Sós-Turán Theorem we get that here $\mathrm{ex}(n, \mathcal{L}) = O(n^{2-c})$ for some $c = c(\mathcal{L})$. We shall call these cases **degenerate extremal graph problems** and find them among the most interesting problems in extremal graph theory. One special case is when $\mathcal{L}$ contains a tree (or a forest). These cases could be called **very degenerate**. Observe, that if a problem is **non-degenerate**, then $T_{n,2}$ contains no excluded subgraphs. Therefore $\mathrm{ex}(n, \mathcal{L}) \geq \left[\frac{n^2}{4}\right]$.

Structural Results

The structure of the extremal graphs is also almost determined by $p(\mathcal{L})$, and is very similar to that of $T_{n,p}$ This is expressed by the following results of Erdős and Simonovits [52], [54], [150]:

Theorem 5.2 (The Asymptotic Structure Theorem). *Let $\mathcal{L}$ be a family of forbidden graphs with subchromatic number p. If S_n is any graph in $\mathrm{EX}(n, \mathcal{L})$, then it can be obtained from $T_{n,p}$ by deleting and adding $o(n^2)$ edges. Furthermore, if $\mathcal{L}$ is finite, then the minimum degree*

$$d_{min}(S_n) = \left(1 - \frac{1}{p}\right) n + o(n).$$

The structure of extremal graphs is fairly stable, in the sense that the almost-extremal graphs have almost the same structure as the extremal graphs (for $\mathcal{L}$ or for K_{p+1}). This is expressed in our next result:

Theorem 5.3 (The First Stability Theorem). *Let $\mathcal{L}$ be a family of forbidden graphs with subchromatic number $p \geq 2$. For every $\varepsilon > 0$, there exist a $\delta > 0$ and n_ε such that, if G_n contains no $L \in \mathcal{L}$, and if, for $n > n_\varepsilon$,*

$$e(G_n) > \mathrm{ex}(n, \mathcal{L}) - \delta n^2,$$

then G_n can be obtained from $T_{n,p}$ by changing at most εn^2 edges.

These theorems are interesting on their own and also widely applicable.

In the remainder of this section we formulate a sharper variant of the stability theorem.

One can ask whether further information on the structure of forbidden subgraphs yields better bounds on $\mathrm{ex}(n, \mathcal{L})$ and further information on the structure of extremal graphs. At this point, we need a definition.

Let $\mathcal{L}$ be a family of forbidden subgraphs, and let $p = p(\mathcal{L})$ be its subchromatic number. The **decomposition** $\mathcal{M}$ of $\mathcal{L}$ is the family of graphs M with the property that, for some $L \in \mathcal{L}$ L contains M as an induced subgraph and $L - V(M)$ is $(p-1)$–colorable.

In other words, for $r = v(L)$, $L \subseteq M \times K_{p-1}(r, \dots, r)$, and M is minimal with this property. The following result is due to Simonovits [150], (see also [54]). In case of $\mathcal{L} = \{K_p\}$ the family $\mathcal{M}$ consists of one graph K_2.

Theorem 5.4 (The Decomposition Theorem, [150]). *Let $\mathcal{L}$ be a forbidden family of graphs with $p(\mathcal{L}) = p$ and decomposition $\mathcal{M}$. Then every extremal graph $S_n \in \mathrm{EX}(n,\mathcal{L})$ can be obtained from a suitable $K_p(n_1,\dots,n_p)$ by changing $O(\mathrm{ex}(n,\mathcal{M})+n)$ edges. Furthermore, $n_j = (n/p) + O(\mathrm{ex}(n,\mathcal{M})/n) + O(1)$, and*

$$d_{\min}(S_n) = \left(1-\frac{1}{p}\right)n + O(\mathrm{ex}(n,\mathcal{M})/n) + O(1).$$

It follows from this theorem that, with $m = \lceil n/p \rceil$, $\mathrm{ex}(n,\mathcal{L}) = e(T_{n,p}) + O(\mathrm{ex}(m,\mathcal{M})+n)$. If $\mathrm{ex}(n,\mathcal{M}) > cn$, then $O(\mathrm{ex}(m,\mathcal{M}))$ is sharp: put edges into the first class of a $T_{n,p}$ so that they form a $G_m \in \mathrm{EX}(m,\mathcal{M})$; the resulting graph contains no L, and has $e(T_{n,p}) + \mathrm{ex}(m,\mathcal{M})$ edges.

A second stability theorem can be established using the methods of [150]. To formulate it, we introduce some new terms. Consider a partition $S_1, S_2, \dots, S_p$ of the vertex-set of G_n, and the complete p-partite graph $H_n = K(s_1,\dots,s_p)$ corresponding to this partition of $V(G_n)$, where $s_i = |S_i|$. An edge is called an **extra edge** if it is in G_n but not in H_n, and is a **missing edge** if it is in H_n but not in G_n. For given p and G_n, the partition $S_1, S_2, \dots, S_p$ is called **optimal** if the number of missing edges is minimum. Finally, for a given vertex v, let $a(v)$ and $b(v)$ denote the numbers of missing and extra edges at v.

Theorem 5.5 (The Second Stability Theorem). *Let $\mathcal{L}$ be a forbidden family of graphs with $p(\mathcal{L}) = p$ and decomposition $\mathcal{M}$, and let $k > 0$. Suppose that G_n contains no $L \in \mathcal{L}$,*

$$e(G_n) \geq \mathrm{ex}(n,\mathcal{L}) - k \cdot \mathrm{ex}(n,\mathcal{M}),$$

and let $S_1,\dots,S_p$ be the optimal partition of G_n, $G_i := G(S_i)$. Then

(i) G_n can be obtained from $\times G_i$ by deleting $O(\mathrm{ex}(n,\mathcal{M})+n)$ edges;

(ii) $e(G_j) = O(\mathrm{ex}(n,\mathcal{M})) + O(n)$, and $v(G_j) = (n/p) + O(\sqrt{\mathrm{ex}(n,\mathcal{M})} + \sqrt{n})$;

(iii) for any constant $c > 0$, the number of vertices v in G_i with $a(v) > cn$ is only $O(1)$, and the number of vertices with $b(v) > cn$ is only $O(\mathrm{ex}(n,\mathcal{M})/n) + O(1)$;

(iv) let $L \in \mathcal{L}$, with $v(L) = r$, and let A_i be the set of vertices v in S_i for which $b(v) < (n/2pr)$; if $M \times K_{p-1}(r_1,\dots,r) \subseteq L$, then the graph $G(A_i)$ contains no M.

The constant k of the condition cannot be seen in (i)-(iv): it is hidden in the constants of the $O(.)$'s This theorem is useful also in applications. The deepest part is the first part of (iii). This implies (iv), which in turn implies all the other statements. A proof is sketched in [77], where the theorem was needed.

We conclude this section with the theorem characterizing those cases where $T_{n,p}$ is the extremal graph.

Theorem 5.6 (Simonovits, [153]). *The following statements are equivalent:*

(a) The minimum chromatic number in $\mathcal{L}$ is $p+1$ but there exists (at least one) $L \in \mathcal{L}$ with an edge e such that $\chi(L-e) = p$. (Colour-critical edge.)

(b) There exists an n_0 such that for $n > n_0(\mathcal{L})$, $T_{n,p}$ is extremal.

(c) There exists an n_0 such that for $n > n_0(\mathcal{L})$, $T_{n,p}$ is the **only** *extremal graph.*

The Product Conjecture

When I started working in extremal graph theory, I formulated (and later slightly modified) a conjecture on the structure of extremal graphs in non-degenerate cases. The meaning of this conjecture is that all the non-degenerate extremal graph problems can be reduced to degenerate extremal graph problems.

Conjecture 5.7 (Product structure). *Let $\mathcal{L}$ be a family of forbidden graphs and $\mathcal{M}$ be the decomposition family of $\mathcal{L}$. If no trees and forests occur in $\mathcal{M}$, then all the extremal graphs S_n for $\mathcal{L}$ have the following structure: $V(S_n)$ can be partitioned into $p = p(\mathcal{L})$ subsets $V_1, \ldots, V_p$ so that V_i is* **completely joined** *to V_j for every $1 \le i < j \le p$.*

This implies that each S_n is the product of p graphs G_i, where each G_i is extremal for some degenerate family $\mathcal{L}_{i,n}$. The meaning of this conjecture is that (almost) all the non-degenerate extremal graph problems can be reduced to degenerate extremal graph problems.

One non-trivial illustration of this conjecture is the Octahedron theorem:

Theorem 5.8 (Erdős–Simonovits [82]). *Let $O_6 = K_3(2,2,2)$ (i.e. O_6 is the graph defined by the vertices and edges of the octahedron.) If S_n is an extremal graph for O_6 for $n > n_0(O_6)$, then $S_n = H_m \otimes H_{n-m}$ for some $m = \frac{1}{2}n + o(n)$. Further, H_m is an extremal graph for C_4 and H_{n-m} is extremal for P_3.*

Remark 5.9. The last sentence of this theorem is an easy consequence of that S_n is the product of two other graphs of approximately the same size.

Remark 5.10. In [82] some generalizations of the above theorem can also be found. Thus e.g., the analogous product result holds for all the forbidden graphs $L = K_{p+1}(2, t_2, \ldots, t_p)$ and $L = K_{p+1}(3, t_2, \ldots, t_p)$.

Probably the octahedron theorem can be extended to all graphs $L = K_{p+1}(t_1, t_2, \ldots, t_p)$ and even to more general cases. On the other hand, in [155] counterexamples are constructed to the product-conjecture if we allow trees or forests in the decomposition family. In this case, when the decomposition contains trees, both cases can occur: the extremal graphs may be non-products and also they may be products. Turán's theorem itself is a product-case, where the decomposition family contains $K_2 = P_2$.

Szemerédi Lemma on Regular Partitions of Graphs

There are many important tools in Extremal Graph Theory that became quite standard to use over the last 20 years. One of them is the Szemerédi Regularity Lemma.

Let G be an arbitrary graph, $X, Y \subset V(G)$ be two disjoint vertex-sets and let $d(X,Y)$ denote the edge-density between them:

$$d(X,Y) = \frac{e(X,Y)}{|X| \cdot |Y|}.$$

Regularity lemma. [168] *For every $\varepsilon > 0$, and every integer κ there exists a $k_0(\varepsilon, \kappa)$ such that for every G_n $V(G_n)$ can be partitioned into sets $V_0, V_1, \ldots, V_k$ – for some $\kappa < k < k_0(\varepsilon, \kappa)$ – so that $|V_0| < \varepsilon n$, each $|V_i| = m$ for $i > 0$ and* *for all but at most* $\varepsilon \cdot \binom{k}{2}$ *pairs* (i,j)*, for every* $X \subseteq V_i$ *and* $Y \subseteq V_j$*, satisfying* $|X|, |Y| > \varepsilon m$*, we have*

$$|d(X,Y) - d(V_i, V_j)| < \varepsilon.$$

The applications of Szemerédi's Regularity Lemma are plentiful and are explained in details in [117], so here we shall describe it only very briefly.

One feature of the Regularity Lemma is that – in some sense – it allows us to handle a deterministic graph as if it were a (generalized) random one. One can easily prove for random graphs the existence of various subgraphs and the Regularity Lemma often helps us to ensure the existence of the same subgraphs when otherwise that would be far from trivial.

One example of this is the Erdős–Stone Theorem. Knowing Turán's theorem, the Szemerédi Lemma immediately implies the Erdős–Stone theorem. In the previous section we have mentioned a few improvements of the original Erdős–Stone theorem. The proof of the Chvátal–Szemerédi version also uses the Regularity Lemma as its main tool. Joining the work of Thomason, [170] [171] Fan Chung, Graham, Wilson, [35] [34] and others, V. T. Sós and I used the Regularity Lemma to give a transparent description of the so-called **quasi-random** graph sequences, [158] that was generalized by Fan K. Chung to hypergraphs [33].

The Regularity Lemma can be generalized in various ways. One of these generalizations states that if the edges of G_n are r–coloured for some fixed r, then we can partition the vertices of the graph so that the above Regularity Lemma remains true in all the colours simultaneously. This is what we used among others in proving some Turán–Ramsey type theorems [69], [70], [71] but it has also many other applications.

Generalized Regularity Lemma. *For every* $\varepsilon > 0$*, and integers* r, κ *there exists a* $k_0(\varepsilon, \kappa, r)$ *such that for every graph* G_n *the edges of which are* r*–coloured, the vertex set* $V(G_n)$ *can be partitioned into sets* $V_0, V_1, \ldots, V_k$ *– for some* $\kappa < k < k_0(\varepsilon, \kappa, r)$ *– so that* $|V_0| < \varepsilon n$*,* $|V_i| = m$ *(is the same) for every* $i > 0$*, and for all but at most* $\varepsilon\binom{k}{2}$ *pairs* (i,j)*, for every* $X \subseteq V_i$ *and* $Y \subseteq V_j$ *satisfying* $|X|, |Y| > \varepsilon m$*, we have*

$$|d_\nu(X,Y) - d_\nu(V_i, V_j)| < \varepsilon \qquad \text{simultaneously for} \qquad \nu = 1, \ldots, r,$$

where $d_\nu(X,Y)$ *is the edge-density in colour* ν*.*

As I mentioned, we describe the various applications of Szemerédi Lemma in more details in some other place [117]. Here I mention only that it was extended to hypergraphs by Frankl and Rödl [96], see also Chung, [33]. Prömel and Steger also use a hypergraph version of the Regularity lemma in "induced extremal graph problems" [141]. Algorithmic versions were found by Alon, Duke, Leffman and Rödl, Yuster, [4], and in some sense it was extended to sparse graphs by Rödl and independently, by Kohayakawa [116].

I should also mention some new variants due to Komlós, see, for example, [117], [118].

6. Turán–Ramsey Problems

V. T. Sós has a survey [162] on

– application of Turán type theorems to distance-distribution, initiated by Erdős and Turán,

– Turán–Ramsey type theorems initiated by her,

– and the connection of these fields.

These fields belong to Extremal Graph Theory and are strongly influenced by Paul Erdős. I will touch on these fields only very briefly.

These problems were partly motivated by applications of graph theory to distance distribution. Turán theorem combined with some geometrical facts can provide us with estimates on the number of short distances in various geometrical situations. Thus they can be applied in some estimates in analysis, probability theory, and so on. It was Erdős who first pointed out this possibility of applying Graph Theory to distance distribution theorems [44] and later Turán in [175] initiated investigating these problems more systematically. This work culminated in 3 joint papers of Erdős, Meir, Sós and Turán [73], [74], [75].

The structure of the extremal graphs in Turán type theorems seems to be too regular. So we arrive at the question: How do the upper bounds in extremal graph theorems improve if we exclude graphs very similar to the Turán graphs? Basically this was what motivated V. T. Sós [160] in initiating a new field of investigation. Erdős joined her and they have proved quite a few nice results, see e.g., [86], [87].

Let $\alpha_p(G)$ denote the maximum cardinality of vertices in G such that the subgraph spanned by these vertices contains no K_p.

General Problem. Assume that $L_1, \dots, L_r$ are given graphs, and G_n is a graph on n vertices the edges of which are coloured by r colours $\chi_1, \dots, \chi_r$, and

$$\begin{cases} \text{for } \nu = 1, \dots, r \text{ the subgraph of colour } \chi_\nu \text{ contains no } L_\nu \\ \text{and } \alpha_p(G_n) \le m. \end{cases}$$

What is the maximum of $e(G_n)$ under these conditions?

Originally the general problem was investigated only for $p = 2$,[8] and one breakthrough was the Szemerédi–Bollobás–Erdős theorem:

Theorem 6.1 (Szemerédi [169]). *If (G_n) is a sequence of graphs not containing K_4 and the stability number $\alpha(G_n) = o(n)$, then*

$$e(G_n) \le \frac{1}{8}n^2 + o(n^2). \tag{6.1}$$

Erdős asks if the $o(n^2)$ error term is necessary: Is it true that in Theorem 6.1 the stronger

$$e(G_n) \le \frac{n^2}{8}$$

also holds?

Theorem 6.2 (Bollobás–Erdős [16]). *(6.1) is sharp.*

Many estimates concerning general and various special cases of this field are proved in [69], [70], [71]. Here we mention just one result:

Theorem 6.3 (Erdős–Hajnal–Simonovits–Sós–Szemerédi [71]). *(a) For any integers $p > 1$ and $q > p$ if $\alpha_p(G_n) = o(n)$ and $K_q \not\subseteq G_n$, then*

$$e(G_n) \le \frac{1}{2}\left(1 - \frac{p}{q-1}\right)\binom{n}{2} + o(n^2).$$

(b) For $q = pk + 1$ this upper bound is sharp.

[8] In all the papers quoted but in [71] we consider this special case.

One of the most intriguing open problems of the field is (among many other very interesting questions)

Problem 6.4. *Assume that (G_n) is a sequence of graphs not containing the Octahedron graph $K(2,2,2)$. If $\alpha(G_n) = o(n)$, does it follow that $e(G_n) = o(n^2)$?*

I conclude this section with a slightly different result of Ajtai, Erdős, Komlós and Szemerédi. Let t be the average degree of G_n. Turán's theorem guarantees an independent set of size $\frac{n}{t+1}$.

Theorem 6.5 ([1]). *If the number of $K_3 \subseteq G_n$ is $o(n^3)$, then*

$$\alpha(G_n) > c\frac{n}{t}\log t \qquad \text{for} \qquad t = \frac{2e(G_n)}{n}.$$

The nice feature of the above theorem is that it says: if the number of triangles in G_n is $o(n^3)$, then the size of the maximum independent set jumps by a log-factor. This is sharp: $\frac{n}{t}\log t$ is achieved for random graphs.

Theorem 6.5 can also be interpreted as follows: excluding the triangles (or assuming that there are only few triangles in our graph) leads to randomlike behaviour.

7. Cycles in Graphs

Cycles play central role in graph theory. Many results provide conditions to ensure the existence of some cycles in graphs. Among others, the theory of Hamiltonian cycles (and paths) constitute an important part of graph theory. The Handbook of Combinatorics contains a chapter by A. Bondy [21] giving a lot of information on ensuring cycles via various types of conditions. Also, the book of Walther and Voss [178] and the book of Voss [177] contain many relevant results. Below we shall approach the theory of degenerate extremal graph problems (see Section 8) through extremal graph problems with forbidden cycles. Of course, one of the simplest extremal graph problems is when $\mathcal{L}$ is the family of all cycles. If we exclude them, the considered graphs will be all the trees and forests; the extremal graphs are the trees.

Remark 7.1. Describing walks and cycles in graphs is perhaps one of those parts of extremal graph theory, where algebraic methods may come in more often than in other extremal problems. So here occasionally, and very superficially, I will speak of Margulis graphs, Ramanujan graphs and Cayley graphs. I feel, these topics are very important, not only because of expander graphs but also because they provide new methods to construct nice graphs in extremal graph theory. Hoping that the Handbook of Combinatorics will sooner or later appear, I warmly recommend Noga Alons' chapter: Tools from Higher Algebra [2], which provides a lot of interesting and useful information – among others – on topics I had to describe very shortly.

The Long Cycle Problem

One problem posed by Turán was the extremal problem of cycles of length m. If we exclude all the odd cycles, the extremal graph will be the Turán graph $T_{n,2}$. What are the extremal graphs if family $\mathcal{L}_m$ of excluded graphs is the family of cycles of length at least m. The answer is given by the Erdős–Gallai theorem:

Theorem 7.2 (Erdős and Gallai [66]). *Let* $\mathcal{L}_m = \{C_k \;:\; k \geq m\}$. *Then*
(i) $\frac{m-1}{2}n - \frac{1}{2}m^2 < \text{ex}(n, \mathcal{L}_m) \leq \frac{m-1}{2}n$ *and*
(ii) the connected graphs G_n *whose 2-connected blocks are* K_{m-1}*'s are extremal.*

Graphs described in (ii) do not exist for all n, but we get asymptotically extremal graphs for all n, by taking those graphs in which one 2-connected component has size at most $m-1$ and all the other blocks are complete $m-1$–graphs.
The following theorem is the twin of the previous one's.

Theorem 7.3 (Erdős and Gallai [66]).

$$\text{ex}(n, P_m) \leq \frac{m-2}{2}n.$$

The union of $\lfloor \frac{n}{m-1} \rfloor$ *vertex disjoint* K_{m-1} *(and one smaller* K_q*) shows that this is sharp:* $\text{ex}(n, P_m) = \frac{m-2}{2}n + O(m^2)$.

This theorem has a sharper form, proved by Faudree and Schelp [92]. They needed the sharper form to prove some Ramsey theorems on paths.
These theorems can also be used to deduce the existence of Hamilton paths and cycles. Thus, for example, Theorem 7.2 implies Dirac's famous result:

Theorem 7.4 (Dirac). *If the minimum degree of* G_{2k} *is at least* k, *then* G_{2k} *is Hamiltonian.*

Erdős and T. Sós observed that the same estimates hold both for the path P_m and the star $K_2(1, m-1)$ and these being two extremes among the trees of m vertices, they conjectured that:

Conjecture 7.5 (Erdős–T. Sós). *For any tree* T_m,

$$\text{ex}(n, T_m) = \frac{m-2}{2}n + O(1).$$

Some asymptotical approximations of this conjecture were proved by
Ajtai, Komlós and Szemerédi, (unpublished), also, the conjecture is proved in its sharp form for some special families of trees, like caterpillars.

The Case of Excluded C_{2k}

Since the odd cycles are 3-chromatic colour-critical, one can apply Theorem 5.6 to them to get

$$\text{ex}(n, C_{2k+1}) = \left[\frac{n^2}{4}\right] \qquad \text{if} \qquad n > n_0(k).$$

The case of even cycles is much more fascinating. The upper bound would become trivial if we assumed that G_n is (almost) regular and contains no cycles of length $\le 2k$. The difficulty comes from that we exclude *only* C_{2k}.

Theorem 7.6 (Erdős, Bondy–Simonovits [23]).

$$\mathrm{ex}(n, C_{2k}) < ckn^{1+1/k} + o(n^{1+1/k}),$$

Theorem 7.7 (Bondy–Simonovits [23]). *If* $e(G_n) > 100kn^{1+1/k}$, *then*

$$C_{2\ell} \subseteq G_n \qquad \text{for every integer} \qquad \ell \in [k, kn^{1/k}].$$

Erdős stated Theorem 7.6 in [51] without proof and conjectured Theorem 7.7, which we proved. The upper bound on the cycle-length is sharp: take a G_n which is the union of complete graphs.

Let us return to Theorem 7.6. Is it sharp? Finite geometrical (and other) constructions show that for $k = 2, 3, 5$ YES. (Singleton, [159] Benson, [7] Wenger, [179] ...). Unfortunately, nobody knows if this is sharp for C_8, or for other C_{2k}'s.

Faudree and I sharpened Theorem 7.6 in another direction:

Definition 7.8 (Theta-graph). $\Theta(k,p)$ *is the graph consisting of* p *vertex-independent paths of length* k *joining two vertices* x *and* y.

Clearly, $\Theta(k,p)$ is a generalization of C_{2k}. We have proved

Theorem 7.9 (Faudree–Simonovits [93]). $\mathrm{ex}(n, \Theta(k,p)) < c_{k,p} \cdot n^{1+1/k}$.

The Erdős–Rényi Theorem [77] shows that Theorem 7.9 is sharp in the sense that

$$\mathrm{ex}(n, \Theta(k,p)) > c^*_{k,p} n^{1+\frac{1}{k}+\frac{1}{kp}} \qquad \text{as} \qquad n \to \infty.$$

One could ask if there are other global ways to state that if a graph has many edges then it has many cycles of different length. Erdős and Hajnal formulated such a conjecture, which was proved by A. Gyárfás, J. Komlós and E. Szemerédi. Among others, they proved

Theorem 7.10 (Gyárfás–Komlós–Szemerédi [105]). *If* $d_{min}(G) \ge \delta$ *and* $\ell_1, \ldots, \ell_m$ *are the cycle-lengths of* G, *then*

$$\sum \frac{1}{\ell_i} \ge c_1 \log \delta.$$

The meaning of this is as follows: If we regard all the graphs with minimum degree δ and try to minimize the sum of the reciprocals of the cycle-lengths, two candidates should first be checked. One is the union of disjoint $K_{\delta+1}$'s, the other is the union of disjoint complete bipartite graphs $K(\delta,\delta)$'s. In the first case we get $\log \delta + O(1)$, in the second one $\frac{1}{2}\log\delta + O(1)$. The above theorem asserts that these cases minimize $\sum \frac{1}{\ell_i}$.

Some graph theorists could be surprised by measuring the density of cycle lengths this way. Yet, whenever we want to express that something is nearly linear, then in number theory we tend to use this measure. Thus, e.g. the famous $3000 problem of Erdős asks for the following sharpening of Szemerédi's theorem on Arithmetic Progressions [167]:

Conjecture 7.11 (Erdős). *Prove that if $A = \{a_1 < a_2 < \cdots\}$ is an infinite sequence of positive integers and*

$$\sum \frac{1}{a_i} = \infty,$$

then for every k, A contains a k-term arithmetic progression.

Very Long Cycles

We know that a graph with minimum degree 3 contains a cycle of length at most $2\log_2 n$. The other extreme is when (instead of short cycles) we wish to ensure very long cycles. We may go much beyond the Erdős–Gallai theorem if we increase the connectivity and put an upper bound on the maximum degree.

Theorem 7.12 (Bondy–Entringer [22]). *Let $f(n,d)$ be the largest integer k such that every 2-connected G_n with maximum degree d contains a cycle of length at least k. Then*

$$4\log_{d-1} n - 4\log_{d-1}\log_{d-1} n - 20 < f(n,d) < 4\log_{d-1} n + 4.$$

[9]

Clearly, the connectivity is needed, otherwise – as we have seen – the Erdős–Gallai theorem is sharp.

Increasing the connectivity to 3 we can ensure longer cycles:

Theorem 7.13 (Bondy–Simonovits [24]). *If G_n is 3-connected and the minimum degree of G_n is d, the maximum degree is D, then G_n contains a cycle of length at least $e^{c\sqrt{\log n}}$ for some $c = c(d,D)$.*

We conjectured that $e^{c\sqrt{\log n}}$ can be improved to n^c. Bill Jackson, Jackson and Wormald succeded in proving this:

Theorem 7.14 (B. Jackson [109], [112]). *If G_n is 3-connected and the minimum degree of G_n is d, the maximum degree is D, then G_n contains a cycle of length at least n^c for some $c = c(d,D)$.*

Increasing the connectivity higher does not help in getting longer cycles:

Theorem 7.15 (Jackson, Parson [110]). *For every $d > 0$ there are infinitely many $d+2$-regular d-connected graphs without cycles longer than n^γ for some $\gamma = \gamma_d < 1$.*

We close this topic with an open problem:

Conjecture 7.16 (J. A. Bondy). *There exists a constant $c > 0$, such that every cyclically 4-connected 3-regular graph G_n contains a cycle of length at least cn.*

---

[9] Here $\log_{d-1} x$ means log base $d-1$ and not the iterated log.

Erdős–Pósa Theorem

The following question of Gallai is motivated partly by Menger Theorem. If G is a graph

(*) not containing two independent cycles,

how many vertices are needed to represent all the cycles?

K_5 satisfies (*) and we need at least 3 vertices to represent all its cycles. Bollobás [9] proved that in all the graphs satisfying (*) there exist 3 vertices the deletion of which results in a tree (or forest). More generally,

> Let $RC(k)$ denote the minimum t such that if a graph G contains no $k+1$ independent cycles, then one can delete t vertices of G ruining all the cycles of the graph. Determine $RC(k)$!

Erdős and Pósa [76] proved the existence of two positive constants, c_1 and c_2 such that

$$c_1 k \log k \leq RC(k) \leq c_2 k \log k. \tag{7.1}$$

This theorem is strongly connected to the following extremal graph theoretical question:

> Assume that G_n is a graph in which the minimum degree is D. Find an upper bound on the girth of the graph.

Here the usual upper bound is

$$\approx \frac{2\log n}{\log(D-1)}. \tag{7.2}$$

The proof is easy: Assume that the girth is g and let $k = \lfloor \frac{g-1}{2} \rfloor$. Take a vertex x and denote the set of vertices having distance t from x by X_t. Then for $t \leq k$ we have $|X_t| \geq (D-1)|X_{t-1}|$. Therefore

$$n = v(G_n) \geq 1 + D + D(D-1) + D(D-1)^2 + \ldots + D(D-1)^k.$$

This implies (7.2).

These things are connected to many other parts of Graph Theory, in some sense even to the Robertson–Seymour theory. Below I shall try to convince the reader that the Gallai problem is strongly connected to the girth problem.

In [151] I gave a short proof of the upper bound of (7.1). My proof goes as follows (sketch!):

Let G_n be an arbitrary graph not containing $k+1$ independent circuits. Let H_m be a maximal subgraph of G_n all whose degrees are 2,3 or 4. Then one can immediately see that the ramification vertices of H_m, i.e. the vertices of degree 3 or 4 represent all the cycles of G_n.[10] Let μ be the number of these vertices. Replacing the hanging chains[11] by single edges, we get an H_μ each degree of which is 3 or 4. So one can easily find a cycle $C^{(1)}$ of length $\leq c_3 \log \mu$ in H_μ. Applying this to $H_\mu - C^{(1)}$ (but first cleaning up the resulting low degrees) we get another short cycle $C^{(2)}$. This cleaning up is where we have to use that the degrees are bounded from above. Iterating this (and using that the degrees are bounded from above) one can find $c_4\mu/\log\mu$ vertex–independent cycles in G_n. Since $k \leq c_4\mu/\log\mu$, therefore $\mu \leq c_5 k \log k$.

The Erdős–Pósa theorem is strongly connected with the girth problem. If, e.g. we had shorter circuits in graphs with degrees 3 and 4 then the above proof would give better upper bound on $RC(k)$ – but that is ruled out.

[10] The vertices of degree 4 are not really needed ...

[11] paths all whose inner vertices have degree 2

The Margulis Graphs and the Lubotzky–Phillips–Sarnak Graphs

Sometimes we insist on finding **constructions** in certain cases when the randomized methods work easily. Often finding explicit constructions is very difficult. A good example of this is the famous case of the Ramsey 2–colouring, where Erdős offered $... for finding a construction of a graph of n vertices not containing complete graphs or independent sets of at least $c \log n$ vertices. (See Frankl, Wilson [97])

Another similar case is the **girth** problem discussed above, with one exception. Namely, in the girth problem Margulis, [131], [132] and Lubotzky–Phillips–Sarnak [126], [127], succeeded in constructing regular graphs G_n of (arbitrary high) but fixed degree d and girth at least $c_d \log n$. The original random-graph existence proof is due to Erdős and Sachs [79].

These graphs are Cayley graphs. Below (skipping many details)

(a) first we explain, why should one try Cayley graphs of **non-commutative groups**,

(b) then we give a sketch of the description of the first, simpler Margulis graph and

(c) finally we list the main features of the Lubotzky–Phillips–Sarnak graph.

(a) Often cyclic graphs are used in the constructions. Cyclic graphs are the graphs where a set $A_n \subseteq [1, n]$ is given, the vertices of our graphs are the residue classes Z_i (mod n), and Z_i is joined to Z_j if $|i - j| \in A_n$ (or $|n + i - j| \in A_n$). One such well known graph is Q_p (the Paley graph) obtained by joining Z_i to Z_j if their difference is a quadratic non-residue. The advantage of such graphs is that they have great deal of fuzzy (randomlike) structure. From the point of view of the short cycles they are not the best: they have many short even cycles.

(b) Given an arbitrary group $\mathcal{G}$ and some elements $g_1, \dots, g_t \in \mathcal{G}$, these elements generate a Cayley graph on $\mathcal{G}$: we join each $a \in \mathcal{G}$ to the elements $ag_1, \dots, ag_t$. This is a digraph. If we are interested in ordinary graphs, we choose $g_1, \dots, g_t$ so that whenever g is one of them, then also $g^{-1} \in \{g_1, \dots, g_t\}$. Thus we get an undirected graph. Still, if $\mathcal{G}$ is commutative, then this Cayley graph will have many even cycles. For example, $a, ag_1, ag_1g_2, ag_1g_2g_1^{-1}, ag_1g_2g_1^{-1}g_2^{-1}$ is (mostly) a C_4 for commutative groups and a P_4 for non-commutative groups. So, if we wanted to obtain Cayley graphs with large girth, we have better to start with non-Abelian groups. This is what Margulis did in [131]:

Let X denote the set of all 2×2 matrices with integer entries and with determinant 1. Pick the following two matrices:

$$A = \begin{pmatrix} 1 & 2 \\ 0 & 1 \end{pmatrix} \quad \text{and} \quad B = \begin{pmatrix} 1 & 0 \\ 2 & 1 \end{pmatrix}.$$

It is known that they are independent in the sense that there is no non-trivial multiplicative relation between them. So, if we take the 4 matrices A, B, A^{-1} and B^{-1}, they generate an infinite Cayley graph which is a 4-regular tree. If we take everything mod p, then it is easy to see that the tree collapses into a graph of $n \approx p^3$ vertices, in which the shortest cycle has length at least $c \log p$ for some constant $c > 0$. This yields a sequence of 4–regular graphs X_n with girth $\approx c^* \log n$ for some $c^* \approx 0.91...$ Margulis also explains, how the above graphs can be used in constructing certain (explicit) error-correcting codes. Margulis has also generalized this construction (in the same paper) to arbitrary even degrees.

Theorem 7.17 (Margulis, [131]). *For every $\varepsilon > 0$ we have infinitely many values of r, and for each of them an infinity of regular graphs X_j of degree $2r$ with* *girth*

$$g(X_j) > \left(\frac{4}{9} - \varepsilon\right) \frac{\log v(X_j)}{\log r}.$$

(c) The next breakthrough was due to Margulis [132] and to Lubotzky, Phillips and Sarnak [126]. The graph of Lubotzky, Phillips and Sarnak was obtained *not* for extremal graph purposes. The authors, investigating the extremal spectral gap of d-regular graphs, constructed graphs where the difference between the first and second eigenvalues is as large as possible. Graphs with large spectral gaps are good expanders, and this was perhaps the primary interest in [126] or in [132]. As the authors of [126] remarked, Noga Alon turned their attention to the fact that their graphs can be "used" also for many other, classical purposes.

Definition 7.18. *Let X be a connected k–regular graph. Denote by $\lambda(X)$ the second largest eigenvalue (in absolute value) of the adjacency matrix of X.*

Definition 7.19. *A k–regular graph on n vertices, $X = X_{n,k}$, will be called a* **Ramanujan graph**, *if $\lambda(X_{n,k}) \le 2\sqrt{k-1}$.*

I do not have the place here to go into details, but the basic idea is that random graphs have roughly the spectral gap[12] required above and vice versa: if the graph has a large spectral gap, then it may be regarded in some sense, as if it were a random graph. So the Ramanujan graphs provide near-extremum in some problems, where random graphs are near-extremal. (See also [3], [35])

Let p, q be distinct primes congruent to 1 mod 4. The Ramanujan graph $X^{p,q}$ of [126] is a $p+1$–regular Cayley graph of $PSL(2, \mathbb{Z}_q)$ if the Legendre symbol $(\frac{p}{q}) = 1$ and of $PGL(2, \mathbb{Z}_q)$ if $(\frac{p}{q}) = -1$. (Here $\mathbb{Z}_q$ is the field of integers mod q.)

Theorem 7.20 (Alon, quoted in [126]). *Let $X_{n,k} = X^{p,q}$ be a non-bipartite Ramanujan graph; $(\frac{p}{q}) = 1$, $k = p+1$, $n = q(q^2-1)/2$. Then the independence number*

$$\alpha(X_{p,q}) \le \frac{2\sqrt{k-1}}{k} n.$$

Corollary 7.21 ([126]). *If $X_{n,k}$ is a non-bipartite Ramanujan graph, then*

$$\chi(X_{n,k}) \ge \frac{k}{2\sqrt{k-1}}.$$

Margulis, Lubotzky, Phillips and Sarnak have constructed Ramanujan graphs which are $p+1$–regular, and

(a) **bipartite** with $n = q(q^2-1)$ vertices, satisfying

$$\text{girth}(X_{n,p+1}) \ge \frac{4}{3}\frac{\log n}{\log p} - O(1) \qquad \text{and} \qquad \text{diam}(X_{n,p+1}) \le \frac{2}{3}\frac{\log n}{\log p} + 3.$$

Further, they constructed **non-bipartite Ramanujan graphs** with $n = q(q^2 - 1)/2$ vertices, and with the same diameter estimate and with

[12] = difference between the largest and second largest eigenvalues

$$\text{girth}(X_{n,p+1}) \geq \frac{2}{3}\frac{\log n}{\log p} + O(1), \qquad \alpha(X_{n,p+1}) \leq \frac{2\sqrt{p}}{p+1}n, \qquad \chi(X_{n,p+1}) \geq \frac{p+1}{2\sqrt{p}}.$$

Putting $p = const$ or $p \approx n^c$ we get constructions of graphs the existence of which were known earlier only via random graph methods. As a matter of fact, they are better than the known "random constructions", showing that

$$\text{ex}(n, C_{2k}) > c_k n^{1+\frac{4}{3k+25}}.$$

8. Further Degenerate Extremal Graph Problems

We have already seen the most important degenerate extremal graph problems. Unfortunately we do not have as many results in this field as we would like to. Here we mention just a few of them.

Topological Subgraphs

Given a graph L, we may associate with it all its topologically equivalent forms. Slightly more generally, let $\mathcal{T}(L)$ be the set of graphs obtained by replacing some edges of L by "hanging chains", i.e., paths, all inner vertices of which are of degree 2.

Problem 8.1. *Find the maximum number of edges a graph G_n can have without containing subgraphs from $\mathcal{T}(L)$.*

Denote the topological complete p-graphs by $< K_p >$. G. Dirac [39] have proved that every G_n of $2n-2$ edges contains a $< K_4 >$. This is sharp: Dirac gave a graph G_n of $2n-3$ edges and not containing $< K_4 >$. Erdős and Hajnal pointed out that there exist graphs G_n of $cp^2 n$ edges and not containing $< K_p >$. (This can be seen, e.g. by taking $[n/q]$ vertex-disjoint union $K(q,q)$'s for $q = \binom{p/2}{2}$.) Mader [128] showed that

Theorem 8.2. *For every integer $p > 0$ there exists a $D = D(p)$ such that if the minimum degree of G is at least $D(p)$, then G contains a $< K_p >$.*

More precisely,

Theorem 8.3. *There exists a constant $c > 0$ such that if $e(G) > tn$, then G contains a $< K_p >$ for $p = [c\sqrt{\log t}]$.*

Corollary 8.4. *For every L, $\text{ex}(n, \mathcal{T}(L)) = O(n)$.*

Conjecture 8.5 (Erdős – Hajnal – Mader [68], [128]). *If $e(G_n) > tn$, then G_n contains a $< K_p >$ with $p \geq c\sqrt{t}$.*

Mader's result was improved by Komlós and Szemerédi to almost the best:

Theorem 8.6 ([119]). *There is a positive* c_1 *such that if* $e(G_n) > tn$, *then* G_n *contains a* $< K_p >$ *with*

$$p > c_1 \frac{\sqrt{t}}{(\log t)^6}.$$

Very recently, improving some arguments of Alon and Seymour, Bollobás and Thomason completely settled Mader's problem:

Theorem 8.7 ([20]). *There is a positive* c_1 *such that if* $e(G_n) > tn$, *then* G_n *contains a* $< K_p >$ *with*

$$p > c_1\sqrt{t}.$$

Their proof-method was completely different from that of Komlós and Szemerédi. Komlós and Szemerédi slightly later also obtained a proof of Theorem 8.7 along their original lines [[120]].

Recursion Theorems

Recursion theorems could be defined for ordinary graphs and hypergraphs, for ordinary degenerate extremal problems and non-degenerate extremal graph problems, for supersaturated graph problems, ... However, here we shall restrict our considerations to ordinary degenerate extremal graph problems. In this case we have a bipartite L and a procedure assigning an L' to L. Then we wish to deduce upper bounds on $\mathrm{ex}(n, L')$, using upper bounds on $\mathrm{ex}(n, L)$. To illustrate this, we start with two trivial statements.
Claim. Let L be a bipartite graph and L' be a graph obtained from L by attaching a rooted tree T to L at one of its vertices.[13] Then

$$\mathrm{ex}(n, L') = \mathrm{ex}(n, L) + O(n).$$

Claim. Let L be a bipartite graph and L' be a graph obtained by taking two vertex-disjoint copies of L. Then (again)

$$\mathrm{ex}(n, L') = \mathrm{ex}(n, L) + O(n).$$

The proofs are trivial.

One of the problems Turán asked in connection with his graph theorem was to find the extremal numbers for the graphs of the regular (Platonic) polytopes. For the tetrahedron the answer is given by Turán Theorem (applied to K_4). The question of the Octahedron graph is solved by Theorem 5.8, the problems of the Icosahedron and Dodecahedron can be found in Section 12, ([152], [153]). On the cube-graph we have

Theorem 8.8 (Cube Theorem, Erdős–Simonovits, [83]).

$$\mathrm{ex}(n, Q_8) = O(n^{8/5}).$$

[13] This means that we take vertex-disjoint copies of L and T, a vertex $x \in V(T)$ and a vertex $y \in V(L)$ and identify x and y.

We conjecture that the exponent 8/5 is sharp. Unfortunately we do not have any "reasonable" lower bound.

The above theorem and many others follow from a recursion theorem:

Theorem 8.9 (Recursion Theorem, [83]). *Let L be a bipartite graph, coloured in BLUE and RED and $K(t,t)$ be also coloured in BLUE and RED. Let L^* be the graph obtained from these two (vertex-disjoint) graphs by joining each vertex of L to all the vertices of $K(t,t)$ of the other colour. If $\mathrm{ex}(n,L) = O(n^{2-\alpha})$ and*

$$\frac{1}{\beta} - \frac{1}{\alpha} = t,$$

then $\mathrm{ex}(n,L^*) = O(n^{2-\beta})$.

Applying this recursion theorem with $t = 1$ and $L = C_6$ we obtain the Cube-theorem. Another type of recursion theorem was proved by Faudree and me in [93].

Regular subgraphs

Let $\mathcal{L}_{r-reg}$ denote the family of r–regular graphs. Erdős and Sauer posed the following problem [61]:

> *What is the maximum number of edges in a graph G_n not containing any k–regular subgraph?*

Since $K(3,3)$ is 3-regular, one immediately sees that $\mathrm{ex}(n,\mathcal{L}_{3-reg}) = O(n^{5/3})$. Using the Cube Theorem one gets a better upper bound, $\mathrm{ex}(n,\mathcal{L}_{3-reg}) = O(n^{8/5})$. Erdős and Sauer conjectured that for every $\varepsilon > 0$ there exists an $n_0(k,\varepsilon)$ such that for $n > n_0(k,\varepsilon)$ $\mathrm{ex}(n,\mathcal{L}_{k-reg}) \le n^{1+\varepsilon}$. Pyber proved the following stronger theorem.

Theorem 8.10 (Pyber, [142]). *For every k, $\mathrm{ex}(n,\mathcal{L}_{k-reg}) = 50k^2 n\log n$.*

The proof is based on a somewhat similar but much less general theorem of Alon, Friedland and Kalai [5]. For further information, see e.g. Noga Alon, [2]

One more theorem

We close this section with an old problem of Erdős solved not so long ago by Füredi. Let $F(k,t)$ be the bipartite graph with k vertices $x_1,\dots,x_k$ and $\binom{k}{2}t$ further vertices in groups U_{ij} of size t, where all the vertices of $\cup U_{ij}$ are independent and the t vertices of U_{ij} are joined to x_i and x_j $(1 \le i < j \le k)$. Erdős asked for the determination of $\mathrm{ex}(n,F(k,t))$ for $t = 1$. For $t = 1$ and $k = 2$ this is just C_4, so the extremal number is $O(n^{3/2})$. Erdős also proved (and it follows from [83] as well) that $\mathrm{ex}(n,F(3,1)) = O(n^{3/2})$.

Theorem 8.11 (Füredi [100]). $\mathrm{ex}(n,F(k,t)) = O(n^{3/2})$.

9. Supersaturated Graphs

Rademacher Type Theorems

Almost immediately after Turán's result, Rademacher proved the following nice theorem (unpublished, see [47]) :

Theorem 9.1 (Rademacher Theorem). *If* $e(G_n) > \left[\frac{n^2}{4}\right]$ *then* G_n *contains at least* $[\frac{n}{2}]$ *triangles.*

This is sharp: adding an edge to (the smaller class of) $T_{n,2}$ we get $[\frac{n}{2}]$ K_3's. Erdős generalized this result by proving the following two basic theorems [47] :

Theorem 9.2. *There exists a positive constant* $c_1 > 0$ *such that if* $e(G_n) > \left[\frac{n^2}{4}\right]$, *then* G_n *contains an edge* e *with at least* $c_1 n$ *triangles on it.*

Theorem 9.3 (Generalized Rademacher Theorem). *There exists a positive constant* $c_2 > 0$ *such that if* $0 < k < c_2 n$ *and* $e(G_n) > \left[\frac{n^2}{4}\right] + k$, *then* G_n *contains at least* $k\left[\frac{n}{2}\right]$ *copies of* K_3.

(Lovász and I proved that $c_2 = \frac{1}{2}$ [124]. For further results see Moser and Moon [136], Bollobás, [11], [12], and [124], [125].) Erdős also proved the following theorem, going into the other direction.

Theorem 9.4 (Erdős [58]). *If* $e(G_n) = \left[\frac{n^2}{4}\right] - \ell$ *and* G_n *contains at least one triangle, then it contains at least* $[\frac{n}{2}] - \ell - 1$ *triangles.*

(Of course, we may assume that $0 \le \ell \le [\frac{n}{2}] - 3$.)

The General Case

Working on multigraph and digraph extremal problems, Brown and I needed some generalizations of some theorems of Erdős [50], [56].The results below are direct generalizations of some theorems of Erdős. To avoid proving the theorems in a setting narrower than what might be needed later, Brown and I formulated our results in the "most general, still reasonable" form.

Definition 9.5 (Directed multi-hypergraphs [31]). *A directed* (r,q)*–multi-hypergraph has a set* V *of vertices, a set* $\mathcal{H}$ *of directed hyperedges, i.e. ordered* r*–tuples, and a multiplicity function* $\mu(\mathbf{H}) \le q$ *(the multiplicity of the ordered hyperedge)* $\mathbf{H} \in \mathcal{H}$.

We shall return to the multigraph and digraph problems later, here I formulate only some simpler facts. The extremal graph problems directly generalize to directed multi-hypergraphs with bounded hyper-edge-multiplicity:

Given a family $\mathcal{L}$ of excluded directed (r,q)–multi-hypergraphs, we may ask the maximum number of directed hyperedges (counted with multiplicity) a directed (r,q)–multi-hypergraph $\mathbf{G}_n$ can have without containing forbidden sub-multi-hypergraphs from $\mathcal{L}$. The maximum is again denoted by $\mathrm{ex}(n,\mathcal{L})$.

Let $\mathbf{L}$ be a directed (r,q)–multi-hypergraph and $\mathbf{L}[t]$ be obtained from $\mathbf{L}$ by replacing each vertex v_i of $\mathbf{L}$ by a set X_i of t independent vertices, and forming a directed multihyperedge $(y_1,\dots,y_r)$ of multiplicity μ if $y_1 \in X_{i_1},\dots,y_r \in X_{i_r}$ and the corresponding $(v_{i_1},\dots,v_{i_r})$ is a directed hyperedge of multiplicity μ in $\mathbf{L}$.

Theorem 9.6 (Brown–Simonovits, [31]).

$$\mathrm{ex}(n,\mathbf{L}[t]) - \mathrm{ex}(n,\mathbf{L}) = o(n^r).$$

Again, the influence of Erdős is very direct: the above theorem is a direct generalization of his result in [56].

Theorem 9.7 (Brown–Simonovits, [31]). *Let $\mathcal{L}$ be an arbitrary family of (r,q)-hypergraphs, and $\gamma = \lim \frac{\mathrm{ex}(n,\mathcal{L})}{n^r}$, as $n \to \infty$. There exists a constant $c_2 = c_2(\mathcal{L},\varepsilon)$ such that, if*

$$e(\mathbf{G}_n) \ge (\gamma+\varepsilon)n^r$$

and n is sufficiently large, then there exists some $\mathbf{L} \in \mathcal{L}$ for which $\mathbf{G}_n$ contains at least $c_2 n^{v(\mathbf{L})}$ copies of this $\mathbf{L}$.

10. Typical K_p-Free Graphs: The Erdős-Kleitman-Rothschild Theory

Erdős, Kleitman and Rothschild [72] started investigating the following problem:

How many labelled graphs not containing L exist on n vertices?

Denote this number by $M(n,L)$. We have a trivial lower bound on $M(n,L)$: take any fixed extremal graph S_n and take all the $2^{\mathrm{ex}(n,L)}$ subgraphs of it:

$$M(n,L) \ge 2^{\mathrm{ex}(n,L)}.$$

In some sense it is irrelevant if we count *labelled* or *unlabelled* graphs. The number of labelled graphs is at most $n!$ times the number of unlabelled graphs and $\mathrm{ex}(n,L) \ge \left[\frac{n^2}{4}\right]$ for all non-degenerate cases, (and $\mathrm{ex}(n,L) \ge cn^{1+\alpha}$ for all the *non-tree-non-forest* cases). So, if we are satisfied with rough estimates, we may say: counting only labelled graphs is not a real restriction here.

Strictly speaking, this problem is not an extremal graph problem, neither a supersaturated graph problem. However, the answer to the question shows that this problem is in surprisingly strong connection with the corresponding extremal graph problem.

Theorem 10.1 (Erdős-Kleitman-Rothschild [72]). *The number of K_p-free graphs on n vertices and the number of $p-1$-chromatic graphs on n vertices are in logarithm asymptotically equal: For every $\varepsilon(n) \to 0$ there exists an $\eta(n) \to 0$ such that if $M(n, K_p, \varepsilon)$ denotes the number of graphs of n vertices and with at most εn^p subgraphs K_p, then*

$$\mathrm{ex}(n, K_p) \le \log M(n, K_p, \varepsilon) \le \mathrm{ex}(n, K_p) + \eta n^2.$$

In other word, we get "almost all of them" by simply taking all the $(p-1)$-chromatic graphs.

More generally, Erdős, Frankl and Rödl [65] proved that if $\chi(L) > 2$, then

$$M(n, L) = 2^{\mathrm{ex}(n,L)+o(\mathrm{ex}(n,L))}.$$

The corresponding question for bipartite graphs is unsolved. Even for the simplest non trivial case, i.e. for C_4 the results are not satisfactory. This is not so surprising. All these problems are connected with random graphs, where for low edge-density the problems often become much more difficult. Kleitman and Winston [115] showed that

$$M(n, C_4) \le 2^{cn\sqrt{n}},$$

but the best value of the constant c is unknown. Erdős conjectured that

$$M(n, L) = 2^{(1+o(1))\mathrm{ex}(n,L)}.$$

Then the truth should be, of course

$$M(n, C_4) = 2^{((1/2)+o(1))n\sqrt{n}}.$$

I finish with a recent problem of Erdős.

Problem 10.2 (Erdős). *Determine or estimate the number of* **maximal** *triangle-free graphs on n vertices.*

Some explanation. In the Erdős–Kleitman–Rothschild case the number of bipartite graphs was large enough to give a logarithmically sharp estimate. Here $K(a, n-a)$ are the maximal *bipartite* graphs, their number is negligible. This is why the situation becomes less transparent.

11. Induced Subgraphs

One could ask, why do we always speak of *not necessarily induced* subgraphs. What if we try to exclude *induced* copies of L? If we are careless, we immediately run into a complete nonsense. If L is not a complete graph and we ask:

> *What is the maximum number of edges a G_n can have without having an induced copy of L?*

the answer is the trivial $\binom{n}{2}$ and the only extremal graph is K_n. So let us give up this question for a short while and try to attack the corresponding *counting problem* which turned out in the previous section to be in a strong connection with the extremal problem.

How many labelled graphs not containing induced copies of L are on n vertices?

Denote this number by $M^*(n,L)$. Prömel and Steger succeeded in describing $M^*(n,L)$. They started with the case of C_4 and proved that almost all G_n not containing an induced C_4 have the following very specific structure. They are **split graphs**, which means that they are obtained by taking a K_p and $(n-p)$ further independent points and joining them to K_p arbitrarily. (Trivially, these graphs contain no induced C_4's.)

Theorem 11.1 (Prömel–Steger, [139]). *If $\mathbf{S}_n^*$ is the family of split graphs, then*

$$\frac{M^*(n,C_4)}{|\mathbf{S}_n^*|} \to 1 \qquad \text{as} \qquad n \to \infty.$$

This implies – by a result of Prömel [138] that

Corollary 11.2. *There exist two constants, $c_{even} > 0$ and $c_{odd} > 0$, such that*

$$\frac{M^*(n,C_4)}{2^{n^2/4+n-(1/2)n\log n}} \to \begin{cases} c_{even} & \text{for even } n, \\ c_{odd} & \text{for odd } n. \end{cases}$$

Can one generalize this theorem to arbitrary excluded induced subgraphs? To answer this question, first Prömel and Steger generalized the notion of chromatic number.

Definition 11.3. *Let $\tau(L)$ be the largest integer k for which there exists an integer $j \in [0, k-1]$ such that no $k-1$-chromatic graph in which j colour–classes are replaced by cliques contains L as an induced subgraph.*

Clearly, if $\sigma(L)$ denotes the clique covering number, (= the minimum number of complete subgraphs of L to cover all the vertices of L) then

Lemma 11.4 (Prömel–Steger). $\chi(L), \sigma(L) \le \tau(L) \le \chi(L) + \sigma(L)$.

Now, taking a $T_{n,p}$ for $p = \tau(L) - 1$ and replacing j appropriate classes of it (in the above definition) by complete graphs and then deleting arbitrary edges of the $T_{n,p}$ we get graphs not having induced L's:

$$M^*(n,H) \ge 2^{(1-\frac{1}{\tau-1})\binom{n}{2}+o(n^2)}.$$

Theorem 11.5 (Prömel–Steger). *Let H be a fixed nonempty subgraph with $\tau \ge 3$. Then*

$$M^*(n,H) = 2^{(1-\frac{1}{\tau-1})\binom{n}{2}+o(n^2)}.$$

Definition 11.6. *Given a sample graph L, call G_n "good" if there exists a* **fixed** *subgraph $U_n \subseteq \overline{G_n}$ (= the complementary graph of G_n) such that whichever way we add some edges of G_n to U_n, the resulting U' contains no induced copies of L. $\mathrm{ex}^*(n,H)$ denotes the maximum number of edges such a G_n can have.*

Example 11.1. In case of C_4, any bipartite graph $G(A, B)$ is "good", since taking all the edges in A, no edges in B and some edges from $G(A, B)$ we get a U_n not containing C_4 as an induced subgraph.

Theorem 11.7 (Prömel–Steger [140]).

$$\mathrm{ex}^*(n, L) = \left(1 - \frac{1}{\tau - 1}\right)\binom{n}{2} + o(n^2).$$

Thus Prömel and Steger convincingly showed that there is a possibility to generalize ordinary extremal problems and the corresponding counting problems to induced subgraph problems. For further information, see [140] [141].

12. The Number of Disjoint Complete Graphs

There are many problems where instead of ensuring many K_{p+1}'s (see Section 9) we would like to ensure many **edge-disjoint** or **vertex-disjoint** copies of K_{p+1}. Let us start with the case of **vertex-disjoint** copies.

If G_n is a graph from which one can delete $s - 1$ vertices so that the resulting graph is p-chromatic, then G_n cannot contain s vertex-disjoint copies of K_{p+1}. This is sharp: let $H_{n,p,s} = T_{n-s+1,p} \otimes K_{s-1}$. Then $H_{n,p,s}$ has the most edges among the graphs from which one can delete $s-1$ vertices to get a graph of chromatic number at most p. Further:

Theorem 12.1 (Moon [135]). *Among all the graph not containing s vertex-independent K_{p+1}'s $H_{n,p,s}$ has the most edges, assumed that $n > n_0(p, s)$.*

This theorem was first proved by Erdős and Gallai for $p = 1$, then for K_3 by Erdős [48], and then it was generalized for arbitrary p by J. W. Moon, and finally, a more general theorem was proved by me [153]. This more general theorem contained the answer to Turán's two "Platonic" problem: it guaranteed that $H_{n,2,6}$ is the only extremal graph for the dodecahedron graph and $H_{n,3,3}$ for the icosahedron, if n is sufficiently large.

We get a slightly different result, if we look for edge-independent complete graphs. Clearly, if one puts k edges into the first class of $T_{n,p}$, then one gets k edge-independent K_p's as long as $k < cn$. One would conjecture that this is sharp. As long as k is fixed, the general theorems of [153] provide the correct answer. If we wish to find the maximum number of edge-independent copies of K_{p+1} for

$$e(G_n) = \left(1 - \frac{1}{p}\right)\binom{n}{2} + k,$$

for $k = k(n) \to \infty$, the problem changes in character, see e.g. recent papers of Győri [106], [108]. We mention just one theorem here:

Theorem 12.2 (Győri [107]). *Let $e(G_n) = e(T_{n,p}) + k$, $(p \geq 3)$, where $k \leq 3\lfloor \frac{n+1}{p} \rfloor - 5$. The G_n contains k edge-independent K_{p+1}'s, assumed that $n > n_0(p)$.*

For $p = 2$ (for triangles) the result is different in style, see [106].

13. Extremal Graph Problems Connected to Pentagonlike Graphs

A lemma of Erdős asserts that each graph G_n can be turned into a bipartite graph by deleting at most half of its edges. (Above: Theorem 3.6) The proof of this triviality is as follows. Take a bipartite $H_n \subseteq G_n$ of maximum number of edges. By the maximality, each $x \in V(G_n)$ having degree $d(x)$ in G_n must have degree $\geq \frac{1}{2}d(x)$ in H_n. Summing the degrees in both graphs we get $e(H_n) \geq \frac{1}{2}e(G_n)$. This estimate is sharp for random graphs of edge probability $p > 0$, in asymptotical sense. Now, our first question is if this estimate can be improved in cases when we know some extra information on the structure of the graph, say, excluding triangles in G_n. The next theorem asserts that this is not so. Let $D(G_n)$ denote the minimum number of edges one has to delete from G_n to turn it into a bipartite graph.

Theorem 13.1 (Erdős, [53]). *For every $\varepsilon > 0$ there exists a constant $c = c_\varepsilon > 0$ such that for infinitely many n, there exists a G_n for which $K_3 \not\subseteq G_n$, $e(G_n) > c_\varepsilon n^2$, and*

$$D(G_n) > \left(\frac{1}{2} - \varepsilon\right) e(G_n).$$

Conjecture 13.2. *If $K_3 \not\subseteq G_n$, then one can delete (at most) $n^2/25$ edges so that the remaining graph is bipartite.*

Let us call a graph G_n **pentagonlike** if its vertex-set V can be partitioned into $V_1, \dots, V_5$ so that $x \in V_i$ and $y \in V_j$ are joined iff $i - j \equiv \pm 1 \mod 5$. The pentagonlike graph $Q_n := C_5[n/5]$ shows that, if true, this conjecture is sharp. The conjecture is still open, in spite of the fact that good approximations of its solutions were obtained by Erdős, Faudree, Pach and Spencer. This conjecture is proven for $e(G_n) \geq \frac{n^2}{5}$ (see below) and the following (other) weakening is also known, [64]:

Theorem 13.3. *If $K_3 \not\subseteq G_n$ then*

$$D(G_n) \leq \frac{n^2}{18 + \delta}.$$

for some (explicite) constant $\delta > 0$.

In fact, Erdős, Faudree, Pach, and Spencer [64] proved that

Theorem 13.4. *For every triangle-free graph G with n vertices and m edges*

$$D(G_n) \leq \max\left\{\frac{1}{2}m - \frac{2m(2m^2 - n^3)}{n^2(n^2 - 2m)}\ ,\ m - \frac{4m^2}{n^2}\right\} \tag{13.1}$$

Since the second term of (13.1) decreases in $[\frac{1}{8}n^2, \frac{1}{2}n^2]$, and its value is exactly $\frac{1}{25}n^2$ for $m = \frac{1}{5}n^2$, therefore (13.1) implies that if $e(G_n) > \frac{1}{5}n^2$, and $K_3 \not\subseteq G_n$, then $D(G_n) \leq \frac{1}{25}n^2$. It is trivial that if $e(G_n) \leq \frac{2}{25}n^2$, then $D(G_n) \leq \frac{1}{25}n^2$. However, the general conjecture is still open: it is unsettled in the middle interval $\frac{2n^2}{25} < e(G_n) < \frac{n^2}{5}$.

The next theorem of Erdős, Győri and myself [67] states that if $e(G_n) > \frac{1}{5}n^2$, then the pentagon-like graphs need the most edges to be deleted to become bipartite. (This is sharper than the earlier results, since it provides also information on the near-extremal structure.)

Theorem 13.5. *If $K_3 \not\subseteq G_n$ and $e(G_n) \geq \frac{n^2}{5}$, then there is a pentagonlike graph H_n^* with at least the same number of edges: $e(G_n) \leq e(H_n^*)$, for which $D(G_n) \leq D(H_n^*)$.*

14. Problems on the Booksize of a Graph

We have already seen a theorem of Erdős, stating that if a graph has many edges, then it has an edge e with cn triangles on it. Such configurations are usually called books. The existence of such edges is one of the crucial tools Erdős used in many of his graph theorems. Still, it was a longstanding open problem, what is the proper value of this constant c above. Without going into details we just mention three results:

Theorem 14.1 (Edwards [41, 42]). *If $e(G_n) > \left[\frac{n^2}{4}\right]$, then G_n has an edge with $[n/6]+1$ triangles containing this edge.*

This is sharp. The theorem would follow if we knew that there exists a $K_3 = (x,y,z)$ for which the sum of the degrees, $d(x)+d(y)+d(z) > \frac{3n}{2}$. Indeed, at least $\frac{n}{6}$ vertices would be joined to the same pair, say, to xy. An other paper of Edwards contains results of this type, but only for $e(G_n) > \frac{1}{3}n^2$. Let $\Delta_r = \Delta_r(G_n)$ denote the maximum of the sums of the degrees in a $K_r \in G_n$. (For instance, in a random graph R_n $\Delta_r(R_n) \approx r \cdot \frac{2e(R_n)}{n}$.)

Theorem 14.2 (Edwards [41]). *If $\frac{1}{r}\Delta_r > \left(1 - \frac{1}{r+1}\right)n$, $n \geq 1$, then*

$$\frac{1}{r+1}\Delta_{r+1} \geq \frac{2e(G_n)}{n}.$$

This theorem says that if G_n has enough edges to ensure a K_{r+1}, then it also contains a K_{r+1} whose vertex-degree-sum is as large as it should be by averaging. Erdős, Faudree and Győri have improved Theorem 14.1 if we replace the edge-density condition by the corresponding degree-condition. Among others, they have shown that

Theorem 14.3 (Erdős–Faudree-Győri [63]). *There exists a $c > 0$ such that if the minimum degree of G_n is at least $[n/2]+1$, then G_n contains an edge with $[n/6]+cn$ triangles containing this edge.*

15. Digraph/Multigraph Extremal Graph Problems

We have already seen supersaturated extremal graph theorems on multi-digraphs. Here we are interested in simple asymptotically extremal sequences for digraph extremal problems.

Multigraph or digraph extremal problems are closely related and in some sense the digraph problems are the slightly more general ones. So we shall restrict ourselves to digraph extremal problems. A digraph extremal problem means that some q is given and we consider the class of digraphs where loops are excluded and any two vertices may be joined by at most q arcs in one direction and by at most q arcs of the opposite direction. This applies to both the excluded graphs and to the graphs on n vertices the edges of which should be maximized. So our problem is:

> Fix the multiplicity bound q described above. A family $\mathcal{L}$ of digraphs is given and $\mathrm{ex}(n, \mathcal{L})$ denotes the maximum number of arcs a digraph $\mathbf{D}_n$ can have under the condition that it contains no $\mathbf{L} \in \mathcal{L}$ and satisfies the multiplicity condition. Determine or estimate $\mathrm{ex}(n, \mathcal{L})$.

The Digraph and Multigraph Extremal graph problems first occur in a paper of Brown and Harary [30]. The authors described fairly systematically all the cases of small forbidden multigraphs or digraphs. Next Erdős and Brown extended the investigation to the general case, finally I joined the "project". Our papers [26], [28] and [29] describes fairly well the situation $q = 1$ for digraphs (which is roughly equivalent with $q = 2$ for multigraphs). We thought that our results can be extended to all q but Sidorenko [147] and then Rödl and Sidorenko [144] ruined all our hopes. One of our main results was in a somewhat simplified form:

Theorem 15.1. *Let $q = 1$ and $\mathcal{L}$ be a given family of excluded digraphs. Then there exists a matrix $A = (a_{ij})$ of r rows and columns, depending only on $\mathcal{L}$, such that there exists a sequence $(\mathbf{S}_n)$ of asymptotically extremal graphs for $\mathcal{L}$ whose vertex-set V can be partitioned into $V_1, \dots, V_r$ so that for $1 \le i < j \le r$, a $v \in V_i$ is joined to a $v' \in V_j$ by an arc of this direction, iff the corresponding matrix-element $a_{ij} = 2$; further, the subdigraphs spanned by the V_i's are either independent sets or tournaments, depending on whether $a_{ii} = 0$ or 1.*

One crucial tool in our research was a **density** notion for matrices. We associated with every matrix A a quadratic form and maximized it over the standard simplex:

$$g(A) = \max \left\{ uAu^T \ : \ \sum u_i = 1, \ u_i \ge 0. \right\}$$

The matrices are used to characterize some generalizations of graph sequences like $(T_{n,p})_{n>n_0}$ of the general theory for ordinary graphs, and $g(A)$ measures the edge-density of these structures: replaces $(1-\frac{1}{p})$ of the Erdős–Stone–Simonovits theorem.

Definition 15.2. *A matrix A is called* **dense** *if for every submatrix B' of symmetric position $g(B') < g(B)$. In other words, B is minimal for $g(B) = \lambda$.*

We conjectured that – as described below – the numbers $g(B)$ are of **finite multiplicity** and **well ordered** if the matrices are dense:

Conjecture 15.3. *If q is fixed, then for each λ there are only finitely many dense matrices B with $g(B) = \lambda$. Further, if (B_n) is a sequence of matrices of bounded integer entries then $(g(B_n))$ cannot be strictly monotone decreasing.*

One could wonder how one arrives at such conjectures, but we do not have the space to explain that here. Similar matrices (actually, multigraph extremal problems) occur when one attacks Turán–Ramsey problems, see [69], [70], [71].

Our conjecture was disproved by Sidorenko and Rödl [144]. As a consequence, while we feel that the case $q = 1$ (i.e. the case of digraphs where any two points can be joined only by one arc of each direction) is sufficiently well described, for $q > 1$ the problem today seems to be fairly hopeless. Multidigraphs have also been considered by Katona in [113], where he was primarily interested in continuous versions of Turán-type extremal problems.

16. Erdős and Nassredin

Let me finish this paper with an anecdote. Nassredin, the hero of many middle-east jokes, stories (at least this is how we know it in Budapest), once met his friends who were eager to listen to his speech. "Do you know what I wish to speak about" Nassredin asked them. "No, we don't" they answered. "Then why should I speak about it" said Nassredin and left.[14] Next time the friends really wanted to listen to the clever and entertaining Nassredin. So, when Nassredin asked the audience "Do you know what I want to speak about", they answered: "YES, we do" . "Then why should I speak about it" said Nassredin again and went home. The third time the audience decided to be more clever. When Nassredin asked them "Do you know what I will speak about", half of the people said "YES" the other half said "NO". Nassredin probably was lasy to speak: "Those who know what I wanted to tell you should tell it to the others" he said and left again.

I am in some sense in Nassredin's shoes. How could I explain on 30 or 50 pages the influence of Erdős on Extremal Graph Theory to people who do not know it. And why should I explain to those who know it. Yet I think, Nassredin did not behave in the most appropriate way. So I tried – as I promised – to illustrate on some examples this enourmous influence of Paul. I do not think it covered half the topics and I have not tried to be too systematic.

Long Live Paul Erdős!

References

1. M. Ajtai, P. Erdős, J. Komlós and E. Szemerédi: On Turán's theorem for sparse graphs, Combinatorica, **1**(4) (1981) 313–317.
2. N. Alon: Tools from Higher Algebra, Chapter 32 of Handbooks of Combinatorics (ed. Graham, Lovász, Grötschel), 1995.
3. N. Alon: Eigenvalues and expanders, Combinatorica **6,** (1986) 83–96.
4. N. Alon, R. Duke, H. Leffman, V. Rödl and R. Yuster: Algorithmic aspects of the regularity lemma, FOCS, **33** (1993) 479–481, Journal of Algorithms, **16**(1) (1994) 80–109.
5. N. Alon, S. Friedland and A. Kalai: Regular subgraphs of almost regular graphs, J. Combinatorial Theory, Series B **37** (1984) 79–91.
6. N. Alon and J. Spencer: The Probabilistic Method, Wiley Interscience, 1992.

[14] I am not saying I follow his logic, but this is how the story goes.

7. C. Benson: Minimal regular graphs of girth eight and twelve, Canad. J. Math., **18** (1966) 1091–1094.
8. C. Berge and M. Simonovits: The colouring numbers of the direct product of two hypergraphs, Lecture Notes in Math., **411**, Hypergraph Seminar, Columbus, Ohio 1972 (1974) 21–33.
9. B. Bolloás: Graphs without two independent cycles (in Hungarian), Mat. Lapok, **14** (1963) 311–321.
10. B. Bollobás: Extremal Graph Theory, Academic Press, London, (1978).
11. B. Bollobás: Relations between sets of complete subgraphs, Proc. Fifth British Combinatorial Conf. Aberdeen, (1975) 79–84.
12. B. Bollobás: On complete subgraphs of different orders, Math. Proc. Cambridge Philos. Soc., **79** (1976) 19–24.
13. B. Bollobás: Random Graphs, Academic Press, (1985).
14. B. Bollobás: Extremal graph theory with emphasis on probabilistic methods, Conference Board of Mathematical Sciences, Regional Conference Series in Math, No62, AMS (1986).
15. B. Bollobás and P. Erdős: On the structure of edge graphs, Bull. London Math. Soc., **5** (1973) 317–321.
16. B. Bollobás and P. Erdős: On a Ramsey-Turán type problem, Journal of Combinatorial Theory, (B) **21** (1976) 166–168.
17. B. Bollobás, P. Erdős and M. Simonovits: On the structure of edge graphs II, Journal of London Math. Soc. (2), **12** (1976) 219–224.
18. B. Bollobás and Y. Kohayakawa: An extension of the Erdős–Stone theorem, Combinatorica, **14**(3) (1994) 279–286.
19. B. Bollobás and A. Thomason: Large dense neighbourhoods in Turán's theorem, Journal of Combinatorial Theory, (B) **31** (1981) 111–114.
20. B. Bollobás and A. Thomason: Topological subgraphs, European Journal of Combinatorics.
21. J. A. Bondy, Basic Graph Theory: Paths and Circuits, Manuscript (1992), to appear as a chapter in Handbook of Combinatorics, (eds Graham, Grötschel, Lovász)
22. J. A. Bondy and R. C. Entringer: Longest cycles in 2-connected graphs with prescribed maximum degree, Can. J. Math, **32** (1980) 987–992.
23. J. A. Bondy and M. Simonovits: Cycles of even length in graphs, Journal of Combinatorial Theory, **16**B (2) April (1974) 97–105.
24. J. A. Bondy and M. Simonovits: Longest cycles in 3-connected 3-regular graphs, Canadian Journal Math. XXXII (4) (1980) 987–992.
25. W. G. Brown: On graphs that do not contain a Thomsen graph, Canad. Math. Bull., **9** (1966) 281–285.
26. W. G. Brown, P. Erdős and M. Simonovits: Extremal problems for directed graphs, Journal of Combinatorial Theory, B **15**(1) (1973) 77–93.
27. W. G. Brown, P. Erdős and M. Simonovits: On multigraph extremal problems, Problèmes Combin. et Theorie des graphes, (ed. J. Bermond et al.), Proc. Conf. Orsay 1976 (1978) 63–66.
28. W. G. Brown, P. Erdős, and M. Simonovits: Inverse extremal digraph problems, Proc. Colloq. Math. Soc. János Bolyai **37** Finite and Infinite Sets, Eger (Hungary) 1981 Akad. Kiadó, Budapest (1985) 119–156.
29. W. G. Brown, P. Erdős and M. Simonovits: Algorithmic Solution of Extremal digraph Problems, Transactions of the American Math Soc., **292/2** (1985) 421–449.
30. W. G. Brown and F. Harary: Extremal digraphs, Combinatorial theory and its applications, Colloq. Math. Soc. J. Bolyai, **4** (1970) I. 135–198; MR 45 #8576.

31. W. G. Brown and M. Simonovits: Digraph extremal problems, hypergraph extremal problems, and densities of graph structures. Discrete Mathematics, **48** (1984) 147–162.
32. De Caen and L. Székely: The maximum edge number of 4-and 6-cycle free bipartite graphs, Sets, Graphs, and Numbers, Proc. Colloq. Math. Soc. János Bolyai **60** (1992) 135–142.
33. F. R. K. Chung: Regularity lemmas for hypergraphs and quasi-randomness, Random Structures and Algorithms, Vol. **2**(2) (1991) 241–252.
34. F. R. K. Chung and R. L. Graham: Quasi-random hypergraphs, Random Structures and Algorithms, **1** (1990) 105–124.
35. F. K. Chung, R. L. Graham and R. M. Wilson: Quasi-random graphs, Combinatorica, **9**(4) (1989) 345–362.
36. V. Chvátal: On finite polarized partition relations, Canad. Math. Bull., **12** (1969) 321–326.
37. V. Chvátal and E. Szemerédi: On the Erdős-Stone theorem, Journal of the London Mathematics Society, Ser. 2, **23** (1981) 207–214.
38. K. Clarkson, H. Edelsbrunner, L. Guibas, M. Sharir and E. Welzl: Combinatorial complexity bounds for arrangements of curves and spheres, Discrete Computational Geometry, **55** (1990) 99–160.
39. G. Dirac: Some theorems on abstract graphs, Proc. London Math. Soc. (3), 2 (1952), 69–81.
40. G. Dirac: Extensions of Turán's theorem on graphs, Acta Math., **14** (1963) 417–422.
41. C. S. Edwards: Complete subgraphs with degree-sum of vertex-degrees, Combinatorics, Proc. Colloq. Math. Soc. János Bolyai **18** (1976), 293–306.
42. C. S. Edwards: The largest degree-sum for a triangle in a graph, Bull. London Math. Soc., **9** (1977) 203–208.
43. P. Erdős: On sequences of integers no one of which divides the product of two others and related problems, Mitt. Forsch. Institut Mat. und Mech. Tomsk **2** (1938) 74–82.
44. P. Erdős: Neue Aufgaben 250. Elemente der Math., **10** (1955) p114.
45. P. Erdős: Graph Theory and Probability, Canad. Journal of Math., **11** (1959) 34–38.
46. P. Erdős: Graph Theory and Probability, II. Canad. Journal of Math., **13** (1961) 346–352.
47. P. Erdős: On a theorem of Rademacher-Turán, Illinois J. Math., **6** (1962) 122–127. (Reprinted in [57].)
48. P. Erdős: Über ein Extremalproblem in Graphentheorie, Arch. Math. (Basel), **13** (1962) 222–227.
49. P. Erdős: On a problem in graph theory, Math. Gazette, **47** (1963) 220–223.
50. P. Erdős: On extremal problems of graphs and generalized graphs, Israel J. Math, **2**(3) (1964) 183–190.
51. P. Erdős: Extremal problems in graph theory, Theory of Graphs and its Appl., (M. Fiedler ed.) Proc. Symp. Smolenice, 1963), Acad. Press, NY (1965) 29–36.
52. Erdős: Some recent results on extremal problems in graph theory (Results), Theory of Graphs (International Symposium, Rome, 1966), Gordon and Breach, New York and Dunod, Paris, (1967), 117–130, MR 37, #2634.
53. P. Erdős: On bipartite subgraphs of graphs (in Hungarian), Matematikai Lapok, (1967) pp283–288.
54. P. Erdős: On some new inequalities concerning extremal properties of graphs, Theory of Graphs, Proc. Coll. Tihany, Hungary (eds. P. Erdős and G. Katona) Acad. Press. N. Y. (1968) 77–81.

55. P. Erdős: On some applications of graph theory to number theoretic problems, *Publ. Ramanujan Inst.*, **1** (1969) 131–136. (Sharpness of [43].)
56. P. Erdős: On some extremal problems on r-graphs, Discrete Mathematics **1**(1), (1971) 1–6.
57. P. Erdős: The Art of Counting, Selected Writings (in Combinatorics and Graph Theory, ed. J. Spencer), The MIT Press, Cambridge, Mass., (1973) MR**58**#27144
58. P. Erdős: On the number of triangles contained in certain graphs, Canad. Math. Bull., **7**(1) January, (1974) 53–56.
59. P. Erdős: Problems and results in combinatorial analysis, Theorie Combinatorie, Proc. Conf. held at Rome, 1973, Roma, Acad. Nazionale dei Lincei (1976) 3–17.
60. P. Erdős: Paul Turán 1910–1976: His work in graph theory, J. Graph Theory, **1** (1977) 96–101.
61. P. Erdős: On the combinatorial problems which I would most like to see solved, Combinatorica **1,** (1981), 25–42.
62. P. Erdős: On some of my favourite problems in various branches of combinatorics, Fourth Czechoslovakian Symposium on Combinatorics, Graphs and Complexity, J. Nesetril, M Fiedler (editors) (1992), Elsevier Science Publisher B. V.
63. P. Erdős, R. Faudree, and E. Győri: On the booksize of graphs with large minimum degree, Studia. Sci. Math. Hungar., **30** (1995) 1–2.
64. P. Erdős, R. Faudree, J. Pach, J. Spencer: How to make a graph bipartite, Journal of Combinatorial Theory, (B) **45** (1988) 86–98.
65. P. Erdős, P. Frankl and V. Rödl: The asymptotic number of graphs not containing a fixed subgraph and a problem for hypergraphs having no exponent, Graphs and Combinatorics, **2** (1986) 113–121.
66. P. Erdős and T. Gallai: On maximal paths and circuits of graphs, Acta Math. Acad. Sci. Hungar., **10** (1959) 337–356.
67. P. Erdős, E. Győri and M. Simonovits: How many edges should be deleted to make a triangle-free graph bipartite, Sets, Graphs, and Numbers, Proc. Colloq. Math. Soc. János Bolyai **60** 239–263.
68. P. Erdős, A. Hajnal: On complete topological subgraphs of certain graphs, Annales Univ. Sci. Budapest, **7** (1964) 143–149. (Reprinted in [57].)
69. P. Erdős, A. Hajnal, V. T. Sós, E. Szemerédi: More results on Ramsey-Turán type problems, Combinatorica, **3**(1) (1983) 69–82.
70. P. Erdős, A. Hajnal, M. Simonovits, V. T. Sós, and E. Szemerédi: Turán-Ramsey theorems and simple asymptotically extremal structures, Combinatorica, **13** (1993) 31–56.
71. P. Erdős, A. Hajnal, M. Simonovits, V. T. Sós and E. Szemerédi: Turán-Ramsey theorems for K_p–stability numbers, Combinatorics, Probability and Computing, **3** (1994) 297–325. (Proc. Cambridge Conf on the occasion of 80th birthday of P. Erdős, 1994.)
72. P. Erdős, D. J. Kleitman and B. L. Rothschild: Asymptotic enumeration of K_n–free graphs, Theorie Combinatorie, Proc. Conf. held at Rome, 1973, Roma, Acad. Nazionale dei Lincei 1976, vol II, 19–27.
73. P. Erdős, A. Meir, V. T. Sós and P. Turán: On some applications of graph theory I. Discrete Math., **2** (1972) (3) 207–228.
74. P. Erdős, A. Meir, V. T. Sós and P. Turán: On some applications of graph theory II. Studies in Pure Mathematics (presented to R. Rado) 89–99, Academic Press, London, 1971.
75. P. Erdős, A. Meir, V. T. Sós and P. Turán: On some applications of graph theory III. Canadian Math. Bulletin, **15** (1972) 27–32.

76. P. Erdős and L. Pósa: On independent circuits contained in a graph, Canadian J. Math., **17** (1965) 347–352. (Reprinted in [57].)
77. Erdős and A. Rényi: On the evolution of random graphs, Magyar Tud. Akad. Mat. Kut. Int. Közl., **5** (1960) 17–65. (Reprinted in [89] and in [57].)
78. P. Erdős, A. Rényi and V. T. Sós: On a problem of graph theory, Studia Sci. Math. Hung., **1** (1966) 215–235.
79. P. Erdős and H. Sachs: Reguläre Graphen gegebener taillenweite mit minimalen Knotenzahl, Wiss. Z. Univ. Halle-Wittenberg, Math-Nat. R. 12(1963) 251–258.
80. P. Erdős, A. Sárközy and V. T. Sós: On product representation of powers, I, Reprint of the Mathematical Inst. of Hung. Acad. Sci, 12/1993., submitted to European Journal of Combinatorics.
81. P. Erdős and M. Simonovits: A limit theorem in graph theory, Studia Sci. Math. Hungar., **1** (1966) 51–57. (Reprinted in [57].)
82. P. Erdős and M. Simonovits: An extremal graph problem, Acta Math. Acad. Sci. Hung., **22**(3–4) (1971) 275–282.
83. P. Erdős and M. Simonovits: Some extremal problems in graph theory, Combinatorial Theory and its Appl., Proc. Colloq. Math. Soc. János Bolyai **4** (1969) 377–384. (Reprinted in [57].)
84. P. Erdős and M. Simonovits: Compactness results in extremal graph theory, Combinatorica, **2**(3) (1982) 275–288.
85. P. Erdős and M. Simonovits: Supersaturated graphs and hypergraphs, Combinatorica, **3**(2) (1983) 181–192.
86. P. Erdős, V. T. Sós: Some remarks on Ramsey's and Turán's theorems, Combin. Theory and Appl. (P. Erdős et al eds) Proc. Colloq. Math. Soc. János Bolyai **4** Balatonfüred (1969), 395–404.
87. P. Erdős and V. T. Sós: On Ramsey-Turán type theorems for hypergraphs Combinatorica **2,** (3) (1982) 289–295.
88. P. Erdős, V. T. Sós: Problems and results on Ramsey–Turán type theorems Proc. Conf. on Comb., Graph Theory, and Computing, Congr. Num., **26** 17–23.
89. P. Erdős and J. Spencer: Probabilistic Methods in Combinatorics, Acad. Press, NY, 1974; MR52 #2895.
90. P. Erdős and A. H. Stone: On the structure of linear graphs, Bull. Amer. Math. Soc., **52** (1946) 1089–1091.
91. P. Erdős and G. Szekeres: A combinatorial problem in geometry, Compositio Math., **2** (1935) 463–470.
92. R. J. Faudree and R. H. Schelp: Ramsey type results, Proc. Colloq. Math. Soc. János Bolyai **10** Infinite and Finite Sets, Keszthely, 1973, 657–665.
93. R. J. Faudree and M. Simonovits: On a class of degenerate extremal graph problems, Combinatorica, **3**(1) (1983) 83–93.
94. R. J. Faudree and M. Simonovits: Ramsey problems and their connection to Turán type extremal problems, Journal of Graph Theory, Vol **16**(1) (1992) 25–50.
95. P. Frankl and V. Rödl: Hypergraphs do not jump, Combinatorica, **4**(4) (1984).
96. P. Frankl and V. Rödl: The Uniformity lemma for hypergraphs, Graphs and Combinatorics, **8**(4) (1992) 309–312.
97. P. Frankl and R. M. Wilson: Intersection theorems with geometric consequences, Combinatorica **1,** (4) (1981) 357–368.
98. Z. Füredi: Graphs without quadrilaterals, Journal of Combinatorial Theory, (B) **34** (1983) 187–190.
99. Z. Füredi: Turán type problems, Surveys in Combinatorics, (A. D. Keedwell, ed.) Cambridge Univ. Press, London Math. Soc. Lecture Note Series, **166** (1991) 253–300.

100. Z. Füredi: On a Turán type problem of Erdős, Combinatorica, **11**(1) (1991) 75–79.
101. Z. Füredi: New asymptotics for bipartite Turán numbers, submitted to Journal of Combinatorial Theory, (B) (see also the abstracts of invited lectures at ICM Zürich, 1994)
102. Z. Füredi and Á. Seress: Maximal triangle-free graphs with restrictions on the degrees, Journal of Graph Theory, **18**(1) 11–24.
103. R. K. Guy: A problem of Zarankiewicz, Proc. Coll. Theory of Graphs, (Tihany, 1966), (eds: Erdős, Katona) Akad. Kiadó, Budapest, 1968, 119–150.
104. R. K. Guy and S. Znám: A problem of Zarankiewicz, Recent Progress in Combinatorics, (eds: J. A. Bondy, R. Murty) Academic Press, New York, 1969, 237–243.
105. A. Gyárfás, J. Komlós and E. Szemerédi: On the distribution of cycle length in graphs, Journal of Graph Theory, **8**(4) (1984) 441–462.
106. E. Győri: On the number of edge-disjoint triangles in graphs of given size, Proc. Colloq. Math. Soc. János Bolyai **52** 7th Hungarian Combinatorial Coll. (Eger) North Holland (1987) 267–276.
107. E. Győri: On the number of edge-disjoint cliques in graphs, Combinatorica **11,** (1991) 231–243.
108. E. Győri: Edge-disjoint cliques in graphs, Proc. Colloq. Math. Soc. János Bolyai **60** (Proc. Coll. dedicated to the 60th birthday of A. Hajnal and V. T. Sós, Budapest, 1991), 357–363.
109. B. Jackson: Longest cycles in 3-connected cubic graphs, Journal of Combinatorial Theory, (B) **41** (1986) 17–26.
110. B. Jackson and T. D. Parson: On r–regular, r–connected non-hamiltonian graphs, Bull. Australian Math. Soc., **24** (1981) 205–220.
111. B. Jackson and T. D. Parson: Longest cycles in r–regular r–connected graphs, Journal of Combinatorial Theory, (B) **32** (3) (1982) 231–245.
112. B. Jackson, N. C. Wormald: Longest cycles in 3-connected graphs of bounded maximum degree, *Graphs, Matrices, Designs*, Lecture Notes in Pure and Applied Math, Marcel Dekker Inc., (1993) 237–254.
113. Gy. Katona: Continuous versions of some extremal hypergraph problems, Proc. Colloq. Math. Soc. János Bolyai **18** Combinatorics, (Keszthely, 1976) II. 653–678, MR 80e#05071.
114. G. Katona, T. Nemetz and M. Simonovits: A new proof of a theorem of P. Turán and some remarks on a generalization of it, (In Hungarian), Mat. Lapok, XV. 1–3 (1964) 228–238.
115. D. J. Kleitman and K. J. Winston: On the number of graphs without 4-cycles, Discrete Mathematics **41,** (1982), 167–172.
116. Y. Kohayakawa: The Regularity Lemma of Szemerédi for Sparse Graphs, (Manuscript, August, 1993)
117. J. Komlós and M. Simonovits: Szemerédi regularity lemma and its application in graph theory, to appear in Paul Erdős is 80, Proc. Colloq. Math. Soc. János Bolyai (1995)
118. J. Komlós, V. T. Sós: Regular subgraphs in graphs, manuscript, (1992)
119. J. Komlós and E. Szemerédi: Topological cliques in graphs, Combinatorics, Probability and Computing 3 (1994), 247-256.
120. J. Komlós and E. Szemerédi: Topological cliques in graphs II, Combinatorics, Probability and Computing, to appear.
121. T. Kővári, V. T. Sós, P. Turán: On a problem of Zarankiewicz, Colloq. Math., **3** (1954), 50–57.

122. F. Lazebnik and V. A. Ustimenko: New examples of graphs without small cycles and of large size, European Journal of Combinatorics, **14**(5) (1993) 445–460.
123. F. Lazebnik, V. A. Ustimenko, and A. J. Woldar: New constructions of bipartite graphs on m, n vertices with many edges and without small cycles, to appear in JCT(B).
124. L. Lovász and M. Simonovits: On the number of complete subgraphs of a graph I. Proc. Fifth British Combin. Conf. Aberdeen (1975) 431–442.
125. L. Lovász and M. Simonovits: On the number of complete subgraphs of a graph II. Studies in Pure Math (dedicated to the memory of P. Turán), Akadémiai Kiadó+Birkhäuser Verlag, (1983) 459–495.
126. A. Lubotzky, R. Phillips, and P. Sarnak: Ramanujan Conjecture and explicite construction of expanders, (Extended Abstract), Proc. STOC 1986, 240–246
127. A. Lubotzky, R. Phillips, and P. Sarnak: Ramanujan graphs, Combinatorica, 8(3) 1988, 261–277.
128. W. Mader: Hinreichende Bedingungen für die Existenz von Teilgraphen die zu einem vollständigen Graphen homöomorph sind, Math. Nachr. **53** (1972), 145–150.
129. W. Mader: Homomorphieeigenschaften und mittlere Kantendichte von Graphen, Math. Annalen **174** (1967), 265–268.
130. W. Mantel: Problem 28, Wiskundige Opgaven, **10** (1907) 60–61.
131. G. A. Margulis: Explicit constructions of graphs without short cycles and low density codes, Combinatorica, **2**(1) (1982) 71–78.
132. G. A. Margulis: Arithmetic groups and graphs without short cycles, 6th Internat. Symp. on Information Theory, Tashkent 1984, Abstracts, Vol. 1, 123–125 (in Russian).
133. G. A. Margulis: Some new constructions of low-density parity-check codes, convolution codes and multi-user communication, 3rd Internat. Seminar on Information Theory, Sochi (1987), 275–279 (in Russian)
134. G. A. Margulis: Explicit group theoretic construction of group theoretic schemes and their applications for the construction of expanders and concentrators, Journal of Problems of Information Transmission, 1988 pp 39–46 (translation from Problemy Peredachi Informatsii, **24**(1) 51–60 (January–March 1988)
135. J. W. Moon: On independent complete subgraphs in a graph, Canad. J. Math., **20** (1968) 95–102. also in: International Congress of Math. Moscow, (1966), vol 13.
136. J. W. Moon and Leo Moser: On a problem of Turán, MTA, Mat. Kutató Int. Közl., **7** (1962) 283–286.
137. J. Pach and P. Agarwal: Combinatorial Geometry, Courant Institute Lecture Notes, New York University, 1991 (200 pages) to appear in ...
138. H. J. Prömel: Asymptotic enumeration of ℓ-colorable graphs, TR Forshungsinstitute für Diskrete Mathematik, (1988) Bonn, Report No 88???-OR
139. H. J. Prömel and A. Steger: Excluding induced subgraphs: Quadrilaterals, Random Structures and Algorithms, **2**(1) (1991) 55–71.
140. H. J. Prömel and A. Steger: Excluding induced subgraphs II TR Forshungsinstitute für Diskrete Mathematik, (1990) Bonn, Report No 90642-OR to appear in Discrete Applied Math.
141. H. J. Prömel and A. Steger: Excluding induced subgraphs III A general asymptotics, Random Structures and Algorithms, **3**(1) (1992) 19–31.
142. L. Pyber, Regular subgraphs of dense graphs, Combinatorica, **5**(4) (1985) 347–349.
143. F. P. Ramsey: On a problem of formal logic, Proc. London Math. Soc. 2nd Series, **30** (1930) 264–286.

144. V. Rödl and A. Sidorenko: On jumping constant conjecture for multigraphs, (Manuscript, 1993)
145. I. Z. Ruzsa and E. Szemerédi: Triple systems with no six points carrying three triangles, *Combinatorics* (Keszthely, 1976), (1978), **18**, Vol. II., 939–945. North-Holland, Amsterdam-New York.
146. G. N. Sárközy: Cycles in bipartite graphs and an application in number theory, (1993), TR DIMACS forthcoming in Journal of Graph Theory, ( )
147. A. F. Sidorenko: Boundedness of optimal matrices in extremal multigraph and digraph problems, Combinatorica, **13**(1) (1993) 109–120.
148. A. F. Sidorenko: Extremal estimates of probability measures and their combinatorial nature Math. USSR - Izv 20 (1983) N3 503–533 MR 84d: 60031. (=Translation) Original: Izvest. Acad. Nauk SSSR. ser. matem. 46(1982) N3 535–568.
149. A. F. Sidorenko: What do we know and what we do not know about Turán Numbers, Manuscript, (submitted to Graphs and Combinatorics, 1992, Febr.)
150. M. Simonovits: A method for solving extremal problems in graph theory, Theory of graphs, Proc. Coll. Tihany, (1966), (Ed. P. Erdős and G. Katona) Acad. Press, N.Y., (1968) 279–319.
151. M. Simonovits: A new proof and generalizations of a theorem of Erdős and Pósa on graphs without $k+1$ independent circuits, Acta Math. Acad. Sci. Hungar., **18**(1–2) (1967) 191–206.
152. M. Simonovits: The extremal graph problem of the icosahedron, Journal of Combinatorial Theory, **17**B (1974) 69–79.
153. M. Simonovits: Extremal graph problems with symmetrical extremal graphs, additional chromatic conditions, Discrete Math., **7** (1974) 349–376.
154. M. Simonovits: On Paul Turán's influence on graph theory, J. Graph Theory, **1** (1977) 102–116.
155. M. Simonovits: Extremal graph problems and graph products, Studies in Pure Math. (dedicated to the memory of P. Turán) Akadémiai Kiadó+Birkhauser Verlag (1982)
156. M. Simonovits: Extremal Graph Theory, Selected Topics in Graph Theory, (ed. by Beineke and Wilson) Academic Press, London, New York, San Francisco, 161–200. (1983)
157. M. Simonovits: Extremal graph problems, Degenerate extremal problems and Supersaturated graphs, Progress in Graph Theory (Acad Press, ed. Bondy and Murty) (1984) 419–437.
158. M. Simonovits and V. T. Sós: Szemerédi's partition and quasi-randomness, Random Structures and Algorithms, Vol **2**, No. 1 (1991) 1–10.
159. R. Singleton: On minimal graphs of maximum even girth, Journal of Combinatorial Theory **1** (1966), 306–332.
160. V. T. Sós: On extremal problems in graph theory, Proc. Calgary International Conf. on Combinatorial Structures and their Application, (1969) 407–410.
161. V. T. Sós: Some remarks on the connection between graph-theory, finite geometry and block designs Theorie Combinatorie, Acc. Naz.dei Lincei (1976) 223–233
162. V. T. Sós: Survey on Turán–Ramsey Problems, manuscript, to be published.
163. J. Spencer, E. Szemerédi and W. T. Trotter: Unit distances in the Euclidean plane, Graph Theory and Cominatorics, Proc. Cambridge Combin. Conf. (ed B. Bollobás) Academic Press (1983) 293–304.
164. F. Sterboul: A class of extremal problems, Recent Advances in Graph Theory, Proc. Conf. Praga, 1974, 493–499.

165. F. Sterboul: On the chromatic number of the direct product of hypergraphs, Lecture Notes in Math., **411**, Hypergraph Seminar, Columbus, Ohio 1972 (1974),
166. T. Szele: Combinatorial investigations, Matematikai és Physikai Lapok, **50** (1943) 223–256.
167. E. Szemerédi: On a set containing no k elements in an arithmetic progression, Acta Arithmetica, **27** (1975) 199–245.
168. E. Szemerédi: On regular partitions of graphs, Problemes Combinatoires et Théorie des Graphes (ed. J. Bermond et al.), CNRS Paris, 1978, 399–401.
169. E. Szemerédi: On graphs containing no complete subgraphs with 4 vertices (in Hungarian) Mat. Lapok, **23** (1972) 111–116.
170. A. Thomason: Random graphs, strongly regular graphs and pseudo-random graphs, in Surveys in Combinatorics, 1987 (Whitehead, ed.) LMS Lecture Notes Series 123, Cambridge Univ. Press, Cambridge, 1987, 173–196
171. A. Thomason: Pseudo-random graphs, in Proceedings of Random graphs, Poznan, 1985, (M. Karonski, ed.), Annals of Discrete Math., **33** (1987) 307–331.
172. Collected papers of Paul Turán: Akadémiai Kiadó, Budapest, 1989. Vol 1–3, (with comments of Simonovits on Turán's graph theorem).
173. P. Turán: On an extremal problem in graph theory, Matematikai Lapok, **48** (1941) 436–452 (in Hungarian), (see also [174], [172]).
174. P. Turán: On the theory of graphs, Colloq. Math., **3** (1954) 19–30, (see also [172]).
175. P. Turán, Applications of graph theory to geometry and potential theory, Proc. Calgary International Conf. on Combinatorial Structures and their Application, (1969) 423–434 (see also [172]).
176. P. Turán: A Note of Welcome, Journal of Graph Theory, **1** (1977) 7–9.
177. H.-J, Voss: Cycles and bridges in graphs, Deutscher Verlag der Wissenschaften, Berlin, Kluwer Academic Publisher, (1991)
178. H. Walter and H.-J, Voss: Über Kreise in Graphen, (Cycles in graphs, book, in German) VEB Deutscher Verlag der Wissenschaften, Berlin, 1974.
179. R. Wenger: Extremal graphs with no C^4, C^6 and C^{10}, Journal of Combinatorial Theory, (B) **52** (1991) p.113–116.
180. K. Zarankiewicz: Problem P101, Colloq. Math, **2** (1951) 301.
181. A. A. Zykov: On some properties of linear complexes, Mat Sbornik, **24** (1949) 163–188, Amer. Math. Soc. Translations, **79** 1952.

Ramsey Theory in the Work of Paul Erdős

R. L. Graham[1] and J. Nešetřil[2]

[1] AT&T Bell Laboratories, Murray Hill, NJ 07974
[2] Charles University, Praha, The Czech Republic

Summary. Ramsey's theorem was not discovered by P. Erdős. But perhaps one could say that Ramsey theory was created largely by him. This paper will attempt to demonstrate this claim.

1. Introduction

Ramsey's theorem was not discovered by Paul Erdős. This was barely technically possible: Ramsey proved his theorem in 1928 (or 1930, depending on the quoted source) and this is well before the earliest Erdős publication in 1932. He was then 19. At such an early age four years makes a big difference. And also at this time Erdős was not even predominantly active in combinatorics. The absolute majority of the earliest publications of Erdős is devoted to number theory, as can be seen from the following table:

	1932	1933	1934	1935	1936	1937	1938	1939
all papers	2	0	5	10	11	10	13	13
number theory	2	0	5	9	10	10	12	13

The three combinatorial exceptions among his first 8^2 papers published in 8 years are 2 papers on infinite Eulerian graphs and the paper [1] by Erdős and G. Szekeres. Thus, the very young P. Erdős was not a driving force of the development of Ramsey theory or Ramsey-type theorems in the thirties. That position should be reserved for Issac Schur who not only proved his sum theorem [2] in 1916 but, as it appears now [3], also conjectured van der Waerden's theorem [4], proved an important extension, and thus put it into a context which inspired his student R. Rado to completely settle (in 1933) the question of monochromatic solutions of linear equations [5]. This result stands apart even after 60 years.

Yet, in retrospect, it is fair to say that P. Erdős was responsible for the continuously growing popularity of the field. Ever since his pioneering work in the thirties he proved, conjectured and asked seminal questions which together, some 40 to 50 years later, formed Ramsey theory. And for Erdős, Ramsey theory was a constant source of problems which motivated some of the key pieces of his combinatorial research.

It is the purpose of this note to partially justify these claims, using a few examples of Erdős' activity in Ramsey theory which we will discuss from a contemporary point of view.

In the first section we cover paper [1] and later development in a great detail. In Section 2, we consider the development based on Erdős' work related to bounds on various Ramsey functions. Finally, in Section 3 we consider his work related to structural extensions of Ramsey's theorem.

No mention will be made of his work on infinite extensions of Ramsey's theorem. This is covered in these volumes by the comprehensive paper of A. Hajnal.

2. The Erdős-Szekeres Theorem

F. P. Ramsey discovered his theorem [6] in a sound mathematical context (of the decision problem for a class of first-order formulas). But since the time of Dirichlet the "Schubfach principle" and its extensions and variations played a distinguished role in mathematics. The same holds for the other early contributions of Hilbert [19], Schur [2] and van der Waerden [4].

Perhaps because of this context Ramsey's theorem was never regarded as a puzzle and/or a combinatorial curiosity only. Thanks to Erdős and Szekeres [1] the theorem found an early application in a quite different context, namely, plane geometry:

Theorem 2.1 **([1]).** *Let n be a positive integer. Then there exists a least integer $N(n)$ with the following property: If X is a set of $N(n)$ points in the plane in general position (i.e. no three of which are collinear) then X contains an n-tuple which forms the vertices of a convex n-gon.*

One should note that (like in Ramsey's original application in logic) this statement does not involve any coloring (or partition) and thus, by itself, fails to be of "Ramsey type". Rather it fits to a more philosophical description of Ramsey type statements as formulated by Mirsky:

> "There are numerous theorems in mathematics which assert, crudely speaking, that every system of a certain class possesses a large subsystem with a higher degree of organization than the original system."

It is perhaps noteworthy to list the main features of the paper. What a wealth of ideas it contains!

I. It is proved that $N(4) = 5$ and this is attributed to Mrs. E. Klein. This is tied to the social and intellectual climate in Budapest in the thirties which has been described both by Paul Erdős and Szekeres on several occasions (see e.g. [7]), and with names like the Happy End Theorem.

II. The following two questions related to statement of Theorem 2.1 are explicitly formulated:

(a) Does the number $N(n)$ exist for every n?
(b) If so, estimate the value of $N(n)$.

It is clear that the estimates were considered by Erdős from the very beginning. This is evident at several places in the article.

III. The first proof proves the existence of $N(n)$ by applying Ramsey's theorem for partitions of quadruples. It is proved that $N(n) \leq r(2, 4, \{5, n\})$. This is still a textbook argument. Another proof based on Ramsey's theorem for partitions of triples was found by A. Tarsi (see [8]). So far no proof has emerged which is based on the graph Ramsey theorem only.

IV. The authors give "a new proof of Ramsey's theorem which differs entirely from the previous ones and gives for $m_i(k, \ell)$ slightly smaller limits". Here $m_i(k, \ell)$ denotes the minimal value of $|X|$ such that every partition of i-element subsets of X into two classes, say α and β, each k-element contains an i-element subset of class α or each ℓ-element subset contains an i-element subset of class β.

Thus, $m_i(k, \ell)$ is the Ramsey number for 2-partitions of i-element subsets. These numbers are denoted today by $r(2, i, \{k, \ell\})$ or $r_i(k, \ell)$. The proof is close to the standard textbook proofs of Ramsey's theorem. Several times P. Erdős attributed it to G. Szekeres.

Erdős and Szekeres explicitly state that $(r_2(k{+}1,\ell{+}1) =)m_2(k{+}1,\ell{+}1) \leq \binom{k+\ell}{2}$ and this value remained for 50 years essentially the best available upper bound for graph Ramsey numbers until the recent (independent) improvements by Rödl and Thomason. The current best upper bound (for $k = \ell$) is [9] essentially

$$\binom{2k}{k}/\sqrt{k}\ .$$

V. It is not as well known that [1] contains yet another proof of the graph theoretic formulation of Ramsey's theorem (in the above notation, $i = 2$) which is stated for its particular simplicity. We reproduce its formulation here.

Theorem. *In an arbitrary graph let the maximum number of independent points be k; if the number of points is $N \geqq m(k,\ell)$ then three exists in our graph a complete graph of order ℓ.*

Proof. For $\ell = 1$, the theorem is trivial for any k, since the maximum number of independent points is k and if the number of points is $(k+1)$, there must be an edge (complete graph of order 1).

Now suppose the theorem proved for $(\ell - 1)$ with any k. Then at least $\frac{N-k}{k}$ edges start from one of the independent points. Hence if

$$\frac{N-k}{k} \geqq m(k,\ell-1)\ ,$$

i.e.,

$$N \geqq k \cdot m(k,\ell-1) + k\ ,$$

then, out of the end points of these edges we may select, in virtue of our induction hypothesis, a complete graph whose order is at least $(\ell - 1)$. As the points of this graph are connected with the same point, they form together a complete graph of order ℓ.

This indicates that Erdős and Szekeres were well aware of the novelty of the approach to Ramsey's theorem. Also this is the formulation of Ramsey's problem which motivated some of the key pieces of Erdős' research. First an early use of the averaging argument and then the formulation of Ramsey's theorem in a "high off-diagonal" form: If a graph G has a bounded clique number (for example, if it is triangle-free) then its independence number has to be large. The study of this phenomenon led Erdős so key papers [10], [11], [12] which will be discussed in the next section in greater detail.

VI. The paper [1] contains a second proof of Theorem 2.1. This is a more geometrical proof which yields a better bound

$$N(n) \leq \binom{2n-4}{n-2} + 1$$

and it is conjectured (based on the exact values of $N(n)$ for $n = 3,4,5$) that $N(n) = 2^{n-2} + 1$. This is still an unsolved problem. The second proof (which 50 years later very nicely fits to a computational geometry context) is based on yet another Ramsey-type result.

Theorem 2.2 (the ordered pigeon hole principle). *Let m, n be positive integers. Then every set of $(m-1)(n-1)+1$ distinct integers contains either a monotone increasing n-set or monotone decreasing m-set.*

The authors of [1] note that the same problem was considered by R. Rado. The stage has been set.

The ordered pigeon-hole principle has been generalized in many different directions (see e.g., [13], [14]).

All this is contained in this truly seminal paper. Viewed from a contemporary perspective, the Erdős-Szekeres paper did not solve any well-known problem at the time and did not contribute to Erdős' instant mathematical fame (as a number theorist). But the importance of the paper [1] for the later development of combinatorial mathematics cannot be overestimated. To illustrate this development is one of the aims of this paper.

Apart from the problem of a good estimation of the value of N there is a peculiar structural problem related to [1]:

Call a set $Y \subseteq X$ an *n-hole* in X if Y is the set of vertices of a convex n-gon which does not contain other points in X.

Problem. Does there always exist $N^*(n)$ such that if X is any set of at least $N^*(n)$ points in the plane in general position then X contains an n-hole.

It is easy to prove that $N^*(n)$ exists for $n \leq 5$ (see Harborth (1978) where these numbers are determined). Horton (1983) showed that $N^*(7)$ does not exist. Thus only the existence of $N^*(6)$ is an open problem (see [15], [16] for recent related problems).

3. Estimating Ramsey numbers

Today it seems that the first question in this area which one might be tempted to consider is the problem of determining the actual sizes of the sets which are guaranteed by Ramsey's theorem (and other Ramsey-type theorems). But one should try to resist this temptation since it is "well-known" that Ramsey numbers (of all sorts) are difficult to determine and even good asymptotic estimates are difficult to find.

It seems that these difficulties were known to both Erdős and Ramsey. But Erdős considered them very challenging and addressed this question in several of his key articles. In many cases his estimations obtained decades ago are still the best available. Not only that, his innovative techniques became standard and whole theories evolved from his key papers.

Here is a side comment which may partly explain this success: Erdős was certainly one of the first number theorists who took an interest in combinatorics in the contemporary sense (being preceded by isolated events, for example, V. Jarnik's work on the minimum spanning tree problem and the Steiner problem see e.g. [17], [18]. Incidentally Jarnik was one of the first coauthors of Erdős). Together with Turán, Erdős brought to the "slums of topology" not only his brilliance but also his expertise and "good taste". It is our opinion that these facts profoundly influenced further development of the whole field. Thus it is not perhaps surprising that if one would isolate a single feature of Erdős' contribution to Ramsey theory then it is perhaps his continuing emphasis on estimates of various Ramsey-related questions. From the large number of results and papers we decided to cover several key articles and comment on them from a contemporary point of view.

I. 1947 paper [10]. In a classically clear way, Erdős proved

$$2^{k/2} \leq r(k) < 4^k \tag{3.1}$$

for every $k \geq 3$.

His proof became one of the standard textbook examples of the power of the probabilistic method. (Another example perhaps being the strikingly simple proof of Shannon of the existence of exponentially complex Boolean functions.)

The paper [10] proceeds by stating (3.1) in an inverse form: Define $A(n)$ as the greatest integer such that given any graph G of n vertices, either it or its complementary graph contains a complete subgraph of order $A(n)$. Then for $A(n) \geq 3$,

$$\frac{\log n}{2\log 2} < A(n) < \frac{2\log n}{\log 2} .$$

Despite considerable efforts over many years, these bounds have been improved only slightly (see [9], [20]). We commented on the upper bound improvements above. The best current lower bound is

$$r(n) \geq (1+O(1))\frac{\sqrt{2}e}{n}2^{n/2}$$

which is twice the Erdős bound (when computed from his proof).

The paper [10] was one of 23 papers which Erdős published within 3 years in the *Bull. Amer. Math. Soc.* and already here it is mentioned that although the upper bound for $r(3,n)$ is quadratic, the present proof does not yield a nonlinear lower bound. That had to wait for another 10 years.

II. The 1958 paper [11] — Graph theory and probability. The main result of this paper deals with graphs, circuits, and chromatic number and as such does not seem to have much to do with Ramsey theory.

Yet the paper starts with the review of bounds for $r(k,k)$ and $r(3,k)$ (all due to Erdős and Szekeres). Ramsey numbers are denoted as in most older Erdős papers by symbols of $f(k)$, $f(3,k)$, $g(k)$. He then defines analogously the function $h(k,\ell)$ as "the least integer so that every graph of $h(k,\ell)$ vertices contains either a closed circuit of k or fewer lines or the graph contains a set of ℓ independent points. Clearly $h(3,\ell)=f(3,\ell)$".

The main result of [11] is that $h(k,\ell) > \ell^{1+1/2k}$ for any fixed $k \geq 3$ and ℓ sufficiently large. The proof is one of the most striking early use of the probabilistic method. Erdős was probably aware of it and this may explain (and justify) the title of the paper. It is also proved that $h(2k+1,\ell) < c\ell^{1+1/k}$ and this is proved by a variant of the greedy algorithm by induction on ℓ. Now after this is claimed, it is *remarked* that the above estimation (3.1) leads to the fact that there exists a graph G with n vertices which contain no closed circuit of fewer than k edges and such that its chromatic number is $> n^{\epsilon}$.

This side remark is in fact perhaps the most well known formulation of the main result of [11]:

Theorem 3.1. *For every choice of positive integers k, t and ℓ there exists a k-graph G with the following properties:*

(1) The chromatic number of $G > t$.
(2) The graph of $G > \ell$.

This is one of the few true combinatorial classics. It started in the 40's with Tutte [21] and Zykov [26] for the case $k=2$ and $\ell=2$ (i.e., for triangle-free graphs). Later, this particular case was rediscovered and also conjectured several times [22], [23]. Kelly and Kelly [23] proved the case $k=2$, $\ell \leqq 5$, and conjectured the general statement for graphs. This was settled by Erdős in [11] and the same probabilistic method has been applied by Erdős and Hajnal [27] to yield the general result.

Erdős and Rado [29] proved the extension of $k=2$, $\ell=2$ to transfinite chromatic numbers while Erdős and Hajnal [28] gave a particularly simple construction of triangle-free graphs, so called shift graphs $G = (V, E)$: $V = \{(i,j);\ 1 \le i < j \le n\}$ and $E = \{(i,j),(i',j');\ i < j = i' < j'\}$. G_n is triangle-free and $\chi(G_n) = \lceil \log n \rceil$.

For many reasons it is desirable to have a constructive proof of Theorem 3.1. This has been stressed by Erdős on many occasions. This appeared to be difficult (see [25]) and a construction in full generality was finally given by Lovász [30]. A simplified construction has been found in the context of Ramsey theory by Nešetřil and Rödl [31]. The graphs and hypergraphs with the above properties (i), (ii) are called *highly chromatic (locally) sparse graphs*, for short.

Their existence could be regarded as one of the true paradoxes of finite set theory and it has always been felt that this result is one of the central results in combinatorics.

Recently it has been realized that sparse and complex graphs may be used in theoretical computer science for the design of fast algorithms. However, what is needed there is not only a construction of these "paradoxical" structures but also their reasonable size. In one of the most striking recent developments, a program for constructing complex sparse graphs has been successfully carried out. Using several highly ingenious constructions which combine algebraic and topological methods it has been shown that there are complex sparse graphs, the size of which in several instances improves the size of random objects. See Margulis [32], Alon [34] and Lubotzky et al. [33].

Particularly, it follows from Lubotzky et al. [33] that there are examples of graphs with girth ℓ, chromatic number t and the size at most $t^{3\ell}$. A bit surprisingly, the following is still open:

Problem. Find a primitive recursive construction of highly chromatic locally sparse k-uniform hypergraphs. Indeed, even triple systems (i.e., $k = 3$) present a problem.

III. $r(3,n)$ **[12].** The paper [12] provides the lower bound estimate on the Ramsey number $r(3,n)$.

Using probabilistic methods Erdős proved

$$r(3,n) \ge \frac{n^2}{\log^2 n} \tag{3.2}$$

(while the upper bound $r(3,n) \le \binom{n+1}{2}$ follows from [1]).

The estimation of Ramsey numbers $r(3,n)$ was Erdős' favorite problem for many years. We find it already in his 1947 paper [10] where he mentioned that he cannot prove the nonlinearity of $r(3,n)$. Later he stressed this problem (of estimating $r(3,n)$) on many occasions and conjectured various forms of it. He certainly felt the importance of this special case. How right he was is clear from the later developments, which read as a saga of modern combinatorics. And as isolated as this may seems, the problem of estimating $r(3,n)$ became a cradle of many methods and results, perhaps far exceeding the original motivation.

In 1981 Ajtai, Komlós and Szemerédi in their important paper [35] proved by a novel method

$$r(3,n) \le c\frac{n^2}{\log n} . \tag{3.3}$$

This bound and their method of proof has found many applications. The Ajtai, Komlós and Szemerédi proof was motivated by yet another Erdős problem from combinatorial number theory.

In 1941 Erdős and Turán [37] considered problem of dense Sidon sequences (or B_2-sequences). An infinite sequence $S = a_1 < a_2 < \cdots$ of natural numbers is called *Sidon sequence* if all pairwise sums $a_i + a_j$ are distinct. Define

$$f_S(n) = \max\{x : a_x \leq n\}$$

and for a given n, let $f(n)$ denote the maximal possible value of $f_S(n)$. In [37], Erdős and Turán prove that for *finite* Sidon sequences $f(n) \sim n^{1/2}$ (improving Sidon's bound of $n^{1/4}$; Sidon's motivation came from Fourier analysis [38]). However for every *infinite* Sidon sequence S growth of $f_S(n)$ is a more difficult problem and as noted by Erdős and Turán,

$$\underline{\lim} f_S(n)/n^{1/2} = 0 \, .$$

By using a greedy argument it was shown by Erdős [36] that $f_S(n) > n^{1/3}$. (Indeed, given k numbers $x_1 < \cdots < x_k$ up to n, each triple $x_i < x_j < x_k$ kills at most 3 other numbers x, $x_i + x_j = x_k + x$, $x_i + x_k = x_j + x$ and $x_j + x_k = x_i + x$ and thus if $k + 3\binom{k}{3} < ck^2 < n$ we can always find a number $x < n$ which can be added to S.) Ajtai, Komlós and Szemerédi proved using a novel "random construction" the existence of an infinite Sidon sequence S such that

$$f_S(n) > c \cdot (n \log n)^{1/3} \, .$$

An analysis of independent sets in triangle-free graphs is the basis of their approach and this yields as a corollary the above mentioned upper bound on $r(3,n)$. (The best upper bound for $f_S(n)$ is of order $c \cdot (n \log n)^{1/2}$.) It should be noted that the above Erdős-Turán paper [37] contains the following still unsolved problem: Let $a_1 < a_2 < \cdots$ be an arbitrary sequence. Denote by $f(n)$ the number of representations of n as $a_i + a_j$. Erdős and Turán prove that $f(n)$ cannot be a constant for all sufficiently large n and conjectured that if $f(n) > 0$ *for* all sufficiently large n then $\limsup f(n) = \infty$. This is still open. Erdős provided a multiplicative analogue of this conjecture (i.e., for the function $g(n)$, the number of representation of n as $a_i a_j$); this is noted already in [37]). One can ask what this has to do with Ramsey theory. Well, not only was this the motivation for [35] but a simple proof of the fact that $\limsup g(n) = \infty$ was given by Nešetřil and Rödl in [39] just using Ramsey's theorem.

We started this paper by listing the predominance of Erdős's first works in number theory. But in a way this is misleading since the early papers of Erdős stressed elementary methods and often used combinatorial or graph-theoretical methods. The Erdős-Turán paper is such an example and the paper [40] even more so.

The innovative Ajtai-Komlós-Szemerédi paper was the basis for a further development (see, e.g., [41]) and this in turn led somewhat surprisingly to the recent remarkable solution of Kim [42], who proved that the Ajtai-Komlós-Szemerédi bound is up to a constant factor, the best possible, i.e.,

$$r(n,3) > c\frac{n^2}{\log n} \, .$$

Thus $r(n,3)$ is the only nontrivial infinite family of (classical) Ramsey numbers with known asymptotics.

IV. Constructions. It was realized early by Erdős the importance of finding explicit constructions of various combinatorial objects whose existence he justified by probabilistic methods (e.g., by counting). In most case such constructions have not yet found but yet even constructions producing weaker results (or bounds) formed an important line of research. For example, the search for an explicit graph of size (say) $2^{n/2}$ which would demonstrate this Ramsey lower bound has been so far unsuccessful. This is not an entirely satisfactory situation since it is believed that such graphs share many properties with random graphs and thus they could be good candidates for various lower bounds, for example, in theoretical computer science for lower bounds for various measures of complexity. (See the papers [43] and [44] which discuss properties of pseudo- and quasi-random graphs.)

The best constructive lower bound for Ramsey numbers $r(n)$ is due to Frankl and Wilson. This improves on an earlier construction of Frankl [46] who found a first constructive superpolynomial lower bound.

The construction of Frankl-Wilson graphs is simple:

Let p be a prime number, put $q = p^3$. Define the graph $G_p = (V, E)$ as follows:

$$V = \binom{[q]}{p^2 - 1} = \{F \subseteq \{1, \ldots, p^3\} : |F| = p^2 - 1\} ,$$

$$\{F, F'\} \in E \text{ iff } |F \cap F'| \equiv -1 (\text{mod } q) .$$

The graph G_p has $\binom{p^3}{p^2-1}$ vertices. However, the Ramsey properties of the graph G_p are not trivial to prove: It follows only from deep extremal set theory results due to Frankl and Wilson [45] that neither G_p nor its complement contain K_n for $n \geq \binom{p^3}{p-1}$. This construction itself was motivated by several extremal problems of Erdős and in a way (again!) the Frankl-Wilson construction was a byproduct of these efforts.

We already mentioned earlier the developments related to Erdős paper [11]. The constructive version of bounds for $r(3, n)$ led Erdős to geometrically defined graphs. An early example is Erdős-Rogers paper [47] where they prove that there exists a graph G with ℓ^{1+c_k} vertices, which contains no complete k-gon, but such that each subgraph with ℓ vertices contains a complete $(k-1)$-gon.

If we denote by $h(k, \ell)$ the minimum integer such that every graph of $h(k, \ell)$ vertices contains either a complete graph of k vertices or a set of ℓ points not containing a complete graph with $k - 1$ vertices, then

$$h(k, \ell) \leq r(k, \ell) .$$

However, for every $k \geq 3$ we still have $h(k, \ell) > \ell^{1+c_k}$.

This variant of the Ramsey problem is due to A. Hajnal. The construction of the graph G is geometrical: the vertices of G are points on an n-dimensional sphere with unit radius, and two points are joined if their Euclidean distance exceeds $\sqrt{2k/(k-1)}$.

Graphs defined by distances have been studied by many people (e.g., see [48]). The best constructive lower bound on $r(3, n)$ is due to Alon [49] and gives $r(3, n) \geq cn^{3/2}$. See also a remarkable elementary construction [50] giving a weaker result.

4. Ramsey Theory

It seems that the building of a theory *per se* was never Erdős's preference. He is a life long problem solver, problem poser, admirer of mathematical miniatures and beauties. THE BOOK is an ideal. Instead of developing the whole field he seemed always to prefer consideration of particular cases. However, many of these cases turned out to be key cases and somehow theories emerged.

Nevertheless, one can say that Erdős and Rado systematically investigated problems related to Ramsey's theorem with a clear vision that here was a new basis for a theory. In their early papers [51], [52] they investigated possibilities of various extensions of Ramsey's theorem. It is clear that these papers are a result of a longer research and understanding of Ramsey's theorem.

As if these two papers summarized what was known, before Erdős and Rado went on with their partition calculus projects reflected by the grand papers [53] and [54]. But this is beyond the scope of this paper. [51] contains an extension of Ramsey's theorem for colorings by an infinite number of colors. This is the celebrated Erdős-Rado canonization lemma:

Theorem 4.1 ([51]). *For every choice of positive integers p and n there exists $N = N(p,n)$ such that for every set X, $|X| \geq N$, and for every coloring $c : \binom{X}{p} \to \mathbb{N}$ (i.e., a coloring by arbitrarily many colors) there exists an n-element subset Y of X such that the coloring c restricted to the set $\binom{Y}{p}$ is "canonical".*

Here a coloring of $\binom{Y}{p}$ is said to be canonical if there exists an ordering $Y = y_1 < \cdots < y_n$ and a subset $w \subseteq \{1,\ldots,p\}$ such that two n-sets $\{z_1 < \cdots < z_p\}$ and $\{z'_1 < \cdots < z'_p\}$ get the same color if and only if $z_i = z'_i$ for exactly $i \in w$. Thus there are exactly 2^p canonical colorings of p-tuples. The case $w = \phi$ corresponds to a monochromatic set while $w = \{1,\ldots,p\}$ to a coloring where each p-tuple gets a different color (such a coloring is sometimes called a "rainbow" or totally multicoloring).

Erdős and Rado deduced Theorem 4.1 from Ramsey's theorem. For example, the bound $N(p,n) \leq r(2p, 2^{2p}, n)$ gives a hint as to how to prove it. One of the most elegant forms of this argument was published by Rado [55] in one of his last papers.

The problem of estimating $N(p,n)$ was recently attacked by Lefman and Rödl [56] and Shelah [57]. One can see easily that Theorem 4.1 implies Ramsey's theorem (e.g., $N(p,n) \geq r(p,n-2,n)$) and the natural question arises as to how many exponentiations one needs. In [56] this was solved for graphs ($p = 2$) and Shelah [57] solved recently this problem in full generality: $N(p,n)$ is the lower function of the same height $r(p,4,n)$ i.e., $(p-1)$ exponentiations.

The Canonization Lemma found many interesting applications (see, e.g., [58]) and it was extended to other structures. For example, the canonical van der Waerden theorem was proved by Erdős and Graham [59].

Theorem 4.2 ([59]). *For every coloring of positive integers one can find either a monochromatic or a rainbow arithmetic progression of every length. (Recall: a rainbow set is a set with all its elements colored differently.)*

This result was extended by Lefman [60] to all regular systems of linear equations (see also [78]).

One of the essential parts of the development of the "new Ramsey theory" age was the stress on various structural extensions and structure analogies of the original results. A key role was played by Hales-Jewett theorem (viewed as a combinatorial axiomatization of van der Waerden's theorem), Rota's conjecture (the vector-space analogue of Ramsey's theorem), Graham-Rothschild parameter sets, all dealing with new structures. These questions and results displayed the richness of the field and attracted so much attention to it.

It seems that one of the significant turns appeared in the late 60's when Erdős, Hajnal and Galvin started to ask questions such as "which graphs contain a monochromatic triangle in any 2-coloring of its edges". Perhaps the essential parts of this development can be illustrated with this particular example.

We say that a graph $G = (V,E)$ is t-Ramsey for the triangle (i.e., K_3) if for every coloring of E by t-colors, one of the colors contains a triangle. Symbolically we denote this by $G \to (K_3)_t^2$. This is a variant of the Erdős-Rado partition arrow. Ramsey's theorem gives us $K_6 \to (K_3)_2^2$ (and $K_{r(2,t,3)} \to (K_3)_t^2$). But there are other essentially different examples. For example, a 2-Ramsey graph for K_3 need not contain K_6. Graham [62] constructed the unique minimal graph with this property: The graph $K_3 + C_5$ (triangle and pentagon completely joined) is the smallest graph G with $G \to (K_3)_2^2$ which does not contain a K_6. Yet $K_3 + C_5$ contains K_5 and subsequently van Lint, Graham and Spencer constructed a graph G not containing even a K_5, with $G \to (K_3)_2^2$. Until recently, the smallest example was due to Irving [63] and had 18 vertices. Very recently, two more constructions appeared by Erickson [64] and Bukor [65] who found examples with 17 and 16 vertices (both of them use properties of Graham's graph).

Of course, the next question which was asked is whether there exists a K_4-free graph G with $G \to (K_3)_2^2$. This question proved to be considerably harder and it is possible to say that it has not yet been solved completely satisfactorily.

The existence of a K_4-free graph G which is t-Ramsey for K_3 was settled by Folkman [66] ($t = 2$) and Nešetřil and Rödl [67]. The proofs are complicated and the graphs constructed are very large. Perhaps just to be explicit Erdős [68] asked whether there exists a K_4-free graph G which arrows triangle with $< 10^{10}$ vertices. This question proved to be very accurate and it was finally shown by Spencer [69] that there exists such a graph with 3×10^8 vertices. Of course, it is possible that such a graph exists with only 100 vertices!

The proof of this statement is probabilistic. The probabilistic methods were not only applied to get various bounds for Ramsey numbers. Recently, the Ramsey properties of the Random Graph $K(n,p)$ were analyzed by Rödl and Ruciński and the threshold probability for p needed to guarantee $K(n,p) \to (K_3)_t^2$ with probability tending to 1 as $n \to \infty$, was determined (see [70]).

Many of these questions were answered in a much greater generality and this seems to be a typical feature for the whole area. On the other side these more general statements explain the unique role of original Erdős problem. Let us be more specific. We need a few definitions: An ordered graph is a graph with a linearly ordered set of vertices. Isomorphism of ordered graphs means isomorphism preserving orderings. If A, B are ordered graphs (for now we will find it convenient to denote graphs by $A, B, C, \ldots$) then $\binom{B}{A}$ will denote the set of all induced subgraphs of B which are isomorphic to A. We say that a class $\mathcal{K}$ of graphs is *Ramsey* if for every choice of ordered graphs A, B from $\mathcal{K}$ there exists $C \in \mathcal{K}$ such that $C \to (B)_2^A$. Here, the notation $C \to (B)_2^A$ means: for every coloring $c : \binom{C}{A} \to \{1,2\}$ there exists $B' \in \binom{C}{B}$ such that the set $\binom{B'}{A}$ is monochromatic (see, e.g., [71].) Similarly we say that a class $\mathcal{K}$ of graphs is *canonical* if for every choice of ordered graphs A, B from $\mathcal{K}$ there exists $C \in \mathcal{K}$ with the following property: For every coloring $c : \binom{C}{A} \to \mathbb{N}$ there exists $B' \in \binom{C}{B}$ such that the set $\binom{B'}{A}$ has a canonical coloring.

Denote by $Forb(K_k)$ the class of all K_k-free graphs. Now we have the following

Theorem 4.3. *For a hereditary class $\mathcal{K}$ of graphs the following statements are equivalent:*

1. *$\mathcal{K}$ is Ramsey*
2. *$\mathcal{K}$ is canonical*
3. *$\mathcal{K}$ is a union of the following 4 types of classes: the class $Forb(K_k)$, the class of complements of graphs from $Forb(K_k)$, the class of Turán graphs (i.e., complete multipartite graphs) and the class of equivalences (i.e., complements of Turán graphs).*

(1. $\Leftrightarrow$ 3. is proved in [71], 2. $\Rightarrow$ 1. is easy, and one can prove 1. $\Rightarrow$ 2. directly along the lines of Erdős-Rado proof of canonization lemma.) Thus, as often in Erdős' case, the triangle-free graphs were not just any case but rather the typical case.

From today's perspective it seems to be just a natural step to consider. Ramsey properties of geometrical graphs. This was initiated in a series of papers by Erdős, Graham, Montgomery, Rothschild, Spencer and Straus, [72], [73], [74]. Let us call a finite configuration C of points in $\mathbb{E}^n$ *Ramsey* if for every r there is an $N = N(r)$ is that in every r-coloring of the points of $\mathbb{E}^N$, a monochromatic congruent copy of C is always formed. For example, the vertices of a unit simplex in $\mathbb{E}^n$ is Ramsey (with $N(r) = n(r-1)+n$), and it is not hard to show that the Cartesian product of two Ramsey configurations is also Ramsey. More recently, Frankl and Rödl [75] showed that any simplex in $\mathbb{E}^n$ is Ramsey (a simplex is a set of $n+1$ points having a positive n-volume).

In the other direction, it is known [72] that any Ramsey configuration must lie on the surface of a sphere (i.e., be "spherical"). Hence, 3-collinear points do not form a Ramsey configuration, and in fact, for any such set C_3, $\mathbb{E}^N$ can always be 16-colored so as to avoid a monochromatic congruent copy of C_3. It is not known if the value 16 can be reduced (almost certainly it can). The major open question is to characterize the Ramsey configurations. It is natural to conjecture that they are exactly the class of spherical sets. Additional evidence of this was found by Kříž [76] who showed for example, that the set of vertices of any regular polygon is Ramsey. A fuller discussion of this interesting topic can be found in [77].

5. Adventures in Arithmetic Progressions

Besides Ramsey's theorem itself the following result provided constant motivation for Ramsey Theory:

Theorem 5.1 (van der Waerden [79]). *For every choice of positive integers k and n, there exists a least $N(k,n) = N$ such that for every partition of the set $\{1,2,\ldots,N\}$ into k classes, one of the classes always contains an arithmetic progression with n terms.*

The original proof of van der Waerden (which developed through discussions with Artin and Schreier — see [80] for an account of the discovery) and which is included in an enchanting and moving book of Khinchine [81] was until recently essentially the only known proof. However, interesting modifications of the proof were also found, the most important of which is perhaps the combinatorial formulation of van der Waerden's result by Hales and Jewett [82].

The distinctive feature of van der Waerden's proof (and also of Hales-Jewett's proof) is that one proves a more general statement and then uses double induction. Consequently, this procedure does not provide a primitive recursive upper bound for the size of N (in van der Waerden's theorem). On the other hand, the best bound (for n prime) is (only!) $W(n+1) \geq n2^n$, n prime (due to Berlekamp [83]). Thus, the question of whether such a huge upper bound was also necessary, was and remains to be one of the main research problems in the area. In 1988, Shelah [84] gave a new proof of both van der Waerden's and the Hales-Jewett's theorem which provided a primitive recursive upper bound for $N(k,n)$. However the bound is still very large being of the order of fifth function in the Ackermann hierarchy — "tower of tower functions". Schematically,

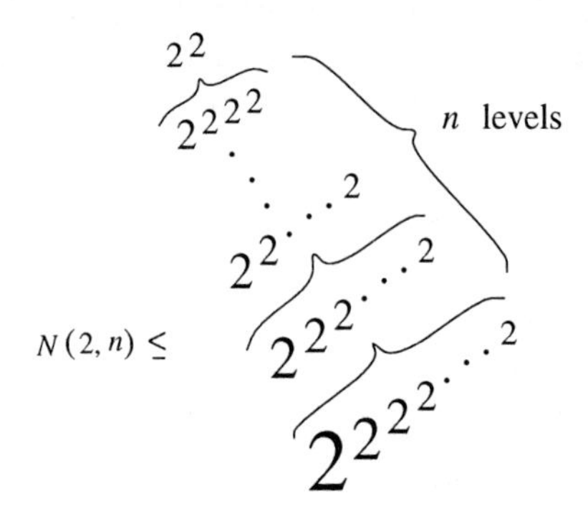

Even for the solution of the modest looking conjecture $N(2,n) \leq 2^{\cdot^{\cdot^{2}}} \Big\} n$, the first author of this paper is presently offering \$1000.

The discrepancy between the general upper bound for van der Waerden numbers and the known values is the best illustrated for the first nontrivial value: while $N(2,3)=9$ the Shelah proof gives the stack of 2's of height 2^{16}.

These observations are not new and were considered already in the Erdős and Turán 1936 paper [85]. For the purpose of improving the estimates for the van der Waerden numbers, they had the idea of proving a stronger — now called a *density* — statement. They considered (how typical!) the particular case of 3-term arithmetic progressions and for a given positive integer N, defined $r(N)$ (their notation) to denote the maximum number elements of a sequence of numbers $\leq N$ which does not contain a 3-term arithmetic progression. They observed the subadditivity of function $r(N)$ (which implies the existence of a limiting value of $r(N)/N$) and proved $r(N) \leq \left(\frac{3}{8}+\epsilon\right) N$ for all $N \geq N(\epsilon)$.

After that they remarked that probably $r(N) = o(N)$. And in the last few lines of their short paper they define numbers $r_\ell(N)$ to denote the maximum number of integers less than or equal to N such that no ℓ of them form an arithmetic progression. Although they do not ask explicitly whether $r_\ell(N) = o(N)$ (as Erdős did many times since), this is clearly in their mind as they list consequences of a good upper bound for $r_\ell(N)$: long arithmetic progressions formed by primes and a better bound for the van der Waerden numbers.

As with the Erdős-Szekeres paper [1], the impact of the modest Erdős-Turán note [85] is hard to overestimate. Thanks to its originality, both in combinatorial and number theoretic contexts, and to Paul Erdős' persistence, this led eventually to beautiful and difficult research, and probably beyond Erdős' expectations, to a rich general theory. We wish to briefly mention some key points of this development.

Good lower estimates for $r(N)$ were obtained soon after by Salem and Spencer [86] and Behrend [87] which still gives the best bound. These bounds recently found a surprising application in a least expected area, namely in the fast multiplication of matrices (Coppersmith, Winograd [88]).

The upper bounds and $r(N) = o(N)$ appeared to be much harder. In 1953 K. Roth [89] proved $r_3(N) = o(N)$ and after several years of partial results, E. Szemerédi in 1975 [91] proved the general case

$$r_\ell(N) = o(N) \quad \text{for every} \quad \ell\ .$$

This is generally recognized as the single most important Erdős solved problem, the problem for which he has paid the largest amount. By now there are more expensive problems (see Erdős' article in these volumes) but they have not yet been solved. And taking inflation into account, possibly none of them will ever have as an expensive solution. Szemerédi's proof changed Ramsey theory in at least two aspects. First, several of its pieces, most notably so called Regularity Lemma, proved to be useful in many other combinatorial situations (see e.g., [90], [91], [92]). Secondly, perhaps due to the complexity of Szemerédi's combinatorial argument, and the beauty of the result itself, an alternative approach was called for. Such an approach was found by Hillel Furstenberg [93], [94] and developed further in many aspects in his joint work with B. Weiss, Y. Katznelson and others. Let us just mention two results which in our opinion best characterize the power of this approach: In [95] Furstenberg and Katznelson proved the density version of Hales-Jewett theorem. More recently, Bergelson and Leibman [96] proved the following striking result (conjectured by Furstenberg):

Theorem 5.2 ([96]). *Let $p_1, \ldots, p_k$ be polynomials with rational coefficients taking integer values on integers and satisfying $p_i(0) = 0$ for $i = 1, \ldots, k$. Then every set X of integers of positive density contains for every choice of numbers $\nu_1, \ldots, \nu_k$, a subset*

$$\mu + p_1(d)\nu_1,\ \mu + p_2(d)\nu_2, \ldots, \mu + p_k(d)\nu_k$$

for some μ and $d > 0$.

Choosing $p_i(x) = x$ and $\nu_i = i$ we get the van der Waerden theorem. Already, the case $p_i(x) = x^2$ and $\nu_i = i$ was open for several years (this gives long arithmetic progressions in sets of positive density with their differences being some square).

For none of these results are there known combinatorial proofs. Instead, they are all proved by a blend of topological dynamics and ergodic theory methods, proving countable extensions of these results. For this part of Ramsey theory this setting seems to be most appropriate. However, it is a long way from the original Erdős-Turán paper.

Let us close this section with a more recent example. In 1983 G. Pisier formulated (in a harmonic analysis context) the following problem: A set of integers $x_1 < x_2 < \cdots$ is said to be *independent* if all finite subsums of distinct elements are distinct. Now let X be an infinite set and suppose for some $\epsilon > 0$ that every finite subset $Y \subseteq X$ contains a subsubset Z of size $\geq \epsilon|Z|$ which is independent. Is it then true that X is a finite union of independent sets?

Despite many efforts and partial solutions the problem is still open. It was again Paul Erdős who quickly realized the importance of the Pisier problem and Erdős, Nešetřil and Rödl recently [97], [98] studied "Pisier type problems". For various notions of an independence relation, the following question was considered: Assume that an infinite set X satisfies for some $\epsilon > 0$, some hereditary density condition (i.e., we assume that every finite set Y contains an independent subsubset of size $\geq \epsilon|Y|$). Is it then true that X can be partitioned into finitely many independent sets?

Positive instances (such as collinearity, and linear independence) as well as negative instances (such as Sidon sets) were given in [97], [98]. Also various "finitization versions" and analogues of the Pisier problem were answered in the negative. But at present the original Pisier problem is still open. In a way one can consider Pisier type problems as dual to the density results in Ramsey theory: One attempts to prove a positive Ramsey type statement under a strong (hereditary) density condition. This is exemplified in [98] by the following problem which is perhaps a fitting conclusion to this paper surveying 60 years of Paul Erdős' service to Ramsey theory.

The Anti-Szemerédi Problem [98]

Does there exist a set X of positive integers such that for some $\epsilon > 0$ the following two conditions hold simultaneously:

(1) For every finite $Y \subseteq X$ there exists a subset $Z \subseteq X$, $|Z| \geq \epsilon|Y|$, which does not contain a 3-term arithmetic progression;

(2) Every finite partition of X contains a 3-term arithmetic progression in one of its classes.

References

1. P. Erdős and G. Szekeres, A combinatorial problem in geometry, Composito Math. 2 (1935), 464–470.
2. I. Schur, Über die Kongruens $x^m + y^m = z^m (\text{mod } p)$, Jber. Deutch. Math. Verein 25 (1916), 114–117.
3. I. Schur, Gesammelte Abhandlungen (eds. A. Brauer, H. Rohrbach), 1973, Springer.
4. B. L. van der Waerden, Berver's einer Baudetschen Vermutung, Nieuw. Arch. Wisk. 15 (1927), 212–216.
5. R. Rado, Studien zur Kombinatorik, Math. Zeitschrift 36 (1933), 242–280.
6. F. P. Ramsey, On a problem of formal logic, Proc. London Math. Soc. 48 (1930), 264–286.
7. P. Erdős, Art of Counting, MIT Press.
8. R. L. Graham, B. L. Rothschild, and J. Spencer, Ramsey theory, Wiley, 1980, 2nd edition, 1990.
9. A. Thomason, An upper bound for some Ramsey numbers, J. Graph Theory 12 (1988), 509–517.
10. P. Erdős, Some remarks on the theory of graphs, Bull. Amer. Math. Soc. 53 (1947), 292–294.
11. P. Erdős, Graph theory and probabilities, Canad. J. Math. 11 (1959), 34–38.
12. P. Erdős, Graph theory and probability II, Canad. J. Math. 13 (1961), 346–352.
13. V. Chvátal and J. Komlos, Some combinatorial theorems on monotonicity, Canad. Math. Bull. 14, 2 (1971).
14. J. Nešetřil and V. Rödl, A probabilistic graph theoretical method, Proc. Amer. Math. Soc. 72 (1978), 417–421.
15. P. Valtr, Convex independent sets and 7-holes in restricted planar point sets, Discrete Comput. Geom. 7 (1992), 135–152.
16. J. Nešetřil and P. Valtr, A Ramsey-type result in the plane, Combinatorics, Probability and Computing 3 (1994), 127–135.
17. V. Jarnik and M. Kössler, Sur les graphes minima, contenant n points donnés, Čas. Pěst. Mat. 63 (1934), 223–235.
18. P. Hell and R. L. Graham, On the history of the minimum spanning tree problem, Annals of Hist. Comp. 7 (1985), 43–57.
19. D. Hilbert, Über die irreduribilität ganzer rationaler funktionen mit ganssahligen koeffieienten, J. Reine und Angew. Math. 110 (1892), 104–129.
20. J. H. Spencer, Ramsey's theorem — a new lower bound, J. Comb. Th. A, 18 (1975), 108–115.
21. Blanche Descartes, A three colour problem, Eureka 9 (1947), 21, Eureka 10 (1948), 24. (See also the solution to Advanced problem 1526, Amer. Math. Monthly 61 (1954), 352.)

22. G. A. Dirac, The structure of k-chromatic graphs, Fund. Math. 40 (1953), 42–55.
23. J. B. Kelly and L. M. Kelly, Paths and Circuits in critical graphs, Amer. J. Math. 76 (1954), 786–792.
24. J. Mycielski, Sur le coloriage des graphes, Collog. Math. 3 (1955), 161–162.
25. J. Nešetřil, Chromatic graphs without cycles of length ≤ 7, Comment. Math., Univ. Carolina (1966).
26. A. A. Zykov, On some properties of linear complexes, Math. Sbornik 66, 24 (1949), 163–188.
27. P. Erdős and A. Hajnal, On chromatic number of set systems, Acta Math. Acad. Sci. Hungar. 17 (1966), 61–99.
28. P. Erdős and A. Hajnal, Some remarks on set theory IX, Mich. Math. J. 11 (1964), 107–112.
29. P. Erdős and R. Rado, A construction of graphs without triangles having preassigned order and chromatic number, J. London Math. Soc. 35 (1960), 445–448.
30. L. Lovász, On the chromatic number of finite set-systems, Acta Math. Acad. Sci. Hungar. 19 (1968), 59–67.
31. J. Nešetřil and V. Rödl, A short proof of the existence of highly chromatic graphs without short cycles, J. Combin. Th. B, 27 (1979), 525–52?.
32. G. A. Margulis, Explicit constructions of concentrators, Problemy Peredachi Informatsii 9, 4 (1975), 71–80.
33. A. Lubotzky, R. Phillips, and P. Sarnak, Ramanujan Graphs, Combinatorica 8(3) (1988), 261–277.
34. N. Alon, Eigenvalues, geometric expanders, sorting in sounds and Ramsey theory, Combinatorica 3 (1986), 207–219.
35. M. Ajtai, J. Komlós and E. Szemerédi, A dense infinite Sidon sequence, European J. Comb. 2 (1981), 1–11.
36. P. Erdős, Problems and results in additive number theory, Colloque sur la Theorie des Numbres, Bruxelles (1955), 127–137.
37. P. Erdős and P. Turán, On a problem of Sidon in additive number theory and on some related problems, J. London Math. Soc. 16 (1941), 212–215.
38. S. Sidon, Ein Satz über trigonometrische Polynome und seine Anwendungen in der Theorie der Fourier-Reihen, Math. Ann. 106 (1932), 539.
39. J. Nešetřil and V. Rödl, Two proofs in combinatorial number theory, Proc. Amer. Math. Soc. 93, 1 (1985), 185–188.
40. P. Erdős, On sequences of integers no one of which divides the product of two others and on some related problems, Izv. Nanc. Ise. Inset. Mat. Mech. Tomsrk 2 (1938), 74–82.
41. N. Alon and J. Spencer, Probabilistic methods, Wiley, New York, 1992.
42. J. H. Kim, The Ramsey number $R(3,t)$ has order of magnitude $t^2/\log t$, Random Structures and Algorithms (to appear).
43. F. R. K. Chung, R. L. Graham and R. M. Wilson, Quasirandom graphs, Combinatorica 9 (1989), 345–362.
44. A. Thomason, Random graphs, strongly regular graphs and pseudorandom graphs, In: Survey in Combinatorics, Cambridge Univ. Press (1987), 173–196.
45. P. Frankl and R. M. Wilson, Intersection theorems with geometric consequences, Combinatorica 1 (1981), 357–368.
46. P. Frankl, A constructive lower bound fo Ramsey numbers, Ars Combinatorica 2 (1977), 297–302.
47. P. Erdős and C. A. Rogers, The construction of certain graphs, Canad. J. Math. (1962), 702–707.
48. D. Preiss and V. Rödl, Note on decomposition of spheres with Hilbert spaces, J. Comb. Th. A 43 (1) (1986), 38–44.

49. N. Alon, Explicit Ramsey graphs and orthonormal labellings, Electron. J. Combin. 1, R12 (1994), (8pp).
50. F. R. K. Chung, R. Cleve, and P. Dagum, A note on constructive lower bound for the Ramsey numbers $R(3,t)$, J. Comb. Theory 57 (1993), 150–155.
51. P. Erdős and R. Rado, A combinatorial theorem, J. London Math. Soc. 25 (1950), 249–255.
52. P. Erdős and R. Rado, Combinatorial theorems on classifications of subsets of a given set, Proc. London Math. Soc. 3 (1951), 417–439.
53. P. Erdős and R. Rado, A partition calculus in set theory, Bull. Amer. Math. Soc. 62 (1956), 427–489.
54. P. Erdős, A. Hajnal and R. Rado, Partition relations for cardinal numbers, Acta Math. Hungar. 16 (1965), 93–196.
55. R. Rado, Note on canonical partitions, Bull. London Math. Soc. 18 (1986), 123–126. Reprinted: Mathematics of Ramsey Theory (ed. J. Nešetřil and V. Rödl), Springer 1990, pp. 29–32.
56. H. Lefman and V. Rödl, On Erdős-Rado numbers, Combinatorica (1995).
57. S. Shelah, Finite canonization, 1994 (preprint) (to appear in Comm. Math. Univ. Carolinae).
58. J. Pelans and V. Rödl, On coverings of infinite dimensional metric spaces. In Topics in Discrete Math., vol. 8 (ed. J. Nešetřil), North Holland (1992), 75–81.
59. P. Erdős and R. L. Graham, Old and New Problems and Results in Combinatorial Number Theory, L' Enseignement Math. 28 (1980), 128 pp.
60. H. Lefman, A canonical version for partition regular systems of linear equations, J. Comb. Th. A 41 (1986), 95–104.
61. P. Erdős, J. Nešetřil and V. Rödl, Selectivity of hypergraphs, Colloq. Math. Soc. János Bolyai 37 (1984), 265–284.
62. R. L. Graham, On edgewise 2-colored graphs with monochromatic triangles and containing no complete hexagon, J. Comb. Th. 4 (1968), 300.
63. R. Irving, On a bound of Graham and Spencer for a graph-coloring constant, J. Comb. Th. B, 15 (1973), 200–203.
64. M. Erickson, An upper bound for the Folkman number $F(3,3,5)$, J. Graph Th. 17 (6), (1993), 679–68.
65. J. Bukor, A note on Folkman number $F(3,3,5)$, Math. Slovaka 44 (4), (1994), 479–480.
66. J. Folkman, Graphs with monochromatic complete subgraphs in every edge coloring, SIAM J. Appl. Math. 18 (1970), 19–24.
67. J. Nešetřil and V. Rödl, Type theory of partition properties of graphs, In: Recent Advances in Graph Theory (ed. M. Fiedler), Academia, Prague (1975), 405-412.
68. P. Erdős, Problems and result on finite and infinite graphs, In: Recent Advances in Graph Theory (ed. M. Fiedler), Academia, Prague (1975), 183–192.
69. J. Spencer, Three hundred million points suffice, J. Comb. Th. A 49 (1988), 210–217.
70. V. Rödl and A. Ruciński, Threshold functions for Ramsey properties, J. Amer. Math. Soc. (1995), to appear.
71. J. Nešetřil, Ramsey Theory, In: Handbook of Combinatorics, North Holland (1995), to appear.
72. P. Erdős, R. L. Graham, P. Montgomery, B. L. Rothschild, J. H. Spencer, and E. G. Straus, Euclidean Ramsey Theorem, J. Combin. Th. (A) 14 (1973), 341–363.
73. P. Erdős, R. L. Graham, P. Montgomery, B. L. Rothschild, J. H. Spencer, and E. G. Straus, Euclidean Ramsey Theorems II, In A. Hajnal, R. Rado and V.Sós, eds., Infinite and Finite Sets I, North Holland, Amsterdam, 1975, pp. 529–557.

74. P. Erdős, R. L. Graham, P. Montgomery, B. L. Rothschild, J. H. Spencer, and E. G. Straus, Euclidean Ramsey Theorems III, In A. Hajnal, R. Rado and V. Sós, eds., Infinite and Finite Sets II, North Holland, Amsterdam, 1975, pp. 559–583.
75. P. Frankl and V. Rödl, A partition property of simplices in Euclidean space, J. Amer. Math. Soc. 3 (1990), 1–7.
76. I. Kříž Permutation groups in Euclidean Ramsey theory, Proc. Amer. Math. Soc. 112 (1991), 899–907.
77. R. L. Graham, Recent trends in Euclidean Ramsey theory, Disc. Math. 136 (1994), 119–127.
78. W. Deuber, R. L. Graham, H. J. Prömel and B. Voigt, A canonical partition theorem for equivalence relations on $\mathbb{Z}^t$, J. Comb. Th. (A) 34 (1983), 331–339.
79. B. L. van der Waerden, Beweis einer Baudetschen Vermutung, Nieuw Arch. Wisk. 15 (1927), 212–216.
80. B. L. van der Waerden, How the Proof of Baudet's Conjecture was found, in Studies in Pure Mathematics (ed. L. Mirsky), Academic Press, New York, 1971, pp. 251–260.
81. A. J. Khinchine, Drei Perlen der Zahlen Theorie, Akademie Verlag, Berlin 1951 (reprinted Verlag Harri Deutsch, Frankfurt 1984).
82. A. W. Hales and R. I. Jewett, Regularity and positional games, Trans. Amer. Math. Soc. 106 (1963), 222-229.
83. E. R. Berlekamp, A construction for partitions which avoid long arithmetic progressions, Canad. Math. Bull 11 (1968), 409–414.
84. S. Shelah, Primitive recursive bounds for van der Waerden numbers, J. Amer. Math. Soc. 1 (1988), 683–697.
85. P. Erdős and P. Turán, On some sequences of integers, J. London Math. Soc. 11 (1936), 261–264.
86. R. Salem and D. C. Spencer, On sets of integers which contain no three terms in arithmetic progression, Proc. Nat. Acad. Sci. 28 (1942), 561–563.
87. F. A. Behrend, On sets of integers which contain no three in arithmetic progression, Proc. Nat. Acad. Sci. 23 (1946), 331–332.
88. D. Coppersmith and S. Winograd, Matrix multiplication via arithmetic progressions, J. Symb. Comput. 9 (1987), 251–280.
89. K. Roth, On certain sets of integers, J. London Math. Soc. 28 (1953), 104–109.
90. V. Chvátal, V. Rödl, E. Szemerédi, and W. Trotter, The Ramsey number of graph with bounded maximum degree, J. Comb. Th. B, 34 (1983), 239–243.
91. J. Nešetřil and V. Rödl, Partition theory and its applications, in Surveys in Combinatorics, Cambridge Univ. Press, 1979, pp. 96–156.
92. V. Rödl and A. Ruciński, Threshold functions for Ramsey properties, J. Amer. Math. Soc. (1995), to appear.
93. H. Furstenberg, Ergodic behavior of diagonal measures and a theorem of Szemerédi on arithmetic progressions, J. Anal. Math. 31 (1977), 204–256.
94. H. Furstenberg, Recurrence in Ergodic Theory and Combinatorial Number Theory, Princeton Univ. Press, Princeton, 1981.
95. H. Furstenberg and Y. Katznelson, A density version of the Hales-Jewett theorem, J. Analyze Math. 57 (1991), 61–85.
96. V. Bergelson and A. Leibman, Polynomial extensions of van der Waerden's and Szemerédi's theorems, J. Amer. Math. Soc. (1995), to appear.
97. P. Erdős, J. Nešetřil, and V. Rödl, On Pisier Type Problems and Results (Combinatorial Applications to Number Theory). In: Mathematics of Ramsey Theory (ed. J. Nešetřil and V. Rödl), Springer Verlag (1990), 214–231.
98. P. Erdős, J. Nešetřil, and V. Rödl, On Colorings and Independent Sets (Pisier Type Theorems) (preprint).

Memories on Shadows and Shadows of Memories

G. O. H. Katona

Mathematical Institute, Hungarian Academy of Sciences, Budapest Pf. 127, 1364, Hungary

I am one of the very few mathematicians who knew Paul's aunt Irma before I knew him. She and my grandmother were neighbors during World War II. Aunt Irma was one of the few Jews in Budapest who survived the holocaust. This is how I met her since I was raised from this time by my grandmother and my aunt. Aunt Irma must have had good memories about my grandmother since they kept a good relationship, she regularly visited my grandmother even after her move to another place. She learned about my interest in mathematics and suggested I meet her nephew *Pali* who happened to be a mathematician. "Of course" I had never heard of him, but I was very glad to meet an old "real mathematician". He was a very respectful old man (46!). I immediately understood that I was seeing an extraordinary personality.

He gave me three problems to solve. I remember one of them.

How many numbers can we choose from the set $\{1, 2, \ldots, 2n\}$ *without having a number and its proper divisor in the set?*

He probably had in his mind that I should call him the next day with the solutions. I took the problems very seriously. But only with my pace. It was summer. The summer between high school and university studies. I had to do so much. But I regularly returned to the problems, I solved all of them by September and reported the solutions to him.

Next time when Aunt Irma visited us forwarded the following message:

He is probably very talented in other fields but not so much in mathematics.

It did not touch me deeply. I had a lot of self-confidence. I won the National Olympiad in mathematics for high school students and was a member of the Hungarian team at the first International Olympiad. But I should have taken his opinion more seriously! At least concerning my pace and assertiveness. Is it too late?

I started my studies. Beside my regular classes, I attended several special lectures and seminars. Turán's seminar proved to be the most important for me. In spite of his negative (perhaps non-positive) opinion on my abilities, Uncle Paul (in Hungarian: Pali bácsi (=Pauli baachy)) did not forget about me. He handed me a reprint of the famous Erdős-Ko-Rado paper [1], saying that there are some open problems in it. I was in my second year of studies, I had a joint result with my friend Domokos Szász, but no one was too much interested in it because the problem was posed by ourselves. So I started to read the paper which was not without difficulties: I did not know English. The influence of this paper on my mathematical life was decisive.

Let me remind the reader what the main theorem of this paper was.

If we have a family of k element subsets of an n-element set, $2k \leq n$ and any two sets meet then the size of the family is at most $\binom{n-1}{k-1}$.

This theorem became one of the centers of my interest. Much later, in 1971, I found an elegant proof of it [5]. The open problem I started to work on was the following.

Determine the largest family of subsets of an n-element set if any two of the sets meet in at least l elements.

I spent all my free time in a period of 3 or 4 months thinking on this problem. Knowing my pace, this is not so much! Let me show you what I observed.

For sake of simplicity consider only the case when $l = 2$ and n is even. The conjectured optimal family for this case was the family of all subsets having at least $\frac{n+2}{2}$ elements. A family satisfying the conditions cannot contain the empty set or a one-element set. The total number of the 2-element and $n-1$-element sets is maximum when all $n-1$-element ones are chosen. The total number of the 3-element and $n-2$-element sets is maximum when all $n-2$-element ones are chosen. In general, it seemed that the total number of the i-element and $n-i+1$-element sets was maximum when all $n-i+1$-element ones are chosen. That is, if we have m sets of size i in the family then they push out at least m sets of size $n-i+1$. A set of size i pushes out exactly the complements of its $i-1$-element subsets. So, it would be enough to prove that the number of $i-1$-element subsets of i-element members of the family is at least m.

Let us repeat this more formally. Let $\mathcal{A}$ be the family satisfying the condition of the problem and let $\mathcal{A}_i$ be the subfamily of its i-element members. The *shadow* (this name was introduced later by someone else) of $\mathcal{A}_i$ is

$$\sigma(\mathcal{A}_i) = \{B : \ |B| = i-1, B \subset A \in \mathcal{A}\}.$$

We need $|\mathcal{A}_i| \leq |\sigma(\mathcal{A}_i)|$. After several months of work I managed to prove a somewhat stronger statement:

If any two members of the family $\mathcal{A}_i$ meet in at least 2 elements then

$$\frac{\binom{2i-2}{i-1}}{\binom{2i-2}{i}} \leq \frac{|\sigma(\mathcal{A}_i)|}{|\mathcal{A}_i|}. \tag{1}$$

The left hand side is $\frac{i}{i-1}$, therefore (1) implies that the size of the shadow is larger than the size of the original family. This gives the solution of the problem. The case of general l is much the same ([3]).

I was extremely happy and reported the result to Uncle Paul (that time Professor). He seemed to be satisfied and mentioned some consequences of "my theorem". Moreover, he invented an unusual reward for me. He invited me for lunch in a very nice hotel. The name of the hotel was Red Star (Vörös

Csillag). It was on the top of a larger hill. Its restaurant was partially open air and had a fantastic view on Budapest. Of course, his mother (Anyuka) was with us. I was on one hand very proud on the other hand embarrassed. This was the first time I ate in a restaurant. Uncle Paul probably did understand my embarrassment and helped me to choose the food.

Equation (1) determined the minimum of the ratio of the size of the shadow and the original family. It is easy to see that this estimate is sharp when the size of the family is $\binom{2i-2}{i}$. However it is not sharp if the number of sets is different. It was disturbing that I could not determine the minimum of $|\sigma(\mathcal{A}_i)|$ when $|\mathcal{A}_i|$ was given. However I did not see any nice construction that could be conjectured to be the optimum. Then I realized that the problem makes sense without the intersection-property, too.

From this time (about 1962) I concentrated on this problem and after two or three years I found a complicated inductional solution. I presented my theorem in 1965 in a lecture organized by the János Bolyai Mathematical Society. This became my best known result [4]. Let me formulate it in a special case, only.

The minimum of $|\sigma(\mathcal{A}_i)|$ under the condition that $|\mathcal{A}_i| = \binom{a}{i}$ for some fixed integer a is $\binom{a}{i-1}$. The optimal construction is the family of all i-element subsets of an a-element set.

Although it was not asked or conjectured by Erdős, it was an indirect consequence of Uncle Paul giving me his reprint. Later Branko Grünbaum called my attention to the fact that Kruskal [6] proved the same theorem earlier. His motivation was coding, therefore the combinatorial world was not aware of his result. Both (very different) proofs were quite lengthy. Many authors tried to find simpler proofs. Probably the shortest one is due to Frankl [2].

A young student from Greifswald (that time German Democratic Republic) spent a few months in Budapest in 1990 as an exchange student. I suggested to him to think about the following problem. What is the minimum size of the shadow if $|\mathcal{A}_i|$ is fixed, like before, and the family $\mathcal{A}_i$ has a system of distinct representatives, that is, one can find elements in each member of $\mathcal{A}_i$ in such a way that these elements are distinct. (More precisely, I asked a somewhat different problem and he found this nicer variant.) After 3 years of working on it he found the solution what I would formulate here only in a special case.

Theorem (Leck, [7]). *Suppose that $|\mathcal{A}_i|$ is fixed, is of the form $m(i+1)$ and $\mathcal{A}_i$ has a system of distinct representatives. The minimum of $|\sigma(\mathcal{A}_i)|$ is $m\binom{i+1}{2}$ under these conditions. The optimal construction is a disjoint union of m $(i+1)$-element sets containing all i-element subsets.*

The underground stream of problems started by Uncle Paul has reached Uwe Leck. But it is my responsibility to find a reward for him. Should I take him to Hotel Red Star? Its name is Golf now. And Uwe Leck has since become a "rich (West-)German". I do not think lunch in a restaurant would be a great experience for him. Any suggestions?

References

1. Erdős, P., Chao Ko and Rado, R., Intersection theorems for systems of finite sets, *Quart. J. Math. Oxford Ser (2)* **12** (1961) 313–320.
2. Frankl, P., A new short proof of the Kruskal-Katona theorem, *Discrete Math.* **48** (1984) 327–329.
3. Katona, G., Intersection theorems for systems of finite sets, *Acta Math. Hungar.* **15** (1964) 329–337.
4. Katona, G., A theorem on finite sets, *Theory of Graphs*, Akadémiai Kiadó, Budapest, 1968.
5. Katona, G. O. H., A simple proof of the Erdős-Chao Ko-Rado theorem, *J. Combin. Theory Ser B* **13** (1972) 183–184.
6. Kruskal, J. B., The number of simplices in a complex, *Mathematical Optimization Technics*, Univ. California Press, Berkeley, 1963.
7. Leck, U., On the minimum size of the shadow of set systems with a SDR, Preprint No. A 93-97, Freie Universität Berlin, Fachb. Math. Serie A, Math., 1993.

Applications of the Probabilistic Method to Partially Ordered Sets

William T. Trotter

Department of Mathematics, Arizona State University, Tempe, Arizona 85287, U.S.A.

This paper is dedicated to Paul Erdős with appreciation for his impact on mathematics and the lives of mathematicians all over the world.

Summary. There are two central themes to research involving applications of probabilistic methods to partially ordered sets. The first of these can be described as the study of random partially ordered sets. Among the specific models which have been studied are: random labelled posets; random t-dimensional posets; and the transitive closure of random graphs. A second theme concentrates on the adaptation of random methods so as to be applicable to general partially ordered sets. In this paper, we concentrate on the second theme. Among the topics we discuss are fibers and co-fibers; the dimension of subposets of the subset lattice; the dimension of posets of bounded degree; and fractional dimension. This last topic leads to a discussion of ramsey theoretic questions for probability spaces.

1. Introduction

Probabilistic methods have been used extensively throughout combinatorial mathematics, so it no great surprise to see that researchers have applied these techniques with great success to finite partially ordered sets. One central theme to this research is to define appropriate definitions of a *random poset*, and G. Brightwell's excellent survey article [1] provides a summary of work in this direction.

A second theme involves the application of random methods to more general classes of posets. After this brief introductory section, we present four examples of this theme. The first example is quite elementary and involves fibers and co-fibers, concepts which generalize the notions of chains and antichains. The prinicpal result here is an application of random methods to provide a non-trivial upper bound on the minimum size of fibers.

Our second example is more substantial. It involves the dimension of subposets of the subset lattice, an instance in which many of the classic techniques and results pioneered by Paul Erdős play major roles. The third example involves an application of the Lovász Local Lemma and leads naturally to the the investigation of the dimension of a random poset of height two.

Our last example involves fractional dimension for posets—an area where there are many attractive open problems. This topic leads to natural questions involving ramsey theory for probability spaces.

1991 *Mathematics Subject Classification.* 06A07, 05C35.

Key words and phrases. Partially ordered set, poset, graph, random methods, dimension, fractional dimension, chromatic number

Research supported in part by the Office of Naval Research.

The remainder of this section is a very brief condensation of key ideas and notation necessary for the remaining five sections. In this article, we consider a *partially ordered set* (or *poset*) $\mathbf{P} = (X, P)$ as a discrete structure consisting of a set X and a reflexive, antisymmetric and transitive binary relation P on X. We call X the *ground set* of the poset $\mathbf{P}$, and we refer to P as a *partial order* on X. The notations $x \leq y$ in P, $y \geq x$ in P and $(x, y) \in P$ are used interchangeably, and the reference to the partial order P is often dropped when its definition is fixed throughout the discussion. We write $x < y$ in P and $y > x$ in P when $x \leq y$ in P and $x \neq y$. When $x, y \in X$, $(x, y) \notin P$ and $(y, x) \notin P$, we say x and y are *incomparable* and write $x \| y$ in P.

Although we are concerned almost exclusively with *finite* posets, i.e., those posets with finite ground sets, we find it convenient to use the familiar notation $\mathbb{R}$, $\mathbb{Q}$, $\mathbb{Z}$ and $\mathbb{N}$ to denote respectively the reals, rationals, integers and positive integers equipped with the usual orders. Note that these four infinite posets are *total* orders; in each case, any two distinct points are comparable. Total orders are also called *linear* orders, or *chains*. We use $\mathbf{n}$ to denote an n–element chain with the points labelled as $0 < 1 < \ldots < n-1$.

A subset $A \subseteq X$ is called an *antichain* if no two distinct points in A are comparable. We also use $\mathbf{P} + \mathbf{Q}$ to denote the disjoint sum of $\mathbf{P}$ and $\mathbf{Q}$.

In the remainder of this article, we will assume that the reader is familiar with the basic concepts for partially ordered sets, including maximal and minimal elements, chains and antichains, sums and cartesian products, comparability graphs and Hasse diagrams. For additional background information on posets, the reader is referred to the author's monograph [23], the survey article [14] on dimension by Kelly and Trotter and the author's survey articles [21], [22], [25] and [26]. Another good source of background information on posets is Brightwell's general survey article [2].

2. Fibers and Co-Fibers

The classic theorem of Dilworth [4] asserts that a poset $\mathbf{P} = (X, P)$ of width n can be partitioned into n chains. Also, a poset of height h can be partitioned into h antichains. For graph theorists, these results can be translated into the simple statement that comparability graphs are perfect. Against this backdrop, researchers have devoted considerable energy to generalizations of the concepts of chains and antichains. Here is one such example.

Let $\mathbf{P} = (X, P)$ be a poset. Lonc and Rival [18] called a subset $A \subseteq X$ a *co-fiber* if it intersects every non-trivial maximal chain in $\mathbf{P}$. Let $\mathrm{cof}(\mathbf{P})$ denote the least m so that $\mathbf{P}$ has a co-fiber of cardinality m. Then let $\mathrm{cof}(n)$ denote the maximum value of $\mathrm{cof}(\mathbf{P})$ taken over all n–element posets. In any poset, the set A_1 consisting of all maximal elements which are not minimal elements and the set A_2 of all minimal elements which are not maximal are both co-fibers. As $A_1 \cap A_2 = \emptyset$, it follows that $\mathrm{cof}(n) \leq \lfloor n/2 \rfloor$. On the other hand, the fact that $\mathrm{cof}(n) \geq \lfloor n/2 \rfloor$ is evidenced by a height 2 poset with $\lfloor n/2 \rfloor$ minimal elements each of which is less than all $\lceil n/2 \rceil$ maximal elements. So $\mathrm{cof}(n) = \lfloor n/2 \rfloor$ (this argument appears in [18]).

Dually, a subset $B \subseteq X$ is called a *fiber* if it intersects every non-trivial maximal antichain. Let $\mathrm{fib}(\mathbf{P})$ denote the least m so that $\mathbf{P}$ has a fiber of cardinality m. Then let $\mathrm{fib}(n)$ denote the maximum value of $\mathrm{fib}(\mathbf{P})$ taken over all n-element posets. Trivially, $\mathrm{fib}(n) \geq \lfloor n/2 \rfloor$, and Lonc and Rival asked whether equality holds.

In [6], Duffus, Sands, Sauer and Woodrow showed that if $\mathbf{P} = (X, P)$ is an n-element poset, then there exists a set $F \subseteq X$ which intersects every 2-element maximal antichain so that $|F| \leq \lfloor n/2 \rfloor$. However, B. Sands then constructed a 17-point poset in which the smallest fiber contains 9 points. This construction was generalized by R. Maltby [19] who proved that for every $\epsilon > 0$, there exist a n_0 so that for all $n > n_0$ there exists an n-element poset in which the smallest fiber has at least $(8/15 - \epsilon)n$ points.

From above, there is no elementary way to see that there exists a constant $\alpha > 0$ so that $\text{fib}(n) < (1 - \alpha)n$. However, this is an instance where random methods provided real insights into the truth. In the remainder of this paper, we use the notation $[n]$ to denote the n–element set $\{1, 2, \ldots, n\}$ (No order is implied on $[n]$, except for the natural order on positive integers).

Theorem 2.1. *Let* $\mathbf{P} = (X, P)$ *be a poset with* $|X| = n$. *Then* X *contains a fiber of cardinality at most* $4n/5$. *Consequently,* $\text{fib}(n) \leq 4n/5$.

Proof. Let $C \subseteq X$ be a maximum chain. Then $X - C$ is a fiber. So we may assume that $|C| < n/5$. Label the points of C as $x_1 < x_2 < \ldots < x_t$, where $t = |C| < n/5$. Next we define two different partitions of $X - C$. First, for each $i \in [t]$, set $U_i = \{x \in X - C : i \text{ is the least integer for which } x \| x_i\}$. Then set $D_i = \{x \in X - C : i \text{ is the largest integer for which } x \| x_i\}$.

Then for each subset $S \subseteq [t-1]$, define

$$B(S) = C \cup (\cup\{D_i : i \in S\}) \cup (\cup\{U_{i+1} : i \notin S\})$$

Note that for each $i \in [t-1]$, the maximality of C implies that $D_i \cap U_{i+1} = \emptyset$.

Claim 1. For every subset $S \subseteq [t-1]$, $B(S)$ is a fiber.

Proof. Let $S \subseteq [t-1]$ and let A be a non-trivial maximal antichain. We show that $A \cap B(S) \neq \emptyset$. This intersection is nonempty if $A \cap C \neq \emptyset$, so we may assume that $A \cap C = \emptyset$. Now the fact that C is a maximal chain implies that every point of C is comparable with one or more points of A. However, no point of C can be greater than one point of A and less than another point of A. Also, x_1 can only be less than points in A, and x_t can only be greater than points in A. It follows that $t \geq 2$ and that there is an integer $i \in [t-1]$ and points $a, a' \in A$ for which $x_i < a$ in P and $x_{i+1} > a'$ in P. Clearly, $a' \in D_i$ and $a \in U_{i+1}$. If $i \in S$, then $D_i \subset B(S)$, and if $i \notin S$, then $U_{i+1} \subset B(S)$. In either case, we conclude that $A \cap B(S) \neq \emptyset$.

Claim 2. The expected cardinality of $B(S)$ with all subsets $S \subseteq [t-1]$ equally likely is $t + 3(n-t)/4$.

Proof. Note that $C \subseteq B(S)$, for all S. For each element $x \in X - C$, let i and j be the unique integers for which $x \in D_i$ and $x \in U_j$. Then $j \neq i + 1$. It follows that the probability that x belongs to $B(S)$ is exactly $3/4$.

To complete the proof of the theorem, we note that there is some $S \subseteq [t-1]$ for which the fiber $B(S)$ has at most $t + 3(n-t)/4$ points. However, $t < n/5$ implies that $t + 3(n-t)/4 < 4n/5$.

The preceding theorem remains an interesting (although admittedly elementary) illustration of applying random methods to *general* partially ordered sets. Characteristically, it shows that an n-point poset has a fiber containing at most $4n/5$ points without actually producing the fiber. Furthermore, this is also an instance in which the constant provided by random methods can be improved by another approach.

The following result is due to Duffus, Kierstead and Trotter [5].

Theorem 2.2. (Duffus, Kierstead and Trotter) *Let* $\mathbf{P} = (X, P)$ *be a poset and let* $\mathcal{H}$ *be the hypergraph of non-trivial maximal antichains of* $\mathbf{P}$. *Then the chromatic number of* $\mathcal{H}$ *is at most* 3. □

Theorem 2.2 shows that $\text{fib}(n) \leq 2n/3$, since whenever $X = B_1 \cup B_2 \cup B_3$ is a 3-coloring of the hypergraph $\mathcal{H}$ of non-trivial maximal antichains, then the union of any two of $\{B_1, B_2, B_3\}$ is a fiber. Quite recently, Lonc [17] has obtained the following intersesting result providing a better upper bound for posets with small width.

Theorem 2.3. (Lonc) *Let* $\mathbf{P} = (X, P)$ *be a poset of width* 3 *and let* $|X| = n$. *Then* $\mathbf{P}$ *has a fiber of cardinality at most* $11n/18$. □

I am still tempted to assert that $\lim_{n\to\infty} \text{fib}(n)/n = 2/3$.

3. Dimension Theory

When $\mathbf{P} = (X, P)$ is a poset, a linear order L on X is called a *linear extension* of P when $x < y$ in L for all $x, y \in X$ with $x < y$ in P. A set $\mathcal{R}$ of linear extensions of P is called a *realizer* of $\mathbf{P}$ when $P = \cap\mathcal{R}$, i.e., for all x, y in X, $x < y$ in P if and only if $x < y$ in L, for every $L \in \mathcal{R}$. The minimum cardinality of a realizer of $\mathbf{P}$ is called the *dimension* of $\mathbf{P}$ and is denoted $\dim(\mathbf{P})$.

It is useful to have a simple test to determine whether a family of linear extensions of P is actually a realizer. The first such test is just a reformulation of the definition. Let $\text{inc}(\mathbf{P}) = \text{inc}(X, P)$ denote the set of all incomparable pairs in $\mathbf{P}$. Then a family $\mathcal{R}$ of linear extensions of P is a realizer of P if and only if for every $(x, y) \in \text{inc}(X, P)$, there exist distinct linear extensions $L, L' \in \mathcal{R}$ so that $x > y$ in L and $y > x$ in L'.

Here is a more useful test. Call a pair $(x, y) \in X \times X$ a *critical pair* if:

1. $x \| y$ in P;
2. $z < x$ in P implies $z < y$ in P, for all $z \in X$; and
3. $w > y$ in P implies $w > x$ in P, for all $w \in X$.

The set of all critical pairs of $\mathbf{P}$ is denoted $\text{crit}(\mathbf{P})$ or $\text{crit}(X, P$. Then it is easy to see that a family $\mathcal{R}$ of linear extensions of P is a realizer of P if and only if for every critical pair (x, y), there is some $L \in \mathcal{R}$ with $x > y$ in L. We say that a linear order L on X *reverses* (x, y) if $x > y$ in L. So the dimension of a poset is just the minimum number of linear extensions required to reverse all critical pairs.

For each $n \geq 3$, let $\mathbf{S}_n$ denote the height 2 poset with n minimal elements $a_1, a_2, \ldots, a_n$, n maximal elements $b_1, b_2, \ldots, b_n$ and $a_i < b_j$, for $i, j \in [n]$ and $j \neq i$. The poset $\mathbf{S}_n$ is called the standard example of an n–dimensional poset. Note that $\dim(\mathbf{S}_n)$ is at most n, since $\text{crit}(\mathbf{S}_n) = \{(a_i, b_i) : i \in [n]\}$ and n linear extensions suffice to reverse the n critical pairs in $\text{crit}(\mathbf{S}_n)$. On the other hand, $\dim(\mathbf{S}_n) \geq n$, since no linear extension can reverse more than one critical pair.

4. The Dimension of Subposets of the Subset Lattice

For integers k, r and n with $1 \leq k < r < n$, let $\mathbf{P}(k, r; n)$ denote the poset consisting all k–element and all r–element subsets of $\{1, 2, \ldots, n\}$ partially ordered by inclusion. For simplicity, we use $\dim(k, r; n)$ to denote the dimension of $\mathbf{P}(k, r; n)$.

Historically, most researchers have concentrated on the case $k = 1$. In a classic 1950 paper in dimension theory, Dushnik [7] gave an exact formula for $\dim(1, r; n)$, when $r \geq 2\sqrt{n}$.

Theorem 4.1. (Dushnik) *Let n, r and j be positive integers with $n \geq 4$ and $2\sqrt{n} - 2 \leq r < n - 1$. If j is the unique integer with $2 \leq j \leq \sqrt{n}$ for which*

$$\lfloor \frac{n - 2j + j^2}{j} \rfloor \leq k < \lfloor \frac{n - 2(j-1) + (j-1)^2}{j-1} \rfloor,$$

then $\dim(1, r; n) = n - j + 1$. □

No general formula for $\dim(1, r; n)$ is known when r is relatively small in comparison to n, although some surprisingly tight estimates have been found. Here is a very brief overview of this work, beginning with an elementary reformulation of the problem. When L is a linear order on X, $S \subset X$ and $x \in X - S$, we say $x > S$ in L when $x > s$ in L, for every $s \in S$.

Proposition 4.2. $\dim(1, r; n)$ *is the least t so that there exist t linear orders $L_1, L_2, \ldots, L_t$ of $[n]$ so that for every r-element subset $S \subset [n]$ and every $x \in [n] - S$, there is some $i \in [t]$ for which $x > S$ in L_i.* □

Spencer [20] used this proposition to estimate $\dim(1, 2; n)$. First, he noted that by the Erdős/Szekeres theorem, if $n > 2^{2^t}$ and $\mathcal{R}$ is any set of t linear orders on $[n]$, then there exists a 3–element set $\{x, y, z\} \subset [n]$ so that for all $L \in \mathcal{R}$, either $x < y < z$ in L or $x > y > z$ in L. Thus $\dim(1, 2; n) > t$ when $n > 2^{2^t}$. On the other hand, if $n \leq 2^{2^t}$, then there exists a family $\mathcal{R}$ of t linear orders on $[n]$ so that for every 3–element subset $S \subset [n]$ and every $x \in S$, there exists some $L \in \mathcal{R}$ so that either $x < S - \{x\}$ in L or $x > S - \{x\}$ in L. Then let $\mathcal{S}$ be the family of $2t$ linear orders on X determined by adding to $\mathcal{R}$ the duals of the linear orders in $\mathcal{R}$. Clearly, the $2t$ linear orders in $\mathcal{S}$ satisfy the requirements of Proposition 4.2 when $r = 2$, and we conclude:

Theorem 4.3. (Spencer) *For all $n \geq 4$,*

$$\lg \lg n < \dim(1, 2; n) \leq 2 \lg \lg n.$$

□

Spencer [20] then proceeded to determine a more accurate upper bound for $\dim(1, 2; n)$ using a technique applicable to larger values of r. Let t be a positive integer, and let $\mathcal{F}$ be a family of subsets of $[t]$. Then let r be an integer with $1 \leq r \leq t$. We say $\mathcal{F}$ is *r-scrambling* if $|\mathcal{F}| \geq r$ and for every sequence $(A_1, A_2, \ldots, A_r)$ of r distinct sets from $\mathcal{F}$ and for every subset $B \subseteq [r]$, there is an element $\alpha \in [t]$ so that $\alpha \in A_\beta$ if and only if $\beta \in B$. We let $M(r, t)$ denote the maximum size of a r-scrambling family of subsets of $[t]$. Spencer then applied the Erdős/Ko/Rado theorem to provide a precise answer for the size of $M(2, t)$.

Theorem 4.4. (Spencer) $M(2, t) = \binom{t-1}{\lfloor \frac{t-2}{2} \rfloor}$, *for all $t \geq 4$.* □

As a consequence, Spencer observed that

$$\lg \lg n < \dim(1, 2; n) \leq \lg \lg n + (\frac{1}{2} + o(1)) \lg \lg \lg n.$$

Almost 20 years later, Füredi, Hajnal, Rödl and Trotter [13] were able to show that the upper bound in this inequality is tight, i.e.,

$$\dim(1,2;n) = \lg\lg n + (\frac{1}{2} + o(1))\lg\lg\lg n.$$

For larger values of r, Spencer used random methods to produce the following bound.

Theorem 4.5. (Spencer) *For every $r \geq 2$, there exists a constant $c = c_r > 1$ so that $M(r,t) > c^t$.*

Proof. Let p be a positive integer and consider the set of all sequences of length p whose elements are subsets of $[t]$. There are 2^{pt} such sequences. The number of such sequences which fail to be r-scrambling is easily seen to be at most

$$\binom{p}{r} 2^r (2^r - 1)^t 2^{(p-r)t}.$$

So at least one of these sequences is a r-scrambling family of subsets of $[t]$ provided $\binom{p}{r}2^r(2^r-1)^t 2^{(p-r)t} < 2^{pt}$. Clearly this inequality holds for $p > c^t$ where $c = c_r \sim e^{\frac{1}{r2^r}}$ is a constant larger than 1.

Here's how the concept of scrambling families is used in provide upper bounds for $\dim(q,r;n)$.

Theorem 4.6. (Spencer) *If $p = M(r,t)$ and $n = 2^p$, then* $\dim(1,r;n) \leq t$.

Proof. Let $\mathcal{F}$ be an r-scrambling family of subsets of $[t]$, say $\mathcal{F} = \{A_1, A_2, \ldots, A_p\}$ where $p = M(r,t)$. Then set $n = 2^p$ and let $Q_1, Q_2, \ldots, Q_n$ be the subsets of $[p]$. For each $\alpha \in [t]$, define a linear order L_α on the set $[n]$ by the following rules. Let x and y be distinct integers from $[n]$ and let $u = \min((Q_x - Q_y) \cup (Q_y - Q_x))$. Set $x > y$ in L_α if either

1. $\alpha \in A_u$ and $u \in Q_x - Q_y$, or
2. $\alpha \notin A_u$ and $u \in Q_y - Q_x$.

It is not immediately clear why L_α is a linear order on $[n]$ for each $\alpha \in [t]$, but it is easy to check that this is so. Now let S be an r–element subset of $[n]$ and let $x \in [n] - S$. We must show that $x > S$ in L_α for some $\alpha \in [t]$. For each $y \in S$, let $u_y = \min((Q_x - Q_y) \cup (Q_y \cup Q_x))$ and then consider the family $\{A_{u_y} : y \in S\}$. Since $\mathcal{F}$ is a r-scrambling family of subsets of $[t]$, there exists some $\alpha \in [t]$ such that $\alpha \in A_{u_y}$ if and only if $u_y \in Q_x$. It follows from the definition of L_α that $x > S$ in L_α.

By paying just a bit of attention to constants, the preceding results of Spencer actually yield the following upper bound on $\dim(1,r;n$.

Theorem 4.7. (Spencer) *For all $r \geq 2$,* $\dim(1,r;n) \leq (1+o(1))\frac{1}{\lg e} r 2^r \lg\lg n$. □

Of course, this bound is only meaningful if r is relatively small in comparison to n, but in this range, it is surprisingly tight. The following lower bound is a quite recent result due to Kierstead.

Theorem 4.8. (Kierstead) *If* $2 \leq r \leq \lg\lg n - \lg\lg\lg n$, *then*

$$\frac{(r+2-\lg\lg n + \lg\lg\lg n)^2 \lg n}{32\lg(r+2-\lg\lg n + \lg\lg\lg n)} \leq \dim(1,r;n).$$

□

We will return to the issue of estimating $\dim(1,r;n)$ in the next section.

5. The Dimension of Posets of Bounded Degree

Given a poset $\mathbf{P} = (X, P)$ and a point $x \in X$, define the *degree* of x in $\mathbf{P}$, denoted $\deg_{\mathbf{P}}(x)$, as the number of points in X which are comparable to x. This is just the degree of the vertex x in the associated comparability graph. Then define $\Delta(\mathbf{P})$ as the maximum degree of $\mathbf{P}$. Finally, define $\text{Dim}(k)$ as the maximum dimension of a poset $\mathbf{P}$ with $\Delta(\mathbf{P}) \leq k$. Rödl and Trotter were the first to prove that $\text{Dim}(k)$ is well defined. Their argument showed that $\text{Dim}(k) \leq 2k^2 + 2$. It is now possible to present a very short argument for this result by first developing the following idea due to Füredi and Kahn [12].

For a poset $\mathbf{P} = (X, P)$ and a point $x \in X$, let $U(x) = \{y \in X : y > x \text{ in } P\}$ and let $U[x] = U(x) \cup \{x\}$. Dually, let $D(x) = \{y \in X : y < x \text{ in } P\}$ and $D[x] = D(x) \cup \{x\}$. The following proposition admits an elementary proof. In fact, something more can be said, and we will comment on this in the next section.

Proposition 5.1. (Füredi and Kahn) *Let $\mathbf{P} = (X, P)$ be a poset and let L be any linear order on X. Then there exist a linear extension L' of P so that if (x, y) is a critical pair and $x > D[y]$ in L, then $x > y$ in L', so that $x > D[y]$ in L'.* □

Theorem 5.2. (Rödl and Trotter) *If $\mathbf{P} = (X, P)$ is a poset with $\Delta(\mathbf{P}) \leq k$, then* $\dim(\mathbf{P}) \leq 2k^2 + 2$.

Proof. Define a graph $\mathbf{G} = (X, E)$ as follows. The vertex set X is the ground set of $\mathbf{P}$. The edge set E contains those two element subsets $\{x, y\}$ for which $U[x] \cap U[y] \neq \emptyset$. Clearly, the maximum degree of a vertex in $\mathbf{G}$ is at most k^2. Therefore, the chromatic number of $\mathbf{G}$ is at most k^2+1. Let $t = k^2+1$ and let $X = X_1 \cup X_2 \cup \ldots \cup X_t$ be a partition of X into subsets which are independent in $\mathbf{G}$. Then for each $i \in [t]$, let L_i be any linear order on X with $X_i > X - X_i$ in L_i. Finally, define L_{t+i} to be any linear order on X so that:

1. $X_i > X - X_i$ in L_{t+i}, and
2. The restriction of L_{t+i} to X_i is the dual of the restriction of L_i to X_i.

We claim that for every critical pair $(x, y) \in \text{crit}(\mathbf{P})$, if $x \in X_i$, then either $x > D[y]$ in L_i or $x > D[y]$ in L_{t+i}. This claim follows easily from the observation that any two points of $D[y]$ are adjacent in $\mathbf{G}$ so that $|D[y] \cap X_i| \leq 1$.

Füredi and Kahn [12] made a dramatic improvement in the upper bound for in $\text{Dim}(k)$ by applying the Lovász Local Lemma [9]. We sketch their argument which begins with an application of random methods to provide an upper bound for $\dim(1, r; n)$. In this sketch, we make no attempt to provide the best possible constants.

Theorem 5.3. (Füredi and Kahn) *Let r and n be integers with $1 < r < n$. If t is an integer such that*

$$n\binom{n-1}{r}\left(\frac{r}{r+1}\right)^t < 1, \tag{5.1}$$

then $\dim(1, r; n) \leq t$. *In particular,* $\dim(1, r; n) \leq r(r+1)\log(n/r)$.

Proof. Let t be an integer satisfying the inequality given in the statement of the theorem. Then let $\{L_i : i \in [t]\}$ be a sequence of t random linear orders on X. The expected number of pairs (x, S) where S is an r–element subset of $[n]$, $x \in [n] - S$ and there is no $i \in [t]$ for which $x > S$ in L_i is exactly what the left hand side of this inequality is calculating. It follows that this quantity is less than one, so the probability that there are no such pairs is positive. This shows that $\dim(1, r; n) \leq t$. The estimate $\dim(1, r; n) \leq r(r+1)\log(n/r)$ follows easily.

Theorem 5.4. (Füredi and Kahn) *If* $\mathbf{P} = (X, P)$ *is a poset for which* $\Delta(\mathbf{P}) \leq k$, *then* $\dim(\mathbf{P}) \leq 100k\log^2 k$, *i.e.*, $\mathrm{Dim}(k) \leq 100k\log^2 k$.

Proof. The inequality $\dim(\mathbf{P}) \leq 100k\log^2 k$ follows from the preceding theorem if $k \leq 1000$, so we may assume that $k > 1000$. Set $m = \lceil k/\log k \rceil$ and $r = \lceil 9\log k \rceil$. Using the Lovász Local Lemma, we see that there exists a partition $X = Y_1 \cup Y_2 \ldots \cup Y_m$, with $|D[x] \cap Y_i| \leq r$, for every $x \in X$. Now fix $i \in [m]$, let $q = rk + 1$ and let $s = \dim(1, r; q)$. We construct a family $\mathcal{R}_\rangle = \{\mathcal{L}_{\rangle,|} : | \in [\in \int]\}$ as follows.

Let $\mathbf{G}$ be the graph on X defined in the proof of Theorem 5.2. Then let $\mathbf{G}_i$ be the subgraph induced by Y_i. Now it is easy to see that any point of Y_i is adjacent to at most rk other points in Y_i in the graph $\mathbf{G}_i$. It follows that the chromatic number of $\mathbf{G}_i$ is at most $rk+1$. Let $Y_i = Y_{i,1} \cup \ldots \cup Y_{i,q}$ be a partition into subsets each of which is independent in $\mathbf{G}_i$. Then let $\mathcal{R} = \{\mathcal{M}_| : | \in [\int]\}$ be a family of linear orders of $[q]$ so that for every r–element subset $S \subset [q]$ and every $x \in [q] - S$, there is some $j \in [s]$ for which $x > S$ in M_j.

Then for each $j \in [s]$, define $L_{i,j}$ as any linear order for which:

1. $Y_i > X - Y_i$ in $L_{i,j}$ and
2. if $a < b$ in M_j, then $Y_{i,a} < Y_{i,b}$ in $L_{i,j}$.

. Finally, for each $j \in [s]$, define $L_{i,s+j}$ as any linear order for which:

1. $Y_i > X - Y_i$ in $L_{i,s+j}$,
2. if $a < b$ in M_j, then $Y_{i,a} < Y_{i,b}$ in $L_{i,j}$, and
3. if $a \in [q]$, then the restriction of $L_{i,s+j}$ to $Y_{i,a}$ is the dual of the restriction of $L_{i,j}$ to $Y_{i,a}$.

.

Next we claim that if (x, y) is a critical pair and $x \in Y_i$, then there is some $j \in [2s]$ so that $x > D[y]$ in $L(i,j)$. To see this observe that any two points in $D[y]$ are adjacent in $\mathbf{G}$ so at most r points in $D[y]$ belong to Y_i, and all points of $D[y] \cap Y_i$ belong to distinct subsets in the partition of Y_i into independent subsets. Let $x \in Y_{i,j_0}$. Then there exists some $j \in [s]$ so that $j_0 > j$ in M_j whenever $j \neq j_0$ and $D[y] \cap Y_{i,j} \neq \emptyset$. It follows that either $x > D[y]$ in $L(i,j)$ or $x > D[y]$ in $L(i, s+j)$.

Finally, we note that $s = \dim(1, r; q) \leq r(r+1)\log(q/r)$, so that $\dim(\mathbf{P}) \leq 100k\log^2 k$ as claimed.

There are two fundamentally important problems which leap out from the preceding inequality limiting the dimension of posets of bounded degree, beginning with the obvious question: Is the inequality $\mathrm{Dim}(k) = O(k\log^2 k)$ best possible? However, the details of the proof also suggest that the inequality could be improved if one could provide a better upper bound than $\dim(1, \log k; k) = O(\log^3 k)$. Unfortunately, the second approach will not yield much as Kierstead [15] has recently provided the following lower bound.

Theorem 5.5. (Kierstead) *If* $\lg\lg n - \lg\lg\lg n \leq r \leq 2^{\lg^{1/2} n}$, *then*

$$\frac{(r + 2 - \lg\lg n + \lg\lg\lg n)^2 \lg n}{32\lg(r + 2 - \lg\lg n + \lg\lg\lg n)} \leq \dim(1, r; n) \leq \frac{2k^2\lg^2 n}{\lg^2 k}. \tag{5.2}$$

□

As a consequence, it follows that $\dim(1, \log k; k) = \Omega(\log^3 k/\log\log k)$. So the remaining challenge is to provide better lower bounds on $\mathrm{Dim}(k)$. Random methods seem to be our best hope. Here is a sketch of the technique used by Erdős, Kierstead and Trotter [8] to show that $\mathrm{Dim}(k) = \Omega(k\log k)$.

For a fixed positive integer n, consider a random poset $\mathbf{P}_n$ having n minimal elements $a_1, a_2, \ldots, a_n$ and n maximal elements $b_1, b_2, \ldots, b_n$. The order relation is defined by setting $a_i < b_j$ with probability $p = p(n)$; also, events corresponding to distinct min-max pairs are independent.

Erdős, Kierstead and Trotter then determine estimates for the expected value of the dimension of the resulting random poset. The arguments are far too complex to be conveniently summarized here, as they make non-trivial use of correlation inequalities. However, the following theorem summarizes the lower bounds obtained in [8].

Theorem 5.6. (Erdős, Kierstead and Trotter)

1. *For every $\epsilon > 0$, there exists $\delta > 0$ so that if*

$$\frac{\log^{1+\epsilon} n}{n} < p \leq \frac{1}{\log n},$$

then

$$\dim(\mathbf{P}) > \delta\, pn \log pn, \textit{ for almost all } \mathbf{P}.$$

2. *For every $\epsilon > 0$, there exist $\delta, c > 0$ so that if*

$$\frac{1}{\log n} \leq p < 1 - n^{-1+\epsilon},$$

then

$$\dim(\mathbf{P}) > \max\{\delta n, n - \frac{cn}{p \log n}\}\ , \textit{ for almost all } \mathbf{P}.$$

□

The following result is then an easy corollary.

Corollary 5.7. (Erdős, Kierstead and Trotter) *For every $\epsilon > 0$, there exists $\delta > 0$ so that if*

$$n^{-1+\epsilon} < p \leq \frac{1}{\log n},$$

then

$$\dim(\mathbf{P}) > \delta \Delta(\mathbf{P}) \log n, \textit{ for almost all } \mathbf{P}.$$

□

Summarizing, we now know that

$$\Omega(k \log k) = D(k) = O(k \log^2 k). \tag{5.3}$$

It is the author's opinion that the upper bound is more likely to be correct and that the proof of this assertion will come from investigating the dimension of a slightly different model of random height 2 posets. For integers n and k with k large but much smaller than n, we consider a poset with n minimal points and n maximal points. However, the comparabilities come from taking k random matchings.

The techniques used by Erdős, Kierstead and Trotter in [8] break down when $p = o(\log n/n)$. But this is just the point at which we can no longer guarantee that the maximum degree is $O(pn)$.

6. Fractional Dimension and Ramsey Theory for Probability Spaces

In many instances, it is useful to consider a fractional version of an integer valued combinatorial parameter, as in many cases, the resulting LP relaxation sheds light on the original problem. In [3], Brightwell and Scheinerman proposed to investigate fractional dimension for posets. This concept has already produced some interesting results, and many appealing questions have been raised. Here's a brief sketch of some questions with immediate connections to random methods.

Let $\mathbf{P} = (X, P)$ be a poset and let $\mathcal{F} = \{\mathcal{M}_\infty, \ldots, \mathcal{M}_\sqcup\}$ be a multiset of linear extensions of P. Brightwell and Scheinerman [3] call $\mathcal{F}$ a *k–fold realizer* of P if for each incomparable pair (x, y), there are at least k linear extensions in $\mathcal{F}$ which reverse the pair (x, y), i.e., $|\{i : 1 \leq i \leq t, x > y \text{ in } M_i\}| \geq k$. The *fractional dimension* of $\mathbf{P}$, denoted by $\text{fdim}(\mathbf{P})$, is then defined as the least real number $q \geq 1$ for which there exists a k–fold realizer $\mathcal{F} = \{M_1, \ldots, M_t\}$ of P so that $k/t \geq 1/q$ (it is easily verified that the least upper bound of such real numbers q is indeed attained and is a rational number). Using this terminology, the *dimension* of $\mathbf{P}$ is just the least t for which there exists a 1–fold realizer of P. It follows immediately that $\text{fdim}(\mathbf{P}) \leq \dim(\mathbf{P})$, for every poset $\mathbf{P}$.

Note that the standard example of an n–dimensional poset also has fractional dimension n. Brightwell and Scheinerman [3] proved that if $\mathbf{P}$ is a poset and $|D(x)| \leq k$, for all $x \in X$, then $\text{fdim}(\mathbf{P}) \leq k + 2$. They conjectured that this inequality could be improved to $\text{fdim}(\mathbf{P}) \leq k + 1$. This was proved by Felsner and Trotter [10], and the argument yielded a much stronger conclusion, a result with much the same flavor as Brooks' theorem for graphs.

Theorem 6.1. (Felsner and Trotter) *Let k be a positive integer, and let $\mathbf{P}$ be any poset with $|D(x)| \leq k$, for all $x \in X$. Then $\text{fdim}(\mathbf{P}) \leq k+1$. Furthermore, if $k \geq 2$, then $\text{fdim}(\mathbf{P}) < k+1$ unless one of the components of $\mathbf{P}$ is isomorphic to $\mathbf{S}_{k+1}$, the standard example of a poset of dimension $k+1$.* □

We do not discuss the proof of this result here except to comment that it requires a strengthening of Proposition 5.1, and to note that it implies that the fractional dimension of the poset $\mathbf{P}(1, r; n)$ is $r+1$. Thus a poset can have large dimension and small fractional dimension. However, there is one elementary bound which limits dimension in terms of fractional dimension.

Theorem 6.2. *If $\mathbf{P} = (X, P)$ is a poset with $|X| = n$ and $\text{fdim}(\mathbf{P}) = q$, then $\dim(\mathbf{P}) \leq (2 + o(1))q \log n$.*

Proof. Let $\mathcal{F}$ be a multi-realizer of P so that $\text{Prob}_{\mathcal{F}}[x > y] \geq 1/q$, for every critical pair $(x, y) \in \text{crit}(\mathbf{P})$. Then take t to be any integer for which

$$n(n-1)(1 - 1/q)^t < 1.$$

Then let $\{L_1, \ldots, L_t\}$ be a sequence of length t in which the linear extensions in $\mathcal{F}$ are equally likely to be chosen. Then the expected number of critical pairs which are not reversed is less than one, so the probability that we have a realizer of cardinality t is positive.

Felsner and Trotter [10] derive several other inequalities for fractional dimension, and these lead to some challenging problems as to the relative tightness of inequalities similar to the one given in Theorem 6.1. However, the subject of fractional dimension has produced a number of challenging problems which are certain to require random methods in their solutions. Here is two such problems, one of which has recently been solved.

A poset $\mathbf{P} = (X, P)$ is called an *interval order* if there exists a family $\{[a_x, b_x] : x \in X\}$ of non-empty closed intervals of $\mathbb{R}$ so that $x < y$ in P if and only if $b_x < a_y$ in $\mathbb{R}$. Fishburn [11] showed that a poset is an interval order if and only if it does not contain $\mathbf{2} + \mathbf{2}$ as a subposet. The interval order $\mathbf{I}_n$ consisting of all intervals with integer endpoints from $\{1, 2, \ldots, n\}$ is called the *canonical interval order*.

Although posets of height 2 can have arbitrarily large dimension, this is not true for interval orders. For these posets, large height is a prerequisite for large dimension.

Theorem 6.3. (Füredi, Hajnal, Rödl and Trotter) *If* $\mathbf{P} = (X, P)$ *is an interval order of height* n, *then*

$$\dim(\mathbf{P}) \leq \lg\lg n + (1/2 + o(1)) \lg\lg\lg n. \tag{6.1}$$

□

The inequality in the preceding theorem is best possible.

Theorem 6.4. (Füredi, Hajnal, Rödl and Trotter) *The dimension of the canonical interval order satisfies*

$$\dim(\mathbf{I}_n) = \lg\lg n + (1/2 + o(1)) \lg\lg\lg n. \tag{6.2}$$

□

Although interval orders may have large dimension, they have bounded fractional dimension. Brightwell and Scheinerman [3] proved that the dimension of any finite interval order is less than 4, and they conjectured that for every $\epsilon > 0$, there exists an interval order with dimension greater than $4 - \epsilon$. We believe that this conjecture is correct, but confess that our intuition is not really tested. For example, no interval order is known to have fractional dimension greater than 3.

Motivated by the preceding inequalities and the known bounds on the dimension and fractional dimension of interval orders and the posets $\mathbf{P}(1, r; n)$, Brightwell asked whether there exists a function $f : \mathbb{Q} \to \mathbb{R}$ so that if $\mathbf{P} = (X, P)$ is a poset with $|X| = n$ and $\operatorname{fdim}(\mathbf{P}) = q$, then $\dim(\mathbf{P}) \leq f(q) \lg\lg n$. If such a function exists, then the family $P(1, r; n)$ shows that we would need to have $f(q) = \Omega(2^q)$.

But we will show that there is no such function. The argument requires some additional notation and terminology. Fix integers n and k with $1 \leq k < n$. We call an ordered pair (A, B) of k-element sets a (k, n)-*shift pair* if there exists a $(k+1)$-element subset $C = \{i_1 < i_2 < \cdots < i_{k+1}\} \subseteq \{1, 2, \ldots, n\}$ so that $A = \{i_1, i_2, \ldots, i_k\}$ and $B = \{i_2, i_3, \ldots, i_{k+1}\}$. We then define the (k, n)-*shift graph* $\mathbf{S}(k, n)$ as the graph whose vertex set consists of all k-element subsets of $\{1, 2, \ldots, n\}$ with a k-element set A adjacent to a k-element set B exactly when (A, B) is a (k, n)-shift pair. Note that the $(1, n)$ shift graph $\mathbf{S}(1, n)$ is just a complete graph. It is customary to call a $(2, n)$-shift graph just a shift graph; similarly, a $(3, n)$-shift graph is called a *double shift graph*. The formula for the chromatic number of the $(2, n)$-shift graph $\mathbf{S}(2, n)$ is a folklore result of graph theory: $\chi(\mathbf{S}(2, n)) = \lceil \lg n \rceil$. Several researchers in graph theory have told me that this result is due to Andras Hajnal, but Andras says that it is not. In any case, it is an easy exercise.

The following construction exploits the properties of the shift graph to provide a negative answer for Brightwell's question.

Theorem 6.5. *For every* $m \geq 3$, *there exists a poset* $\mathbf{P} = (X, P)$ *so that*

1. $|X| = m^2$*;*
2. $\dim(X,P) \geq \lg m$*; and*
3. $\mathrm{fdim}(X,P) \leq 4$*.*

Proof. The poset $\mathbf{P} = (X,P)$ is constructed as follows. Set $X = \{x(i,j) : 1 \leq i,j \leq m\}$, so that $|X| = m^2$. The partial order P is defined by first defining $x(i,j_1) < x(i,j_2)$ in P, for each $i \in [m]$ whenever $1 \leq j_1 < j_2 \leq m$. Furthermore, for each $i \in [m]$, $x(i_1,j_1) < x(i_2,j_2)$ in P if and only if $(i_2 - i_1) + (j_2 - j_1) > m$.

We now show that $\dim(X,P) \geq \lg m$. Note first that for each i, j with $1 \leq i < j \leq m$, $x(i,j-i) \| x(j,m)$. Let $\dim(X,P) = t$, and let $\mathcal{R} = \{\mathcal{L}_\infty, \mathcal{L}_\in, \ldots, \mathcal{L}_\sqcup\}$ be a realizer of P. For each i,j with $1 \leq i < j \leq m$, choose an integer $\phi(\{i,j\}) = \alpha \in \{1,2,\ldots,t\}$ so that $x(i,j-i) > x(j,m)$ in L_α. We claim that ϕ is a proper coloring of the $(2,m)$ shift graph $\mathbf{S}(1,m)$ using t colors, which requires that $\dim(X,P) = t \geq \chi(\mathbf{S}(2,m) = \lceil \lg m \rceil$. To see that ϕ is a proper coloring, let i, j and k be integers with $1 \leq i < j < k \leq m$, let $\phi(\{i,j\}) = \alpha$ and let $\phi(\{j,k\}) = \beta$. If $\alpha = \beta$, then $x(i,j-i) > x(j,m)$ in L_α and $x(j,k-j) > x(k,m)$ in L_α. Also, $x(j,m) > x(j,k-j)$ in P. However, since $(k-i)+(m-j+i) > m$, it follows that $x(k,m) > x(i,j-i)$ in P, so that $x(k,m) > x(i,j-i)$ in L_α. Thus,

$$x(i,j-i) > x(j,m) > x(j,k-j) > x(k,m) > x(i,j-i) \text{ in } P \tag{6.3}$$

The inequalities in equation 6.3 cannot all be true. The contradiction shows that ϕ is a proper coloring of the shift graph $\mathbf{S}(2,m)$ as claimed. In turn, this shows that $\dim(X,P) \geq \lceil \lg m \rceil$.

Finally, we show that $\mathrm{fdim}(X,P) \leq 4$. For each element $x \in X$, let p_1 and p_2 be the natural projection maps defined by $p_(x) = i$ and $p_2(x) = j$ when $x = x(i,j)$. Next, we claim that for each subset $A \subset [m]$, there exists a linear extension $L(A)$ of P so that $x > y$ in $L(A)$ if:

1. $x \| y$ in P;
2. $p_1(x) \in A$ and $p_1(y) \notin A$.

To show that such linear extensions exist, we use the alternating cycle test (see Chapter 2 of [23]). Let $A \subseteq [m]$, and let $S(A) = \{(x,y) \in X \times X : x\|y \text{ in } P, p_1(x) \in A \text{ and } p_1(y) \notin A\}$. Now suppose that $\{(u_k,v_k) : 1 \leq k \leq p\} \subseteq S(A)$ is an alternating cycle of length p, i.e., $u_k \| v_k$ and $u_k \leq v_{k+1}$ in P, for all $k \in [p]$ (subscripts are interpreted cyclically). Let $k \in [m]$. Then $p_1(u_k) \in A$ and $p_1(v_{k+1}) \notin A$. It follows that $u_k < v_{k+1}$ in P, for each $k \in [p]$. It follows that $p_1(v_{k+1} - p_1(u_k) + p_2(v_{k+1}) - p_2(u_k) > m$. Also, we know that $m \geq p_1(v_k) - p_1(u_k) + p_2(v_k) - p_2(u_k)$. Thus $p_1(v_{k+1}) + p_2(v_{k+1}) > p_1(v_k) + p_2(v_k)$. Clearly, this last inequality cannot hold for all $k \in [p]$. The contradiction shows that $S(A)$ cannot contain any alternating cycles. Thus the desired linear extension $L(A)$ exists.

Finally, we note that if we take $\mathcal{F} = \{\mathcal{L}(\mathcal{A}) : \mathcal{A} \subseteq [\Updownarrow]\}$ and set $s = |\mathcal{F}|$, then $x > y$ in at least $s/4$ of the linear extensions in $\mathcal{F}$, whenever $x\|y$ in P. To see this, observe that there are exactly $2^s/4$ subsets of $[m]$ which contain $p_1(x)$ but do not contain $p_1(y)$. This shows that $\mathrm{fdim}(X,P) \leq 4$ as claimed. It also completes the proof of the theorem.

Now we turn our attention to the double shift graph. If $\mathbf{P} = (X,P)$ is a poset, a subset $D \subseteq X$ is called a *down set*, or an *order ideal*, if $x \leq y$ in P and $y \in D$ always imply that $x \in P$. The following result appears in [13] but may have been known to other researchers in the area.

Theorem 6.6. *Let n be a positive integer. Then the chromatic number of the double shift graph $\mathbf{S}(3,n)$ is the least t so that there are at least n down sets in the subset lattice $\mathbf{2}^t$.* □

The problem of counting the number of down sets in the subset lattice $\mathbf{2}^t$ is a classic problem and is traditionally called Dedekind's problem. Although no closed form expression is known, relatively tight asymptotic formulas have been given. For our purposes, the estimate provided by Kleitman and Markovsky [16] suffices. Theorem 6.6, coupled with the estimates from [16] permit the following surprisingly accurate estimate on the chromatic number $\chi(\mathbf{S}(3,n))$ of the double shift graph [13].

Theorem 6.7. (Füredi, Hajnal, Rödl and Trotter)

$$\chi(\mathbf{S}(3,n)) = \lg\lg n + (1/2 + o(1))\lg\lg\lg n.$$

□

Now that we have introduced the double shift graph, the following elementary observation can be made [13].

Proposition 6.8. *For each* $n \geq 3$, $\dim(1,2;n) \geq \chi(\mathbf{S}(3,n))$, *and* $\dim(\mathbf{I}_n) \geq \chi(\mathbf{S}(3,n))$. □

Although the original intent was to investigate questions involving the fractional dimension of posets, Trotter and Winkler [27] began to attack a ramsey theoretic problem for probability spaces which seems to have broader implications. Fix an integer $k \geq 1$, and let $n \geq k+1$. Now suppose that Ω is a probability space containing an event E_S for every k-element subset $S \subset \{1,2,\ldots,n\}$. We abuse terminology slightly and use the notation $\text{Prob}(S)$ rather than $\text{Prob}(E_S)$.

Now let $f(\Omega)$ denote the minimum value of $\text{Prob}(A\overline{B})$, taken over all (k,n)-shift pairs (A,B). Note that we are evaluating the probability that A is true and B is false. Then let $f(n,k)$ denote the maximum value of $f(\Omega)$ and let $f(k)$ denote the limit of $f(n,k)$ as n tends to infinity.

Even the case $k=1$ is non-trivial, as it takes some work to show that $f(1) = 1/4$. However, there is a natural interpretation of this result. Given a sufficiently long sequence of events, it is inescapable that there are two events, A and B with A occurring before B in the sequence, so that

$$\text{Prob}(A\overline{B}) < \frac{1}{4} + \epsilon.$$

The $\frac{1}{4}$ term in this inequality represents coin flips. The ϵ is present because, for finite n, we can always do slightly better than tossing a fair coin.

For $k=2$, Trotter and Winkler [27] show that $f(2) = 1/3$. Note that this is just the fractional chromatic number of the double shift graph. This result is also natural and comes from taking a random linear order L on $\{1,2,\ldots,n\}$ and then saying that a 2-element set $\{i,j\}$ is *true* if $i < j$ in L. Trotter and Winkler conjecture that $f(3) = 3/8$, $f(4) = 2/5$, and are able to prove that $\lim_{k\to\infty} f(k) = 1/2$. They originally conjectured that $f(k) = k/(2k+2)$, but they have since been able to show that $f(5) \geq \frac{27}{64}$ which is larger than $\frac{5}{12}$.

As an added bonus to this line of research, we are beginning to ask natural (and I suspect quite important) questions about patterns appearing in probability spaces.

References

1. G. R. Brightwell, Models of random partial orders, in *Surveys in Combinatorics 1993*, K. Walker, ed., 53–83.

2. G. R. Brightwell, Graphs and partial orders, *Graphs and Mathematics*, L. Beineke and R. J. Wilson, eds., to appear.
3. G. R. Brightwell and E. R. Scheinerman, Fractional dimension of partial orders, *Order* **9** (1992), 139–158.
4. R. P. Dilworth, A decomposition theorem for partially ordered sets, *Ann. Math.* **51** (1950), 161–165.
5. D. Duffus, H. Kierstead and W. T. Trotter, Fibres and ordered set coloring, *J. Comb. Theory Series A* **58** (1991) 158–164.
6. D. Duffus, B. Sands, N. Sauer and R.Woodrow, Two coloring all two-element maximal antichains, *J. Comb. Theory Series A* **57** (1991) 109–116.
7. B. Dushnik, Concerning a certain set of arrangements, *Proc. Amer. Math. Soc.* **1** (1950), 788–796.
8. P. Erdős, H. Kierstead and W. T. Trotter, The dimension of random ordered sets, *Random Structures and Algorithms* **2** (1991), 253–275.
9. P. Erdős and L. Lovász, Problems and results on 3-chromatic hypergraphs and some related questions, in *Infinite and Finite Sets*, A. Hajnal et al., eds., North Holland, Amsterdam (1975) 609–628.
10. S. Felsner and W. T. Trotter, On the fractional dimension of partially ordered sets, *Discrete Math.* **136** (1994), 101–117.
11. P. C. Fishburn, Intransitive indifference with unequal indifference intervals, *J. Math. Psych.* **7** (1970), 144–149.
12. Z. Füredi and J. Kahn, On the dimensions of ordered sets of bounded degree, *Order* **3** (1986) 17–20.
13. Z. Füredi, P. Hajnal,V. Rödl, and W. T. Trotter, Interval orders and shift graphs, in *Sets, Graphs and Numbers*, A. Hajnal and V. T. Sos, eds., Colloq. Math. Soc. Janos Bolyai **60** (1991) 297–313.
14. D. Kelly and W. T. Trotter, Dimension theory for ordered sets, in *Proceedings of the Symposium on Ordered Sets*, I. Rival et al., eds., Reidel Publishing (1982), 171–212.
15. H. A. Kierstead, The order dimension of 1-sets versus k-sets, J. Comb. Theory Series A, to appear.
16. D. J. Kleitman and G. Markovsky, On Dedekind's problem: The number of isotone boolean functions, II, *Trans. Amer. Math. Soc.* **213** (1975), 373–390.
17. Z. Lonc, Fibres of width 3 ordered sets, *Order* **11** (1994) 149–158.
18. Z. Lonc and I. Rival, Chains, antichains and fibers, *J. Comb. Theory Series A* **44** (1987) 207–228.
19. R. Maltby, A smallest fibre-size to poset-size ratio approaching 8/15, *J. Comb. Theory Series A* **61** (1992) 331–332.
20. J. Spencer, Minimal scrambling sets of simple orders, *Acta Math. Acad. Sci. Hungar.* **22**, 349–353.
21. W. T. Trotter, Graphs and Partially Ordered Sets, in *Selected Topics in Graph Theory II*, R. Wilson and L. Beineke, eds., Academic Press (1983), 237–268.
22. W. T. Trotter, Problems and conjectures in the combinatorial theory of ordered sets, *Annals of Discrete Math.* **41** (1989), 401–416.
23. W. T. Trotter, *Combinatorics and partially Ordered Sets: Dimension Theory*, The Johns Hopkins University Press, Baltimore, Maryland (1992).
24. W. T. Trotter, Progress and new directions in dimension theory for finite partially ordered sets, in *Extremal Problems for Finite Sets*, P. Frankl, Z. Füredi, G. Katona and D. Miklós, eds., Bolyai Soc. Math. Studies **3** (1994), 457–477.

25. W. T. Trotter, Partially ordered sets, in *Handbook of Combinatorics*, R. L. Graham, M. Grötschel, L. Lovász, eds., to appear.
26. W. T. Trotter, Graphs and Partially Ordered Sets, to appear.
27. W. T. Trotter and P. Winkler, Ramsey theory and sequences of random variables, in preparation.

A Bound of the Cardinality of Families not Containing Δ-Systems

A. V. Kostochka*

Institute of Mathematics, Siberian Branch, Russian Academy of Sciences

Dedicated to Professor Paul Erdős on the occasion of his 80th birthday

Summary. P.Erdős and R.Rado defined a Δ-system as a family in which every two members have the same intersection. Here we obtain a new upper bound of the maximum cardinality $\varphi(n)$ of an $n-$uniform family not containing any Δ-system of cardinality 3. Namely, we prove that for any $\alpha > 1$, there exists $C = C(\alpha)$ such that for any n,

$$\varphi(n) \leq C\ n!\alpha^{-n}.$$

1. Introduction

P. Erdős and R. Rado [2] introduced the notion of a Δ-system. They called a family $\mathcal{H}$ of finite sets a Δ-*system* if every two members of $\mathcal{H}$ have the same intersection.

Let $\varphi(n)$(respectively, $\gamma(n)$) denote the the maximum cardinality of an $n-$uniform family (respectively, intersecting $n-$uniform family) not containing any Δ-system of cardinality 3.

P. Erdős and R. Rado [2] proved that

$$2^n n! > \varphi(n) \geq 2^n$$

and conjectured that

$$\varphi(n) < c^n \quad \text{for some absolute constant } c.$$

The best published upper bound for $\varphi(n)$ is due to J. Spencer [3]:

$$\varphi(n) < e^{cn^{3/4}} n!\ .$$

Z. Füredi and J. Kahn (see [1]) proved that

$$\varphi(n) < e^{c\sqrt{n}} n!\ .$$

The aim of the present paper is to prove

Theorem 1.1. *For any integer $\alpha > 1$, there exists $C = C(\alpha)$ such that for any n,*

$$\varphi(n) \leq C\ n!\alpha^{-n}.$$

* This work was partly supported by the grant 93-011-1486 of the Russian Foundation of Fundamental Research and by the grant RPY000 of International Science Foundation.

We follow here the ideas of J. Spencer [3]. In particular, in the course of proofs some inequalities are true if n is large in comparison with α. And we choose C so that the statement of the theorem holds for smaller n.

Remark. Certainly, since the statement is true for any constant α, it is also true for some function on n tending to infinity. In fact, along the lines of the proof of the theorem one can prove that there exists a positive constant C such that

$$\varphi(n) \leq C\, n! \left(\frac{30 \log\log\log n}{\log\log n} \right)^n .$$

(All the logarithms throughout the paper are taken to the base e.) It is enough to change slightly Lemma 2.2 and to take in Section 3. $k = \lceil \frac{n}{\log n} (\log\log n)^3 \rceil$.

2. Preliminary lemmas

Call a family $\mathcal{F}$ of sets a $(3, n, k)-$*family* if it is an $n-$uniform family not containing any Δ-system of cardinality 3 such that the cardinality of the intersection of each two members of $\mathcal{F}$ is at most $n - k$.

Lemma 2.1. *For any $(3, n, k)-$family $\mathcal{F}$,*

$$|\mathcal{F}| \leq 2^{n-k+1} \frac{n!}{k!}.$$

Proof. We use induction on $n - k$. Obviously, any $(3, k, k)-$family has at most two members. Hence the lemma is true for $n - k = 0$.

Let the lemma be valid for $n - k \leq m - 1$ and $\mathcal{F}$ be a $(3, m + k, k)-$family. Choose in $\mathcal{F}$ two edges A_1 and A_2 with minimum cardinality of intersection and denote $Z = A_1 \cup A_2$. Then each $A \in \mathcal{F}$ has a non-empty intersection with Z.

For any $x \in Z$, let $\mathcal{F}(x) = \{A \in \mathcal{F} \mid x \in A\}$, $\widetilde{\mathcal{F}}(x) = \{A \setminus \{x\} \mid A \in \mathcal{F}(x)\}$. Then for any $x \in Z$, $\widetilde{\mathcal{F}}(x)$ is a $(3, m + k - 1, k)-$family. Thus,

$$|\mathcal{F}| \leq \sum_{x \in Z} |\widetilde{\mathcal{F}}(x)| \leq |Z| 2^m \frac{(m - 1 + k)!}{k!} \leq 2^{m+1} \frac{(m + k)!}{k!}. \quad \square$$

Lemma 2.2. *If $\gamma(k) \leq C\; k! \alpha^{-k} e^{-\alpha}$ for any $k \leq n$, then $\varphi(n) \leq C\; n! \alpha^{-n}$.*

Proof. Let $\mathcal{F}$ be a $(3, n, 1)-$family with $|\mathcal{F}| = \varphi(n)$ and $A \in \mathcal{F}$. For any $X \subset A$, the family $\mathcal{F}(A, X) = \{B \setminus X \mid B \in \mathcal{F} \& \; B \cap A = X\}$ is a $(3, n - |X|, 1)-$family. Moreover, if for some $X \subset A$, there exist two disjoint sets $B_1 \setminus X, B_2 \setminus X \in \mathcal{F}(A, X)$ then A, B_1 and B_2 form a Δ-system. Hence $\mathcal{F}(A, X)$ is an intersecting family and

$$\varphi(n) \leq \sum_{X \subset A} |\mathcal{F}(A, X)| \leq \sum_{i=0}^{n} \binom{n}{i} \gamma(n-i) \leq C\, n! \alpha^{-n} e^{-\alpha} \sum_{i=0}^{n} \frac{\alpha^i}{i!} < C\, n! \alpha^{-n}. \; \square$$

¿From now on, we suppose that for each $m \leq n - 1$,

$$\gamma(m) \leq C\; m! \alpha^{-m} e^{-\alpha}, \tag{2.1}$$

$$\varphi(m) \leq C\; m! \alpha^{-m}. \tag{2.2}$$

In view of Lemma 2.2 it is enough to show that (2.1) holds for $m = n$.

The following observation from [3] will be used throughout the paper. Let $B_1, \ldots, B_t$ be disjoint finite sets and $\mathcal{F}$ be a $(3, n, 1)$–family such that $|A \cap B_i| \geq b_i$ for each $A \in \mathcal{F}$. Then

$$|\mathcal{F}| \leq \binom{|B_1|}{b_1} \cdot \ldots \cdot \binom{|B_t|}{b_t} \varphi(n - b_1 - \ldots - b_t). \tag{2.3}$$

Lemma 2.3. *Let $0 < r \leq k \leq n/2$ and for any members $A_1, \ldots, A_r$ of a $(3, n, 1)$–family $\mathcal{F}$,*

$$|A_1 \cup \ldots \cup A_r| \leq rn - kr^2/2. \tag{2.4}$$

Then

$$|\mathcal{F}| \leq C \frac{n!}{k!}.$$

Proof. For $r = 1$ the lemma is valid since (2.4) is impossible as $r = 1$. Suppose that the lemma is true for $r \leq s - 1$ and $|\mathcal{F}| > Cn!/k!$. By the induction hypothesis there exist $A_1, \ldots, A_{s-1} \in \mathcal{F}$ such that for the set $B = A_1 \cup \ldots \cup A_{s-1}$ we have $|B| > (s-1)n - k(s-1)^2/2$. If the lemma does not hold for $\mathcal{F}$ then for any $A \in \mathcal{F}$,

$$|A \cap B| > n + ((s-1)n - k(s-1)^2/2) - (sn - ks^2/2) = k(s - 1/2),$$

and there is an $i, 1 \leq i \leq s - 1$ such that $|A \cap A_i| > k$. Thus by (2.3),

$$|\mathcal{F}| \leq (s-1) \binom{n}{k+1} \varphi(n - k - 1) \leq$$

$$(s-1) \binom{n}{k+1} C\,(n-k-1)!\alpha^{-n+k+1} = C\,n!\alpha^{-n+k+1} \frac{s-1}{(k+1)!} < C \frac{n!}{k!}. \qquad \square$$

Lemma 2.4. *Let $\xi \geq 2$, $1 \leq t < s \leq n$ and $\mathcal{F}$ be a $(3, s, 1)$–family with $|\mathcal{F}| \geq C\, s!\xi^{-s}$. Then there exist $\mathcal{F}' \subset \mathcal{F}$ and X such that*
1) $|X| = s - t$;
2) for any $A \in \mathcal{F}'$, $A \supset X$;
3) $|\mathcal{F}'| \geq C\, t!\beta^{-t}$, where $\beta = (4\xi)^{s/t}$.

Proof. CASE 1. *For any $A \in \mathcal{F}$, $|\{B \in \mathcal{F} \mid |B \cap A| \geq s - t\}| \leq Ct!2^s\beta^{-t} - 1$.* Then a simple greedy algorithm gives us a $(3, s, t+1)$–subfamily of $\mathcal{F}$ with cardinality at least

$$\frac{|\mathcal{F}|}{Ct!2^s\beta^{-t}} = \frac{s!(4\xi)^s}{t!(2\xi)^s} = 2^s \frac{s!}{t!}.$$

But the existence of such a big $(3, s, t+1)$–family contradicts Lemma 2.1.

CASE 2. *There exists $A \in \mathcal{F}$ such that $|\{B \in \mathcal{F} \mid |B \cap A| \geq s - t\}| \geq \lfloor Ct!2^s\beta^{-t} \rfloor$.* Then for some $X \subset A$ with $|X| = s - t$ we have

$$|\{B \in \mathcal{F} \mid B \cap A \supset X\}| \geq \lfloor Ct!2^s\beta^{-t} \rfloor \binom{s}{s-t}^{-1} > Ct!\beta^{-t}.$$

This is the family we need. $\square$

3. Main construction

Let $\mathcal{F}$ be an intersecting $(3,n,1)-$family with $|\mathcal{F}|=\gamma(n)$. The idea is to find a (not too large) family of collections of disjoint (and considerably small) sets such that each member of $\mathcal{F}$ intersects each set from some collection and then apply (2.3). We put

$$y=\lfloor n/3\alpha\rfloor,\ m=3\alpha-1,\ k=\lceil\frac{n}{\log n}\log\log n\rceil, r=\lfloor\log\log n\rfloor.$$

Lemma 3.1. *For all $s=0,1,...,m$ and for $i_0=1$ and any $i_1,...,i_s\in\{1,...,r\}$ there are subfamilies $\mathcal{F}(1,i_1,...,i_s)$ of the family $\mathcal{F}$ and sets $X(i_1,...,i_s)$ and $Z(1,i_1,...,i_{s-1})$ such that for any $s=1,...,m$ and for any $i_1,...,i_s,i_s'\in\{1,...,r\}$,*
1) $\mathcal{F}(1,i_1,...,i_s)\subset\mathcal{F}(1,i_1,...,i_{s-1})$;
2) for all $A\in\mathcal{F}(1,i_1,...,i_s)$,

$$A\supset X(i_1)\cup X(i_1,i_2)\cup\ldots\cup X(i_1,i_2,\ldots,i_s);$$

3) the sets $X(i_1),X(i_1,i_2),\ldots,X(i_1,i_2,\ldots,i_s)$ are pairwise disjoint;
4) $|X(i_1,i_2,\ldots,i_s)|=y$;
5) $|Z(1,i_1,i_2,\ldots,i_{s-1})|\le kr(r+1)/2$;
6) $X(i_1,i_2,\ldots,i_{s-1},i_s)\cap X(i_1,i_2,\ldots,i_{s-1},i_s')\subset Z(1,i_1,i_2,\ldots,i_{s-1})$;
7) $|\mathcal{F}(1,i_1,...,i_s)|\ge C(n-sy)!\xi_s^{sy-n}$, where

$$\xi_s=\left((2\alpha)^{\frac{n}{n-sy}}8^{\frac{n+(n-y)+\ldots+(n-(s-1)y)}{n-sy}}\right)=\left((2\alpha)^n8^{ns-s(s-1)y/2}\right)^{\frac{1}{n-sy}}.$$

Proof. We use induction on s. Put $\mathcal{F}(1):=\mathcal{F}$, $\xi_0:=2\alpha$.

STEP s $(0\le s<m)$. We have at hand $\mathcal{F}(1,i_1,...,i_s)$ for any $i_1,...,i_s\in\{1,...,r\}$ and if $s>0$ we also have sets $X(i_1,...,i_s)$ and $Z(1,i_1,...,i_{s-1})$ as needed. Consider

$$\begin{aligned}\widetilde{\mathcal{F}}&=\widetilde{\mathcal{F}}(1,i_1,...,i_s)\\&=\{A\setminus(X(i_1)\cup X(i_1,i_2)\cup\ldots\cup X(i_1,i_2,\ldots,i_s))\mid A\in\mathcal{F}(1,i_1,...,i_s)\}.\end{aligned}$$

According to the statements of the lemma, $\widetilde{\mathcal{F}}$ is a $(3,n-sy,1)$-family. Note that

$$(ns-s(s-1)y/2)/(n-sy)\le ns/(n-(m-1)y)\le\frac{mn}{n-(n-2y)}<(3\alpha)^2.$$

Hence $\xi_s\le(2\alpha)^{3\alpha}8^{9\alpha^2}$ and due to Statement 7) of the lemma, we can use Lemma 2.4. This Lemma 2.4 provides that there exists X_1 of cardinality y and $\mathcal{H}_1\subset\widetilde{\mathcal{F}}$ with $|\mathcal{H}_1|\ge C(n-(s+1)y)!\beta^{(s+1)y-n}$ (where $\beta=(4\xi_s)^{\frac{n-sy}{n-(s+1)y}}$) such that any $A\in\mathcal{H}_1$ contains X_1. We put $Z_1:=\emptyset$.

Suppose that $(3,n-sy,1)$-families $\mathcal{H}_1,\ldots,\mathcal{H}_l$ and sets $X_1,\ldots,X_l,Z_1,\ldots,Z_l$ are constructed and that for each $1\le j\le l,\ 1\le j'\le l,\ j\ne j'$,
(i) $|X_j|=y$;
(ii) $|Z_l|\le kl(l-1)/2$;
(iii) $X_j\cap X_j'\subset Z_l$;
(iv) for any $A\in\mathcal{H}_j$, $X_j\subset A$;
(v) $|\mathcal{H}_j|\ge C(n-(s+1)y)!\xi_{s+1}^{(s+1)y-n}$.

If $l<r$ then we construct $\mathcal{H}_{l+1},X_{l+1}$ and Z_{l+1} as follows. Remark that for each $A\in\widetilde{\mathcal{F}}$, we have $|A|=n-sy>lk$ and the number of $A\in\widetilde{\mathcal{F}}$ with $|A\cap(X_1\cup\ldots\cup X_l)|\ge lk$ does not exceed (by (2.3))

$$\binom{|X_1 \cup \ldots \cup X_l|}{lk} \varphi(n - sy - lk) \leq \binom{ly}{lk} C(n - sy - lk)! \alpha^{sy+lk-n} \leq$$

$$\left(\frac{eln}{3\alpha lk}\right)^{lk} C(n - sy - lk)! \alpha^{sy+lk-n} \leq \frac{C(n - sy)!}{\alpha^{n-sy} k^{lk}} .$$

But for large n we have

$$k^k \geq \left(\frac{n}{\log n}\right)^{\frac{n \log\log n}{\log n}} \geq e^{0.5 n \log\log n} > (2\xi_m)^n .$$

Hence for the family $\mathcal{H}' := \{A \in \widetilde{\mathcal{F}} \mid |A \cap (X_1 \cup \ldots \cup X_l)| < lk\}$ we have $|\mathcal{H}'| \geq |\widetilde{\mathcal{F}}| - C(n - sy)!(2\xi_s)^{sy-n} \geq C(n - sy)!(2\xi_s)^{sy-n}$. Then by Lemma 2.4 there exist $\mathcal{H}_{l+1} \subset \mathcal{H}'$ and X_{l+1} with $|X_{l+1}| = y$ such that each $A \in \mathcal{H}_{l+1}$ contains X_{l+1} and $|\mathcal{H}_{l+1}| \geq C(n - (s+1)y)!(\beta)^{(s+1)y-n}$, where

$$\beta = (4 \times 2\xi_s)^{\frac{n-sy}{n-(s+1)y}} = \left(8 \times \left((2\alpha)^n 8^{ns-s(s-1)y/2}\right)^{\frac{1}{n-sy}}\right)^{\frac{n-sy}{n-(s+1)y}} =$$

$$\left((2\alpha)^n 8^{n(s+1)-s(s+1)y/2}\right)^{\frac{1}{n-(s+1)y}} = \xi_{s+1} .$$

By definition of $\mathcal{H}'$,

$$|X_{l+1} \cap (X_1 \cup \ldots \cup X_l)| < lk.$$

Putting $Z_{l+1} := Z_l \cup (X_{l+1} \cap (X_1 \cup \ldots \cup X_l))$, we have $|Z_{l+1}| \leq |Z_l| + lk \leq kl(l+1)/2$ and conditions (i)-(v) are fulfilled for $l+1$. Thus we can proceed till $l = r$.

After constructing $\mathcal{H}_r, X_r$ and Z_r we put for $j = 1, \ldots, r$, $X(i_1, i_2, ..., i_s, j) := X_j$ and

$$\mathcal{F}(1, i_1, ..., i_s, j) := \{A \cup X(i_1) \cup X(i_1, i_2) \cup \ldots \cup X(i_1, i_2, ..., i_s, j) \mid A \in \mathcal{H}_j\},$$

and

$$Z(1, i_1, ..., i_s) := Z_r .$$

By construction, the statements 1)-7) of the lemma will be fulfilled for $s+1$. □

Lemma 3.2. *For all $s = 0, 1, ..., m+1$ and for any $i_1, ..., i_s \in \{1, ..., r\}$ there are sets $X(i_1, ..., i_s)$ and $Z(1, i_1, ..., i_{s-1})$ and for any $i_1, ..., i_{m+1} \in \{1, ..., r\}$ there are sets $A(i_1, ..., i_{m+1}) \in \mathcal{F}$ such that*
1) the sets $X(i_1), X(i_1, i_2), \ldots, X(i_1, i_2, \ldots, i_{m+1})$ are pairwise disjoint;
2) $|X(i_1, i_2, \ldots, i_s)| = y$ if $1 \leq s \leq m$;
3) $|X(i_1, i_2, \ldots, i_{m+1})| = n - my$;
4) $A(i_1, ..., i_{m+1}) = X(i_1) \cup X(i_1, i_2) \cup \ldots \cup X(i_1, i_2, \ldots, i_{m+1})$;
for any $s = 1, ..., m$ and for any $i_1, ..., i_s, i'_s \in \{1, ..., r\}$
5) $X(i_1, i_2, \ldots, i_{s-1}, i_s) \cap X(i_1, i_2, \ldots, i_{s-1}, i'_s) \subset Z(1, i_1, i_2, \ldots, i_{s-1})$;
6) $|Z(1, i_1, i_2, \ldots, i_{s-1})| \leq kr(r+1)/2$.

Proof. For $s = 0, 1, ..., m$ and for any $i_1, ..., i_s \in \{1, ..., r\}$, consider $\mathcal{F}(1, i_1, ..., i_s)$, $X(i_1, ..., i_s)$ and $Z(1, i_1, ..., i_{s-1})$ from Lemma 3.1.

Now, for an arbitrary $(m+1)-$tuple $(1, i_1, ..., i_m)$, consider

$$\mathcal{H} = \mathcal{H}(1, i_1, ..., i_m) :=$$

$$\{A \setminus (X(i_1) \cup X(i_1, i_2) \cup \ldots \cup X(i_1, i_2, \ldots, i_m)) \mid A \in \mathcal{F}(1, i_1, ..., i_m)\}.$$

By construction, $\mathcal{H}$ is a $(3, n - my, 1)$−family and by Lemma 3.1, $|\mathcal{H}| \geq C(n - my)!\xi_m^{my-n}$.

Recall that $m = 3\alpha - 1$, and $n - my \geq y \geq n/(3\alpha) - 1$. For large n,

$$\xi_m^{n-my} = (2\alpha)^n 8^{nm-m(m-1)y/2} \leq (2\alpha)^n 8^{nm} \leq (2\alpha 8^{3\alpha-1})^n < k!$$

Hence $|\mathcal{H}| > C(n - my)!/k!$ and by Lemma 2.3 (note that $0 < r < k < (n/3\alpha - 1)/2 \leq (n - my)/2$) there exist $A_1, \ldots, A_r \in \mathcal{H}$ such that

$$|A_1 \cup \ldots \cup A_r| > r(n - my) - kr^2/2. \tag{3.1}$$

Let us denote

$$Z(1, i_1, ..., i_m) := \bigcup_{1 \leq j < h \leq r} A_j \cap A_h,$$

and for $j = 1, \ldots, r$, $X(i_1, i_2, ..., i_s, j) := A_j$ and

$$A(i_1, ..., i_m, j) = X(i_1) \cup X(i_1, i_2) \cup \ldots \cup X(i_1, i_2, \ldots, i_m) \cup A_j.$$

In view of (3.1), $|Z(1, i_1, ..., i_m)| \leq kr^2/2$. Now, by Lemma 3.1 and the construction, all the statements of the lemma are fulfilled. □

Lemma 3.3. *For each $A \in \mathcal{F}$, there exist $s, 0 \leq s \leq m$ and $i_1, ..., i_s \in \{1, ..., r\}$ such that*

$$A \bigcap X(i_1, i_2, \ldots, i_s, j) \neq \emptyset \qquad \forall j \in \{1, ..., r\}. \tag{3.2}$$

Proof. Assume that for some $B \in \mathcal{F}$ for each $s, 0 \leq s \leq m$ and each $i_1, ..., i_s \in \{1, ..., r\}$, there exists $j^*(1, ..., i_s)$ such that

$$B \bigcap X(i_1, i_2, \ldots, i_s, j^*(1, ..., i_s)) = \emptyset.$$

Let further $q_0 = 1$ and for $s = 1, \ldots, m + 1$,

$$q_s = j^*(q_0, \ldots, q_{s-1}).$$

Then B has empty intersection with every member of the sequence $X(q_1)$, $X(q_1, q_2), \ldots, X(q_1, q_2, \ldots, q_{m+1})$. But this means that B is disjoint from $A(q_1, q_2, \ldots, q_{m+1})$, a contradiction to the definition of $\mathcal{F}$. □

Completion of the proof of the theorem. Consider

$$Z := \bigcup_{s=1}^{m+1} \bigcup_{(1,i_1,\ldots,i_{s-1})} Z(1, i_1, ..., i_{s-1}).$$

Clearly, $|Z| \leq (1 + r + r^2 + \ldots + r^m)kr(r + 1)/2 \leq kr^{m+2} = kr^{3\alpha+1}$.

Let $\mathcal{E} = \{A \in \mathcal{F} \mid A \cap Z \neq \emptyset\}$ and

$$\mathcal{H}(1, i_1, ..., i_s) = \{A \in \mathcal{F} \setminus \mathcal{E} \mid \quad A \bigcap X(i_1, i_2, \ldots, i_s, j) \neq \emptyset \quad \forall j \in \{1, ..., r\}\}.$$

By Lemma 3.2, for each $s, 0 \leq s \leq m$ and $i_1, ..., i_s \in \{1, ..., r\}$ the sets $X(i_1, i_2, \ldots, i_s, j) \setminus Z$ for distinct j are disjoint. Hence by Lemma 3.3, we can write $\mathcal{F}$ in the form

$$\mathcal{F} = \mathcal{E} \bigcup \left(\bigcup_{s=0}^{m} \bigcup_{(1,i_1,\dots,i_s)} \mathcal{H}(1, i_1, \dots, i_s) \right).$$

Let us estimate

$$|\mathcal{E}| \leq |Z| \varphi(n-1) \leq kr^{3\alpha+1}\, C(n-1)! \alpha^{1-n}.$$

Note that each $A \in \mathcal{H}(1, i_1, \dots, i_s)$ should intersect each of r disjoint sets $X(i_1, i_2, \dots, i_s, 1) \setminus Z, X(i_1, i_2, \dots, i_s, 2) \setminus Z, \dots, X(i_1, i_2, \dots, i_s, r) \setminus Z$. The cardinalities of these sets for $s < m$ are at most y and for $s = m$ are less than $2y$. Consequently by (2.3),

$$|\mathcal{H}(1, i_1, \dots, i_s)| \leq (2y)^r \varphi(n-r) \leq \left(\frac{2n}{3\alpha}\right)^r C(n-r)! \alpha^{r-n} \leq \left(\frac{2+o(1)}{3}\right)^r Cn! \alpha^{-n}.$$

Thus,

$$|\mathcal{F}| \leq kr^{3\alpha+1}\, C(n-1)! \alpha^{1-n} + \frac{r^{m+1}-1}{r-1} \left(\frac{2+o(1)}{3}\right)^r Cn! \alpha^{-n} <$$

$$C\, n! \alpha^{-n} e^{-\alpha} \left(\frac{1.5 (\log \log n)^{3\alpha+2} e^{\alpha} \alpha}{\log n} + \frac{(\log \log n)^{3\alpha} e^{\alpha}}{(1.5 - o(1))^{\lfloor \log \log n \rfloor}} \right).$$

But for large n the expression in big parentheses is less than one. So, the theorem is proved.

Acknowledgement. The author is very indebted to M.Axenovich and D.G.Fon-Der-Flaass for many creative conversations.

References

1. P.Erdős, Problems and results on set systems and hypergraphs, *Extended Abstract, Conf.on Extremal Problems for Finite Sets,1991, Visegrad, Hungary,* 1991, 85-92.
2. P.Erdős and R.Rado, Intersection theorems for systems of sets, *J.London Math. Soc.* **35**(1960), 85-90.
3. J.Spencer, Intersection theorems for systems of sets, *Canad. Math. Bull.* **20**(1977), 249-254.

Arrangeability and Clique Subdivisions

Vojtěch Rödl[1*] and Robin Thomas[2**]

[1] Department of Mathematics and Computer Science, Emory University, Atlanta, GA 30322
[2] School of Mathematics, Georgia Institute of Technology, Atlanta, GA 30332

Summary. Let k be an integer. A graph G is *k-arrangeable* (concept introduced by Chen and Schelp) if the vertices of G can be numbered $v_1, v_2, \ldots, v_n$ in such a way that for every integer i with $1 \leq i \leq n$, at most k vertices among $\{v_1, v_2, \ldots, v_i\}$ have a neighbor $v \in \{v_{i+1}, v_{i+2}, \ldots, v_n\}$ that is adjacent to v_i. We prove that for every integer $p \geq 1$, if a graph G is not p^8-arrangeable, then it contains a K_p-subdivision. By a result of Chen and Schelp this implies that graphs with no K_p-subdivision have "linearly bounded Ramsey numbers," and by a result of Kierstead and Trotter it implies that such graphs have bounded "game chromatic number."

The Theorem

In this paper graphs are finite, may have parallel edges, but may not have loops. We begin by defining the concept of admissibility, introduced by Kierstead and Trotter [6].

Let G be a graph, let $M \subseteq V(G)$, and let $v \in M$. A set $A \subseteq V(G)$ is called an *M-blade* with center v if either

(i) $A = \{a\}$ and $a \in M$ is adjacent to v, or
(ii) $A = \{a, b\}$, $a \in M$, $b \in V(G) - M$, and b is adjacent to both v and a.

An *M-fan with center v* is a set of pairwise disjoint *M-blades* with center v. Let k be an integer. A graph G is *k-admissible* if the vertices of G can be numbered $v_1, v_2, \ldots, v_n$ in such a way that for every $i = 1, 2, \ldots, n$, G has no $\{v_1, v_2, \ldots, v_i\}$-fan with center v_i of size $k+1$.

As pointed out in [6] the concepts of arrangeability and admissibility are asymptotically equivalent in the sense that if a graph is k-arrangeable, then it is $2k$-admissible, and if it is k-admissible, then it is $(k^2 - k + 1)$-arrangeable.

Let p be an integer. A graph G has a K_p-subdivision if G contains p distinct vertices $v_1, v_2, \ldots, v_p$ and $\binom{p}{2}$ paths P_{ij} $(i, j = 1, 2, \ldots, p, i < j)$ such that P_{ij} has ends v_i and v_j, and if a vertex of G belongs to both P_{ij} and $P_{i'j'}$ for $(i,j) \neq (i',j')$, then it is an end of both. The following is our main result.

Theorem 1. *Let $p \geq 1$ be an integer. If a graph G is not $\frac{1}{2}p^2(p^2+1)$-admissible, then it has a K_p-subdivision.*

We first prove Theorem 1, and then discuss its applications. For the proof, we need the following result of Komlós and Szemerédi [7].

Theorem 2. *Let $p \geq 1$ be an integer. If a simple graph on n vertices has at least $\frac{1}{2}p^2 n$ edges, then it has a K_p-subdivision.*

---

* Supported in part by NSF under Grant No. DMS-9401559.
** Supported in part by NSF under Grant No. DMS-9303761, and by ONR under Contract No. N00014-93-1-0325.

We first prove a lemma.

Lemma 3. *Let $p \geq 1$ be an integer, let G be a graph, and let M be a non-empty subset of $V(G)$. If for every $v \in M$ there is an M-fan in G with center v of size $\frac{1}{2}p^2(p^2+1)$, then G has a K_p-subdivision.*

Proof. Let p, G, and M be as stated in the lemma, and for $v \in M$ let F_v be a fan in G with center v of size $\frac{1}{2}p^2(p^2+1)$. We may assume that G is minimal subject to $M \subseteq V(G)$ and the existence of all F_v $(v \in M)$. Let $|M| = m$, let e_1 be the number of edges of G with both ends in M, and let e_2 be the number of edges of G with one end in M and the other in $V(G) - M$. Then from the existence of the fans F_v for $v \in M$ we deduce that $2e_1 + e_2 \geq \frac{1}{2}p^2(p^2+1)m$.

We claim that if $|V(G) - M| \geq p^2 m - e_1$, then G has a K_p-subdivision. Indeed, by our minimality assumption for every $w \in V(G) - M$ there exist vertices $u, v \in M$ such that $\{u, w\} \in F_v$. For $w \in V(G) - M$ let us denote by $e(w)$ some such pair of vertices. Let J be the graph obtained from G by deleting $V(G) - M$ and for every $w \in V(G) - M$ adding an edge between the vertices in $e(w)$. Then $|E(J)| \geq p^2 m$, and since every pair of vertices is joined by at most two (parallel) edges, J has a simple subgraph J' on the same vertex-set with at least $\frac{1}{2}p^2 m$ edges. By Theorem 2 J' has a K_p-subdivision L. Every edge of L that does not belong to G joins two vertices u, v with $\{u, v\} = e(w)$ for some $w \in V(G) - M$. By replacing each such edge by the edges uw, vw we obtain a K_p-subdivision in G. This proves our claim, and so we may assume that $|V(G) - M| \leq p^2 m - e_1$.

Now $|V(G)| \leq (p^2+1)m - e_1$, and

$$\begin{aligned} |E(G)| &\geq e_1 + e_2 = 2e_1 + e_2 - e_1 \geq \frac{1}{2}p^2(p^2+1)m - e_1 \\ &\geq \frac{1}{2}p^2((p^2+1)m - e_1) \geq \frac{1}{2}p^2|V(G)|, \end{aligned}$$

and hence G has a K_p-subdivision by Theorem 2, as required.

Proof of Theorem 1. Let p be an integer, and let G be a graph on n vertices with no K_p-subdivision. We are going to show that G is $\frac{1}{2}p^2(p^2+1)$-admissible by exhibiting a suitable ordering of $V(G)$. Let $i \in \{0, 1, \ldots, n\}$ be the least integer such that there exist vertices $v_{i+1}, v_{i+2}, \ldots, v_n$ with the property that for all $j = i, i+1, \ldots, n$, G has no $(V(G) - \{v_{j+1}, v_{j+2}, \ldots, v_n\})$-fan with center v_j of size $p^2(\frac{1}{2}p^2+1)+1$. We claim that $i = 0$. Indeed, otherwise by Lemma 3 applied to $M = V(G) - \{v_{i+1}, v_{i+2}, \ldots, v_n\}$ there exists a vertex v_i with no M-fan with center v_i of size $\frac{1}{2}p^2(p^2+1)$, and so the sequence $v_i, v_{i+1}, \ldots, v_n$ contradicts the choice of i. Hence $i = 0$, and $v_1, v_2, \ldots, v_n$ is the desired enumeration of the vertices of G.

Applications

We now mention two applications of Theorem 1. Let $\mathcal{G}$ be a class of graphs. We say that $\mathcal{G}$ has *linearly bounded Ramsey numbers* if there exists a constant c such that if $G \in \mathcal{G}$ has n vertices, then for every graph H on at least cn vertices, either H or its complement contain a subgraph isomorphic to G. The class of all graphs does not have linearly bounded Ramsey numbers, but some classes do. Burr and Erdös [3] conjectured the following.

Conjecture 4. Let t be an integer, and let $\mathcal{G}$ be the class of all graphs whose edge-sets can be partitioned into t forests. Then $\mathcal{G}$ has linearly bounded Ramsey numbers.

Chvátal, Rödl, Szemerédi and Trotter [5] proved that for every integer d, the class of graphs of maximum degree at most d has linearly bounded Ramsey numbers, and Chen and Schelp [4] extended that to the class of k-arrangeable graphs. Chen and Schelp also showed that every planar graph has arrangeability at most 761, a bound that has been subsequently lowered to 10 by Kierstead and Trotter [6]. From Chen and Schelp's result and Theorem 1 we deduce

Corollary 5. *For every integer $p \geq 1$, the class of graphs with no K_p-subdivision has linearly bounded Ramsey numbers.*

For the second application we need to introduce the following two-person game, first considered by Bodlaender [2]. Let G be a graph, and let t be an integer, both fixed in advance. The game is played by two players Alice and Bob. Alice is trying to color the graph, and Bob is trying to prevent that from happening. They alternate turns with Alice having the first move. A move consists of selecting a previously uncolored vertex v and assigning it a color from $\{1, 2, \ldots, t\}$ distinct from the colors assigned previously (by either player) to neighbors of v. If after $|V(G)|$ moves the graph is (properly) colored, Alice wins, otherwise Bob wins. More precisely, Bob wins if after less than $|V(G)|$ steps either player cannot make his or her next move. The *game chromatic number* of a graph G is the least integer t such that Alice has a winning strategy in the above game. Kierstead and Trotter [6] have shown the following.

Theorem 6. *Let k and t be positive integers. If a k-admissible graph has chromatic number t, then its game chromatic number is at most $kt + 1$.*

They have also shown that planar graphs have admissibility at most 8, and hence planar graphs have game chromatic number at most 33 by Theorem 6 and the Four Color Theorem [1]. From Theorems 1, 2 and 6 we deduce

Corollary 7. *Let p be a positive integer. Then every graph with no K_p-subdivision has game chromatic number at most $\frac{1}{2}p^4(p^2 + 1)$.*

Added in proof: Bollobás and Thomason proved Theorem 1 with the constant 1/2 replaced by 22 in [B. Bollobás and A. G. Thomason, Highly Linked Graphs, manuscript 1994].

References

1. K. Appel and W. Haken, Every planar map is four colorable, *Contemporary Mathematics* **98**, Providence, RI, 1989.
2. H. L. Bodlaender, On the complexity of some coloring games, Proceedings of the WG 1990 Workshop on Graph Theoretical Concepts in Computer Science, to appear.
3. S. A. Burr and P. Erdös, On the magnitude of generalized Ramsey numbers, *Infinite and Finite Sets*, Vol. 1, A. Hajnal, R. Rado and V. T. Sós, eds., Colloq. Math. Soc. Janos Bolyai, North Holland, Amsterdam/London, 1975.
4. G. Chen and R. H. Schelp, Graphs with linearly bounded Ramsey numbers, *J. Comb. Theory Ser B*, to appear.

5. V. Chvatál, V. Rödl, E. Szemerédi and W. T. Trotter, The Ramsey number of a graph of bounded degree, *J. Comb. Theory Ser B.* **34** (1983), 239–243.

6. H. A. Kierstead and W. T. Trotter, Planar graph coloring with an uncooperative partner, manuscript 1992.

7. E. Szemerédi, Colloquium at Emory University, Atlanta, GA, April 22, 1994.

A Finite Partition Theorem with Double Exponential Bound

Saharon Shelah*

[1] Institute of Mathematics, The Hebrew University, Jerusalem, Israel
[2] Rutgers University, Department of Mathematics, New Brunswick, NJ USA

Dedicated to Paul Erdős

Summary. We prove that double exponentiation is an upper bound to Ramsey's theorem for colouring of pairs when we want to predetermine the order of the differences of successive members of the homogeneous set.

The following problem was raised by Jouko Vaananen for model theoretic reasons (having a natural example of the difference between two kinds of quantifiers, actually his question was a specific case), and propagated by Joel Spencer: Is there for any n, c an m such that

(∗) for every colouring f of the pairs from $\{0, 1, \ldots, m-1\}$ by 2 (or even c) colours, there is a monochromatic subset $\{a_0, \ldots, a_{n-1}\}$, $a_0 < a_1 < \ldots$ such that the sequence $\langle a_{i+1} - a_i : i < n-1 \rangle$ is with no repetition and is with any pregiven order.

Noga Alon [Al] and independently Janos Pach proved that for every n, c there is such an m as in (∗); Alon used van der Waerden numbers (see [Sh:329]) (so obtained weak bounds).
Later Alon improved it to iterated exponential (Alan Stacey and also the author have later and independently obtained a similar improvement). We get a double exponential bound. The proof continues [Sh:37]. Within the "realm" of double exponential in c, n we do not try to save.
We thank Joel Spencer for telling us the problem, and Martin Goldstern for help in proofreading.

Notation. Let ℓ, k, m, n, c, d belong to the set $\mathbb{N}$ of natural numbers (which include zero). A sequence η is $\langle \eta(0), \ldots, \eta(\ell g \eta - 1) \rangle$, also ρ, ν are sequences. $\eta \triangleleft \nu$ means that η is a proper initial subsequence of ν. We consider sequences as graphs of functions (with domain of the form $n = \{0, \ldots, n-1\}$) but $\eta \cap \nu$ means the largest initial segment common to η and ν. $\eta^\frown\langle s \rangle$ is the sequence $\langle \eta(0), \ldots, \eta(\ell g \eta - 1), s \rangle$ (of length $\ell g(\eta) + 1$).

Let ${}^{\ell}m := \left\{ \eta : \ell g(\eta) = \ell, \text{ and } \mathrm{Rang}(\eta) \subseteq \{0, \ldots, m-1\} \right\}$,

${}^{\ell>}m = \bigcup_{k<\ell} {}^{k}m.$

For $\nu \in {}^{\ell>}m$ we write $[\nu]_{{}^{\ell}m}$ (or $[\nu]$ if ℓ and m are clear from the context) for the set $\{\eta \in {}^{\ell}m : \nu \triangleleft \eta\}$.
$[A]^n = \{w \subseteq A : |w| = n\}$.
Intervals $[a, b), (a, b), [a, b)$ are the usual intervals of integers. The proof is similar to [Sh:37] but sets are replaced by trees.

---

* I thank Alice Leonhardt for excellent typing.
Research supported by the United States-Israel Binational Science Foundation
Publication number 515; Latest Revision 02/95
Done Erdős 80th birthday meeting in Keszthely, Hungary; July 1993

1. Definition

$r = r(n,c)$ is the first number m such that:

$(*)^{n,c}_m$ <u>for every</u> $f : [\{0,\dots,m-1\}]^2 \to \{0,\dots,c-1\}$ and linear order $<^*$ on $\{0,\dots,n-2\}$ <u>we can find</u> $a_0 < \cdots < a_{n-1} \in [0,m)$ such that:

(a) $f{\restriction}\{\{a_i,a_j\} : 0 \le i < j < n\}$ is constant

(b) the numbers $b_\ell := a_{\ell+1} - a_\ell$ (for $\ell < n-1$) are with no repetitions and are ordered by $<^*$, i.e. $i <^* j \Rightarrow b_i < b_j$

We will find a double exponential bound for $r = r(n,c)$, specifically, $r = r(n,c) \le 2^{(c(n+1)^3)^{nc}}$ (so our bound is double exponential in n and in c).

This is done in conclusion 6. Alon conjectures that the true order of magnitude of $r(n,c)$ is single exponential, and Alon and Spencer have proved this for the case where the sequence $\langle a_{i+1} - a_i : i < n-1\rangle$ is monotone.

2. Definition

We say S is an (ℓ, m^*, m, u)-tree if:

(a) $u \subseteq \{0,1,\dots,\ell-1\}$ and $m^* \ge m$

(b) $S \subseteq {}^{\ell\ge}(m^*)$

(c) S is closed under initial segments

(d) if $\nu \in S$ is $\triangleleft$-maximal <u>then</u> $\ell g(\nu) = \ell$

(e) if $\nu \in S$ and $\ell g(\nu) \in u$ <u>then</u> for m numbers $j < m^*$ we have $\nu^\frown\langle j\rangle \in S$

3. Claim

Suppose $k, \ell, m, p, m^* \in \mathbb{N}$ satisfy

$$(*)_{k,\ell,m,p,m^*} \qquad m^* \ge p^k m^{\ell k+1},$$

<u>then</u> for every $i(*) < \ell$, and $u \subseteq [0,\ell-1)$, and $|u| \le p$ and $f : [{}^\ell(m^*)]^2 \to \{0,1\}$ <u>there is</u> $T \subseteq {}^{\ell\ge}(m^*)$ closed under initial segments, $(T,\triangleleft) \cong ({}^{\ell\ge}m, \triangleleft)$ satisfying

$\bigoplus$ for every $\eta_1,\dots,\eta_k \in T \cap {}^\ell(m^*)$ with $\langle \eta_1{\restriction}i(*),\dots,\eta_k{\restriction}i(*)\rangle$ pairwise distinct we can find a set S such that:

(a) S is an $(\ell, m^*, m, u \setminus i(*))$-tree

(b) if $j \in [1,k)$ and $\nu \in S \cap {}^\ell(m^*)$ then $f(\{\eta_j,\eta_k\}) = f(\{\eta_j,\nu\})$

(c) $\nu \in S \cap {}^\ell(m^*) \Rightarrow \eta_k{\restriction}i(*) \triangleleft \nu$

Remark 3.1. (1) With minor change we can demand in $\bigoplus$ "for any $i(*) < \ell$".

(2) We could use here f with range $\{0,\dots,c-1\}$, and in claim 3 get a longer sequence $\langle \nu_\ell : \ell < n^*\rangle$ such that $f(\{\nu_{\ell_1},\nu_{\ell_2}\})$ depends just on $\nu_{\ell_1} \cap \nu_{\ell_2}$, then use a partition theorem on such colouring.

Proof. For each $\eta \in {}^{\ell>}(m^*)$ choose randomly a set $A_\eta \subseteq [0,m^*)$, $|A_\eta| = m$, $A_\eta = \{x^\eta_1,\dots,x^\eta_m\}$ (pairwise distinct, chosen by order) (not all are relevant, some can be fixed).

We define $T = \{\eta : \eta \in {}^{\ell\ge}(m^*)$, and $i < \ell g(\eta) \Rightarrow \eta(i) \in A_{\eta{\restriction}i}\}$.
We have a natural isomorphism h from ${}^{\ell\ge}m$ onto T:

$$h(\nu) = \eta \Leftrightarrow \bigwedge_{i<\ell g\nu} \eta(i) = x^{\eta\restriction i}_{\nu(i)}.$$

Our problem is to verify $\bigoplus$, we prove that the probability that it fails is < 1, this suffices.
We can represent it as:

$(*)_1$ if $\nu_1, \dots, \nu_k \in {}^{\ell}m$ and $\nu_1\restriction i(*), \dots, \nu_k\restriction i(*)$ distinct, then for $h(\nu_1), \dots, h(\nu_k)$ there is S as required there.

So it suffices to prove that for any given such $i(*) < \ell$ and $\nu_1, \dots, \nu_k$ the probability of failure is $< \dfrac{1}{\binom{|{}^{\ell}m|}{k}} = \dfrac{1}{\binom{m^\ell}{k}}$ as it suffices the demand in $(*)$, to hold for the minimal suitable $i(*)$. W.l.o.g. we may assume $u = u \setminus i(*)$. For this we can assume x^{ρ}_{j} are fixed whenever $\neg[h(\nu_k)\restriction i(*)) \trianglelefteq \rho]$ or $\ell g(\rho) \notin u$.
Let $Y = \{\eta \in {}^{\ell}(m^*)$:Prob$[h(\nu_k) = \eta] \neq 0\}$, so $|Y| = m^{|u|}$.
So $h(\nu_1), \dots, h(\nu_{k-1})$ are determined. Now $h(\nu_1), \dots, h(\nu_{k-1})$ and f induces an equivalence relation E on Y:

$$\eta' E \eta'' \text{ iff } \bigwedge_{j=1}^{k-1} f(\{h(\nu_j), \eta'\}) = f(\{h(\nu_j), \eta''\}).$$

The number of classes is $\leq 2^{k-1}$, let them be $A_1, \dots, A_{2^{k-1}}$ (they are pairwise disjoint, some may be empty).

We call A_j large if there is S as required in clauses (a) and (c) of $\bigoplus$ such that $(\forall \rho)[\rho \in S \cap {}^{\ell}(m^*) \Rightarrow \rho \in A_j]$.

It is enough to show that the probability of $h(\nu_k)$ belonging to a non-large equivalence class is $< \frac{1}{\binom{m^\ell}{k}}$, hence it is enough to prove:

$(*)_2$ A_j not large $\Rightarrow$ Prob$(h(\nu_k) \in A_j) < \frac{1}{\binom{m^\ell}{k} \times 2^k}$.

So assume A_j is not large. Let $Y^* := \{\eta\restriction i : \eta \in Y \text{ and } i \leq \ell\}$.

$$\text{Let } Z_j := \left\{ \eta \in Y^* : \begin{array}{ll} \text{there is } S \subseteq Y^* \text{ satisfying} & \\ (a)' & S \text{ is an } (\ell, m^*, m, u \backslash \ell g(\eta))\text{-tree} \\ (b)' & \nu \in S \cap {}^{\ell}(m^*) \Rightarrow \eta \trianglelefteq \nu \\ (c)' & \text{for every } \nu \in S \cap {}^{\ell}(m^*) \text{ we have } \nu \in A_j \end{array} \right\}.$$

Let $Z^*_j := \{\eta \in Z_j$: there is no $\eta' \triangleleft \eta, \eta' \in Z_j\}$.
Clearly $h(\nu_k\restriction i(*)) \notin Z_j$, (as A_j is not large) hence $h(\nu_k\restriction i(*)) \notin Z^*_j$.
Clearly

$(*)_3$ for $\eta \in Y^* \setminus Z_j$ such that $\ell g(\eta) \in u$ we have

$$|\{i : \eta^\frown\langle i\rangle \in Z^*_j \text{ (or even } \in Z_j)\}| < m.$$

But if $\nu \triangleleft \eta$, $\nu \in Z_j$ then $\eta \notin Z^*_j$. Hence

$(*)_4$ for $\eta \in Y^*$ such that $\ell g(\eta) \in u$ we have

$$|\{i : \eta^\frown\langle i\rangle \in Z^*_j\}| < m.$$

Now

$(*)_5$ if $\eta \in A_j$ (hence $\eta \in Y$; remember that A_j is not large) then $\eta \in Z_j$, hence $\bigvee_{j \in u} \eta \restriction (j+1) \in Z_j^*$.

So

$$\begin{aligned}\mathrm{Prob}(h(\nu_k) \in A_j) &\leq \mathrm{Prob}\Big(\bigvee_{j\in u} [h(\nu_k \restriction (j+1)) \in Z_j^*]\Big) \\ &\leq \sum_{j \in u} \mathrm{Prob}(h(\nu_k \restriction (j+1) \in Z_j^*)) \\ &< |u| \times \frac{m}{m^*}.\end{aligned}$$

(first inequality by $(*)_5$, second inequality trivial, last inequality by $(*)_4$ above). So it suffices to show:

$$|u| \times \frac{m}{m^*} \leq \frac{1}{\binom{m^\ell}{k} \times 2^k}$$

equivalently

$$m^* \geq |u| \times m \times \binom{m^\ell}{k} \times 2^k$$

as $\binom{m^\ell}{k} \leq m^{\ell k}/k!$, and $|u| \leq p$ by the hypothesis $(*)_{k,\ell,m,p,m^*}$ we finish. $\square_2$

4. Lemma

Assume

(a) $\rho_1, \ldots, \rho_n \in {}^{n-1}2$ *are distinct, for* $\ell \in \{2, \ldots, n\}$, *we have* $r_\ell \in \{1, \ldots, \ell-1\}$ *such that* $\ell g(\rho_\ell \cap \rho_{r_\ell}) = \ell - 1$ *and* $r \in \{1, \ldots, \ell-1\} \setminus \{r_\ell\} \Rightarrow \ell g(\rho_\ell \cap \rho_r) < \ell - 1$

(b) $f : [{}^\ell m]^2 \to [0, c) = \{0, \ldots, c-1\}$

(c) $m = 2^{(n+1)^{(c+1)n}}$.

Then we can find $\eta_1, \ldots, \eta_n \in {}^\ell m$ such that:

(α) $f \restriction [\{\eta_1, \ldots, \eta_n\}]^2$ is a constant function

(β) $\langle \ell g(\eta_{i+1} \cap \eta_{r_{i+1}}) : i = 1, \ldots, n-1\} \rangle$ is a sequence with no repetitions ordered just like $\langle \ell g(\rho_{i+1} \cap \rho_{r_{i+1}}) : i = 1, n-1\} \rangle$; also:

$$\begin{aligned}\eta_{i+1}(\ell g(\eta_{i+1} \cap \eta_{r_{i+1}})) &< \eta_{r_{i+1}}(\ell g(\eta_i \cap \eta_{r_{i+1}})) \\ &\Leftrightarrow \rho_{i+1}(\ell g(\rho_{i+1} \cap \rho_{r_{i+1}})) < \rho_{r_{i+1}}(\ell g(\rho_{i+1} \cap \rho_{r_{i+1}})).\end{aligned}$$

Remark 4.1. 1) Note that if $\Gamma \subseteq {}^{n-1}2$, $|\Gamma| = n$ and the set $\{\rho_1 \cap \rho_2 : \rho_1 \neq \rho_2$ are from $\Gamma\}$ has no two distinct members with the same length then we can list Γ as $\langle \rho_1, \ldots, \rho_n \rangle$ as required in clause (a) of Lemma 4.

2) So if $<^*$ is a linear order on $\{1, \ldots, n-1\}$ then we can find distinct $\rho_1, \ldots, \rho_n \in {}^{n-1}2$ as in clause (a) of Lemma 4. and a permutation σ of $\{1, \ldots, n\}$ such that: for $i \neq j \in \{1, \ldots, n-2\}$ we have

$$i <^* j \quad \text{iff} \quad \ell g(\rho_{\sigma(i)} \cap \rho_{\sigma(i+1)}) > \ell g(\rho_{\sigma(j)} \cap \rho_{\sigma(j+1)}).$$

(E.g. use induction on n.)

Proof. Let us define $\langle m_j : 2 \leq j \leq cn \rangle$ by induction on j:

$$m_2 = n^c, \qquad m_{j+1} = n^c m_j^{n^c(n+1)+1}.$$

Check that $m_j \leq 2^{(n+1)^{(c+1)j}}$, so in particular $m_{(cn)} \leq m$. Now we claim that for any number $d \in [1, c]$ the following holds:

$\bigotimes_d$ Assume $q \in [0, n^c)$ such that q divisible by n^d and $q + n^d \leq cn$, $u = [q, q+n^d)$, T^* is an $(\ell, m, m_{(q+dn)}, u)$-tree and f is a function from $[T^* \cap {}^\ell m]^2$ with range of cardinality d. Then we can find $\eta_1, \ldots, \eta_n \in T^* \cap ({}^\ell m)$ such that clauses (α) and (β) of the conclusion of Lemma 4. hold.

This suffices: use $q = 0, d = c$. We prove this by induction on d. If $d = 1$, trivial as only one colour occurs. For $d + 1 > 1$, without loss of generality Rang$(f) = [0, d]$, let $f' : [{}^\ell m]^2 \to \{0, 1\}$ be $f'(\{\eta', \eta''\}) =$ Min$\{f(\{\eta', \eta''\}, 1\}$. Let for $j < n$, $u_j := [q + n^d j, q + n^d j + n^d)$. By downward induction on $j \in [1, n]$ we try to define T_j such that:

(i) T_j is an $(\ell, m, m_{(q+nd+j)}, \bigcup_{i<j} u_i)$-tree

(ii) $T_j \subseteq T_{j+1}$

(iii) for every $j \in [1, n-1]$ and $\eta_1, \ldots, \eta_j \in T_j \cap {}^\ell m$ with $\langle \eta_1 \restriction (q + n^d j), \ldots, \eta_j \restriction (q + n^d j) \rangle$ pairwise distinct, we can find $\eta' \neq \eta'' \in T_{j+1} \cap {}^\ell m$ such that:

$$\eta_j \restriction (q + n^d j) \lhd \eta', \eta''$$
$$\ell g(\eta' \cap \eta'') \in u_j$$
$$f'(\{\eta', \eta''\}) = 0$$
$$\bigwedge_{t \in [1,j)} \left[0 = f'(\{\eta_t, \eta_j\}) \Rightarrow 0 = f'(\{\eta_t, \eta'\}) = f'(\{\eta_t, \eta''\}) \right].$$

<u>This suffices</u> as then we can choose by induction on $j = 1, \ldots, n$ a sequence $\nu_j \in T_j \cap ({}^\ell m)$ such that (after reordering) the set $\{\nu_1, \ldots, \nu_n\}$ will serve as $\{\eta_1, \ldots, \eta_n\}$ of $\otimes_d$ (with the constant colour being zero). Let us do it in detail.

By induction on $j = 1, \ldots, n$ we choose $\nu_1^j, \ldots, \nu_{j-1}^j$ such that:

(a) $\nu_1^j, \ldots, \nu_j^j$ are distinct members of $T_j \cap ({}^\ell m)$

(b) $f \restriction [\{\nu_1^j, \ldots, \nu_j^j\}]^2$ is constantly zero

(c) for $\ell = \{2, \ldots, j\}$ we have $\ell g(\nu_\ell^j \cap \nu_{r_\ell}^j) \in u_{\ell-1}$

For $j = 1$ no problem. In the induction step, i.e. for $j + 1$, we apply the condition (iii) above with $\langle \nu_\ell^j : \ell \in [1, j], \ell \neq q_{j+1} \rangle ^\frown \langle \nu_{q_{j+1}}^j \rangle$ here standing for $\eta_1, \ldots, \eta_j$ there (we want $\nu_{q_{j+1}}^j$ be the last), the condition $\langle \eta_\ell \restriction (q + n^d j) : \ell = 1, \ldots, j \rangle$ with no repetition follows by clause (c), so we get η', $\eta'' \in T_{j+1} \cap ({}^\ell m)$ as there. W.l.o.g. $\eta'(\ell g(\eta' \cap \eta'')) < \eta''(\ell g(\eta' \cap \eta''))$.

We now define ν_ℓ^{j+1} for $\ell = 1, \ldots, j + 1$:

If $\ell \in \{1, \ldots, j + 1\} \setminus \{j + 1, r_{j+1}\}$ then $\nu_\ell^{j+1} = \nu_\ell^j$ (remember $T_j \subseteq T_{j+1}$).
If $\rho_{j+1}(\ell g(\rho_{j+1} \cap \rho_{r_{j+1}})) < \rho_{r_{j+1}}(\ell g(\rho_{j+1} \cap \rho_{r_{j+1}}))$
then $\nu_{j+1}^{j+1} = \eta'$ and $\nu_{r_{j+1}}^{j+1} = \eta''$.
If $\rho_{r_{j+1}}(\ell g(\rho_{j+1} \cap \rho_{r_{j+1}})) < \rho_{j+1}(\ell g(\rho_{j+1} \cap \rho_{r_{j+1}}))$
then $\nu_{j+1}^{j+1} = \eta''$ and $\nu_{r_{j+1}}^{j+1} = \eta'$.

Now check. (Note T_0, u_0 could be omitted above.)

Carrying the Inductive Definition. For $j = n$, trivial: let $T_n = T^*$ (given in $\otimes_d$).
For $j \in [2, n)$, where $T_{j+1}, \ldots, T_n$ are already defined, we apply Claim 2 with

$$m_{q+nd+j+1}, \quad m_{q+nd+j}, \quad n^d(j+1), \quad j, \quad q+n^dj, \quad [q+n^dj, q+n^dj+n^d), \quad n^dj$$

here standing for

$$m^*, \quad m, \quad \ell, \quad k, \quad i(*), \quad u, \quad p$$

there. (I.e. the tree in claim 2 is replaced by one isomorphic to it, levels outside $\bigcup_{i<j} u_i$ can be ignored.)
So we need to check $m_{q+nd+j+1} \geq n^dj(m_{q+nd+j})^{n^d(j+1)j+1}$, which holds by the definition of the m_i's (as $d \leq c-1$). <u>But</u> we require above more than in Claim 2 (preferring the colour 0). But if it fails for T_j then for some $\eta_1, \ldots, \eta_j$ in T_j we have S as in $\oplus$ of Claim 2, with no $\eta^1 \neq \eta^2 \in S \cap {}^{\ell}m$ such that $f'(\{\eta', \eta''\}) = 0$. On S we can apply our induction hypothesis on d — allowed as the original f misses a colour (the colour zero) when restricted to S. $\square_3$

5. Fact

Let $\langle \eta_i : i < (2m-1)^{\ell} \rangle$ enumerate ${}^{\ell}(2m-1)$ in lexicographic order. Let

$$A = \{0, 2, \ldots, 2m-2\} \subsetneq [0, 2m-1), \text{ so } |A| = m,$$

$$B := \{i < (2m-1)^{\ell} : \eta_i \in {}^{\ell}A\}.$$

Let

$$low_{m,\ell}(k) = (2m-1)^{\ell-k-1}, \text{ and}$$
$$high_{m,\ell}(k) = (2m-1)^{\ell-k} - 1.$$

Then:
(0) if $i \neq j$ are in B, $|\eta_i \cap \eta_j| = k$ then:

$$i < j \text{ iff } \eta_i(k) < \eta_j(k).$$

(1) If $i < j$ are in B, $|\eta_i \cap \eta_j| = k$, then:

$$low_{m,\ell}(k) \leq j - i \leq high_{m,\ell}(k)$$

(2) If $i_1 < j_1$, $i_2 < j_2$ are all in B, and $|\eta_{i_1} \cap \eta_{j_1}| \neq |\eta_{i_2} \cap \eta_{j_2}|$, then:

$$j_1 - i_1 < j_2 - i_2 \quad \text{iff} \quad |\eta_{i_1} \cap \eta_{j_1}| > |\eta_{i_2} \cap \eta_{j_2}|.$$

Proof. (0) Check.
(1) Let $\nu := \eta_i \cap \eta_j$ and $k := |\nu|$. We are looking for upper and lower bounds of the cardinality of the set (the order is lexicographic)

$$C := \{\eta \in {}^{\ell}(2m-1) : \eta_i \leq \eta < \eta_j\}.$$

Clearly each $\eta \in C$ satisfies $\nu \triangleleft \eta$. Moreover, since $\eta_j \in {}^\ell m$, each element $\eta \in C$ must satisfy $\eta(k) \leq \eta_j(k) < 2m-1$. Hence

$$C \subseteq \bigcup_{s<2m-1} [\nu^\frown\langle s\rangle]_{({}^\ell(2m-1))} \setminus \{\eta_j\}$$

so we get $|C| \leq (2m-1)\cdot(2m-1)^{\ell-k-1} - 1 = (2m-1)^{\ell-k} - 1$.
For $k = \ell - 1$ the lower bound claimed in (1) is trivial, so assume $k \leq \ell - 2$. Let $\nu' := (\eta_i{\restriction}k)^\frown\langle\eta_i(k)+1\rangle$ (note: as $\eta_i(k) < \eta_j(k)$ are both in A, necessarily $\nu'(k) < \eta_j(k)$). Then we have

$$C \supseteq \bigcup_{s<2m-1} [\nu'^\frown\langle s\rangle]_{({}^\ell(2m-1))} \cup \{\eta_i\},$$

so $|C| \geq (2m-1)^{\ell-k-1} + 1$.
Proof of (2): Check that $high_{m,\ell}(k+1) < low_{m,\ell}(k)$, and use (1). $\square_4$

Remark. Also $low_{m,\ell}(k) = (2m-1)^{\ell-k-1} + 1$ is O.K. but with the present bound we can use only η_i with $\eta_i(\ell-1) < m$, $B = \{i : \eta_i{\restriction}(\ell-1) \in {}^\ell A\}$. So $(2m-1)^\ell$ can be replaced by $m(2m-1)^{\ell-1}$.

6. Conclusion

If $f : [0,(2m-1)^\ell)^{[2]} \to [0,c)$, $\ell := nc$, $m := 2^{(n+1)^{(c+1)n}}$, then we can find $a_0 < \ldots < a_{n-1} < m$ such that $f{\restriction}[\{a_1,\ldots,a_{n-1}\}]^2$ constant and $\langle a_{i+1} - a_i : i < n-1\rangle$ in any pregiven order

Proof. As in fact 5, let $\langle \eta_i : i < (2m-1)^\ell\rangle$ enumerate ${}^\ell(2m-1)$ in lexicographic order, and let $\overline{B} := \{i < (2m-1)^\ell : \eta_i \in {}^\ell m\}$ and for $i \in \overline{B}$ define η_i' by $\eta_i'(j) = 2\cdot\eta_i(j)$ for $j < \ell$. So η_i' is in the set B from Fact 5. Define a function $f' : [{}^\ell m]^2 \to c$ by requiring $f'(\{\eta_i', \eta_j'\}) = f(\{i,j\})$ for all $i,j \in B$. Now the conclusion follows from Lemma 4., Remark 4.1(2) and Fact 5., particularly clause (2). $\square_5$

References

[Al] Noga Alon. Notes.

[Sh:329] Saharon Shelah. *Primitive recursive bounds for van der Waerden numbers.* Journal of the American Mathematical Society, **1**:683–697, 1988.

[GrRoSp] R. Graham, B. Rotchild, and J. Spencer. *Ramsey Theory.* Wiley – Interscience Series in Discrete Mathematics. Wiley, New York, 1980.

[Sh:37] Saharon Shelah. *A two-cardinal theorem.* Proceedings of the American Mathematical Society, **48**:207–213, 1975.

VI. Geometry

Introduction

Erdős' love for geometry and elementary or discrete geometry in particular, dates back to his beginnings. The Erdős-Szekeres paper has been influential and certainly helped to create discrete geometry as we know it today. But Erdős also put geometry to the service to other branches, giving definition to various geometrical graphs and proving bounds on their chromatic and independence numbers. We are happy to include papers by Moshe Rosenfeld, Pavel Valtr, Janos Pach, Jiří Matoušek and, in particular, a paper by Miklós Laczkovich and Imre Ruzsa on the number of homothetic sets. While the paper of Peter Fishburn is closely related to Erdős' favorite theme, the papers of N. G. de Bruijn (on Penrose tiling) and J. Aczél and L. Losonczi (on functional equations) cover broader related aspects. It is perhaps fitting to complement this introduction by a few related Erdős problems in his own words:

Let $x_1, \dots, x_n$ be n points in the plane, not all on a line, and join every two of them. Thus we get at least n distinct lines. This follows from Gallai-Sylvester but also from a theorem of de Bruijn and myself.

My most striking contribution to geometry is no doubt my problem on the number of distinct distances. This can be found in many of my papers on combinatorial and geometric problems.

Hickerson, Pach and I proved that on the unit sphere one can find n points for which the distance $\sqrt{2}$ can occur among $n^{1/3}$ pairs. Perhaps this is best possible. In fact, there are $n^{1/3}$ points at distance 1 from every other point. For every $0 < \alpha < 2$ there are n points so that for every point there are $\log^* n$ other points at distance α — again we do not know if this is best possible.

Purdy and I proved (using an idea of Kárteszi) that there are n points in the plane with no three on a line for which the unit distance occurs at least $cn \log n$ times. We have no nontrivial upper bound. If the points are in 3-space the unit distance can occur $n^{4/3}$ times (Hickerson, Pach and myself) but if we also assume that no four are on a plane, we can do no better than $cn \log n$.

Szekeres and I proved, that if $\binom{2n-4}{n-2} + 1$ points are given in a plane no three on a line, then we can always select among them n points which are the vertices of a convex n-gon. Probably $2^{n-2} + 1$ is the correct value — we proved that 2^{n-2} is not enough. This problem (which was due to E. Klein, i.e., Mrs. Szekeres) had a great influence.

Extension of Functional Equations

J. Aczél[1*] and L. Losonczi[2]

[1] University of Waterloo, Department of Pure Mathematics, Waterloo, Ontario N2L 3G1, Canada
[2] Kuwait University, Department of Mathematics, P.O.Box 5969, Safat 13060, Kuwait

1. Introduction

Extension theorems are common in various areas of mathematics. In topology continuous extensions of continuous functions are studied. In functional analysis one is interested mainly in linear extensions of linear operators preserving continuity or some other properties like bounds or norm. In algebra extensions of homomorphisms and isomorphisms are investigated. The latter can be considered as extensions of functional equations.

In the area of functional equations one is interested in extending functional equations, i.e., either extending functions satisfying a functional equation (or a system of equations) on a "restricted" set to functions which satisfy the same equation (or system of equations) on the "maximal" set or showing that given functions satisfying a functional equation (or a system of equations) on a restricted set satisfy the same equation (or system) on a "larger" set.

The first extension theorem for the Cauchy functional equation is in the 1965 joint paper [AE 65] of Erdős with Aczél who proved that, if a function $f :]0, \infty[\to \mathbb{R}$ satisfies the Cauchy functional equation for all positive values of x and y, that is

$$f(x+y) = f(x) + f(y) \quad (x, y \in]0, \infty[),$$

then there exists a unique function $F : \mathbb{R} \to \mathbb{R}$ such that $F(x) = f(x)$ if $x \in]0, \infty[$ (i.e. F is an extension of f) and F satisfies the Cauchy equation for all values of x, y:

$$F(x+y) = F(x) + F(y) \quad (x, y \in \mathbb{R}). \tag{1.1}$$

In the same paper they showed that there exist no Hamel bases of the set of nonnegative numbers, i.e., there does not exist any set B' whose elements are nonnegative and any nonnegative number is representable as a linear combination of a finite number of elements of B' with nonnegative rational coefficients in a unique way.

---

* Research supported by the Natural Sciences and Engineering Research Council of Canada grants nr. OGP002972 and CPG0164211
1991 Mathematics Subject Classification: 39 B 22, 39 B 52
Key words and phrases: functional equation, extension of functional equation

In 1960 in the problems section of Colloquium Mathematicum Erdős [E 60] raised the following problem. If $f(x+y) = f(x) + f(y)$ is satisfied for almost all (x, y) in the plane (in the sense of plane Lebesgue measure) does there exist a function F satisfying the Cauchy equation everywhere (i.e. satisfying (1.1)) such that $f(x) = F(x)$ almost everywhere (in the sense of the linear Lebesgue measure)?

The Aczél–Erdős theorem and the problem of Erdős (solved independently by Jurkat [J 65] and de Bruijn [B 66]) became the starting points of two different directions of the extension theory of functional equations, which by now is quite extensive (Kuczma [K 78] quoted already in 1978 more than 100 related papers; see also [PR 95]).

The aim of this paper is to show how the above result and problem of Paul Erdős influenced the development of the extension theory of functional equations.

2. Extensions of homomorphisms

Let G, H be (multiplicatively written) groups and S be a subsemigroup of S. Let further $f : S \to H$ be a homomorphism of S into H. Under what conditions can f be extended in a unique way to a homomorphism of G? With $G = H =$ the additive group of $\mathbb{R}$ and S=the additive group of positive real numbers, we get back to the Aczél– Erdős extension problem. The general problem was treated by Aczél et al. [ABDKR 71] by proving several theorems which under different assumptions on S gave a solution. The first one is closely related to the extension theorem in [AE 65].

Suppose that for every element $x \in G$, different from the unit element, either $x \in S$ or $x^{-1} \in S$ (or both). Then every homomorphism $f : S \to H$ can be extended uniquely to a homomorphism $F : G \to H$ of G into H.

This can be proved by defining F by

$$F(x) := \begin{cases} E & \text{if } x = e \\ f(x) & \text{if } x \in S \\ f(x^{-1})^{-1} & \text{if } x^{-1} \in S \end{cases}$$

where e, E are the unit elements of G, H, respectively.

A unique extension also exists when S generates the Abelian group G.

Indeed, in this case $G = S \cdot S^{-1} = \{xy^{-1} \mid x \in S, y \in S\}$ and by

$$F(xy^{-1}) := f(x)f(y)^{-1} \quad (x, y \in S)$$

F is well defined on G (because $xy^{-1} = uv^{-1}$ $(x, y, u, v \in S)$ implies $xv = uy$) and supplies the unique extension of f.

A more elaborate result is the following.

Suppose that S is a subsemigroup of the group G such that

$$G = S \cdot S^{-1} \cdot S \cdot S^{-1} = \{xy^{-1}uv^{-1} \mid x, y, u, v \in S\} \tag{2.1}$$

and f is a homomorphism of S into H. The map f can be extended to a homomorphism of G into H if and only if

$$x, y, z, u, v, w \in S, xy^{-1}z = uv^{-1}w \textit{ implies } f(x)f(y)^{-1}f(z) = f(u)f(v)^{-1}f(w).$$

When the extension exists, then it is unique.

There are further results in this vein in [ABDKR 71]. Martin [M 77] generalized these results to the case when G can be represented in a product form similar to (2.1) but with an arbitrary number of factors.

Necessary and sufficient conditions for the extension were found by Osondu [O 78] in the case when S generates G:

Let S be a subsemigroup of a group G which is generated by S and f a homomorphism of S into some group H. Then f can be extended to a homomorphism of G into H if and only if f satisfies the following condition.

For every positive integer n and for every $s_i \in S$, $\epsilon_i \in \{-1, 1\}$ $(i = 1, 2, \ldots, n)$

$$\prod_{i=1}^{n} s_i^{\epsilon_i} = e \Longrightarrow \prod_{i=1}^{n} f(s_i)^{\epsilon_i} = E$$

holds, where e, E are the unit elements of G, H respectively.

In this result the necessary and sufficient condition seems to be too "close" to the statement itself. A more detached condition may be desirable.

3. Extensions of the Cauchy and Pexider equation and their applications

We introduce some terminology (see Daróczy–Losonczi [DaL 67]). Let D be a nonempty subset of $\mathbb{R}^2$. We write

$$\begin{aligned} D_x &:= \{x \mid \exists y : (x, y) \in D\}, \\ D_y &:= \{y \mid \exists x : (x, y) \in D\}, \\ D_{x+y} &:= \{x + y \mid (x, y) \in D\}. \end{aligned}$$

Then D_x, D_y are the projections of D onto the x-axis, y-axis, respectively, and D_{x+y} is the projection of D, parallel to the line $x+y = 0$, onto the x-axis. Let further $D' := D_x \cup D_y \cup D_{x+y}$. We say that a function f is *additive on the set D* (or satisfies the Cauchy equation on D) if $f : D' \to \mathbb{R}$ and

$$f(x + y) = f(x) + f(y) \quad ((x, y) \in D) \tag{3.1}$$

holds. A function $F : \mathbb{R} \to \mathbb{R}$ is called an *(additive) extension* of a function f additive on D (we say also that F extends the Cauchy equation from D to $\mathbb{R}^2$) if F satisfies the Cauchy equation on $\mathbb{R}^2$ and $F(x) = f(x)$ for $x \in D'$.

We saw above that there exists a unique (additive) extension from $\mathbb{R}_+^2$ ($\mathbb{R}_+$ is the set of positive reals) to $\mathbb{R}^2$ (take $S = \mathbb{R}_+$). As Daróczy and Losonczi [DaL 67] showed, *there always exists a unique (additive) extension from a neighborhood of* $(0,0)$ *to* $\mathbb{R}^2$. They took the neighborhood to be circular but it turned out [A 83] that "hexagonal neighborhoods" of $(0,0)$

$$H_r := \{(x,y) \mid x, y, x+y \in]-r, r[\}$$

are more convenient. Indeed, if

$$f(x+y) = f(x) + f(y) \text{ for } (x,y) \in H_r \tag{3.2}$$

then define for *any* $t \in \mathbb{R}$

$$F(t) := nf\left(\frac{t}{n}\right) \quad (n \in \mathbb{N}, \frac{t}{n} \in]-r, r[).$$

This definition is unambiguous since, by (3.2),

$$nf\left(\frac{t}{n}\right) = nmf\left(\frac{t}{nm}\right) = mnf\left(\frac{t}{mn}\right) = mf\left(\frac{t}{m}\right) \text{ for } \frac{t}{n}, \frac{t}{m} \in]-r, r[.$$

Clearly $F(x) = f(x)$ for $x \in]-r, r[$ (choose $n = 1$). The function F, so defined, is additive on $\mathbb{R}^2$ since, for arbitrary $u, v \in \mathbb{R}$, there exists an $n \in \mathbb{N}$ such that $\frac{u}{n}, \frac{v}{n}, \frac{u+v}{n} \in]-r, r[$, thus, again by (3.2),

$$F(u+v) = nf\left(\frac{u}{n} + \frac{v}{n}\right) = nf\left(\frac{u}{n}\right) + nf\left(\frac{v}{n}\right) = F(u) + F(v).$$

This $F : \mathbb{R} \to \mathbb{R}$, extending (3.2) from H_r to $\mathbb{R}^2$, is unique because, if also $\bar{F} : \mathbb{R} \to \mathbb{R}$ were additive on $\mathbb{R}^2$ and would satisfy $\bar{F}(x) = f(x)$ for $x \in]-r, r[$ then, for arbitrary $t \in \mathbb{R}$ choosing $n \in \mathbb{N}$ again so that $\frac{t}{n} \in]-r, r[$, we get

$$\bar{F}(t) = \bar{F}\left(n\frac{t}{n}\right) = n\bar{F}\left(\frac{t}{n}\right) = nf\left(\frac{t}{n}\right) = F(t) \text{ for all } t \in \mathbb{R}$$

as asserted.

However there may not exist an additive extension even in some very simple cases. Let, for instance f be defined by

$$f(x) = \begin{cases} 2x+1 & \text{for } x \in]2,3[\\ 2x+3 & \text{for } x \in]4,5[. \\ 2x+4 & \text{for } x \in]6,8[\end{cases} \tag{3.3}$$

This function is additive on $D =]2,3[\times]4,5[$ but no additive extension to $\mathbb{R}^2$ exists. Indeed, such an extension F would be continuous on $]2,3[$, thus (see e.g. [A 66]) there would exist a constant c such that $F(x) = cx$ for all $x \in \mathbb{R}$ which contradicts $F(x) = f(x)$ for $x \in D' =]2,3[\cup]4,5[\cup]6,8[$. Nevertheless, we see from (3.3), that $F(x) = 2x$ is "nearly" an extension of this f: it differs from f on D_x, D_y and D_{x+y} just by a constant each and is additive everywhere. Such functions are called *quasi–extensions.*

To be exact, if

$$f(x+y) = f(x) + f(y) \quad ((x,y) \in D) \tag{3.1}$$

holds and there exist constants α, β and a function F, additive on $\mathbb{R}^2$, such that

$$\begin{aligned} f(x) &= F(x) + \alpha && \text{for all } x \in D_x, \\ f(y) &= F(y) + \beta && \text{for all } y \in D_y, \\ f(z) &= F(z) + \alpha + \beta && \text{for all } z \in D_{x+y} \end{aligned} \tag{3.4}$$

then F is an (additive) quasi–extension of f (from D to $\mathbb{R}^2$). These formulas are reminiscent of the general solution

$$\begin{aligned} f(x) &= F(x) + \alpha \\ g(y) &= F(y) + \beta \\ h(z) &= F(z) + \alpha + \beta \end{aligned} \tag{3.5}$$

(F is an arbitrary additive function on $\mathbb{R}^2$, α and β are arbitrary constants) of the Pexider equation

$$h(x+y) = f(x) + g(y) \quad ((x,y) \in \mathbb{R}^2).$$

Since in (3.3) or (3.4) (and also in (3.1)) the three occurences of f are anyway defined on possibly different intervals, this is conducive to consider (3.1) as a *Pexider equation*

$$h(x+y) = f(x) + g(y) \quad ((x,y) \in D) \tag{3.6}$$

with f, g, h defined on D_x, D_y, D_{x+y}, respectively. As it turns out, these have an extension (not only quasi–extension) to $\mathbb{R}^2$ from the above set $D =]2,3[\times]4,5[$ and from every open connected set (region) D.

In fact, the basic result concerning quasi–extensions of the Cauchy equation and extensions of the Pexider equation is that *from any open connected set (region) D in $\mathbb{R}^2$ there exists a unique quasi–extension of the Cauchy equation* (3.1) *to* $\mathbb{R}^2$ (Daróczy–Losonczi [DaL 67]) *and a unique extension of the Pexider equation* (3.6) *to* $\mathbb{R}^2$ (Rimán [R 76], Aczél [A 85], Radó–Baker [RB 87]). The former follows from the latter. Both can be proved by taking for D first a hexagon, like H_r above, but its centre shifted from the origin to (a,b), showing, for equation (3.6), by substitutions that

$$h(u+v+a+b) - h(a+b) = f(u+a) - f(a) + g(v+b) - g(b), \tag{3.7}$$

where (u,v) is in the hexagon H_r around the origin. This implies (putting $v = 0$ or $u = 0$, respectively) that

$$h(t+a+b)-h(a+b)=f(t+a)-f(a)=g(t+b)-g(b) \quad (t\in]-r,r[).$$

Defining the function Φ by equating $\Phi(t)$ to this common value on $]-r,r[$, (3.7) shows that Φ is additive on H_r so, by what we proved above, has a unique (additive) extension F to $\mathbb{R}^2$. The functions given by

$$\begin{aligned}\tilde{f}(x) &= F(x)-F(a)+f(a)\\ \tilde{g}(y) &= F(y)-F(b)+g(b)\\ \tilde{h}(z) &= F(z)-F(a+b)+h(a+b)\end{aligned} \tag{3.8}$$

($x,y,z\in\mathbb{R}$), and only these, extend the Pexider equation from the (a,b)-centred hexagon to $\mathbb{R}^2$. The proof that *the Pexider equation has a unique extension from any open connected region D to $\mathbb{R}^2$* can be completed by covering D by a sequence of hexagons such that each hexagon is contained in D and any two consecutive hexagons have nonempty intersection and by applying the above extension process to each hexagon. Of course, (3.8) has the form (3.5) of the general solution of Pexider's equation on $\mathbb{R}^2$.

The general solution of the Pexider equation (3.6) on D is obtained by replacing, in (3.8), $-F(a)+f(a)$ and $-F(b)+g(b)$ by two arbitrary constants and $-F(a+b)+h(a+b)$ by their sum, and restricting $\tilde{f},\tilde{g},\tilde{h}$ to D_x, D_y, D_{x+y} respectively. The function F, additive on $\mathbb{R}^2$, which we have just determined, is also the unique quasi–extension of f in (3.1) from D to $\mathbb{R}^2$, in the sense (3.4). If two of the sets $D_x\cap D_y$, $D_x\cap D_{x+y}$, $D_y\cap D_{x+y}$ are nonempty, then the quasi–extension is an extension of the Cauchy equation.

We note that the concept of extension and quasi–extension can be defined similarly for more general (in particular some closed) domains, ranges (groups, vector spaces) and equations. In their paper [RB 87], quoted above, Radó and Baker had a real or complex topological vector space as domain of $\tilde{f},\tilde{g},\tilde{h}$ and an Abelian group as range. We mention a few further results.

Let G and H be Abelian divisible groups and let X be a subset of G having the properties

(i) *for any $x\in X$ and for all rational numbers $\lambda\in]0,1[$ we have $\lambda x\in X$;*

(ii) *for any pair $(x,y)\in X^2$ there exists an integer n, which may depend upon x and y, such that $(x+y)/n\in X$.*

Suppose that $f: X\to H$ is additive on $D:=\{(x,y)\mid x\in X, y\in X, x+y\in X\}$. Then there exists an additive extension $F: G\to H$ of f. Moreover F is unique if, and only if, the subgroup generated by X is G. (Ng [N 74], Dhombres–Ger [DG 78]).

Székelyhidi [S 72] found the general representation of functions which are *additive on an open set $D\subset\mathbb{R}^2$*. He also proved [S 81] an extension theorem for the equation $\Delta_y^{n+1}f(x)=0$ where Δ_y is the usual difference operator defined by $\Delta_y f(x):=f(x+y)-f(x)$. Let again $D\subset\mathbb{R}^2$, n a positive integer and for $k=0,1,\ldots,n+1$ let

$$D_k:=\{x+(n+1-k)y\mid (x,y)\in D\}.$$

The result by Székelyhidi is the following. *Let $D \subset \mathbb{R}^2$ be an open and connected set with $(0,0) \in D$. Let n be a positive integer and $f : \cup_{k=0}^{n+1} D_k \to \mathbb{R}$ be a function such that*

$$\Delta_y^{n+1} f(x) = \sum_{k=0}^{n+1} \binom{n+1}{k} (-1)^k f[x + (n+1-k)y] = 0 \quad ((x,y) \in D).$$

Then there exists an extension $F : \mathbb{R} \to \mathbb{R}$ of f such that $\Delta_y^{n+1} F(x) = 0$.

The Levi–Cività equation

$$f(x+y) = \sum_{k=1}^{n} g_k(x) h_k(y) \tag{3.9}$$

is a common generalization of the Cauchy and Pexider equation and the sine and cosine equations.

Recently the second author [L 85a], [L 90] proved an extension theorem which corresponds to the Aczél–Erdős result for equation (3.9).

Let the functions $f, g_k, h_k :]0, \infty[\to \mathbb{C}$ $(k = 1, \ldots, n)$ satisfy the functional equation (3.9) *for all positive x, y and assume that the functions $g_1, \ldots, g_n$ and $h_1, \ldots, h_n$ are linearly independent on $]0, \infty[$. Then there exists a unique set of functions $F, G_k, H_k : \mathbb{R} \to \mathbb{C}$ $(k = 1, \ldots, n)$ such that*

$$F(x+y) = \sum_{k=1}^{n} G_k(x) H_k(y) \qquad (x, y \in \mathbb{R})$$

and

$$\begin{aligned} F(x) &= f(x) \\ G_k(x) &= g_k(x) \qquad (x \in]0,\infty[;\ k = 1, \ldots, n) \\ H_k(x) &= h_k(x) \end{aligned}$$

hold. If f, g_k, h_k $(k = 1, \ldots, n)$ are continuous or measurable on $]0, \infty[$ then so are F, G_k, H_k $(k = 1, \ldots, n)$ on $\mathbb{R}$. The proof is based on the following observation. It is known ([A 66] pp. 201–203) that the functions $F_1, \ldots, F_n$, defined on a set A with values in an arbitrary field, are linearly independent on A if, and only if, there exist elements $a_1, \ldots, a_n \in A$ (necessarily distinct) such that $\det(F_i(a_j))_{i,j=1}^n \neq 0$. Thus by the linear independence of the functions $g_1, \ldots, g_n$ there exist $x'_1, \ldots, x'_n \in]0, \infty[$ such that $\det(g_i(x'_j))_{i,j=1}^n \neq 0$. Without loss of generality we may suppose that $x'_1 = \min\{x'_1, \ldots, x'_n\}$. Let $x_k = x'_k - x'_1$ $(k = 1, \ldots, n)$ and substitute $x + x_k$ for x $(k = 1, \ldots, n)$ in (3.9). We obtain a system of equations which can be written as

$$\mathcal{F}(x+y) = \mathcal{G}(x)\mathcal{H}(y) \qquad (x, y \in]0, \infty[) \tag{3.10}$$

where, for $x \in]0, \infty[$ $(x_1 = 0)$,

$$\begin{aligned}
\mathcal{F}(x) &= \Big(f(x+x_1),\ldots,f(x+x_n)\Big)^T \\
\mathcal{G}(x) &= \begin{pmatrix} g_1(x+x_1) & \ldots & g_n(x+x_1) \\ \vdots & & \vdots \\ g_1(x+x_n) & \ldots & g_n(x+x_n) \end{pmatrix} \\
\mathcal{H}(x) &= \Big(h_1(x),\ldots,h_n(x)\Big)^T
\end{aligned}$$

and T denotes transposition. Extending the Pexider equation (3.10) we obtain an extension of (3.9).

4. Almost everywhere additive functions

The problem of Erdős, mentioned in the introduction, was inspired by Hartman [H 61] who proved that, if $f(x+y) = f(x) + f(y)$ is satisfied for $(x,y) \in A \times A$, where the complement of A has (linear) measure zero, then there exists a function $F : \mathbb{R} \to \mathbb{R}$ satisfying the Cauchy equation everywhere and $F(x) = f(x)$ almost everywhere on $\mathbb{R}$. Actually Hartman's paper [H 61] appeared later than [E 60] as a partial solution to the Erdős problem [E 60]. Jurkat [J 65] and de Bruijn [B 66] (independently) solved the problem in the affirmative, showing that, *if $f(x+y) = f(x) + f(y)$ holds for all $(x,y) \in D$ where the Lebesgue measure of $\mathbb{R}^2 \setminus D$ is zero, then there exists a function $F : \mathbb{R} \to \mathbb{R}$ satisfying the Cauchy equation everywhere and $F(x) = f(x)$ a.e., that is, for all $x \in D_1$ where $\mathbb{R} \setminus D_1$ is of Lebesgue measure zero.*

Is this F an (additive) extension of f (in the sense of Daróczy–Losonczi)?

The answer is *no.* Counter example (by V. Zinde–Walsh, simplified by C.T. Ng and J. Aczél, see [A 80], [AD 89]) :

$$D := \{(x,y) \mid x, y, x+y \neq 1, 2\} \cup \{(1,1)\} ,$$

i.e., D is obtained from $\mathbb{R}^2$ by removing from it the lines $x = 1,\ x = 2,\ y = 1,\ y = 2,\ x+y = 1,\ x+y = 2$ except the point $(1,1)$. Clearly $\mathbb{R}^2 \setminus D$ is of Lebesgue measure zero and $D' = \mathbb{R}$. Let

$$f(x) = \begin{cases} 0 & \text{if } x \in \mathbb{R} \setminus \{1,2\}, \\ 1 & \text{if } x = 1, \\ 2 & \text{if } x = 2. \end{cases}$$

This function is additive on D (we have $f(1+1) = f(2) = 2 = 1+1 = f(1) + f(1)$ and elsewhere, for $(x,y) \in D$, we get $f(x) = f(y) = f(x+y) = 0$) but f is obviously not additive on $D' = \mathbb{R}$.

Questions similar to Erdős's problem can be asked for other functional equations and inequalities.

Let $I \subset \mathbb{R}$ be an open interval. A function $f : I \to \mathbb{R}$ is called *almost convex* if

$$f\left(\frac{x+y}{2}\right) \leq \frac{f(x)+f(y)}{2} \qquad ((x,y) \in I \times I \setminus M),$$

where $M \subset I \times I$ is of planar Lebesgue measure zero. Kuczma [K 70] proved that, *if $f : I \to \mathbb{R}$ is almost convex, then there exists a unique convex function $F : I \to \mathbb{R}$ such that $F(x) = f(x)$ a.e. in I.*

Ger [G 71], [G 75] noticed that "equality almost everywhere" can also be introduced in an axiomatic way.

A nonempty family $\mathcal{I}^k$ of subsets of the k-dimensional euclidean space $\mathbb{R}^k$ is called *a Linearly Independent Proper Ideal* (abbreviated as LIPI) if

(a) $A, B \in \mathcal{I}^k$ implies $A \cup B \in \mathcal{I}^k$,

(b) $A \in \mathcal{I}^k, B \subset A$ implies $B \in \mathcal{I}^k$,

(c) $\mathbb{R}^k \notin \mathcal{I}^k$,

(d) for every real number $\alpha \neq 0, \beta \in \mathbb{R}^k$, and $A \in \mathcal{I}^k$ we have $\alpha A + \beta \in \mathcal{I}^k$.

It can be easily seen that the family $\mathcal{L}_0^k$ of all subsets of $\mathbb{R}^k$ with Lebesgue measure zero is a LIPI. Similarly, the families $\mathcal{F}^k$ and $\mathcal{L}_f^k$ of all sets of the first category in $\mathbb{R}^k$ and of all sets having finite Lebesgue measure in $\mathbb{R}^k$, are also LIPIs.

We say that that two LIPIs $\mathcal{I}^2$ and $\mathcal{I}^1$ are *conjugate* if for every $M \in \mathcal{I}^2$ there exists a set $U \in \mathcal{I}^1$ such that, for every $x \notin U$, the set

$$V_x := \{y \mid (x,y) \in M\}$$

belongs to $\mathcal{I}^1$. In view of Fubini's theorem, the LIPIs $\mathcal{L}_0^2, \mathcal{F}^2, \mathcal{L}_f^2$ and $\mathcal{L}_0^1, \mathcal{F}^1, \mathcal{L}_f^1$ are, respectively, conjugate.

Ger [G 71] proved the following result. *Let $\mathcal{I}^2$ and $\mathcal{I}^1$ be conjugate linearly invariant proper ideals. If $f : \mathbb{R} \to \mathbb{R}$ satisfies the equation $\Delta_y^{n+1} f(x) = 0$ for all $(x,y) \in \mathbb{R}^2 \setminus M$ where $M \in \mathcal{I}^2$ then there exists exactly one function $F : \mathbb{R} \to \mathbb{R}$ and a set $U \in \mathcal{I}^1$ such that $\Delta_y^{n+1} F(x) = 0$ holds for all $(x,y) \in \mathbb{R}^2$ (i.e., F is a polynomial function) and $F(x) = f(x)$ for all $x \notin U$.*

Ger [G 75] found similar results for the Mikusiński equation

$$f(x+y)\,[f(x+y) - f(x) - f(y)] = 0$$

and for the Pexider equation

$$f(x+y) = g(x) + h(y).$$

5. Some applications

The above extension theorems have important and amusing applications. An example is that of "aggregated allocations" see [AKNW 83], [ANW 84], [AW 81].

Suppose that there is a certain amount s of quantifiable goods (raw materials, energy, money for scientific projects, grants etc.) is to be allocated (completely) to $m(\geq 3)$ projects. For this purpose a committee of n assessors is formed. The i–th assessor recommends that the amount x_{ij} should be allocated to the j–th project. If he (she) can count, then $\sum_{j=1}^{m} x_{ij} = s$. Now the recommendations should be aggregated into a consensus allocation by a chairman or external authority. Again the allocated consensus amounts should *add up to s*. It is supposed (this is a restriction) that the aggregated allocation for the j–th project depends only on the recommended allocations to *that* project, that both the individual and the aggregated allocations be *nonnegative* (!) and, if all assessors recommend rejection, then the consensus allocation to that project should be 0 ('*consensus on rejection*'). If we write the individual allocations for the j–th project as a vector $\mathbf{x}_j = (x_{1j}, x_{2j}, \ldots, x_{nj})$ and denote the aggregated allocation by $f_j(\mathbf{x}_j)$ then these conditions mean the following:

$$f_j : [0,s]^n \to [0,s], \quad f_j(\mathbf{0}) = 0 \quad (j = 1, 2, \ldots, m)$$

and

$$\sum_{j=1}^{m} \mathbf{x}_j = s\mathbf{1} \Longrightarrow \sum_{j=1}^{m} f_j(\mathbf{x}_j) = s \tag{5.1}$$

where $\mathbf{1} = (1, 1, \ldots, 1)$, $s\mathbf{1} = (s, s, \ldots, s)$, and $\mathbf{0} = 0\mathbf{1} = (0, 0, \ldots, 0)$. With $\mathbf{x}_1 = s\mathbf{1}, \mathbf{x}_2 = \ldots = \mathbf{x}_m = \mathbf{0}$ we get $f_1(s\mathbf{1}) = s$ and, since the subscript 1 has no privileged role, $f_j(s\mathbf{1}) = s$ for all $j = 1, 2, \ldots, m$ ('*consensus on overwhelming merit*'). Now we substitute into (5.1) $\mathbf{x}_1 = \mathbf{z} = (z_1, z_2, \ldots, z_n)$, $\mathbf{x}_3 = s\mathbf{1} - \mathbf{z}, \mathbf{x}_2 = \mathbf{x}_4 = \ldots = \mathbf{x}_m = \mathbf{0}$:

$$f_1(\mathbf{z}) = s - f_3(s\mathbf{1} - \mathbf{z}) \quad (\mathbf{z} \in [0,s]^n).$$

The substitution of $\mathbf{x}_1 = \mathbf{x} = (x_1, x_2, \ldots, x_n)$, $\mathbf{x}_2 = \mathbf{y} = (y_1, y_2, \ldots, y_n)$, $\mathbf{x}_3 = s\mathbf{1} - \mathbf{x} - \mathbf{y}, \mathbf{x}_4 = \ldots = \mathbf{x}_m = \mathbf{0}$ gives

$$f_1(\mathbf{x}) + f_2(\mathbf{y}) = s - f_3(s\mathbf{1} - \mathbf{x} - \mathbf{y}) = f_1(\mathbf{x} + \mathbf{y}), \tag{5.2}$$

whenever $\mathbf{x}, \mathbf{y}, \mathbf{x} + \mathbf{y}$ are all in $[0,s]^n$. Putting here $\mathbf{x} = \mathbf{0}$ we get $f_2(\mathbf{y}) = f_1(\mathbf{0}) + f_2(\mathbf{y}) = f_1(\mathbf{y})$ for all $\mathbf{y} \in [0,s]^n$. Again the subscripts 1,2 have no privileged role, so

$$f_1 = f_2 = \ldots = f_m = f \quad \text{on } [0,s]^n$$

and (5.2) reduces to the Cauchy equation

$$f(\mathbf{x} + \mathbf{y}) = f(\mathbf{x}) + f(\mathbf{y}) \quad ((\mathbf{x}, \mathbf{y}) \in D), \tag{5.3}$$

where $D := \{(\mathbf{x}, \mathbf{y}) \mid \mathbf{x}, \mathbf{y}, \mathbf{x} + \mathbf{y} \in [0,s]^n\}$. All solutions of (5.3) can be extended from D (even though it is closed) to $\mathbb{R}^n \times \mathbb{R}^n$. Applying the known result that every n–place additive function, nonnegative on an n–dimensional interval, is of the form of an inner product $< \mathbf{a}, \mathbf{x} >$ ($\mathbf{a}$ is a constant vector with nonnegative components a_i), we have

$$f_j(x_1, x_2, \ldots, x_n) = \sum_{i=1}^{n} a_i x_i \quad (a_i > 0, \sum_{i=1}^{n} a_i = 1;\ j = 1, 2, \ldots, m).$$

We have found that the *weighted arithmetic mean* is the general solution of our allocation problem.

Many applications concern *local solutions* of functional equations. The functional equation

$$f(x + y - xy) + f(xy) = f(x) + f(y) \quad (x, y \in]0, 1[, [0, 1], \text{ or } \mathbb{R})$$

was introduced by Hosszú. The most complicated case is the one where $f :]0, 1[\to \mathbb{R}$ and the equation holds for $x, y \in]0, 1[$. The *general solution* is (Lajkó [La 74])

$$f(x) = A(x) + b \quad (x \in]0, 1[)$$

where $A : \mathbb{R} \to \mathbb{R}$ is an arbitrary additive function (on $\mathbb{R}^2$) and b is an arbitrary constant. The proof is based on the fact that the function $x \to f(x + 1/2) - f(1/2)$ is additive on the "triangle–like" set

$$D = \{(x, y) \mid -\frac{1}{2} < x < 0, \frac{1}{2} + \frac{1}{2x - 1} < y < \frac{1}{2} + 1\}.$$

A is obtained as the quasi–extension of this function.

The extension theorem for the Levi–Cività equation (together with the extension theorem of Daróczy–Losonczi) can be applied to the solution of *functional equations of sum form* (see [L 85a], [L 91]).

Kiesewetter [Ki 65] proved that the "arctan equation"

$$f(x) + f(y) = f\left(\frac{x + y}{1 - xy}\right) \quad (x, y \in \mathbb{R}, 1 - xy \neq 0) \tag{5.4}$$

has no continuous solutions except $f(x) = 0$ $(x \in \mathbb{R})$. Crstici, Muntean and Vornicescu [CMV 83] found the local solutions of the arctan equation valid on the set $D_1 := \{(x, y) \in \mathbb{R}^2 \mid xy < 1\}$ satisfying various regularity conditions (continuity at one point or boundedness or measurability on an interval etc). The general local solution on D_1 was found by Muntean and Vornicescu [MV 83].

The second author proved [L 85b], [L 90] that *the general solution of* (5.4) *is the function*

$$f(x) = A(\arctan x) \quad (x \in \mathbb{R})$$

where $A : \mathbb{R} \to \mathbb{R}$ *is an arbitrary additive function (on* $\mathbb{R}^2$*) periodic with period* π*.*

This clearly implies $f(\tan r\pi) = 0$ for all rational numbers r, explaining Kiesewetter's result. In [L 85b], [L 90], [MV 83] again the extension theorem of Daróczy and Losonczi was the main tool of the proof.

Finally we mention that the result of de Bruijn and Jurkat can be applied to find all bounded multiplicative linear functionals on the complex Banach algebra of Lebesgue integrable functions (under convolution) [A 80].

References

[A 66] J. Aczél, *Lectures on Functional Equations and Their Applications,* Academic Press, New York–London, 1966 [Mathematics in Science and Engineering, Vol. 19].

[A 80] J. Aczél, *Some good and bad characters I have known and where they led. (Harmonic analysis and functional equations),* In: 1980 Seminar on Harmonic Analysis. [Canad. Math. Soc. Conf. Proc., Vol. 1]. Amer Math. Soc., Providence, RI, 1981, pp. 177–187.

[A 83] J. Aczél, *Diamonds are not the Cauchy extensionist's best friend,* C. R. Math. Rep. Acad. Sci. Canada **5** (1983), 259–264.

[A 85] J. Aczél, *28. Remark,* Report of Meeting. The Twenty–second International Symposium on Functional Equations (December 16–December 22, 1984, Oberwolfach, Germany). Aequationes Math. **29** (1985), p. 101.

[ABDKR 71] J. Aczél, J. A. Baker, D. Ž. Djoković, Pl. Kannappan and F. Radó, *Extensions of certain homomorphisms of subsemigroups to homomorphisms of groups,* Aequationes Math. **6** (1971), 263–271.

[AD 89] J. Aczél and J. Dhombres, *Functional Equations in Several Variables,* Cambridge University Press, Cambridge–New York–New Rochelle–Melbourne–Sydney, 1989.

[AE 65] J. Aczél and P. Erdős, *The non–existence of a Hamel–basis and the general solution of Cauchy's functional equation for nonnegative numbers,* Publ. Math. Debrecen **12** (1965), 259–265.

[AKNW 83] J. Aczél, Pl. Kannappan, C. T. Ng and C. Wagner, *Functional equations and inequalities in 'rational group decision making',* In: General Inequalities 3 (Proc. Third Internat. Conf. on General Inequalities, Oberwolfach, 1981). Birkhäuser, Basel–Boston–Stuttgart, 1983, pp. 239–243.

[ANW 84] J. Aczél, C. T. Ng and C. Wagner, *Aggregation theorems for allocation problems,* SIAM J. Alg. Disc. Meth. **5** (1984), 1–8.

[AW 81] J. Aczél and C. Wagner, *Rational group decision making generalized: the case of several unknown functions,* C. R. Math. Rep. Acad. Sci. Canada **3** (1981), 139–142.

[B 66] N. G. de Bruijn, *On almost additive functions,* Colloquium Math. **15** (1966), 59–63.

[CMV 83] B. Crstici, I. Muntean and N. Vornicescu, *General solution of the arctangent functional equation,* Anal. Numér. Théor. Approx. **12** (1983), 113–123.

[DaL 67] Z. Daróczy and L. Losonczi, *Über die Erweiterung der auf einer Punktmenge additiven Funktionen,* Publ. Math. Debrecen **14** (1967), 239–245.

[DG 78] J. Dhombres and R. Ger, *Conditional Cauchy equations,* Glas. Mat. Ser. III. **13 (33)** (1978), 39–62.

[E 60] P. Erdős, *P 310,* Colloquium Math. **7** (1960), 311.

[G 71] R. Ger, *On almost polynomial functions,* Colloquium Math. **24** (1971), 95–101.

[G 75] R. Ger, *On some functional equations with a restricted domain,* Fundamenta Math. **89** (1975), 95–101.

[H 61] S. Hartman, *A remark about Cauchy's equation,* Colloquium Math. **8** (1961), 77–79.

[J 65] W. B. Jurkat, *On Cauchy's functional equation,* Proc. Amer. Math. Soc. **16** (1965), 683–686.

[Ki 65] H. Kiesewetter, *Über die arc tan–Funktionalgleichung, ihre mehrdeutigen stetigen Lösungen und eine nichtstetige Gruppe,* Wiss. Z. Friedrich–Schiller–Univ. Jena Math.–Natur. **14** (1965), 417–421.

[K 70] M. Kuczma, *Almost convex functions,* Colloquium Math. **21** (1970), 279–284.

[K 78] M. Kuczma, *Functional equations on restricted domains,* Aequationes Math. **18** (1978), 1–35.

[La 74] K. Lajkó, *Applications of extensions of additive functions,* Aequationes Math. **11** (1974), 68–76.

[L 85a] L. Losonczi, *An extension theorem,* Aequationes Math. **28** (1985), 293–299.

[L 85b] L. Losonczi, *Remark 32: The general solution of the arc tan equation,* In: Proc. Twenty–third Internat. Symp. on Functional Equations (Gargnano, Italy, June 2–11, 1985). Univ. of Waterloo, Centre for Information Theory, Waterloo, Ont., 1985, pp.74–76.

[L 90] L. Losonczi, *Local solutions of functional equations,* Glasnik Mat. **25 (45)** (1990), 57–67.

[L 91] L. Losonczi, *An extension theorem for the Levi–Cività functional equation and its applications,* Grazer Math. Ber. **315** (1991), 51–68.

[M 77] S. C. Martin, *Extensions and decompositions of homomorphisms of semigroups,* Manuscript, University of Waterloo, Ont., 1977.

[MV 83] I. Muntean and N. Vornicescu, *On the arctangent functional equation,* (Roumanian), Seminarul "Theodor Angheluţa", Cluj–Napoca, 1983, pp. 241–246.

[N 74] C. T. Ng, *Representation for measures of information with the branching property,* Inform. and Control **25** (1974), 45–56.

[O 78] K. E. Osondu, *Extensions of homomorphisms of a subsemigroup of a group,* Semigroup Forum **15** (1978), 311–318.

[PR 95] L. Paganoni and J. Rätz, *Conditional functional equations and orthogonal additivity,* Aequationes Math. **50** (1995), 134–141.

[RB 87] F. Radó and J. A. Baker, *Pexider's equation and aggregation of allocations,* Aequationes Math. **32** (1987), 227–239.

[R 76] J. Rimán, *On an extension of Pexider's equation,* Zbornik Radova Mat. Inst. Beograd N. S. **1(9)** (1976), 65–72.

[S 72] L. Székelyhidi, *The general representation of an additive function on an open point set,* (Hungarian), Magyar Tud. Akad. Mat. Fiz. Oszt. Közl. **21** (1972), 503–509.

[S 81] L. Székelyhidi, *An extension theorem for a functional equation,* Publ. Math. Debrecen **28** (1981), 275–279.

Remarks on Penrose Tilings

N. G. de Bruijn

Department of Mathematics and Computing Science, Technological University Eindhoven, Eindhoven, The Netherlands

1. Introduction

1.1. This paper will cover some details on Penrose tilings presented in lectures over the years but never published in print before. The main topics are: (i) the characterizability of Penrose tilings by means of a local rule that does not refer to arrows on the edges of the tiles, and (ii) the fact that the Ammann quasigrid of the inflation of a Penrose tiling is topologically equivalent to the pentagrid that generates the original tiling.

The fact that any Penrose tiling is the topological dual of the Ammann quasigrid of the inflation was first noticed by Socolar and Steinhardt ([9]). They presented it in the equivalent form that the topological dual of the Ammann quasigrid of a Penrose tiling is the *deflation* of that tiling.

The Ammann quasigrid of the inflation of a Penrose tiling can also be defined as the union of the central lines of the stacks of that tiling. Therefore I refer to that union as the *central grid* of the tiling. It can be defined independently of the original notion of Ammann grid. Actually the definition of the central grid can also be given for other kinds of tilings where there is no obvious definition of an Ammann grid and no obvious definition of deflation.

The paper is intended to be readable more or less independently of previous ones, at least in the sense that all relevant notions will be explained in the paper itself.

1.2. My geometric terminology will use the notion of shapes and 1-shapes in the plane. A *shape* is an equivalence class of figures under similarity transform. Similarity includes multiplications (with respect to a point), shifts and rotations, but no reflections with respect to a line. In terms of complex numbers, this similarity transform just means linear transformation. A *1-shape* is defined similarly, but without multiplications. That means that similarity is replaced by congruence. In other words, figures with the same 1-shape have the same shape and the same size.

Throughout this paper I shall use the two 1-shapes V and W, pictured in Figure 1.1. They are rhombs, with all edges having unit length. The acute angles of V are 36°, and those of W are 72°. V will be called the *thin rhomb* and W the *thick rhomb*. The word "rhomb" will always refer to a V or a W.

The *arrowed rhombs* V_a and W_a (the subscript a stands for "arrowed") are obtained from V and W by putting single and double arrows on the edges in the way depicted in Figure 1.2. They will be called the *Penrose rhombs*.

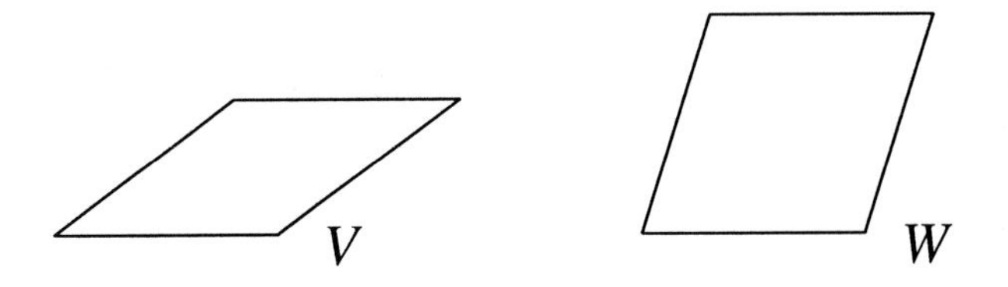

Fig. 1.1. The 1-shapes V (thin rhomb) and W (thick rhomb).

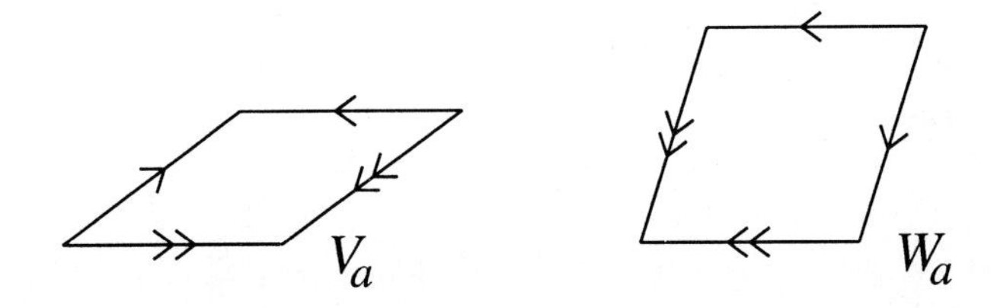

Fig. 1.2. Penrose's arrowed rhombs V_a (thin) and W_a (thick).

1.3. A *Penrose tiling* is a tiling of the plane by V_a's and W_a's with the property that two tiles always have either nothing, or a vertex, or a full edge in common; in the latter case they are called *direct neighbors* and it is required that along the common edge they have the same kind of arrow in the same direction. Figure 1.3 shows a fragment of such a tiling.

Various procedures for obtaining all Penrose tilings are known, like

(i) Penrose's use of deflation, in combination with a (non-constructive) selection argument (see [4] for an exposition).

(ii) the use of deflation, in combination with "updown-generation" (see [5]).

(iii) forming the dual of a "pentagrid" and providing the arrows afterwards (see [1]).

In [5] the relation between (ii) and (iii) was studied in detail.

Somewhat more complicated, but nevertheless interesting and promising, are ways to take the Ammann bars (see [7, 9]) as the basis of the Penrose tilings.

1.4. One of the topics to be treated in the present paper is the question how one can determine whether a tiling of the plane by V's and W's can be provided with arrows so as to form a Penrose tiling with V_a's and W_a's. It

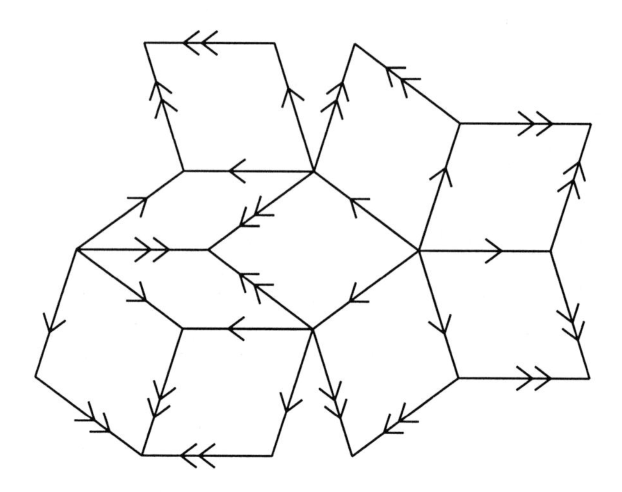

Fig. 1.3. A piece of a Penrose tiling.

will be shown (section 2.6) that this can be settled by inspecting, for every rhomb in the unarrowed tiling, the figure formed by its four direct neighbors.

Section 2.9 will pay attention to the question of a tiling give by the *vertices* only.

2. Arrow-free characterization of Penrose tilings

2.1. The term *unarrowed rhomb tiling* will mean a tiling of the plane with V's and W's, with the condition that neighboring tiles have a full edge in common.

Any Penrose tiling turns into an unarrowed rhomb tiling by omitting the arrows on the edges.

For any unarrowed rhomb tiling, any way to attach an arrow to every edge such that it becomes a Penrose tiling, will be called an *arrowing* of the tiling, and whenever such an arrowing exists the tiling is called *arrowable.* Not every unarrowed rhomb tiling is arrowable. A simple counterexample is the doubly periodic tiling by thick rhombs only.

It will be shown that the condition for an unarrowed rhomb tiling to be arrowable can be put into a form that does not refer to arrows any more.

2.2. Consider a figure formed by an unarrowed rhomb t and four neighboring rhombs t_1, t_2, t_3, t_4, each one of these having one edge in common with t. It is required that there is no overlap between any two of these rhombs (apart from possible common edges). Let me call such a figure a *cross*.

If to a cross formed by t, t_1, t_2, t_3, t_4 one adds a number of further rhombs $t_5, \ldots, t_k$ which all have a vertex in common with t, such that there is no overlap between any two of $t, t_1, \ldots, t_k$, and such that the areas around the vertices of t are entirely covered, then that figure will be called an *extended cross*.

2.3. An *arrowing* of a cross or an extended cross is a way to attach an arrow (either single or double) to every edge in the figure such that each one of its rhombs becomes a Penrose rhomb.

Not every cross can be arrowed. And a cross that can be arrowed is not necessarily extendable to an extended cross that can be arrowed.

A cross will be called *perfect* if it is extendable to an extended cross that permits an arrowing.

2.4. It is not hard to make a list of all possibilities for perfect crosses. Starting with a thin rhomb t, trying all possible sets of direct neighbors t_1, t_2, t_3, t_4, and investigating whether they have at least one arrowable extension, one finds 6 possibilities, pairwise related by 180° rotation, leading to 3 different 1-shapes P_1, P_2, P_3. Similarly, if starting from a thick rhomb, one gets to 8 possibilities, which are pairwise related by 180° rotation, so there are 4 different 1-shapes P_4, P_5, P_6, P_7.

For the notion "1-shape of a perfect cross" I shall also use the term *neighborhood pattern*. Figure 2.1 gives them all.

P_1 and P_2 form a pair, in the sense that the mirror image of P_1 has the same 1-shape as P_2. In the same sense P_5 and P_6 form a pair. The others, P_3, P_4 and P_7, are symmetric: the axis of symmetry is the short diagonal of t in P_3 and the long diagonal of t in P_4 and P_7 .

2.5. Each one of the seven neighborhood patterns can be arrowed in exactly one way. These arrowed patterns are called $P_{1a}, \ldots, P_{7a}$, and are shown in Figure 2.2.

2.6. At this stage it becomes possible to answer the question which unarrowed rhomb tilings can be arrowed.

I shall say that an unarrowed rhomb tiling satisfies the *cross condition* if for each one of its rhombs the cross formed by that rhomb and its four direct neighbors is one of the seven 1-shapes $P_1, \ldots, P_7$.

Theorem 2.1. *(i) If an unarrowed rhomb tiling of the plane is arrowable then it satisfies the cross condition.*

(ii) If an unarrowed rhomb tiling of the plane satisfies the cross condition then it can be arrowed in exactly one way.

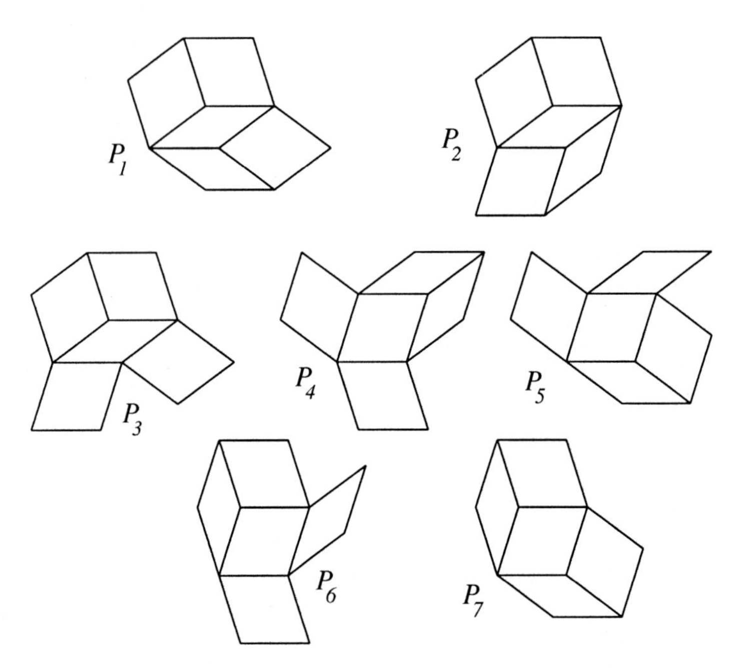

Fig. 2.1. The neighborhood patterns $P_1, \ldots, P_7$.

Proof. (i) This is little more than the fact that the list of neighborhood patterns provided in section 2.4 is exhaustive. If a cross is a part of an arrowable rhomb tiling, then it also has an extension to an extended cross that lies in that tiling, and any arrowing of the whole rhomb tiling implies an arrowing of that extended cross.

(ii) Start from an unarrowed rhomb tiling that satisfies the cross condition. So if t is one of the rhombs in the tiling then it is the central rhomb of a cross of one of the types $P_1, \ldots, P_7$, and that uniquely defines an arrowing of t according to Figure 2.2. So the arrows on t are imposed by the cross of t. They will be called the *imposed arrows* of t.

In this way an arrowing is prescribed for every rhomb in the tiling, but it is the question whether neighbors get the same imposed arrow on their common edge. To that end it has to be shown that if in the tiling two rhombs s, t have an edge in common, then along that edge the arrow of s (imposed by the cross of s) is the same as the arrow of t (imposed by the cross of t). This can be checked without having the full tiling available.

If one of s, t is thin and the other one is thick, then uniqueness of arrowing holds already for the figure consisting of that s and t only. That arrowing on that figure is imposed by the cross of s as well as by the one of t, so in particular the imposed arrow on the common edge is the same in both cases.

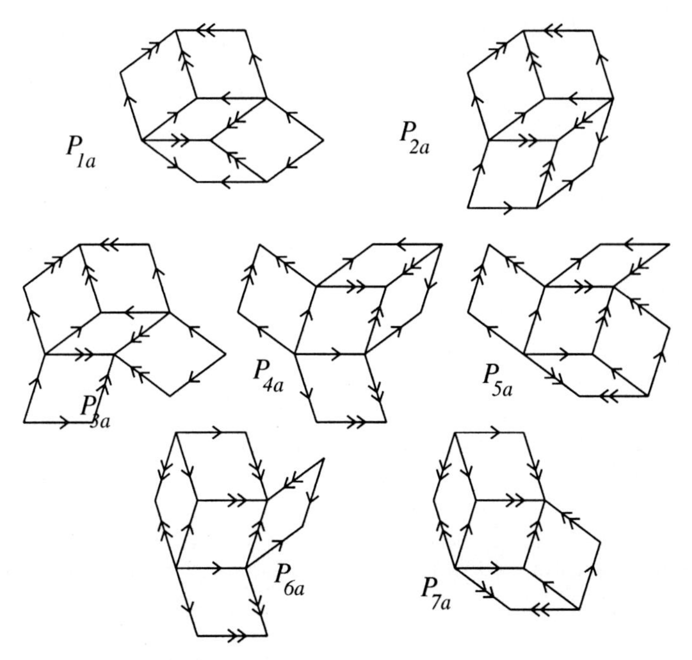

Fig. 2.2. The arrowed neighborhood patterns $P_{1a}, \ldots, P_{7a}$.

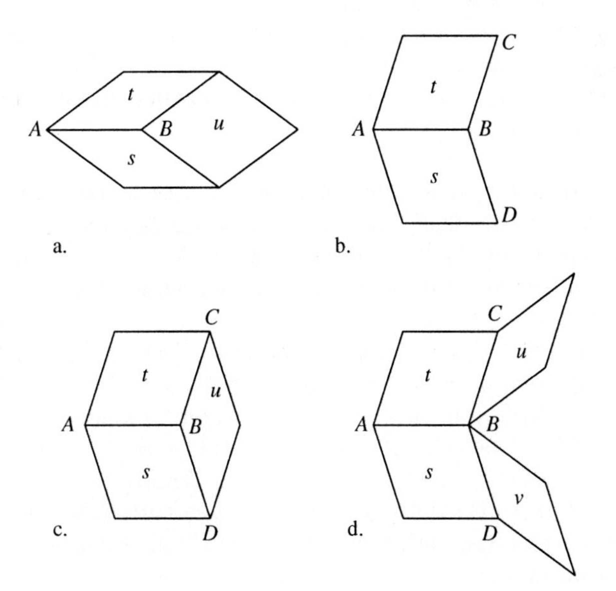

Fig. 2.3. The arrowing of s imposed by the cross of s and the arrowing of t imposed by the cross of t lead to the same arrow on the common edge.

The next case is that s and t are both thin. The cases where the cross of a thin rhomb contains a second thin rhomb are the P_1 and P_2 of Figure 2.1. In both cases s and t have a thick rhomb u as a common neighbor in the way depicted in Figure 2.3a (possibly with s and t interchanged). Just by observing the pair s, u it is seen that the cross of s imposes a double arrow on the common edge of s and t, pointing to the right. With the pair t, u the same result is obtained for the imposed arrow of t.

The remaining case is that both s and t are thick. This is shown in Figure 2.3b (possibly with s and t interchanged). It is a part of a tiling in which the cross condition holds for both s and t. Since the cross of s contains t it cannot contain a thick rhomb u pasted to the edge BC: such a configuration formed by s, t, u does not occur in any of the crosses of s shown in Figure 2.1. The same argument shows that the tiling does not contain a thick rhomb pasted to BD.

The only possibilities to paste thin rhombs to BC and BD are shown in Figures 2.3c and 2.3d.

In Figure 2.3c, the only possible arrowing of the pair s, u has a double arrow from B to A, whence that double arrow is imposed by the cross of s. The same argument applied to the pair t, u leads to the same imposed arrow of t.

In Figure 2.3d the pair t, u shows that the imposed arrow of t on AB is a single one, pointing to the right. Inspection of the pair s, v shows the same imposed arrow of s on AB.

This completes the proof of the theorem.

2.7. As a corollary of the theorem of section 2.6 it can be noted that an unarrowed rhomb tiling can be arrowed in at most one way. This does not require the whole theorem: uniqueness of arrowing holds already for all arrowable extended crosses.

2.8. Every arrowable tiling contains infinitely many copies of each one of $P_1, \ldots, P_7$. Inspection of an arbitrary Penrose tiling shows occurrences of all these perfect crosses, and it is known that any finite configuration occurring in any Penrose tiling occurs infinitely often in any other one.

2.9. An unarrowed rhomb tiling is completely determined by its set of vertices, irrespective of whether the tiling is arrowable or not. In order to see this, it suffices to check that in an unarrowed rhomb tiling any two vertices have distance 1 if and only if they are connected by an edge of a rhomb of the tiling.

It is also easy to see that two vertices in an unarrowed rhomb tiling have distance beteen 0 and 1 if and only if they are the end-points of the short diagonal of a thin rhomb.

The shortest distance exceeding 1 is 2 sin 36° (=1.17557). It is the length of the short diagonal in a thick rhomb, but this distance may occur in a different way between vertices in a pair of adjacent thin rhombs.

3. Deflation and inflation

3.1. This section gives some information about inflation and deflation of Penrose tilings. It is intended as a kind of motivating background, but will not be used in the rest of the paper.

Penrose's idea of inflation and deflation was very central in the discovery of his arrowed rhomb tilings. *Deflation* is a certain way to get from a Penrose tiling to a new tiling with smaller pieces ($\frac{1}{2}(-1+\sqrt{5})$ times the original ones), and *inflation* is the inverse operation, leading to a tiling with bigger pieces. For a description of the process see [1, 6, 7, 8]. It is somewhat easier to describe the deflation as an operation on rhombus *halves*. The half thin rhomb is obtained by cutting a thin rhomb along its short diagonal, the half thick rhomb by cutting a thick rhomb along its long diagonal. For these rhomb halves the deflation is a matter of *subdivision* (see [5]). This is shown in Figure 3.1.

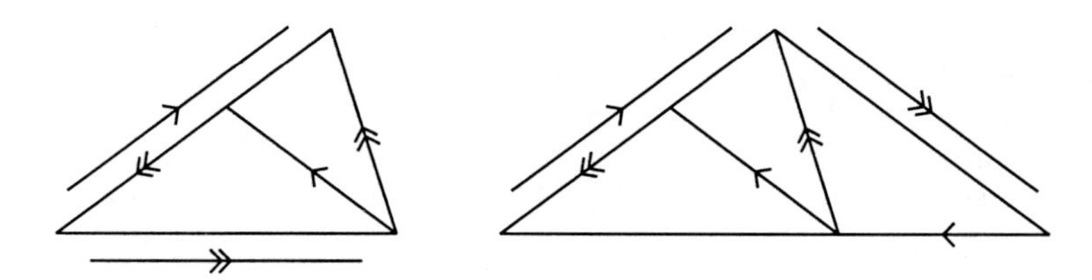

Fig. 3.1. The deflation of the half thin and the half thick rhomb. The full figure on the left is a half thin rhomb, with arrows drawn at the outside. It is subdivided into a half thin and a half thick rhomb, with arrows indicated along the edges themselves. These two small pieces belong to the deflation. Similarly the full figure on the right is a half thick rhomb, and here the deflation has three pieces: a half thin rhomb and two half thick rhombs. There are of course two shapes of half thin rhombs, which are each other's mirror image, and similarly there are two shapes of half thick rhombs. For the mirror images of the half rhombs shown in the figure one has to take the mirror image subdivision.

When this subdivision operation is carried out for every half thin rhomb and every half thick rhomb of a Penrose tiling, the smaller pieces fit together to a Penrose tiling with smaller pieces.

In section 2.3 of [5] it is explained that the idea of inflation comes naturally by observing how in an arrowed rhombus tiling pieces can be grouped together. Those groups form the pieces of the inflation.

3.2. If deflation is applied to a tiling of a *finite* portion of the plane, and if the result is enlarged by a factor $\frac{1}{2}(1+\sqrt{5})$ afterwards, it becomes a tiling of a bigger part of the plane with pieces of the original size. Infinite repetition of this process and application of an infinite selection principle leads, albeit in an inconstructive way, to Penrose tilings of the full plane (see [4, 6, 8]). But the idea of inflation and deflation can also be used in a completely different and very constructive way for the production of *all* infinite Penrose tilings. The method is *updown generation*, a process that is controlled by taking an arbitrary infinite path in a particular finite automaton (see [5]).

4. Duality

4.1. The key method in [1] was the production of tilings by means of their topological dual. I spend a few words here as a short independent introduction.

4.2. In any Penrose tiling the rhombs are arranged in what I shall call *stacks*. Consider an arbitrary edge e of an arbitrary rhomb r of the tiling. Two rhombs of the tiling are called *e-neighbors* if they have a common edge that is parallel to e. Two rhombs of the tiling are called *e-related* if they can be connected by a chain of e-neighbors. The stack generated by r and e is the set of all rhombs which are e-related to r.

Every rhomb of the tiling belongs to exactly two stacks, and any two stacks have exactly one rhomb in common, unless their e's are parallel.

The idea of a stack can at once be generalized to tilings by parallelotopes in higher dimensional spaces. In that case there can be stacks of various dimensions. In the two-dimensional case the stacks are one-dimensional, but in three dimensions there are two-dimensional stacks that can be considered as unions of one-dimensional stacks.

4.3. Consider a Penrose tiling and a stack generated by some r and e. In any rhomb of the stack connect the mid-points of the edges parallel to e by a straight line segment. These line segments form an infinite broken line that can be called the *back bone* of the stack.

The union of the back bones of all stacks is called the *skeleton* of the tiling. In a skeleton one has *curves* (the back bones), *points* (the intersection points of the curves) and *mazes* (the connected components of the complement of the union of all curves). These curves, points and mazes can be deformed topologically without disturbing their relation to the tiling. On any one of the curves the order of the intersection points with other curves keeps reflecting the order in which the corresponding stack is intersected by other stacks. The topological deformations do not disturb the duality between skeleton and tiling. In that duality curves correspond to stacks, points to rhombs, mazes to rhomb vertices.

4.4. If some topological deformation of a skeleton is given, one is not yet able to reconstruct the tiling. First one has to know, for any curve of the skeleton, the direction and the length of the edge e that was used in the discussion of the corresponding stack. Once all these are known, the Penrose tiling is determined up to a parallel shift in the plane. That is, the Penrose tiling deprived of its arrows. But the latter is no great loss: since one started from a Penrose tiling in the first place, one knows that a consistent system of arrows exists, and section 2.7 takes care of the uniqueness of the arrowing.

4.5. In [1] it was shown that the skeletons are topologically equivalent to so-called pentagrids, but this had to be taken with a grain of salt in so-called singular cases. The pentagrids are parametrized by real numbers $\gamma_0, \ldots, \gamma_4$ satisfying $\gamma_0 + \ldots + \gamma_4 = 0$. If j is one of the numbers $0, \ldots, 4$ and $\zeta = e^{2\pi i/5}$ then the set of all complex numbers z satisfying $\mathrm{Re}(z\zeta^{-j}) + \gamma_j \in \mathcal{Z}$ is a grid of parallel lines in the complex plane. The superposition of these five grids is called the *pentagrid* generated by $\gamma_0, \ldots, \gamma_4$. The pentagrid is called *singular* if it contains three lines passing through one point.

Key results of [1] are the following. Any non-singular pentagrid is the dual of a Penrose tiling. The singular pentagrids do not have a dual in the ordinary sense, but by infinitesimal perturbations they turn into grids that do have a dual. There can be various perturbations with different effect, and the result is that a singular grid corresponds either to 2 or to 10 different Penrose tilings. With these extra arrangements for the singular cases, *all* Penrose tilings are obtained from pentagrids.

The perturbations will get some more attention in section 6.2 below.

4.6. The idea of producing a tiling as a topological dual of the superposition of a number of parallel grids can be extended to general classes of tilings of spaces of arbitrary dimension by means of parallelotopes. See [2] for a proof that under fairly general conditions the dual covers the whole space uniquely.

4.7. It can be concluded from section 4.5 that the skeleton of a Penrose tiling is topologically equivalent to some regular pentagrid or to some perturbed singular pentagrid. But remarkably, there is always a second and completely different straight line grid with the topological structure of the skeleton. This matter will be treated in sections 5 and 6, where it is also shown that there are not just these two: there are infinitely many straight line grids with the same topology. The pentagrid can be deformed continuously in such a way that it always keeps the same topology and always remains a superposition of five straight line grids, with the central grid as final result. And one can even get beyond that.

5. Ammann bars and central bars.

5.1. An attractive way to introduce the *Ammann bars* of a Penrose tiling is to consider both the thin and the thick arrowed rhomb as a billiard table and to choose particular billiard ball tracks on them. These tracks can already be shown on the rhomb halves, displayed in Figure 5.1. The tracks have been chosen in such a way that when fitting pieces together the segments combine to infinite straight lines. These are called the Ammann bars after their discoverer R. Ammann (see [7, 9]).

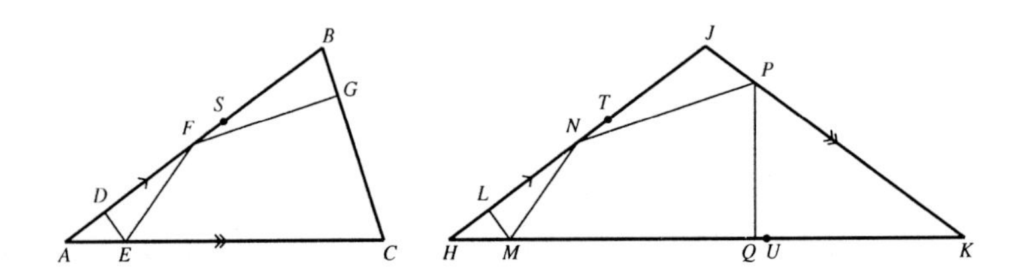

Fig. 5.1. The billiard ball tracks on the arrowed half rhombs. The angles at D, G, L and Q are 90°. The lines EF and FG form equal angles with AB, which is expressed by saying that the billiard ball bounces at the edge AB. At E, M, N and P there are similar cases of bouncing. Finally $AD = HL$, $AF = HN$ and $JP = AE = HM$. In the mirror images of these half rhombs one of course takes the mirror image tracks.

The grid formed by these bars, called *Ammann quasigrid* in [9], has some of the properties of the *pentagrid* used to produce the Penrose tiling in [1]. Both are superpositions of five parallel grids, making angles which are multiples of 72°. In the pentagrid the parallel grids are all equidistant, but in the Ammann quasigrid they show two different distances. On the other hand, the Ammann quasigrid looks much simpler than the pentagrid since its meshes show only a small finite number of shapes. A pentagrid contains infinitely many different mesh shapes, and arbitrarily small ones.

It was observed by J.E.S. Socolar and P.J. Steinhardt (see [9]) that dualization of the Ammann quasigrid of a Penrose tiling leads exactly to the deflation of that tiling.

A quite simple suggestive argument can be given for this statement, although it might not be easy to turn it into a formal proof. It is explained by Figure 5.2. The straight line billiard track segments can be continuously deformed into the segments in Figure 5.3. The points D,E,G,L,M,P can stay where they were, but F, N and Q move. In the mirror images of these half rhombs the corresponding things are done correspondingly. If in each rhomb of a Penrose tiling the original billiard ball track is deformed continuously into the new track, then the union of all segments shows the picture of a grid deforming itself by bending its bars, all the time keeping the same topology. In the final situation (the one of Figure 5.2) each mesh contains exactly one vertex of the deflation, and that is exactly the one produced by the dualization.

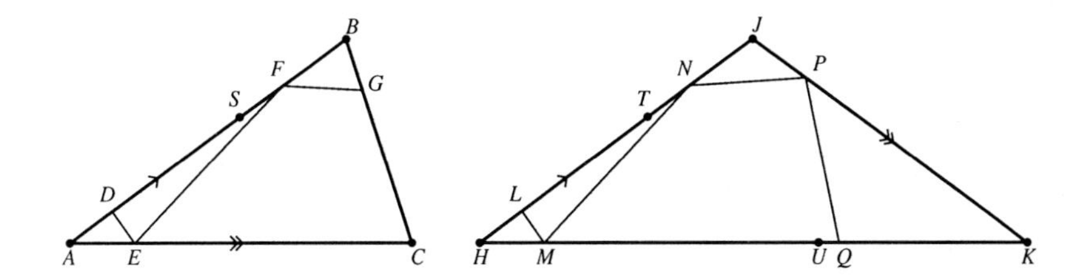

Fig. 5.2. The deformed billiard ball tracks on the half thin and the half thick rhomb. In the mirror images of these half rhombs mirror image tracks have to be taken. The vertices of the deflation in the above half rhombs are A,B,C,S,H,K,J,U,T. Having these tracks in all half rhombs of a Penrose tiling results in a set of mazes where each maze contains exactly one deflation vertex, and in the duality that vertex is exactly the vertex corresponding to that maze.

5.2. The Socolar-Steinhardt phenomenon can also be expressed like this: the Ammann bars of the *inflation* of a Penrose tiling form a quasigrid that is the topological dual of that tiling itself. Let me call those Ammann bars of the inflation the *central* bars of the tiling itself. It will be shown that they can be defined independently of the billiard ball construction, and that a proof for the duality can be given without reference to the argument of section 5.1.

The name "central" was chosen because of the the central position of these bars in the stacks of the tiling (see section 6.4).

The central bars can be found by comparing Figures 3.1 and 5.1. The double arrows of the deflation are from S to A, from C to B, from T to H and from U to J. These are all intersected orthogonally by the billiard ball track. The distance from the intersection point to the end-point of the double arrow is just one fourth of the length of the arrow. This shows that the central bars of a Penrose tiling have to contain the segments EF, GH, TU, VW indicated in Figure 5.3.

It has to be admitted that some of the rhombs contain further pieces of central bars, but in order to build the full central grid it suffices to consider those pieces of Figure 5.3. Since

$$EB = GB = TP = VP, \quad CF = AH = RU = RW \tag{5.2.1}$$

it is obvious that in a stack the segments in the direction of the stack fit together to a full straight line. It is easy to show directly that (5.2.1) implies $EB = GB = TP = VP = \frac{1}{4}$, $CF = AH = RU = RW = \frac{1}{4}(2+\sqrt{5})$, and therefore the central bars can be introduced without any reference to the billiard ball tracks.

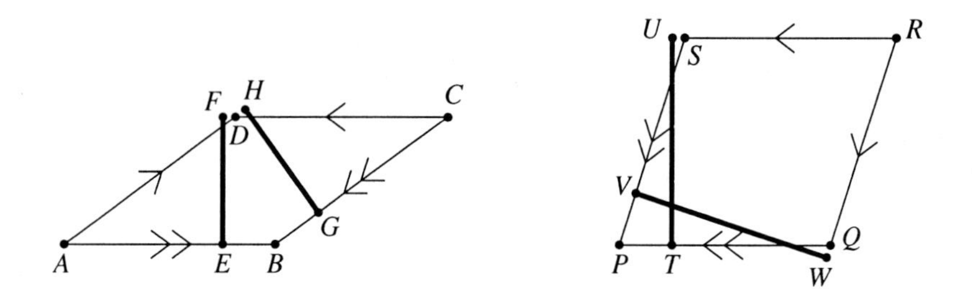

Fig. 5.3. The basic segments of the central bars. In the thin rhomb $ABCD$ (with double arrows AB and CB) the segments are EF and GH, in the thick rhomb $PQRS$ (with double arrows QP and SP) they are TU and VW. There are right angles at E, G, T and V. The points F, H, U, W lie on the extensions of the segments CD, AD, RS, RQ, respectively. The lengths of the bar segments are $EF = GH = \sin 36°$, $TU = VW = \sin 72°$. The positions are determined by $EB = GB = TP = VP = 0.25$. Moreover $DF = DH = SU = QW = (-2+\sqrt{5})/4 = 0.059017$.

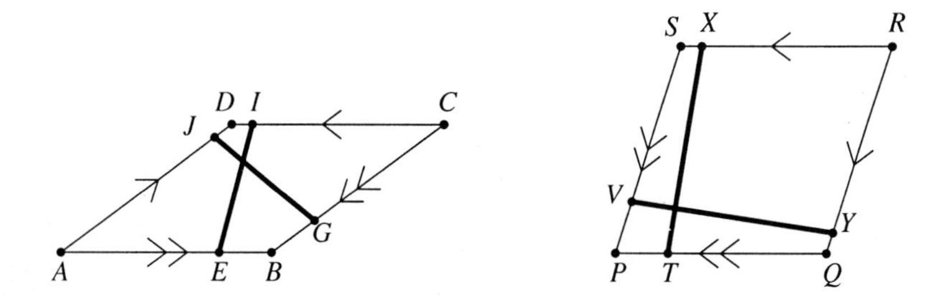

Fig. 5.4. The segments of the deformed bars EI, GJ in the thin rhomb and TX, VY in the thick rhomb, determined by $DI = DJ = SX = QY = 0.1$. Any maze formed by these deformed bars contains exactly one vertex of the Penrose tiling, and that is exactly the vertex corresponding to the maze in the duality.

The segments of Figure 5.3 can be transformed into those of Figure 5.4 by continuous deformation. Taking the segments of Figure 5.4 in all rhombs of a Penrose tiling one gets something that is topologically equivalent to the skeleton of the tiling (that was defined in section 4.3 by connecting mid-points of the edges).

The transition from Figure 5.3 to Figure 5.4 does not affect the topology of the grid. This is not completely trivial, since (in contrast to what what was described in section 5.1) the operations take place partly *outside* the rhombs. So it has to be made sure that the corresponding operations in neighboring rhombs do not interfere.

In section 6 the topological equivalence of pentagrid and central grid will be shown in quite a different way.

6. Algebraic proof of the topological equivalence

6.1. The various grids to be considered are generalizations of the one described in section 4.5. They are characterized by real numbers $\omega_{j,k}$, where j runs through the set $\{0,1,2,3,4\}$ and k through the set $\mathcal{Z}$ of all integers. The grid determined by these ω's is the superposition of $\Gamma_{\omega,0},\ldots,\Gamma_{\omega,4}$, where each $\Gamma_{\omega,j}$ is is a grid of parallel lines in the complex plane. The line with index k in $\Gamma_{\omega,j}$ is the set of all complex numbers z for which $\mathrm{Re}(z\zeta^{-j})=\omega_{j,k}$ (as before, $\zeta=e^{2\pi i/5}$).

So the pentagrid described in section 4.5 with its five real parameters $\gamma_0,\ldots,\gamma_4$ satisfying $\gamma_0+\ldots+\gamma_4=0$ is the case where $\omega_{j,k}=k-\gamma_j$.

6.2. In [1] it was indicated how the pentagrid defines a Penrose tiling. Singular grids cannot be used directly: they first have to get an infinitesimal deformation in order to admit proper dualization (see [1], section 12).

A few words may be devoted to the meaning of the word "infinitesimal" here. If in a singular pentagrid a grid line, a vertical one, say, is moved over a small finite distance to the right, then triple intersection points may have been avoided in a big finite range only. For a given line in a pentagrid and a given large positive number R there exists a small positive ε such that within the circular disk given by $|z|<R$ all triple intersections are avoided by shifting the line over a distance δ with $0<\delta<\varepsilon$. But outside the range R such shifts may cause other intersection points, and alter the topology. The topology of the grid obtained by an infinitesimal shift to the right can be defined as the limit of the topology obtained within $|z|<R$ by means of sufficiently small shifts of the grid line.

But of course, perturbation of a pentagrid is not just a perturbation of a single line, but of the set of the five γ's. I now introduce an index p that describes which infinitesimal perturbation has to be taken. It has ten possible values: $0,\ldots,9$, but in the majority of the singular cases there are only two that have a different effect, and in the regular cases all ten have the same effect. In [1], section 9, it was explained that the complex number $\xi=\sum_{j=0}^{4}\gamma_j\zeta^{2j}$ is an essential parameter for the Penrose tiling. It is related to the real numbers μ_j, to be used in this section, given by

$$\mu_j = \gamma_j - (\gamma_{j-1} + \gamma_{j+1})\tau^{-1},$$

for $j = 0, \ldots, 4$, where, as before, $\tau = 2\text{Re}\ \zeta = \frac{1}{2}(-1+\sqrt{5})$, and j is taken mod 5 (so $\gamma_5 = \gamma_0$, etc.). The μ's can be derived from ξ by $\text{Re}(\xi\zeta^{-2j}) = (1-\frac{1}{2}\tau)\mu_j$.

It was shown in [1], section 11 that a pentagrid is singular if and only if for one of the j's there is an α of the form $(1-\zeta)(n_0+n_1\zeta+n_2\zeta^2+n_3\zeta^3+n_4\zeta^4)$ with integers $n_0, \ldots, n_4$ such that $\text{Re}((\xi-\alpha)\zeta^{-j}) = 0$. It is not hard to show that this condition is equivalent to the following one: for at least one value of j there is an integer k such that $(k-\mu_j)\tau$ is an integer.

Now consider infinitesimal perturbations of ξ, given by a positive infinitesimal dw and a perturbation parameter p (one of the values $0, \ldots, 9$). The perturbation of ξ is $d\xi = e^{p\pi i/5}dw$. And perturbations of the γ's that produce this $d\xi$ can be taken as (cf. [1], formula (9.2)): $d\gamma_j = (2/5)\text{Re}(\zeta^{-2j}d\xi)$. The perturbations of the μ's turn out to be

$$d\mu_j = (1-\frac{1}{2}\tau)^{-1}\text{Re}(e^{(p-4j)\pi i/5})dw.$$

6.3. The lines of a pentagrid correspond to stacks of the Penrose tiling, and for each stack there is a central bar according to section 5. These central bars can be evaluated in terms of the μ's. This will be explained in section 6.4. The result is that the line with index k in the j-th subgrid of the pentagrid leads to a central bar given by $\text{Re}(z\zeta^{-j}) = \beta_{j,k}$, where, at least in the case of a non-singular grid, $\beta_{j,k} = (1+\frac{1}{2}\tau^{-1})(k+\tau\lceil(k-\mu_j)\tau\rceil-\frac{1}{2}\tau)$. ($\lceil x \rceil$ is the usual notation for the least integer $\geq x$). In singular cases the value of $\lceil(k-\mu_j)\tau\rceil$ can be affected by perturbation of μ_j if $(k-\mu_j)\tau$ is an integer. The effect can be described by means of the notations $\lceil x\rceil_+$ and $\lceil x\rceil_-$, intended as $\lceil x+dw\rceil$ and $\lceil x-dw\rceil$, respectively, where dw is a positive infinitesimal. This just means

$$\lceil x\rceil_+ = \lfloor x\rfloor + 1, \quad \lceil x\rceil_- = \lceil x\rceil$$

for all real values of x. With this notation the result for the central bar becomes

$$\beta_{j,k} = (1+\frac{1}{2}\tau^{-1})(k+\tau\lceil(k-\mu_j)\tau\rceil_{\varphi(p,j)} - \frac{1}{2}\tau), \tag{6.3.1}$$

where $\varphi(p,j)$ stands for $+$ or $-$ according to whether $\text{Re}(e^{(p-4j)\pi i/5})$ is < 0 or > 0.

In the non-singular cases the $\varphi(p,j)$ can be ignored, of course. But it should be noted that it is not so easy to see from the γ's whether a case is singular or not. If one does not want to bother one might just take the $\varphi(0,j)$ as a standard, but it is dangerous to omit the $\varphi(p,j)$ altogether. In the case $\gamma_0 = \cdots = \gamma_4 = 0$ with its ten different perturbations the formula $\beta_{j,k} = (1+\frac{1}{2}\tau^{-1})(k+\tau\lceil k\tau\rceil - \frac{1}{2}\tau)$ would definitely *not* represent the central bars of a Penrose tiling.

6.4. Here are some details about the derivation of the expression (6.3.1) for the central bar of the stack corresponding to a given line of a pentagrid. For simplicity, non-singularity will be assumed. Moreover, the (unessential) restriction is made that $j = 0$, which makes it possible to talk in terms of left and right. Moreover, the letter j becomes available for other purposes.

So the grid line is (with some integer k) $\mathrm{Re}(z) = k - \gamma_0$. The meshes directly to the left and directly to the right of this line produce the vertices of the rhombs of the stack. According to [1], section 5, the vertices are derived from the meshes as follows. Take any point z in the mesh, and form the integers $K_j(z) = \lceil \mathrm{Re}(z\zeta^{-j}) + \gamma_j \rceil$; then the vertex is $\sum_{j=0}^{4} K_j(z)\zeta^j$. With a real variable t the points of the grid line are represented as $(k - \gamma_0) + it$. So the vertices corresponding to the meshes directly to the left of the line are given by $M(t) = k + \sum_{j=1}^{4} \lceil u_j \rceil \zeta^j$, where $u_j = \mathrm{Re}(((k - \gamma_0) + it)\zeta^{-j}) + \gamma_j$. It will be shown that the real part of $M(t)$ takes only four different values if t varies. That real part is $k + (\lceil u_1 \rceil + \lceil u_4 \rceil)\mathrm{Re}(\zeta) + (\lceil u_2 \rceil + \lceil u_3 \rceil)\mathrm{Re}(\zeta^2)$. Obviously $\lceil u_1 \rceil + \lceil u_4 \rceil = \lceil u_1 + u_4 \rceil + q$, where q is either 0 or 1, and similarly $\lceil u_2 \rceil + \lceil u_3 \rceil = \lceil u_2 + u_3 \rceil + r$, with r either zero or 1. With the abbreviation

$$A = \lceil (k - \gamma_0)\tau + \gamma_1 + \gamma_4 \rceil \tau/2 - \lceil -(k - \gamma_0)\tau^{-1} + \gamma_2 + \gamma_3 \rceil \tau^{-1}/2$$

this leads to

$$\mathrm{Re}(M(t)) = k + A + q\mathrm{Re}(\zeta) + r\mathrm{Re}(\zeta^2).$$

It gives the four different horizontal coordinates of the left end-points of the horizontal edges in the stack. The right end-points are obtained by adding 1. So all the vertices involved here are lying on eight vertical lines. The figure formed by those eight lines has the central bar as an axis of symmetry, and that is why the name "central" was chosen. Since $\mathrm{Re}(\zeta) + \mathrm{Re}(\zeta^2) = -\frac{1}{2}$ the average of the eight horizontal coordinates is $k + A + \frac{1}{4}$, whence the central bar can be represented as $\mathrm{Re}(z) = k + A + \frac{1}{4}$.

From the arrowing of the horizontal vertices in the stack it is easy to derive that this central bar cuts the doubly arrowed horizontal edges at a point $\frac{1}{4}$ from the end, exactly in accordance with Figure 5.3.

These calculations did not yet use the condition that $\sum \gamma_j = 0$. Tilings without that condition were mentioned in [3] in connection with a riffle shuffle card trick, and it is actually the riffle shuffle arrangement in a stack that guarantees that there are only eight different horizontal coordinates.

But $\sum \gamma_j = 0$ gives a simplification:

$$-(k - \gamma_0)\tau^{-1} + \gamma_2 + \gamma_3 = \mu_0\tau - k\tau - k,$$

and since non-singularity is assumed, this cannot be an integer. If x is not an integer one has $-\lceil -x \rceil = \lceil x \rceil - 1$, which leads to the following formula for the central bar:

$$\mathrm{Re}(z) = (1 + \frac{1}{2}\tau^{-1})(k + \tau\lceil (k - \mu_0)\tau \rceil - \frac{1}{2}\tau).$$

For the singular cases the result (6.3.1) can now be derived by obvious limit operations.

6.5. The topological equivalence of pentagrid and central grid will now be established directly, on the basis of (6.3.1). Figure 6.1 presents an illustration. It looks reasonable to enlarge the pentagrid by a factor 5/2 when comparing it to the central grid (it is the same factor as in [1], section 5, formula above (5.3)), but for the topology of the grids such factors make no difference at all.

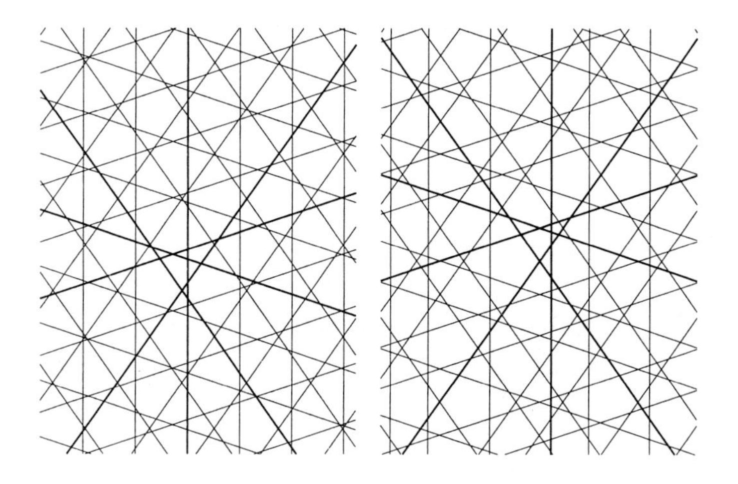

Fig. 6.1. On the left there is a piece of the pentagrid with parameters $\gamma_0 = 0.2$, $\gamma_1 = 0.4$, $\gamma_2 = 0.3$, $\gamma_3 = -0.8$, $\gamma_4 = -0.1$, on the right a corresponding piece of the central grid with the same parameters. The scale of the picture on the left is 5/2 times the one on the right. In order to facilitate the comparison of the topologies, the lines with $k_j = 0$ in the j-th subgrid are thicker than the others.

The method is as follows. Consider two grids of the form of section 6.1, given as $\mathrm{Re}(z\zeta^{-j}) = \alpha_{j,k}$ and $\mathrm{Re}(z\zeta^{-j}) = \beta_{j,k}$. It is assumed that for each j the $\alpha_{j,k}$ and the $\beta_{j,k}$ increase with k. And it is assumed that nowhere in the grids three lines pass through a point. In the topological correspondence the k-th line of the j-th subgrid of the α-grid will correspond to the k-th line of the j-th subgrid of the β-grid, for all j and k.

If p, q, r are straight lines forming a triangle, then they determine an orientation: seen from the inside of the triangle the circular order p, q, r is either clockwise or counter-clockwise. This gives the criterion for the topological equivalence in the grid: for any three lines in the α-grid the orientation has to be the same as for the corresponding three lines in the β-grid.

There are two kinds of triangles here. One is of the kind formed with values $j, j-1, j+1$, the other one with $j, j+2, j+3$ (j taken mod 5). The orientation is a simple matter of determinants. The result is as follows. The topological equivalence of the two grids is guaranteed if for all j and for all integers p, q, r the combination $\alpha_{j+1,p} + \alpha_{j-1,q} - \tau\alpha_{j,r}$ has the same sign as $\beta_{j+1,p} + \beta_{j-1,q} - \tau\beta_{j,r}$, and $\alpha_{j+2,p} + \alpha_{j-2,q} + \tau^{-1}\alpha_{j,r}$ has the same sign as $\beta_{j+2,p} + \beta_{j-2,q} + \tau^{-1}\beta_{j,r}$.

Since no three lines pass through a point none of these expressions is zero.

It is this very explicit condition that has to be verified for the regular pentagrid and the corresponding central grid. It will be established in sections 6.7-6.9 as a theorem on inequalities that can be understood independently of the previous sections. In order to get rid of the factor $(1 + \frac{1}{2}\tau^{-1})$ those sections work with θ instead of β, where $\beta = (1+\frac{1}{2}\tau^{-1})\theta$. And for simplicity, sections 6.7-6.9 restrict themselves to $j = 0$; the other cases are completely similar.

6.6. If the conditions mentioned in section 6.5 are satisfied for α and β then they are obviously also satisfied for α and $\omega(\lambda)$, where $0 \le \lambda \le 1$, and $\omega(\lambda)$ is obtained by linear interpolation:

$$\omega(\lambda)_{j,k} = (1-\lambda)\alpha_{j,k} + \lambda\beta_{j,k}.$$

This means that the the α-grid can be deformed continuously into the β-grid without ever changing the topology. It is even possible to push the λ beyond 1. The lower bound $\frac{1}{2}\tau$ in theorem 6.1 (section 6.7) has the effect that the topology of the $\omega(\lambda)$-grid remains the same over the interval $0 \le \lambda < (1 - \frac{1}{2}(1+\frac{1}{2}\tau^{-1})\tau)^{-1}$.

Section 4.5 discussed infinitesimal perturbations for the interpretation of the dual of a singular pentagrid. The same thing can now be achieved with the $\omega(\lambda)$, with small positive but not infinitesimal values of λ.

6.7. The sections 6.7-6.9 do not make use of anything said in the previous sections.

Let $\gamma_0, \dots, \gamma_4$ be real numbers with $\gamma_0 + \dots + \gamma_4 = 0$. For $j = 0, \dots, 4$, $k_j \in \mathcal{Z}$, the real numbers $\alpha_{j,k}$ and $\theta_{j,k}$ are defined by

$$\alpha_{j,k} = k - \gamma_j, \quad \theta_{j,k} = k + \tau\lceil (k-\mu_j)\tau \rceil - \frac{1}{2}\tau,$$

where $\mu_j = \gamma_j - (\gamma_{j-1} + \gamma_{j+1})\tau^{-1}$, $\gamma_5 = \gamma_0$ and $\tau = \frac{1}{2}(-1+\sqrt{5})$.

Let $k_0, \dots, k_4$ be integers, and abbreviate

$$H = \alpha_{1,k_1} + \alpha_{4,k_4} - \tau\alpha_{0,k_0}, \quad L = \theta_{1,k_1} + \theta_{4,k_4} - \tau\theta_{0,k_0},$$

$$K = \alpha_{2,k_2} + \alpha_{3,k_3} + \tau^{-1}\alpha_{0,k_0}, \quad M = \theta_{2,k_2} + \theta_{3,k_3} + \tau^{-1}\theta_{0,k_0}.$$

With these abbreviations the following theorem can be proved:

Theorem 6.1. *If $H \neq 0$ then $L/H \geq \frac{1}{2}\tau$. If $K \neq 0$ then $M/K \geq \frac{1}{2}\tau$.*

6.8. Here is the proof of the first part of the theorem. H and L can be expressed like this:

$$H = k_1 + k_4 - \tau k_0 + \tau\mu_0,$$

$$L = k_1 + k_4 - \tau k_0 + \tau\lceil(k_1 - \mu_1)\tau\rceil + \tau\lceil(k_4 - \mu_4)\tau\rceil - \tau^2\lceil(k_0 - \mu_0)\tau\rceil - \tau + \tfrac{1}{2}\tau^2.$$

If x and y are real numbers then $\lceil x\rceil + \lceil y\rceil - \lceil x + y\rceil$ is either 0 or 1. So

$$\lceil(k_1 - \mu_1)\tau\rceil + \lceil(k_4 - \mu_4)\tau\rceil = \lceil(k_1 + k_4)\tau - (\mu_1 + \mu_4)\tau\rceil + p$$

with $p = 0$ or $p = 1$. Moreover

$$(k_0 - \mu_0)\tau = k_1 + k_4 - H,$$

and since $(\mu_1 + \mu_4)\tau = -\mu_0$,

$$(k_1 + k_4)\tau - (\mu_1 + \mu_4)\tau = H\tau^{-1} + k_0 - k_1 - k_4.$$

So there is a simple expression for L in terms of H and p:

$$L = \tau\lceil H\tau^{-1}\rceil - \tau^2\lceil -H\rceil - \tau + \frac{1}{2}\tau^2 + p\tau.$$

Now first assume $H > 0$. Use $p \geq 0$, $-\lceil -H\rceil \geq 0$, and note that if $0 < c < 1$ then $\lceil x\rceil \geq 1 - c + cx$ for all $x > 0$. The special case $x = H\tau^{-1}$, $c = \frac{1}{2}\tau$ leads to $L/H \geq \frac{1}{2}\tau$.

Next assume $H < 0$. Use $p \leq 1$, $\lceil -H\rceil \geq 1$, and note that if $0 < c < 1$ then $\lceil x\rceil \leq cx + c$ for all $x < 0$. With $x = H\tau^{-1}$, $c = \frac{1}{2}\tau$ it follows that $L/H \geq \frac{1}{2}\tau$.

6.9. The proof of the second part of the theorem is similar. K and M can be expressed as follows: $K = k_2 + k_3 + k_0\tau^{-1} - \tau\mu_0$ and

$$M = k_2 + k_3 + k_0\tau^{-1} + \tau\lceil(k_2 - \mu_2)\tau\rceil + \tau\lceil(k_3 - \mu_3)\tau\rceil + \lceil(k_0 - \mu_0)\tau\rceil - \tau - \frac{1}{2}.$$

Note that

$$\lceil(k_2 - \mu_2)\tau\rceil + \lceil(k_3 - \mu_3)\tau\rceil = \lceil(k_2 + k_3)\tau - (\mu_2 + \mu_3)\tau\rceil + q$$

where q is either 0 or 1. Moreover $(k_0 - \mu_0)\tau = K - k_0 - k_2 - k_3$, and since $\mu_2 + \mu_3 = \mu_0\tau$,

$$(k_2 + k_3)\tau - (\mu_2 + \mu_3)\tau = K\tau - k_0.$$

So M can be expressed in terms of K and q:

$$M = \tau\lceil K\tau\rceil + \lceil K\rceil + \tau q - \tau - \frac{1}{2}.$$

If $K > 0$ it can be used that $\lceil K\rceil \geq 1$, whence $M \geq \tau\lceil K\tau\rceil - \tau + \frac{1}{2}$. If $0 < c < 1$ then $\lceil x\rceil \geq 1 - c + cx$ for all $x > 0$. With $x = K\tau$, $c = \frac{1}{2}\tau^{-1}$ this leads to $M/K \geq \frac{1}{2}\tau$. If $K < 0$ one can use the inequalities $q \geq 0$, $\lceil K\rceil \leq 0$ and $\lceil x\rceil \leq cx + c$ $(x < 0,\ 0 < c < 1)$ with $x = K\tau$, $c = \frac{1}{2}\tau^{-1}$. This again leads to $M/K \geq \frac{1}{2}\tau$, and that finishes the proof of theorem 6.1.

References

1. N. G. de Bruijn, Algebraic theory of Penrose's non-periodic tilings of the plane. Kon. Nederl. Akad. Wetensch. Proc. Ser. A 84 (=Indagationes Mathematicae 43), 38-52 and 53-66 (1981). Reprinted in: P. J. Steinhardt and Stellan Ostlund: The Physics of Quasicrystals, World Scientific Publ., Singapore, New Jersey, Hong Kong.
2. N.G. de Bruijn, Dualization of multigrids. In: Proceedings of the International Workshop Aperiodic Crystals, Les Houches 1986. Journal de Physique, Vol.47, Colloque C3, supplement to nr. 7, July 1986, pp. 9-18.
3. N. G. de Bruijn, A riffle shuffle card trick and its relation to quasicrystal theory. Nieuw Archief Wiskunde (4) 5 (1987) 285-301.
4. N. G. de Bruijn, Symmetry and quasisymmetry. In: Symmetrie in Geistes- und Naturwissenschaft. Herausg. R. Wille. Springer Verlag 1988, pp. 215-233.
5. N. G. de Bruijn, Updown generation of Penrose tilings. Indagationes Mathematicae, N.S., 1, pp. 201-219 (1990).
6. Martin Gardner, Mathematical games. Extraordinary nonperiodic tiling that enriches the theory of tiles. Scientific American 236 (1) 110-121 (Jan. 1977).
7. Branko Grünbaum and G.C. Shephard. Tilings and patterns. New York, W.H. Freeman and Co. 1986.
8. R. Penrose. Pentaplexity. Mathematical Intelligencer vol 2 (1) pp. 32-37 (1979).
9. J. E. S. Socolar and P. J. Steinhardt. Quasicrystals. II. Unit cell configurations. Physical Rev. B Vol. 34 (1986), 617-647. Reprinted in: P.J. Steinhardt and Stellan Ostlund: The Physics of Quasicrystals, World Scientific Publ., Singapore, New Jersey, Hong Kong.

Distances in Convex Polygons

Peter Fishburn

AT&T Bell Laboratories, Murray Hill, NJ 07974

Summary. One of Paul Erdős's many continuing interests is distances between points in finite sets. We focus here on conjectures and results on intervertex distances in convex polygons in the Euclidean plane. Two conjectures are highlighted. Let $t(x)$ be the number of different distances from vertex x to the other vertices of a convex polygon C, let $T(C) = \Sigma t(x)$, and take $T_n = \min\{T(C) : C \text{ has } n \text{ vertices}\}$. The first conjecture is $T_n = \binom{n}{2}$. The second says that if $T(C) = \binom{n}{2}$ for a convex n-gon, then the n-gon is regular if n is odd, and is what we refer to as bi-regular if n is even. The conjectures are confirmed for small n.

1. Introduction

Let $n_2 = \lfloor n/2 \rfloor$, the integer part of $n/2$ for $n \geq 3$. We begin with three conjectures about every convex n-gon in $\mathbb{R}^2$.

C1. Its vertices determine at least n_2 different distances.
C2. Some vertex has at least n_2 distances to the other vertices.
C3. Each of at least n_2 vertices has at least n_2 distances to the other vertices.

C1 was stated by Erdős in 1946 [3] and settled affirmatively by Altman in 1963 [1]. C2, another old conjecture of Erdős, is open. C3 was suggested by recent work with Erdős. It might be false, but we have no evidence of this.

My purpose here is to explore two conjectures related to C1–C3. We preface them with results for C1 and C2. Throughout, a convex polygon is unique up to similarity transformations (translation, uniform rescaling, rotation around a point, reflection around a line). R_n denotes the regular n-gon, and $R_{n+1} - 1$ is the n-gon whose vertices are n of the vertices of R_{n+1}. With $d(x, y)$ the distance between x and y, a *length-k run* in a convex n-gon is a sequence $x_0, x_1, \ldots, x_k$ of successively adjacent vertices for which

$$d(x_0, x_1) < d(x_0, x_2) < \cdots < d(x_0, x_k) \ .$$

Theorem 1.1. *C1 is true. Suppose convex n-gon C has exactly n_2 different intervertex distances. If n is odd then $C = R_n$. If n is even and $n \geq 8$ then $C \in \{R_n, R_{n+1} - 1\}$.*

Altman [1] proves all but the final sentence, which is proved in [8]. At $n = 6$, a third hexagon besides R_6 and $R_7 - 1$ has exactly three intervertex distances (A_6, defined below); at $n = 4$, two other quadrilaterals besides R_4 and $R_5 - 1$ have exactly two intervertex distances.

The next theorem, from [6], gives the best result known toward C2. It implies that some vertex has at least $n/3$ different distances to the others. It is not known if there is a $\lambda > 1$ such that every convex n-gon ($n \geq 4$) has a vertex with at least $\lfloor \lambda n/3 \rfloor$ distances to the other vertices.

Theorem 1.2. *For every $n \geq 4$, every convex n-gon has a run of length $\lfloor (n+3)/3 \rfloor$, and some convex n-gons have no run of length $\lfloor (n+3)/3 \rfloor + 1$.*

The proof uses the following observation in Moser [12].

Lemma 1.1 (Moser's). *Every convex n-gon has a circumscribed circle with a sub-semicircular arc ending in vertices such that the region enclosed by this arc and the chord between its endpoints contains at least $\lfloor (n-1)/3 \rfloor$ other vertices. Beginning from either endpoint, these vertices produce a run of length at least $\lfloor (n+2)/3 \rfloor$.*

Figure 1.1 illustrates Moser's lemma and some special polygons, including the bi-regular polygons A_4, A_6 and A_8.

Convex $2k$-gon A_{2k} is called *bi-regular* because it is formed from two regular polygons. A_4 is composed of two equilateral triangles with a common side. To form A_{2k} for $k \geq 3$, begin with a fixed copy R_k^0 of R_k with maximum intervertex distance 1. On each ray from the center of R_k^0 that bisects a side, add a vertex that has maximum distance 1 to those of R_k^0. The vertices of A_{2k} are the k of R_k^0 and the added k on the perpendicular bisectors of the sides of R_k^0.

Let $t(x)$ be the number of distances from vertex x to the other vertices of a convex n-gon C. The t-sequence of C, which we view cyclically, is $(t(x_1), t(x_2), \ldots, t(x_n))$ with $x_1, x_2, \ldots, x_n$ the vertices in clockwise or counterclockwise succession. Let $T(C) = t(x_1) + t(x_2) + \cdots + t(x_n)$. We focus on

$$T_n = \min\{T(C) : C \text{ is a convex } n\text{-gon}\} .$$

For odd n, $T(R_n) = n[(n-1)/2] = \binom{n}{2}$ with t-sequence $((n-1)/2, \ldots, (n-1)/2)$. For even n,

$$R_n \text{ has } t\text{-sequence } (n/2, n/2, \ldots, n/2);$$

$$A_n \text{ has } t\text{-sequence } (n/2-1, n/2, n/2-1, n/2, \ldots, n/2-1, n/2),$$

so

$$T(R_n) = n^2/2 \text{ and } T(A_n) = \binom{n}{2} < n^2/2 .$$

Since we have no intimations of smaller values for $T(C)$, we consider two conjectures. The first was introduced in [4].

C4. $T_n = \binom{n}{2}$ for all $n \geq 3$.

C5. For all $n \geq 3$, if $T(C) = \binom{n}{2}$ for a convex n-gon C, then $C = R_n$ if n is odd, and $C = A_n$ if n is even.

There is no apparent implication between these conjectures. If C4 is true, polygons other than R_n and A_n might realize T_n. If C5 holds, there might be other polygons that have $T(C) < \binom{n}{2}$. Because C4 implies that the minimum average $t(x)$ is $(n-1)/2$, it strengthens C2 but is apparently independent of C3. And C5 does not obviously imply either C2 or C3.

Suppose the preceding conjectures are all true. When n is odd, R_n is the unique realizer for T_n and, by Theorem 1.1, for the minimum n_2 of C1. But it clearly does not minimize the number of vertices with $t(x) \geq n_2$ in regard to C3 for $n \geq 5$. When n is even and $n \geq 8$, the realizers R_n and $R_{n+1} - 1$ for n_2 in C1 differ from the unique realizer A_n for T_n. Moreover, since A_n has exactly $n/2$ vertices with $t(x) \geq n/2$, it gives the tightest possible realization of C3.

Evidence for C4 and C5 is provided in the next two sections. Section 2 sketches proofs which show that C4 is true for $n \leq 8$, and Section 3 outlines proofs which verify C5 for $n \leq 7$. Although some of the proofs are case-intensive, it is hoped that their ideas will encourage refinements or extensions that will settle the conjectures fully.

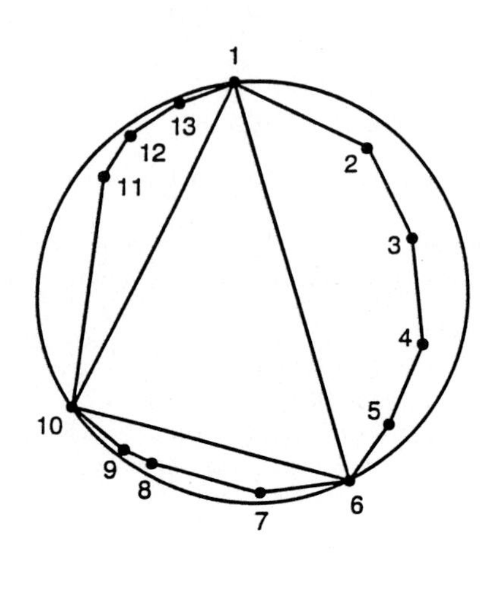
1
2
3
4
5
6
7
8
9
10
11
12
13
Moser's lemma
for n = 13

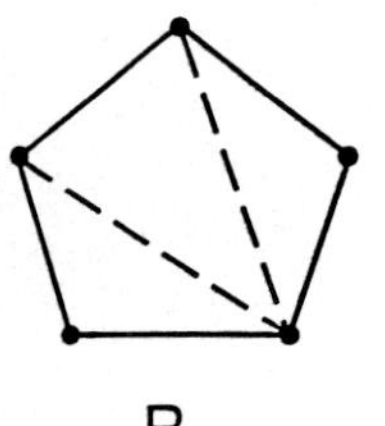
R_5

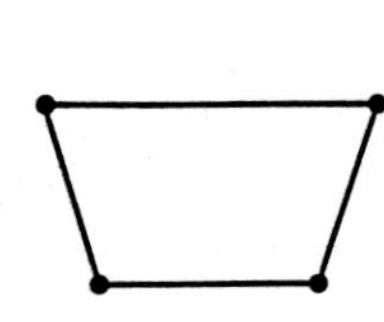
R_5 - 1

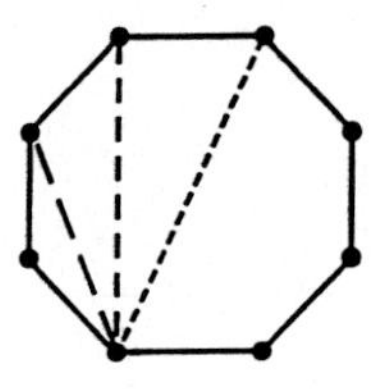
R_8

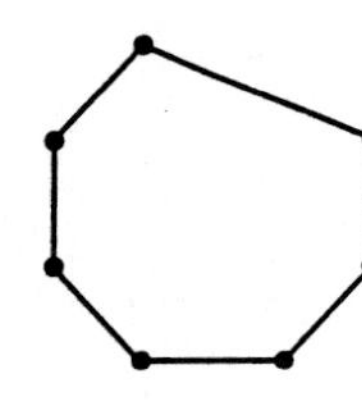
R_8 - 1

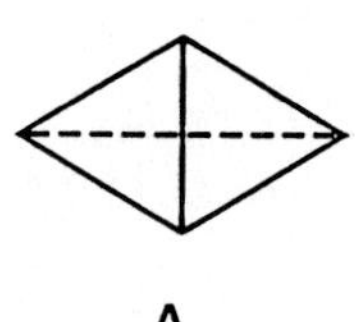
A_4

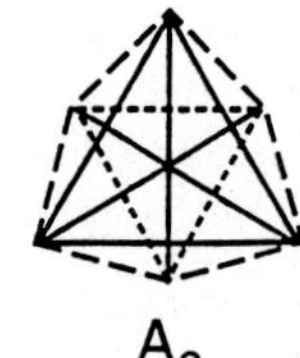
A_6

A_8

Fig. 1.1.

Section 4 offers further remarks on C2–C5, then discusses a set of distance problems for convex polygons motivated by the conjecture of Erdős and Moser which says that the maximum number of pairs of vertices $\{x, y\}$ in a convex n-gon for which $d(x, y) = 1$ is bounded above by cn for some constant c.

2. Evidence for C4

Theorem 2.1. *C4 is true for $n \leq 8$.*

Since $t(x) \geq 1$ for all x, $T_3 \geq 3$, and $T_3 = 3$ is uniquely realized by R_3. When $t(x) = 1$ for some x, the other $n-1$ vertices lie on a subsemicircular arc with center at x. Equal spacing along this arc yields the following lemma.

Lemma 2.1. *Suppose $t(x) = 1$. If y is adjacent to x for convex n-gon C then $t(y) \geq n - 2$. Moreover,*

$$T(C) \geq (3n^2 - 8n + 9)/4 \text{ if } n \text{ is odd ;}$$
$$T(C) \geq (3n^2 - 8n + 8)/4 \text{ if } n \text{ is even .}$$

It follows that the minimizing t-sequence for $n = 4$ is $(1, 2, 1, 2)$. Thus $T_4 = 6$, which is uniquely realized by A_4.

For $n \geq 5$, Lemma 2.1 implies that $T(C) > \binom{n}{2}$ when $t(x) = 1$ for some vertex, so we assume henceforth that min $t(x) \geq 2$. The minimizing t-sequence at $n = 5$ is then $(2, 2, 2, 2, 2)$, with $T_5 = 10$ realized by R_5. The next section shows that R_5 is the only pentagon for which $T(C) = 10$.

The next lemma will be used to verify $T_6 = 15$.

Lemma 2.2. *If x and y are adjacent vertices of convex n-gon C and $t(x) = t(y) = 2$, then $n \leq 5$.*

Proof. Let the hypotheses of the lemma hold. Assume without loss of generality that $x = (0, 0)$, $y = (1, 0)$, and the other vertices of C lie above the abscissa. We have $d(x, y) = 1$. Let d_x and d_y be the second intervertex distances for x and y respectively. Since the other vertices lie at intersection points of the x-centered circles of radii 1 and d_x and the y-centered circles of radii 1 and d_y, $n \leq 6$. To obtain $n = 6$, all four intersections above the abscissa must occur, so assume this. Then $\max\{d_x, d_y\} < 2$.

It is easily seen that convexity is violated at $n = 6$ if either $d_x = d_y$ or $\min\{d_x, d_y\} < 1$. The latter is illustrated in Figure 2.1A. Assume henceforth that $1 < d_x < d_y < 2$ as pictured in Figure 2.1B. Point q has $d(x, q) = d(y, q) = 1$. The other three intersection points are p, z and v.

We prove that convexity is violated at $n = 6$ by showing that q lies below line segment vp. With d_x fixed, the best chance for convexity occurs when d_y is near 2 [z near $(-1, 0)$] since this moves v down the d_x circle. We prove that vp intersects the perpendicular bisector $\perp$ of xy above q when $d_y = 2$.

Fix $d_y = 2$ and let $d = d_x$. Let s be p's horizontal component. Since $\cos\theta = (1 + d^2 - 1)/2d = d/2$ and $\cos\theta = s/d$, $s = d^2/2$. Then p's vertical component is $[d^2 - (d^2/2)^2]^{1/2}$. Similar computations for v give

$$p = (d^2/2, [d^2 - (d^2/2)^2]^{1/2})$$
$$v = (-(3 - d^2)/2,\ [d^2 - \{(3 - d^2)/2\}^2]^{1/2}) .$$

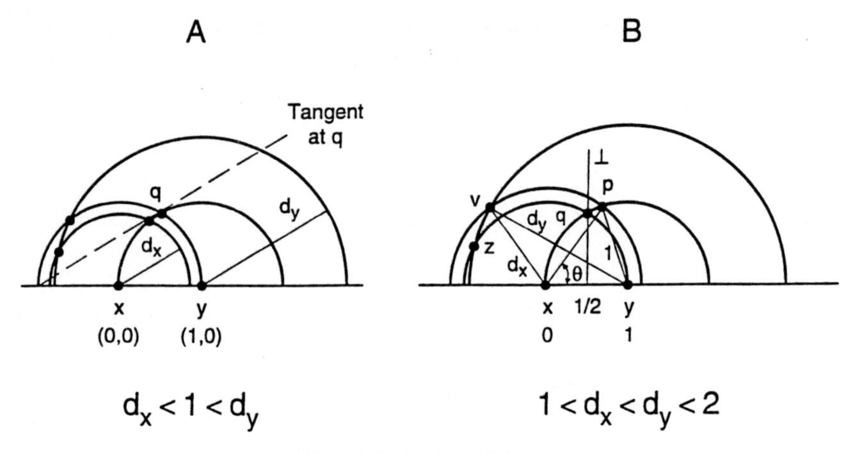

Fig. 2.1. $t(x) = t(y) = 2$.

Line segment $\alpha p + (1-\alpha)v$ has first component $\alpha(d^2/2) - (1-\alpha)(3-d^2)/2$, which equals $1/2$ for intersection with $\perp$ when $\alpha = (4-d^2)/3$. Let h be the height at which $\alpha p + (1-\alpha)v$ intersects $\perp$. Then

$$h = \{d(4-d^2)^{3/2} + (d^2-1)[4d^2 - (3-d^2)^2]^{1/2}\}/6 .$$

Since $q = (1/2, \sqrt{3}/2)$, our claim that vp intersects 1 above q is

$$(4-d^2)^{3/2}d + (d^2-1)[4d^2 - (3-d^2)^2]^{1/2} > 3\sqrt{3} .$$

Let $u = d^2 - 1$, $0 < u < 3$. Then, after squaring both sides of the preceding inequality, it reduces to

$$(3-u)^{3/2}\sqrt{8+7u-u^2} > (9-8u+u^2)\sqrt{u} .$$

This holds when $u \geq 2$ since its right side is then negative. Suppose $u < 2$. We square sides and cancel identical terms to get $216 - 27u > 81u$, i.e., $u < 2$, so the inequality holds for all $0 < u < 3$. ∎

It follows from $\min t(x) \geq 2$ and Lemma 2.2 that the smallest possible $T(C)$ at $n = 6$ occurs uniquely for the t-sequence $(2,3,2,3,2,3)$. Since this is realized by $A_6, T_6 = 15$.

Consider $n = 7$. Since R_7 has t-sequence $(3,3,\ldots,3)$, $T_7 \leq 21$. If adjacent vertices never have t-counts of 2 and 3 but $t(x) = 2$ for some x, then $T \geq 22$, as with $(2,4,2,4,3,3,4)$. Hence $T \leq 20$ is conceivable only if there are adjacent vertices with $t(x) = 3$ and $t(y) = 2$. These t-counts are possible at $n = 7$, but only under special circumstances.

Lemma 2.3. *Suppose $t(x) = 3$ and $t(y) = 2$ for adjacent vertices of C with $d(x,y) = 1$. Let $d_1 < d_2$ be the second and third distances for x; let d_y be the second distance for y. Then $n \leq 7$, and $n \leq 6$ if $\min\{d_1, d_y\} < 1$.*

Proof (outline). Lemma 2.2 requires $n \leq 5$ if we omit the third x distance. Since $t(y) = 2$, at most two more vertices are added by x's third distance, so $n \leq 7$. If $d_y < 1$, straightforward geometric arguments yield a convexity violation if $n > 6$. If $d_y > 1$ but $d_1 < 1$, convexity also forces $n \leq 6$. When $1 < d_2$ for this case, we use the argument in the latter part of the proof of Lemma 2.2. ∎

Figure 2.2 illustrates $1 < \min\{d_1, d_y\}$ for Lemma 2.3. The top diagram has all possible circle intersections ($n \leq 8$). As drawn, q must be removed for convexity at $n = 7$, but p could be removed instead if d_y were a bit smaller. By increasing d_y we lose intersection point p, but can still have a convex heptagon with the seven remaining points.

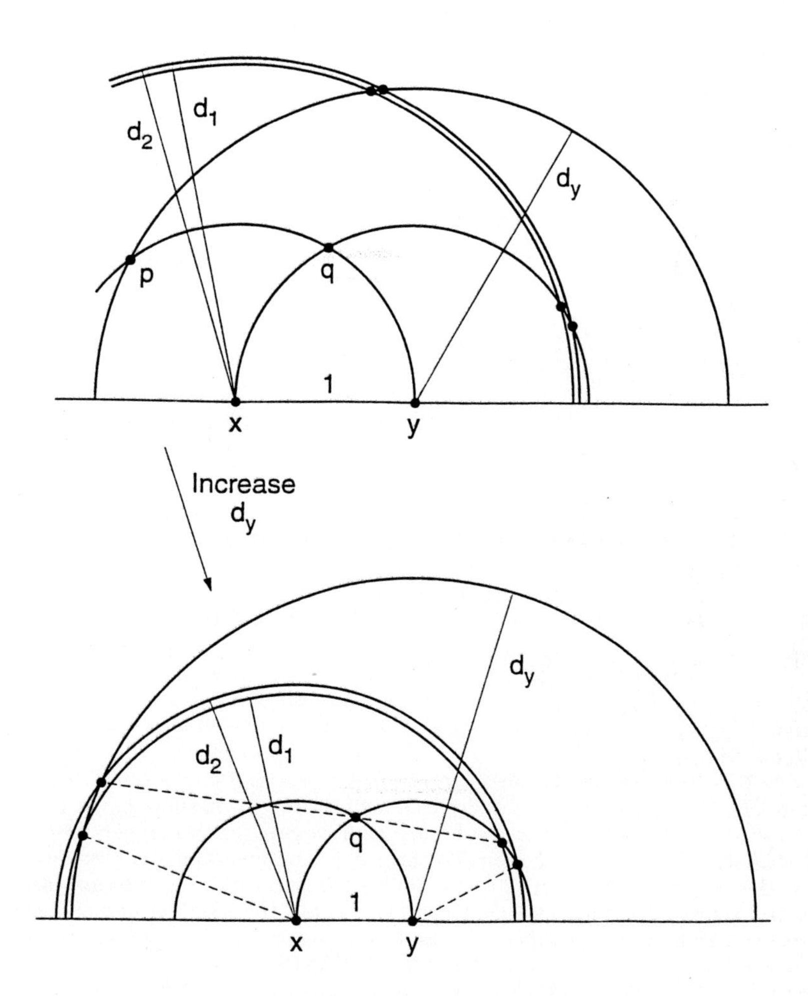

Fig. 2.2. $t(x) = 3,\ t(y) = 2.$

Modulo minor changes in d_1, d_2 and d_y, Figure 2.2 shows the only ways we can have a convex heptagon when there are contiguous t-counts 2 and 3. To have $T \leq 20$, the five vertices besides x and y must have a t sum no greater than 15, or an average of $t \leq 3$ per vertex. It is easily checked that no other vertex has $t = 2$, and some clearly have $t \geq 4$, so in fact $T(C) \geq 22$ for these heptagons. We conclude that $T_7 = 21$, which is attainable only with t-sequence $(3, 3, \ldots, 3)$.

Finally, consider $n = 8$ where A_8 with t-sequence $(3, 4, 3, 4, 3, 4, 3, 4)$ gives $T_8 \leq 28$. By Lemmas 2.1 and 2.2, the minimum sum of adjacent t-counts for a convex octagon is 6, with $(t(x), t(y))$ either $(4, 2)$ or $(3, 3)$. In the first case, addition of a d_3 circle centered at x on Figure 2.2 will give a $T(C)$ for $n = 8$ at least in the low 30's. A similar analysis for $(3, 3)$ yields a similar result. I omit details. It follows that $T_8 = 28$.

At the present time I know of no convex octagon for which $28 < T(C) < 32$. We have $T(R_8) = 32$, but R_8 is not the only octagon with $T = 32$. An example is O_1 on Figure 6 in [9]. Its t-sequence is $(2, 5, 4, 5, 2, 5, 4, 5)$.

3. Evidence for C5

Theorem 3.1. *C5 is true for $n \leq 7$.*

The preceding section verifies this for $n \leq 4$ and shows for $n \in \{5, 6, 7\}$ that the unique t-sequences that have $T(C) = \binom{n}{2}$ are

$$\begin{array}{ll} (2,2,2,2,2) & \text{for } n = 5 \\ (2,3,2,3,2,3) & \text{for } n = 6 \\ (3,3,3,3,3,3,3) & \text{for } n = 7\,. \end{array}$$

These t-sequences are assumed throughout this section. We also let m denote the number of different intervertex distances in a convex polygon.

Suppose $n = 5$. If $m = 2$, we have R_5 by Theorem 1.1. If $m = 3$, Theorem 2 and Figure 2 in [9] show that $t(x) \geq 3$ for some vertex. Suppose $m = 4$. Denote the four distances by d_1 through d_4 and label each vertex with its two distances. Suppose without loss of generality that one is labeled d_1d_2 and another d_1d_3. Then each of the other three vertices has (d_1 or d_2) and (d_1 or d_3), and to obtain a fourth distance one of these has label d_1d_4, so each of the remaining two also has (d_1 or d_4). Then, because $t(x) = 2$ for all vertices, all five have label d_1, and this yields the contradiction that one of the other labels is used on only one vertex. A similar contradiction holds when $m > 4$, so R_5 is the only pentagon with t-sequence $(2, 2, 2, 2, 2)$.

Let $n = 6$. Label the vertices 1 through 6 in succession clockwise and let jk denote $d(j, k)$. By Theorem 1.2 there is a run of length 3. Assume for definiteness that $12 = a < 13 = b < 14 = c$: see Figure 3.1A, where circled numbers give the t-sequence. By Moser's construction [12] there is a minimum-diameter circumscribed circle that contains three vertices that give a triangle with each interior angle no greater than $\pi/2$: see Figures 3.1B and 3.1C. If one of the resulting subsemicircular sectors contains two of the other vertices, its end vertices both have $t \geq 3$, a contradiction to the t-sequence pattern $(2, 3, 2, 3, 2, 3)$. We therefore have the arrangement of Figure 3.1C, where all three vertices of the noted triangle have either $t = 2$ or $t = 3$. If they have $t = 3$, we get a contradiction, for each of other three vertices has $t = 2$, which would force us back to Figure 3.1B or to the conclusion that some vertex lies outside the circle. As a consequence, Moser's construction implies the arrangement shown in Figure 3.1A.

Because the sectors outside the sides of Moser's triangle are subsemicircular, $21 < 26$ and $23 < 24$. Since $t(2) = 2$, we have $23 = 21 = a$ and $26 = 24$. By analogy, $43 = 45$, $42 = 46$, $65 = 61$ and $64 = 62$. Since $26 = 24 = 46$, Moser's triangle is

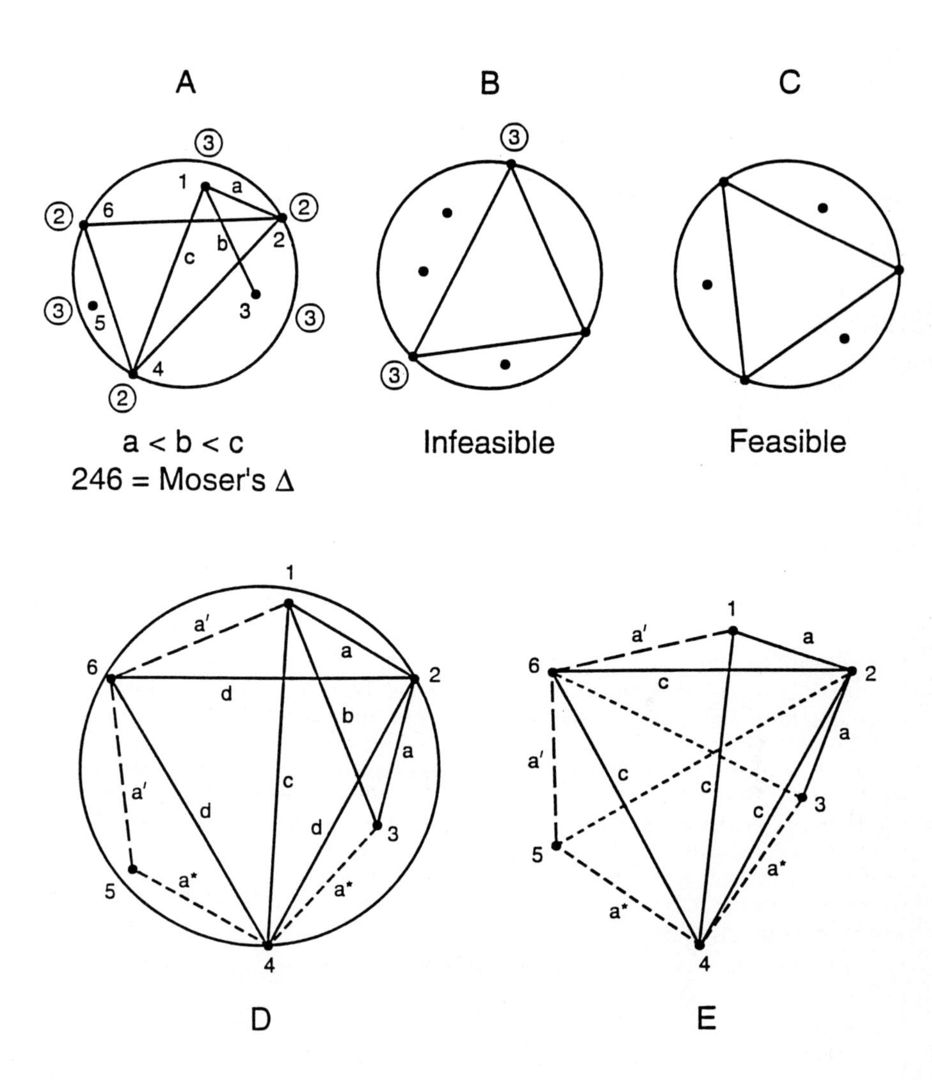

Fig. 3.1. $n = 6,\ (t(1), \ldots, t(6)) = (3, 2, 3, 2, 3, 2)$.

equilateral: see Figure 3.1D. Let $d = 26$. We have $43 = 45 < d = 42 = 46$, $c = 41$ and $t(4) = 2$. A simple geometric argument implies that $c = d$: if $c = 43$ with 1 as pictured, 3 would lie outside the circle. This brings us to Figure 3.1E. For reasons similar to those just given for $c = d$, we have $25 = 63 = c$. Therefore triangles 124 and 326 are congruent, so the line through 1 and 3 is parallel to the line through 6 and 4, and as a consequence $a' = a^*$. Similarly, $a' = a$, so all sides of the hexagon have the same length. These and the six c diagonals define A_6.

My proof that R_7 is the only convex heptagon C with t-sequence $(3, 3, \ldots, 3)$ is unreasonably long. A shorter proof is needed. The main steps in mine are as follows:

1. Show that $C = R_7$ if all sides are equally long.
2. Prove that $d(x, z) > \max\{d(x, y), d(y, z)\}$ whenever x, y and z are consecutive vertices.

Label the points 1 through 7 clockwise in such a way that Moser's lemma for $n = 7$ gives 1 through 4 in a subsemicircular sector of a circumscribed circle. Set $12 = a < 13 = b < 14 = c$.

3. Prove that $17 = a$, $16 = b$ and $15 = c$.
4. Apply the result just proved in a sequence of steps (the first is $43 < 42 < 41 \Rightarrow 45 = 43, 42 = 46, 41 = 47 = c$) to conclude that $C = R_7$.

The main labor involves eliminating other possibilities in step 3.

4. Discussion

Conjecture C4, which claims that T_n equals $\binom{n}{2}$, evolved from work on C2, and in turn suggested conjectures C3 and C5. It has been featured here because its appealing form and global character may suggest approaches that prove it as well as its parent C2.

We noted earlier that, in view of A_n, C3 is formulated as tightly as possible for even n. This is not true for odd n. At least four vertices of a convex pentagon have $t \geq 2$, and at least five vertices of a convex heptagon have $t \geq 3$. A small challenge is to give a convincingly tight alternative to C3 for odd n.

Another set of conjectures for distances in convex polygons is based on multiplicity vectors. The *multiplicity vector* of n-gon C is $r(C) = (r_1(C), r_2(C), \ldots, r_m(C))$ where m is the number of different intervertex distances and $r_i(C)$ is the number of times the ith most-frequent distance occurs, ties resolved arbitrarily. Thus $r_1(C) \geq r_2(C) \geq \cdots \geq r_m(C) \geq 1$ and $\Sigma r_i(C) = \binom{n}{2}$.

Let $r_i(n) = \max_C \{r_i(C) : C \text{ is a convex } n\text{-gon}\}$. Erdős and Moser [7] conjecture that $r_1(n) < cn$ for some constant c. The best general bounds we are aware of are $2n - 7 \leq r_1(n) \leq \pi n(2\log_2 n - 1)$, due to Edelsbrunner and Hajnal [2] and Füredi [11] respectively.

It is known [5] that $r_2(n) = n$ for $5 \leq n \leq 8$ and that $r_2(25) > 25$. We do not know the smallest n at which the second most-frequent distance can exceed n, nor do we have a very good idea of the growth rate of $r_2(n)/n$.

It is not known if $r_3(n) > n$ for some n.

We conjecture [4, 5] that $\Sigma[r_i(C)]^2$ is maximized uniquely over all convex n-gons at $C = R_n$, except for $n \in \{4, 6, 8\}$. The nonregular maximizers for the exceptional cases are identified in [9].

Many years ago Danzer (see [1]) disproved an Erdős conjecture [3] by constructing a convex 9-gon in which each vertex has three others equidistant from it. Fishburn and Reeds [10] constructs a convex 20-gon in which each vertex has three others distance 1 from it. Erdős and Fishburn [4] conjecture that there is no convex n-gon in which every vertex has distance 1 to four other vertices. If true, then $r_1(n) \leq 3n - 6$.

References

1. E. Altman, On a problem of P. Erdős, Amer. Math. Monthly 70 (1963) 148–157.
2. H. Edelsbrunner and P. Hajnal, A lower bound on the number of unit distances between the vertices of a convex polygon, J. Combin. Theory A 56 (1991) 312–316.
3. P. Erdős, On sets of distances of n points, Amer. Math. Monthly 53 (1946) 248–250.
4. P. Erdős and P. Fishburn, Multiplicities of interpoint distances in finite planar sets, Discrete Appl. Math. (to appear).
5. P. Erdős and P. Fishburn, Intervertex distances in convex polygons, Discrete Appl. Math. (to appear).
6. P. Erdős and P. Fishburn, A postscript on distances in convex n-gons, Discrete Comput. Geom. 11 (1994) 111–117.
7. P. Erdős and L. Moser, Problem 11, Canad. Math. Bull. 2 (1959) 43.
8. P. Fishburn, Convex polygons with few intervertex distances, DIMACS report 92–18 (April 1992), AT&T Bell Laboratories, Murray Hill, NJ.
9. P. Fishburn, Convex polygons with few vertices, DIMACS report 92–17 (April 1992), AT&T Bell Laboratories, Murray Hill, NJ.
10. P. C. Fishburn and J. A. Reeds, Unit distances between vertices of a convex polygon, Comput. Geom.: Theory and Appls. 2 (1992) 81–91.
11. Z. Füredi, The maximum number of unit distances in a convex n-gon, J. Combin. Theory A 55 (1990) 316–320.
12. L. Moser, On the different distances determined by n points, Amer. Math. Monthly 59 (1952) 85–91.

The Number of Homothetic Subsets

M. Laczkovich[1*] and I. Z. Ruzsa[2*]

[1] Department of Analysis, Eötvös Loránd University, Múzeum krt. 6-8, Budapest, H-1088 Hungary
[2] Mathematical Institute of the Hungarian Academy of Sciences, Budapest, Pf. 127, H-1364 Hungary

Summary. We investigate the maximal number $S(P,n)$ of subsets of a set of n elements homothetic to a fixed set P. Elekes and Erdős proved that $S(P,n) > cn^2$ if $|P| = 3$ or the elements of P are algebraic. For $|P| \geq 4$ we show that $S(P,n) > cn^2$ if and only if every quadruple in P has an algebraic cross ratio. Moreover, there is a sequence S_n of numbers such that $S(P,n) \asymp S_n$ whenever $|P| = 4$ and the cross ratio of P is transcendental.

AMS subject classification: primary 52C10, secondary 05D99.

Let $\mathbf{C}$ denote the set of complex numbers. We say that the sets $A, B \subset \mathbf{C}$ are homothetic, and write $A \sim B$, if $B = a \cdot A + b = \{ax + b : x \in A\}$ with suitable $a, b \in \mathbf{C}$, $a \neq 0$. If P and A are finite subsets of $\mathbf{C}$ then let

$$s(P,A) \stackrel{\text{def}}{=} |\{X \subset A : X \sim P\}|,$$

where $|H|$ denotes the cardinality of H. In [2] G. Elekes and P. Erdős investigated the behaviour of the sequence

$$S(P,n) \stackrel{\text{def}}{=} \max_{|A|=n} s(P,A).$$

It is easy to see that $S(P,n) \leq 2n(n-1)$ holds for every finite P and $n \in \mathbf{N}$. Elekes and Erdős proved that the order of magnitude of $S(P,n)$ is close to n^2 for every P; namely

$$S(P,n) \geq c \cdot n^{2 - b \cdot \log^{-a} n}$$

holds for every $n \geq |P|$ with positive constants a, b, and c depending on P but not on n. They also showed that

$$S(P,n) \geq c \cdot n^2 \quad (n \geq |P|) \tag{1}$$

if $|P| = 3$ or if the elements of P are algebraic, and asked whether or not this is true for every finite P. In this paper we answer this question in the negative, and characterize the sets satisfying (1).

Our characterization will be given in terms of the cross ratio and projective equivalence. The cross ratio of the distinct complex numbers a, b, c, d is defined by

$$(a;b;c;d) \stackrel{\text{def}}{=} \frac{c-a}{c-b} : \frac{d-a}{d-b}.$$

* Supported by Hungarian National Foundation for Scientific Research, Grant T 7582.

A map $f : A \to B$ ($A, B \subset \mathbf{C}$) is said to be projective if it preserves the cross ratios of the quadruples of A. Two sets are projective equivalent if there is a projective bijection between them. Note that P and P' are projective equivalent whenever $|P| = |P'| \le 3$. Our main result is the following.

Theorem 1. *For every finite P the following are equivalent.*

(i) *There is a positive constant c such that* (1) *holds.*
(ii) *The cross ratio of every quadruple of P is algebraic.*
(iii) *There is a set P' such that P and P' are projective equivalent, and the elements of P' are algebraic.*

First we prove the implication (ii)$\Longrightarrow$(iii). Suppose that the set $P = \{a_1, \ldots, a_k\}$ satisfies (ii), where $a_1, \ldots, a_k$ are distinct complex numbers and $k \ge 4$ (if $k \le 3$ then the statement is obvious). Let $f(x) = (px+q)/(rx+s)$ ($x \in \mathbf{C}$), where p, q, r, s are chosen such that $ps - qr \neq 0$, $ra_i + s \neq 0$ for $i = 1, \ldots, k$, and $f(a_i)$ is algebraic for $i = 1, 2, 3$. Let $P' = f(P)$. Since f is projective, P and P' are projective equivalent. For the same reason, the cross ratio of the numbers $f(a_1), f(a_2), f(a_3), f(a_i)$ is algebraic for every $4 \le i \le k$. Since the first three of these numbers are algebraic, so is $f(a_i)$. Thus the elements of P' are algebraic, and hence P satisfies (iii).

The implication (iii)$\Longrightarrow$(i) is an immediate consequence of the following theorem and of the result of Elekes and Erdős stating that (1) holds if the elements of P are algebraic.

Theorem 2. *Let $|P| = |P'| = k$, and suppose that P and P' are projective equivalent. Then there are positive constants c_1 and c_2 depending only on k such that*

$$c_1 \cdot S(P,n) \le S(P',n) \le c_2 \cdot S(P,n) \tag{2}$$

for every $n \ge k$.

Lemma 1. *For every $Q \subset \mathbf{C}$, $|Q| = k$ and $n \ge k$ we have $S(Q,kn) \le k^{2k} S(Q,n)$.*

Proof. Let $E \subset \mathbf{C}$ be such that $|E| = kn$ and $s(Q,E) = S(Q,kn)$. Let N denote the number of pairs (C,D) such that $C \subset D \subset E$, $C \sim Q$ and $|D| = n$. Clearly, $N = S(Q,kn)\binom{kn-k}{n-k}$. On the other hand, as each D contains at most $S(Q,n)$ C's, we have $N \le \binom{kn}{n} S(Q,n)$. This gives

$$\frac{S(Q,kn)}{S(Q,n)} \le \frac{\binom{kn}{n}}{\binom{kn-k}{n-k}} = \frac{kn(kn-1)\ldots(kn-k+1)}{n(n-1)\ldots(n-k+1)} < k^{2k}.$$

Proof (of Theorem 2). The statement of the theorem is obvious if $k = 1$ or $k = 2$, so that we may assume $k \ge 3$. Let a, b, c be distinct elements of P, let f be a projective bijection from P onto P', and let $f(a) = a'$, $f(b) = b'$, $f(c) = c'$. We can suppose that $a = a' = 0$ and $b = b' = 1$, since otherwise we replace P by $(P-a)/(b-a)$ and P' by $(P'-a')/(b'-a')$. This replacement will not affect the projective equivalence of the sets P and P', nor will it change the values of $S(P,n)$ and $S(P',n)$.

Let $A \subset \mathbf{C}$ be such that $|A| = n$ and $s(P,A) = S(P,n)$. If $X \subset A$ and $X \sim P$ then there are elements $x, y \in X$ such that $X = \{x + p(y-x) : p \in P\}$. Therefore, if T denotes the set of all pairs $(x,y) \in A \times A$ satisfying $x + p(y-x) \in A$ for all $p \in P$, then we have $|T| \ge S(P,n)$. Let $\lambda = c(c'-1) \cdot (c'(c-1))^{-1}$, and let μ be a number to be fixed later. We put

$$B = \{x + p'(\lambda y + \mu - x) : (x,y) \in T,\ p' \in P'\}.$$

If $x \neq \lambda y + \mu$ then the set $U_{x,y} = \{x + p'(\lambda y + \mu - x) : p' \in P'\}$ is similar to P'. We can select the number μ in such a way that $x \neq \lambda y + \mu$ for every $(x, y) \in A \times A$, and the sets $U_{x,y}$ $((x, y) \in A \times A)$ are distinct. Fixing such a μ it follows that

$$s(P', B) \geq |T| \geq S(P, n). \tag{3}$$

For every $p \in P$ we denote $p' = f(p)$,

$$A_p = \{x + p(y - x) : (x, y) \in T\}, \quad B_p = \{x + p'(\lambda y + \mu - x) : (x, y) \in T\},$$

and

$$\phi_p(x + p(y - x)) = x + p'(\lambda y + \mu - x) \ \ ((x, y) \in T).$$

First we show that ϕ_p is a well-defined map from A_p onto B_p. To this end we have to prove that if $(x, y) \in T$, $(u, v) \in T$ and

$$x + p(y - x) = u + p(v - u), \tag{4}$$

then

$$x + p'(\lambda y + \mu - x) = u + p'(\lambda v + \mu - u). \tag{5}$$

If $p = p' = 0$ or $p = p' = 1$ then the implication (4) $\Longrightarrow$ (5) is clear. If $p = c$, $p' = c'$, then (4) implies $(1 - c)x + cy = (1 - c)u + cv$. Therefore, by the definition of λ,

$$x + c'(\lambda y + \mu - x) = (1 - c')x + \lambda c' y + c'\mu = \frac{1 - c'}{1 - c}\big((1 - c)x + cy\big) + c'\mu =$$

$$\frac{1 - c'}{1 - c}\big((1 - c)u + cv\big) + c'\mu = (1 - c')u + \lambda c' v + c'\mu = u + c'(\lambda v + \mu - u),$$

which gives (5). Finally suppose $p \in P \setminus \{0, 1, c\}$. Since f is projective, we have $(0; 1; c; p) = (0; 1; c'; p')$ and thus $\lambda = p(p' - 1) \cdot (p'(p - 1))^{-1}$. Then a computation identical to the one above shows that (4) implies (5) in this case as well. This proves that the map ϕ_p is well-defined, and it is clear that $\phi_p(A_p) = B_p$.

It follows from the definition of T that $A_p \subset A$, and hence $|A_p| \leq n$. Thus we have $|B_p| \leq n$, and then $|B| \leq \sum_{p \in P} |B_p| \leq kn$. Combining with (3), this gives $S(P', kn) \geq S(P, n)$. Then, by Lemma 1, we obtain $S(P', n) \geq k^{-2k} S(P, n)$. Interchanging the roles of P and P' we get the other inequality of (2). ■

In order to prove the implication (i)$\Longrightarrow$(ii) of Theorem 1, we may assume that $|P| = 4$. Indeed, if (i) holds for P then it holds for every four-element subset of P as well, and if (ii) holds for every four-element subset of P, then it also holds for P.

Let α be a transcendental number, and denote

$$S_n = S(\{0, 1, 2, \alpha\}, n) \ \ (n \geq 4).$$

The value of S_n does not depend on the choice of α. Indeed, if β is another transcendental number, then there is a field-automorphism σ of $\mathbf{C}$ such that $\sigma(\alpha) = \beta$. For every $X \subset \mathbf{C}$, $|X| = 4$ we have $X \sim \{0, 1, 2, \alpha\}$ if and only if $\sigma(X) \sim \{0, 1, 2, \beta\}$ and this easily implies that $S(\{0, 1, 2, \alpha\}, n) = S(\{0, 1, 2, \beta\}, n)$ for every $n \geq 4$.

Theorem 3. *There are positive absolute constants c_1 and c_2 such that for every quadruple $P \subset \mathbf{C}$, if the cross ratio of the elements of P is transcendental, then*

$$c_1 \cdot S_n \leq S(P, n) \leq c_2 \cdot S_n$$

for every $n \geq 4$.

Proof. Let $P = \{a,b,c,d\}$ and $(a;b;c;d) = \alpha$. We put $\beta = 2/(2-\alpha)$ and $P' = \{0,1,2,\beta\}$. Since $(0;1;2;\beta) = \alpha$, the sets P and P' are projective equivalent. Consequently, by Theorem 2, there are absolute constants $c_1, c_2 > 0$ such that $c_1 \cdot S(P',n) \le S(P,n) \le c_2 \cdot S(P',n)$ for every $n \ge 4$. Also, since β is transcendental, we have $S(P',n) = S_n$ by the remark preceding the theorem. ∎

Now, in order to prove the implication (i)$\Longrightarrow$(ii) of Theorem 1, it is enough to show that $S_n = o(n^2)$ as $n \to \infty$. Indeed, suppose this is true, and let $P \subset \mathbf{C}$ be a four-element set for which (i) holds but (ii) does not. Then, by Theorem 3, $c_2 \cdot S_n \ge S(P,n) \ge c \cdot n^2$ for every $n \ge 4$, a contradiction.

The rest of the paper will be devoted to the proof of $S_n = o(n^2)$. In the sequel we denote

$$S(n,c) = S(\{0,1,2,c\},n) \quad (c \in \mathbf{C} \setminus \{0,1,2\},\ n \in \mathbf{N}).$$

Lemma 2. *For every n we have $S(n,c) \ge S_n$ for all, but a finite number of $c \in \mathbf{C}$.*

Proof. Let α be transcendental and let $A = \{a_1, \ldots, a_n\} \subset \mathbf{C}$ be such that $s(\{0,1,2,\alpha\},A) = S_n$. This means that there is a set I of quadruples of indices such that $|I| = S_n$ and for every $(i,j,k,m) \in I$ we have

$$a_i - 2a_j + a_k = 0 \quad \text{and} \quad (\alpha - 1)a_i - \alpha a_j + a_m = 0. \tag{6}$$

Let

$$V = \left\{ \sum_{i=1}^{n} r_i \cdot a_i : \quad r_i \in \mathbf{Q}(\alpha), \quad i = 1, \ldots, n \right\},$$

then V is a linear space over the field $\mathbf{Q}(\alpha)$. Let $B = \{b_1, \ldots, b_d\}$ be a basis of V, and let

$$a_i = \sum_{j=1}^{d} r_{ij} \cdot b_j \quad (i = 1, \ldots, n) \tag{7}$$

be representations with $r_{ij} \in \mathbf{Q}(\alpha)$ for every $i = 1, \ldots, n$ and $j = 1, \ldots, d$. Since the coefficients of a_i, a_j, a_k, a_m in the equations (6) belong to $\mathbf{Q}(\alpha)$, and $b_1, \ldots, b_d$ are linearly independent over $\mathbf{Q}(\alpha)$, it follows that substituting the representations (7) into the equations (6) we obtain identities. In other words, for every choice of the variables $x_1, \ldots, x_d$, the numbers

$$c_i = \sum_{j=1}^{d} r_{ij} \cdot x_j \quad (i = 1, \ldots, n) \tag{8}$$

will satisfy the equations

$$c_i - 2c_j + c_k = 0 \quad \text{and} \quad (\alpha - 1)c_i - \alpha c_j + c_m = 0 \tag{9}$$

for every $(i,j,k,m) \in I$. The right-hand sides of (8), as linear forms, are different, because for $x_j = b_j$ they have different values (namely $a_1, \ldots, a_n$). Then $x_1, \ldots, x_d$ can be chosen to be integers such that the values of the corresponding $c_1, \ldots, c_n$ are different. Indeed, for every $i_1 \ne i_2$, the set $\{(x_1, \ldots, x_d) : c_{i_1} = c_{i_2}\}$ is a hyperplane, and $\mathbf{Z}^d$ cannot be covered by finitely many hyperplanes. Clearly, if the $x_j's$ are integers then $c_i \in \mathbf{Q}(\alpha)$ for every $i = 1, \ldots, n$.

We have proved that there are distinct elements $c_i \in \mathbf{Q}(\alpha)$ satisfying the equations (9). For every $i = 1, \ldots, n$ there is a rational function R_i with rational coefficients such that $c_i = R_i(\alpha)$. Thus we have

$$R_i(\alpha) - 2R_j(\alpha) + R_k(\alpha) = 0 \text{ and } (\alpha - 1)R_i(\alpha) - \alpha R_j(\alpha) + R_m(\alpha) = 0$$

for every $(i,j,k,m) \in I$. Since α is transcendental, the rational functions $R_i(t) - 2R_j(t) + R_k(t)$ and $(t-1)R_i(t) - tR_j(t) + R_m(t)$ must be identically zero for every $(i,j,k,m) \in I$. Therefore, whenever the numbers $R_1(c), \dots, R_n(c)$ are defined (that is, c is not a root of the denominator of $R_1 \cdot \ldots \cdot R_n$), and are distinct, then the set $A_c = \{R_1(c), \dots, R_n(c)\}$ contains S_n subsets similar to $\{0,1,2,c\}$. The rational functions R_i are distinct, since they have different values at α. Then $|A_c| = n$ for all but a finite number of c's, and for such a c we have $S(n,c) \geq S(\{0,1,2,c\}, A_c) \geq S_n$. ■

We remark that, in fact, $S(n,c) = S_n$ holds for all but finitely many $c \in \mathbf{C}$. As we shall not need this result, we omit the proof. What we need is the fact that $S_n \leq S(n,k)$ for every $k \in \mathbf{N}$, $k > k_0(n)$. This implies

$$\limsup_{n\to\infty} \frac{S_n}{n^2} \leq \limsup_{n\to\infty} \left(\limsup_{k\to\infty} \frac{S(n,k)}{n^2} \right), \tag{10}$$

and hence, in order to prove $S_n = o(n^2)$, it is enough to show that the right-hand side of (10) is zero. In the next theorem we prove somewhat more.

Theorem 4.

$$\lim_{\substack{n\to\infty \\ k\to\infty}} \frac{S(n,k)}{n^2} = 0.$$

Let $z, q_1, \dots, q_d \in \mathbf{Z}$ and $X_1, \dots, X_d \in \mathbf{N}$. We shall say that the set

$$R(z, q_1, \dots, q_d; X_1, \dots, X_d) = \left\{ z + \sum_{i=1}^{d} x_i q_i : x_i \in \mathbf{Z},\ 0 \leq x_i < X_i \ (i = 1, \dots, d) \right\}$$

is a *d-dimensional arithmetical progression of size* $\prod_{i=1}^d X_i$. We shall need the following theorem proved by G. A. Freiman in [3, 4, 5] and I. Z. Ruzsa in [6].

Theorem A. *If $A \subset \mathbf{Z}$, $|A| = n$, and $|A+A| \leq cn$, then A is contained in a d-dimensional arithmetical progression of size not exceeding $c'n$, where d and c' only depend on c.*

Let $A \subset \mathbf{Z}$ be a finite set, let $G = (A, E)$ be a graph, and put $S = \{a + b : (a,b) \in E\}$. The following result was proved by A. Balog and E. Szemerédi in [1].

Theorem B. *If $|A| = n$, $|E| \geq c_1 n^2$ and $|S| \leq c_2 n$, then there is a subset $A' \subset A$ such that $|A'| \geq c_3 n$ and $|A' + A'| \leq c_4 n$, where c_3 and c_4 only depend on c_1 and c_2.*

As Balog and Szemerédi remark, these two theorems can be combined to obtain the following result: if $|A| = n$, $|E| \geq c_1 n^2$ and $|S| \leq c_2 n$, then there is a d-dimensional arithmetical progression R of size not exceeding $c_5 n$ such that $|R \cap A| \geq c_6 n$, where d, c_5, c_6 depend only on c_1 and c_2. In the next lemma we prove a slight improvement of this result.

Lemma 3. *If $|A| = n$, $|E| \geq c_1 n^2$ and $|S| \leq c_2 n$, then there is d-dimensional arithmetical progression R of size not exceeding $c_{11} n$ such that the subgraph of G induced by the set $R \cap A$ contains at least $c_{12} n^2$ edges. Here the constants d, c_{11} and c_{12} depend only on c_1 and c_2.*

Proof. In the sequel $c_7, c_8, \dots$ will denote constants depending only on c_1 and c_2. Since $|E| \geq c_1 n^2$, there is a subset $A_1 \subset A$ such that in the subgraph of G induced by A_1 the degree of every point is greater than $c_1 n/2$. (Indeed, delete one by one each point of degree at most $c_1 n/2$. In this way we cannot remove all points of A, and the set of remaining points will have the required property.) Then $|A_1| \geq c_1 n/2$. Applying Theorem B to the graph induced by A_1, we obtain a subset $A' \subset A_1$ such that $|A'| \geq c_7 n$ and $|A' + A'| \leq c_8 n$. Then we apply Theorem A to obtain a d-dimensional arithmetical progression $R = R(z, q_1, \dots, q_d; X_1, \dots, X_d)$ such that $A' \subset R$ and the size of R does not exceed $c_9 n$.

Let $c_{10} = c_1 c_7/8$ and let B be the set of those points of A that are connected to at least $c_{10} n$ points of A'. Since there are at least $(c_7 n \cdot c_1 n/2)/2 = 2c_{10} n^2$ edges starting from the points of A', we have $|B| \geq c_{10} n$. Let $\mathcal{R} = \{R + b : b \in B\}$. Since $A' \subset R$, each set $R + b$ ($b \in B$) contains at least $c_{10} n$ elements of the form $a + b$, where $a \in A'$ and $(a, b) \in E$. These elements belong to S and, by assumption, $|S| \leq c_2 n$. This implies that any pairwise disjoint subsystem of $\mathcal{R}$ contains at most c_2/c_{10} sets. Let $\{R + b_i : i = 1, \dots, k\}$ be a maximal disjoint subsystem of $\mathcal{R}$. Then $k \leq c_2/c_{10}$ and for every $b \in B$ there is an $i \leq k$ such that $(R + b) \cap (R + b_i) \neq \emptyset$. This implies $b \in (R - R) + b_i$, and hence $B \subset \cup_{i=1}^{k} [(R - R) + b_i]$. Let $b_{k+1} = z$ and $H = \cup_{i=1}^{k+1} [(R - R) + b_i]$. Then $A' \cup B \subset H$ as $A' \subset R \subset (R - R) + b_{k+1}$. Thus the subgraph of G induced by $H \cap A$ contains at least $c_{12} n^2$ edges ($c_{12} = c_{10}^2/2$), since every point of B is connected to at least $c_{10} n$ points of A'. Let $q_{d+j} = b_j$ for $j = 1, \dots, k+1$, and put

$$R' = \left\{ \sum_{i=1}^{d+k+1} x_i q_i : -X_i \leq x_i < X_i \ (1 \leq i \leq d) \text{ and } 0 \leq x_i < 2 \ (d < i \leq d+k+1) \right\}.$$

Then $H \subset R'$, and in order to complete the proof of the lemma it is enough to note that R' is a $d + k + 1$-dimensional arithmetical progression of size not exceeding $2^{k+1} \prod_{i=1}^{d} (2X_i) \leq 2^{d+k+1} c_9 n = c_{11} n$. ■

Lemma 4. *Let $n, k \in \mathbf{N}$, $c > 0$, and suppose that $S(n, k) > cn^2$. Then there are positive constants d, c', c'' depending only on c, and there exists a d-dimensional arithmetical progression $R = R(z, q_1, \dots, q_d; X_1, \dots, X_d)$ such that the gcd $(q_1, \dots, q_d) = 1$, the size of R is at most $c'n$, and for a suitable integer u the set*

$$\{(a, b) \in R \times R : 2a + k(b - a) = u\} \tag{11}$$

contains at least $c''n$ distinct pairs.

Proof. Since $S(n, k) > cn^2$, there is a set $A \subset \mathbf{C}$ such that $|A| = n$ and A contains more than cn^2 subsets similar to $\{0, 1, 2, k\}$. Repeating the argument of the proof of Lemma 2 with $\mathbf{Q}$ instead of $\mathbf{Q}(\alpha)$, we can see that A can be chosen to be a subset of $\mathbf{Q}$. Multiplying by a suitable integer, we may assume that $A \subset \mathbf{Z}$.

Let E denote the set of all pairs (a, b) such that each number a, b, $(a + b)/2$ and $a + \frac{k}{2}(b - a)$ belongs to A; then $|E| > cn^2$. Let $S = \{a + b : (a, b) \in E\}$, then $S \subset 2A$, and hence $|S| \leq n$. Therefore, by Lemma 3, there are positive constants d, c', c'' depending only on c and there exists a d-dimensional arithmetical progression $R = R(z, q_1, \dots, q_d; X_1, \dots, X_d)$ of size not exceeding $c'n$ such that $A \cap R$ contains at least $c''n^2$ edges from E. Let $F = \{(a, b) \in E : a, b \in R\}$. Then $|F| \geq c''n^2$ and $|\{2a + k(b - a) : (a, b) \in F\}| \leq |2 \cdot A| = n$. Consequently, there exists an element $u \in A$ such that the set defined in (11) contains at least $c''n$ distinct pairs. If $(q_1, \dots, q_d) = 1$, then the proof is complete. Otherwise let $(q_1, \dots, q_d) = m$, and replace R by $R(z, q_1/m, \dots, q_d/m; X_1, \dots, X_d)$. It is clear that this modified R satisfies the requirements. ■

Lemma 5. *Let $R = R(z, q_1, \dots, q_d; X_1, \dots, X_d)$ be a d-dimensional arithmetical progression with $(q_1, \dots, q_d) = 1$, and denote $s = \prod_{i=1}^{d} X_i$ and $M = \min_{1 \le i \le d} X_i$. Then for every positive integer k, the number of elements of R in any residue class mod k is at most $(s/M) + (s/k)$.*

Proof. For $i = 1, ..., d$ define

$$k_i = \frac{(k, q_1, ..., q_{i-1})}{(k, q_1, ..., q_i)}, \quad k_1 = \frac{k}{(k, q_1)}.$$

We have $k_1 ... k_d = k$. Now consider the k numbers

$$y_1 q_1 + ... + y_d q_d, \quad 0 \le y_i \le k_i - 1.$$

We show that they are all incongruent modulo k. Indeed, suppose that

$$y_1 q_1 + ... + y_d q_d \equiv y_1' q_1 + ... + y_d' q_d \pmod{k}.$$

Let j be the largest subscript for which $y_j \ne y_j'$. Since the first $j - 1$ summands are divisible by $(q_1, ..., q_{j-1})$, we have

$$(k, q_1, ..., q_{j-1}) | q_j (y_j - y_j'),$$

hence

$$k_j = \frac{(k, q_1, ..., q_{j-1})}{(k, q_1, ..., q_j)} \Bigg| \, y_j - y_j',$$

which contradicts the assumption $0 \le y_j, y_j' \le k_j - 1$, $y_j \ne y_j'$.

Let R^* be the set of those elements of R that are $\equiv a \pmod{k}$. The sets

$$R^* + y_1 q_1 + ... + y_d q_d, \quad 0 \le y_i \le k_i - 1$$

lie in different residue classes modulo k, hence they are disjoint, and they are contained in $R(z, q_1, ..., q_d, X_1 + k_1 - 1, ..., X_d + k_d - 1)$, consequently

$$k|R^*| \le \prod (X_i + k_i - 1)$$

(we recall that $k = \prod k_i$), or

$$|R^*| \le \frac{s}{k} \prod \left(1 + \frac{k_i - 1}{X_i}\right) \le \frac{s}{k} \prod \left(1 + \frac{k_i - 1}{M}\right).$$

To complete the proof we show that the inequality

$$\prod_{i=1}^{d} \left(1 + \frac{k_i - 1}{M}\right) \le 1 + \frac{(\prod k_i) - 1}{M}$$

holds for arbitrary real numbes $k_i \ge 1$, $M \ge 1$. For $d = 2$ this inequality asserts that

$$\left(1 + \frac{k_1 - 1}{M}\right) \left(1 + \frac{k_2 - 1}{M}\right) \le 1 + \frac{k_1 k_2 - 1}{M}.$$

After multiplying by M and rearranging this becomes

$$\frac{(k_1 - 1)(k_2 - 1)}{M} \le (k_1 - 1)(k_2 - 1),$$

which is true if all the variables are ≥ 1. The case for $d \ge 3$ now easily follows by an induction on d. ■

Now we turn to the proof of Theorem 4. Suppose that $\limsup_{n,k\to\infty} S(n,k)/n^2 > 0$. Then there is a constant $c > 0$ such that for every K there are integers $n, k > K$ with $S(n,k) > cn^2$. Thus, by Lemma 4, there are positive constants d, c', c'' such that the following statement holds:

(∗) for every K there is a d-dimensional arithmetical progression

$$R_K = R(z^K, q_1^K, \ldots, q_d^K; X_1^K, \ldots, X_d^K)$$

and there are integers $n, k > K$ such that $(q_1^K, \ldots, q_d^K) = 1$, the size of R_K is at most $c'n$, and for a suitable integer u the set

$$\{(a,b) \in R_K \times R_K : 2a + k(b-a) = u\}$$

contains at least $c''n$ distinct pairs.

We prove that this is impossible. Let d be the smallest positive integer such that (∗) holds for suitable positive constants c' and c''. If $2a + k(b-a) = u$ then $2a \equiv u \pmod{k}$ and hence, if R_K, n, k are as in (∗) then $2 \cdot R_K$ contains at least $c''n$ elements in one of the residue classes mod k. This implies that R_K contains at least $c''n/2$ elements in one of the residue classes mod k. Let $M_K = \min_{1\le i\le d} X_i^K$. Then, by Lemma 5, we have

$$c''/2 \le (c'/M_K) + (c'/k). \tag{12}$$

Let $C = 4c'/c''$. If $K > C$, then $k > K$ implies $c'/k < c''/4$ and thus (12) gives $M_K < C$. By rearranging the indices we may assume that $X_d^K < C$ holds for every $K > C$. This implies $d > 1$. Indeed, if $d = 1$ then we have $|R_K| = |X_1^K| < C$ for $K > C$. This gives $|R_K \times R_K| \le C^2$, which contradicts (∗) for $K > C^2/c''$.

We complete the proof by showing that (∗) also holds for $d-1$ instead of d, if we replace c'' by c''/C^2. Since d was minimal, this will provide the contradiction we were looking for.

For every K we put $R_K' = R(z^K, q_1^K, \ldots, q_{d-1}^K; X_1^K, \ldots, X_{d-1}^K)$. Also, for every $a \in R_K$ we choose a representation $a = z^K + \sum_{i=1}^d x_i q_i^K$ with $0 \le x_i < X_i$ $(i = 1, \ldots, d)$ and define $a' = z^K + \sum_{i=1}^{d-1} x_i q_i^K$. In this way we have defined a map $a \mapsto a'$ from R_K into R_K'.

Let $K > C$, and let R_K, n, k and u be as in (∗). Obviously, $a - a' \in \{i \cdot q_d^K : 0 \le i \le X_d^K - 1\}$ for every $a \in R_K$ and hence the set

$$P = \{2a' + k(b' - a') : (a,b) \in R_K \times R_K,\ 2a + k(b-a) = u\}$$

contains at most $(X_d^K)^2$ distinct elements. Since $K > C$, we have $X_d^K < C$ and thus $|P| \le C^2$. This implies that for a suitable u' the set

$$\{(a',b') \in R_K' \times R_K' : 2a' + k(b'-a') = u'\}$$

contains at least $(c''/C^2)n$ distinct pairs. Therefore, replacing c'' by c''/C^2, the $d-1$-dimensional arithmetical progression R_K' and the integers n, k, u' will satisfy the statement of (∗) apart from the condition that $(q_1^K, \ldots, q_{d-1}^K) = 1$. But this condition can also be fulfilled if we replace R_K' by

$$R(z^K, q_1^K/m, \ldots, q_{d-1}^K/m; X_1^K, \ldots, X_{d-1}^K),$$

where $m = (q_1^K, \ldots, q_{d-1}^k)$. ∎

References

1. A. Balog and E. Szemerédi, A statistical theorem of set addition, Combinatorica 14 (1994), 263–268.
2. G. Elekes and P. Erdős, Similar configurations and pseudogrids, preprint.
3. G. A. Freiman, Foundations of a structural theory of set addition (in Russian), Kazan Gos. Ped. Inst., Kazan, 1966.
4. G. A. Freiman, Foundations of a structural theory of set addition, Translation of Mathematical Monographs vol. 37, Amer. Math. Soc., Providence, R. I., USA, 1973.
5. G. A. Freiman, What is the structure of K if $K + K$ is small?, Lecture Notes in Mathematics 1240, Springer, Berlin-New York 1987, pp. 109-134.
6. I. Z. Ruzsa, Generalized arithmetical progressions and sumsets, Acta Math. Sci. Hung. 65 (1994), 379–388.

On Lipschitz Mappings onto a Square

Jiří Matoušek

Department of Applied Mathematics, Charles University, Malostranské nám. 25, 118 00 Praha 1, Czech Republic

Summary. Recently Preiss [4] proved that every subset of the plane of a positive Lebesgue measure can be mapped onto a square by a Lipschitz map. In this note we give an alternative proof of this result, based on a well-known combinatorial lemma of Erdős and Szekeres. The validity of an appropriate generalization of this lemma into higher dimensions remains as an open problem.

1. Introduction

The following problem was posed by Laczkovich [3]: Let $E \subseteq \mathbb{R}^d$ $(d \geq 2)$ be a set with positive Lebesgue measure $\lambda^d(E) > 0$. Does there exist a Lipschitz mapping $f : \mathbb{R}^d \to Q = [0,1]^d$, such that $f(E) = Q$? Recently Preiss [4] answered this question affirmatively for $d = 2$:

Theorem 1.1. [4] *Let $E \subseteq \mathbb{R}^2$ be a set with $\lambda^2(E) > 0$. There exists a Lipschitz mapping f of the plane onto the square $Q = [0,1]^2$, such that $f(E) = Q$.*

In this note we give a somewhat different proof of this result[1] based on a well-known combinatorial lemma due to Erdős and Szekerés. By an additional trick, we also prove the following "absolute constant" version:

Theorem 1.2. *There exists a constant $c > 0$ such that for any E in the plane with Lebesgue measure $\lambda^2(E) = 1$ there exists a 1-Lipschitz mapping f of the plane onto the square $Q = [0,c]^2$, such that $f(E) = Q$.*

The value of c obtained from our proof is quite small. Clearly some improvement is possible, but it seems that our method is not suitable for obtaining the best possible value of c. It is easy to see that one cannot hope to push c arbitrarily close to 1 (e.g., consider E being a disk of unit area; its diameter $2/\sqrt{\pi}$ is smaller than the diagonal of the unit square, and so there is no 1-Lipschitz mapping onto the unit square).

Laczkovich's question for $d > 2$ remains open; a combinatorial problem related to an attempt to generalize our proof is stated in the end of section 2..

2. Proof of Theorem 1.1

The metric in $\mathbb{R}^2$, implicitly considered in Theorem 1.1, is the usual Euclidean metric. It appears more convenient to work with the maximum metric d_∞ defined by

[1] Preiss in fact proves a slightly stronger statement, namely that f can be taken such that $f(\mathbb{R}^2 \setminus E)$ is countably rectifiable (i.e. it can be covered by a countable set of Lipschitz curves). In order to keep this note technically simple, we will not prove this strengthening here, although our method also provides it with some extra care.

$$d_\infty((x_1,y_1),(x_2,y_2)) = \max(|x_1-x_2|,|y_1-y_2|)\,.$$

This metric will be used throughout the rest of this paper. Clearly this modification does not affect the validity of Theorem 1.1.

We begin the proof by a simple lemma, which is essentially contained in [4]. Let $a > 0, w > 0$ be real numbers and $\varphi : [0,a] \to [0,a]$ be a 1-Lipschitz real function. Let Q denote the square $[0,a]^2$. We partition Q into three regions defined as follows (see Fig. 2.1): L (resp. S, resp. U) is the set of points $(x,y) \in Q$ with $y < \varphi(x) - w$ (resp. $\varphi(x) - w \le y \le \varphi(x) + w$, resp. $y > \varphi(x) + w$). We define a

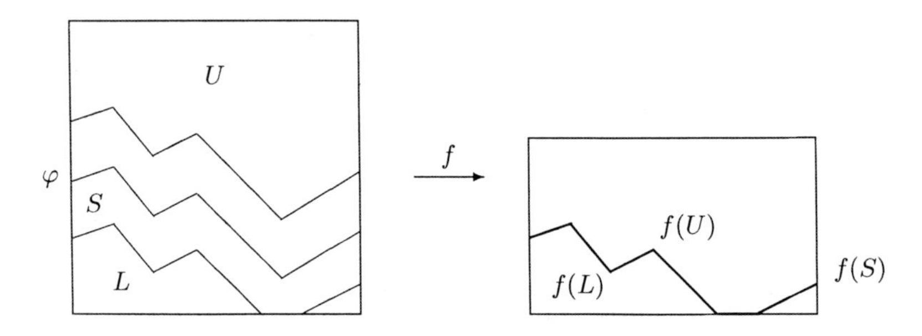

Fig. 2.1. Contraction along φ

mapping $f = f_{\varphi,w} : Q \to [0,1] \times [0, 1-2w]$ (called the *contraction of Q along φ by w*) as follows:

$$f(x,y) = \begin{cases} (x, \min(y, 1-2w)) & \text{for } (x,y) \in L \\ (x, \max(y-2w, 0)) & \text{for } (x,y) \in U \\ (x, \max(\min(\varphi(x) - w, 1-2w), 0)) & \text{for } (x,y) \in S. \end{cases}$$

Lemma 2.1. *With a, w, φ as above, $f = f_{w,\varphi}$ is a 1-Lipschitz mapping.*

Proof. Let $p_1 = (x_1,y_1)$ and $p_2 = (x_2,y_2)$ be two points in Q and $q_1 = (u_1,v_1)$ and $q_2 = (u_2,v_2)$ their f-images, resp., and suppose that $v_1 \ge v_2$. We must show $d_\infty(p_1,p_2) \ge d_\infty(q_1,q_2)$. First, if $|u_1-u_2| \ge |v_1-v_2|$, then $d_\infty(q_1,q_2) = |u_1-u_2| = |x_1-x_2| \le d_\infty(p_1,p_2)$, as f does not alter the x-coordinate. On the other hand, suppose that $|u_1-u_2| < |v_1-v_2| = v_1-v_2$; then $q_1 \in f(S) \cup f(L)$ implies $q_2 \in f(L)$, as $f(S)$ is a graph of the 1-Lipschitz function $x \mapsto \max(0, \min(1-2w, \varphi(x)-w))$. Therefore f decreases the y-coordinate of p_1 by at least as much as the y-coordinate of p_2, and we have $d_\infty(q_1,q_2) = v_1 - v_2 \le y_1 - y_2 \le d_\infty(p_1,p_2)$. □

We also note that the mapping f in the above lemma does not increase the measure of sets.

Let $E \subseteq \mathbb{R}^2$ be as in Theorem 1.1. By the Density Theorem for the Lebesgue measure, one can choose a square Q_0 with side a_0 and a compact subset $K \subseteq E \cap Q_0$ with $\lambda^2(K) \ge 0.99\lambda^2(Q_0)$. For notational convenience, let us assume $Q_0 = [0,a_0]^2$. Let $f_0 : \mathbb{R}^2 \to Q_0$ be a 1-Lipschitz retraction of the plane onto Q_0.

We derive Theorem 1.1 from the following lemma.

Lemma 2.2. *Let $Q = [0,a]^2$ be an axis-parallel square, let $K \subseteq Q$ be compact with $\lambda^2(K) = \lambda^2(Q)(1-\varepsilon)$, $0 < \varepsilon \le 0.01$. Then one can find a 1-Lipschitz mapping g of Q onto $Q' = [0,a']^2$ such that*

(i) $a' \geq a(1 - \sqrt{\varepsilon})$,
(ii) $d_\infty(p, g(p)) \leq a\sqrt{\varepsilon}$ *for any* $p \in Q$,
(iii) $\lambda^2(K') \geq \lambda^2(Q')(1 - 0.9\varepsilon)$, *where* $K' = g(K)$.

Assuming this Lemma, the proof of Theorem 1.1 is finished by an inductive construction as follows. Suppose that we have already constructed a square $Q_i = [0, a_i]^2$ and a 1-Lipschitz mapping f_i of $\mathbb{R}^2$ onto Q_i in such a way that $\lambda^2(K_i) \geq \lambda^2(Q_i)(1 - \varepsilon_i)$, where $K_i = f_i(K)$. We apply Lemma 2.2 with $Q = Q_i$, $a = a_i$, $K = K_i$, and we get a 1-Lipschitz mapping $g_i : Q_i \to Q_{i+1} = [0, a_{i+1}]^2$ with properties as in the Lemma. We set $f_{i+1} = g_i \circ f_i$ and continue by the next step of the construction.

From the Lemma we get $\varepsilon_{i+1} \leq 0.9\varepsilon_i$ and $a_{i+1} \geq a_i(1 - \sqrt{\varepsilon_i})$. We thus have $\varepsilon_i \leq 0.01(0.9)^i$, $a_i \geq a_0 \prod_{j=0}^{i-1}(1 - 0.95^j/10)$, and straightforward estimates give $a = \lim_{i\to\infty} a_i \geq 0.1a_0$.

For every point $p \in \mathbb{R}^2$, we set $f(p) = \lim_{i\to\infty} f_i(p)$. The condition (ii) of Lemma 2.2 guarantees that the limit exists. The mapping f is clearly 1-Lipschitz, and its image is contained in the square $Q = [0, a]^2$. The set $f(K)$ is compact and it is easily seen that it is also dense in Q, thus $f(K) = Q$. The proof of Theorem 1.1 is finished by rescaling f so that it maps K onto $[0, 1]^2$.

Proof (of Lemma 2.2). Let G be the set of squares with side a/n from an $n \times n$ grid covering Q. Let $B_0 \subseteq G$ be the set of squares $s \in G$ with $\lambda^2(s \setminus K) \geq \frac{15}{16}\lambda^2(s)$. Choose n so large that the squares of B_0 contain at least 1/2 of the measure of $Q \setminus K$; this is possible by elementary measure theoretic considerations. Then $|B_0| \geq \varepsilon n^2/2$.

The required mapping g will again be constructed inductively, this time in a finite number t of steps. Let us explain the first step of the construction. Let $\varphi_0 : [0, a] \to [0, a]$ be a suitable 1-Lipschitz function; its choice is the heart of the proof. The requirement on φ_0 is that either the graph of φ_0 (i.e. the set $\{(x, \varphi_0(x));\ x \in [0, a]\}$) or the graph of φ_0 rotated by $\pi/2$ (i.e. the set $\{(\varphi_0(y), y);\ y \in [0, a]\}$) contain the centers of possibly many squares of B_0, namely at least $\sqrt{|B_0|}$ such centers. We will explain the construction assuming that the first case occurs (the graph of φ_0 contains at least $\sqrt{|B_0|}$ centers). The second case is handled symmetrically, by exchanging the role of the coordinate axes.

Let $D_0 \subseteq B_0$ be the squares whose centers are contained in the graph of φ_0, $|D_0| \geq \sqrt{|B_0|}$. We let $\bar{g}_0 : Q \to Q$ be the contraction of Q along φ_0 by a/n (see the definition above Lemma 2.1). We define an auxiliary mapping $h_0 : [0, a] \times [0, a(n - 2)/n] \to [0, a(n-2)/n]^2$, acting as the identity on the first $n-2$ columns of the grid and contracting the last two columns to a vertical segment, and we set $g_0 = h_0 \circ \bar{g}_0$. Thus, the range of g_0 is the square $Q_1 = [0, a(n - 2)/n]^2$.

It is easily checked that the mapping g_0 maps each square of the grid G into some square of G in Q_1. All squares in D_0 are contracted to pieces of Lipschitz curves.

We define $B_1 = \{s \in G;\ \exists s' \in B_0 : g_0(s') \subseteq s \text{ and } \lambda^2(g_0(s')) > 0\}$. Thus, we have $|B_1| \leq |B_0| - |D_0|$, so at least $|D_0|$ squares of B_0 are "killed" by g_0.

Similarly we define mappings $g_1, g_2, \ldots$. Having defined the mapping g_{i-1}, we put $B_i = \{s \in G;\ \exists s' \in B_{i-1} : g_{i-1}(s') \subseteq s \text{ and } \lambda^2(g_{i-1}(s')) > 0\}$ and we construct an appropriate 1-Lipschitz mapping $g_i : Q_i \to Q_{i+1}$, where $Q_i = [0, a(n - 2i)/n]^2$. The mapping g_i kills at least $\sqrt{|B_i|}$ squares of B_i, in the sense that $|B_i| - |B_{i+1}| \geq \sqrt{|B_i|}$.

Let t be the first index such that $|B_0| - |B_t| \geq \varepsilon n^2/5$. For $i < t$, we thus have $|B_0| - |B_i| < \varepsilon n^2/5$ (and therefore $|B_i| > \varepsilon n^2/4$). In particular, we get $\varepsilon n^2/5 >$

$|B_0| - |B_{t-1}| \geq \sum_{i=0}^{t-2} \sqrt{|B_i|} \geq (t-1)n\sqrt{\varepsilon}/2$, which implies that $t < \frac{2}{5}n\sqrt{\varepsilon} + 1 \leq n\sqrt{\varepsilon}/2$ (since we may assume that $n\sqrt{\varepsilon}$ is large).

By composing the mappings $g_0, \ldots, g_{t-1}$, we obtain a 1-Lipschitz mapping g of Q onto $Q' = [0, a']^2$, with $a' = a(1 - 2t/n) \geq a(1 - \sqrt{\varepsilon})$. Under this mapping, at least $\varepsilon n^2/5$ of the squares of B_0 have images of measure 0, and these squares contain at least $\frac{15}{16}(a^2/n^2)(\varepsilon n^2/5) = \frac{3}{16}\varepsilon a^2$ of the measure of $Q \setminus K$. Hence $\lambda^2(g(Q \setminus K)) \leq \frac{13}{16}\varepsilon a^2$ and $\lambda^2(g(K)) \geq \lambda^2(Q' \setminus g(Q \setminus K)) \geq a'^2 - \frac{13}{16}\varepsilon a^2$. Using the assumption $\varepsilon \leq 0.01$, we have $a' \geq a(1 - \sqrt{\varepsilon}) \geq 0.9a$, and thus $\lambda^2(g(K)) \geq a'^2(1 - \frac{10}{9} \cdot \frac{13}{16}\varepsilon) \geq a'^2(1 - 0.9\varepsilon)$ as required.

It remains to show how to choose the mappings φ_i, whose graphs or rotated graphs pass through sufficiently many center points of the squares of B_i. Call a set $D \subseteq \mathbb{R}^2$ *1-Lipschitz in the y-coordinate* (resp. *in the x-coordinate*) if $|y_1 - y_2| \leq |x_1 - x_2|$ (resp. $|x_1 - x_2| \leq |y_1 - y_2|$) for any two points $(x_1, y_1), (x_2, y_2) \in D$. It is easy to see that if D is a finite set which is 1-Lipschitz in the y-coordinate, then there is a 1-Lipschitz function whose graph contains D. Hence it suffices to establish the following

Lemma 2.3. *Let P be an m point set in the plane, $k = \lceil \sqrt{m} \rceil$. Then there exists either a k point subset of P 1-Lipschitz in the x-coordinate or a k point subset of P 1-Lipschitz in the y-coordinate.*

Proof. Call a set C in the plane a *nondecreasing chain* if for any two points $(x_1, y_1), (x_2, y_2) \in C$ with $x_1 \leq x_2$ we have $y_1 \leq y_2$; a *nonincreasing chain* is defined analogously with an opposite inequality for the y-coordinates. A lemma due to Erdős and Szekeres [2] can be stated as follows: Any m point set in the plane contains a k point nondecreasing chain or a k point nonincreasing chain. To see the relevance to the above lemma, rotate the coordinate system by $\pi/4$; then nondecreasing chains become precisely 1-Lipschitz subsets in the x-coordinate and nonincreasing chains become 1-Lipschitz subsets in the x-coordinate. □

For a direct application of the above method for Laczkovich's problem in dimension d, one would need the following: Given an m-point set P in $\mathbb{R}^d$, find a subset S of size $cm^{1-1/d}$ ($c > 0$ a constant) which is 1-Lipschitz in the x_i-coordinate for some $i = 1, 2, \ldots, d$. Here a subset $S \subseteq P$ is C-Lipschitz in the x_i-coordinate if for any two points $a = (a_1, \ldots, a_d), b = (b_1, \ldots, b_d) \in S$ one has $|a_i - b_i| \leq C \max_{i \neq j}(|a_j - b_j|)$. As noted by Gabor Tardos [5], already for $d = 3$ such a large 1-Lipschitz subset need not exist. Indeed, let $P \subset \mathbb{R}^3$ be the set $\{(i, j, i + i);\ i, j = 1, 2, \ldots \sqrt{m}\}$; it is not difficult to check that there is no subset of size exceeding $O(\sqrt{m})$ which is 1-Lipschitz in one of the coordinates. Still, it is possible that there always exists a subset of size $cm^{1-1/d}$ which is C-Lipschitz in one of the coordinates, C a constant. This problem looks interesting in its own right, although it is not completely clear whether a positive answer would solve Laczkovich's problem.

3. Proof of Theorem 1.2

In the above proof of Theorem 1.1, the only part where an arbitrarily large part of the measure of E is wasted is the initial application of the Density Theorem for the choice of the square Q_0 with $\lambda^2(Q_0 \cap E) > 0.99\lambda^2(Q_0)$. In the rest of the proof, $Q_0 \cap E$ is then mapped onto a square with side at most 10 times smaller than the side of Q_0. Thus, it suffices to prove the following lemma:

Lemma 3.1. *For any $\varepsilon_0 \in (0,1)$ and any $E \subset \mathbb{R}^2$ with $\lambda^2(E) = 1$, there exists a 1-Lipschitz mapping $f : \mathbb{R}^2 \to Q = [0,c]^2$ with $\lambda^2(f(E)) \geq (1-\varepsilon_0)\lambda^2(Q)$, where $c = c(\varepsilon_0) > 0$ only depends on ε_0.*

Proof. The proof is probabilistic, using the so-called Second Moment Method, see e.g., [1].

Fix $c = \sqrt{\varepsilon_0}/7$. For simplicity, we will construct a 2-Lipschitz mapping onto $[0,c]^2$; the Lemma is then obtained by taking $c/2$ instead of c and rescaling. Choose an integer N large enough so that some axis-parallel square S with side $3cN$ contains at least $9/10$ of the measure of E. Let G be an $N \times N$ grid consisting of $3c \times 3c$ axis-parallel squares; the squares of G are denoted by s_{ij}, $i = 1, 2, \ldots N$ denoting the row index and $j = 1, 2, \ldots, N$ the column index. Let M_{ij} denote the middle $c \times c$ square in s_{ij}, and set $M = \bigcup_{i,j=1}^{N} M_{ij}$. Since 9 translational copies of the set M cover the square S, it is possible to place the grid G in such a way that $\lambda^2(M \cap E) \geq 1/10$; we assume that such a placement has been fixed.

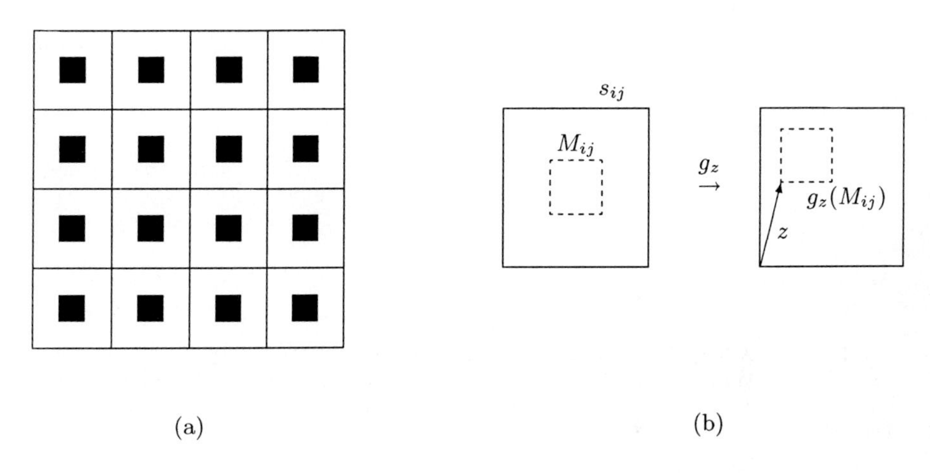

Fig. 3.1. (a) The grid G and the set M; (b) the mapping g_z.

Consider a grid square s_{ij}. Let $z = z_{ij} \in [0, 2c]^2$ be a vector. We define a mapping $g_z : s_{ij} \to s_{ij}$ as follows. We require that g_z is the identity map on the boundary ∂s_{ij} of s_{ij}, and on M_{ij}, g_z acts as the translation by the vector $z - (c,c)$ (see Fig. 3.1). With respect to the d_∞ metric, g_z is a 2-Lipschitz mapping on $\partial s_{ij} \cup M_{ij}$. We extend it to a 2-Lipschitz mapping $s_{ij} \to s_{ij}$ (this is possible since for any metric space Y and any $X \subseteq Y$, a C-Lipschitz mapping from X to the plane with the d_∞ metric can be extended to a C-Lipschitz mapping defined on Y, see e.g., [6]).

For later use, we observe that if z is randomly chosen from the uniform probability distribution on $[0, 2c]^2$, then for any fixed $x \in M_{ij}$, we have

$$\mathrm{Prob}(x \in g_z(E \cap M_{ij})) \geq \frac{\lambda^2(E \cap M_{ij})}{4c^2} \tag{3.1}$$

(since for each $y \in E \cap M_{ij}$, $g_{x-y+(c,c)}$ sends y to x).

Consider $Z = \{(z_{ij};\ i,j = 1,2,\ldots,N);\ z_{ij} \in [0,2c]^2\}$ with the product probability measure, and let $\mathbf{z} \in Z$ be a random element. Let $\bar{g}_{\mathbf{z}} : G \to G$ be the 2-Lipschitz mapping whose restriction on each s_{ij} is $g_{z_{ij}}$. Further we define a 1-Lipschitz mapping $f_1 : G \to [0,3c]^2$ corresponding to a "folding" of the grid G (as if G were a piece of paper and we wanted to fold it to one little square). Formally, if $p \in s_{ij}$ is a point in G with displacement (x,y) from the lower left corner of s_{ij}, we set

$$f_1(p) = \begin{cases} (x,y) & \text{for } i,j \text{ odd,} \\ (x, 3c-y) & \text{for } i \text{ odd, } j \text{ even,} \\ (3c-x, y) & \text{for } i \text{ even, } j \text{ odd,} \\ (3c-x, 3c-y) & \text{for } i,j \text{ even.} \end{cases}$$

Finally, we let f_0 be the retraction of the plane onto the grid G, and $f_2 : [0,3c]^2 \to Q$ be the mapping translating the middle $c \times c$ square of $[0,3c]^2$ onto $Q = [0,c]^2$ and contracting the rest of $[0,3c]^2$ onto the boundary of Q. Both f_0, f_2 are 1-Lipschitz. We put $f_{\mathbf{z}} = f_2 \circ f_1 \circ \bar{g}_{\mathbf{z}} \circ f_0$. This is a 2-Lipschitz map.

Fix any point $x \in [0,c]^2$, and let X_{ij} be the 0/1 indicator variable for the event $x \in f_{\mathbf{z}}(E \cap M_{ij})$ (this is a random variable depending on the choice of $\mathbf{z}$), and set $X = \sum_{i,j=1}^{N} X_{ij}$. From (3.1), we get for the expectation of each X_{ij}

$$\mathrm{E}\, X_{ij} \geq \frac{1}{4c^2} \lambda^2(E \cap M_{ij}),$$

and so

$$\mathrm{E}\, X = \sum_{i,j=1}^{N} \mathrm{E}\, X_{ij} \geq \frac{1}{40c^2} > \frac{1}{\varepsilon_0}.$$

As the variables X_{ij} are independent, we have

$$\mathrm{Var}\, X = \sum_{i,j=1}^{N} \mathrm{Var}\, X_{ij} \leq \sum_{i,j=1}^{N} \mathrm{E}\, X_{ij}^2 = \sum_{i,j} \mathrm{E}\, X_{ij} = \mathrm{E}\, X.$$

From Chebyshev Inequality we thus get

$$\mathrm{Prob}(X = 0) \leq \mathrm{Prob}(|X - \mathrm{E}\, X| \geq \mathrm{E}\, X) \leq \frac{\mathrm{Var}\, X}{(\mathrm{E}\, X)^2} \leq \frac{1}{\mathrm{E}\, X} < \varepsilon_0.$$

The event $X \neq 0$ means $x \in f_{\mathbf{z}}(M \cap E)$, where x is our fixed point of Q. Considering all x's simultaneously and using Fubini's Theorem, we get that there exists a $\mathbf{z}_0 \in Z$ such that $\lambda^2(f_{\mathbf{z}_0}(E \cap M)) \geq (1-\varepsilon_0)\lambda^2(Q)$. This concludes the proof of Lemma 3.1. □

Acknowledgment. I would like to thank David Preiss for explaining me his proof, useful discussions and for his hospitality during my visit. Thanks also to Luděk Zajíček for valuable comments to the presentation and noticing several mistakes in it.

References

1. N. Alon, J. Spencer, P. Erdős: *The probabilistic method.* Cambridge Univ. Press 1992.
2. P. Erdős, G. Szekeres: A combinatorial problem in geometry. *Compositio Math.* 2(1935) 463–470.
3. M. Laczkovich: Paradoxical decompositions using Lipschitz functions, *Real Analysis Exchange* 17(1991–92), 439-443.
4. D. Preiss, manuscript, 1992.
5. G. Tardos, private communication, April 1993.
6. J. H. Wells, L. R. Williams: *Embeddings and extensions in analysis*, Springer-Verlag 1975.

A Remark on Transversal Numbers

János Pach

Mathematical Institute, Hungarian Academy of Sciences, Budapest Pf. 127, 1364, Hungary

"What does the Hungarian parrot say?"
"Log. Log log log log ... "
(Riddle. Folklore.)

1. Introduction

In his classical monograph published in 1935, Dénes König [K] included one of Paul Erdős's first remarkable results: an infinite version of the Menger theorem. This result (as well as the König–Hall theorem for bipartite graphs, and many related results covered in the book) can be reformulated as a statement about transversals of certain hypergraphs.

Let H be a hypergraph with vertex set $V(H)$ and edge set $E(H)$. A subset $T \subseteq V(H)$ is called a *transversal* of H if it meets every edge $E \in E(H)$. The *transversal number*, $\tau(H)$, is defined as the minimum cardinality of a transversal of H. Clearly, $\tau(H) \geq \nu(H)$, where $\nu(H)$ denotes the maximum number of pairwise disjoint edges of H. In the above mentioned examples, $\tau(H) = \nu(H)$ holds for the corresponding hypergraphs. However, in general it is impossible to bound τ from above by any function of ν, without putting some restriction on the structure of H.

One of Erdős's closest friends and collaborators, Tibor Gallai (who is also quoted in König's book) once said: "I don't care for bounds involving log n's and loglog n's. I like exact answers. But Paul has always been most interested in asymptotic results." In fact, this quality of Erdős has contributed a great deal to the discovery and to the development of the "probabilistic method" (see [ES], [AS]).

The search for "exact answers" (e.g. to the perfect graph conjecture of Berge [Be]) has revealed some important connections between transversal problems and linear programming that led to the deeper understanding of the König–Hall–Menger–type theorems. It proved to be useful to introduce another parameter, the *fractional transversal number* of a hypergraph, defined by

$$\tau^\star(H) = \min_t \sum_{x \in V(H)} t(x) \quad ,$$

where the minimum is taken over all non–negative functions $t : V(H) \longrightarrow \mathbb{R}$ with the property that

$$\sum_{x \in E} t(x) \geq 1 \qquad \text{for every } E \in E(H).$$

Obviously, $\tau(H) \geq \tau^\star(H) \geq \nu(H)$, and $\tau^\star(H)$ can be easily calculated by linear programming. (See [Lo] and [S].)

At the same time, the probabilistic (or shall we say, asymptotic) approach has also led to many exciting discoveries about extremal problems related to transversals (e.g. Ramsey–Turán–type theorems, property B). It was pointed out by Vapnik and Chervonenkis [VC] that in some important families of hypergraphs a relatively small set of randomly selected vertices will, with high probability, be a transversal. They defined the *dimension* of a hypergraph as the size of the largest subset $A \subseteq V(H)$ with the property that for every $B \subseteq A$ there exists an edge $E \in E(H)$ such that $E \cap A = B$. Adapting the original ideas of [VC] and [HW], it was shown in [KPW] (see also [PA]) that

$$\tau(H) \leq (1 + o(1)) \dim(H)\tau^\star(H) \log \tau^\star(H), \tag{1.1}$$

as $\tau^\star \longrightarrow \infty$, and that this bound is almost tight.

Ding, Seymour and Winkler [DSW] have introduced another parameter of a hypergraph, closely related to its dimension. They defined $\lambda(H)$ as the size of the largest collection of edges $\{E_1, \ldots, E_k\} \subseteq E(H)$ with the property that for every pair (E_i, E_j), $1 \leq i \neq j \leq k$, there exists a vertex x such that $x \in E_i \cap E_j$ but $x \notin E_h$ for any $h \neq i, j$. Combining (1.1) with Ramsey's theorem, they showed that

$$\tau(H) \leq \ 6\lambda^2(H) \ (\lambda(H) + \nu(H)) \binom{\lambda(H) + \nu(H)}{\lambda(H)}^2 \tag{1.2}$$

holds for every hypergraph H.

As far as we know, Haussner and Welzl [HW] were the first to recognize that (1.1) has a wide range of interesting geometric applications, due to the fact that a large variety of hypergraphs defined by geometric means have low Vapnik–Chervonenkis dimensions.

The aim of this note is to illustrate the power of this approach by two examples. In Section 2 we show that (1.2) easily implies some far–reaching generalizations of results of Erdős and Szekeres [ES1], [ES2]. In Section 3, we use (1.2) to extend and to give alternative proofs for some old results of Gyárfás and Lehel (see [G], [GL], [L]) bounding the transversal numbers of box hypergraphs.

2. Covering with boxes

Given two points $p, q, \in \mathbb{R}^d$, let $\mathrm{Box}[p, q]$ be defined as the smallest box containing p and q, whose edges are parallel to the axes of the coordinate system. The following theorem settles a conjecture of Bárány and Lehel [BL], who established the first non–trivial result of this kind.

Theorem 2.1. *Any finite (or compact) set $P \subseteq \mathbb{R}^d$ contains a subset with at most $2^{2^{d+2}}$ elements, $\{p_i | 1 \le i \le 2^{2^{d+2}}\}$, such that*

$$P \subseteq \bigcup_{i,j=1}^{2^{2^{d+2}}} \mathrm{Box}[p_i,\, p_j]\,.$$

Proof. Let H be a hypergraph on the vertex set

$$V(H) = \{\mathrm{Box}[p,q] \mid p,q \in P\}\,,$$

defined as follows. Associate with each point $r \in P$ the set

$$E_r = \{\mathrm{Box}[p,q] | r \in \mathrm{Box}[p,q]\},$$

and let $E(H) = \{E_r | r \in P\}$.

Clearly, $E_p \cap E_q \neq \emptyset$ for any $p, q, \in P$, because $\mathrm{Box}[p,\, q] \in E_p \cap E_q$. Hence, $\nu(H) = 1$.

According to a well–known lemma of Erdős and Szekeres [ES1], any sequence of $k^2 + 1$ real numbers contains a monotone subsequence of length $k+1$. By repeated application of this statement, we obtain that any set of $2^{2^{d-1}} + 1$ points in $\mathbb{R}^d$ has three elements $p_i,\, p_j,\, p_k$ with $p_k \in \mathrm{Box}[p_i,\, p_j]$. This immediately implies that

$$\lambda(H) \le 2^{2^{d-1}}\,.$$

Indeed, for any family of more than $2^{2^{d-1}}$ edges $E_{p_1}, \dots E_{p_\lambda} \in E(H)$, one can choose three distinct indices i, j, k with $p_k \in \mathrm{Box}\,[p_i,\, p_j]$, which yields that

$$E_{p_i} \cap E_{p_j} \subseteq E_{p_k}\,.$$

Thus, we can apply (1.2) to obtain

$$\tau(H) \le 6\lambda^2(H)\,(\lambda(H)+1)^3 < 2^{2^{d+2}}\,,$$

and the result follows. □

As was shown in [BL], the bound $2^{2^{d+2}}$ in Theorem 2.1 is nearly optimal. In fact, the above argument yields a slightly stronger result.

Theorem 2.2. *Let $P \subseteq \mathbb{R}^d$ by any compact set, and let $\mathcal{B}$ be any family of boxes in parallel position with the property that for any $(\nu+1)$–element subset $P' \subseteq P$ there is a box $B \in \mathcal{B}$ which covers at least two points of P' $(d\,, \nu \ge 1)$. Then one can choose at most $\binom{2^{2^d}+\nu}{\nu}^5$ members of $\mathcal{B}$ such that their union will cover P.*

In [ES2], Erdős and Szekeres proved the following.

Lemma 2.1. *Every set $P \subseteq \mathbb{R}^2$ with at least 2^k elements contains three points p_1, p_2, p_3 such that $\sphericalangle\, p_1p_2p_3 \ge \pi\left(1 - \frac{1}{k}\right)$.*

Our next result, which improves a theorem of Bárány [B], can be regarded as a generalization of Lemma 2.1.

Theorem 2.3. *Let d be a positive integer, $\epsilon > 0$. Every finite (or compact) set $P \subseteq \mathbb{R}^d$ has a subset of at most $2^{(c/\epsilon)^{d-1}}$ elements, $P' = \{p_1, p_2, \ldots\}$, with the property that for any $p \in P \backslash P'$ there exist $p_i, p_j \in P'$ satisfying*

$$\sphericalangle p_i p p_j \geq \pi - \epsilon \ .$$

(Here $c \leq 8$ is a constant.)

Proof. For $d \geq 2, \epsilon > 0$ fixed, let us cover the unit hemisphere centered at $O \in \mathbb{R}^d$ with $(4/\epsilon)^{d-1}$ spherical $(d-1)$–dimensional simplices $S_1, S_2, \ldots$ such that the diameter of each S_i is at most $\epsilon/2$. Let $h_{t1}, \ldots, h_{td}$ denote the hyperplanes induced by O and the $((d-2)$–dimensional) facets of S_t. A parallelotope whose facets are parallel to $h_{t1}, \ldots, h_{td}$, respectively, is called a *box of type t*. The smallest box of type t containing $p, q \in \mathbb{R}^d$ will be denoted by $\text{Box}_t[p, q]$.

For any $p, q \in \mathbb{R}^d$, choose an index t such that pq is parallel to Os for some $s \in S_t$, and let $\text{Box}[p, q] = \text{Box}_t[p, q]$. Notice that if $r \in \text{Box}\ [p, q]$ then $\sphericalangle prq \geq \pi - \epsilon$.

Just like in the previous proof, define a hypergraph H by

$$\begin{aligned} V(H) &= \{\text{Box}[p, q] \mid p, q \in P\}, \\ E(H) &= \{E_r | r \in P\}, \end{aligned}$$

where $E_r = \{\text{Box}[p, q] | r \in \text{Box}[p, q]\}$, and observe that it is sufficient to bound the transversal number of H. Clearly, $\nu(H) = 1$.

By the definition of $\lambda(H)$, one can select $\lambda(H) = \lambda$ elements $p_1, \ldots, p_\lambda \in P$ with the property that any two of them are enclosed in a box of some type, which does not cover any other p_k. More precisely, for every $1 \leq i < j \leq \lambda$ there exists $t(i, j) \leq (4/\epsilon)^{d-1}$ such that

$$\{p_1, \ldots, p_\lambda\} \cap \text{Box}_{t(i,j)}\ [p_i, p_j] = \{p_i, p_j\} \ .$$

Obviously, p_i and p_j are two antipodal vertices of $\text{Box}_{t(i,j)}\ [p_i, p_j]$, and every box has 2^{d-1} pairs of antipodal vertices. Let us color the segments $p_i p_j (1 \leq i < j \leq \lambda)$ with $(4/\epsilon)^{d-1} 2^{d-1}$ colors according to the value of $t(i, j)$ and to the particular position of the diagonal $p_i p_j$ within $\text{Box}_{t(i,j)}\ [p_i, p_j]$. It is easy to see that the segments of the same fixed color form a bipartite subgraph of the complete graph K_λ on the vertex set $p_1, \ldots, p_\lambda$. Hence the chromatic number of K_λ,

$$\lambda(H) \leq 2^{(4/\epsilon)^{d-1} 2^{d-1}} \ ,$$

and the result follows from (1.2).

It is not hard to see that the bound in Theorem 2.3 is asymptotically tight, apart from the exact value of c (see [B], [EF]).

Combining these observations with an analogue of Turán's theorem for hypergraphs (see e.g. [?]), we immediately obtain the following result related to a problem of Conway, Croft Erdős and Guy [CC].

Corollary 2.1. *There exists a constant $c > 0$ such that, for any set of n distinct points $p_1, \dots p_n \in \mathbb{R}^d$, the number of triples $i < j < k$ for which $\sphericalangle p_i p_j p_k > \pi - \epsilon$, is at least $\lfloor n^3 / 2^{(c/\epsilon)^{d-1}} \rfloor$. Moreover, apart from the value of c, this bound cannot be improved.*

Finally, we mention another straightforward generalization of Lemma 2.1.

Theorem 2.4. *Let P be any set of at least $k^{(c/\epsilon)^{d-1}}$ points in $\mathbb{R}^d$, where c is a suitable constant. Then one can find $p_0, \dots, p_k \in P$ such that they are "almost collinear", i.e., $\sphericalangle p_{i-1} p_i p_{i+1} > \pi - \epsilon$ for every i $(1 \le i < k)$.*

3. Gallai–type theorems

Many problems in geometric transversal theory were motivated by the following famous question of Gallai. Given a family of pairwise intersecting disks in the plane, what is the smallest number of needles required to pierce all of them? (The answer is three. See [D], [DGK], [GPW], [E].).

First we show that (1.2) implies the following result of Gyárfás and Lehel.

Theorem 3.1 (GL). *For any positive integers k and ν, there exists a number $f = f(k, \nu)$ with the following property. Let H be any finite family of subsets of $\mathbb{R}$ such that each of them can be obtained as the union of at most k intervals. If H has no $\nu+1$ pairwise disjoint members, then all of its members can be pierced by at most f points.*

Proof. In order to apply (1.2), we have to bound $\lambda(H)$. Let $E_1, \dots, E_\lambda$ be some members (edges) of H such that, for any $i < j$, $E_i \cap E_j$ has a point x_{ij} which does not belong to any other $E_h (h \ne i, j)$. Write each E_i $(1 \le i \le \lambda)$ as the union of k intervals,

$$E_i = I_{i1} \cup \dots \cup I_{ik} \, .$$

If $x_{ij} \in I_{ip} \cap I_{jq}$ for some $i < j$, then (E_i, E_j) is called a pair of type (p, q). (A pair may have several different types.)

It is easy to check that there are no four edges E_i such that all $\binom{4}{2} = 6$ pairs determined by them are of the same type. Thus,

$$\lambda < R_{k^2}(4) \, ,$$

where $R_s(t)$ denotes the smallest number R such that any complete graph of R vertices, whose edges are colored with s colors, has a monochromatic complete subgraph of t vertices. Hence, the theorem is true with

$$f(k,\nu) \le 6 \binom{R_{k^2}(4)+\nu}{\nu}^5 . \qquad \square$$

Theorem 3.1 does not generalize to subsets of the plane that can be obtained as the union of k axis–parallel rectangles. Indeed, let $H = \{E_i | 1 \le i \le n\}$, where

$$E_i = \left\{(x,y) \in \mathbb{R}^2 | 0 \le x, y \le n \text{ and } \min(|x-i|, |y-i|) \le \frac{1}{4}\right\}.$$

Then $\nu(H) = 1$, while $\lambda(H) = \tau(H) = n$.

However, one can easily establish the following.

Theorem 3.2. *Let F be a family of open domains in the plane such that each of them is bounded by a closed Jordan curve, and any two of them share at most two boundary points. Furthermore, let H be a finite set system, whose every element can be obtained by taking the union of at most k members of F. If H has no $\nu+1$ pairwise disjoint elements, then all of its elements can be pierced by at most $g(k,\nu)$ points (where g depends only on k and ν).*

Proof. Pick λ elements (edges) of H,

$$E_i = I_{i1} \cup \ldots \cup I_{ik} \qquad (I_{ip} \in F\ , 1 \le i \le \lambda, 1 \le p \le k)\ ,$$

and suitable points

$$x_{ij} \in (E_i \cap E_j) \setminus \cup_{h \ne i,j}\ E_h\ \ ,$$

as in the previous proof. After defining the type of a pair (E_i, E_j), $i < j$ in exactly the same way as above, now one can argue that there are no 6 edges E_i such that all the $\binom{6}{2} = 15$ pairs determined by them have the same type (p,q). Assume, for contradiction, that e.g. $E_1, \ldots, E_6$ satisfy this condition for some $p \ne q$. Then any I_{ip} $(1 \le i \le 3)$ and any I_{jq} $(4 \le j \le 6)$ have a common interior point (x_{ij}) which is not covered by any other E_k $(k \ne i,j)$. We can conclude (by tedious case analysis) that there exist pairwise disjoint connected open subsets $I'_{ip} \subseteq I_{ip}$ $(1 \le i \le 3)$, $I'_{jq} \subseteq I_{jq}$ $(4 \le j \le 6)$ such that every I'_{ip} and I'_{jq} share a common boundary segment. This contradicts Kuratowski's theorem on planar maps. The case $p = q$ can be treated similarly.

Thus, $\lambda < R_{k^2}(6)$ and the result follows. We could also apply Theorem 1.1 of Sharir [Sh] to deduce $\lambda < R_{k^2}(c)$ with a much larger constant $c > 6$.

Theorem 3.2 can be applied to the family F_C of all homothetic copies of a convex set C in the plane. The special case when C is a convex polygon with a bounded number of sides was settled by Gyárfás [G]. (An easy compactness argument shows that C does not need to be strictly convex.)

For any hypergraph H and for any integer $t \ge 1$, let $\nu_t(H)$ denote the maximum number of (not necessarily distinct) edges of H such that every vertex is contained in at most t of them. Furthermore, let $\lambda_t(H)$ be the size of the largest collection of edges $\{E_i | i \in I\} \subseteq E(H)$ with the property that for any t–tuple $J \subseteq I$ there exists $x_J \in V(H)$ such that

$$x_J \in \left(\bigcap_{i \in J} E_i \right) \setminus \left(\bigcup_{i \notin J} E_i \right) .$$

Clearly, $\nu_1(H) = \nu(H)$ and $\lambda_2(H) = \lambda(H)$.

Ding, Seymour and Winkler [DSW] have established an upper bound for $\tau(H)$ in terms of $\nu_t(H)$ and $\lambda_{t+1}(H)$, for any fixed $t \geq 1$. Applying their result with $t = 2$, we obtain the following generalization of Theorem 3.1 for the plane.

Theorem 3.3. *Let H be a finite family of open sets in the plane such that*
(i) every member of H is bounded by at most k closed Jordan curves;
(ii) any two distinct members of H have at most ℓ boundary points in common.
Assume that among any $\nu + 1$ members of H there are three with non–empty intersection. Then all members of H can be pierced by at most $f(k, \ell, \nu)$ points, where f does not depend on H.

In higher dimensions we obtain e.g. the following result.

Theorem 3.4. *Let H be a finite family of not necessarily connected polyhedra in $\mathbb{R}^d$ ($d \geq 2$). Assume that every member of H has at most k vertices, and that among any $\nu + 1$ members of H there are $d + 1$ whose intersection is non–empty. Then all members of H can be pierced by at most $g(d, k, \nu)$ points, where g does not depend on H.*

The special case of Theorem 3.4, when every member of H is the union of a bounded number of axis–parallel boxes, was proved by Lehel [L].

References

[AK] N. Alon and D. J. Kleitman: Piercing convex sets and the Hadwiger–Debrunner (p, q)–problem, Adv. Math. 96 (1992), 103–112.

[AS] N. Alon and J. Spencer: The Probabilistic Method, J. Wiley Interscience, New York, 1991.

[B] I. Bárány: An extension of the Erdős–Szekeres theorem on large angles, Combinatorica 7 (1987), 161–169.

[BL] I. Bárány and J. Lehel: Covering with Euclidean boxes, European J. Combinatorics 8 (1987), 113–119.

[Be] C. Berge: Graphs and Hypergraphs, North–Holland, 1982.

[CC] J. H. Conway, H. T. Croft, P. Erdős and M. J. T. Guy: On the distribution of values of angles determined by coplanar points, J. London Math. Soc. (2) 19 (1979), 137–143.

[D] L. Danzer: Zur Lösung des Gallaischen Problems über Kreisscheiben in der euklidischen Ebene, Studia Sci. Math. Hung. 21 (1986), 111-134.

[DGK] L. Danzer, B. Grünbaum and V. Klee: Helly's theorem and its relatives, in: Convexity, Proc. Symp. Pure Math., Vol. 7, Amer. Math. Soc., Providence, 1963, 100–181.

[DSW] G. Ding, P. Seymour and P. Winkler: Bounding the vertex cover number of a hypergraph, Combinatorica 14 (1994), 23–34.

[E] J. Eckhoff: Helly, Radon and Carathéodory type theorems, in: Handbook of Convex Geometry (P. Gruber, J. Wills, eds.), North–Holland, Amsterdam, 1993, 389–448.

[EF] P. Erdős and Z. Füredi: The greatest angle among n points in the d–dimensional Euclidean space, Annals of Discrete Mathematics 17 (1983), 275–283.

[ES] P. Erdős and J. Spencer: Probabilistic Methods in Combinatorics, Akadémiai Kiadó, Budapest and Academic Press, New York, 1974.

[ES1] P. Erdős and G. Szekeres: A combinatorial problem in geometry, Compositio Math. 2 (1935), 463–470.

[ES2] P. Erdős and G. Szekeres: On some extremum problems in elementary geometry, Ann. Univ. Sci. Budapest. Eötvös, Sect. Math. III–IV (1960–61), 53–62.

[GPW] J. Goodman, R. Pollack and R. Wenger: Geometric transversal theory, in: New Trends in Discrete and Computational Geometry (J. Pach, ed.), Springer–Verlag, Berlin, 1993, 163–198.

[GRS] R. Graham, B. Rothschild and J. Spencer: Ramsey Theory, J. Wiley and Sons, New York, 1980.

[G] A. Gyárfás: A Ramsey–type theorem and its applications to relatives of Helly's theorem, Periodica Math. Hung. 3 (1973), 261–270.

[GL] A. Gyárfás and J. Lehel: A Helly–type problem in trees, Coll. Math. Soc. J. Bolyai 4. Combinatorial Theory and its Applications, North–Holland, Amsterdam, 1969, 571–584.

[HW] D. Haussler and E. Welzl: ϵ–nets and simplex range queries, Discrete Comput. Geometry 2 (1987), 127–151.

[KPW] J. Komlós, J. Pach and G. Woeginger: Almost tight bounds for epsilon–nets, Discrete Comput. Geometry 7 (1992), 163–173.

[K] D. König: Theory of Finite and Infinite Graphs, Birkhäuser Verlag, Basel – Boston, 1990.

[L] J. Lehel: Gallai–type results for multiple boxes and forests, Europ. J. Combinatorics 9 (1988), 113–120.

[Lo] L. Lovász: Normal hypergraphs and the perfect graph conjecture, Discrete Math. 2 (1972), 253–267.

[PA] J. Pach and P. Agarwal: Combinatorial Geometry, DIMACS Technical Report 41–51, 1991. To be published by J. Wiley and Sons, New York.

[R] F. P. Ramsey: On a problem of formal logic, Proc. London Math. Soc. 30 (1930), 264–286.

[S] A. Schrijver: Linear and Integer Programming, J. Wiley and Sons, New York, 1986.

[Sh] M. Sharir: On k–sets in arrangements of curves and surfaces, Discrete Comput. Geom. 6 (1991), 593–613.

[VC] V. N. Vapnik and A. Ya. Chervonenkis: On the uniform convergence of relative frequencies of events to their probabilities, Theory Probab. Appl. 16 (1971), 264–280.

In Praise of the Gram Matrix

Moshe Rosenfeld

Department of Computer Science, Pacific Lutheran University, Tacoma, WA 98447

Summary. We use the Gram matrix to prove that the largest number of points in R^d such that the distance between all pairs is an odd integer (the square root of an odd integer) is $\leq d+2$ and we characterize all dimensions d for which the upper bound is attained. We also use the Gram matrix to obtain an upper bound for the smallest angle determined by sets of n lines through the origin in R^d.

1. Introduction

An $n \times n$ symmetric matrix M is positive semi-definite if $\langle Mx, x \rangle \geq 0$ for all vectors $x \in E^n$. Equivalently, M is positive semi-definite if it is symmetric and its eigenvalues are nonnegative. The Gram Matrix (Grammian) of the set of vectors $\{u_1, \ldots, u_n\} \subset E^d$ is the $n \times n$ matrix M defined by $M = (\langle u_i, u_j \rangle)$. This matrix is a positive semi definite matrix and its rank is the dimension of the subspace spanned by $\{u_1, \ldots, u_n\}$. It is well known that if N is a positive semi definite matrix of order $n \times n$ and rank d, then N is the Grammian of a set of n vectors $\{u_1, \ldots, u_n\} \subset E^d$. In other words, $N = (\langle u_i, u_j \rangle)$. This bi-directional relation between vectors, their inner products and the Grammian has provided a powerful tool for solving problems in combinatorics and combinatorial geometry, it plays a central role in the extensive work of Seidel on equiangular lines, it was used in [4] to solve a problem of L. Lovàsz where they proved that $\sqrt[3]{2n^2} \geq \max \| \sum_{i=1}^{n} u_i \| \geq c \frac{\sqrt[3]{n^2}}{\sqrt{\ln n}}$, the max taken over all families of n almost orthogonal unit vectors in R^n that is all n tuples of vectors such that among any three vectors there is at least one orthogonal pair, it was used in [9] to answer a question of P. Erdös showing that the maximum size of a set of almost orthogonal lines in R^d is $2d$ and the list can go on and on. One could fill a large volume exploring the many applications of the gram matrix. The work in this note was motivated by the following simple and attractive problem that appeared in the Nov. 1993 54th W. L. Putnam Mathematical Competition (problem B-5): Can four points in the plane have pairwise odd integral distances? The answer is NO and the Gram matrix provides a very short proof of that. Surprisingly, we cannot do much better if we relax the distance requirement and permit points to have distances whose square is an odd integer. We prove that if the square of all distances among pairs of n points in R^d is an odd integer then $n \leq d+2$. We show that the upper bound is attained if and only if $d = 2 \mod 4$. A slight modification of this proof yields also an alternative proof to a theorem of R. L. Graham, B. L. Rothschild and E. G. Strauss [3] that $d+2$ points in R^d with odd integral distances exist if and only if $d = 14 \mod 16$. We conclude by using the Gram matrix to obtain an upper bound for the smallest angle determined by sets of n lines through the origin in R^d. Cases for which this upper bound is attained are discussed and also some related open problems. In Section 2 we discuss distances among points and related open problems and in Section 3 we discuss angles among lines.

2. Odd distances among points in R^d

Theorem 2.1. *Let $p_1, \ldots, p_n$ be n points in R^d such that $\|p_i - p_j\|^2 = 1 \bmod 2$ if $i \neq j$. Then*

$$n \leq \begin{cases} d+2 & \text{if } d = 2 \bmod 4 \\ d+1 & \text{if } d \neq 2 \bmod 4 \end{cases} \quad \text{and this is best possible .}$$

Proof. Let $u_k = p_k - p_1$ and let $M = (\langle u_k, u_j \rangle)$ be the Gram Matrix of the vectors $\{u_k\}$. M is a positive semi-definite matrix of rank $\leq d$ and order $n-1$, hence if $n-1 > d$ $\mathrm{Det}(M) = 0$ and therefore $\mathrm{Det}(2M) = 0$. From $2\langle u_i, u_j \rangle = \|u_i\|^2 + \|u_i\|^2 - \|u_i - u_j\|^2$ and $\|u_i - u_j\|^2 = \|p_i - p_j\|^2 = 1 \bmod 2$ we deduce that $2\langle u_i, u_j \rangle$ is an odd integer. From the above discussion it follows that:

$$A = 2M \bmod 4 = \begin{pmatrix} 2 & a_{1,2} & a_{1,3} & \cdots & a_{1,n+1} \\ a_{1,2} & 2 & a_{2,3} & \cdots & a_{2,n+1} \\ \cdot & \cdot & \cdot & \cdots & \cdot \\ \cdot & \cdot & \cdot & \cdots & \cdot \\ a_{n+1,1} & a_{n+1,2} & \cdot & \cdots & 2 \end{pmatrix} \quad (a_{ij} = a_{ji} = \pm 1) .$$

Let $A_{ij}(x)$ be the matrix derived from the matrix A by replacing a_{ij} and a_{ji} by x. Consider the quadratic $p(x) = \mathrm{Det}(A_{ij}(x)) = ax^2 + bx + c$. Since all entries in A are integers clearly, a, b and c are also integers. Furthermore, since $A_{ij}(x)$ is symmetric we also have $b = 2k$. Hence $p(1) - p(-1) = 4k$. This means that by replacing any pair of symmetric -1's by 1's the value of the determinant mod 4 remains unchanged or $\mathrm{Det}(A) \bmod 4 = \mathrm{Det}(J+I) \bmod 4$ (the matrix J is the square matrix with $J_{ij} = 1$). Clearly 1 is an eigenvalue of $J + I$ with multiplicity $n-2$ and since the sum of the entries in each row is n, n is the remaining eigenvalue. Therefore $\mathrm{Det}(A) \bmod 4 = \mathrm{Det}(J+I) \bmod 4 = n \bmod 4$. As noted above, $\mathrm{Det}(A) = 0$ if $n > d+1$ hence we must have $n = 0 \bmod 4$. If $n \geq d+3$ and if we had n points in R^d with mutually odd integral squared distances then $n = 0 \bmod 4$. But then removing one of the points will yield a set of $n-1$ such points in R^d while $n - 1 \neq 0 \bmod 4$ and $n - 1 > d + 1$ which is impossible. Hence $n \leq d+2$. It remains to show that for all $d = 2 \bmod 4$ it is possible to construct $d+2$ points in R^d with mutually odd integral squared distances. The following construction using the Gram matrix will do the job.

Let $d = 4k+2$ and let A be a matrix of order $4k+3$ with $A_{ii} = 2k+1$ and $A_{ij} = -\frac{1}{2}$ $i \neq j$. Since $A - (2k + \frac{3}{2})I = -\frac{1}{2}J$, $(2k + \frac{3}{2})$ is an eigenvalue of A and since J has rank 1 the multiplicity of this eigenvalue is $4k+2$. The sum of the entries in each row is zero, hence 0 is the remaining eigenvalue of A. This implies that $\mathrm{rank}(A) = 4k+2 = d$ that A is positive semi-definite and can be written as: $A = M \cdot M^{tr}$ where M is a $(4k+3) \times (4k+2)$ matrix. The rows of M will determine $4k+3$ points $\{p_i\}$ in R^d such that: $\|p_i\|^2 = 2k+1$ and $\|p_i - p_j\|^2 = \|p_i\|^2 + \|p_j\|^2 + 1 = 4k+3$. These points together with the origin yield $4k+4 = d+2$ points as claimed. Geometrically, the $d+2$ points consist of the $d+1$ points of a regular simplex with side $\sqrt{4k+3}$ and its center. To see this, recall that the length of the edge of the regular simplex inscribed in the unit sphere in R^d is $\sqrt{2 + 2/d} = \sqrt{\frac{4k+3}{2k+1}}$ if $d = 4k+2$. Hence if we inflate the regular unit simplex by $\sqrt{2k+1}$ the distance from the center to the vertices will be $\sqrt{2k+1}$ and the distance between the other vertices will be $\sqrt{4k+3}$. ■

A slight modification of the above proof yields an alternative proof to the following theorem of Graham, Rothschild and Strauss [3].

Theorem 2.2. *For the existence of $n+2$ points in R^n so that the distance between any two of them is an odd integer it is necessary and sufficient that $(n+2) = 0$ MOD 16.*

Proof. As in the previous proof, let $\{p_0, \ldots, p_{n+1}\}$ be $n+2$ points in R^n with mutual distances $\|p_i - p_j\| = 1 \text{ mod } 2$. Let $u_i = p_i - p_0$, $i = 1, \ldots, n+1$. The matrix $M = (\langle u_i, u_j \rangle)$ has rank $\leq d$ and order $n+1$. Hence if $n \geq d$ we have:

$$\text{Det}(M) = 0 \tag{2.1}$$

$$\text{Det}(M) = 0 \Rightarrow \text{Det}(2M) = 0 \Rightarrow \text{Det}(2M) \text{ Mod } 16 = 0 \,. \tag{2.2}$$

Since $\|u_i - u_j\| = \|p_i - p_j\| = 1$ Mod 2 we have:

$$\|u_i - u_j\|^2 = 1 \text{ Mod } 8 \tag{2.3}$$

$$2\langle u_i, u_j \rangle = \|u_i\|^2 + \|u_j\|^2 - \|u_i - u_j\|^2 = 1 \text{ Mod } 8 \,. \tag{2.4}$$

Since $\|u_i\| = 1 \text{ mod } 2$ we have:

$$2\langle u_i, u_i \rangle^2 = 2\|u_i\|^2 = 2 \text{ Mod } 16 \,. \tag{2.5}$$

From (2.4) and (2.5) we get:

$$2M \text{ mod } 16 = \begin{pmatrix} 2 & a_{1,2} & a_{1,3} & \cdots & a_{1,n+1} \\ a_{1,2} & 2 & a_{2,3} & \cdots & a_{2,n+1} \\ \cdot & \cdot & \cdot & \cdots & \cdot \\ \cdot & \cdot & \cdot & \cdots & \cdot \\ a_{n+1,1} & a_{n+1,2} & \cdot & \cdots & 2 \end{pmatrix}$$
$$= A \text{ (where } a_{ij} = a_{ji} = 1 \text{ or } 9) \,.$$

Again we let $A_{ij}(x)$ be the matrix derived from the matrix A by replacing a_{ij} and a_{ji} by x and consider the quadratic $p(x) = \text{Det}(A_{ij}(x)) = ax^2 + bx + c$. As in Theorem 2.1, a, b and c are integers and $b = 2k$. Hence $p(9) - p(1) = 80a + 16k$. This means that replacing any pair of symmetric 9's by 1's the value of the determinant mod 16 is unchanged, in other words, $\text{Det}(A)$ Mod 16 $= \text{Det}(J+I)$ Mod 16.

Since the eigenvalues of $J+I$ are $(n+2)$ and $(1)^{\{n\}}$ we have:

$$\text{Det}(J+I) \text{ Mod } 16 = (n+2) \text{ Mod } 16 \,.$$

Thus $\text{Det}(M)$ can be 0 only if $(n+2) = 0 \text{ mod } 16$. Again, the Gram matrix can be used to describe a construction of $d+2$ points so that the distance between any two of them is an odd integer. As in Theorem 1, it will suffice to construct a positive semi-definite matrix M of order $d+1$, rank d, with $M_{ii} = (2k_i+1)^2$ and $M_{ii} + M_{jj} - 2M_{ij} = (2k_{ij}+1)^2$. For $d = 16k - 2$ let M be the matrix of order $16k - 1$ shown below.

$$M = \begin{pmatrix} (4k-1)^2 & \frac{(4k-1)^2}{2} & \cdot & \cdot\ \cdot & \frac{(4k-1)^2}{2} \\ \frac{(4k-1)^2}{2} & (6k-1)^2 & \frac{8k^2-8k+1}{2} & \cdot\ \cdot & \frac{8k^2-8k+1}{2} \\ \cdot & \frac{8k^2-8k+1}{2} & \cdot & \cdot\ \cdot & \cdot \\ \cdot & \cdot & \cdot & \cdot\ \cdot & \cdot \\ \cdot & \cdot & \cdot & \cdot\ \cdot & \cdot \\ \frac{(4k-1)^2}{2} & \frac{8k^2-8k+1}{2} & \cdot & \cdot\ \cdot & (6k-1)^2 \end{pmatrix}.$$

A simple but tedious computation shows that $(8k-1)$ times the top row is the sum of the remaining $16k-2$ rows. Hence 0 is an eigenvalue of M. Clearly, $\left((6k-1)^2 - \frac{8k^2-8k+1}{2}\right)$ is an eigenvalue with multiplicity $16k-3$ and using the trace we can see that the remaining eigenvalue is also positive. Hence M is positive semi-definite of rank $16k-2$. Since $2(6k-1)^2-(8k^2-8k+1)=(8k-1)^2$ and $(6k-1)^2+(4k-1)^2-(4k-1)^2=(6k-1)^2$ M is the desired matrix. The matrix M is the Gram matrix of the vectors determined by the $d+2$ points constructed in [3]. ■

Remarks. Unlike the $d+2$ points constructed in Theorem 2.1, this set is a 3-distance set. H. Harbroth (private communication) showed that there is no 2-distance set of $d+2$ points in R^d so that the two distances are odd integers. On the other extreme it might be interesting to construct $d+2$ points so that all distances (squared distances) are distinct odd integers. Both theorems raise some interesting questions already in the plane. If we define a graph whose vertices are the points in the plane and connect two vertices by an edge if their distance (distance squared) is an odd integer then this graph does not contain a K_4 (K_5). Do either of the two graphs have a finite chromatic number? (Clearly if the odd distance-squared graph has a finite chromatic number so does the odd distance graph). This problem is similar to Nelson's classical problem of coloring the plane so that points at distance 1 get distinct colors.

As noted by P. Erdős, Turàn's theorem implies that the maximum number of distances among n points in the plane that are odd integers is $\leq n^2/3$. On the other hand, one can easily construct n points on the line with $n^2/4$ odd distances. Erdős asked whether it is possible to construct n points in the plane so that more than $n^2/4$ of the distances will be an odd integer. This question as well as the similar question with odd squared-distances is also interesting in higher dimensions.

3. Angles among lines in R^d

Let $L=\{L_1,\ldots,L_n\}\subset R^d$ be a set of lines through the origin. Let $\alpha(L)$ denote the smallest angle among all angles determined by pairs of distinct lines from L. Let $\alpha(n,d)=\sup\{\alpha(L)|L=\{L_1,\ldots,L_n\}\subset R^d\}$.

Theorem 3.1. *[1]* $\cos\alpha(n,d)\geq\sqrt{\frac{n-d}{d(n-1)}}$.

[2] For each n and d there is a set of lines $L=\{L_1,\ldots,L_n\}\subset R^d$ *such that* $\cos\alpha(L)=\cos\alpha(n,d)$.

Proof. We assume that $n > d$. Let u_i be a unit vector along the line L_i and let M be the Grammian of $\{u_1, \ldots, u_n\}$. Note that $\cos^2 a(L) = \text{Max } \langle u_i, u_j \rangle^2$. Since M is positive semi-definite of rank $\leq d$ it has at most d nonzero eigenvalues $\{\lambda_1, \ldots, \lambda_d\}$, $\lambda_i \geq 0$. Since the u_i's are unit vectors, $M_{ii} = 1$ and therefore:

$$\text{Trace}(M) = n = \sum_{i=1}^{d} \lambda_i \tag{3.1}$$

$$\text{Trace}(M^2) = \sum_{i=1}^{n} \sum_{j=1}^{n} \langle u_i, u_j \rangle^2 = \sum_{i=1}^{d} \lambda_i^2 . \tag{3.2}$$

From (3.1):

$$\sum_{i=1}^{d} \lambda_i^2 \geq d \left(\frac{n}{d} \right)^2 = \frac{n^2}{d} .$$

From (3.2):

$$\sum_{i=1}^{n} \sum_{j=1}^{n} \langle u_i, u_j \rangle^2 = n + 2 \sum_{i=1}^{n} \sum_{j=i+1} \langle u_i, u_j \rangle^2 \geq \frac{n^2}{d} .$$

Or:

$$n(n-1)\text{Max}(\langle u_i, u_j \rangle^2) \geq \frac{n^2}{d} - n = \frac{n(n-d)}{d} \Rightarrow \text{Max}(\langle u_i, u_j \rangle^2) \geq \frac{n-d}{d(n-1)} .$$

To prove (3.2) note that $\text{Max } \cos^2 \alpha(L) = \text{Max}\{\text{Max}(\langle u_i, u_j \rangle^2)\}$ where the inner maximum is taken over the $\binom{n}{2}$ pairs of distinct vectors and the outer maximum is taken over all n-tuples of unit vectors in E^d (which is a closed bounded set in E^{nd}). Since this is a continuous function the maximum is attained. ■

Remarks. Note that if equality holds in (2.3) then $\langle u_i, u_j \rangle^2 = \frac{n-d}{d(n-1)}$ or the lines form a set of equiangular lines. It is well known that the existence of n equiangular lines in E^d with angle θ is equivalent to the existence of a graph with n vertices so that the smallest eigenvalue of its Seidel matrix is $-\frac{1}{\arccos \theta}$ and has multiplicity $n - d$. Already in E^3 it seems that the determination of the exact value of $\alpha(n, 3)$ may be a very difficult problem. For instance, it is not possible to construct a set L of 5 equiangular lines in E^3 with $\cos \alpha(L) = \frac{1}{\sqrt{6}}$ even though we can construct a set of 5 equiangular lines. Indeed the only way to do it is by taking 5 of the 6 diagonals of the icosahedron. The angle determined by any pair of these lines is: $\arccos \frac{1}{\sqrt{5}}$. Raphael Robinson (private communication) asked whether $\alpha(5, 3) = \alpha(6, 3)$, or more generally if there are integers n for which $\alpha(n, 3) = \alpha(n + 1, 3)$? For spherical caps there are a few values for which this is the case, L. Danzer [6] proved that the widest angle for 11 spherical caps is the same as for 12 $\left(\arccos \frac{1}{\sqrt{5}} = 63°26'5.8''\right)$. He also showed that for 10 spherical caps the widest angle is $\geq 66°8'48.3''$ hence spherical caps cannot be used to prove that $\alpha(5, 3) = \alpha(6, 3)$, for more information see [7] and [8]. Since the maximum number of equiangular lines in E^3 is 6, the bound also will not be attained for $n > 6$. It will also be interesting to know whether there are values of $n \geq 5$ for which the extremal configuration is not unique.

For $n = d+1$ we get $\cos(\alpha(d+1, d) = \frac{1}{d}$ and this is always attained by the $d+1$ lines connecting the center of the regular simplex to its vertices. There are many sporadic cases where the bound is attained. For instance $\alpha(16,6) = \frac{1}{3}$ and there are 16 equiangular lines in R^6 with angle $\arccos\left(\frac{1}{3}\right)$; $\alpha(176,22) = \alpha(276,23) = \frac{1}{5}$ and in both these cases there are appropriate sets of equiangular lines (see [5]). A particularly attractive case is the case of $2d$ lines in R^d. In this case $\cos\alpha(2d,d) = \sqrt{\frac{2d-d}{d(2d-1)}} = \sqrt{\frac{1}{2d-1}}$. This bound is attained for $d = 3, 5$ but not for $d = 4$. Indeed, there is no Seidel Matrix of order 8 with smallest eigenvalue $-\sqrt{7}$. On the other hand for $d = 5$ it is possible to construct 10 lines in R^5 so that the angle between any distinct pair is $\arccos\frac{1}{3}$: to see this consider the Seidel matrix of the Petersen graph. It's eigenvalues are $(3)^{\{5\}}$, $(-3)^{\{5\}}$ hence this matrix can be used to construct 10 equiangular lines in R^5 with angle $\arccos\frac{1}{3}$. Recall that a conference matrix of order n is a symmetric $n \times n$ matrix A with 0's along the diagonal, ± 1 elsewhere satisfying $A^2 = (n-1)I$ ([1]). Its eigenvalues are $\pm\sqrt{n-1}$. Hence if a conference matrix of order $2d$ exists it will yield $2d$ equiangular lines in R^d with angle $\alpha(2d,d)$. A necessary condition for the existence of a conference matrix of order n is that $n = 2$ Mod 4 and $n-1$ be a sum of two squares (J. H. van Lint and J. J. Seidel). There are exactly 4 distinct conference matrices of order 26 (Weisfeiler [10]), there are at least 18 conference matrices of order 50. In each one of these cases $\alpha(2d,d)$ will be attained by distinct configurations. Conference matrices of orders 6, 14, 30, 38, 42, 46 are known to exist hence in all corresponding dimensions the bound for the largest angle is attained.

Acknowledgement: The author is indebted to Branko Grünbaum and Victor Klee for helpful discussions during preparation of this note.

References

1. V. Belevitch, Conference networks and Hadamard matrices, Ann. Soc. Sci. Bruxelles, Ser. I 82 (1968), pp. 13–32.
2. L. Danzer, Finite point-sets on S^2, Discrete Mathematics, Vol. 60 (1986), pp. 3–66.
3. R. L. Graham, B. L. Rothschild and E. G. Strauss, Are there $n+2$ points in E^n with odd integral distances? Amer. Math. Monthly, 81 (1974), pp. 21–25.
4. B. S. Kashin and S. V. Konyagin, On systems of vectors in a Hilbert space, Proceedings of the Steklov Institute of Mathematics (AMS Translation) 1983, Issue 3, pp. 67–70.
5. P. W. H. Lemmens and J. J. Seidel, Equiangular lines J. of algebra 24 (1973), pp. 494–512.
6. J. H. van Lint and J. J. Seidel, Equilateral points sets in elliptic geometry, Proc. Kon. Nederl. Akad. Wet., Ser. A, 69 (1966), pp. 335–348.
7. Raphael M. Robinson, Arrangement of 24 points on a sphere, Math. Annalen 144 (1961), pp. 17–48.
8. Raphael M. Robinson, Finite sets of points on a sphere with each nearest to five others, Math. Annalen 1979 (1969), pp. 296–318.
9. M. Rosenfeld, Almost orthogonal lines in E^d, Applied Geometry and Discrete Mathematics, The "Victor Klee Festschrift," DIMACS Series in Discrete Mathematics and Theoretical Computer Science, Vol. 4 (1991), pp. 489–492.
10. B. Weisfeiler, On construction and identification of graphs, Lecture Notes 558, Springer-Verlag (1976).

On Mutually Avoiding Sets*

Pavel Valtr

[1] Department of Applied Mathematics, Charles University, Malostranské nám. 25, 118 00 Praha 1, Czech Republic
[2] Graduiertenkolleg "Algorithmische Diskrete Mathematik", Fachbereich Mathematik, Freie Universität Berlin, Takustrasse 9, 14195 Berlin, Germany

Summary. Two finite sets of points in the plane are called mutually avoiding if any straight line passing through two points of anyone of these two sets does not intersect the convex hull of the other set. For any integer n, we construct a set of n points in general position in the plane which contains no pair of mutually avoiding sets of size more than $\mathcal{O}(\sqrt{n})$ each. The given bound is tight up to a constant factor, since Aronov *et al.* [1] showed a polynomial-time algorithm for finding two mutually avoiding sets of size $\Omega(\sqrt{n})$ each in any set of n points in general position in the plane.

1. Introduction

Let A and B be two disjoint finite sets of points in the plane such that their union contains no three points on a line. We say that A *avoids* B if no straight line determined by a pair of points of A intersects the convex hull of B. A and B are called *mutually avoiding* if A avoids B and B avoids A. In this note we investigate the maximum size of a pair of mutually avoiding sets in a given set of n points in the plane.

Aronov *et al.* [1] showed that any set of n points in general position in the plane (i.e., no three points lie on a line) contains a pair of mutually avoiding sets, both of size at least $\sqrt{n/12}$. Moreover, they gave an algorithm for finding such a pair of sets in time $\mathcal{O}(n \log n)$. In Section 2 we construct, for any integer n, a set of n points in general position in the plane which contains no pair of mutually avoiding sets of size more than $11\sqrt{n}$ each.

Mutually avoiding sets in a $d-$dimensional space are defined similarly. Any set of n points in general position in $\mathbb{R}^d$ contains a pair of mutually avoiding sets, both of size at least $\Omega(n^{\frac{1}{d^2-d+1}})$ (see [1]). On the other hand, our method described for the planar case in Section 2 yields a construction of sets of n points in R^d with no pair of mutually avoiding sets of size more than $\mathcal{O}(n^{1-1/d})$.

Now we recall some definitions from [1]. A set of line segments, each joining a pair of the given points, is called *a crossing family* if any two line segments intersect in the interior. Two line segments are called *parallel* if they are two opposite sides of a convex quadrilateral. In other words, two line segments are parallel if their end-points form two mutually avoiding sets of size 2. It is an easy observation that any pair of avoiding sets of size s can be rebuilt into s pairwise parallel line segments or into a crossing family of size s. Aronov *et al.* [1] used this observation and the above result on mutually avoiding sets for finding a crossing family of size $\Omega(\sqrt{n})$ and a set of $\Omega(\sqrt{n})$ pairwise parallel line segments.

* This research was supported by "Deutsche Forschungsgemeinschaft", grant We 1265/2–1 and by Charles University grant No. 351.

The result on pairwise parallel line segments was strengthened and extended to a higher dimension by Pach. Pach [8] showed that any set of n points in general position in $\mathbb{R}^d$ contains at least $\Omega(n^{1/d})$ d–dimensional simplices (i.e., $(d+1)$–point subsets) which are pairwise mutually avoiding.

In Section 3 we show a relation between mutually avoiding sets and Erdős' well-known empty-hexagon-problem.

2. Sets with small mutually avoiding subsets

For a finite set P of points in the plane, let $q(P)$ denote the ratio of the maximum distance of any pair of points of P to the minimum distance of any pair of points of P. For example, if P is a square grid $\sqrt{n} \times \sqrt{n}$ then $q(P) = \sqrt{2}(\sqrt{n} - 1)$. In this section we show:

Theorem 2.1. *Let $c > 0$ be a positive constant. Then any set P of n points in the plane satisfying $q(P) \leq c\sqrt{n}$ contains no pair of mutually avoiding sets of size more than $\lceil 2(\sqrt{17} + 1)c\sqrt{n}\, \rceil$ each.*

One of the basic results about covering says that for any integer $n \geq 2$ there is a set P of n points in the plane with $q(P) < c_0\sqrt{n}$, where $c_0 = \sqrt{2\sqrt{3}/\pi} \approx 1.05$ (see [5]). Such a set P can be found as the triangle grid inside a disk of appropriate size. If we slightly perturb the points of P, we obtain a set in general position still satisfying $q(P) < c_0\sqrt{n}$. According to Theorem 2.1 this set contains no pair of mutually avoiding sets of size more than $11\sqrt{n}$. (It is obvious for $n \leq 100$. For $n > 100$ we use the estimation $\lceil 2(\sqrt{17} + 1)c_0\sqrt{n} \rceil < 11\sqrt{n}$.)

Proof (of Theorem 2.1). Let P be a set of n points in the plane satisfying $q(P) \leq c\sqrt{n}$. Without loss of generality, we may and shall assume that the minimum distance in P is 1. Let A and B be two mutually avoiding subsets of the set P. Define Cartesian coordinates so that for some positive constant $d \in (0, \frac{1}{2}c\sqrt{n}\rangle$ all points of $A \cup B$ lie in the closed strip between the two vertical lines $p : x = -d$ and $q : x = d$, and one of the sets A and B, say A, has a point on the line p and a point on the line q. Moreover, let the topmost point b_0 of the set B lie on the x–axis and let the set A lie "above" the set B (i.e., the set A lies above any straight line connecting two points of B). Since b_0 lies on the x–axis, all points of $A \cup B$ lie between the two horizontal lines $r : y = -c\sqrt{n}$ and $s : y = c\sqrt{n}$. Define three points u, v, w as those points in which the line s intersects the line p, the y–axis, and the line q, respectively (see Fig. 2.1).

For each point $b \in B$, let $f(b)$ be that point in which the line segment bv intersects the x–axis. Now we show that, for any two points $b, b' \in B$, the distance $|f(b)f(b')|$ between $f(b)$ and $f(b')$ is greater than $\frac{d}{(\sqrt{17}+1)c\sqrt{n}}$.

If the line bb' is horizontal then $|f(b)f(b')| \geq \frac{1}{2}|bb'| \geq \frac{1}{2} > \frac{d}{(\sqrt{17}+1)c\sqrt{n}}$. If the line bb' is not horizontal then it intersects the line s in some point g outside the segment uw (see Fig. 2.2). Thus $|gv| > d$. Without loss of generality, assume that b is closer to the line s than b'. Let z be that point on $b'v$, for which bz is horizontal.

Now estimate

$$|bz| = \frac{|bb'|}{|gb'|} \cdot |gv| > |bb'| \cdot \frac{|gv|}{|gv| + |vb'|} > 1 \cdot \frac{|gv|}{|gv| + \sqrt{(2c\sqrt{n})^2 + d^2}} >$$

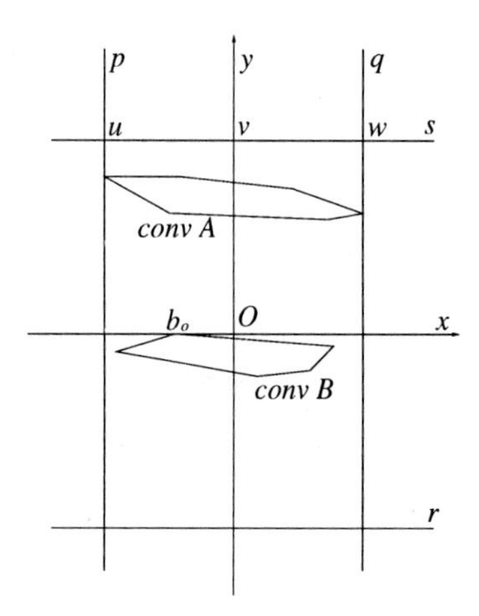

Fig. 2.1. Auxiliary lines and points

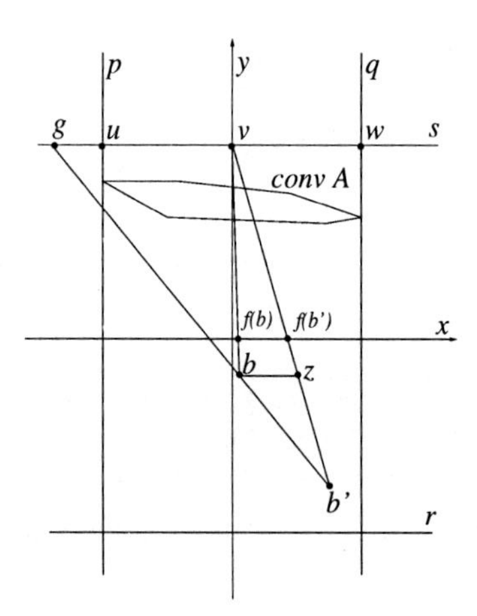

Fig. 2.2. Auxiliary points and triangles

$$> \frac{d}{d+\sqrt{(2c\sqrt{n})^2+d^2}} \geq \frac{d}{\frac{1}{2}c\sqrt{n}+\sqrt{(2c\sqrt{n})^2+(\frac{1}{2}c\sqrt{n})^2}} = \frac{d}{(\frac{1}{2}+\sqrt{\frac{17}{4}})c\sqrt{n}}$$

and

$$|f(b)f(b')| \geq \frac{1}{2}|bz| > \frac{d}{(\sqrt{17}+1)c\sqrt{n}}.$$

Since the points $f(b), b \in B$ are placed on a line segment of length $2d$, the size of B is at most $\lceil 2d/\frac{d}{(\sqrt{17}+1)c\sqrt{n}}\rceil = \lceil 2(\sqrt{17}+1)c\sqrt{n}\rceil$.

3. Relation between mutually avoiding sets and the empty-hexagon problem

Let A be a finite set of points in general position in the plane. A subset S of A of size k is called *convex* if its elements are vertices of a convex k-gon. If S is convex and the interior of the corresponding convex k-gon contains no point of A, then S is called *a k-hole* (or *an empty k-gon*). The classical Erdős-Szekeres Theorem [4] (1935) says that if the size of A is at least $\binom{2k-4}{k-2}+1$ then A contains a convex subset of size k.

Erdős [3] asked whether the following sharpening of the Erdős-Szekeres theorem is true. Is there a least integer $n(k)$ such that any set of $n(k)$ points in general position in the plane contains a k-hole? He pointed out that $n(4) = 5$ and Harborth [6] proved $n(5) = 10$. However, as Horton [7] shows, $n(k)$ does not exist for $k \geq 7$. The question about the existence of $n(6)$ (the empty-hexagon-problem) is still open. After a definition we formulate a conjecture which, if true, would imply that the number $n(6)$ exists.

Definition 3.1. *Let A be a finite set of points in general position in the plane. Let $k \geq 2$, $l \geq 2$. A subset S of A of size $k+l$ is called a (k,l)-set if S is a union of two disjoint sets K and L so that the following three conditions hold:*
(i) $|K| = k, \quad |L| = l$,
(ii) K and L are mutually avoiding,
(iii) the convex hull of S contains no points of $A - S$.

Conjecture 3.1 (Bárány, Valtr). For any two integers $k \geq 2$ and $l \geq 2$, there is an integer $p(k,l)$ such that any set of at least $p(k,l)$ points in general position in the plane contains a (k,l)-set.

If Conjecture 3.1 is true for $k = l = 6$ then the number $n(6)$ exists. It follows from the fact that any (6,6)-set contains a 6-hole (it can be shown that either one of the corresponding sets K and L is a 6-hole or there is a 6-hole containing three points of K and three points of L).

Note that all known constructions of large sets with no 6-hole (see [7], [9]) satisfy Conjecture 3.1 already for rather small integers $p(k,l)$.

We cannot even prove that the numbers $p(k,2), k \geq 5$ exist. (Note that $p(2,2) = 5$ and $p(3,2) = 7$ are the minimum values of $p(k,2), k = 2,3$, for which Conjecture 3.1 holds.) The existence of numbers $p(k,2), k \geq 2$ would imply the following conjecture:

Conjecture 3.2 (Bárány, Valtr). For any integer $k > 0$, there is an integer $R(k)$ such that any set of at least $R(k)$ points in general position in the plane contains $k+2$ points $x, y, z_1, z_2, \ldots, z_k$ such that the k sets $\{x, y, z_i\}, i = 1, \ldots, k$ are 3-holes (i.e., they form empty triangles).

Bárány [2] proved that Conjecture 4 holds for $k \leq 10$.

References

1. B. Aronov, P. Erdős, W. Goddard, D. J. Kleitman, M. Klugerman, J. Pach, and L. J. Schulman, Crossing Families, *Combinatorica* 14 (1994), 127-134; also Proc. Seventh Annual Sympos. on Comp. Geom., ACM Press, New York (1991), 351-356.
2. I. Bárány, personal communication.
3. P. Erdős, On some problems of elementary and combinatorial geometry, Ann. Mat. Pura. Appl. (4) 103 (1975), 99-108.
4. P. Erdős and G. Szekeres, A combinatorial problem in geometry, Compositio Math. 2 (1935), 463-470.
5. L. Fejes Tóth, Regular figures, Pergamon Press, Oxford 1964.
6. H. Harborth, Konvexe Fünfecke in ebenen Punktmengen, Elem. Math. 33 (1978), 116-118.
7. J. D. Horton, Sets with no empty convex 7-gons, Canadian Math. Bull. 26 (1983), 482-484.
8. J. Pach, Notes on Geometric Graph Theory, DIMACS Series in Discrete Mathematics and Theoretical Computer Science, Vol. 6 (1991), 273-285.
9. P. Valtr, Convex independent sets and 7-holes in restricted planar point sets, Discrete Comput. Geom. 7 (1992), 135-152.

VII. Infinity

Introduction

Paul Erdős was always interested in infinity. One of his earliest results is an infinite analogue of (the then very recent) Menger's theorem (which was included in a classical book of his teacher Denes König). Two out of his earliest three combinatorial papers are devoted to infinite graphs. According to his personal recollections, Erdős always had an interest in "large cardinals" although his earliest work on this subject are joint papers with A. Tarski from the end of thirties. These interests evolved over the years into the Giant Triple Paper, with the Partition Calculus forming a field rightly called here Erdősian Set Theory.

We wish to thank András Hajnal for a beautiful paper which perhaps best captures the special style and spirit of Erdős' mathematics. We solicited another two survey papers as well. An extensive survey was written by Peter Cameron on the seemingly simple subject of the infinite random graph, which describes the surprising discovery of Rado and Erdős-Renyi finding many new fascinating connections and applications. The paper by Peter Komjáth deals with another (this time more geometrical) aspect of Erdősian set theory. In addition, the research articles by Shelah and Kříž complement the broad scope of today's set theory research, while the paper of Aharoni looks at another pre-war Erdős conjecture. And here are the masters own words:

I have nearly 50 joint papers with Hajnal on set theory and many with Rado and Fodor and many triple papers and I only state a few samples of our results. The first was my result with Dushnik and Miller.

Let $m \geq \aleph_0$. Then

$$m \to (m, \aleph_0)^2_2 \,.$$

I use the arrow notation invented by R. Rado — in human language: If one colors the pairs of a set of power $m \geq \aleph_0$ by two colors either in color 1 there is a complete graph of power m or in color 2 an infinite complete graph. Hajnal, Rado and I nearly completely settled $m \to (n, q)^2_2$ but the results are very technical and can be found in our joint triple paper and in our book. In one of our joint papers, Hajnal and I proved that a graph G of chromatic number $\aleph_1$ contains all finite bipartite graphs and with Shelah and Hajnal we proved that it contains all sufficiently large odd cycles (Hajnal and Komjáth have sharper results).

Hajnal and I have quite a few results on property B. A family of sets $\{A_\alpha\}$ has property B if there is a set C which meets each of the sets A_α

and contains none of them. This definition is due to Miller. It is now more customary to call the family two chromatic. Here is a sample of our many results: Let $\{A_\alpha\}$ be a family of $\aleph_k$ countable sets with $|A_\alpha \cap A_\beta| \le n$. Then there is a set S which meets each A_α in a set of size $(k+1)n+1$. The result is best possible. If $k \ge \omega$ then there is a set for which $|S \cap A_\alpha| < \aleph_0$. We have to assume the generalized hypothesis of the continuum.

Todorčevič proved with $c = \aleph_1$ that one can color the edges of the complete graph on $|S| = \aleph_1$ with $\aleph_1$ colors so that every $S_1 \subset S$, $|S_1| = \aleph_1$, contains all colors.

We wrote two papers on solved and unsolved problems in set theory. Most of them have been superseded in many cases because undecidability raised its ugly head (according to many: its pretty head). Here is a problem where, as far as we know, no progress has been made. One can divide the triples of a set of power $2^{2^{\aleph_0}}$ into t classes so that every set of power $\aleph_1$ contains a triple of both classes. On the other hand if we divide the triples of a set of power $(2^{2^{\aleph_0}})^+$ into two classes there is always a set of size $\aleph_1$ all whose triples are in the same class. If $S = 2^{2^{\aleph_0}}$ can we divide the triples into two classes so that every subset of size $\aleph_1$ should contain a $K(4)$ of both classes (or more generally a homogeneous subset of size $\aleph_0$?) I offer 500 dollars for clearing up this problem.

Erdős-Galvin-Hajnal problem. Let G have chromatic number $\aleph_1$. Can one color the edges by 2 (or $\aleph_0$, or $\aleph_1$) colors so that if we divide the vertices into $\aleph_0$ classes there always is a class which contains all the colors? Todorčevič proved this if G is the complete graph of $\aleph_1$ vertices — in the general case Hajnal and Komjáth have some results.

The Random Graph

Peter J. Cameron

School of Mathematical Sciences, Queen Mary and Westfield College,
Mile End Road, London E1 4NS, U.K.

Summary. Erdős and Rényi showed the paradoxical result that there is a unique (and highly symmetric) countably infinite random graph. This graph, and its automorphism group, form the subject of the present survey.

1. Introduction

In 1963, Erdős and Rényi [18] showed:

Theorem 1.1. *There exists a graph R with the following property. If a countable graph is chosen at random, by selecting edges independently with probability $\frac{1}{2}$ from the set of* 2*-element subsets of the vertex set, then almost surely (i.e., with probability* 1*), the resulting graph is isomorphic to R.*

This theorem, on first acquaintance, seems to defy common sense — a random process whose outcome is predictable. Nevertheless, the argument which establishes it is quite short. (It is given below.) Indeed, it formed a tailpiece to the paper of Erdős and Rényi, which mainly concerned the much less predictable world of finite random graphs. (In their book *Probabilistic Methods in Combinatorics*, Erdős and Spencer [19] remark that this result "demolishes the theory of infinite random graphs.")

I will give the proof in detail, since it underlies much that follows. The key is to consider the following property, which a graph may or may not have:

(∗) *Given finitely many distinct vertices $u_1, \ldots, u_m$, $v_1, \ldots, v_n$, there exists a vertex z which is adjacent to $u_1, \ldots, u_m$ and nonadjacent to $v_1, \ldots, v_n$.*

Often I will say, for brevity, "z is correctly joined". Obviously, a graph satisfying (∗) is infinite, since z is distinct from all of $u_1, \ldots, u_m, v_1, \ldots, v_n$. It is not obvious that any graph has this property. The theorem follows from two facts:

Fact 1.1. *With probability* 1*, a countable random graph satisfies* (∗)*.*

Fact 1.2. *Any two countable graphs satisfying* (∗) *are isomorphic.*

Proof (of Fact 1.1). We have to show that the event that (∗) fails has probability 0, i.e., the set of graphs not satisfying (∗) is a null set. For this, it is enough to show that the set of graphs for which (∗) fails for some given vertices $u_1, \ldots, u_m, v_1, \ldots, v_n$ is null. (For this deduction, we use an elementary lemma from measure theory: the union of countably many null sets is null. There are only countably many values of m and n, and for each pair of values, only countably many choices of the vertices $u_1, \ldots, u_m, v_1, \ldots, v_n$.) Now we can calculate the probability of this set. Let $z_1, \ldots, z_N$ be vertices distinct from $u_1, \ldots, u_m, v_1, \ldots, v_n$. The probability that any z_i is not correctly joined is $1-\frac{1}{2^{m+n}}$; since these events are independent (for different z_i), the probability that none of $z_1, \ldots, z_N$ is correctly joined is $\left(1-\frac{1}{2^{m+n}}\right)^N$. This tends to 0 as $N \to \infty$; so the event that no vertex is correctly joined does have probability 0.

Note that, at this stage, we know that graphs satisfying $(*)$ exist, though we have not constructed one — a typical "probabilistic existence proof". Note also that "probability $\frac{1}{2}$" is not essential to the proof; the same result holds if edges are chosen with fixed probability p, where $0 < p < 1$. Some variation in the edge probability can also be permitted.

Proof (of Fact 1.2). Let Γ_1 and Γ_2 be two countable graphs satisfying $(*)$. Suppose that f is a map from a finite set $\{x_1, \ldots, x_n\}$ of vertices of Γ_1 to Γ_2, which is an isomorphism of induced subgraphs, and x_{n+1} is another vertex of Γ_1. We show that f can be extended to x_{n+1}. Let U be the set of neighbours of x_{n+1} within $\{x_1, \ldots, x_n\}$, and $V = \{x_1, \ldots, x_n\} \setminus U$. A potential image of x_{n+1} must be a vertex of Γ_2 adjacent to every vertex in $f(U)$ and nonadjacent to every vertex in $f(V)$. Now property $(*)$ (for the graph Γ_2) guarantees that such a vertex exists.

Now we use a model-theoretic device called "back-and-forth". (This is often attributed to Cantor [11], in his characterization of the rationals as countable dense ordered set without endpoints. However, as Plotkin [41] has shown, it was not used by Cantor; it was discovered by Huntington [32] and popularized by Hausdorff [25].)

Enumerate the vertices of Γ_1 and Γ_2, as $\{x_1, x_2, \ldots\}$ and $\{y_1, y_2, \ldots\}$ respectively. We build finite isomorphisms f_n as follows. Start with $f_0 = \emptyset$. Suppose that f_n has been constructed. If n is even, let m be the smallest index of a vertex of Γ_1 not in the domain of f_n; then extend f_n (as above) to a map f_{n+1} with x_m in its domain. (To avoid the use of the Axiom of Choice, select the correctly-joined vertex of Γ_2 with smallest index to be the image of x_m.) If n is odd, we work backwards. Let m be the smallest index of a vertex of Γ_2 which is not in the range of f_n; extend f_n to a map f_{n+1} with y_m in its range (using property $(*)$ for Γ_1).

Take f to be the union of all these partial maps. By going alternately back and forth, we guaranteed that every vertex of Γ_1 is in the domain, and every vertex of Γ_2 is in the range, of f. So f is the required isomorphism.

The graph R holds as central a position in graph theory as $\mathbb{Q}$ does in the theory of ordered sets. It is surprising that it was not discovered long before the 1960s! Partly because of this, a considerable amount of knowledge about R is "folklore". I apologize in advance if I have misattributed, or failed to attribute, any results.

2. Some constructions

Erdős and Rényi did not feel it necessary to give an explicit construction of R; the fact that almost all countable graphs are isomorphic to R guarantees its existence. Nevertheless, such constructions may tell us more about R. Of course, to show that we have constructed R, it is necessary and sufficient to verify condition $(*)$.

I begin with an example from set theory. The downward Löwenheim–Skolem theorem says that a consistent first-order theory over a countable language has a countable model. In particular, there is a countable model of set theory (the *Skolem paradox*).

Theorem 2.1. *Let M be a countable model of set theory. Define a graph M^* by the rule that $x \sim y$ if and only if either $x \in y$ or $y \in x$. Then M^* is isomorphic to R.*

Proof. Let $u_1, \ldots, u_m, v_1, \ldots, v_n$ be distinct elements of M. Let $x = \{v_1, \ldots, v_n\}$ and $z = \{u_1, \ldots, u_m, x\}$. We claim that z is a witness to condition $(*)$. Clearly $u_i \sim z$ for all i. Suppose that $v_j \sim z$. If $v_j \in z$, then either $v_j = u_i$ (contrary to assumption), or $v_j = x$ (whence $x \in x$, contradicting the Axiom of Foundation). If $z \in v_j$, then $x \in z \in v_j \in x$, again contradicting Foundation.

Note how little set theory was actually used: only our ability to gather finitely many elements into a set (a consequence of the Empty Set, Pairing and Union Axioms) and the Axiom of Foundation. In particular, the Axiom of Infinity is not required. Now there is a familiar way to encode finite subsets of $\mathbb{N}$ as natural numbers: the set $\{a_1, \ldots, a_n\}$ of distinct elements is encoded as $2^{a_1} + \ldots + 2^{a_n}$. This leads to an explicit description of R: the vertex set is $\mathbb{N}$; x and y are adjacent if the xth digit in the base 2 expansion of y is a 1 or *vice versa*. This description was given by Rado [42].

The next construction is more number-theoretic. Take as vertices the set $\mathbb{P}$ of primes congruent to 1 mod 4. By quadratic reciprocity, if $p, q \in \mathbb{P}$, then $\left(\frac{p}{q}\right) = 1$ if and only if $\left(\frac{q}{p}\right) = 1$. (Here "$\left(\frac{p}{q}\right) = 1$" means that p is a quadratic residue mod q.) We declare p and q adjacent if $\left(\frac{p}{q}\right) = 1$.

Let $u_1, \ldots, u_m, v_1, \ldots, v_n \in \mathbb{P}$. Choose a fixed quadratic residue a_i mod u_i (for example, $a_i = 1$), and a fixed non-residue b_j mod v_j. By the Chinese Remainder Theorem, the congruences

$$x \equiv 1 \pmod 4, \quad x \equiv a_i \pmod{u_i}, \quad x \equiv b_j \pmod{v_j},$$

have a unique solution $x \equiv x_0 \pmod{4u_1 \ldots u_m v_1 \ldots v_n}$. By Dirichlet's Theorem, there is a prime z satisfying this congruence. So property $(*)$ holds.

A set S of positive integers is called *universal* if, given $k \in \mathbb{N}$ and $T \subseteq \{1, \ldots, k\}$, there is an integer N such that, for $i = 1, \ldots, k$,

$$N + i \in S \qquad \text{if and only if} \qquad i \in T.$$

(It is often convenient to consider binary sequences instead of sets. There is an obvious bijection, under which the sequence σ and the set S correspond when $(\sigma_i = 1) \Leftrightarrow (i \in S)$ — thus σ is the characteristic function of S. Now a binary sequence σ is universal if and only if it contains every finite binary sequence as a consecutive subsequence.)

Let S be a universal set. Define a graph with vertex set $\mathbb{Z}$, in which x and y are adjacent if and only if $|x - y| \in S$. This graph is isomorphic to R. For let $u_1, \ldots, u_m, v_1, \ldots, v_n$ be distinct integers; let l and L be the least and greatest of these integers. Let $k = L - l + 1$ and $T = \{u_i - l + 1 : i = 1, \ldots, m\}$. Choose N as in the definition of universality. Then $z = l - 1 - N$ has the required adjacencies.

The simplest construction of a universal sequence is to enumerate all finite binary sequences and concatenate them. But there are many others. It is straightforward to show that a random subset of $\mathbb{N}$ (obtained by choosing positive integers independently with probability $\frac{1}{2}$) is almost surely universal. (Said otherwise, the base 2 expansion of almost every real number in $[0, 1]$ is a universal sequence.)

Incidentally, it seems to be unknown whether

$$\{n : \text{the } n^{\text{th}} \text{ prime is congruent to 1 mod 4}\}$$

is universal.

Of course, it is possible to construct a graph satisfying $(*)$ directly. For example, let Γ_0 be the empty graph; if Γ_k has been constructed, let Γ_{k+1} be obtained by adding, for each subset U of the vertex set of Γ_k, a vertex $z(U)$ whose neighbour set is precisely U. Clearly, the union of this sequence of graphs satisfies $(*)$.

3. Indestructibility

The graph R is remarkably stable: if small changes are made to it, the resulting graph is still isomorphic to R. Some of these results depend on the following analogue of property $(*)$, which appears stronger but is an immediate consequence of $(*)$ itself.

Proposition 3.1. *Let $u_1, \ldots, u_m, v_1, \ldots, v_n$ be distinct vertices of R. Then the set*

$$Z = \{z : z \sim u_i \text{ for } i = 1, \ldots, m;\ z \not\sim v_j \text{ for } j = 1, \ldots\ n\}$$

is infinite; and the induced subgraph on this set is isomorphic to R.

Proof. It is enough to verify property $(*)$ for Z. So let $u'_1, \ldots, u'_k,\ v'_1, \ldots, v'_l$ be distinct vertices of Z. Now the vertex z adjacent to $u_1, \ldots, u_n, u'_1, \ldots, u'_k$ and not to $v_1, \ldots, v_n,\ v'_1, \ldots, v'_l$, belongs to Z and witnesses the truth of this instance of $(*)$ there.

The operation of *switching* a graph with respect to a set X of vertices is defined as follows. Replace each edge between a vertex of X and a vertex of its complement by a non-edge, and each such non-edge by an edge; leave the adjacencies within X or outside X unaltered. See Seidel [46] for more properties of this operation.

Proposition 3.2. *The result of any of the following operations on R is isomorphic to R:*

(a) deleting a finite number of vertices;
(b) changing a finite number of edges into non-edges or vice versa;
(c) switching with respect to a finite set of vertices.

Proof. In cases (a) and (b), to verify an instance of property $(*)$, we use Proposition 3.1 to avoid the vertices which have been tampered with. For (c), if $U = \{u_1, \ldots, u_m\}$ and $V = \{v_1, \ldots, v_n\}$, we choose a vertex outside X which is adjacent (in R) to the vertices of $U \setminus X$ and $V \cap X$, and non-adjacent to those of $U \cap X$ and $V \setminus X$.

Not every graph obtained from R by switching is isomorphic to R. For example, if we switch with respect to the neighbours of a vertex x, then x is an isolated vertex in the resulting graph. However, if x is deleted, we obtain R once again! Moreover, if we switch with respect to a random set of vertices, the result is almost certainly isomorphic to R.

R satisfies the *pigeonhole principle*:

Proposition 3.3. *If the vertex set of R is partitioned into a finite number of parts, then the induced subgraph on one of these parts is isomorphic to R.*

Proof. Suppose that the conclusion is false for the partition $X_1 \cup \ldots \cup X_k$ of the vertex set. Then, for each i, property $(*)$ fails in X_i, so there are finite disjoint subsets U_i, V_i of X_i such that no vertex of X_i is "correctly joined" to all vertices of U_i and to none of V_i. Setting $U = U_1 \cup \ldots \cup U_k$ and $V = V_1 \cup \ldots \cup V_k$, we find that condition $(*)$ fails in R for the sets U and V, a contradiction.

Indeed, this property is characteristic:

Proposition 3.4. *The only countable graphs Γ which have the property that, if the vertex set is partitioned into two parts, then one of those parts induces a subgraph isomorphic to Γ, are the complete and null graphs and R.*

Proof. Suppose that Γ has this property but is not complete or null. Since any graph can be partitioned into a null graph and a graph with no isolated vertices, we see that Γ has no isolated vertices. Similarly, it has no vertices joined to all others.

Now suppose that Γ is not isomorphic to R. Then we can find $u_1, \ldots, u_m$ and $v_1, \ldots, v_n$ such that $(*)$ fails, with $m+n$ minimal subject to this. By the preceding paragraph, $m+n > 1$. So the set $\{u_1, \ldots, v_n\}$ can be partitioned into two non-empty subsets A and B. Now let X consist of A together with all vertices (not in B) which are not "correctly joined" to the vertices in A; and let Y consist of B together with all vertices (not in X) which are not "correctly joined" to the vertices in B. By assumption, X and Y form a partition of the vertex set. Moreover, the induced subgraphs on X and Y fail instances of condition $(*)$ with fewer than $m+n$ vertices; by minimality, neither is isomorphic to Γ, a contradiction.

Remark 3.1. A different argument gives a weaker conclusion (that one of the parts contains an induced copy of R) but is useful in other situations. We may suppose that there are just two parts X_1, X_2. If X_1 is not isomorphic to R, then there exist $u_1, \ldots, u_m, v_1, \ldots, v_n \in X_1$ such that no "correctly joined" vertex is in X_1. Thus, all such vertices are in X_2, and Proposition 3.1 applies. Indeed, there are many more graphs which satisfy this weaker version of the pigeonhole principle.

Finally:

Proposition 3.5. *R is isomorphic to its complement.*

For property $(*)$ is clearly self-complementary.

4. Graph-theoretic properties

The most important property of R (and the reason for Rado's interest) is that it is *universal*:

Proposition 4.1. *Every finite or countable graph can be embedded as an induced subgraph of R.*

Proof. We apply the proof technique of Fact 1.2; but, instead of back-and-forth, we just "go forth". Let Γ have vertex set $\{x_1, x_2, \ldots\}$, and suppose that we have a map $f_n : \{x_1, \ldots, x_n\} \to R$ which is an isomorphism of induced subgraphs. Let U and V be the sets of neighbours and non-neighbours respectively of x_{n+1} in $\{x_1, \ldots, x_n\}$. Choose $z \in R$ adjacent to the vertices of $f(U)$ and nonadjacent to those of $f(V)$, and extend f_n to map x_{n+1} to z. The resulting map f_{n+1} is still an isomorphism of induced subgraphs. Then $f = \bigcup f_n$ is the required embedding. (The point is that, going forth, we only require that property $(*)$ holds in the target graph.)

In particular, R contains infinite cliques and cocliques. Clearly no finite clique or coclique can be maximal. There do exist infinite maximal cliques and cocliques. For example, if we enumerate the vertices of R as $\{x_1, x_2, \ldots\}$, and build a set S by $S_0 = \emptyset$, $S_{n+1} = S_n \cup \{x_m\}$ where m is the least index of a vertex joined to every vertex in S_n, and $S = \bigcup S_n$, then S is a maximal clique.

Dual to the concept of induced subgraph is that of *spanning subgraph*, using all the vertices and some of the edges. Not every countable graph is a spanning subgraph of R (for example, the complete graph is not). We have the following characterization:

Proposition 4.2. *A countable graph Γ is isomorphic to a spanning subgraph of R if and only if, given any finite set $\{v_1, \ldots, v_n\}$ of vertices of Γ, there is a vertex z joined to none of $v_1, \ldots, v_n$.*

Proof. We use back-and-forth to construct a bijection between the vertex sets of Γ and R, but when going back from R to Γ, we only require that *nonadjacencies* should be preserved.

This shows, in particular, that every infinite locally finite graph is a spanning subgraph (so R contains 1-factors, one- and two-way infinite Hamiltonian paths, etc.). But more can be said.

The argument can be modified to show that, given any non-null locally finite graph Γ, any edge of R lies in a spanning subgraph isomorphic to Γ. Moreover, as in the last section, if the edges of a locally finite graph are deleted from R, the result is still isomorphic to R. Now let $\Gamma_1, \Gamma_2, \ldots$ be given non-null locally finite countable graphs. Enumerate the edges of R, as $\{e_1, e_2, \ldots\}$. Suppose that we have found edge-disjoint spanning subgraphs of R isomorphic to $\Gamma_1, \ldots, \Gamma_n$. Let m be the smallest index of an edge of R lying in none of these subgraphs. Then we can find a spanning subgraph of $R - (\Gamma_1 \cup \ldots \cup \Gamma_n)$containing e_m and isomorphic to Γ_{n+1}. We conclude:

Proposition 4.3. *The edge set of R can be partitioned into spanning subgraphs isomorphic to any given countable sequence of non-null countable locally finite graphs.*

In particular, R has a 1-factorization, and a partition into Hamiltonian paths.

5. Homogeneity and categoricity

We come now to two model-theoretic properties of R. These illustrate two important general theorems, the Engeler–Ryll-Nardzewski–Svenonius theorem and Fraïssé's theorem. The context is first-order logic; so a *structure* is a set equipped with a collection of relations, functions and constants whose names are specified in the language. If there are no functions or constants, we have a *relational structure.* The significance is that any subset of a relational structure carries an induced substructure. (In general, a substructure must contain the constants and be closed with respect to the functions.)

Let M be a relational structure. We say that M is *homogeneous* if every isomorphism between finite induced substructures of M can be extended to an automorphism of M.

Proposition 5.1. *R is homogeneous.*

Proof. In the proof of Fact 1.2, the back-and-forth machine can be started with any given isomorphism between finite substructures of the graphs Γ_1 and Γ_2, and extends it to an isomorphism between the two structures. Now, taking Γ_1 and Γ_2 to be R gives the conclusion.

Fraïssé[21] observed that $\mathbb{Q}$ (as ordered set) is homogeneous, and used this as a prototype: he gave a necessary and sufficient condition for the existence of a homogeneous structure with prescribed finite substructures. Following his terminology, the *age* of a structure M is the class of all finite structures embeddable in M. A class $\mathcal{C}$ of finite structures has the *amalgamation property* if, given $A, B_1, B_2 \in \mathcal{C}$ and embeddings $f_1 : A \to B_1$ and $f_2 : A \to B_2$, there exists $C \in \mathcal{C}$ and embeddings $g_1 : B_1 \to C$ and $g_2 : B_2 \to C$ such that $f_1 g_1 = f_2 g_2$. (Less formally, if the two structures B_1, B_2 have isomorphic substructures A, they can be "glued together" so that the copies of A coincide, the resulting structure C also belonging to the class $\mathcal{C}$.)

Theorem 5.1. *(a) A class $\mathcal{C}$ of finite structures (over a fixed relational language) is the age of a countable homogeneous structure M if and only if $\mathcal{C}$ is closed under isomorphism, closed under taking induced substructures, contains only countably many non-isomorphic structures, and has the amalgamation property.*
(b) If the conditions of (a) are satisfied, then the structure M is unique up to isomorphism.

The countable homogeneous structure M whose age is $\mathcal{C}$ is called the *Fraïssé limit* of $\mathcal{C}$. The class of all finite graphs trivially satisfies Fraïssé's conditions; its Fraïssé limit is R. The Fraïssé limit of a class $\mathcal{C}$ is characterized by a condition generalizing property $(*)$, namely

> *If A and B are members of the age of M with $A \subseteq B$ and $|B| = |A| + 1$, then every embedding of A into M can be extended to an embedding of B into M.*

In the statement of the amalgamation property, when the two structures B_1, B_2 are "glued together", the overlap may be larger than A. We say that the class $\mathcal{C}$ has the *strong amalgamation property* if this doesn't occur; formally, if the embeddings g_1, g_2 can be chosen so that, if $b_1 g_1 = b_2 g_2$, then there exists $a \in A$ such that $b_1 = af_1$ and $b_2 = af_2$. This property is equivalent to others we have met.

Proposition 5.2. *Let M be the Fraïssé limit of the class $\mathcal{C}$, and $G = \mathrm{Aut}(M)$. Then the following are equivalent:*

(a) $\mathcal{C}$ has the strong amalgamation property;
(b) $M \setminus A \cong M$ for any finite subset A of M;
(c) the orbits of G_A on $M \setminus A$ are infinite for any finite subset A of M, where G_A is the setwise stabiliser of A.

See Cameron [6], El-Zahar and Sauer [15].

A structure M is called *$\aleph_0$-categorical* if any countable structure satisfying the same first-order sentences as M is isomorphic to M. (We must specify countability here: the upward Löwenheim–Skolem theorem shows that, if M is infinite, then there are structures of arbitrarily large cardinality which satisfy the same first-order sentences as M.)

Proposition 5.3. *R is $\aleph_0$-categorical.*

Proof. Property $(*)$ is not first-order as it stands, but it can be translated into a countable set of first-order sentences $\sigma_{m,n}$ (for $m, n \in \mathbb{N}$), where $\sigma_{m,n}$ is the sentence

$$(\forall u_1..u_m v_1..v_n)\left(\begin{pmatrix}(u_1 \neq v_1)\&\ldots\& \\ (u_m \neq v_n)\end{pmatrix} \to (\exists z)\begin{pmatrix}(z \sim u_1)\&\ldots\&(z \sim u_m)\& \\ \neg(z \sim v_1)\&\ldots\&\neg(z \sim v_n)\end{pmatrix}\right).$$

Once again this is an instance of a more general result. An *n-type* in a structure M is an equivalence class of n-tuples, where two tuples are equivalent if they satisfy the same (n-variable) first-order formulae. Now the following theorem was proved by Engeler [16], Ryll-Nardzewski [44] and Svenonius [49]:

Theorem 5.2. *For a countable first-order structure M, the following conditions are equivalent:*

(a) M is $\aleph_0$-categorical;
(b) M has only finitely many n-types, for every n;
(c) the automorphism group of M has only finitely many orbits on M^n, for every n.

Note that the equivalence of conditions (a) (axiomatizability) and (c) (symmetry) is in the spirit of Klein's Erlanger Programm. The fact that R satisfies (c) is a consequence of its homogeneity, since $(x_1, \ldots x_n)$ and $(y_1, \ldots, y_n)$ lie in the same orbit of $\mathrm{Aut}(R)$ if and only if the map $(x_i \to y_i)$ $(i = 1, \ldots, n)$ is an isomorphism of induced subgraphs, and there are only finitely many n-vertex graphs.

Remark 5.1. The general definition of an n-type in first-order logic is more complicated than the one given here: roughly, it is a maximal set of n-variable formulae consistent with a given theory. I have used the fact that, in an $\aleph_0$-categorical structure, any n-type is *realized* (i.e., satisfied by some tuple) — this is a consequence of the Gödel–Henkin completeness theorem and the downward Löwenheim–Skolem theorem. See Hodges [28] for more details.

Some properties of R can be deduced from either its homogeneity or its $\aleph_0$-categoricity. For example, Proposition 4.1 generalizes. We say that a countable relational structure M is *universal* (or *rich for its age*, in Fraïssé's terminology [22]) if every countable structure N whose age is contained in that of M (i.e., which is *younger* than M) is embeddable in M.

Theorem 5.3. *If M is either $\aleph_0$-categorical or homogeneous, then it is universal.*

The proof for homogeneous structures follows that of Proposition 4.1, using the analogue of property $(*)$ described above. The argument for $\aleph_0$-categorical structures is a bit more subtle, using Theorem 5.4 and König's Infinity Lemma: see Cameron [7].

6. First-order theory of random graphs

The graph R controls the first-order theory of finite random graphs, in a manner I now describe. This theory is due to Glebskii *et al.* [24], Fagin [20], and Blass and Harary [2]. A property P holds in *almost all finite random graphs* if the proportion of N-vertex graphs which satisfy P tends to 1 as $N \to \infty$. Recall the sentences $\sigma_{m,n}$ which axiomatize R.

Theorem 6.1. *Let θ be a first-order sentence in the language of graph theory. Then the following are equivalent:*

(a) θ holds in almost all finite random graphs;
(b) θ holds in the graph R;
(c) θ is a logical consequence of $\{\sigma_{m,n} : m, n \in \mathbb{N}\}$.

Proof. The equivalence of (b) and (c) is immediate from the Gödel–Henkin completeness theorem for first-order logic and the fact that the sentences $\sigma_{m,n}$ axiomatize R.

We show that (c) implies (a). First we show that $\sigma_{m,n}$ holds in almost all finite random graphs. The probability that it fails in an N-vertex graph is not greater than $N^{m+n}\left(1 - \frac{1}{2^{m+n}}\right)^{N-m-n}$, since there are at most N^{m+n} ways of choosing $m + n$ distinct points, and $\left(1 - \frac{1}{2^{m+n}}\right)^{N-m-n}$ is the probability that no further point is correctly joined. This probability tends to 0 as $N \to \infty$.

Now let θ be an arbitrary sentence satisfying (c). Since proofs in first-order logic are finite, the deduction of θ involves only a finite set Σ of sentences $\sigma_{m,n}$. It follows from the last paragraph that almost all finite graphs satisfy the sentences in Σ; so almost all satisfy θ too.

Finally, we show that not (c) implies not (a). If (c) fails, then θ doesn't hold in R, so $(\neg\theta)$ holds in R, so $(\neg\theta)$ is a logical consequence of the sentences $\sigma_{m,n}$. By the preceding paragraph, $(\neg\theta)$ holds in almost all random graphs.

The last part of the argument shows that there is a *zero-one law*:

Corollary 6.1. *Let θ be a sentence in the language of graph theory. Then either θ holds in almost all finite random graphs, or it holds in almost none.*

It should be stressed that, striking though this result is, most interesting graph properties (connectedness, hamiltonicity, etc.) are not first-order, and most interesting results on finite random graphs are obtained by letting the probability of an edge tend to zero in a specified manner as $N \to \infty$, rather than keeping it constant (see Bollobás [3]). Nevertheless, we will see a recent application of Theorem 6.1 later.

7. Measure and category

When the existence of an infinite object can be proved by a probabilistic argument (as we did with R in Section 1), it is often the case that an alternative argument using the concept of Baire category can be found. In this section, I will sketch the tools briefly. See Oxtoby [39] for a discussion of measure and Baire category.

In a topological space, a set is *dense* if it meets every nonempty open set; a set is *residual* if it contains a countable intersection of open dense sets. The *Baire category theorem* states:

Theorem 7.1. *In a complete metric space, any residual set is non-empty.*

(The analogous statement for probability is that a set which contains a countable intersection of sets of measure 1 is non-empty. We used this to prove Fact 1.1.)

The simplest situation concerns the space $2^{\mathbb{N}}$ of all infinite sequences of zeros and ones. This is a probability space, with the "coin-tossing measure" — this was the basis of our earlier discussion — and also a complete metric space, where we define $d(x, y) = \frac{1}{2^n}$ if the sequences x and y agree in positions $0, 1, \ldots, n-1$ and disagree in position n. Now the topological concepts translate into combinatorial ones as follows. A set S of sequences is open if and only if it is *finitely determined*, i.e., any $x \in S$ has a finite initial segment such that all sequences with this initial segment are in S. A set S is dense if and only if it is *always reachable*, i.e., any finite sequence has a continuation lying in S. Now it is a simple exercise to prove the Baire category theorem for this space, and indeed to show that a residual set is dense and has cardinality $2^{\aleph_0}$. We will say that "almost all sequences have property P (in the sense of Baire category)" if the set of sequences which have property P is residual.

We can describe countable graphs by binary sequences: take a fixed enumeration of the 2-element sets of vertices, and regard the sequence as the characteristic function of the edge set of the graph. This gives meaning to the phrase "almost all graphs (in the sense of Baire category)". Now, by analogy with Fact 1.1, we have:

Fact 7.1. *Almost all countable graphs (in the sense of either measure or Baire category) have property $(*)$.*

The proof is an easy exercise. In fact, it is simpler for Baire category than for measure — no limit is required!

In the same way, almost all binary sequences (in either sense) are universal (as defined in Section 2).

A binary sequence defines a path in the binary tree of countable height, if we sart at the root and interpret 0 and 1 as instructions to take the left or right branch at any node. More generally, given any countable tree, the set of paths is a complete metric space, where we define the distance between two paths to be $\frac{1}{2^n}$ if they first split apart at level n in the tree. So the concept of Baire category is applicable. The combinatorial interpretation of open and dense sets is similar to that given for the binary case.

For example, the age of a countable relational structure M can be described by a tree: nodes at level n are structures in the age which have point set $\{0, 1, \ldots, n-1\}$, and nodes X_n, X_{n+1} at levels n and $n+1$ are declared to be adjacent if the induced structure of X_{n+1} on the set $\{0, 1, \ldots, n-1\}$ is X_n. A path in this tree uniquely describes a structure N on the natural numbers which is younger than M, and conversely. Now Fact 7.1 generalizes as follows:

Proposition 7.1. *If M is a countable homogeneous relational structure, then almost all countable structures younger than M are isomorphic to M.*

It is possible to formulate analogous concepts in the measure-theoretic framework, though with more difficulty. But the results are not so straightforward. For example, almost all finite triangle-free graphs are bipartite (a result of Erdős, Kleitman and Rothschild [17]); so the "random countable triangle-free graph" is almost surely bipartite. (In fact, it is almost surely isomorphic to the "random countable bipartite graph", obtained by taking two disjoint countable sets and selecting edges between them at random.)

A structure which satisfies the conclusion of Proposition 7.1 is called *ubiquitous* (or sometimes *ubiquitous in category*, if we want to distinguish measure-theoretic or other forms of ubiquity). Thus the random graph is ubiquitous in both measure and category. See Bankston and Ruitenberg [1] for further discussion.

8. The automorphism group

From the homogeneity of R (Proposition 5.1), we see that it has a large and rich group of automorphisms: the automorphism group $G = \mathrm{Aut}(R)$ acts transitively on the vertices, edges, non-edges, etc. — indeed, on finite configurations of any given isomorphism type. In the language of permutation groups, it is a rank 3 permutation group on the vertex set, since it has three orbits on ordered pairs of vertices, viz., equal, adjacent and non-adjacent pairs. Much more is known about G; this section will be the longest so far.

First, the cardinality:

Proposition 8.1. $|\mathrm{Aut}(R)| = 2^{\aleph_0}$.

This is a special case of a more general fact. The automorphism group of any countable first-order structure is either at most countable or of cardinality $2^{\aleph_0}$, the first alternative holding if and only if the stabilizer of some finite tuple of points is the identity.

The normal subgroup structure was settled by Truss [51]:

Theorem 8.1. $\mathrm{Aut}(R)$ *is simple.*

Truss proved a stronger result: if g and h are two non-identity elements of $\mathrm{Aut}(R)$, then h can be expressed as a product of five conjugates of g or g^{-1}. (This clearly implies simplicity.)

Truss also described the cycle structures of all elements of $\mathrm{Aut}(R)$.

A countable structure M is said to have the *small index property* if any subgroup of $\mathrm{Aut}(M)$ with index less than $2^{\aleph_0}$ contains the stabilizer of some (finite) tuple of points of M. Hodges *et al.* [29] showed:

Theorem 8.2. *R has the small index property.*

The exact significance of this property will not be explained here. However, it is related to the question of the reconstruction of a structure from its automorphism group. For example, Theorem 8.2 has the following consequence:

Corollary 8.1. *Let Γ be a graph with fewer than $2^{\aleph_0}$ vertices, on which* $\mathrm{Aut}(R)$ *acts transitively on vertices, edges and non-edges. Then Γ is isomorphic to R (and the isomorphism respects the action of* $\mathrm{Aut}(R)$*).*

Another field of study concerns small subgroups. To introduce this, we reinterpret the last construction of R in Section 2. Recall that we took a universal set $S \subseteq \mathbb{N}$, and showed that the graph $\Gamma(S)$ with vertex set $\mathbb{Z}$, in which x and y are adjacent whenever $|x - y| \in S$, is isomorphic to R. Now this graph admits the "shift" automorphism $x \mapsto x + 1$, which permutes the vertices in a single cycle. Conversely, let g be a cyclic automorphism of R. We can index the vertices of R by integers so that g is the map $x \mapsto x + 1$. Then, if $S = \{n \in \mathbb{N} : n \sim 0\}$, we see that $x \sim y$ if and only if $|x - y| \in S$, and that S is universal. A short calculation shows that two cyclic automorphisms are conjugate in $\mathrm{Aut}(R)$ if and only if they give rise to the same set S. Since there are $2^{\aleph_0}$ universal sets, we conclude:

Proposition 8.2. *R has $2^{\aleph_0}$ non-conjugate cyclic automorphisms.*

(Note that this gives another proof of Proposition 8.1.)

Almost all subsets of $\mathbb{N}$ are universal — this is true in either sense discussed in Section 7. The construction preceding Proposition 8.2 shows that graphs admitting a given cyclic automorphism correspond to subsets of $\mathbb{N}$; so almost all "cyclic graphs" are isomorphic to R. What if the cyclic permutation is replaced by an arbitrary permutation or permutation group? The general answer is unknown:

Conjecture 8.1. Given a permutation group G on a countable set, the following are equivalent:

(a) some G-invariant graph is isomorphic to R;
(b) a random G-invariant graph is isomorphic to R with positive probability.

A random G-invariant graph is obtained by listing the orbits of G on the 2-subsets of the vertex set, and deciding randomly whether the pairs in each orbit are edges or not. We cannot replace "positive probability" by "probability 1" here. For example, consider a permutation with one fixed point x and two infinite cycles. With probability $\frac{1}{2}$, x is joined to all or none of the other vertices; if this occurs, the graph is not isomorphic to R. However, almost all graphs for which this event does not occur are isomorphic to R. It can be shown that the conjecture is true for the group generated by a single permutation; and Truss' list of cycle structures of automorphisms can be re-derived in this way.

Another interesting class consists of the *regular* permutation groups. A group is *regular* if it is transitive and the stabilizer of a point is the identity. Such a group G can be considered to act on itself by right multiplication. Then any G-invariant graph is a *Cayley graph* for G; in other words, there is a subset S of G, closed under inverses and not containing the identity, so that x and y are adjacent if and only if $xy^{-1} \in S$. Now we can choose a *random Cayley graph* for G by putting inverse pairs into S with probability $\frac{1}{2}$. It is not true that, for every countable group G, a random Cayley graph for G is almost surely isomorphic to R. Necessary and sufficient conditions can be given; they are somewhat untidy. I will state here a fairly general sufficient condition.

A *square-root set* in G is a set

$$\sqrt{a} = \{x \in G : x^2 = a\};$$

it is *principal* if $a = 1$, and *non-principal* otherwise.

Proposition 8.3. *Suppose that the countable group G cannot be expressed as the union of finitely many translates of non-principal square-root sets and a finite set. Then almost all Cayley graphs for G are isomorphic to R.*

This proposition is true in the sense of Baire category as well. In the infinite cyclic group, a square-root set has cardinality at most 1; so the earlier result about cyclic automorphisms follows. See Cameron and Johnson [9] for further details.

Next, we discuss a recent theorem of Thomas [50]. The concept of a *closed* group of permutations will be considered further in the next section. For now, it suffices to say that a group is closed if and only if it is the full automorphism group of some first-order structure. An *anti-automorphism* of R is an isomorphism from R to its complement; a *switching automorphism* maps R to a graph equivalent to R by switching. The concept of a *switching anti-automorphism* should be clear.

Theorem 8.3. *There are exactly five closed permutation groups which contain* $\mathrm{Aut}(R)$, *viz.:* $A = \mathrm{Aut}(R)$; *the group D of automorphisms and anti-automorphisms of R; the group S of switching automorphisms of R; the group B of switching automorphisms and anti-automorphisms of R; and the symmetric group.*

Remark 8.1. The set of all graphs on a given vertex set is a $\mathbb{Z}_2$-vector space, where the sum of two graphs is obtained by taking the symmetric difference of their edge sets. Now complementation corresponds to adding the complete graph, and switching to adding a complete bipartite graph. Thus, it follows from Theorem 8.3 that, if G is a closed supergroup of $\mathrm{Aut}(R)$, then the set of all images of R under G is contained in a coset of a subspace $W(G)$ of this vector space. (For example, $W(B)$ consists of all complete bipartite graphs and all unions of at most two complete graphs.) Moreover, these subspaces are invariant under the symmetric group. It is remarkable that the combinatorial proof leads to this algebraic conclusion.

I will sketch, very briefly, a proof of the theorem. (I am grateful to Dima Fon-Der-Flaass for discussions about this.) First, a remark. Let U and V be finite disjoint sets of vertices of R. We saw in Proposition 3.1 that the set Z of vertices satisfying the conclusion of (*) induces a graph isomorphic to R. It follows from the homogeneity of R that the pointwise stabiliser of $U \cup V$ in $\mathrm{Aut}(R)$ acts homogeneously on Z. In topological language, the induced permutation group is a dense subgroup of $\mathrm{Aut}(Z)$.

Let G be a closed supergroup of $A = \mathrm{Aut}(R)$. We call two finite graphs *equivalent* if some (or any) copy of the first in R can be mapped to a copy of the second by some element of G. Let X be a finite subset of R. The pointwise stabilizer $A_{(X)}$ has $2^{|X|}$ orbits outside X, each orbit Z corresponding to the subset of X consisting of vertices adjacent to that orbit. The orbits of $G_{(X)}$ are unions of these orbits. We distinguish two cases:

(a) there exists a non-empty set X and an $A_{(X)}$-orbit Z which is fixed by $G_{(X)}$;
(b) no such X and Z exist.

(Note that the groups A and D fall under case (a), the others under (b).) I will discuss case (a); the arguments in case (b) are somewhat similar.

Now assume that G is a counterexample to the theorem. Let G be n-transitive but not $(n+1)$-transitive, and assume that n is as small as possible. Note that $n > 1$: for if G is not 2-transitive, it preserves the A-orbits on pairs, which are the edges and non-edges of R. Let X and Z be as in the definition of case (a). By our earlier remark, the closure A^* of the permutation group induced by $A_{(X)}$ on Z is isomorphic to A. Now G^*, the closure of ${G_{(X)}}^Z$, is a closed supergroup of A^*. A combinatorial argument shows that G^* is not n-transitive. By choice of n, G^* must be isomorphic to A, D, S or B. But also G^* falls under case (a), so $G^* \cong A$ or D. By enlarging X by two points of Z if necessary, we may assume that $G^* \cong A$. (No anti-automorphism can fix two points.) An equivalent statement is that, if $g \in G$ induces an isomorphism on X, then it induces an isomorphism on Z. From here, a simple combinatorial argument shows that $G_{(X)} \leq A$; that is, if $g \in G$ induces an isomorphism on X, then $g \in A$.

To complete the proof, we must show that, for every $g \in G$, there is a subgraph of R isomorphic to X on which g induces an isomorphism or anti-isomorphism. This uses a famous Ramsey-type theorem of Deuber [13] and Nešetřil and Rödl [37]: *given a finite graph U, there is a finite graph V with the property that, for any 2-colouring of the edges of V, there is a monochrome induced subgraph isomorphic to U.* By iterating, we obtain W with the property that, given 2-colourings of both edges and non-edges of W, there is an induced subgraph isomorphic to U in which both the edges and the non-edges are monochromatic. We apply this theorem with U the union of X (constructed above), a graph not equivalent to a complete graph, and a graph not equivalent to a null graph. Let W be chosen thus, and assume that W is embedded in R. Now give an edge e of W the colour c_1 if eg is an edge, and c_2 if eg is a non-edge; and 2-colour the non-edges of W similarly. We find a copy of U in R for which both edges and non-edges are monochromatic; in other words, Ug is complete or null or isomorphic to U or its complement. We chose U so that Ug cannot be complete or null; so g induces an isomorphism or anti-isomorphism on U (and hence on X), and we are done.

Here is a recent application due to Cameron and Martins [10], which draws together several threads from earlier sections. Though it is a result about finite random graphs, the graph R is inextricably involved in the proof.

Let $\mathcal{F}$ be a finite collection of finite graphs. For any graph Γ, let $\mathcal{F}(\Gamma)$ be the hypergraph whose vertices are those of Γ, and whose edges are the subsets which induce graphs in $\mathcal{F}$. To what extent does $\mathcal{F}(\Gamma)$ determine Γ?

Theorem 8.4. *Given $\mathcal{F}$, one of the following possibilities holds for almost all finite random graphs Γ:*

(a) $\mathcal{F}(\Gamma)$ determines Γ uniquely;
(b) $\mathcal{F}(\Gamma)$ determines Γ up to complementation;
(c) $\mathcal{F}(\Gamma)$ determines Γ up to switching;
(d) $\mathcal{F}(\Gamma)$ determines Γ up to switching and/or complementation;
(e) $\mathcal{F}(\Gamma)$ determines only the number of vertices of Γ.

I sketch the proof in the first case, that in which $\mathcal{F}$ is not closed under either complementation or switching. We distinguish two first-order languages, that of graphs and that of hypergraphs (with relations of the arities appropriate for the graphs in $\mathcal{F}$). Any sentence in the hypergraph language can be "translated" into the graph language, by replacing "E is an edge" by "the induced subgraph on E is one of the graphs in $\mathcal{F}$".

By the case assumption and Theorem 8.3, we have $\mathrm{Aut}(\mathcal{F}(R)) = \mathrm{Aut}(R)$. Now by Theorem 5.2, the edges and non-edges in R are 2-types in $\mathcal{F}(R)$, so there is a formula $\phi(x,y)$ (in the hypergraph language) such that $x \sim y$ in R if and only if $\phi(x,y)$ holds in $\mathcal{F}(R)$. If ϕ^* is the "translation" of ϕ, then R satisfies the sentence

$$(\forall x,y)((x \sim y) \leftrightarrow \phi^*(x,y)).$$

By Theorem 6.1, this sentence holds in almost all finite graphs. Thus, in almost all finite graphs, Γ, vertices x and y are joined if and only if $\phi(x,y)$ holds in $\mathcal{F}(\Gamma)$. So $\mathcal{F}(\Gamma)$ determines Γ uniquely.

By Theorem 8.3, $\mathrm{Aut}(\mathcal{F}(R))$ must be one of the five possibilities listed; in each case, an argument like the one just given shows that the appropriate conclusion holds.

To conclude, here is an interesting (non-closed) permutation group containing $\mathrm{Aut}(R)$. A permutation is *almost an automorphism* of a graph Γ if it maps edges to edges and non-edges to non-edges with finitely many exceptions. Now the "almost-automorphisms" of Γ form a group $\mathrm{AAut}(\Gamma)$.

Proposition 8.4. $\mathrm{AAut}(R)$ *is highly transitive (i.e., n-transitive for every n).*

Proof. Take vertices $x_1, \ldots, x_n$ and $y_1, \ldots y_n$. Let R_1 and R_2 be obtained by deleting all edges of R within the sets $\{x_1, \ldots, x_n\}$ and $\{y_1, \ldots, y_n\}$ respectively. By Proposition 3.2(b), R_1 and R_2 are isomorphic to R. By homogeneity, there is an isomorphism from R_1 to R_2 mapping x_i to y_i for $i = 1, \ldots, n$. This map is almost an automorphism of R.

9. Topological aspects

There is a natural way to define a topology on the vertex set of R: we take as a basis for the open sets the set of all finite intersections of vertex neighbourhoods. It can be shown that this topology is homeomorphic to $\mathbb{Q}$ (using the characterization of $\mathbb{Q}$ as the unique countable, totally disconnected, topological space without isolated points, due to Sierpiński [48], see also Neumann [38]). Thus:

Proposition 9.1. $\mathrm{Aut}(R)$ *is a subgroup of the homeomorphism group of* $\mathbb{Q}$.

This is related to a theorem of Mekler [36]:

Theorem 9.1. *A countable permutation group G is embeddable in the homeomorphism group of $\mathbb{Q}$ if and only if the intersection of the supports of any finite number of elements of G is empty or infinite.*

Here, the *support* of a permutation is the set of points it doesn't fix. Now of course $\mathrm{Aut}(R)$ is not countable; yet it does satisfy Mekler's condition. (If x is moved by each of the automorphisms $g_1, \ldots, g_n$, then the infinitely many vertices joined to x but to none of $xg_1, \ldots, xg_n$ are also moved by these permutations.)

More attention has been given to a natural topology on the symmetric group $\mathrm{Sym}(X)$ on a countable set X. Here, a basis for the open sets consists of all sets of the form

$$\{g \in \mathrm{Sym}(X) : x_i g = y_i \text{ for } i = 1, \ldots, n\},$$

where $(x_1, \ldots, x_n)$ and $(y_1, \ldots, y_n)$ are n-tuples of distinct points of X, and $n \in \mathbb{N}$. In other words, a basis of neighbourhoods of the identity consists of the stabilizers of all finite tuples. Now, as mentioned earlier, a subgroup G of $\mathrm{Sym}(X)$ is closed if and only if it is the automorphism group of some first-order structure. It is possible to formulate the small index property in topological terms: M has the small index property if and only if every subgroup of $\mathrm{Aut}(M)$ with index less than $2^{\aleph_0}$ is open in $\mathrm{Aut}(M)$.

The topology on the symmetric group can be derived from a metric, so that the group is a complete metric space. (Take the set on which the group acts to be $\mathbb{N}$; then let the distance between permutations g and h be $\frac{1}{2^n}$ if g and h agree, and also g^{-1} and h^{-1} agree, on $0, 1, \ldots, n-1$, but one of these pairs disagrees on n.) Now a closed subgroup is thus also a complete metric space. So we can apply the concept of Baire category to $\mathrm{Aut}(R)$, for example, and ask: is there a "typical" automorphism? Truss [53] showed the following nice result.

Theorem 9.2. *There is a conjugacy class which is residual in* $\mathrm{Aut}(R)$. *Its members have infinitely many cycles of each finite length, and no infinite cycles.*

Members of the residual conjugacy class (which is, of course, unique) are called *generic automorphisms* of R. I outline the argument. Each of the following sets of automorphisms is residual:

(a) those with no infinite cycles;
(b) those automorphisms g with the property that, if Γ is any finite graph and f any isomorphism between subgraphs of Γ, then there is an embedding of Γ into R in such a way that g extends f.

(Here (a) holds because the set of automorphisms for which the first n points lie in finite cycles is open and dense.) In fact, (b) can be strengthened; we can require that, if the pair (Γ, f) extends the pair (Γ_0, f_0) (in the obvious sense), then any embedding of Γ_0 into R such that g extends f_0 can be extended to an embedding of Γ such that g extends f. Then a residual set of automorphisms satisfy both (a) and the strengthened (b); this is the required conjugacy class.

Another way of expressing this result is to consider the class $\mathcal{C}$ of finite structures each of which is a graph Γ with an isomorphism f between two induced subgraphs (regarded as a binary relation). This class satisfies Fraïssé's hypotheses, and so has a Fraïssé limit M. It is not hard to show that, as a graph, M is the random graph R; arguing as above, the map f can be shown to be a (generic) automorphism of R.

More generally, Hodges *et al.* [29] showed that there exist "generic n-tuples" of automorphisms of R, and used this to prove the small index property for R; see also Hrushovski [30]. The group generated by a generic n-tuple of automorphisms is, not surprisingly, a free group; all its orbits are finite, so it is unlike the kinds of subgroups we considered in Section 8.

10. Some other structures

As we have seen, R has several properties of a general kind: for example, homogeneity, $\aleph_0$-categoricity, universality, ubiquity. Much effort has gone into studying, and if possible characterizing, structures of other kinds with these properties. (For example, they are all shared by the ordered set $\mathbb{Q}$.)

Note that, of the four properties listed, the first two each imply the third, and the first implies the fourth. Moreover, a homogeneous structure over a finite relational language is $\aleph_0$-categorical, since there are only finitely many isomorphism types of n-element structure for each n. Thus, homogeneity is in practice the strongest condition, most likely to lead to characterizations.

A major result of Lachlan and Woodrow [35] determines the countable homogeneous graphs.

Theorem 10.1. *A countable homogeneous graph is isomorphic to one of the following:*

(a) the disjoining union of m complete graphs of size n, where $m, n \leq \aleph_0$ and at least one of m and n is $\aleph_0$;
(b) complements of (a);
(c) the Fraïssé limit of the class of K_n-free graphs, for fixed $n \geq 3$;
(d) complements of (c);
(e) the random graph R.

Other classes in which the homogeneous structures have been determined include finite graphs (Gardiner [23]), tournaments (Lachlan [34] — surprisingly, there are just three), digraphs (Cherlin [12] (there are uncountably many, see Henson [27]), and posets (Schmerl (1979)). In the case of posets, Droste [14] has characterizations under weaker assumptions.

For a number of structures, properties of the automorphism group, such as normal subgroups, small index property, or existence of generic automorphisms, have been established.

The only case in which an analogue of Theorem 8.3 is known is $\mathbb{Q}$. The next result follows from a theorem of Cameron [4].

Theorem 10.2. *There are just five closed permutation groups containing the group* $\mathrm{Aut}(\mathbb{Q})$ *of order-preserving permutations of $\mathbb{Q}$, viz.:* $\mathrm{Aut}(\mathbb{Q})$*; the group of order preserving or reversing permutations; the group of permutations preserving a cyclic order; the group of permutations preserving or reversing a cyclic order; and* $\mathrm{Sym}(\mathbb{Q})$.

However, there is no analogue of Theorem 8.4 in this case, since there is no Glebskii–Blass–Fagin–Harary theory for ordered sets. ($\mathbb{Q}$ is dense, but no finite ordered set is dense.)

Since my paper with Paul Erdős concerns sum-free sets (Cameron and Erdős [8]), it is appropriate to discuss their relevance here. Let H_n be the Fraïssé limit of the class of K_n-free graphs, for $n \geq 3$ (see Theorem 10.1). These graphs were first constructed by Henson [26], who also showed that H_3 admits cyclic automorphisms but H_n does not for $n > 3$. We have seen how a subset S of $\mathbb{N}$ gives rise to a graph $\Gamma(S)$ admitting a cyclic automorphism: the vertex set is $\mathbb{Z}$, and $x \sim y$ if and only if $|x - y| \in S$. Now $\Gamma(S)$ is triangle-free if and only if S is *sum-free* (i.e., $x, y \in S \Rightarrow x + y \notin S$). It can be shown that, for almost all sum-free sets S (in the sense of Baire category), the graph $\Gamma(S)$ is isomorphic to H_3; so H_3 has $2^{\aleph_0}$ non-conjugate cyclic automorphisms. However, the analogue of this statement for measure is false; and, indeed, random sum-free sets have a rich and surprising structure which is not well understood (Cameron [5]). For example, the probability that $\Gamma(S)$ is bipartite is approximately 0.218.

Universality of a structure M was defined in a somewhat introverted way in Section 5: M is universal if every structure younger than M is embeddable in M. A more general definition would start with a class $\mathcal{C}$ of structures, and say that $M \in \mathcal{C}$ is *universal for* $\mathcal{C}$ if every member of $\mathcal{C}$ embeds into M. For a survey of recent results on this sort of universality, for various classes of graphs, see Komjath and Pach [33]. Two negative results, for the classes of locally finite graphs and of planar graphs, are due to De Bruijn (see Rado [42]) and Pach [40] respectively.

Added in proof: 1. Recent results of Shelah and Spencer [47] and Hrushovski [31] suggest that there are interesting countable graphs which "control" the first-order theory of finite random graphs whose edge-probabilities tend to zero in specified ways. See Wagner [54], Winkler [55] for surveys of this.

2. Thomas (personal communication) has found all closed supergroups of the automorphism group of the random k-uniform hypergraph.

References

1. P. Bankston and W. Ruitenberg (1990), Notions of relative ubiquity for invariant sets of relational structures, *J. Symbolic Logic* **55**, 948–986.
2. A. Blass and F. Harary (1979), Properties of almost all graphs and complexes, *J. Graph Theory* **3**, 225–240.
3. B. Bollobás (1985), *Random Graphs*, Academic Press, London.
4. P. J. Cameron (1976), Transitivity of permutation groups on unordered sets, *Math. Z.* **148**, 127–139.
5. P. J. Cameron (1987), On the structure of a random sum-free set, *Probab. Thy. Rel. Fields* **76**, 523–531.
6. P. J. Cameron (1990), *Oligomorphic Permutation Groups, London Math. Soc. Lecture Notes* **152**, Cambridge Univ. Press, Cambridge.
7. P. J. Cameron (1992), The age of a relational structure, pp. 49–67 in *Directions in Infinite Graph Theory and Combinatorics* (ed. R. Diestel), *Topics in Discrete Math.* **3**, North-Holland, Amsterdam.
8. P. J. Cameron and P. Erdős (1990), On the number of sets of integers with various properties, pp. 61–79 in *Number Theory* (ed. R. A. Mollin), de Gruyter, Berlin.
9. P. J. Cameron and K. W. Johnson (1987), An investigation of countable B-groups, *Math. Proc. Cambridge Philos. Soc.* **102**, 223–231.
10. P. J. Cameron and C. Martins (1993), A theorem on reconstructing random graphs, *Combinatorics, Probability and Computing* **2**, 1–9.
11. G. Cantor (1895), Beiträge zur Begründung der transfiniten Mengenlehre, *Math. Ann.* **46**, 481–512.
12. G. L. Cherlin (1987), Homogeneous directed graphs, *J. Symbolic Logic* **52**, 296. (See also *Memoirs Amer. Math. Soc.*, to appear.)
13. W. Deuber (1975), Partitionstheoreme für Graphen, *Math. Helvetici* **50**, 311–320.
14. M. Droste (1985), Structure of partially ordered sets with transitive automorphism groups, *Mem. Amer. Math. Soc.* **57**.
15. M. El-Zahar and N. W. Sauer (to appear), Ramsey-type properties of relational structures, *Discrete Math.*
16. E. Engeler (1959), Äquivalenz von n-Tupeln, *Z. Math. Logik Grundl. Math.* **5**, 126–131. [?]
17. P. Erdős, D. J. Kleitman and B. L. Rothschild (1977), Asymptotic enumeration of K_n-free graphs, pp. 19–27 in *Colloq. Internat. Teorie Combinatorie*, Accad. Naz. Lincei, Roma.
18. P. Erdős and A. Rényi (1963), Asymmetric graphs, *Acta Math. Acad. Sci. Hungar.* **14**, 295–315.

19. P. Erdős and J. Spencer (1974), *Probabilistic Methods in Combinatorics*, Academic Press, New York/Akad. Kiadó, Budapest.
20. R. Fagin (1976), Probabilities on finite models, *J. Symbolic Logic* **41**, 50–58.
21. R. Fraïssé (1953), Sur certains relations qui généralisent l'ordre des nombres rationnels, *C. R. Acad. Sci. Paris* **237**, 540–542.
22. R. Fraïssé (1986), *Theory of Relations*, North-Holland, Amsterdam.
23. A. Gardiner (1976), Homogeneous graphs, *J. Combinatorial Theory* (B) **20**, 94–102.
24. Y. V. Glebskii, D. I. Kogan, M. I. Liogon'kii and V. A. Talanov (1969), Range and degree of realizability of formulas in the restricted predicate calculus, *Kibernetika* **2**, 17–28.
25. F. Hausdorff (1914), *Grundzügen de Mengenlehre*, Leipzig.
26. C. W. Henson (1971), A family of countable homogeneous graphs, *Pacific J. Math.* **38**, 69–83.
27. C. W. Henson (1972), Countable homogeneous relational structures and $\aleph_0$-categorical theories, *J. Symbolic Logic* **37**, 494–500.
28. W. A. Hodges (1993), *Model Theory*, Cambridge Univ. Press, Cambridge.
29. W. A. Hodges, I. M. Hodkinson, D. Lascar and S. Shelah (1993), The small index property for ω-stable, ω-categorical structures and for the random graph, *J. London Math. Soc.* (2) **48**, 204–218.
30. E. Hrushovski (1992), Extending partial isomorphisms of graphs, *Combinatorica* **12**, 411–416.
31. E. Hrushovski (1993), A new strongly minimal set, *Ann. Pure Appl. Logic* **62**, 147–166.
32. E. V. Huntington (1904), The continuum as a type of order: an exposition of the model theory, *Ann. Math.* **6**, 178–179.
33. P. Komjáth and J. Pach (1992), Universal elements and the complexity of certain classes of graphs, pp. 255–270 in *Directions in Infinite Graph Theory and Combinatorics* (ed. R. Diestel), *Topics in Discrete Math.* **3**, North-Holland, Amsterdam.
34. A. H. Lachlan (1984), Countable homogeneous tournaments, *Trans. Amer. Math. Soc.* **284**, 431–461.
35. A. H. Lachlan and R. E. Woodrow (1980), Countable ultrahomogeneous undirected graphs, *Trans. Amer. Math. Soc.* **262**, 51–94.
36. A. H. Mekler (1986), Groups embeddable in the autohomeomorphisms of $\mathbb{Q}$, *J. London Math. Soc.* (2) **33**, 49–58.
37. J. Nešetřil and V. Rödl (1978), The structure of critical Ramsey graphs, *Colloq. Internat. C.N.R.S.* **260**, 307–308.
38. P. M. Neumann (1985), Automorphisms of the rational world, *J. London Math. Soc.* (2) **32**, 439–448.
39. J. C. Oxtoby (1971), *Measure and Category*, Springer, Berlin.
40. J. Pach (1981), A problem of Ulam on planar graphs, *Europ. J. Combinatorics* **2**, 351–361.
41. J. Plotkin (to appear), Who put the back in back-and-forth?
42. R. Rado (1964), Universal graphs and universal functions, *Acta Arith.* **9**, 393–407.
43. R. Rado (1967), Universal graphs, in *A Seminar in Graph Theory* (ed. F. Harary and L. W. Beineke), Holt, Rinehard & Winston, New York.
44. C. Ryll-Nardzewski (1959), On the categoricity in power $\aleph_0$, *Bull. Acad. Polon. Sci. Ser. Math.* **7**, 545–548.
45. J. H. Schmerl (1979), Countable homogeneous partially ordered sets, *Algebra Universalis* **9**, 317–321.

46. J. J. Seidel (1977), A survey of two-graphs, pp. 481–511 in *Colloq. Internat. Teorie Combinatorie*, Accad. Naz. Lincei, Roma.
47. S. Shelah and J. Spencer (1988), Zero-one laws for sparse random graphs, *J. American. Math. Soc.* **1**, 97–115.
48. W. Sierpiński (1920), Une propriété topologique des ensembles dénombrables denses en soi, *Fund. Math.* **1**, 11–16.
49. L. Svenonius (1955), $\aleph_0$-categoricity in first-order predicate calculus, *Theoria* **25**, 82–94.
50. S. Thomas (1991), Reducts of the random graph, *J. Symbolic Logic* **56**, 176–181.
51. J. K. Truss (1985), The group of the countable universal graph, *Math. Proc. Cambridge Philos. Soc.* **98**, 213–245.
52. J. K. Truss (1989), Infinite permutation groups, I, Products of conjugacy classes, *J. Algebra* **120**, 454–493; II, Subgroups of small index, *ibid.* **120**, 494–515.
53. J. K. Truss (1992), Generic automorphisms of homogeneous structures, *Proc. London Math. Soc.* (3) **65**, 121–141.
54. F. O. Wagner (1994), Relational structures and dimensions, *Automorphisms of First-Order Structures* (ed. R. Kaye and H. D. Macpherson), Oxford University Press, pp. 153–180.
55. P. Winkler (1993), Random structures and zero-one laws, *Finite and Infinite Combinatorics in Sets and Logic* (ed. N. W. Sauer, R. E. Woodrow and B. Sands), NATO Advanced Science Institutes Series, Kluwer Academic Publishers, pp. 399–420.

Paul Erdős' Set Theory

András Hajnal

[1] Mathematical Institute of the Hungarian Academy of Sciences, Budapest, Hungary
[2] University of Calgary, Mathematics Department, Calgary, AB, Canada
[3] Center of Discrete Mathematics and Theoretical Computer Science, Rutgers University, Piscataway, New Jersey 08855, USA

1. An apology

Paul Erdős has published more than one hundred research papers in set theory. It is my rough estimate that these contain more than one thousand theorems, many having an interest in their own right. Although most of his problems and results have a combinatorial flavour, and the subject now known as "combinatorial set theory " is one he helped to create, it is also true to say that his work has had a very important impact upon the direction of research in many parts of present day set theory. Whole theories have developed out of basic questions which he formulated.

This (relatively) short note does not, and is not intended to, give a methodical survey of set theory or combinatorial set theory or even of Erdős' work in set theory. I shall simply write about some of the ideas as I learned them during our cooperation over many years, some of the highlights, and some of the outstanding results. In many cases it will not be possible to give a detailed discussion of the present day status of some of the problems I shall mention. If the reader considers that my own name occurs too frequently in this note, I can only offer the excuse that we have published more than fifty joint papers, mostly in set theory, and I probably know these papers better than the rest of his work.

2. Early days and some philosophy

Paul was a child mathematical prodigy, and he started to discover outstanding original results in number theory as a first year undergraduate. We are familiar with the names of mathematicians who influenced his early work in number theory and analysis, Pál Veress, Fejér, Davenport, Mordell to mention just a few. But this is not true for set theory. Paul told me that he learned the basics of set theory from his father, a well educated high-school teacher, and he soon became fascinated with "Cantor's paradise". However, he discovered set theory as a subject for research by himself.

Paul, who has always refrained from seriously formulating any kind of philosophy, was (and still is) the ultimate Platonist. $\aleph_{\omega_{\omega+1}+1}$ existed for him just as surely as 3, the smallest odd prime. He was driven by the same compulsive search for "truth" whether he was thinking about inaccessible cardinals or twin primes. Moreover, he could switch from one subject to the other in an instant. All questions which admit a relevant answer in finite combinatorics should also be asked and answered in set theory, and vice-versa. A large part of his greatness lies in the fact that he really did find the relevant questions. It was this attitude which led him to his first encounter with (actual) infinity.

In 1931, as a first year undergraduate student attending the graph theory course of Dénes König, he proved a generalization of Menger's theorem for infinite graphs. This only appeared in 1936 at the end of König's book on graph theory. In 1936 he wrote a paper jointly with Tibor Gallai and Endre Vázsonyi having a similar character; they gave necessary and sufficient conditions for an infinite graph to have an Euler line [1], [2].

The next paper I have to mention [ESz], about finite combinatorics, was written with George Szekeres in 1935. They re-discovered the finite version of Ramsey's theorem and proved a fundamental Ramsey-type result of finite character: *If G is a graph having $\binom{k+\ell-2}{k-1}$ vertices ($k, \ell \geq 3$), then either G contains a complete graph on k vertices or there is an independent set of ℓ vertices.* From then on he always had in mind possible generalizations of Ramsey's theorem, and so became the creator of both finite and infinite Ramsey theory.

3. Infinite Ramsey theory — early papers

The Ramsey theorem is about partitions of (finite) k-element subsets of ω (the set of non-negative integers), and in the mid-1930's Erdős began to speculate about partitioning the countable sets. He corresponded with Richard Rado in Cambridge, England about this problem and Rado proved the first theorem saying that "nothing can be said in this case". The result appeared only much later, in the early 1950's, in a sequence of joint papers by them [9],[12],[13],[18]. I will return to this later.

Erdős' first real set-theoretic result appeared in a paper of Dushnik and Miller [DM]. The theorem, now known as the Erdős; Dushnik, Miller theorem, says that *for an infinite cardinal κ, if a graph on κ vertices does not contain an infinite complete subgraph, then there is an independent set of vertices of size κ.* This was the first "unbalanced" generalization of Ramsey's theorem. Once the result is formulated, the verification for regular κ is a fairly easy exercise. Erdős proved it for singular κ, and his proof, which required a good technical knowledge of the set theory of those days, is included in the Dushnik-Miller paper.

Soon after, in 1942, he proved in [3] the basic theorems of infinite Ramsey theory. Let $[X]^r$ denote the set of r-element subsets of X, and let $f : [X]^r \to \gamma$ be an r-partition of γ colours on X. A set $H \subseteq X$ is *homogeneous* for f in the colour $\nu < \gamma$ if $f(Y) = \nu$ for all $Y \in [H]^r$. More than 10 years later in joint work with Paul, [12], Richard Rado introduced the partition symbol

$$\kappa \to (\lambda_\nu)^r_{\nu<\gamma}$$

to denote the following assertion: *for any r-partition $f : [\kappa]^r \to \gamma$ there are $\nu < \gamma$ and $H \subseteq \kappa$ such that $|H| = \lambda_\nu$ and H is homogeneous for colour ν.* The negation of this is denoted by replacing the arrow by a crossed arrow, $\not\to$. If $\lambda_\nu = \lambda$ for all $\nu < \gamma$, the notation $\kappa \to (\lambda)^r_\gamma$ is used; this is called the "balanced" partition symbol.

Using this notation, Ramsey's theorem states

$$\omega \to (\omega)^r_k \text{ for } 1 \le r, k < \omega.$$

The Erdős; Dushnik, Miller theorem says, for any infinite cardinal κ,

$$\kappa \to (\kappa, \aleph_0)^2.$$

The result of Rado I did not state says, for every κ,

$$\kappa \not\to (\aleph_0)^{\aleph_0}_2.$$

Using these notations, the results proved by Erdős in the 1942 paper are the following:

(i) $(2^\lambda)^+ \to (\lambda^+)^2_\lambda$.
(ii) $2^\lambda \not\to (3)^2_\lambda$.
(iii) Assuming the generalized continuum hypothesis (CGH in what follows) $\aleph_{\alpha+2} \to (\aleph_{\alpha+2}, \aleph_{\alpha+1})^2$.
(iv) $2^\lambda \not\to (\lambda^+)^2_2$.

He attributes (iv) to Sierpiński, who proved it for $\lambda = \aleph_0$. He also mentions that the obvious $\not\to$ relation (ii) was pointed out to him by Gödel in conversation.

Of course, the partition symbol was not used in that early paper. In fact Erdős was always slightly resistant to its use. Later, when forced, he did sometimes write the symbol, but I have never seen him read it. When we were discussing such relations he frequently asked me in a complaining voice to "state it in human language".

The observant reader would already have noted that (iii) was an attempt to find the right generalization of the Erdős; Dushnik, Miller theorem. We will return to this topic later.

4. His "Remarks"

As yet we are still in 1943, and in that year two more significant papers appeared. First I want to say a few words about [5], "Some remarks on Set Theory". I quote the first sentence: "This paper contains a few disconnected results on the theory of sets." The "Remarks" became a series. Eleven of them appeared in all. The fifth and sixth were written jointly with Géza Fodor, the seventh to ninth and eleventh with me, and the tenth with Michael Makkai. Several of these papers contain set-theoretical results about Euclidean spaces, Hamel bases and other objects that are familiar to analysts and combinatorialists. The editors and I have decided that this typical Erdős genre deserves a separate treatment, and this will be given by Péter Komjáth in this volume. However, I cannot resist mentioning the first theorem in the first of these papers, the Erdős–Sierpiński duality principle. This generalization of an earlier result of Sierpiński states: *Assuming the continuum hypothesis* (CH) *there is a surjective map* $f : \mathbb{R} \to \mathbb{R}$ *which interchanges sets of Lebesge measure zero and sets of first category.*

Stating this theorem allows me an opportunity to say something about his attitude towards the generalized continuum hypothesis (GCH) and mathematical logic in general. It should be remembered that in 1943 Gödel's proof of the consistency of GCH was quite new. Erdős always knew and appreciated and applied these results. He was happy to have them as a more-or-less justified tool to prove new theorems, and if he could not solve a set theory problem he always tried to solve it assuming GCH. On the other hand, later on he was always uneasy and disappointed if one of his favourite problems turned out to be independent, and he would remark "independence has raised its ugly head".

5. Large cardinals— the Erdős–Tarski paper

The cardinal κ has the property $P(\kappa)$ if there is a field of sets which contains a family of λ pairwise disjoint sets for every $\lambda < \kappa$, but which does not contain such a family of size κ. Much to their surprise, Erdős and Tarski [4] proved that *for limit cardinals* κ, $P(\kappa)$ *holds iff* κ *is an uncountable inaccessible.* First of all it was surprising that such a seemingly harmless problem should involve inaccessible cardinals which, in those days, had "hardly been born". The second surprise was, and this is explicitly mentioned in the paper, that the negation of $P(2^{\aleph_0})$ could not be proved in ZFC since it was generally believed that it would be proved consistent that $2^{\aleph_0}$ is inaccessible. (Indeed, this was one of the first corollaries of Cohen's method.)

In their paper, they formulated several properties of inaccessible cardinals and they mentioned quite a few connections between these properties in footnotes (without proofs). For example, they knew that if κ is measurable then it has the tree property and that this implies that $\kappa \to (\kappa)^r_\lambda$ holds for all

$\lambda < \kappa$ and $r < \omega$. Of course, they also knew that the simplest Ramsey-type theorem $\kappa \to (\kappa)^2_2$ is false if κ is not strongly inaccessible. Later, in 1960 when Tarski, using a result of Hanf, proved that "small" inaccessibles are not measurable, and the theory of large cardinals was created, it became necessary to publish these classical proofs. A new Erdős –Tarski paper (written by Donald Monk) appeared in 1961.

It is of interest to note that Erdős and Tarski made an historical mistake in [4]. It was vaguely speculated that it may turn out to be at least consistent that all strongly inaccessible cardinals are measurable. This probably postponed the discovery of the true situation for almost twenty years. I cannot say how seriously Tarski believed this (I did try to ask him), but Erdős quite happily accepted this hypothesis in the spirit I described in §2. In our joint work from 1956 to 1960 we investigated every combinatorial property of strongly inaccessible cardinals under the assumption that they are measurable, although we did always mention that this was an assumption. However, as was so often the case for Paul, in the end this turned out to be quite fortunate. I will come back to this in §9 and §10.

6. Set mappings and compactness

Erdős visited Hungary in 1948 for the first time after the Second World War. Very likely it was during this visit that he recalled an old problem of Paul Turán. Let f be a set mapping on a set X, i.e. $f : X \to \mathbb{P}(X)$ (the power set of X) such that $x \notin f(x)$. We say that f is of order λ if $|f(x)| < \lambda$ for $x \in X$. A subset $S \subset X$ is independent for f if for all $x, y \in S$, $y \notin f(x)$. For combinatorialists, a set mapping of type λ is just a loop-free digraph having out degrees $< \lambda$. Turán was interested in the case when $X = \mathbb{R}$ and $f(x)$ is finite, i.e. f is of order ω, and he asked if there is a free set of power $2^{\aleph_0}$. A young Hungarian, Dezső Lázár who was killed during the war, proved this and he also proved that *if f is of order $\lambda < \kappa \geq \omega$ then there is a free set of size κ, provided κ is a regular cardinal.* Ruziewicz conjectured that this is true for any $\kappa \geq \omega$.

Erdős proved this conjecture assuming GCH in the second of his "Remarks" in 1950 [10]. It remained an open question for ten more years if GCH is really needed. In 1960, I proved the result in ZFC in [H1] where more history of the problem can be found.

Typically, even before proving the result he conjectured that if λ is an infinite cardinal and f is a set mapping of order λ on any set X, then X is the union of λ independent sets, i.e. the digraph has chromatic number at most λ. This was proved by Géza Fodor [F]. Erdős investigated the problem for finite λ in a paper with N.G. de Bruijn [11]. A little reflection will convince the reader that if the underlying set is finite then it is the union $2\lambda-1$ independent sets (and this is best possible). To show that this is true for arbitrary X they proved that *for any $k < \omega$ if every finite subgraph of a graph G has chromatic number of at most k then G also has chromatic number at most k.*

The reader may say that this is a consequence of either Tychonov's theorem on the product of compact spaces or Gödel's compactness theorem, but this is how compactness was introduced to infinite combinatorics. Let me point out that it would be quite difficult to find a proof of the set mapping theorem not using compactness. I do not know of any.

Erdős continued the investigation of set mappings with Géza Fodor in [19] and [21]. I want to mention one of their theorems, which later proved to be useful in applications.

Assume f is a set mapping of order $\lambda < \kappa \geq \omega$ on κ. Let $\tau < \kappa$ and let X_α $(\alpha < \tau)$ be a sequence of subsets of κ each of size κ. There is a set S free for f which meets each X_α in a set of size κ. For singular κ their proof used GCH, but my method yields this in ZFC as well.

7. A partition calculus in set theory

The partition calculus was developed by Erdős and Rado in the early 1950's. The long paper [18] contains all the results they had proved up until then. Their first discovery was that the partition relation $\kappa \to (\lambda_\nu)^r_{\nu<\gamma}$ made sense for order types as well as cardinals, or even a mixture of these. This led them to a great variety of new problems, some simple and some difficult, but requiring different methods. Let me mention just a few of these. For what countable ordinals α does $\alpha \to (\alpha, 3)^2$ hold? For what $\alpha < \omega_1$ does $\lambda \to (\alpha)^2_k$ hold, where k is finite and λ is the order type of the reals? They proved the pleasing result that $\eta \to (\aleph_0, \eta)^2$, where η is the order type of the rationals, but for what other countable types is this true? They also noticed that the proof of the Erdős; Dushnik, Miller theorem in the special case $\kappa = \omega_1$ actually gives the slightly stronger fact that $\omega_1 \to (\omega_1, \omega+1)^2$, and then a natural question is, what about $\omega_1 \to (\omega_1, \omega+2)^2$?

They proved a great many partial results and isolated the most important problems. We cannot collect all their results and problems here, instead I shall discuss some of the important new discoveries. One of these is the *positive stepping up lemma* which can be stated as follows: *If κ is a cardinal and $\kappa \to (\lambda_\nu)^r_{\nu<\gamma}$ holds, then $(2^{<\kappa})^+ \to (\lambda_\nu+1)^{r+1}_{\nu<\gamma}$.* Here μ^+ denotes the smallest cardinal greater than μ, and $2^{<\kappa} = \sum\{2^\mu : \mu < \kappa \text{ a cardinal}\}$. Let $\exp_n(\kappa)$ denote the n-times iterated exponentiation (i.e. $\exp_0(\kappa) = \kappa$ and $\exp_{n+1}(\kappa) = 2^{\exp_n(\kappa)}$.) Since $2^{<\kappa^+} = 2^\kappa$, and $\kappa^+ \to (\kappa^+)^1_\kappa$ just expresses the fact that κ^+ is a regular cardinal, we obtain by induction the

Erdős-Rado Theorem:

$(\exp_n(\kappa))^+ \to (\kappa^+)^{n+1}_\kappa$.

In particular, we have $(2^\kappa)^+ \to (\kappa^+)^2_\kappa$, $(2^{2^\kappa})^+ \to (\kappa^+)^3_\kappa$ etc. Quite often in the literature only the first of these (case $n = 2$) is referred to as the Erdős-Rado theorem, but we have seen that this was already proved in 1943. There can be very few theorems in set theory which have received so many "simplified" proofs as the Erdős-Rado theorem. But in the early 1950's there was no pressing-down lemma, and elementary substructures and chains had not been introduced into set theory. Erdős and Rado used the so-called *ramification method*. Let me outline this in a simple case. Let $f : [X]^2 \to \gamma$ be a 2-partition of X with γ colours. Pick an $x_0 \in X$. The rest of the points can be split into γ parts according to the colour of $f(\{x_0, y\})$. Repeat this in each part and continue transfinitely. We get a tree or ramification system as they called it. At the α stage we will have $|\gamma|^{|\alpha|}$ parts. If X has large enough cardinality, the tree we get will have large height. The points picked along a branch of this tree will form a *prehomogeneous set*, i.e. the colour of a pair $\{x_\alpha, x_\beta\}$ will depend only upon the point, say x_α, which was chosen first. This method is fairly hard to write down formally, but it is really quite intuitive. It was elaborated in great detail in the Erdős-Rado papers and was used for several years to obtain positive partition relations for the case when the underlying set has regular cardinality. Although most of the important proofs have been streamlined to "linear" ones, there is really no algorithm for this translation, and the intuition behind ramification serves as a good tool to obtain new results.

They also discovered polarized partition relations. The symbol

$$\binom{\kappa}{\lambda} \to \binom{\kappa_\nu}{\lambda_\nu}^{1,1}_{\nu<\gamma}$$

means the following: whenever $f : \kappa \times \lambda \to \nu$ is a colouring with γ colours, then there are $\nu < \gamma$, $K \in [\kappa]^{\kappa_\nu}$ and $L \in [\lambda]^{\lambda_\nu}$ such that $K \times L$ is homogeneous for f in the colour ν. For combinatorialists, this is just Ramsey for complete bipartite graphs, and the reader can easily formulate the generalization to s-partite graphs.

However, this is not just formalism. It turned out that quite a few problems about polarized partitions are basic questions in set theory. As an illustration, they proved

$$\binom{\aleph_1}{\aleph_0} \to \left(\begin{matrix}\aleph_1 \\ \aleph_0\end{matrix} , \begin{matrix}\aleph_0 \\ \aleph_0\end{matrix}\right)^{1,1}.$$

In "human language" this says: *if A_α $(\alpha < \omega_1)$ are arbitrary subsets of ω, then either the intersection of $\aleph_1$ of them is infinite, or the union of $\aleph_0$ of them has an infinite complement.* They attributed the negative relation

$$CH \Rightarrow \binom{\aleph_1}{\aleph_0} \not\to \left(\begin{matrix}\aleph_1 \\ \aleph_0\end{matrix} , \begin{matrix}\aleph_1 \\ \aleph_0\end{matrix}\right)^{1,1}.$$

to Sierpiński who, of course, proved this in a different context.

There is one more important partition relation I should mention, and this realized Paul's old wish to have a Ramsey theorem for something more than just k-element sets. They introduced the symbol $\kappa \to (\lambda)^{<\omega}_{\gamma}$ to denote the following statement: *for every sequence $f_n : [\kappa]^n \to \gamma$ of n -partitions of κ with γ colours, there is a subset $H \subset \kappa$ of cardinality λ which is simultaneously homogeneous for each f_n.* They only proved that $2^{\aleph_0} \not\to (\aleph_0)^{<\omega}_2$ with a clever ad-hoc construction, and they asked if $\kappa \to (\aleph_0)^{<\omega}_2$ can be true for any cardinal κ? We turn back to the discussion of this important symbol in §9 and §10.

There is one more type of partition theorem which I should have mentioned earlier. In [9] they proved the first *canonical Ramsey theorem*, and this also led to a long sequence of investigations, improvements and generalizations. Just to give the flavour of what this is all about, I will state just one special case. *Let $f : [\omega]^2 \to \gamma$ be a 2-partition of ω with any number of colours. Then there is an infinite subset $H \subseteq \omega$ such that either H is homogeneous for some colour, or all pairs in H have different colours, or H is prehomogeneous (the colour of a pair depends only on the least element), or H is endhomogeneous (the colour of a pair depends only on the largest element).*

8. My first encounter with Paul

Erdős visited Hungary in 1955 for the first time since the country had become a member of the Eastern block. He was an Hungarian citizen traveling on an Hungarian passport, and he could not have returned earlier if he wanted to leave again. But in the "liberalized" atmosphere of 1956 the Academy was allowed to elect him as a member and the government granted him a special diplomatic type of passport which allowed him to come and go whenever he wished. This was a great opportunity for young Hungarian mathematicians who had heard of him only by name (of course, it was impossible for us to travel to the West before 1956).

At that time I was a graduate student of László Kalmár in Szeged (a small town on the south-eastern border of Hungary). Paul travelled around the country in 1956 and came to visit the mathematics department at the University of Szeged. He had already corresponded with Géza Fodor who was then a young assistant professor in the department. I was introduced to him as "a promising young man" studying set theory, and soon we were left alone in Professor Kalmár's office sitting in two enormous armchairs facing each other over a coffee table. I thought he was very old — he was 43 years old and I was 25. I felt very honoured, and a little embarrassed, to be left alone with this famous man. I did not know then that he had met most of his young collaborators in a similar way. He first asked me what were my interests in set theory. I was then writing my thesis on a subject which later was called relative constructibility, and I was quite proud of it. So I started to explain my results with some enthusiasm. He listened to me very politely, and when I had finished he asked “and are you interested in normal set theory as well?” Of course, we were not on first-name terms then and so the question was phrased in a very polite form of Hungarian that is used for addressing a stranger. But it was clear that it was a genuine inquiry and he meant no harm.

Earlier I had thought of a problem when I heard about all the set-mapping results from Géza. What if we investigate set-mappings of more variables, and I asked if there would be large free subsets in this case too? To tell the truth, as a student of Kalmár, I was trained to think like a logician. So what I had in mind was this: if X is a large set and to every finite subset $V \subset X$ we associate a countable $F(V) \subset X$ such that $F(V) \cap V = \emptyset$, is there a large independent set? I even had the vague idea that if $F(V)$ is the Skolem hull of V in some structure, then an “independent set” would really deserve its name.

Luckily, Paul liked that one! It started a furious activity and the conversation became more fluent and colloquial. The first thing I learned from him (and this took quite a while) was that he would not start to think about the general case. He first wanted to know what happened for set mappings defined on pairs. He proved several lemmas and some partial results and stated a few conjectures. He then suddenly remembered that there was something else he had to do and he called Géza. Quite close to the mathematics building in Szeged stands a rather ugly cathedral built in the 1930’s with two high towers. It turned out that he “must” climb the three hundred odd stairs to the top! Géza had earlier agreed to accompany him, and he then gently began to persuade me to come along too. I had by then lived for two years in Szeged, and I had never had the slightest difficulty in resisting any pressure to visit the tower, especially since the surrounding countryside is absolutely flat and so there was not very much to see. However, much to my own surprise, I could not resist this invitation. Climbing those stairs more results and conjectures were formulated by him while, at the same time, he was complaining that he felt a little dizzy.

That day ended with dinner at Kalmár’s house where the conversation continued mainly about set mappings, but sometimes interrupted with some of his comments on “Sam and Joe”. When we parted, it was almost as from an old friend — there was a joint-paper half ready, which could be completed by correspondence.

9. Our first joint paper

The notation used for discussing set-mapping problems is not standardized as in the case of partition relations. A set-mapping of *order* λ and *type* μ on κ is a function $f : [\kappa]^{\mu} \to [\kappa]^{<\lambda}$ such that $f(x) \cap x = \emptyset$ for all $x \in \mathrm{dom}(f)$, and a set $S \subseteq \kappa$ is *free* if $f(x) \cap S = \emptyset$ for all $x \in [S]^{\mu}$. I shall denote by

Free$(\kappa, \lambda, \mu, \nu)$ the following assertion: *For every set-mapping of order λ and type μ on κ there is a free set of cardinality ν.* Likewise, Free$(\kappa, \lambda, < \mu, \nu)$ denotes the corresponding assertion when $f : [\kappa]^{<\mu} \to [\kappa]^{<\lambda}$.

Our very first result was to prove that Free$(\exp_{n-1}(\lambda)^+, \lambda, n, \lambda^+)$ holds for every infinite λ. The proof of this uses the Erdős-Rado theorem which I had learned during my first conversation with Paul. I also learned that it was not known if the Erdős-Rado theorem is best possible, for example it was not known then if

$$2^{2^{\aleph_0}} \not\to (\aleph_1)_2^3$$

holds. Assuming GCH we could prove $\neg$ Free$(\aleph_1, 2, 2, \aleph_1)$ so that our theorem is best possible for $n = 2$, but for $n > 2$ any progress seemed to lie far in the future. I will return to this in the section on the negative stepping-up method we developed with Rado, but let me say now that there are only consistency results to show that the theorem is best possible.

I was therefore surprised when shortly afterwards I received a letter from Paul (who was visiting Israel) claiming that Free$(\kappa, \lambda, n, \kappa)$ holds for $n < \omega$ and any uncountable limit cardinal $\kappa > \lambda$. The proof used GCH and I realized that it only worked for singular κ (the real theorem is for κ a singular strong limit cardinal, i.e. $2^\tau < \kappa$ for $\tau < \kappa$.) We now know that this is a special case of a general "canonization" theorem proved later with Rado (see 10.1), but which Paul discovered in at least two other interesting contexts before the general theorem was formulated. I wrote to tell him that I could not see how the proof works for regular limit κ. He replied by return of post that I was right, but the theorem was true since we may use the "measure hypothesis" from Erdős-Tarski, and he wrote down a proof that Free$(\kappa, \lambda, n, \kappa)$ holds for finite n and $\lambda < \kappa$, an uncountable measurable cardinal. I have to say that during my studies I had read the Erdős-Tarski paper, but either I skipped the footnotes or I did not recognize the significance of the remarks. However, after reading the letter, I did understand the strength of the hypothesis, and the same day proved that it implies Free$(\kappa, \lambda, < \omega, \kappa)$. Later, when the paper was actually written, we realized that the proof actually gave the stronger result that

$$\kappa \to (\kappa)_\lambda^{<\omega}$$

holds for $\lambda < \kappa$ if $\kappa > \omega$ is a measurable. This is perhaps one of our best-known joint theorems, and I will say more about this in the next section.

This brings me to our first joint oversight. Although our joint paper with Rado did not appear until 1965, I already had a weak form of the negative stepping-up lemma in 1957 (in fact it was because of this that we decided to write the triple paper on partition relations even though Erdős and Rado had already obtained a great many new unpublished results.)

I told Paul that the negative stepping-up gives us that

$$\kappa \not\to (\omega)_2^{<\omega} \;\Rightarrow\; 2^\kappa \not\to (\omega)_2^{<\omega}.$$

He immediately pointed out that it is easy "to go through" singular cardinals, and we put into the paper the remark that $\kappa \not\to (\omega)_2^{<\omega}$ holds for all κ less than the first strongly inaccessible cardinal $\kappa_0 > \omega$. We only realized later [35], after we learned the Hanf-Tarski results, that this almost trivially implies that

$$\kappa_0 \not\to (\omega+1)_2^{<\omega},$$

and therefore, by our theorem, it follows that κ_0 is not measurable. Unfortunately, this argument is not very strong, we could never make it work beyond the first fixed point in the sequence of inaccessibles.

10. Erdős cardinals and the strength of $\kappa \to (\lambda)_2^{<\omega}$

The real strength of the statement $\kappa \to (\lambda)_2^{<\omega}$ was discovered by the Berkeley school in the early 1960's. Dana Scott first proved that the existence of a measurable cardinal contradicts Gödel's axiom of constructibility (V=L). Soon afterwards Gaiffman and Rowbottom proved that it also implies that ω has only countably many constructible sets, and more generally Rowbottom proved that $\kappa \to (\omega_1)_2^{<\omega}$ implies that ω_1 is inaccessible in L.

Rowbottom also generalized our theorem with Paul. According to the Kiesler-Tarski paper [KT] it was Dana Scott who introduced the notion of a normal measure (a κ-complete $0-1$ measure on κ satisfying the pressing down lemma, i.e. if f is regressive on a subset of measure one, then it is constant on a set of measure one.) It was known that if κ is measurable then it carries a normal measure, and Rowbottom proved that, if there is a normal measure on κ and $f : [\kappa]^n \to \lambda$ $(n < \omega, \lambda < \kappa)$, then there is a homogeneous set of measure one. Of course, this immediately yields our theorem, and the proof actually becomes easier.

I happened to spend 1964 at Berkeley with Tarski's group and gave a course of lectures on Erdős-Rado set theory. I could not have been too successful as a lecturer, more than twenty people attended the first lecture, and in the end I was left with an audience of three — two students, Reinhardt and Silver, and a young assistant professor Donald Monk. I told them everything I knew about ordinary partition theorems and the little I knew about $\kappa \to (\lambda)_2^{<\omega}$. Silver apparently got interested and his thesis [Si], which appeared in 1966, contained some fantastic discoveries.

First he realized that the real strength of $\kappa \to (\lambda)_2^{<\omega}$ is that it yields, for any given structure on κ , a set of indiscernibles having order type λ (i.e a set of ordinals such that any two similarly ordered n-tuples satisfy the same formulas.) Using this he proved that the smallest κ satisfying $\kappa \to (\omega)_2^{<\omega}$ must be very large, for example there must be many weakly compact cardinals less than κ. He showed that, for $\alpha < \omega_1$, if $\kappa \to (\alpha)_2^{<\omega}$ holds, then it is true in L, and finally he proved that if $\kappa \to (\omega_1)_2^{<\omega}$ holds, then $O^{\#}$ exists, which means that L is very small, and this expresses the real strength of $\kappa \to (\omega_1)_2^{<\omega}$. I am not willing to write down the technicalities of this here.

Let me remind the reader that one of the few consequences of the axiom of constructibility which Gödel himself had noticed was that there is an uncountable analytic complement (the complement of a continuous image of a Borel set) which has no perfect subset. It was Solovay who proved that if $\kappa \to (\omega_1)_2^{<\omega}$ holds for some κ, then such a set cannot exist. This was the first application in descriptive set theory. It is stated in Descriptive Set Theory, the 1978 book of Moschovakis, that all applications of the existence of measurable cardinals in descriptive set theory come from a κ satisfying $\kappa \to (\omega_1)_2^{<\omega}$. All this shows that such cardinals deserve a special name, and the story I have written down shows that they are quite rightly called Erdős cardinals.

11. The early sixties. A long chapter

After 1956, Paul came home to visit his mother every year. He usually spent some months in Budapest where I also lived at that time. When he was at home I went to work with him at their apartment two or three times a week. Mrs Erdős was not only a devoted mother to Paul, but she was also an efficient secretary and would keep a record of his publications and look after his papers. When I visited them she would make us coffee and then leave us alone to "work". Our meetings had no prepared agenda, sometimes we went through earlier proofs, sometimes we had to read a manuscript or proof sheets, but the main point of our conversations was always the discovery of new problems and to start thinking about them. Paul was fantastically fast in both making and understanding proofs and finding the new questions. Though I usually made some notes, they were never quite satisfactory. We both needed to rely on our memories. This was quite a good fit, he always remembered the theorems and then I could scrape together the old proofs. I think now that these visits were real highlights in my life.

Now I have to change strategy. I cannot continue telling the results paper by paper, and in any case they were not proved in the order of publication. Starting around 1957 or 58, we agreed to write a triple paper with Rado on the partition calculus and the three of us set aside everything which we thought belonged there. Already in 1960 I visited Rado in Reading to work on the triple paper, carrying with me an almost completed manuscript. So, in this long chapter I will open subsections about the results of these years with an indication of where they appeared.

11.1 Canonization

Let $f : [X]^r \to \gamma$ be an r partition of length γ of X. Let $\langle Y_\alpha : \alpha < \varphi \rangle$ be a sequence of disjoint subsets of X, $Y = \bigcup_{\alpha<\varphi} Y_\alpha$. For a subset $v \in [Y]^r$

there is a number $s(v) \leq r$ an increasing sequence $\alpha(v) = \langle \alpha_i(v) : i < s(v) \rangle$ of ordinals and a sequence $r(v) = \langle r_i(v) : i < s(v) \rangle$ of integers, defining the position of v in the partition $Y = \bigcup_{\alpha<\varphi} Y_\alpha$, so that $\sum_{i<s(v)} r_i(v) = r$ and $|v \cap Y_{\alpha_i(v)}| = r_i(v)$ for $i < s(v)$. Two r-element sets $v, v' \in [Y]^r$ have the same position if $\alpha(v) = \alpha(v')$ and $r(v) = r(v')$. f is *canonical with respect to the sequence* $(Y_\alpha : \alpha < \varphi)$ if for any two v, v' having the same position

$$f(v) = f(v')$$

The "canonization" theorem of [43] tells us that: *there is an integer k_r such that whenever $\langle X_\alpha : \alpha < \varphi \rangle$ is a sequence of subsets of X with fast enough increasing cardinalities,*

$$|X_\alpha| > \exp_{k_r}(|\bigcup_{\beta<\alpha} X_\beta|),$$

then there is a disjointed sequence $Y_\alpha \subset X_\alpha$ $(\alpha < \varphi)$ also with fast increasing cardinalities, $|Y_\alpha| > |\bigcup_{\beta<\alpha} X_\beta|$, such that $f : [X]^r \to \gamma$ is canonical with respect to the sequence $\langle Y_\alpha : \alpha < \varphi \rangle$, provided $\gamma, \varphi < |X_0|$.

One corollary of this is the following. Assume κ is a singular strong limit cardinal then $\kappa \to (\kappa, \lambda_\nu)^2_{\nu<\gamma}$ iff $\mathrm{cf}(\kappa) \to (\mathrm{cf}(\kappa), \lambda_\nu)^2_{\nu<\gamma}$. The 'only if' part comes easily using a canonical partition, and the 'if' part uses the canonization theorem. The reader should remember the Erdős; Dushnik, Miller theorem $\kappa \to (\kappa, \aleph_0)^2$. Now the above result tells us, at least with GCH, for which singular cardinals κ the relation $\kappa \to (\kappa, \aleph_1)^2$ holds. For example, $\aleph_{\omega_1} \not\to (\aleph_{\omega_1}, \aleph_1)^2$ but $\aleph_{\omega_2} \to (\aleph_{\omega_2}, \aleph_1)^2$. By now I do not really have to tell the reader that this is the form discovered by Paul.

It would be nice to have a necessary and sufficient condition for the case of arbitrary singular κ. We knew that for a singular κ, say with $\mathrm{cf}(\kappa) = (2^{\aleph_0})^+$, to have $\kappa \to (\kappa, \aleph_1)^2$ it is necessary to have $\lambda^{\aleph_0} < \kappa$ for $\lambda < \kappa$. We repeatedly asked if this is sufficient. It was proved by Shelah and Stanley in the eighties that this is consistently false [SS1].

When preparing the material of our book with Attila Máté and Richard Rado, [100] where we tried to discuss the ordinary partition relation is ZFC, we isolated the following problem. Assume there is an increasing sequence of integers $n_k : k < \omega$ such that

$$\aleph_\omega < 2^{\aleph_{n_0}} < \ldots < 2^{\aleph_{n_k}} < \ldots$$

Does it follow that $2^{<\aleph_\omega} \to (\aleph_\omega)^2_2$ holds? Clearly our canonization does not work in this case. Shelah proved this with a new type of canonization theorem [S1], and parts of his results are given in the book. Further uses of "canonization" will be mentioned later.

One last remark. It is interesting to see how combinatorial ideas do pop up in different topics. When Shelah obtained with such miraculous speed his celebrated result on the bound in Van der Waerden's theorem, he was already the best expert on canonization, and one of the main lemmas in his proof is indeed a (finite) canonization theorem.

11.2 Square brackets

Sierpiński proved $2^{\aleph_0} \not\to (\aleph_1)^2_2$ by well ordering the continuum and defining a partition of the pairs into two classes, so that a pair ordered in the same way in both the natural ordering and the well-ordering belongs to the first class.

Paul told me that he formulated a generalization of this in 1956 with the following question. Can one split the pairs of reals into three classes so that every subset of size $\aleph_1$ (or $2^{\aleph_0}$) contains a pair from each class?

He soon proved this assuming CH. We discovered that whenever a partition relation fails, one can ask for a corresponding weaker property, and in [43] we introduced the following square bracket relation

$$\kappa \to [\lambda_\nu]^r_{\nu<\gamma}.$$

This means that for every $f : [\kappa]^r \to \gamma$ there is a $\nu < \gamma$ and a subset $H \subset \kappa$, $|H| = \lambda_\nu$ so that f does not take the value ν on the r-tuples of H.

It is worthwhile to formulate separately the negation of the "balanced" form of this (when all the λ_ν are equal). Thus $\kappa \not\to [\lambda]^r_\gamma$ means that there is an $f : [\kappa]^r \to \gamma$ such that all subsets of size λ are completely inhomogeneous i.e. f takes all possible values on the r-tuples of any set of size λ.

Clearly we needed some test cases. We proved that $2^\kappa = \kappa^+$ implies $\kappa^+ \not\to [\kappa^+]^2_{\kappa^+}$, and only later did we realize that this was also known to Sierpiński in a different context. But probably the nicest result was the following:

If κ is a strong limit cardinal of cofinality ω, then $\kappa \to [\kappa]^2_3$.

(Note that $\kappa \not\to (\kappa, (cf(\kappa))^+)^2$ and $\kappa \not\to [\kappa]^2_2$ is a trivial corollary.) This follows from the "canonization" theorem of the previous section. Indeed it gives a stronger result. Under the above conditions on κ, for every $f : [\kappa]^2 \to \gamma$, $\gamma < \kappa$, there is a set $H \in [\kappa]^\kappa$ such that f takes at most two different values on the pairs of H.

So we introduced a third symbol, the strong square bracket. Let γ, δ be cardinals. $\kappa \to [\lambda]^r_{\gamma,\delta}$ ($\kappa \to [\lambda]^r_{\gamma,<\delta}$) means that for every r-partition $f : [\kappa]^r \to \gamma$ with γ-colors, there is a subset $H \subset \kappa$ of size λ such that f takes at most δ (fewer than δ) values on the r-tuples of H.

So the above theorem says that $\kappa \to [\kappa]^2_{\gamma,2}$ for singular strong limit κ of cofinality ω and $\gamma < \kappa$. We used this symbol to ask if $\aleph_2 \to [\aleph_1]^2_{\aleph_1,\aleph_0}$? Paul thought this was an old question of Ulam, but later we discovered that it is equivalent to a well-known model theoretical conjecture of C.C. Chang.

In §12, I will discuss the effect of our 1967 problem paper [67], but this is a good place to write down the present status of some of the square bracket problems stated in that paper. Let me begin with an innocent but very nice result of Fred Galvin

$$\eta \to [\eta]^2_3.$$

Many generalizations of this were published later. Galvin and Shelah proved $2^{\aleph_0} \not\to [2^{\aleph_0}]^2_{\aleph_0}$ and $\mathrm{cf}(2^{\aleph_0}) \not\to [\mathrm{cf}(2^{\aleph_0})]^2_{\aleph_0}$ in 1968 [**GS**], they also proved some weak results like $\aleph_1 \not\rightarrow [\aleph_1]^2_4$ and $\aleph_1 \not\rightarrow [\aleph_1]^3_{\aleph_1}$ for the case when the underlying set has cardinality $\aleph_1$.

Again we had a false feeling. Although we did not state it explicitly, we clearly believed that $\aleph_1 \not\rightarrow [\aleph_1]^2_{\aleph_1}$ could not be proved in ZFC. But in 1987 Stevo Todorčević proved us wrong [T1]. He proved in ZFC that

$$\kappa^+ \not\rightarrow [\kappa^+]^2_{\kappa^+}$$

holds for every regular κ. This was extended by Todorčević and Shelah for more successors, and inaccessibles. This was certainly one of the most significant discoveries in set theory in the eighties requiring entirely new methods. I will come back to this for a moment in the next section. However, this still leaves open the question whether

$$2^{\aleph_0} \not\rightarrow [\aleph_1]^2_3$$

holds? Shelah [S2] proved that $2^{\aleph_0} \rightarrow [\aleph_1]^2_3$ is really consistent with ZFC. In his model $2^{\aleph_0}$ is quite large so it is still possible, but quite unlikely, that

$$2^{\aleph_0} = \aleph_2 \Rightarrow \aleph_2 \not\rightarrow [\aleph_1]^2_3.$$

11.3 Jónsson algebras–negative relations with infinite exponents

A Jónsson algebra is an infinite algebra A with countably many finitary operations such that all proper subalgebras have cardinality strictly less than $|A|$. The question is, for what infinite cardinals κ is there a Jónsson algebra of cardinality κ ? I mention this here because of a connection with the square brackets. As pointed out by Shelah much later in the game, there is a Jónsson algebra on κ iff $\kappa \not\rightarrow [\kappa]^{<\omega}_{\kappa}$ holds.

I heard the problem from Tarski in 1964 and when I returned to Hungary and met Paul, we immediately had some remarks about this which we published in [45]. First we proved that if there is a Jónsson algebra on κ, then there is also one on κ^+, and hence there is one on $\aleph_n$ for $n < \omega$. We also proved that $2^\kappa = \kappa^+$ implies that there is a Jónsson algebra on κ^+ since we knew that $2^\kappa = \kappa^+ \Rightarrow \kappa^+ \not\rightarrow [\kappa^+]^2_{\kappa^+}$.

I must also mention that it was already proved by Kiesler and Rowbottom that there is a Jónsson algebra on every κ if $V = L$ [KR].

It was a metatheorem for the two of us because of Rado's theorem that "nothing is true for infinite exponents". So we proved already in [24] that Free$(\kappa, 2, \aleph_0, \aleph_0)$ fails for every κ and in [43] we strengthened Rado's result to $\kappa \not\rightarrow [\aleph_0]^{\aleph_0}_{2^{\aleph_0}}$. In this spirit we also proved in the Jónsson algebra paper that there is an infinitary Jónsson algebra on every κ, in other words

$$\kappa \not\rightarrow [\kappa]^{\aleph_0}_{\kappa}$$

This became one of our best used theorems. Kunen used it for a simple proof of his famous theorem that there is no nontrivial elementary embedding of the universe into itself (disproving the hoped for existence of Reinhardt cardinals) and Solovay used it in his proof that GCH holds at every singular strong limit cardinal above a strongly compact cardinal.

Our "metatheorem" is not quite true since Foreman and Magidor [FM] recently proved that it is consistent that $\aleph_3 \to [\aleph_2]^{\aleph_0}_{\aleph_2}$. It goes without saying that Erdős always assumed the axiom of choice, and I would not even mention this except that it happens that partition relations with infinite exponents may be true if we do not assume the axiom of choice, and indeed they became an important tool of set theory e.g. in investigations concerning the Axiom of Determinacy and its consequences to descriptive set theory.

Back to the previous chapter, Shelah [S5] recently published quite a few theorems extending the class of cardinals κ for which there is a Jónsson algebra and then later proving the stronger result $\kappa \not\to [\kappa]^2_\kappa$. I presently do not know of any instance of the result where $\kappa^+ \not\to [\kappa^+]^{<\omega}_{\kappa^+}$ is true but $\kappa^+ \not\to [\kappa^+]^2_{\kappa^+}$ is not known.

11.4 Negative stepping-up

This result published in [43] says that, if $r \geq 2$, $\kappa \geq \omega$ and $\kappa \not\to (\lambda_\nu)^r_{\nu<\gamma}$, then

$$2^\kappa \not\to (\lambda_\nu + 1)^{r+1}_{\nu<\gamma}$$

provided the sequence λ_ν *satisfies certain simple conditions.* The simplest of these is that two of them are infinite and one of them is regular. There are about six more conditions to cover relevant cases. These conditions become less restrictive as r grows, and there is no condition at all for $r \geq 5$.

But even the one just stated tells us that the Erdős-Rado theorem of §7 is best possible, i.e.

$$\exp_{n-1}(\kappa) \not\to (\kappa^+)^n_2$$

for $n \geq 2$, since the result $2^\kappa \not\to (\kappa^+)^2_2$ can be lifted by induction on n.

Let me state another example. We know that if $\kappa \not\to (\kappa)^2_2$ then $\kappa \not\to (\kappa, 4)^3$. This should imply $2^\kappa \not\to (\kappa, 5)^4$, but to get this, a special argument is needed say if κ is singular.

Maybe the negative stepping up is true without any conditions at all on the λ_ν, but to the best of my knowledge this is still wide open. There are cases where we do not know what happens without GCH for $n = 2$. Let me explain this with an example. It is very easy to see that $\aleph_\omega^{\aleph_0} \not\to (\aleph_{\omega+1}, (\aleph_0)_{\aleph_0})^2$, but this should still be true if the $\aleph_0$ entries are replaced by 3's, and indeed we did prove this with GCH

$$\aleph_{\omega+1} \not\to (\aleph_{\omega+1}, (3)_{\aleph_0})^2.$$

To stick my neck out again, it seems inconceivable to prove this in ZFC, but no consistency proofs are known in the other direction.

Now a trivial canonization lifts this say to the first singular cardinal with cofinality $\aleph_{\omega+1}$ i.e. to $\aleph_{\omega_{\omega+1}} \not\rightarrow (\aleph_{\omega_{\omega+1}}, (3)^2_{\aleph_0})$ and this should be stepped up to

$$\aleph_{\omega_{\omega+1}+1} \not\rightarrow (\aleph_{\omega_{\omega+1}}, (4)_{\aleph_0})^3.$$

Unfortunately in this case, for $r = 2$, only one of the entries is infinite and even that is singular. So we had nothing to cover this case and it was stated as one of our open problems for a long time. I thought Shelah and Stanley had a proof of this $\not\rightarrow$ from GCH, but I understand that it is still open.

A more significant problem is that our result suggests a negative stepping-up, for square brackets and set mappings as well.

It was recognized early in the game that for square brackets this is consistently false even assuming GCH. For example, $2^{\aleph_0} = \aleph_1 \Rightarrow \aleph_1 \not\rightarrow [\aleph_1]^2_{\aleph_1}$, but it is very easy to see that $\aleph_2 \not\rightarrow [\aleph_1]^3_{\aleph_1}$ implies $\aleph_2 \not\rightarrow [\aleph_1]^2_{\aleph_1,\aleph_0}$, the negation of Chang's conjecture, which was proved to be consistent in an early paper of Silver.

Stevo Todorčević worked out stepping-up methods from combinatorial principles known to hold in L, which do give the stepping-up for square brackets and for set mappings in most cases. See [T2] and also [HK1] for more history.

Let me conclude this chapter with two more interesting recent results of Todorčević which show the present direction of research in this area. He proved that $\aleph_2 \rightarrow [\aleph_1]^3_{\aleph_1}$ is equivalent to Chang's conjecture in ZFC (without assuming CH), and $\aleph_2 \not\rightarrow [\aleph_1]^3_{\aleph_0}$ is true in ZFC [T3]. See also [57]. This is a very deep result, but Erdős had a hand in initiating of this type of theorem as well, in [81] we remarked that the stepping-up method yields $2^{\aleph_1} \not\rightarrow [\aleph_1]^3_4$ in ZFC.

11.5 Polarized partition relations

While working on the triple paper [43], we had to draw the line somewhere, and we decided that we will only include results for polarized partitions of the form

$$\begin{pmatrix} \kappa \\ \lambda \end{pmatrix} \rightarrow \begin{pmatrix} \kappa_0 \\ \lambda_0 \end{pmatrix}, \begin{pmatrix} \kappa_1 \\ \lambda_1 \end{pmatrix}^{1,1}$$

and we gave a number of results assuming GCH. Some of the results inherent in the methods were only stated in the second problem paper [81]. But the simplest problem we isolated was if

$$2^{\aleph_0} = \aleph_1 \Rightarrow \begin{pmatrix} \aleph_2 \\ \aleph_1 \end{pmatrix} \rightarrow \begin{pmatrix} \mu & \nu \\ \aleph_1 & \aleph_1 \end{pmatrix}^{1,1}$$

holds for $\aleph_0 \leq \mu, \nu \leq \aleph_1$.

One of the first results proved after our problem list became public was due to Karel Prikry [P]. I state a special case:

$$\begin{pmatrix} \aleph_2 \\ \aleph_0 \end{pmatrix} \not\to \begin{pmatrix} \aleph_0 \\ \aleph_1 \end{pmatrix}^{1,1} \text{ or even } \begin{pmatrix} \aleph_2 \\ \aleph_0 \end{pmatrix} \not\to \begin{bmatrix} \aleph_0 \\ \aleph_1 \end{bmatrix}^{1,1}_{\aleph_1}$$

is consistent with ZFC and GCH.

Later, Richard Laver [L] proved that relative to a very large cardinal it is consistent with GCH that there is an ω_1 complete ideal I on ω_1 having the following strong saturation property: Given $F \subset I^+$, the complement of I, $|F| = \aleph_2$ (i.e. $\aleph_2$ large subsets of $\aleph_1$) there is an $F^1 \subset F$, $|F^1| = \aleph_2$, such that the intersection of any countably many sets in F^1 is in I^+. This easily yields

$$\begin{pmatrix} \aleph_2 \\ \aleph_1 \end{pmatrix} \to \begin{pmatrix} \aleph_1 \\ \aleph_1 \end{pmatrix}^{1,1}_2 \text{ even } \begin{pmatrix} \aleph_2 \\ \aleph_1 \end{pmatrix} \to \begin{pmatrix} \aleph_1 \\ \aleph_1 \end{pmatrix}^{1,1}_{\aleph_0},$$

and it was one of the first corollaries of Jensen's morasses, that Prikry's result holds in L.

For lack of space and time, we did not include polarized partitions in the book [100] so there is no comprehensive account in the literature about the recent results. Let me state one problem of the form $\begin{pmatrix} \aleph_{\alpha+1} \\ \aleph_\alpha \end{pmatrix} \to (\,\cdot\,)^{1,1}$ which is unsolved and for which there are no consistency results either. Does GCH imply

$$\begin{pmatrix} \aleph_{\omega_1+1} \\ \aleph_{\omega_1} \end{pmatrix} \to \begin{pmatrix} \aleph_{\omega_1} \\ \aleph_{\omega_1} \end{pmatrix}^{1,1}_2 ?$$

A small hope here is an unpublished remark of Shelah from 1989. *Assume* $\langle \kappa_\alpha : \alpha < \omega_1 \rangle$ *is an increasing sequence of measurable cardinals,* $\kappa = \sup_\alpha \kappa_\alpha$ *and* $2^\kappa = \kappa^+$ *then* $\begin{pmatrix} \kappa^+ \\ \kappa \end{pmatrix} \to \begin{pmatrix} \kappa \\ \kappa \end{pmatrix}^{1,1}_2$ *holds.*

Added in proof (March 1995). In September 1994, Shelah proved the following striking result. Assume $\kappa > cf(\kappa)$, κ is strong limit and $2^\kappa > \kappa^+$. Then

$$\begin{pmatrix} \kappa^+ \\ \kappa \end{pmatrix} \to \begin{pmatrix} \kappa \\ \kappa \end{pmatrix}^{1,1}_\tau \quad \text{holds for } \tau < \kappa .$$

Writing up the second problem paper [81], I realized that our theorem in [43] yielding $\begin{pmatrix} \aleph_2 \\ \aleph_2 \end{pmatrix} \to \begin{pmatrix} \aleph_1 \\ \aleph_1 \end{pmatrix}^{1,1}_2$ from CH can be generalized to give

$$2^{\aleph_0} = \aleph_1 \Rightarrow \begin{pmatrix} \aleph_2 \\ \aleph_2 \end{pmatrix} \to \begin{pmatrix} \aleph_1 \\ \aleph_1, \end{pmatrix}^{1,1}_3$$

but it is consistent with CH that

$$\begin{pmatrix} \aleph_2 \\ \aleph_2 \end{pmatrix} \not\to \begin{pmatrix} \aleph_1 \\ \aleph_1 \end{pmatrix}^{1,1}_4$$

holds. A recent still unpublished result of J. Baumgartner using a new kind of argument, says that assuming CH (but no more of GCH)

$$\begin{pmatrix} \aleph_3 \\ \aleph_2 \end{pmatrix} \rightarrow \begin{pmatrix} \aleph_1 \\ \aleph_1 \end{pmatrix}^{1,1}_{\aleph_0}$$

holds.

11.6 Property B and incompactness

Our second major joint paper [33] is about the following property of families of sets, F: *There is a set B which meets every element of F but does not contain any member of F as a subset.* This means, in a terminology introduced later, that the chromatic number of F is two. Before stating some results I want to tell how we came across property **B**. Property **B** was actually discovered by Felix Bernstein in 1908. He proved that for every $\kappa \geq \omega$, if F is a family of size κ of sets of size κ, then F has this property. (He used it to get a subset $B \subset I\!R$, $|B| = |I\!R \backslash B| = 2^{\aleph_0}$ and such that neither B nor $I\!R \backslash B$ contains a perfect subset of $I\!R$.

In those years I often visited Erdős at the summer house of the Academy in Mátraháza (a summer resort in the mountains), where he used to spend part of the summer with his mother. The place was reserved for members of the Academy and I was still young, so I had to find a place in the village for a couple of days. But I did get decently fed in the summer house during the day time. Usually there were other visitors or regular inhabitants to also work with Paul, and he would do this simultaneously. He led his usual life there, alternately proving, conjecturing, playing chess, ping pong, bridge, or walking to mountain tops. It was his habit to stop playing abruptly, when the rest of us were warming up to the game, and to return to work. In those days he went to bed around ten o-clock, but he woke up early, between four or five in the morning, so it was actually safer for me not to be living too close.

There was a vague plan to write a book on set theory and I arrived with a number of old journals. One of them was the 1937 volume of the Comptes Rendus Varsowie containing a long paper of Tarski, "Ideale in Vollständigen Mengenkörper", in which we wanted to find something for the planned book. Erdős volunteered to look it up. I had something else to do and I left him alone for a while. When I returned, he was excitedly reading. But not Tarski's paper, it was a forgotten paper [Mil] of an American set theorist E.W. Miller, which was next to Tarski's paper in the same volume. (Yes, the same as in Dushnik-Miller.) Miller proved that, for n finite, if F is a family of infinite sets and any two members of F intersect in at most n elements, then F has property **B**. "Reading" meant reading the statements and trying to figure out the proofs. After a while, I gave up and started reading the paper in detail.

The proof was by a cardinal induction on $\kappa = |F|$, the size of the family, and for a given κ, the underlying set was split into the increasing continuous union of κ smaller sets $\{A_\alpha : \alpha < \kappa\}$, each A_α being closed with respect to certain operations. In this case, for each $n+1$ element set, there is at most one set containing it, and the elements of this (possibly non-existent) set were the values of these operations. Then the induction hypothesis was applied to the families $F|A_\alpha$. This is called nowadays, the method of elementary chains. Miller actually proved that F has the stronger property $\mathbf{B}(< \aleph_0)$, i.e. there is a set B which meets each element A of F in a finite set.

Paul's only comment was: "You see there are still things we do not know," and before we actually read all the details, he started to ask questions. What if the sets are only almost disjoint (have finite intersections)? There is a counter-example on the second page of Miller's paper, and I tried to return to the details. "Yes" he said, "but we should then assume that the sets are bigger." So, instead of collecting data for the book, we wrote a long paper.

Let me state a special case of one of the main results: *Assume GCH. If F is a family of very strongly almost disjoint sets of size $\aleph_2$, i.e. $|A \cap B| < \aleph_0$ for $A \neq B \in F$, then F has property* $\mathbf{B}(< \aleph_2)$. More importantly, if F *consists of sets of size $\aleph_1$ just strongly almost disjoint, i.e. $|A \cap B| < \aleph_0$ for $A \neq B$ in F, then F still has property* $\mathbf{B}(< \aleph_1)$ *provided* $|F| \leq \aleph_\omega$.

The reason why the proof broke down for $\aleph_{\omega+1}$ was quite clear. In the generalization of Miller's proof we had to use infinitary operations, and alas $\aleph_\omega^{\aleph_0}$ is greater than $\aleph_\omega$ no matter what we assume.

We both felt that this is a real hard-core problem and we tried to find other methods. In doing so we formulated the following statement $(*)$: *For $\alpha < \aleph_{\omega+1}$ there is a partition of $\alpha = \cup_{n<\omega} S_{\alpha,n}$ into countably many pieces such that $|S_{\alpha,n}| \leq \aleph_n$ and for any $\alpha < \aleph_{\omega+1}$ with $cf(\alpha) = \omega_1$ there is an increasing sequence $(\alpha_\nu : \nu < \omega_1)$ of ordinals $\alpha_\nu \to \alpha$ such that for each $n < \omega$ the sequences $\{S_{\alpha_\nu,n} : \nu < \omega_1\}$ are increasing as well.*

Of course, we could not prove this, but we could deduce from it the theorem for $|F| = \aleph_{\omega+1}$.

Later a young German set theorist W. Donder pointed out that our statement is an easy corollary of Jensen's $\square_{\aleph_\omega}$ and as a corollary of this statement and some obvious generalizations, the theorem for families of sets of size $\aleph_1$ is true in L.

In 1986, in a paper with Juhász and Shelah [HJS], we proved that *it is consistent, relative to a super compact cardinal, that there is a family of size $\aleph_{\omega+1}$ of strongly almost disjoint sets of size $\aleph_1$ not having property* $\mathbf{B}$ *and also GCH holds in the model.*

Paul was of course immediately asking if in Miller's theorem $\mathbf{B}(< \aleph_0)$ can be replaced by $\mathbf{B}(k)$ with some $k < \omega$. Let us consider families F of countably infinite sets, such that for any two $A \neq B \in F$, $|A \cap B| \leq n < \omega$. First we proved that *for countable families F, F has property* $\mathbf{B}(n+1)$ *but not necessarily* $\mathbf{B}(n)$. Then much to our surprise, we proved using GCH that, if

$|F| = \aleph_k$, $k < \omega$ *then* F *must have property* $\mathbf{B}((k+1)n+1)$ *but not necessarily* $\mathbf{B}((k+1)n)$. The reason for the surprise was, that these were strong incompactness results saying that there is a family of size $\aleph_{k+1}$ of countable sets not having property $\mathbf{B}((k+1)n+1)$ but every subfamily of size $\aleph_k$ has this property and such incompactness results were not then in the literature (but we already knew of the Hanf-Tarski result by the time we finished the paper). At the end of the paper we gave a long list of incompactness problems for $\aleph_2$ which were later solved by different authors. Eventually, Paul's persistent interest in these problems led to Shelah's celebrated compactness theorem for singular cardinals [S3].

I just state here one of the problems, the fate of which I will describe in § 10.9. *Does there exist a graph on* $\aleph_2$ *vertices having uncountable chromatic number, such that all subgraphs of size* $\aleph_1$ *are at most* $\aleph_0$ *chromatic?*

Finally, let me mention that due to the interest of Paul, property **B** had an even bigger career in finite combinatorics. But fortunately, this is not the subject of this note.

11.7 Chromatic number

In our paper [42] we discovered r-shift graphs. [42] is "Some remarks on set theory IX". Its subject is a general problem involving reals, so I hope it fits into Komjáth's paper. But I must mention that a few years ago, Fremlin and Talagrand obtained some very interesting results solving most of the problems stated there [FT].

The vertices of the κ, r-shift graph $G(\kappa, r)$, $2 \le r < \omega$ are the r-tuples of κ or rather the increasing sequences $\{\alpha_0, ..., \alpha_{r-1}\}$, $\alpha_0 < ... < \alpha_{r-1} < \kappa$ and we join $\{\alpha_0, \alpha_1, ..., \alpha_{r-1}\}$ and $\{\alpha_1, \alpha_2, ..., \alpha_r\}$. We proved, as a corollary of Ramsey's theorem or the Erdős-Rado theorem, that these graphs have large chromatic number and that they do not contain odd circuits of length less than $r+2$.

It was an early result of finite graph theory that there exist graphs having large chromatic number and not containing a K_3 (see [46] for historical references). I think $G(n, 2)$ is the simplest example of this and we were both surprised that this was not known earlier. Paul was always interested in this problem. He proved in 1959 using his probability method that for all $k < \omega$ and $r < \omega$ there are graphs of chromatic number $\ge k$ and of girth $\ge r$ (not containing circuits of length $< r$). He was always interested in infinitary generalizations and in [22] he proved with Rado that, for $\kappa \ge \omega$ there is a K_3-free graph on κ of chromatic number κ.

These results again suggested the wrong generalization, but this time we were not defeated. In [46] we proved that a graph not containing C_4 (a circuit of length 4) has chromatic number at most $\aleph_0$. In fact, we proved a much stronger result. We defined col(G) the coloring number of G as the smallest cardinal κ such that the vertex set of G has a well ordering such that for each vertex x the number of edges having x as the larger element is smaller that κ. This concept was later introduced in finite combinatorics under a different name as well (G is k-degenerate if $\mathrm{col}(G) \le k+1$) [Bo]. Obviously, $chr(G) \le \mathrm{col}(G)$ and we proved that *if G does not contain a complete bipartite graph $K_{k,\aleph_1}$ for every $k < \omega$ then* $\mathrm{col}(G) \le \aleph_0$. We used the cardinal induction method described in the previous section. Again, the problem arose, what can be said if only larger complete bipartite graphs are excluded? Let me again state a special case of our result. *Assume GCH. If G does not contain a $K_{\aleph_0,\aleph_3}$ then $chr(G) \le \aleph_2$ and if G does not contain a $K_{\aleph_0,\aleph_2}$ then $chr(G) \le \aleph_1$ provided $|G| \le \aleph_\omega$* .

The situation is analogous to the one described in the previous chapter. It is consistent that the second clause of the theorem is true for every G e.g. if $V = L$, and it is consistent (relative to a super compact cardinal) that the result strongly fails for $\aleph_{\omega+1}$, i.e. there exists a graph G on $\aleph_{\omega+1}$ of chromatic number $\aleph_2$ not containing a $K_{\aleph_0,\aleph_0}$. This was shown in our paper with Juhász and Shelah mentioned in 10.6. The construction of this example from the one described there is a combinatorial argument, which uses that in the model we have many instances of $\diamondsuit$ (the diamond principle).

I have to mention that we also introduced generalized Specker graphs to show that for $\kappa \ge \omega$ there are graphs of *size* κ , with chromatic number κ having large odd girth.

There are quite a few generalizations of our theorem for $\mathrm{col}(G) > \aleph_0$ but I do not state these here, instead I offer the references [K1] [HK3] and [101]. Let me mention one typical Erdős question: Does $chr(G) > \aleph_0$ imply that G contains all large odd circuits, say of length $2k+1$ for $k \ge k_0$ for some k_0. Note that this is a typical problem where it is the chromatic number that has to be large as $\mathrm{col}(K_{\aleph_0,\aleph_1}) = \aleph_1$. Later we proved this with Shelah in [79].

Rado asked if the de Bruijn-Erdős compactness theorem for finite chromatic number extends to finite coloring numbers. As the definition of the coloring number involves a well-ordering this can not be expected. Indeed we disproved it, but a surprising result of [46] is that still there is a uniform bound. We proved: *If* $\mathrm{col}(G') \le k$ $(2 \le k < \omega)$ *for every finite subgraph G' of G then* $\mathrm{col}(G) \le 2k-2$, and there is a countable graph to show that this is best possible for each k.

There is an important finite theorem hidden at the end of [46]. We proved there, using the probabilistic method, that *for every r, s, k, there are r-uniform hypergraphs of chromatic number greater than k and girth greater than s*. In fact, defined on some n-element set, they do not contain an independent set of size n^{1-d} for some $d > 0$. This fitted logically into the line of thought of [46] and it did not occur to us that no finite combinatorialist will look at, much less read, a forty page paper full of alephs, to find an interesting probabilistic argument on the thirty fifth page.

11.8 Another miss

We first met Eric Milner in 1958 at the IMC meeting in Edinburgh. He was a former student of Rado and was working in Singapore. Rado interested him in partition problems and he settled one of their problems about countable ordinals [M1]. That was enough to induce Paul to visit and work with him in Singapore in 1960. He returned from there to Budapest with a new interesting problem which I solved and this began a long collaboration between the three of us. The Milners' returned to England in 1961 and Eric joined Rado at Reading. On my way back from Berkeley to Budapest in 1965, I stayed in Reading for a month with them discussing a long half-finished manuscript. Although our long joint papers only appeared a few years later, in 1965 we were already deeply involved in our joint work and we thought it would probably help if all three of us could be together at the same place and at the same time. So it was arranged that Eric should visit us during the summer of 1965 to spend a week at the summer house of the Writer's Union in Szigliget on Lake Balaton. Eric arrived with an interesting question about transversals, and as a consequence, instead of regularly working on manuscripts, we wrote another shorter paper [56] which became our first joint work to appear. As a side issue in that paper we proved the following theorem: *Let* $\lambda > \mathrm{cf}(\lambda) = \kappa > \omega$, *and let* λ_α $(\alpha < \kappa)$ *be an increasing continuous sequence of cardinals cofinal in* λ, *and assume that* $\tau^\kappa < \lambda$ *for* $\tau < \lambda$. *If* S *is a stationary subset of* κ *and* $\mathcal{F} \subset \prod_{\alpha \in S} \lambda_\alpha$ *is an almost disjoint set of transversals (i.e.* $|\{\alpha \in S : f(\alpha) = g(\alpha)\}| < \kappa$ *for* $f \neq g \in \mathcal{F}$), *then* $|\mathcal{F}| \leq \lambda$.

Eric made notes of our results and wrote it up and the paper appeared in 1968. We forgot the whole thing, and the paper seems to have gone unnoticed. Even in 1967 when we wrote the first problems paper with Paul, where our intention was to write down all our interesting problems, we omitted any mention of this. However, it seems we were not the only blind ones. During the summer of 1971 Adrian Matthias organized a large conference on set theory in Cambridge, England. Karel Prikry was one of the invited speakers and he gave a talk on a generalization of Jensen's work on Kurepa families. He discovered the following result: *Under the assumptions of our theorem, if* $\mathcal{H} \subseteq \mathbb{P}(\lambda)$ *is a Kurepa family in the sense that* $|\mathcal{H}|\lambda_\alpha| \leq \lambda_\alpha$ *for* $\alpha \in S$ *(*S *a stationary subset of* κ*), then* $|\mathcal{H}| \leq \lambda$. ($\mathcal{H}|\lambda_\alpha = \{H \cap \lambda_\alpha : H \in \mathcal{H}\}$.)

He told me this result the day before his lecture and it sounded vaguely familiar. But it took me the whole day to realize that this was just our earlier theorem applied to the sets $\mathcal{H}|\lambda_\alpha$ in place of λ_α. I managed to get a copy of our paper to give to Prikry before the lecture. Now there were about one hundred set-theorists in attendance, including all the leading ones, when Karel stated our result in a totally digestible form. But nobody asked, what happens if we replaced λ_α by λ_α^+? I suppose the psychological barrier was too strong. In 1974, just before the ICM in Vancouver, I was visiting Eric again in Calgary (he moved there in 1967), when I received a preprint of Silver's ingenious discovery that : *if* $\lambda > \mathrm{cf}(\lambda) > \omega$ *and if* $2^\tau = \tau^+$ *on a stationary set of cardinals* $\tau < \lambda$, *then* $2^\lambda = \lambda^+$. At the same time I received a preprint from Prikry giving a combinatorial proof of Silver's result. Prikry told me that the instant he saw Silver's manuscript it dawned on him that the only thing needed was to lift our old result with λ_α^+ in place of λ_α. Of course this requires a non-trivial argument. Baumgartner and Jensen also found elementary proofs of Silver's result without remembering our theorem. But the real miss, and so uncharacteristic of Paul, was not to have asked the question!

11.9 Incompactness for the Chromatic Number

In 1966, assuming CH, we solved the problem on chromatic numbers stated in §10.6. We proved in [54] that there is a graph of chromatic number at least $\aleph_1$ on $(2^{\aleph_0})^+$ vertices all of whose subgraphs of cardinality at most $2^{\aleph_0}$ have chromatic number at most $\aleph_0$. This also comes with a story and some advice. During a working session at Paul's apartment, we were talking about something totally unrelated to chromatic number and compactness. In the middle of an attempted proof, we found that the pairs of $\mathbb{R}$ are colored with countably many colors and our proof would be finished if there was a monochromatic increasing path of length 2, IP_2, i.e. a triple $x_1 < x_2 < x_3$ with $\{x_1, x_2\}$ and $\{x_2, x_3\}$ having the same color. Unfortunately there was not, and I tried to get another proof. But Paul started to insist that we should know for what order types θ,

$$\theta \to (IP_2)^2_\omega$$

holds. We parted unsuccessful in both attempts. But on the way home, I could not help thinking about his question. I remembered an old idea of Sierpiński which easily implied that there is a $\not\to$ for every θ of cardinality $|\theta| \le 2^{\aleph_0}$. Then I saw in a flash that this just says that all subgraphs of size $2^{\aleph_0}$ of our shift graph $G((2^{\aleph_0})^+, 2)$ have chromatic number $\le \aleph_0$. As this was a hundred dollar problem, I immediately called Paul when I arrived home. (I think eventually I got only \$50 for it, but with some reason.) The advice is this: just answer his questions, you have time later to ponder if it is important or not.

This was the status of the problem when we published it in [67]. Let me tell some later developments. First Jim Baumgartner proved with a forcing argument that it is consistent that *GCH holds and there is an* $\aleph_2$*-chromatic graph on* $\aleph_2$ *vertices, such that the chromatic number of every subgraph of size* $\aleph_1$ *is at most* $\aleph_0$ [B1].

M. Foreman and R. Laver proved that: *it is consistent with GCH relative to a large cardinal that every graph on* $\aleph_2$, *all of whose* $\aleph_1$ *subgraphs are* $\aleph_0$*-chromatic is at most* $\aleph_1$*-chromatic* [FL].

Finally, Shelah proved, after improving his own results several times, that it is consistent (true in L) that: *for every regular non-weakly compact* κ *there* *is a κ-chromatic graph on κ, all of whose subgraphs of size less that κ are $\aleph_0$-chromatic* [S4].

We also invented an interesting graph in [54]. Let $C(\omega_2, \omega)$ be a graph whose vertices are the elements of ${}^{\omega_2}\omega$, i.e. ω_2-sequence of integers and we join two sequences if they are eventually different. We proved, and these are obvious facts, that *every subgraph of size $\aleph_1$ of $C(\omega_2, \omega)$ has chromatic number $\leq \aleph_0$, and moreover, that every graph of size $\aleph_2$ having this property embeds into $C(\omega_2, \omega)$*. However, the chromatic number of $C(\omega_2, \omega)$ in ZFC is a mystery. Our result implies that CH $\Rightarrow chr(C(\omega_2, \omega)) \geq \aleph_1$ and Péter Komjáth proved this from the weaker assumption $2^{\aleph_0} \leq \aleph_2$, and he proved it consistent with GCH that $chr(C(\omega_2, \omega)) = \aleph_3$ [K2]. On the other hand Foreman proved it is consistent relative to a large cardinal that $chr(C(\omega_2, \omega)) \leq \aleph_1$ [Fo]. It seems to be out of the question with the present methods to prove that our graph is consistently $\aleph_0$-chromatic.

11.10 Decomposition of graphs

I just want to mention our paper [53] which appeared in 1967 but was written about two years earlier. In this paper we raised problems of the following type. Does there exist graphs G not containing a K_λ, for some cardinal λ, such that for every vertex partition or edge partition with few colors say a monochromatic K_τ appears. In present notation: For what λ, τ, γ is there a K_λ-free graph G such that

$$G \to (K_\tau)^1_\gamma \quad \text{or} \quad G \to (K_\tau)^2_\gamma$$

holds? We had some results but I just want to restate two edge partition problems from the paper, one of them finite.

Does there exist a finite K_4-free G with $G \to (K_3)^2_2$? This was solved by Folkmann [Fol] affirmatively, but the question became one of the starting points of structural Ramsey theory (see § 15).

The infinitary problem is the following. Does there exist a K_4-free graph G of cardinality $(2^{\aleph_0})^+$ such that $G \to (K_3)^2_\omega$ holds? As far as I remember, this was the last set theory problem Paul offered a prize for (it was worth \$250.) Shelah later proved this to be consistent, but I will speak about the status of this kind of problem in a more general context in §15.

12. Δ-Systems and More Set Mappings

Δ-systems were introduced in a paper of Erdős and Rado which appeared in 1960 [26]. A family F of sets is a Δ-system if there is a set D, the kernel of F, such that $A \cap B = D$ for all $A \neq B \in F$. The paper set the task to determine $\Delta(\kappa, \lambda) = \delta$ the smallest cardinal for which every family F, of sets of size κ and cardinality δ contains a Δ-system of size $\lambda \geq 3$. As it is well known, for finite κ, the problem is still unsolved. A \$1000 reward is offered by Paul for the proof or disproof of the conjecture that for some $c > 0$

$$\Delta(\kappa, 3) < c^\kappa \quad \text{for} \quad \kappa < \omega$$

However, for $\kappa \geq \omega$, Erdős and Rado settled the problem completely. Although some of the details were only cleared up in their second paper [60] on the subject, the main upper bounds were already obtained in [26]. One of the main results says that *if* $\kappa < \delta = cf(\delta)$, $|F| = \delta \geq \omega$ *and* δ *is inaccessible from* κ, *i.e.* $\sigma^\kappa < \delta$ *for* $\sigma < \delta$ *then* F *contains a* Δ*-system of size* δ. This is probably the most frequently used theorem of set theory, since it is the simplest tool to prove that certain partially ordered sets satisfy certain chain conditions. If $\langle P, \preceq \rangle$ is a partially ordered set $p, q \in P$ are *incompatible* if there is no $r \in P$ with $r \preceq p, q$ and P satisfies the κ*-chain condition* if every subset of pairwise incompatible elements has cardinality $\leq \kappa$. It was already an important element of Cohen's proof of the independence of the continuum hypothesis, that finite $0, 1$ sequences from any index set, ordered by reverse inclusion satisfy the $\aleph_0$-chain condition. Cohen and his early followers did not know the Erdős-Rado theorem and they proved it for the special cases they needed. But soon it was discovered by logicians, and it is invoked almost any time forcing is used.

There is another theorem of Erdős and Specker [30], I should have mentioned in § 5, which is used almost as often in forcing arguments to establish chain conditions as the Δ-system theorem. Assume $f : \kappa \to \mathbb{P}(\kappa)$ is an ordinary set mapping. In § 5 we saw that, if $|f(x)| < \tau < \kappa$ for some cardinal $\tau < \kappa$ then there is a free set of size κ, and κ is the union of τ free sets. Now if κ is a successor cardinal λ^+ then the assumption that the type $\text{tp}(f(x)) < \xi < \lambda^+$ for some fixed $\xi < \lambda^+$ is weaker than the assumption that $|f(x)| < \tau$ for some cardinal $\tau < \kappa$, but by the Erdős-Specker theorem, this still implies the existence of a free set of size κ; however, Fodor's theorem does not apply since the graph induced by f can be λ^+-chromatic. Most of the time that we want to construct or force an object on λ^+ such that each subset of size λ^+ contains a subset of size λ^+ of certain kind, but the whole set is not the union of λ sets of this kind, then the Erdős-Specker result is the first thing to remember.

13. The Unsolved Problems in Set Theory [67]

In 1967 the first major post Cohen conference was held at UCLA. By that time, Cohen's method was generally known and developed, and the aim of the conference was to bring together all experts of set theory and to collect and make public all the fantastic new results available. We were both invited, Paul was there, but I could not make it. (It was the only time I did not get a passport from the Hungarian authorities.) The organizers convinced Paul that instead of mentioning a few interesting problems as the spirit moves him, he should write up all the difficult problems he came across during his work in combinatorial style set theory. He immediately promised that *we* will do it in a joint paper. This time we worked hard and fast. A mimeographed version of the manuscript containing 82 problems (or groups of problems really) was ready in the same year and we sent a copy to everyone we knew and who we thought would be interested. It included all the problems I mentioned in the previous chapters and quite a few more. A large number of (*then*) young mathematicians started to work on these, and produced solutions either by applying the newly developed methods of independence proofs or simply divising new combinatorial methods. The paper only appeared four years later in 1971, and by that time the status of most of the problems had changed. We tried to keep the manuscript up to date by adding remarks, but in 1971 we decided to write a second problem paper [81] which contained the status of the problems up until that time.

It would clearly be impossible to write a similar survey today. In the previous chapters, I tried to show on selected topics how the Erdős problems generated new questions and results and how they became integral parts of modern set theory, and how many of them are still alive. In this chapter I can only mention the status of a few more which I omitted earlier.

I did not finish the story of set mappings of type $< \omega$. Shortly after our problem paper was distributed Jim Baumgartner proved in his thesis [B2] that if $V = L$ then $Free(\kappa, 2, < \omega, \aleph_0)$ is equivalent to $\kappa \to (\aleph_0)_2^{<\omega}$ but on the other hand, it is still open if it is consistent relative to a large cardinal that $Free(\aleph_\omega, 2, < \omega, \aleph_0)$ holds, or more strongly, there is no Jónsson algebra on $\aleph_\omega$. As I already mentioned, $Free(\aleph_\omega, 2, < \omega, \aleph_0)$ was our first joint problem. We already suspected at Kalmár's supper that it will be hard, but probably not quite as hard as it turned out to be.

I am afraid I have mentioned too many problems which led to independence results, so here is a difficult theorem of Shelah and Stanley solving one of our problems in ZFC:

$$(2^{\aleph_0})^+.\omega \to ((2^{\aleph_0})^+.\omega, n)^2$$

for $n < \omega$ [SS2]. It is another matter that they also proved $\omega_3.\omega_1 \to (\omega_3.\omega_1, 3)^2$ to be independent of ZFC and GCH.

Erdős proved with Alaoglu in 1950 in [5] that if κ is smaller than the first weakly inaccessible cardinal greater than $\aleph_0$, then one can not have $\aleph_0$ σ-additive $0, 1$ measures so that every subset of S is measurable with respect to one of them. Erdős attributes the question to Stanislaw Ulam but he got the first result. We asked if $\aleph_0$ can be replaced by $\aleph_1$ here? Prikry proved it to be consistent, but this question became the forerunner of so many questions in the theory of large cardinals that I do not dare to write about later developments in detail.

Instead, here are some evergreen problems from the theory of ordinary partition relations for ordinals.

1. $\omega^\omega \to (\omega^\omega, 3)^2$ was proved by C.C. Chang [C] and $\omega^\omega \to (\omega^\omega, n)^2$ $n < \omega$ was proved by E.C. Milner [M2] and independently by Jean Larson [Lar]). But $\omega^{\omega^2} \to (\omega^{\omega^2}, 3)^2$ or $\omega^{\omega^\alpha} \to (\omega^{\omega^\alpha}, 3)^2$ seem to be as hard as ever. Here of course α^β means ordinal exponentiation.
2. Does there exist an α with $\alpha \to (\alpha, 3)^2$ such that $\alpha \not\to (\alpha, 4)^2$?
3. I proved with Jim Baumgartner in 1970 [BH] that $\Phi \to (\omega)^1_\omega \Rightarrow \forall \alpha < \omega_1 \forall k < \omega\ \Phi \to (\alpha)^2_k$. But for exponents > 2 very little is known. For example, $\omega_1 \to (\alpha, 4)^3$ is still open for $\alpha < \omega_1$. The world record is presently held by Milner and Prikry; they proved this for $\alpha \leq \omega.2 + 1$. See [MP].
4. Is $\omega_2 \to (\alpha)^2_2$ for $\alpha < \omega_2$ consistent with GCH? I proved the consistency of $\omega_2 \not\to (\omega_1 + \omega)^2_2$ and it follows from the existence of Laver's ideal mentioned in §10.5 that $\omega_2 \to (\omega_1.2)^2_2$.
5. It follows from a recent result of Baumgartner, myself and Todorčević [BHT] that GCH$\Rightarrow \omega_3 \to (\omega_2 + \xi)^2_k$ for $\xi < \omega_1$ and $k < \omega$ but $\omega_3 \to (\omega_2 + 2)^2_\omega$ is still open. See [BHT] for many new problems arising from our results.

14. Paradoxical Decompositions

Erdős has twelve major joint papers with Eric Milner, nine of those were written by the three of us. These are from a later period so the results and problems are more technical than the ones I described earlier, it is out of question to give a list of them. I want to speak about one idea which features in quite a few of them.

It was always clear that Ramsey's theorem is a generalization of the pigeonhole principle of Dedekind. When partition relations $\kappa \to (\lambda_\nu)^r_{\nu<\gamma}$ were formally introduced, it became apparent that the pigeonhole principle is just a partition relation for cardinals with exponent $r = 1$. For example, $nk + 1 \to (n+1)^1_k$ for the finite case with k boxes, and $\aleph_0 \to (\aleph_0)^1_k$ for $k < \omega$, and more generally, $\kappa \to (\kappa)^1_\lambda$ for $\lambda < cf(\kappa)$, $\kappa \geq \omega$. It was discovered by Milner and Rado in [MR] which appeared in 1965 that the pigeonhole principle does not work the same way for ordinals. They proved that for any $\kappa \geq \omega$

(i) $\xi \not\to (\kappa^n)^1_{n<\omega}$ if $\xi < \kappa^+$ and as a corollary of this $\xi \not\to (\kappa^\omega)^1_\omega$ for $\xi < \kappa^+$.

Here again α^β denotes ordinal exponentiation. This phenomena, often called the *Milner-Rado paradox*, has to be kept in mind, just because it is so contrary to one's first intuition. When partition relations proliferated it was discovered that this (as almost anything) can be written as a polarized partition relation:

(ii) $\begin{pmatrix} \omega \\ \xi \end{pmatrix} \not\to \begin{pmatrix} 1 & , & \omega \\ \kappa^\omega & , & 1 \end{pmatrix}^{1,1}$ for $\xi < \kappa^+$

and also as a square bracket relation:

(iii) $\xi \not\rightarrow [\kappa^\omega]^1_{\aleph_0, <\aleph_0}$ for $\xi < \kappa^+$.

In [63] we have investigated the polarized partition relation

$$\begin{pmatrix} \kappa \\ \xi \end{pmatrix} \rightarrow \begin{pmatrix} 1 & \delta \\ \sigma & \tau \end{pmatrix}^{1,1} \quad \text{for } \kappa = \omega \text{ and } \kappa = \omega_1,\ \xi < \omega_2.$$

We gave a complete discussion, relying heavily on the form (iii) of the paradox in the case $\kappa = \omega_1$, i.e.

(iv) $\xi \not\rightarrow [\omega_1^\omega]^1_{\aleph_0, <\aleph_0}$ for $\xi < \omega_2$.

When we tried to lift our results to higher cardinals we realized that we would need to generalize (iv) to

(v) $\xi \not\rightarrow [\omega_2^{\omega_1}]^1_{\aleph_1, \aleph_0}$ for $\xi < \omega_3$.

We already discovered in 1967, that this will not be possible in ZFC, but we only wrote down the results which we called the $\aleph_2$ -phenomenon in our 1978 paper [93] relying heavily on other people's results. See [93] for references.

Since this is not so well known, I will write down the $\aleph_2$ -phenomenon as it relates to (v).

A.1 $\xi \not\rightarrow [\omega_2^{\omega_1}]^1_{\aleph_1, \aleph_0}$ *holds for* $\xi < \omega_2^{\omega_2}$

A.2 *If* $2^{\aleph_1} = \aleph_2$ *then for some* $\xi_0 < \omega_3$, $\xi_0 \rightarrow [\omega_2^{\omega_1}]^1_{\aleph_1, \aleph_0}$

A.3 *It is consistent with* $2^{\aleph_1} = \aleph_3$ *that* $\xi \rightarrow [\omega_2^{\omega_1}]^1_{\aleph_1, \aleph_0}$ *holds for* $\xi < \omega_3$

A.4 $\omega_2^{\omega_2} \rightarrow [\omega_2^\omega]^1_{\aleph_1, \aleph_0}$ *and* $\omega_2^{\omega_2} \not\rightarrow [\omega_2^\omega]^1_{\aleph_1, \aleph_0}$ *are both consistent with ZFC and GCH.* (The $\not\rightarrow$ holds e.g. in L while the $\rightarrow$ follows from Chang's conjecture.)

All this happens because a counterexample establishing the $\not\rightarrow$ is really a sequence $\{A^\xi_\alpha : \alpha < \omega_1\} \subset \xi$ such that the order type $\mathrm{tp}(\bigcup_{\beta<\alpha} A_\beta) < \omega_2^{f_\xi(\alpha)+1}$ for a function $f_\xi : \omega_1 \rightarrow \omega_1$ and, for $\zeta < \xi$, f_ζ must be smaller than f_ξ in some well known ordering of these functions. In fact, this was the reason why we asked all the problems 19A-19E in the unsolved problems paper, about the relation of the transversal hypothesis and the Kurepa hypothesis.

Problem 19D was slightly out of the line there. Typically, Paul asked something that was quite new: are there $2^{\aleph_1}$ almost disjoint, stationary subsets of ω_1? It is easy to see the consistency of a 'yes' answer, it is true e.g. in L, however the consistency of a 'no' answer with CH is not completely proved. Foreman, Magidor and Shelah proved in [FMS] that 'no' follows from a consistent set-theoretical principle called Martin's Maximum (MM), but MM implies that $2^{\aleph_0} = \aleph_2$. They also proved it consistent with CH that there is a stationary subset of ω_1 on which the nonstationary ideal is $\aleph_2$-saturated. All these very difficult consistency proofs of course are relative to the existence of some large cardinals.

15. A mistake and its consequences

In §12 of our paper [46] about chromatic numbers, we claimed a false theorem. I just state a special case. Let $\mathcal{H} = (h, H)$ be a 3-uniform hypergraph (i.e. $H \subset [h]^3$) such that every pair $e \in [h]^2$ is in at most countably many elements of H. Then we claimed that the chromatic number of $\mathcal{H}$ is at most $\aleph_0$.

As we know now, this is true if $|h| \leq \aleph_1$ and false for a triple system of cardinality $(2^{\aleph_0})^+$. Now I have to disclose a not so surprising secret. Paul actually wrote up some of our joint papers, but these were the short ones. For the long ones it was my job to prepare the manuscript, but we always read the manuscript and even the proof sheets together. The trouble was that he often got bored with mechanical work like this, and he made up new conjectures and theorems and insisted that we should include them by adding remarks — even to the galley proofs. Taking the responsibility, I think I was the one who overlooked that the cardinal induction method breaks down from $\aleph_1$ to $\aleph_2$ in this case. Anyway, if the theorem was really true, the whole structure of the paper should have been changed but fortunately we did not have time for that.

As usual, I forgot the theorem, but Paul did not. I got a phone call from him from abroad about four years after the paper had appeared. He was trying to tell the proof of it to Bruce Rothschild, and got stuck. They soon discovered a counter-example. Let $(2^{\aleph_0})^+ = \kappa$, $h = [\kappa]^2$, $H = \{\{\{\alpha, \beta\}, \{\beta, \gamma\}, \{\alpha, \gamma\}\} : \alpha < \beta < \gamma < \kappa\}$. Clearly, any two elements of H have at most one element in common, and the chromatic number is at least $\aleph_1$ by the Erdős-Rado theorem. We wrote a triple paper [76] about it. However, Paul got interested in this question: *what kind of finite triple systems must appear in an $\aleph_1$-chromatic triple system?* I think the first question was the 6/3, i.e. are there three triples with empty intersection such that each pair has exactly one point in common? (This question for triple systems made quite a splash in finite combinatorics as well.) Fred Galvin came up with a negative answer. Later Fred spent the academic year 72-73 in Budapest, and the three of us started to work on this problem and we asked the same question for triple systems not containing large independent sets. Unfortunately, we did not find a general answer, maybe there isn't one, every time that we constructed a large chromatic system avoiding concrete finite systems, Paul ingeniously invented new ones for which the construction did not work. The motivation behind this was the following. Clearly there is a cardinal κ with the following property. *If for a finite triple system $\mathcal{H}$ there is a $(> \aleph_0)$-chromatic triple system $\mathcal{K}$ not containing $\mathcal{H}$, then there is one of cardinality at most κ.* Cardinals which satisfy a condition like this are quite often impossible to determine, and such was the case with this problem. For two triples with a common edge the number we found is $(2^{\aleph_0})^+$. With GCH all our examples had cardinality $\aleph_2$ associated with them. We ended up with a concisely written paper almost ninety pages long [85] containing some really good theorems, but which remained relatively unknown. Again, it is not possible to give a list of the results, but I do want to mention one concept and problem from the paper that I really like.

We constructed large chromatic r-uniform hypergraphs by induction on r, and to support the induction from r to $r+1$, we needed the r-tuple system $\mathcal{H}$, to have a stronger property than $chr(\mathcal{H}) > \aleph_0$.

Let $\mathcal{H}_\nu = \langle h, H_\nu \rangle$ $(\nu < \varphi)$, be a system of r-uniform hypergraphs on the same vertex set h. The system has *simultaneous chromatic number* $> \aleph_0$ if, for every partition of the vertex set $h = \bigcup_{n<\omega} h_n$ into $\aleph_0$ parts, there is an $n < \omega$ such that h_n contains "edges" from each $\mathcal{H}_\nu$ for $\nu < \varphi$.

We say that a $(> \aleph_0)$-chromatic $\mathcal{H} = \langle h, H \rangle$ *splits to δ parts*, if there is a disjoint partition $H = \bigcup_{\nu<\delta} H_\nu$ so that the system $\mathcal{H}_\nu = \langle h, H_\nu \rangle$ $(\nu < \delta)$ has simultaneous chromatic number $> \aleph_0$. We proved that quite a few known $(> \aleph_0)$-chromatic graphs split to $\aleph_1$ parts and these served as a basis of our induction process.

In those days, before Todorčević's result, we only knew with CH that $K_{\aleph_1}$ splits to $\aleph_1$-parts. Still, as we did not find anything that does not split, we asked the question: *Is it true that every $(> \aleph_0)$-chromatic graph splits to two (or $\aleph_1$) parts?*

This problem as it stands is still unsolved. With Péter Komjáth I have some unpublished partial results. Here are two of them.

(1) *It is consistent that every $\aleph_1$-chromatic graph splits into $\aleph_1$ parts.*
(2) *It is consistent relative to a measurable cardinal, that there is a $(> \aleph_0)$-chromatic graph which does not split into $\aleph_1$ parts.* (We do not know this for two parts.)

16. Structural Ramsey Theory

As I already mentioned in §10.10, we asked the first questions of the following type: Does there exist a K_4-free graph G such that $G \to (K_3)^2_2$.

The following type of generalization appeared first in Deuber's paper [D]. Let G, H be graphs; H embeds into G if G has an induced subgraph isomorphic to H. With present day partition calculus notation, we say $G \rightarrowtail (H)^2_{\kappa,\lambda}$ if for arbitrary colorings $k : G \to \kappa$, $\ell : [g]^2 \backslash G \to \lambda$ of the edges of G with κ colors and non-edges with λ-colors, there is an induced subgraph $H' \subset G$ isomorphic to H such that k and ℓ are constant on the edges and on the non-edges of H' respectively.

Deuber proved that *for all finite H and $k < \omega$, there is a finite G with* $G \rightarrowtail (H)^2_{k,k}$ and the combination of the two types of questions Paul raised became the starting points of the Něsetřil-Rödl type structural Ramsey theory.

With Erdős and Pósa we proved the first infinitary result of this kind [88]. The paper appeared in the volume of the Keszthely conference held for Paul's sixtieth birthday in 1973, and this volume contains the first Něsetřil

Rödl paper on the subject. The finitary theory developed very fast. The problem was generalized for coloring of substructures of a fixed kind instead of coloring pairs, but fortunately I do not have to give an account of this. I just want to say that this was not done in the infinitary case because here some basic problems are still open.

In the paper with Erdős and Pósa we proved that *for every countable H and $k < \omega$ there is a G ($|G| = 2^{\aleph_0}$) such that $G \rightarrowtail (H)^2_{k,k}$* and asked if this holds true for countably many colors, or for larger H.

We discovered a decade later with Péter Komjáth that *it is consistent to have $|H| = \omega_1$ and $G \not\rightarrowtail (H)^2_{2,1}$ for every G* [HK2] and Shelah proved that *it is consistent that for all H and γ there is a G with $G \rightarrowtail (H)^2_{\gamma,\gamma}$* [S6].

In 1989, [H2] I proved *in ZFC that for all finite H and arbitrary γ there is a G with $G \rightarrowtail (H)^2_{\gamma,\gamma}$* but the problem of countable H and countable γ is open (though the $\rightarrowtail$ is consistent by Shelah's result). The answer may turn out to be to Paul's liking (a theorem in ZFC) but I am sure it will be very difficult.

Shelah generalized his consistency results for K_τ-free H as well, but at this point I feel I have to stop and refer the reader to a recent survey paper of mine on this subject [H3].

17. Applications of partition relations in set theoretical topology

In the last thirty years, set theoretical topology became a major area of research as shown e.g. in the Handbook of Set Theoretical Topology. The reason for this is that the new methods of set theory (forcing, large cardinals) made it possible to study topological spaces for what they are, namely set theoretical objects. The point I want to make is that, although Erdős did not take an active part in most of this, combinatorial set theory which he created is one of the major tools in this development.

This happens not just through the applications of positive theorems. There are of course some famous ones. Being closest to the fire, with István Juhász we showed, for example, as a consequence of $(2^{2^{\aleph_0}})^+ \rightarrow (\aleph_1)^2_4$ that, *every Hausdorff space of cardinality $(2^{2^{\aleph_0}})^+$ has discrete subspaces of size $\aleph_1$*. Also, as a consequence of the canonization theorem of §10 that, *the spread (the supremum of the sizes of discrete subspaces) is attained in a Hausdorff space if this supremum is a singular strong limit cardinal.*

More importantly, there are literally dozens and dozens of examples obtained as strengthenings of negative partition relations which would never have turned up in their present form without a detailed analysis of these relations. Let me try to make this clear with an example. I already mentioned Prikry's consistency proof of

$$\binom{\aleph_2}{\aleph_1} \not\to \binom{\aleph_0}{\aleph_1}^{1,1}_2 .$$

To state it in "human language" (but already a little twisted for my purposes), it means that there is a sequence $\{f_\alpha : \alpha < \omega_2\} \subset {}^{\omega_1}2$ such that for all countable $I \in [\omega_2]^{\aleph_0}$ there is a $\nu(I) < \omega_1$ such that for $\nu(I) < \nu < \omega_1$ there are $\alpha_0, \alpha_1 \in I$ with $f_{\alpha_0}(\nu) = 0$ and $f_{\alpha_1}(\nu) = 1$.

When in [HJ] with Juhász we discovered HFD's (hereditarily finally dense sets) and proved the consistency of the existence of a hereditarily separable space of power $\aleph_2$ (assuming $2^{\aleph_0} = \aleph_1$) we only had to change the last clause of the above statement: there is a $\nu(I) < \omega_1$ such that for every finite set F with $\nu(I) < F < \omega_1$ and every 0,1-function ϵ defined on F there is an $\alpha \in I$ such that for all $\nu \in F$, $f_\alpha(\nu) = \epsilon(\nu)$. And now $\{f_\alpha : \alpha < \omega_2\}$ is a hereditarily separable subspace of cardinality $\aleph_2$ of $D(2)^{\omega_1}$.

18. A final apology

I feel that I should stop at this point. One reason is that this is the hundredth page of my handwritten manuscript, but there are other reasons. Paul has continued to work on set theory, stating new and old problems in the numerous problem papers he published. Our last major set theory paper with Jean Larson [109] appeared in 1993. It would not really be appropriate for me to speculate on the reactions that these latest problems may provoke, for we lack the perspective. It is also true, that his interest in set theory is slightly diminished, he does not like the technical problems which already in the assumptions involve consistency results. But he triumphantly continues to carry the flag of Georg Cantor.

I also have some doubts about my manuscript. It is as if I have been trying to sketch a rain forest, but with only enough time and ability to draw the trunks of what I thought to be the largest trees. Paul's real strength is in the great variety of those hundreds of small questions which he has asked that have given some real insights into so many different topics. I can only admire his inventiveness and thank him for everything he has given us.

Finally, I also wish to thank our old friend Eric Milner for helping me to prepare this paper.

19. Paul Erdős set theory papers

[1] Végtelen gráfok Euler-vonalairól, *Mat. és Fiz. Lapok* 1936, 129-141. (On Euler lines of infinite graphs, in Hungarian with T. Grünwald (Gallai) and E. Weiszfeld (Vázsonyi))

[2] Über Euler-Linien unendlicher Graphen, *Journ. of Math. and Phy.* **17** (1938) 59-75. (with T. Grünwald (Gallai) and E. Vázsonyi)

[3] Some set-theoretical properties of graphs, *Revista de la Univ. Nac. de Tucuman, Ser. A. Mat. y Fiz. Teor*, **3** (1942) 363-367.

[4] On families of mutually exclusive sets, *Annals of Math.* **44** (1943) 315-329. (with A. Tarski)

[5] Some remarks on set theory, *Annals of Math.* **44** (1943) 643-646.

[6] Some remarks on connected sets, *Bull. Amer. Math. Soc.* **50** (1944) 443-446.

[7] On the Hausdorff dimension of some sets in Euclidean space, *Bull. Amer. Math. Soc.* **53** (1946) 107-109. (with N.G. de Bruijn)

[8] On a combinatorial problem, *Akademia Amsterdam*, **10** (1948) 421-423. (with N.G. de Bruijn)

[9] A combinatorial theorem, *Journal London Math. Soc.* **25** (1950) 249-255. (with R. Rado)

[10] Some remarks on set theory II, *Proc. Amer. Math. Soc.* (1950) 127-141.

[11] A colour problem for infinite graphs and a problem in the theory of relation, *Akademia Amsterdam* **13** (1951) 371-373. (with N.G. de Bruijn)

[12] Combinatorial theorems on classifications of subsets of a given set, *Proc. London Math. Soc.* **3** (1951) 257-271.

[13] A problem on ordered sets, *Journ. London Math. Soc.* **28** (1953) 426-238. (with R. Rado)

[14] Some remarks on set theory III, *Michigan Math. Journ.* **2** (1953) 51-57.

[15] Some remarks on set theory IV, *Michigan Math. Journ.* **2** (1953) 169-173.

[16] Partitions of the plane into sets having positive measure in every non-null measurable product set, *Amer. Math. Soc.* **79** (1955) 91-102. (with J.C. Oxtoby)

[17] Some theorems on graphs, *Hebrew Univ. Jerusalem* **10** (1955) 13-16.

[18] A partition calculus in set theory, *Bull. Amer. Math. Soc.* **62** (1956) 427-489. (with R. Rado)

[19] Some remarks on set theory V., *Acta Sci. Math. Szeged* **17** (1956) 250-260. (with G. Fodor)

[20] On a perfect set, *Coll. Math.* **4** (1957) 195-196. (with S. Kakutani)

[21] Some remarks on set theory VI., *Acta Sci. Math. Szeged* **18** (1957) 243-260. (with G. Fodor)

[22] Partition relations connected with the chromatic number of graphs, *Journal London Math. Soc.* **34** (1959) 63-72. (with R. Rado)

[23] A theorem on partial well-ordering of sets of vectors, *Journal London Math. Soc.* **34** (1959) 222-224. (with R. Rado)
[24] On the structure of set mappings, Acta Math. Acad. Sci. Hung. 9 (1958) 111-130. (with A. Hajnal)
[25] On the structure of inner set mappings, *Acta Sci. Math. Szeged* **20** (1959) 359-369. (with G. Fodor)
[26] Intersection theorems for systems of set, *Journal London Math. Soc.* **35** (1960) 85-90. (with R. Rado)
[27] Some remarks on set theory VIII, *Michigan Math. Journal* **7** (1960) 187-191. (with A Hajnal)
[28] Some remarks on set theory VII, *Acta Sci. Math.* **21** (1960) 154-163. (with A. Hajnal)
[29] A construction of graphs without triangles having preassigned order and chromatic number, *J. London Math. Soc.* **35** (1960) 445-448. (with R. Rado)
[30] On a theorem in the theory of relations and a solution of a problem of Knaster, Coll. Math. 8, (1961) 19-21. (with E. Specker)
[31] On some problems involving inaccessible cardinals, *Essays on the Foundation of Hebrew University of Jerusalem* (1961) 50-82. (with A. Tarski)
[32] On the topological product of discrete compact spaces,in: *General topology and its relations to Modern Analysis and Algebra*, Proceedings of *symposium in Prague in September* 1961, 148-151. (with A. Hajnal)
[33] On a property of families of sets, Acta *Math. Acad. Sci. Hung.* **12** (1961) 87-123. (with A. Hajnal)
[34] Some extremal problems on infinite graphs, *Publications of the Math. Inst. of the Hungarian Academy of Science*, **7** Ser. A. (1962) 441-457. (with J. Cipszer and A. Hajnal)
[35] Some remarks concerning our paper "On the structure of set mappings" and Non-existence of two-valued 0,1-measure for the first uncountable inaccessible cardinal, *Act Math. Acad. Sci. Hung.* **13** (1962) 223-226. (with A. Hajnal)
[36] On a classification of denumerable order types and an application to the partition calculus, *Fundamental Math.* **51** (1962) 117-129. (with A. Hajnal)
[37] The Hausdorff measure of the intersection of sets of positive Lebesque measure, *Mathematika London* **10** (1963) 1-9. (with S.J. Taylor)
[38] On some properties of Hamel basis, *Colloq. Math.* **10** (1963) 267-269.
[39] An intersection property of sets with positive measure, *Colloq. Math.* **10** (1963) 75-80. (with R. Kestelman and C.A. Rogers)
[40] An interpolation problem associated with the continuum hypothesis, *Michigan Math. Journal* **11** (1964) 9-10.
[41] On complete topological subgraphs of certain graphs, *Annales Univ. Sci. Bp.* **7** (1964) 143-149. (with A. Hajnal)

[42] Some remarks on set theory IX, *Michigan Math. Journ.* II (1964) 107-127. (with A. Hajnal)
[43] Partition relations for cardinal numbers, Acta *Math. Hung.* **16** (1965) 93-196. (with A. Hajnal and R. Rado)
[44] The non-existence of a Hamel-basis and the general solution of Cauchy's functional equation for non-negative numbers, *Publ. Math. Debreen* **12** (1965) 259-263. (with J. Aczel)
[45] On a problem of B. Jónsson, *Bulletin de l'Academic Polonaise des Sciences* **14** (1966) 19-23. (with A. Hajnal)
[46] On chromatic number of graphs and set-systems, *Acta Math. Acad. Sci. Hung.* **17** (1966) 61-99. (with A. Hajnal)
[47] On the complete subgraphs of graphs defined by systems of set, Acta *Math. Acad. Sci. Hung.* **17** (1966) 159-229. (with A. Hajnal and E.C. Milner)
[48] An example concerning open everywhere discontinuous functions, *Revue Roum. de Math. Pures et Appl.* **11** (1966) 621-622.
[49] Some remarks on set theory X, *Studia Sci. Math. Hung.* **1** (1966) 157-159. (with M. Makkai)
[50] Partition relations and transitivity domains of binary relations *J. London Math. Soc.* **42** (1967) 624-633. (with R. Rado)
[51] Some remarks on chromatic graphs, *Coll. Math.* **16** (1967) 253-256.
[52] Kromatikus gráfokról, (On chromatic graphs, in Hungarian) *Mat. Lapok* **18** (1967) 1-2. (with A. Hajnal)
[53] On a decomposition of graphs, *Acta Math. Acad. Sci. Hung* . **18** (1967) 359-3776. (with A. Hajnal)
[54] On chromatic number of infinite graphs, in: *Theory of graphs*, Proc of the Colloqu. held at Tihany, Hungary, Akadémiai Kiadó, Budapest-Academic Press, New York, 1968, 83-98. (with A. Hajnal)
[55] Egy kombinatorikus problémáról, *Mat. Lapok* **19** (1968) 345-348. (On a combinatorial problem. In Hungarian. with A. Hajnal)
[56] On sets of almost disjoint subsets of a set, *Acta Math. Acad. Sci. Hung.* **20** (1968) 209-218. (with A. Hajnal and E.C. Milner)
[57] A problem on well ordered set, *Acta Math. Acad. Sci. Hung* . **20** (1969) 323-329. (with A. Hajnal and E.C. Milner)
[58] Set mappings and polarized partitions, *Combinatorial theory and its applications*, Balatonfüred, Hungary (1969) 327-363. (with A. Hajnal and E.C. Milner)
[59] Problems and results in chromatic graphs theory (dedicated to the memory of Jon Folkman), *Proof Techniques in Graph Theory*, book ed. by Frank Harary, Academic Press, New York and London, 1969, 27-35.
[60] Intersection theorems for systems of sets, II J. *London Math. Soc.* **44** (1969) 467-479. (with R. Rado)

[61] Problems in combinatorial set theory, in: *Combinatorial Structures and their Applications*, (Proc. Calgary Internat. Conf. Calgary, Alta., (1969) 97-100, Gordon and Breach, New York, 1970.
[62] Some results and problems for certain polarized partition, *Acta Math Acad. Sci. Hung.* **21** (1970) 369-392. (with A. Hajnal)
[63] Polarized partition relations for ordinal numbers, *Studies in pure Mathematics, Academic Press*, (1969) 63-87. (with A. Hajnal and E.C. Milner)
[64] Problems and results in finite and infinite combinatorial analysis, *Ann. of the New York A. Sci.* **175** (1970) 115-124. (with A. Hajnal)
[65] Ordinary partition relations for ordinal numbers, *Periodic Math. Hung.* **1** (1971) 171-185. (with A. Hajnal)
[66] Partition relations for η_α sets, *J. London Math. Soc.* **3** (1971) 193-204. (with E.C. Milner and R. Rado)
[67] Unsolved problems in set theory, in: *Axiomatic Set Theory (Proc. Symp. Pure Math. Vol XIII, Part I, Univ. Calif. Los Angeles, Calif. 1967*) Amer. Math. Soc., (1971) 17-48. (with A. Hajnal)
[68] A theorem in the partition calculus, *Canad. Math. Bull.* **15** (1972) 501-505. (with E.C. Milner)
[69] On Ramsey like theorems, Problems and results, in: *Combinatorics (Proc. Conf. Combinatorial Math. Inst. Oxford, 1972)* Inst. Math. Appl. Southend-on-Sea (1972) 123-140. (with A. Hajnal)
[70] On problems of Moser and Hanson, *Graph theory and applications (Proc. Conf. Western Michigan Univ. Kalamazoo, Mich. 1971) Lecture Notes in Math. Vol.* **303**, Springer, Berlin (1972) 75-79. (with S. Shelah)
[71] Partition relations for η_α and for $\aleph_\alpha$-saturated models, in: *Theory of sets and topology (in honor of Felix Hausdorff, 1868-1942*) VEB Deutsch Verlag Wissensch., Berlin, 1972, 95-108. (with A. Hajnal)
[72] Separability properties of almost disjoint families of sets, *Israel J. Math.* **12** (1972) 207-214. (with S. Shelah)
[73] Simple one point extensions of tournaments, *Matematika* **19** (1972) 57-62. (with A. Hajnal and E.C. Milner)
[74] Some remarks on simple tournaments, *Algebra Universalis* **2** (1972) 238-245. (with E. Fried, A. Hajnal, and E.C. Milner)
[75] Chain conditions on set mappings and free sets, *Acta Sci. Math. Szeged* **34** (1973) 69-79. (with A. Hajnal and A. Máté)
[76] On chromatic number of graphs and set systems in: *Cambridge School in Mathematical Logic (Cambridge, England 1971) Lecture Notes in Math.* Vol **337**, Springer, Berlin, 1973, 531-538. (with A. Hajnal and B.L. Rothschild)
[77] Corrigedum: "A theorem in the partition calculus", *Canad. Math. Bull.* **17** (1974) 305. (with E.C. Milner)
[78] Intersection theorems for systems of sets III, *Collection of articles dedicated to the memory of Hanna Newmann*, IX. *J. Austral. Math. Soc.* **18** (1974) 22-40. (with E.C. Milner and R. Rado)

[79] On some general properties chromatic number in: *Topics in topology (Prac. Colloq. Keszthely, 1972) Colloq. Math. Soc. J. Bolyai,* Vol **8**, North Holland, Amsterdam, 1974, 243-255. (with A. Hajnal and S. Shelah)

[80] The chromatic index of an infinite complete hypergraph: a partition theorem, in: *Hypergraph Seminar, Lecture notes in Math.* Vol **411**, Springer, Berlin (1974) 54-60. (with R. Bonnet)

[81] Unsolved and solved problems in set theory in: Proceedings of the Tarski Symposium (*Proc. Symp. Pure Math. Vol XXV, Univ. of California, Berkeley, Calif. (1971)* Amer. Math. Soc. Providence, RI (1974) 269-287. (with A. Hajnal)

[82] Some remarks on set theory, XI, *Collection of articles dedicated to A. Mostourski on the occasion of his 60th birthday,* III, *Fund. Math.* **81** (1974) 261-265. (with A. Hajnal)

[83] A non-normal box product, in: *Infinite and finite sets, (Colloq. Keszthely, 1973, dedicated to P. Erdős on his 60th birthday), Vol* II, *Colloq. Math. Soc. J. Bolyai,* Vol **10**, North Holland, Amsterdam 91975) 629-631. (with M.E. Rudin)

[84] On maximal almost disjoint families over singular cardinals, in: *Infinite and finite sets, (Colloq. Kenthely, 1973; dedicated to P. Erdős on his 60th birthday),*Vol I, *Colloq. Math. Soc. J. Bolyai,* Vol **10**, North Holland, Amsterdam (1975) 597-604. (with S.H. Hechler)

[85] On set-systems having large chromatic number and not containing prescribed subsystems, in: I*nfinite and finite sets (Colloq. Keszthely 1973; dedicated to P. Erdős on his 60th birthday)*, Vol I, *Colloq. Math. Soc. J. Balyai*, Vol **10**, North Holland, Amsterdam, (1975), 425-513. (with F. Galvin and A. Hajnal)

[86] Problems and results on finite and infinite combinatorial analysis, in: Infinite and finite sets *(Colloq. Keszthely 1973; dedicated to P. Erdős on his 60th birthday)*, Vol I, *Colloq. Math. Soc. J. Bolyai*, Vol **10** North Holland, Amsterdam (1975) 403-424.

[87] Problems and results on finite and infinite graphs, in: *Recent advances in graph theory (Proc. Second Chechoslovak Sympos., Prague* 1974 (loose errata), Academia, Prague (1975) 183-192.

[88] Strong embedding of graphs into coloured graphs, in: *Infinite and finite sets (Colloq. Keszthely 1973; dedicated to P. Erdős on his 60th birthday)* Vol I *Colloq. Math. Soc. J. Bolyai*, Vol 10, North Holland, Amsterdam (1975) 585-595. (with A. Hajnal and L. Pósa)

[89] Families of sets whose pairwise intersection have prescribed cardinals or order types, *Math. Proc. Cambridge Philos. Soc.* **80** (1976) 215-221. (with E.C. Milner and R. Rado)

[90] Partition theorems for subsets of vector spaces, *J. Combinatorial Theory Ser. A* **20** (1976) 279-291 (with M. Cates, N. Hindman, and B.L. Rotheschild)

[91] Corrigenda: "Families of sets whose pairwise intersection have prescribed cardinals or order types" (Math. Proc. Cambridge Philos. Soc. 80 (1976), 215-221) *Math. Proc. Cambridge Philos. Soc.* **81** (1977) 523. (with E.C. Milner and R. Rado)
[92] Embedding theorems for graphs establishing negative partition relations, *Period. Math. Hungar.* **9** (1978) 205-230. (with A. Hajnal)
[93] On set systems having paradoxical covering properties, *Acta Math. Acad. Sci. Hungar.* **31** (1978) 89-124. (with A. Hajnal and E.C. Milner)
[94] On some partition properties of families of set, *Studia Sci. Math. Hungar.* **13** (1978) 151-155. (with G. Elekes and A. Hajnal)
[95] On the density of λ-box products, *General Topology Appl.* **9** (1978) 307-312. (with F.S. Cater and F. Galvin)
[96] Colorful partitions of cardinal numbers, *Canad. J. Math* . **31** (1978) 524-541. (with J. Baunegartuer, F. Galvin, and J. Larson)
[97] Transversals and multitransversals, *J. London Math. Soc* . (2) **20** (1979) 387-395 (with F. Galvin and R. Rado)
[98] On almost bipartite large chromatic graphs, *Annals of Discrete Math.* 12 (1982) 117-123. (with A. Hajnal and E. Szemerédi)
[99] Problems and results on finite and infinite combinatorial analysis II, in: *Logic and algorithmic Int. Symp. Zurich, 1980 Ensiegn. Math.* **30** (1982) 131-144.
[100] Combinatorial set theory partition relations for cardinals, in: *Studies in Logic and the Foundations of Mathematics*, **106** North Holland, Amsterdam, New York 1984 347 pp ISBN 0-444-86157-2 (with A. Hajnal, A. Máté, and R. Rado)
[101] Chromatic number of finite and infinite graphs and hypergraphs (French summary), in: *Special volume on ordered sets and their applications* (L'Arbresle 1982) *Discrete Math.* **53** (1985) 281-285. (with A. Hajnal)
[102] Problems and results on chromatic numbers in finite and infinite graphs, in: *Graph theory with applications to algorithms and computer science (Kalamazoo, Mich.* 1984), Wiley Intersci. Publ., Wiley, New York (1985) 201-213.
[103] Coloring graphs with locally few colors, *Discrete Math* . **59** (1986) 21-34. (with Z. Füredi, A. Hajnal, P. Komjath, V. Rödl, and A. Seress)
[104] My joint work with R. Rado, in: *Surveys in combinatorics 1987 (New Cross 1987), London Math. Soc. Lecture Note Ser.* **123** Cambridge Univ. Press, Cambridge-New York (1987) 53-80.
[105] Some problems on finite and infinite graphs, *Logic and combinatorics* (Arcata, Calif. 1985) *Contemp. Math.* **65** Amer. Math. Soc. (1987) 223-228.
[106] Intersection graphs for families of balls in $\mathbb{R}^n$, *European J. Combin.* 9 (1988) 501-505. (with C.D. Godsil, S.G. Krantz, and T. Parsons)
[107] Countable decomposition of $\mathbb{R}^2$ and $\mathbb{R}^3$, *Discrete Comput. Geom.* **5** (1990) 325-331. (with P. Komjáth)

[108] Some Ramsey type theorems, *Discrete Math.* **87** (1991) 261-269. (with F. Galvin)

[109] Ordinal partition behaviour of finite powers of cardinals, in: *Finite and Infinite Combinatorics in Sets and Logic*, NATO ASI *Series*, *Series* C: *Mathematical and Physical Sciences* Vol **411** (1993) 97-116. (with A. Hajnal and J. Larson)

References

[B1] J.E. Baumgartner, Generic graph constructions, *Journal of Symbolic Logic* **49** (1984) 234-240.

[B2] J.E. Baumgartner, Results and independence proofs in combinatorial set theory, *Doctoral Dissertation, University of California*, Berkeley, Calif. (1970).

[BH] J.E. Baumgartner and A. Hajnal, A proof (involving Martin's Axiom) of a partition relation, *Fund. Math.* **78** (1973) 193-203.

[BHT] J.E. Baumgartner, A. Hajnal, and S. Todorčević, Extension of the Erdős-Rado theorem in: *Finite and Infinite Combinatorics in Sets and Logic*, NATO ASI Series, Series C. **411** (1993) 1-18.

[Bo] B.Bollobás, Extremal Graph Theory, Academic Press (1978), xx+488.

[C] C.C. Chang, A partition theorem for the complete graph on ω^ω, *J. Combinatorial Theory* **12** (1972) 396-452.

[D] W. Deuber, Paritionstheoreme für Graphen, *Math. Helv.* **50** (1975) 311-320.

[DM] B. Dushnik and E.W. Miller, Partially ordered sets, *Amer. J. of Math* **63** (1941) 605.

[ESz] P. Erdős and G. Szekeres, A combinatorial problem in geometry, *Compositio Math* **2** (1935) 463-470.

[F] G. Fodor, Proof of a conjecture of P. Erdős, *Acta Sci. Math.* **14** (1952) 219-227.

[Fol] J. Folkman, Graphs with monochromatic complete subgraphs in every edge coloring, *SIAM J. Appl. Math.* **18** (1970) 19-24.

[Fo] M.D. Foreman, preprint.

[FL] M.D. Foreman and R. Laver: Some downward transfer properties for $\aleph_2$, *Advances in Mathematics,* **67** (1988) 230-238.

[FM] M.D. Foreman and M. Magidor, Large cardinals and definable counterexamples to the continuum hypothesis, *Ohio State Math. Res. Inst. Preprints* **92-40** (1992).

[FMS] M.D. Foreman, M. Magidor and S. Shelah, Martin's Maximum, saturated ideals and non regular ultrafilters, part 1, *Annals of Mathematics* **127** (1988) 1-47.

[FT] D.H. Fremlin and H. Talagrand, Subgraphs of random graphs, *Transactions of the A.M.S.* **291** (1985)

[GS] F. Galvin and S. Shelah, Some counterexamples in the partition calculus, *J. of Combinatorial Theory* **15** (1973) 167-174.

[H1] A. Hajnal, Proof of conjecture of S. Ruziewicz, *Fund. Math.* **50** (1961) 123-128.

[H2] A. Hajnal, Embedding finite graphs into graphs colored with infinitely many colors, *Israel Journ. of Math.* **73** (1991) 309-319.

[H3] A. Hajnal, True embedding partition relations, in: *Finite and Infinite Combinatorics in Sets and Logic, NATO ASI Series, Series C* **411** (1993) 135-152.

[HJ] A. Hajnal and I. Juhász, A consistency result concerning hereditarily α-separable spaces, *Indgationes Math.* **35** (1973) 307-312.

[HJS] A. Hajnal, I Juhász and S. Shelah, Splitting strongly almost disjoint families, *Transactions of the A.M.S.* **295** (1986) 369-387.

[HK1] A. Hajnal and P. Komjáth, Some higher gap examples in combinatorial set theory, *Annals of Pure and Applied Logic* **33** (1988) 283-296.

[HK2] A. Hajnal and P. Komjath, Embedding graphs into colored graphs, *Transaction of the A.M.S.* **307** (1988) (395-409), Corrigendum to "Embedding graphs into colored graphs", *Transactions of the A.M.S.* **322** (1992) 475.

[HK3] A. Hajnal and P. Komjáth, What must and what need not be contained in a graph of uncountable chromatic number? *Combinatorica* **4** (1984) 47-52.

[KR] H.J. Keisler and R. Rowbottom, Constructible sets and weakly compact cardinals, *Amer. Math. Soc. Notices* **12** (1965) 373.

[KT] H.J. Keisler and A. Tarski, From accessible to inaccessible cardinals, *Fund. Math.* **53** (1964) 225-308.

[K1] P. Komjáth, The colouring number, *Proc. London Math. Soc.* **54** (1987) 1-14.

[K2] P. Komjáth, The chromatic number of some uncountable graphs in: *Sets, graphs, and numbers, Colloq. Math. Soc. János Bolyai, Budapest, Hungary, 1991,* 439-444.

[L] R. Laver, An $(\aleph_2, \aleph_2, \aleph_0)$ saturated ideal on ω_1, in *Logic Colloquium 1980, (Prague),* North Holland (1982) 173-180.

[Lar] Jean A. Larson, A short proof of a partition theorem for the ordinal ω^ω, *Ann. Math. Logic* **6** (1973) 129-145.

[Mil] E.W. Miller, On property of families of sets, *Comptes Rendue Varsonie* **30** (1937) 31-38.

[M1] E.C. Milner, Partition relations for ordinal numbers, Canad. J. Math. **21** (1969) 317-334.

[M2] E.C. Milner, Lecture notes on partition relations for ordinal numbers (1972) (unpublished).

[MP] E.C. Milner and K. Prikry, A partition relation for triples using a model of Todorčevič, in: *Directions in Infinite Graph Theory and Combinatorics, Proceedings International Conference, Cambridge* (1989), *Annals of Discrete Math.*

[MR] E.C. Milner and R. Rado, The pigeonhole principle for ordinal numbers, *J. London Math. Soc.* **15** (1965) 750-768.

[P] K. Prikry, On a problem of Erdős, Hajnal, and Rado, *Discrete Math* **2** (1972) 51-59.

[S1] S. Shelah, Canonization theorems and applications, *J. Symbolic Logic* **46** (1981) 345-353.

[S2] S. Shelah, Was Sierpiński right?, *Israel Journal of Math.* **62** (1988) 355-380.

[S3] S. Shelah, A compactness theorem for singular cardinals, free algebra, Whitehead problem and transversals, *Israel J. Math.* **21** (1975) 319-349.

[S4] S. Shelah, Incompactness for chromatic numbers of graphs, in: *A tribute to P. Erdős,* Cambridge University Press (1990).

[S5] S. Shelah, There are Jónsson algebras in many inaccessible cardinals, *Cardinal Arithmetic*, Oxford University Press [Sh 365].

[S6] S. Shelah, Consistency of positive partition theorems for graphs and models, in: *Set theory and Applications, Springer Lect. Notes* **1401** 167-193.

[S7] S. Shelah, Was Sierpiński Right? II, in: *Sets, Graphs and Number; Colloquia Math. Soc. Janos Bolyei* **60** (1991) 607-638.

[SS1] S. Shelah and L. Stanley, A theorem and some consistency results in partition calculus, *AFAL* **36** (1987) 119-152.

[SS2] S. Shelah and L. Stanley, More consistency results in partition calculus, Israel J. of Math **81** (1993) 97-110.

[Si] J.H. Silver, Some applications of model theory in set theory, Doctoral dissertation, University of California, Berkeley, 1966 and *Annals of Math. Logic* **3** (1971) 45-110.

[T1] S. Todorčević, Partitioning pairs of countable ordinals, *Acta Math* **159** (1987) 261-294.

[T2] S. Todorčević, Trees, subtrees and order types, *Annals of Math. Logic* **20** (1981) 233-268.

[T3] S. Todorčević, Some partitions of three-dimensional combinatorial cubes, *J. of Combinatorial Theory (A)* **68** (1994) 410-437.

A Few Remarks on a Conjecture of Erdős on the Infinite Version of Menger's Theorem

Ron Aharoni

Technion, Haifa 32000, Israel

Summary. We discuss a few issues concerning Erdős' conjecture on the extension of Menger's theorem to infinite graphs. A key role is given to a lemma to which the conjecture can probably be reduced. The paper is intended to be expository, so rather than claim completeness of proofs, we chose to prove the reduction only for graphs of size $\aleph_1$. We prove the lemma (and hence the $\aleph_1$ case of the conjecture) in two special cases: graphs with countable out-degrees, and graphs with no unending paths. We also present new versions of the proofs of the (already known) cases of countable graphs and graphs with no infinite paths. A main tool used is a transformation converting the graph into a bipartite graph.

1. Introduction

1.1 The problem

Kőnig's classical book [7] contains a proof of an infinite version of Menger's theorem:

Theorem 1.1. *Given any two disjoint sets, A and B, of vertices in any graph, the minimal cardinality of an $A-B$ separating set of vertices is equal to the maximal number of vertex-disjoint $A-B$ paths.*

The proof was by a very young mathematician, Paul Erdős. In fact, it is not very hard: Let $\mathcal{F}$ be a maximal set (with respect to inclusion) of vertex disjoint paths, and let S be the set of all vertices participating in paths from $\mathcal{F}$. Clearly, then, S is $A-B$-separating. If $\mathcal{F}$ is infinite, then $|S| = |\mathcal{F}|$ and the theorem is proved. If $\mathcal{F}$ is finite then so is S, and then techniques from the finite case can be applied. But this proof shows that Theorem 1.1 is not really the right extension to the infinite case: the separating set is of the right cardinality, but still it is obviously too large. It is not clear when did Erdős form the "right" conjecture, which is:

Conjecture 1.1. For any two vertex sets A and B in a graph, there exists a set $\mathcal{F}$ of disjoint $A-B$ paths and an $A-B$ separating set of vertices S, such that S consists of the choice of precisely one vertex from each path in $\mathcal{F}$.

Definition 1.1. *A pair $(\mathcal{F}, S)$ as in the conjecture is called* orthogonal.

As is well known, one can restrict the considerations to digraphs. The undirected case follows by the customary device of replacing each edge by a pair of oppositely directed edges.

The main developments so far have been a proof of the countable bipartite case [8] (where, of course, it is assumed that A and B are the two sides of the graph); the case of countable graphs containing no infinite paths [9]; the general bipartite case [2]; the case of general graphs containing no infinite paths [1], and the countable case [3].

In [1] it was realized that in the absence of infinite paths, the conjecture can be reduced to the bipartite case. The simple transformation leading to the reduction is one of the main themes of the present paper. We shall follow it a bit further, and see how it can be used in other cases.

In fact, already in [9] it was realized that the main difference between the bipartite case and the general one stems from the possible existence of infinite paths. The reason is that in the construction of $\mathcal{F}$, while trying to reach B from A you may end up with infinite paths instead of $A - B$ paths. This obstacle was overcome in [3] in the countable case, but the proof there relies very heavily on countability.

1.2 Warps, waves and hindrances

A triple $\Gamma = (G, A, B)$, where $G = G(\Gamma)$ is a digraph, and $A = A(\Gamma), B = B(\Gamma)$ are subsets of $V(G)$, is called a *web*. A *warp* is a set of disjoint paths (the term is taken from weaving.) If the initial points of the warp are all in A then it is called A*-starting.*

The initial vertex of a path P (if such a vertex exists) is denoted by $in(P)$, and its terminal vertex by $ter(P)$. The vertex set of P is denoted by $V(P)$ and its edge set by $E(P)$. For a family $\mathcal{P}$ of paths, we write $V[\mathcal{P}], E[\mathcal{P}], in[\mathcal{P}], ter[\mathcal{P}]$ for the vertex set, edge set, initial points set, and terminal points set of $\mathcal{P}$, respectively. A *linkage* is a warp $\mathcal{L}$ such that $in[\mathcal{L}] = A$ and $ter[\mathcal{L}] \subseteq B$. A web is called *linkable* if it contains a linkage.

Given a graph G and a subset R of its vertices, we write $G[R]$ for the subgraph of G induced by R. For any set T we write $G - T = G[V(G) \setminus T]$. By $\Gamma - T$ we denote the web $(G - T, A \setminus T, B \setminus T)$. Given a path P (possibly in a super-web of Γ) we write $\Gamma - P$ for $\Gamma - V(P)$. For an A-starting warp $\mathcal{W}$, we write $\Gamma/\mathcal{W}$ for the web $(G - (V[\mathcal{W}] \setminus ter[\mathcal{W}]), (A \setminus in[\mathcal{W}]) \cup ter[\mathcal{W}], B)$. (So, it is the web obtained by moving the points of $in[\mathcal{W}]$ along the paths of $\mathcal{W}$, to serve as new source points.) For a warp consisting of a single path P we shall write Γ/P for $\Gamma/\{P\}$.

If a warp $\mathcal{W}$ is A-starting and $ter[\mathcal{W}]$ is $A - B$ separating then $\mathcal{W}$ is called a *wave.* The *trivial* wave is the set of all singleton paths of the form $(a), a \in A$. A *hindrance* is a wave $\mathcal{W}$ such that $in[\mathcal{W}] \neq A$. Obviously, a hindrance is a non-trivial wave. A web is called *hindered* if it contains a hindrance. It is called *loose* if it contains no non-trivial wave (and then, of course, it is also unhindered.)

For a wave $\mathcal{W}$ we write $\Gamma \angle \mathcal{W}$ for the web $(G - (V[\mathcal{W}] \setminus ter[\mathcal{W}]), ter[\mathcal{W}], B)$ (the difference between this and $\Gamma/\mathcal{W}$ is that points in $A \setminus in[\mathcal{W}]$ are not taken here as source points.)

Let $\mathcal{W}$ be a wave, and let T be the set of vertices in V which are separated by $ter[\mathcal{W}]$ from B. We write $\Gamma[\mathcal{W}]$ for the web $(G[T], A, ter[\mathcal{W}])$.

If $P_1, P_2, \ldots P_k$ are paths and $x_i, 1 \leq i < k$ are vertices such that $x_i \in V(P_i) \cap V(P_{i+1})$ then $P_1 x_1 P_2 x_2 \ldots x_{k-1} P_k$ denotes the path (if indeed it is a path) obtained by going along P_1 until reaching x_1, then switching to P_2, then switching at x_2 to P_3, and so forth. If the sequence ends with a vertex, rather than a path, the path intended is the one ending at that vertex. So, for example, Px means the part of P up to and including x. The same goes for sequences starting with a vertex. If $x = ter(P) = in(Q)$ then PQ denotes the path PxQ.

A useful operation between waves is the following: let $\mathcal{U}, \mathcal{W}$ be waves. We write $\mathcal{U} \uparrow \mathcal{W}$ for the warp $\{PxQ : P \in \mathcal{U}, Q \in \mathcal{W}, x = ter(P) \in V(Q)$ and $V(xQ) \cap V[\mathcal{U}] = \{x\}\}$. It was proved in [3] that $\mathcal{U} \uparrow \mathcal{W}$ is a wave.

There are two natural orders defined between waves: write $\mathcal{W} \leq \mathcal{U}$ if $\mathcal{U}$ is an extension of $\mathcal{W}$, i.e every path $P \in \mathcal{U}$ is a continuation of some path Q from $\mathcal{W}$

such that $V(P) \setminus V(Q) \cap V[\mathcal{W}] = \emptyset$. As usual, "<" means "$\leq$ and not equal". Note that if $\mathcal{U}$ is obtained from $\mathcal{W}$ by the deletion of some paths then $\mathcal{W} < \mathcal{U}$.

Let $\mathcal{W} \preceq \mathcal{U}$ if $ter[\mathcal{W}]$ is separated from B by $ter[\mathcal{U}]$. It is straightforward to see that both relations are indeed partial orders, and that $\mathcal{W} \leq \mathcal{U}$ implies $\mathcal{W} \preceq \mathcal{U}$. In [3] the following was proved:

Lemma 1.1. *In any web Γ there exists a maximal wave in the order $\leq$. If $\mathcal{W}$ is such a maximal wave then $\Gamma \angle \mathcal{W}$ is loose.*

We shall often use this lemma without explicit mention.

Also the following is easy:

Lemma 1.2. *If $\mathcal{W} \not\preceq \mathcal{U}$ then $\mathcal{U} \uparrow \mathcal{W} \succ \mathcal{U}$.*

In [3] it was shown that Conjecture 1.1 is equivalent to the following conjecture:

Conjecture 1.2. An unhindered web is linkable.

The proof of the equivalence is not hard. One direction is even trivial: assuming Conjecture 1.1, if $\mathcal{F}$ is not a linkage then the warp $\{Ps : P \in \mathcal{F}, s \in S \text{ and } s \in V(P)\}$ is a hindrance.

For the proof of the other direction, let $\mathcal{W}$ be a <-maximal wave in Γ. By Lemma 1.1 the web $\Gamma \angle \mathcal{W}$ is loose. Assuming Conjecture 1.2 there exists then a linkage $\mathcal{L}$ in $\Gamma \angle \mathcal{W}$. Taking the concatenation of $\mathcal{W}$ and $\mathcal{L}$ as $\mathcal{F}$ and $ter[\mathcal{W}]$ as S in Conjecture 1.1 yields then the conjecture.

It is Conjecture 1.2 which we shall try to solve. As already mentioned, the countable case of the conjecture is known [3]. The key lemma there is:

Lemma 1.3. *If Γ is an unhindered countable web then for every $a \in A$ there exists an $a - B$ path P such that $\Gamma - P$ is unhindered.*

(The countable case of Conjecture 1.2 follows immediately: if Γ is unhindered, then by the lemma one can link the vertices a_i of A into B by paths P_i one by one, while keeping the web $\Gamma - \bigcup\{V(P_j) : j < i\}$ unhindered.)

In [4] a slight generalization of the countable case was proved:

Lemma 1.4. *Conjecture 1.2 is true for webs Γ in which there exists an $A - B$ warp $\mathcal{W}$ such that $in[A] \setminus in[\mathcal{W}]$ is countable.*

Note that the condition in Lemma 1.4 is not hereditary, that is, it is not passed on to sub-webs. Hence the proof that Conjecture 1.2 implies Conjecture 1.1 does not show immediately that Conjecture 1.1 is true for webs as in the above lemma. But this was proved in [5]:

Lemma 1.5. *Conjecture 1.1 is true for webs satisfying the condition of Lemma 1.4.*

We believe that Lemma 1.3 is true in all webs:

Conjecture 1.3. Lemma 1.3 is true in any web.

The main purpose of this paper is to prove a reduction of Conjecture 1.1 to Conjecture 1.3. But in order not go into too technical details we shall do it only for webs of size $\aleph_1$.

We shall prove Conjecture 1.3 (for graphs of size $\aleph_1$) in two special cases: webs with countable out-degrees, and webs with no unending paths. We shall therefore be able to prove Conjectures 1.1 and 1.2 for these two classes. In both proofs we shall use extensively the bipartite conversion, both technically and for inspiration. While not absolutely necessary, this will prove to be economical. The countable outdegrees case includes, of course, the case of countable webs, and the proof presented here is therefore also a new proof for this case (although the main ideas are similar to those of the old one, that in [3]). The new proof is more natural when considered in the language of the bipartite conversion.

1.3 The bipartite conversion

With any web Γ we associate a bipartite web $b(\Gamma) = (b(G), b(A), b(B))$ in the following way. Every vertex v in $V \setminus A$ is assigned a vertex v'' in $V(b(G))$, and every vertex $v \in V \setminus B$ is assigned a vertex v'. (So, vertices in $V \setminus (A \cup B)$ are assigned two vertices each.) The edge set $E(b(G))$ is defined as $\{(x', y'') : (x, y) \in E(\Gamma)\} \cup \{(x', x'') : x \in V \setminus (A \cup B)\}$. The "source set" of $b(\Gamma)$, which will be denoted by $b(A)$, is just $\{v' : v \in V \setminus B\}$. The "destination set", $b(B)$, is $\{v'' : v \in V \setminus A\}$.

A linkage $\mathcal{F}$ in Γ corresponds in a natural way to a linkage $b(\mathcal{F}) = \{(x', y'') : (x, y) \in E[\mathcal{F}]\} \cup \{(x', x'') : x \notin V[\mathcal{F}]\}$ in $b(\Gamma)$. Conversely, to a linkage J in $b(\Gamma)$ (linkages in bipartite webs are just matchings, hence they are denoted by capital letters, rather than script letters), there corresponds a warp $w(J)$ in Γ, defined by $E[w(J)] = \{(x, y) : (x', y'') \in J, x \neq y\}$. (In general, backwards transformations from $b(\Gamma)$ to Γ will be denoted by w). We write $\tilde{w}(J)$ for the set of those paths in $w(J)$ which begin at A.

The following lemma is very easy to verify:

Lemma 1.6. *Let J be a linkage in $b(\Gamma)$. Then $in[w(J)] = in[\tilde{w}(J)] = A$. If a path in $w(J)$ has an endpoint then this point belongs to B. Hence, if $\tilde{w}(J)$ contains no undending paths, it is a linkage.*

Let now $\mathcal{W}$ be a wave in Γ. Define $b(\mathcal{W}) = \{(x', y'') : (x, y) \in E[\mathcal{W}]\} \cup \{(x') : x \in ter[\mathcal{W}]\} \cup \{(x', x'') : x \notin V[\mathcal{W}]\}$. It is not hard to see that $b(\mathcal{W})$ is a wave in $b(\Gamma)$. Conversely, a wave I in $b(\Gamma)$ corresponds to a warp $w(I)$ in Γ, defined by $E[w(I)] = \{(x, y) : (x', y'') \in E[I]\}$, and the singleton paths in $w(I)$ are the singletons (a), where $a \in A$ and $(a') \in I$.

Lemma 1.7. *$ter[w(I)]$ is $A - B$-separating. Every path in $w(I)$ either starts in A or is non-starting. Hence, if $w(I)$ contains no non-starting paths, it is a wave in Γ. In the latter case, if I is a hindrance, so is $w(I)$.*

Proof. Let $P = x_1 x_2 \ldots x_n$ be an $A - B$ path in Γ. If P is not met by $ter[w(I)]$ then $(x_1') \notin I$. Since the edge (x_1', x_2'') is covered by $ter[I]$, it follows that $x_2'' \in ter[I]$. Let k be the last index for which $x_k'' \in ter[I]$. If $x_k \notin ter[w(I)]$ then there exists y such that $(x_k', y'') \in I$. Then the edge (x_k', x_{k+1}'') necessarily meets $ter[I]$ at x_{k+1}'', contradicting the choice of k. This shows that $ter[w(I)]$ is $A - B$-separating.

Let now $Q \in w(I)$. If Q is a singleton x, then by the definition of $w(I)$ we have $x \in A$, so Q starts at A. If not, choose a vertex $v = v_0 \in V(Q) \setminus ter(Q)$. Then $v_0' \notin ter[I]$, and hence the edge (v_0', v_0'') is covered at v_0''. Hence there exists an edge $(v_0'', v_1') \in I$, and then $(v_1, v_0) \in E(Q)$. Repeating the same argument for v_1, we obtain an edge $(v_2, v_1) \in E(Q)$. This process is either infinite, in which case Q is non-starting, or terminates when some v_k is in A (and then the edge (v_k'', v_k') just does not exist.) This proves the second part of the lemma.

The fact that if I is a hindrance then so is $w(I)$ is obvious.□

From the above it follows that if Γ does not contain unending paths then it is linkable if and only if $b(\Gamma)$ is, and if it contains no non-starting path then it is hindered if and only if $b(\Gamma)$ is. Since Conjecture 1.2 is true for bipartite graphs [2], this proves:

Theorem 1.2. *Conjecture 1.2, and hence also Conjecture 1.1, are true for graphs with no infinite paths.*

This was proved in [1] in a different way.

2. Safely linking one point

For $x \in V \setminus A$ we write $\Gamma \vdash x$ for the web $w(b(\Gamma) - x'')$. In the language of webs, rather than that of bipartite graphs, it is the web obtained upon adding x to A, and deleting all edges going into x. For a subset X of $V \setminus A$ we denote by $\Gamma \vdash X$ the web $w(b(\Gamma) - \{x'' : x \in X\})$.

A key fact in the discussion to follow is:

Lemma 2.1. *If Γ is unhindered, $v \in V \setminus A$ and $\Gamma \vdash v$ is hindered, then there exists in Γ a wave $\mathcal{U}$ such that v is separated by $ter[\mathcal{U}]$ from B (possibly $v \in ter[\mathcal{U}]$.)*

Proof. Let $\mathcal{H}$ be a hindrance in $\Gamma \vdash v$. If v is hindered in $\mathcal{H}$ (i.e $v \notin in[\mathcal{H}]$), then $\mathcal{H}$ is a hindrance also in Γ. If not, then let u be the terminal point of the path R in $\mathcal{H}$ which starts at v, and let h be a point in A hindered by $\mathcal{H}$. Let $\mathcal{H}' = \mathcal{H} \setminus \{R\}$. If there is no $\mathcal{H}'$-alternating path in Γ starting at h and ending at u then the set of edges in $E[\mathcal{H}']$ participating in $\mathcal{H}'$-alternating paths starting at h forms a hindrance in Γ (see [3] for a detailed proof of this fact.) Hence, we may assume that there exists an $\mathcal{H}'$-alternating path Q starting at h and ending at u. Applying Q to $\mathcal{H}$ (that is, taking the warp whose edge set is $E[\mathcal{H}] \triangle E(Q)$) yields then the desired wave $\mathcal{U}$. □

Corollary 2.1. *If Γ is loose and x is a vertex in $V \setminus A$ from which B is reachable then $\Gamma \vdash x$ is unhindered.*

Lemma 2.2. *If Γ is hindered then so is $\Gamma \vdash v$.*

Proof. Let $\mathcal{H}$ be a hindrance in Γ. If $v \notin V[\mathcal{H}]$ then $\mathcal{H} \cup \{(v)\}$ is a hindrance in $\Gamma \vdash v$. If v lies on some path $H \in \mathcal{H}$ then $\mathcal{H} \setminus \{H\} \cup \{vH\}$ is a hindrance in $\Gamma \vdash v$. □

Lemma 2.3. *If Γ is unhindered and $\mathcal{W}$ is a wave in Γ then for every path $P \in \mathcal{W}$ the web Γ/P is unhindered.*

Proof. Suppose that there exists a hindrance $\mathcal{H}$ in Γ/P. Let $\mathcal{J} = \mathcal{W} \setminus \{P\}$, $p = ter(P)$, let Q be the path in $\mathcal{H}$ starting at p, if such exists, and let $\mathcal{K} = \mathcal{H} \setminus \{Q\}$ ($\mathcal{K} = \mathcal{H}$ if $p \notin in[\mathcal{H}]$). Let $\mathcal{L} = \mathcal{K} \uparrow \mathcal{J}$. Assume first that $p \notin in[\mathcal{H}]$. We claim that then $\mathcal{L}$ is a hindrance in Γ, contrary to the assumption that Γ is unhindered. To see this, it is enough to show that $\mathcal{L}$ is a wave, since obviously $in[\mathcal{L}] \neq A$. But this, in turn, is obvious, since in this case $\mathcal{L}$ is just $\mathcal{H} \uparrow \mathcal{W}$, which is a wave .

Assume next that $p \in in[\mathcal{H}]$. Let $\mathcal{H}' = \mathcal{H} \setminus \{Q\} \cup PQ$ and $\mathcal{N} = \mathcal{H}' \uparrow \mathcal{W}$. We claim that $\mathcal{N}$ is a wave (and then, obviously, a hindrance.) To see this, take an $A - B$ path R. Let t be the last point on R which lies on a path from $V[\mathcal{W}]$. Then, since $\mathcal{W}$ is a wave, $t \in ter[\mathcal{W}]$. Hence $t \notin V(P - ter(P))$. But $\mathcal{N}' = \mathcal{H} \uparrow \mathcal{J}$ is a wave in $\Gamma - (P - ter(P))$, and hence the path tR contains a vertex from $ter[\mathcal{N}']$, and hence also from $ter[\mathcal{N}]$. □

Lemma 2.4. *If Γ is unhindered and $a \in A$ then there exists a non-trivial finite path P starting at a such that Γ/P is unhindered.*

Proof. Let $\mathcal{W}$ be a maximal wave in Γ. If $\mathcal{W}$ contains a non-trivial path P starting at a then, by Lemma 2.3, P is the desired path. If not, then, since Γ is unhindered, there exists a vertex v which is connected to a in $\Gamma \angle \mathcal{W}$ and from which B is reachable. We claim that $P = av$ is the desired path. This will clearly follow if we show that $\Gamma \vdash v$ is unhindered. If this is not the case, then by Lemma 2.1 there exists a wave $\mathcal{U}$ in Γ which separates v from B. Since $\mathcal{W}$ does not separate v from B, $\mathcal{W} \not\leq \mathcal{U}$. But, by Lemma 1.2, this contradicts the maximality of $\mathcal{W}$. □

Let S be an $A-B$-separating set of vertices in Γ. We shall denote by $A-part(S)$ the set of all vertices which are separated by S from B, and by $B-part(S)$ the set of all vertices which are separated by S from A (Note that S is contained in both, and so are possibly other vertices.) We write Γ_S for the web $(G[A-part(S)], A, S)$ and Γ^S for the web $(G[B-part(S)], S, B)$. If $\mathcal{W}$ is a wave we write $\Gamma_{\mathcal{W}}$ for $\Gamma_{ter[\mathcal{W}]}$ and $\Gamma^{\mathcal{W}}$ for $\Gamma^{ter[\mathcal{W}]}$.

Lemma 2.5. *If S is $A-B$-separating and Γ is hindered then at least one of Γ_S or Γ^S is hindered.*

Proof. Let $\mathcal{H}$ be a hindrance in Γ. Let $\mathcal{X}$ be the set of all maximal initial parts of paths in $\mathcal{H}$ which are contained in Γ_S, and let $\mathcal{Y}$ be the set of all paths of the form sH, where $s \in S$, $H \in \mathcal{H}$ and s is the last vertex on H belonging to S.

Suppose that $\mathcal{X}$ is not a hindrance in Γ_S. Then there exists a vertex $s \in S$ which is not separated from A by $ter[\mathcal{X}]$. Then the warp $\mathcal{Y} \cup \{(t) : t \in S \setminus \{s\} \setminus in[\mathcal{Y}]\}$ is a hindrance in Γ^S (since a path from s to B in Γ^S which avoids $S \setminus \{s\} \cup ter[\mathcal{Y}]$ could be combined with a path in Γ_S from A to s which avoids $ter[\mathcal{X}]$ to yield a contradiction to the fact that $\mathcal{H}$ is a hindrance.) □

Lemma 2.6. *If Γ has no unending paths, and if it is unhindered, then for any $a \in A$ there exists an $a-B$ path P such that $\Gamma - P$ is unhindered.*

Proof. By Lemma 2.4 there exists a non-trivial path $P_1 = Q_1$ starting at a such that Γ/Q_1 is unhindered. Apply the lemma again, to Γ/Q_1, to obtain a path $P_2 = Q_1Q_2$ in Γ such that Γ/P_2 is unhindered. Since Γ does not contain undending paths, this process must halt at a certain point, and this can happen only when P_k reaches B, for some k. Then P_k is the desired path. □

The following theorem was (essentially) proved in [3]. But we would like to present here a somewhat different proof. The advantage of the new proof is that it makes better use of the bipartite conversion. It differs from the proof in [3] in that it $\vdash$-s vertices, rather than deletes them. Since it is the $\vdash$-ing operation, and not deleting, which is used in the discussion of the uncountable case in the following sections, this may indicate that there is a better chance of extending the proof to the uncountable case.

Theorem 2.1. *Let Γ be an unhindered web, and let $a \in A$ be such that there are only countably many $a-B$ paths. Then there exists an $a-B$ path P such that $\Gamma - P$ is unhindered.*

Proof. Enumerate the $a-B$ paths as $(P_1, P_2, ...)$. We shall choose inductively webs Γ_i, waves $\mathcal{W}_i$ in Γ_i, vertices v_i warps $\mathcal{U}_i$ and sub-warps $\tilde{\mathcal{U}}_i$ of $\mathcal{U}_i$ $(i < \omega)$, as follows. Let $\Gamma_0 = \Gamma$, $\mathcal{W}_0$ and $\mathcal{U}_0$ the trivial wave in Γ, and $v_0 = a$.

Let $\mathcal{W}_1$ be a maximal wave in Γ_0, and let $\Delta_1 = \Gamma_0 \angle \mathcal{W}_1$. By Lemma 1.1 Δ_1 is loose. Let $\mathcal{U}_1 = \mathcal{W}$ and let $\tilde{\mathcal{U}}_1$ be the set of those paths in $\mathcal{U}_1$ which do not start at a. Now let i_1 be first such that P_{i_1} does not meet $V[\tilde{\mathcal{U}}_1]$, and let v_1 be the first vertex on P_{i_1} which does not belong to $V[\mathcal{U}_1]$. Let (u_1, v_1) be the edge of P_{i_1} preceding v_1. Then $u_1 \in ter[\mathcal{W}_1]$, and since Δ_1 is loose, by Corollary 2.1 $\Gamma_1 \stackrel{def}{=} \Delta_1 \vdash v_1$ is unhindered.

Let $\mathcal{W}_2$ be a maximal wave in Γ_1, and define $\Delta_2 = \Gamma_1 \angle \mathcal{W}_2$. Let $\mathcal{U}_2$ be the concatenation of $\mathcal{W}_1$ and $\mathcal{W}_2$. Clearly, then, $\mathcal{U}_2$ is a wave in Γ_2. Let $\tilde{\mathcal{U}}_2$ be the set of those paths in $\mathcal{U}_2$ which do not start at v_0 or v_1. Let i_2 be first such that P_{i_2} does not meet $V[\tilde{\mathcal{U}}_2]$, and let v_2 be the first vertex on P_{i_2} which does not belong to $V[\mathcal{U}_2]$. Let (u_2, v_2) be the edge on P_{i_2} which precedes v_2. Since $\mathcal{U}_2$ is a wave in Γ_2, $u_2 \in ter[\mathcal{U}_2]$. Hence, by Lemma 2.1 $\Gamma_2 \stackrel{def}{=} \Delta_2 \vdash v_2$ is unhindered.

Continuing in this way, we obtain after ω steps a wave $\mathcal{U}$ in the web $\Gamma_\omega \stackrel{def}{=} \Gamma \vdash \{v_0, v_1, \ldots\}$, which is the concatenation of all warps $\mathcal{U}_i$. It is divided into two parts: $\mathcal{Z}$, which consists of those paths in $\mathcal{U}$ which begin at some v_i, and $\tilde{\mathcal{U}} = \mathcal{U} \setminus \mathcal{Z}$, which is the "limit" of all $\tilde{\mathcal{U}}_i$ (i.e its edge set is the union of the edge sets of all $\tilde{\mathcal{U}}_i$,) and whose starting points are all in $A \setminus \{a\}$.

Assertion 2.1. If some path in $\mathcal{Z}$ ends in B (in particular if some v_i is in B,) then the theorem holds.

Proof. Suppose that some path $Q_0 \in \mathcal{Z}$ ends in B. Let k_0 be such that $Q_0 \in \mathcal{W}_{k_0}$ and let $v_{k_1} = in(Q_0)$ (possibly $Q_0 = (v_{k_1})$.) Let u_{k_1} be as in the construction above. Then $u_{k_1} = ter(Q_1)$ for some path $Q_1 \in \mathcal{W}_{k_1}$, and $in(Q_1) = v_{k_2}$ for some $k_2 < k_1$. Let u_{k_2} be as in the construction above, and take a path $Q_2 \in \mathcal{W}_{k_2}$ which terminates at u_{k_2} and starts at v_{k_3} for some $k_3 < k_2$. Continuing this way we get a sequence of triples (v_{k_i}, u_{k_i}, Q_i), where the k_i are descending. Eventually, $k_j = 0$ for some j.

Let $P = Q_j Q_{j-1} \ldots Q_1 Q_0$. We claim that P is the desired path in the theorem. Assume, to the contrary, that $\Gamma - P$ is hindered.

Write $\Pi = \Gamma \vdash \{v_0, v_1, \ldots, v_{k_i}\}$. Let $\Xi_0 = \Pi_{u_{k_0}}$, and for each $i \leq j$ let $\Xi_i = (\Xi_{i-1})_{u_{k_i}}$ and $\Psi_i = (\Xi_{i-1})^{u_{k_i}}$. By Lemma 2.2 the web $\Pi - P$ is hindered. Since Q_0 is a path in the wave $\mathcal{W}_{k_0}$ in Γ_{k_1}, by Lemma 2.3 Γ_{k_1}/Q_0 is unhindered. Hence also $\Psi_1 - P$, which is equal to $\Gamma_{k_1} - Q$, is unhindered. It thus follows that $\Xi_1 - P$ is hindered.

Again, by Lemma 2.5 this implies that either $\Xi_2 - P$ is hindered or $\Psi_2 - P$ is hindered. By the same argument as above, the second of these possibilities would imply that Γ_{k_2}/Q_1 id hindered, in contradiction to Lemma 2.3. Hence $\Xi_2 - P$ is hindered. Continuing this way we obtain that $\Xi_j - P$, which is just the trivial web Γ_A (i.e the web having A as both its source set and its target set, and no edges,) is hindered, which is obviously false. This proves the assertion. □

We shall show that $\tilde{\mathcal{U}}$ is a wave. Since $a \notin in[\tilde{\mathcal{U}}]$, this will mean that it is actually a hindrance, contradicting the assumption that Γ is unhindered. In order to show that $\tilde{\mathcal{U}}$ is a wave, consider an $A - B$ path T. Since $\mathcal{W}_1$ is a wave in Γ, T meets $ter[\mathcal{W}_1]$. Let x be the last vertex on T which belongs to $ter[\mathcal{W}_i]$ for some i. If $x \notin ter[\mathcal{U}]$ then there exists $j > i$ such that $x \in in[\mathcal{W}_j]$. But since $\mathcal{W}_j$ is a wave in Γ_i, the path xT contains a vertex in $ter[\mathcal{W}_j]$, contrary to the choice of i. Thus $x \in ter[\mathcal{U}]$. It remains to show the impossibility of $x \in ter[\mathcal{Z}]$. Suppose that $x = ter(Q)$ for some $Q \in \mathcal{Z}$. Let $in(Q) = v_m$, and suppose that v_m lies on the path P_j. Then the path $P_j v_m Q x T$ is one of the paths P_k. But, by the choice of x, the part xT of this path does not contain a vertex from any $V[\mathcal{W}_n]$, and hence when the turn of P_k arrived, we would have chosen from it all vertices of xT, one by one, as v_p-s. But then $ter(T) = v_p$ for some p, contradicting Assertion 2.1. □

One case in which the conditions of Theorem 2.1 hold is that in which the outdegrees of all vertices are countable, so for webs with this property the theorem is known for all vertices a. Also, clearly, Lemma 1.3 is an immediate corollary. As already mentioned, the countable case of Conjecture 1.1 follows from this lemma directly. So, the theorem yields a new proof of the countable case of Conjecture 1.1.

3. $\aleph_1$-hindrances

The key to the solution of the uncountable bipartite case of Erdős' conjecture was in the definition of higher order hindrances, i.e κ - hindrances for regular uncountable cardinalities κ. The course of the proof there was proving first that a bipartite web is linkable if it does not contain a hindrance or a κ-hindrance for some regular uncountable cardinal κ [6], and then proving that the existence of a κ-hindrance implies the existence of a hindrance (in the old sense [2].)

The notion of higher order hindrances can be carried over to general webs. Since the aim of this paper is to present general techniques rather than complete proofs, we shall be satisfied with presenting the notion of an $\aleph_1$-hindrance, which means that the proofs will apply only to webs of size $\aleph_1$. The reader may take on faith the fact that the notion and proofs can be extended to general cardinalities, and he is referred to [6, 2] for details in the bipartite case.

For an ordinal $\eta \leq \aleph_1$ an *η-ladder* is a sequence $L = (R_\alpha : \alpha < \eta)$, where each "rung" R_α of the ladder is of one of three possible types (the webs Γ^α mentioned in them are to be defined below)

(1) a vertex $v_\alpha \in V(\Gamma^\alpha) \setminus A(\Gamma^\alpha)$

(2) a hindrance $\mathcal{R}_\alpha$ in Γ^α .

(3) an infinite path R_α which is the concatenation of paths in hindrances $\mathcal{R}_\beta, \beta < \alpha$ appearing in rungs of type (2), and has not appeared as yet as a rung of type (3) (i.e it is not R_γ for any $\gamma < \alpha$, γ of type (3).)

The ordinal η is called the *height* of the ladder. (If special mention of the ladder L is necessary, we shall write $\eta(L)$, a remark applying to all notation referring to ladders.) The set of α-s of type (i) is denoted by Φ_i. The webs Γ^α are defined inductively:

If $\alpha \in \Phi_1$ then $\Gamma^{\alpha+1}$ is defined as $\Gamma^\alpha \vdash v_\alpha$.

If $\alpha \in \Phi_2$ then $\Gamma^{\alpha+1} = \Gamma^\alpha \angle R_\alpha$.

If $\alpha \in \Phi_3$ then $\Gamma^{\alpha+1}$ is obtained from Γ^α by deleting $in(R_\alpha)$ from its $'A'$ - set (for the motivation of this definition see below, after the definition of Γ_α.)

For limit α, Γ^α is defined as the "limit" of the webs $\Gamma^\beta, \beta < \alpha$, in an obvious way: its $'A'$ set is the set of all those vertices which belong to $A(\Gamma^\beta)$ for cofinally many β-s in α, and similarly for the definition of its vertex and edge sets. Its $'B'$ set is just B.

Alongside with the the webs Γ^α we define warps $\mathcal{Y}_\alpha = \mathcal{Y}_\alpha(L)$ for each $\alpha < \eta$. Loosely speaking, $\mathcal{Y}_\alpha$ is the concatenation of all hindrances $\mathcal{R}_\beta, \beta < \alpha$. More precisely, we let

$$E[\mathcal{Y}_\alpha] = \bigcup\{E[\mathcal{R}_\beta] : \beta < \alpha, \beta \in \Phi_2\} \setminus \bigcup\{E[\mathcal{R}_\beta] : \beta < \alpha, \beta \in \Phi_3\}$$

The warp $\mathcal{Y}_\alpha$ is not defined solely by its edge set, because it may contain also singleton paths. It will be defined, however, once the set of its initial points is defined. We let $in[\mathcal{Y}_\alpha]$ be the set of all vertices in $A \cup \{v_\beta : \beta < \alpha\}$ which are not hindered in any hindrance $\mathcal{R}_\beta, \beta \in \Phi_2, \beta < \alpha$ and are not $in(R_\beta)$ for any $\beta \in \Phi_3, \beta < \alpha$. Also write $\mathcal{Y}_\eta = \mathcal{Y}(= \mathcal{Y}(L))$.

The following is easily proved by induction:

Lemma 3.1. *$ter[\mathcal{Y}_\alpha]$ separates A from B.*

From the inductive definition of the webs Γ^α it is easy to see that Γ^α is essentially $\Gamma^{ter[\mathcal{Y}_\alpha]}$, the difference being that the vertex set of the latter may be larger - there may be vertices separated by $ter[\mathcal{Y}_\alpha]$ from A, which are therefore in the vertex set of $\Gamma^{ter[\mathcal{Y}_\alpha]}$, but are deleted in the process of forming Γ^α. But for our purposes we may regard these two webs as being the same.

Let $\Gamma_\alpha = \Gamma_{ter[\mathcal{Y}_\alpha]}$.

For $v \in V \setminus A$ let $\rho(v) = \rho_L(v)$ be the first ordinal ρ for which $v \in V(\Gamma_\rho)$.

Rungs of type (3) appear in ladders since the initial point of an infinite path in $\mathcal{Y}$ has lost hope of being linked to B by $\mathcal{Y}$, while it is separated from B by $B(\Gamma_\alpha)$ (Lemma 3.1.) So, it is a "problematic" vertex, which has to be specially taken care of. This complication does not arise, of course, in the case of webs with no undending paths, and as we shall see in Theorem 3.3 also not in the other main case discussed in this paper, namely that of webs with countable outdegrees.

Notation 3.1. Let $F(L) = \cup\{R_\alpha : \alpha \in \Phi_1\}$. For each $\alpha \in \Phi_2$ let HD_α be the set of points in $A(\Gamma^\alpha)$ which are hindered by R_α. We also write HDA_α for the set of points in A which are the initial points of paths in $\mathcal{Y}_\alpha$ terminating at HD_α.

For a limit ordinal α let $\mathcal{NF}_\alpha$ be the set of all infinite paths in $\mathcal{Y}_\alpha$ starting at A.

Also let $NF_\alpha = in[\mathcal{NF}_\alpha]$.

Definition 3.1. *If L is of height $\aleph_1$ and Φ_2 or Φ_3 are stationary in $\aleph_1$ then L is called an* $\aleph_1$-hindrance.

One of the two main ingredients of the proof of Conjecture 1.1 goes over to general webs:

Theorem 3.1. *If Γ contains an $\aleph_1$-hindrance then it is hindered.*

Proof. In [2] the theorem was proved for bipartite webs. We shall use this case of the theorem, but not directly. One has to go a bit into the proof in order to make it applicable to general webs.

Let $L = (R_\alpha : \alpha < \aleph_1)$ be an $\aleph_1$-hindrance. Assume, first, that $\Phi_2(L)$ is stationary. Then L corresponds to an $\aleph_1$-hindrance $N = (T_\alpha : \alpha < \aleph_1)$ in $b(\Gamma)$. By the main theorem of [2], N gives rise to a hindrance H in $b(\Gamma)$. We would like to show that H can be chosen so that $w(H)$ is a hindrance in Γ, i.e so that it does not contain non-starting paths.

Let $M = b(E[\mathcal{Y}(L)])$, i.e the set of edges in $b(\Gamma)$ corresponding to the edges in paths in the warp $\mathcal{Y}$. Then $M = E[\mathcal{Y}(N)]$, and it is a matching in $b(\Gamma)$. We write $F = F(L)$ and $BF = F(N)$. Obviously $BF = b(F)$

The basic concept in the proof in [2] is that of *popularity.*

Given a set of vertices U in any graph, a set $\mathcal{P}$ of paths in the graph is called *U-joined* if every two members of $\mathcal{P}$ meet only within U. A subset U of $B(b(\Gamma))$ is called *popular* if there exists a U-joined set $\{Z_\beta : \beta \in \Psi\}$ of paths, where each Z_β is an M-alternating path from some vertex u'' of U to some vertex $v' \in HD_\beta(N)$ and $\Psi \subseteq \Phi_2$ is stationary.

By the usual abuse of notation we shall say that a vertex $u'' \in B(b(\Gamma))$ is popular if $\{u''\}$ is popular.

We shall now follow a construction given in [2].

Let U_0 be the set of all unpopular vertices in BF (that is, the set of all unpopular vertices $v''_\alpha, \alpha \in \Phi_1$.) Let K_1 be the set of all vertices x' connected to them, U_1 the set of all unpopular vertices in $M[K_1]$, K_2 the set of all unpopular vertices connected to vertices in U_1, U_2 the set of unpopular vertices in $M[K_2]$, and so forth. Let, eventually, $U = \bigcup\{U_i : i < \omega\}, K = \bigcup\{K_i : i < \omega\}$, and $\Delta = b(\Gamma) - U - K$.

Assertion 3.1. All points in $A(\Delta)$ are connected in $b(\Gamma)$ only to points from $B(\Delta)$.

Proof. This is clear, since by definition K is the set of points connected in $b(\Gamma)$ to U. □

Let $M' = M_{|B(\Delta)}$ and $FR = B(\Delta) \setminus \bigcup M$ ("FR" stands for "free".) Clearly, $FR = BF \setminus U$.

The points in FR were popular in Γ. The main idea of the proof in [2] is that they still remain so in Δ, although Δ is formed from Γ by the removal of vertices. In a somewhat informal manner, this can be put in:

Assertion 3.2. Every point p in FR is still popular in Δ, namely it has a p-joined set of M'-alternating paths to "stationarily many" hindered points belonging to $A(\Delta)$ (i.e., points in $A(\Delta) \setminus \bigcup M'$.)

A little more formal, and stronger, is the following:

Assertion 3.3. Let $p \in FR$ and let $\{X_\alpha : \alpha \in \Psi\}$ be a p-joined set of M-alternating paths, where Ψ is stationary and each path X_α goes from p to HD_α. Then $\{\beta : K \text{ meets } X_\beta\}$ is non-stationary.

In [2] Assertion 3.3 was derived from an even stronger one, which we shall also use here:

Assertion 3.4. The entire set U is unpopular.

To see how Assertion 3.3 follows from Assertion 3.4, assume that the set Θ of β-s for which K meets X_β (say, at a point z_β) is stationary. For each $\beta \in \Theta$ choose a point $u_\beta \in U$ connected to z_β. Then the set of paths $\{u_\beta z_\beta X_\beta : \beta \in \Theta\}$ would show that U is popular.

Another immediate corollary of Assertion 3.4 is:

Assertion 3.5. The set of α-s for which $HD_\alpha \cap A(\Delta) \neq \emptyset$ is stationary.

This would clearly follow also from Assertion 3.2 if we knew that $FR \neq \emptyset$. But this is not guaranteed.

In the bipartite case, Assertion 3.2 is sufficient for the proof of Theorem 3.1. In fact, all that is used is that every point $p \in FR$ has a p-joined set of size $\aleph_1$ of M'-alternating paths to points in $A(\Delta) \setminus \bigcup M'$. Since $|FR| \leq \aleph_1$, these alternating paths can be used to match all points of $B(\Delta)$ one by one into $A(\Delta)$. One then uses Fodor's Lemma and Assertion 3.5 to prove that this matching is strictly into. By Assertion 3.1 this matching provides the desired hindrance in $b(\Gamma)$.

An observation which will be used below is that $\mathcal{Y}$-alternating paths in Γ are in one to one correspondence with M-alternating paths in $b(\Gamma)$, with the natural correspondence: an M-alternating path Q corresponds to a $\mathcal{Y}$-alternating path $w(Q)$ whose edge set is $w(E(Q))$. In fact, this can be used as a definition of a $\mathcal{Y}$- alternating path. (A detail worth noting is that when a $\mathcal{Y}$-alternating path Q lingers for more than one edge on a path from $\mathcal{Y}$, its corresponding path alternates between edges of M and edges of the form (x', x''), which are not in M.)

Write $S = w(FR)$ and $\mathcal{Y}' = w(M')$. Then $\mathcal{Y}'$ consists of fragments of paths from $\mathcal{Y}$. Let $\mathcal{Z}$ be the set of paths in $\mathcal{Y}'$ which start at A. Each $s \in S$ has $\aleph_1$ many $\mathcal{Y}'$-alternating paths meeting only at s. Listing the points of S in a sequence of length at most ω_1, we can link them one by one using their $\mathcal{Y}'$-alternating paths, and sticking to the rule that a path in $\mathcal{Y}$ is never used in alternating paths pertaining to more than one point from S.

The trouble is that if we are not careful in the application of the alternating paths, the warp which results may contain non-starting paths. To overcome this difficulty we prove:

Assertion 3.6. Let $s \in S$ and let $\{X_\beta : \beta \in \Psi\}$ be an s-joined set of $\mathcal{Y}'$-alternating paths, each X_β leading to HD_β, and Ψ being stationary. Then there exists a stationary subset Ψ' of Ψ such that for each $\beta \in \Psi'$ the paths from $\mathcal{Y}'$ which X_β meets are all in $\mathcal{Z}$.

Proof. We shall prove something even stronger, namely that the set of those β-s for which X_β meets a path not in $\mathcal{Z}$ is non-stationary. This set consists of two (possibly overlapping) parts. The first, Ξ_1, consists of β-s for which X_β meets a path $Q_\beta \in \mathcal{Y}$ which does not start at A. The second, Ξ_2, consists of β-s for which X_β meets a path, say Y_β, at a point y_β such that there exists a point $k_\beta \in K$ preceding y_β on Y_β.

For each $\beta \in \Xi_1$ the initial point of Q_β is a vertex $v = R_\alpha$ for some $\alpha < \beta$. Hence, if Ξ_1 were stationary, by Fodor's Lemma $\aleph_1$ many X_β-s would share the same Q_β. This is impossible, since the paths X_β are disjoint and each Q_β is finite.

Assume next that Ξ_2 is stationary. For each $\beta \in \Xi_2$ choose a vertex $u''_\beta \in U$ connected to k'_β and witnessing the fact that $k' \in K$. Consider then the paths $J_\beta = u_\beta k_\beta Y_\beta y_\beta X_\beta$. The paths $b(J_\beta)$ in $b(\Gamma)$ then show that U is popular, contradicting Assertion 3.4. □

Let $(s_\delta : \delta < \nu \leq \aleph_1)$ be a listing of all element of S. Choose inductively for each s_δ a $\mathcal{Z}$-alternating path P_δ from s_δ to some $HD_{\alpha(\delta)}(L)$ which meets only paths in $\mathcal{Z}$ which were not met by any $P_\zeta, \zeta < \delta$, does not contain edges from $E[\mathcal{Y}] \setminus E[\mathcal{Z}]$, and such that $\alpha(\delta) \neq \alpha(\zeta)$ for $\zeta < \delta$. Such choice is possible, by Assertion 3.6. Let $\mathcal{H}$ be the warp resulting from the application of all P_δ-s to $\mathcal{Y}$. Since the paths P_δ correspond (by the correspondence b) to M'-alternating paths in Δ, $ter[\mathcal{H}] = w(B(\Delta))$ and $in[\mathcal{H}] \subseteq w(A(\Delta))$. By Assertion 3.1 this means that $H = b(E[\mathcal{H}])$ is a wave in $b(\Gamma)$. By Fodor's Lemma there exists at least one $\alpha \in \Phi_2$ such that no path P_δ reaches $HD_\alpha(N)$. (In fact, there is no need to invoke Fodor's Lemma for this, just choose the paths P_δ so that this condition is fulfilled.) Hence H is, in fact, a hindrance. Since the paths P_δ only use paths from $\mathcal{Z}$ to alternate on, and distinct ones at that, the paths in $\mathcal{H}$ all start at A. Hence, by Lemma 1.7, $\mathcal{H}$ is the hindrance desired in the theorem.

The case that Φ_3 is stationary is similar, but for simplicity of presentation it is better to work directly in Γ. For each $\alpha \in \Phi_3$ let P_α be the infinite path which constitutes the rung R_α. A vertex $w \in F$ is called *popular* if for a stationary subset Ψ of Φ_3 there exist $\mathcal{Y}$-alternating paths $Q_\psi, \psi \in \Psi$ from w to P_ψ. (We demand that such a path ends when it reaches some vertex in P_ψ). The proof then goes along the same lines as in the case of Φ_2 stationary. □

We conjecture that also the other main ingredient of the proof in the bipartite case goes over to general webs, namely:

Conjecture 3.1. If $|V(\Gamma)| \leq \aleph_1$ and Γ contains no hindrance and no $\aleph_1$-hindrance then Γ is linkable.

We are able to prove the conjecture only in the cases in which Conjecture 1.3 is known. So, we have:

Theorem 3.2. *Conjecture 1.2 (and hence also Conjecture 1.1) is true for webs of size* $\aleph_1$ *with no unending paths or with countable outdegrees.*

Proof. By Lemma 1.5 the theorem holds for webs in which A or B are countable. Hence we may assume that $|A| = |B| = \aleph_1$. Enumerate $V \setminus A$ as $(v_\theta : \theta < \aleph_1)$. Define inductively the rungs R_α of a ladder L as follows. Let $\Gamma^0 = \Gamma$. Suppose that R_δ have been defined for all $\delta < \alpha$. Let Γ^α be defined as in the definition of ladders made above. First priority in the choice of rungs is given to type (3). That is, if there exists an infinite path in $\mathcal{NF}_\alpha$ which is not equal to R_β for any $\beta < \alpha$, then choose such a path to be R_α.

Second priority is given to type (2). So, if there does not exist an infinite path as above, but there exists a hindrance in Γ^α, choose R_α to be such a hindrance.

If no path or hindrance as above exist, choose R_α to be the first v_θ in $V \setminus V(\Gamma_\alpha)$.

We proceed in this definition until all vertices in V have been covered. This, obviously, happens for some $\eta \leq \aleph_1$. For each $v \in V \setminus A$ write $\alpha(v)$ for the first α for which $v \in V(\Gamma_\alpha)$.

Assume first that $\eta(L) < \aleph_1$. Then the warp $\mathcal{W}$ consisting of all paths in $\mathcal{Y}(L)$ starting at A satisfies the condition that $A \setminus in[\mathcal{W}]$ is countable. Moreover, since all points of $V(\Gamma) \setminus A$ appear in the ladder L, $ter[\mathcal{Y}(L)]$ must cover all but countably many vertices in B. It follows that Γ satisfies the condition of Lemma 1.4, and hence also Conjecture 1.2.

Thus we may assume that $\eta(L) = \aleph_1$. By the assumption that Γ is $\aleph_1$-unhindered, there exists a club set Π such that $\Pi \cap (\Phi_2 \cup \Phi_3) = \emptyset$.

The idea is to use Π to link Γ. Informally, this is done as follows: we link one point from A to B by a path P, using Lemma 2.6 and Theorem 2.1. Let π_1 be the first element of Π for which all vertices in P are contained in Γ_{π_1}. We wish to take care of the points in A which are not linked by $\mathcal{Y}_{\pi_1}$ into $B(\Gamma_{\pi_1})$. There are only countably many of those, and we link them one by one to B. At each step we take a possibly larger $\pi_i \in \Pi$ so that Γ_{π_i} contains the new paths. We add to the set of problematic points (those which have to be taken care of) all points in A which are hindered in some $R_\alpha, \pi_i > \alpha \in \Pi_2$, and also NF_α, if α is a limit ordinal. After ω steps we have taken care of all problematic points in Γ_π, where π is the limit of all π_i. Since $\pi \in \Pi$, the web Γ^π is unhindered. Also, $NF_\pi = \emptyset$. Hence we can turn over a new leaf - try to link points in Γ^π. By the club-ness of Π we can join all the partial linkages obtained in this procedure to obtain a linkage of A.

Formally all this is a bit more cumbersome:

Clearly, NF_α is countable for every $\alpha < \aleph_1$. For otherwise the rungs of L are of type (3) for all $\alpha < \beta < \aleph_1$, making L an $\aleph_1$-hindrance. Similarly, HD_α is countable, or else we could choose the rungs R_β to be of type (2) from α onwards.

List the points of A as $a_\theta, \theta < \aleph_1$.

For a subset Z of $V \setminus A$ let $\pi(Z)$ be the minimal ordinal $\pi \in \Pi$ such that $\pi > \rho(z)$ for every $z \in Z$.

By Lemma 2.6 and Theorem 2.1 there exists an $a_0 - B$ path P_0 such that $\Delta_1 = \Gamma - P_0$ is unhindered. Let $\pi_1^1 = \pi(V(P_0))$. Let K_1 be the set of all vertices in $A \setminus \{a_0\}$ which are initial vertices of paths from $V[\mathcal{Y}(L)]$ which are met by P_0, and let

$$D_1 = K_1 \cup HDA_{\pi_1^1}(L) \cup NF_{\pi_1^1} \cup \{a_1\}$$

(For the definition of HDA see Notation 3.1 .) Clearly, D_1 is countable, so we can order it in an ω-sequence d_1. Let v_1 be the first vertex in this list. There exists then a $v_1 - B$ path P_1 such that $\Delta_2 = \Delta_1 - P_1$ is unhindered. Let $\pi_1^2 = \pi(V[\{P_0, P_1\}])$.

Let K_2 be the set of vertices from $A \setminus \{v_0, v_1\}$ (where $v_0 = a_0$) which are initial vertices of paths from $\mathcal{Y}$ which are met by P_0, P_1, and let

$$D_2 = K_2 \cup HDA_{\pi_1^2} \cup (NF_{\pi_1^2} \setminus NF_{\pi_1^2}) \cup \{a_2\}$$

Intersperse D_2 in the even numbered places of d_1, to obtain a list d_2. Let v_2 be the first element in d_2 which does not belong to $\{v_0, v_1\}$. Choose a $v_2 - B$ path P_2 such that $\Gamma_3 = \Gamma_2 - P_1$ is unhindered. We continue in this way, and obtain sequences $(v_i), (\pi_1^i), (P_i)$ of length ω. Let $\pi_1 = \sup\{\pi_1^i : i < \omega\}$. The interspersing of the sets D_i in the sequence of points to be linked to B is done so as to guarantee that each point is reached at some stage (so, for example, D_3 is interspersed in every third place, say, in d_2.) Then, after ω steps, when the dust has settled, the warp N_0 consisting of $\{P_i : i < \omega\}$ together with all paths in $\mathcal{Y}_{\pi_1}$ which connect A to $B(\Gamma_{\pi_1})$ and which do not meet any P_i, links A into $B(\Gamma_{\pi_1})$.

Now repeat the procedure with Γ^{π_1} replacing Γ. This is possible, since $\pi_1 \in \Pi$, and hence Γ^{π_1} is unhindered. This produces a linkage N_2 of $B(\Gamma^{\pi_1})$ into $B(\Gamma^{\pi_2})$ for some $\pi_1 < \pi_2 \in \Pi$.

Continuing this way, we obtain an $\aleph_1$-sequence of ordinals $\pi_\zeta \in \Pi$, and linkages $\mathcal{N}_\zeta$ of $A(\Gamma_{\pi_\zeta})$ into $B(\Gamma_{\pi_{\zeta+1}})$. For limit ζ we just define $\pi_\zeta = \sup\{\pi_\beta : \beta < \zeta\}$. Since Π is closed, $\pi_\zeta \in \Pi$.

At each step ζ, the first a_θ in the list of elements of A which have not been linked as yet to B is not hindered in Γ_{π_ζ}, since this web is not hindered. So, it is the starting point of a path $Q \in \mathcal{Y}_{\pi_\zeta}$. This path is not infinite, since if it is then, since $\Pi \cap \Phi_3 = \emptyset$, $Q \in \mathcal{NF}_\alpha$ for some $\alpha < \pi_\zeta$. But then, by our construction, a_θ has been linked already to B at a previous stage. Let $a'_\theta = ter(Q)$. Then $a'_\theta \in A(\Gamma_{\pi_\zeta})$. We put it first in the first list of vertices to be linked, so that after the ζ-th stage a_θ is linked to B. This guarantees that after $\aleph_1$ steps all vertices of A are linked to B. □

Remark: The true content of the theorem, and what is actually proved, is that Conjecture 3.1 can be reduced to Conjecture 1.3 for webs of size $\aleph_1$. We chose the above presentation since it states absolute results, rather than a result about implication. In fact, both cases mentioned in the theorem have simpler proofs. For webs with no unending path one can construct a ladder as in the proof of the theorem, and then link its 'slices' (defined by the club set Π), and these linkages join up to form a linkage of Γ, since no unending path has been generated. In the case of countable outdegrees one can make do with a simpler kind of $\aleph_1$-hindrances, as follows:

Call a hindrance *degenerate* if all of its paths are singletons. A point in A hindered by a degenerate hindrance is just one from which B is unreachable. An $\aleph_1$-hindrance is called degenerate if all of it rungs $R_\alpha, \alpha \in \Phi_2$, are degenerate. Note that in such an $\aleph_1$-hindrance $\mathcal{Y} = \emptyset$, and hence there are no rungs of type (3). Similar arguments to those above (only simpler) show:

Theorem 3.3. *If $|V(\Gamma)| \leq \aleph_1$, the outdegrees of all vertices in Γ are countable, and Γ does not contain a hindrance or a degenerate $\aleph_1$-hindrance, then it is linkable.*

Of course, this follows from the previous theorems, since not being hindered suffices. What it really means is that the above proof can be simplified in this case. A version which does add new information, however, is this:

Theorem 3.4. *A web of size $\aleph_1$ with countable outdegrees is linkable if and only if every countable subset of A is linkable, and it does not contain a degenerate $\aleph_1$-hindrance.*

4. A remark on necessary conditions for linkability

Conjecture 1.2 will yield (if true) a sufficient condition for linkability in webs. The proof of the bipartite case of the conjecture gave, in fact, more than just a sufficient condition - a necessary and sufficient condition is known in this case.

Podewski and Steffens [8] defined (in somewhat different terms) a special type of waves. A wave $\mathcal{W}$ is called *tight* if there is no warp $\mathcal{U}$ such that $in[\mathcal{U}] = in[\mathcal{W}]$, $ter[\mathcal{U}] \subset ter[\mathcal{W}]$ and $ter[\mathcal{U}] \neq ter[\mathcal{W}]$. An *obstruction* is a tight hindrance. In [8] it was proved that a countable bipartite graph is linkable if and only if it is unobstructed. $\aleph_1$- obstructions are defined by taking the rungs $R_\alpha, \alpha \in \Phi_2$ to be obstructions, rather than hindrances. Again, it was proved [6] that a bipartite graph of size $\aleph_1$ is linkable if and only if it does not contain an obstruction or an $\aleph_1$-obstruction.

It is not even completely clear what should be the formulation of the necessary and sufficient condition for linkability in webs. Transferring the notions of tight waves and obstructions verbatim does not work, as the following example shows:

Let $A^1 = \{a_i^1 : i < \omega\}$ $A^2 = \{a_i^2 : i < \omega\}$, $B = \{b_i : i < \omega\}$ and $C = \{c_i : i < \omega\}$. Also let $A = A^1 \cup A^2$ and $V(G) = A \cup B \cup C$. Let $E(G) = A^2 \times C \cup C \times B \cup \{(a_i^1, c_i) : i < \omega\} \cup \{(a_i^2, b_i) : i < \omega\}$. Then the web $\Delta = (G, A, B)$ does not contain an obstruction in the above sense, and yet it is not linkable. What does exist in Δ is a tight wave $\mathcal{T}$ with a proper subwave $\mathcal{T}'$, which means to say that a proper subset of $ter[\mathcal{T}]$ is $A - B$-separating (take $\mathcal{T} = \{(a_i^2, b_i) : i < \omega\} \cup \{(a_i^1, c_i) : i < \omega\}$). It is possible that this is the right notion for an "obstruction" in the case of webs. It will be interesting (and possibly of value for the proof of the general conjecture) to show that the non-existence of obstructions in this sense is sufficient for linkability in countable webs.

Podewski and Steffens [8] noted that in a bipartite web Γ there exists a maximal linkable subset of $A(\Gamma)$ (the set of initial points of a maximal tight wave is such a set.) The web Δ above provides a counterexample to this statement in the case of countable webs: it is not hard to show that a subset N of A is linkable if and only if $A \setminus N$ is infinite, and clearly there is no maximal subset N of A satisfying this condition.

References

1. R. Aharoni. Menger's theorem for graphs containing no infinite paths. *European J. Combin.*, 4:201–204, 1983.
2. R. Aharoni. Konig's duality theorem for infinite bipartite graphs. *J. London Math. Soc.*, 29:1–12, 1984.
3. R. Aharoni. Menger's theorem for countable graphs. *J. Combin. Th., ser. B*, 43:303–313, 1987.
4. R. Aharoni. Linkability in countable-like webs. *Cycles and rays, G. Hahn et al. ed., Nato ASI series*, pages 1–8, 1990.
5. R. Aharoni and R. Diestel. Menger's theorem for countable source sets. *Comb., Pr. and Comp.*, 3:145–156, 1994.
6. R. Aharoni, C. St. J. A. Nash-Williams, and S. Shelah. A general criterion for the existence of transversals. *Proc. London Math. Soc.*, 47:43–68, 1983.
7. D. Kőnig. *Theorie der Endlischen und unendlischen Graphen.* Chelsea, New York, 1950.

8. K.-P Podewski and K.Steffens. Injective choice functions for countable families. *J. Combin. Theory, ser.B*, 21:40–46, 1976.
9. K.-P Podewski and K.Steffens. Uber Translationen und der Satz von Menger in unendlischen Graphen. *Acta Math. Aca. Sci. Hungar.*, 30:69–84, 1977.

On Order-Perfect Lattices

Igor Kříž[1]

Department of Mathematics, The University of Michigan, Ann Arbor, Michigan, 48109 U.S.A.

Summary. We investigate the property of certain well-founded orderings to have a chain of maximal ordinal length. We show that Heyting algebras and countable modular lattices have this property, but we also present an example of a "well-behaved" lattice which does not have it. We prove a general necessary and sufficient condition for modular lattices to have the property in a hereditary form. This hereditary form is called *order-perfectness*, being analogous to perfectness of finite graphs. Certain well-known theorems of D.H.J. de Jongh, R. Parikh, D. Schmidt, E.C. Milner, N. Sauer, and N. Zaguia turn out to be order-perfectness results.

1. Introduction

In every ordering of a set, there are maximal chains with respect to inclusion. In well-founded orderings (see 2.1), chains have ordinal type. However, different maximal chains may have different ordinal types and a question arises as to whether or not there is a chain of maximal order type. This need not be true in general. It turns out to be interesting to investigate well-founded orderings A such that A itself and also all the sets of the form $\{x \in A | x < a\}$ for an $a \in A$ have a chain of maximal order type. We call such orderings *order-perfect.* A motivation for this terminology arises in Section 2. Of course, each finite ordering is order-perfect. In fact, each $AWPO$ (see 2.1) is order-perfect. This is a special case of Theorem 1.2 from Milner and Sauer [3]. Also, the special case of it for countable orderings is equivalent to a Theorem by D. Schmidt [5].

The aim of this paper is to prove order-perfectness for different kinds of orderings, namely for certain classes of lattices. To indicate the motivation, let us take an example. For a WPO (see 2.1) A, the set $\downarrow A$ of all downward closed subsets is well-founded (the converse is also true). Maximal chains in $\downarrow A$ are in $1-1$-correspondence with linear extensions of A. As is well known (see [2]), for each WPO A there exists a linear extension of maximal ordinal type, and so $\downarrow A$ is order-perfect. This is a special case of our result saying that each well-founded completely distributive lattice is order-perfect.

More generally, we present a necessary and sufficient condition for a well-founded modular lattice to be order-perfect. This condition is true for Heyting algebras (cf. [1]) as well as for countable modular lattices. We do not know if there is a natural common generalization of these two facts. In Section 4 we show that one fairly natural candidate fails (the counter-example is a complete sublattice of $\mathrm{Vect}(\mathbb{R}^{\omega})$).

Problem: Is there a class $\mathcal{C}$ of lattices which can be characterized by equations, includes both Heyting algebras and countable modular lattices and has the property that each well-founded lattice from $\mathcal{C}$ is order-perfect?

To conclude this section, let us remark that we call our main concept order-perfectness to avoid confusion with the work of Milner & Pouzet ([4]) who consider perfectness of infinite comparability graphs in the usual graph-theoretical sense.

[1] The author is an Alfred P. Sloan Fellow and an NSF National Young Investigator

2. Preliminaries. Orderings

Definitions 2.1. *In this paper,* $\oplus$ resp. $+$ *denotes the natural (resp. ordinal) sum of ordinal numbers. For a well-ordered set* K *(isomorphic to an ordinal) a subset* $S \subseteq K$ *is called closed if*

$$(T \subseteq S \;\&\; T \text{ has a supremum in } K) \to \sup T \in S.$$

Now let A *be an ordering. The symbol* $\overline{A}$ *designates the ordering obtained from* A *by adding a new greatest element. A chain (resp. strict, decreasing, strictly decreasing chain) in* A *is a sequence* $(a_\alpha)_{\alpha\in\lambda}$ $(\lambda \in \mathrm{Ord})$ *such that* $\alpha < \beta$ *implies* $a_\alpha \leq a_\beta$ *(resp.* $a_\alpha < a_\beta, a_\alpha \geq a_\beta, a_\alpha > a_\beta$*). The length* $|c|$ *of a chain* $c = (a_\alpha)_{\alpha\in\lambda}$ *is the ordinal type of the set* $\{\alpha \in \lambda \mid \beta < \alpha \to a_\beta < a_\alpha\}$*. A chain* $(a_\alpha)_{\alpha\in\lambda}$ *is continuous, if* $a_{\sup I} = \sup\{a_i \mid i \in I\}$ *for any* I *such that* $I \cup \{\sup I\} \subseteq \lambda$*; in particular, the right hand supremum (the unique least upper bound) is required to exist. An antichain in* A *is a set of pairwise incomparable elements.*

The ordering A is *well-founded* (abb. WF) if it has no infinite strictly decreasing chain. A mapping $\varphi : A \to B$, where A, B are orderings, ir called *monotone* resp. *strictly monotone* if it satisfies $a \leq b \to \varphi(a) \leq \varphi(b)$ (resp. $a < b \to \varphi(a) < \varphi(b)$). For well-founded A we define

$$\begin{aligned}\mu(A) &= \min\{\gamma \mid \exists\varphi : A \to \gamma \text{ strictly monotone}\}\\ \overline{\lambda}(A) &= \sup\{\gamma \mid \exists\varphi : \gamma \to A \text{ strictly monotone}\}.\end{aligned}$$

Observe that $\overline{\lambda}(A) \leq \mu(A)$. We write $\lambda(A)$ instead of $\overline{\lambda}(A)$ if this supremum as attained (i.e. if there is a maximum).

Put $\downarrow A = \{B \subseteq A \mid a \in B \;\&\; b \leq a \to b \in B\}$. Then $\downarrow A$ is ordered by inclusion. For $a \in A$ we denote by $\Downarrow a$ (resp. $\downarrow a$) the induced ordering on $\{b \in A \mid b < a\}$ (resp. $\{b \in A \mid b \leq a\}$). An ordering A is called *order-perfect* if we have

$$\lambda(\Downarrow a) = \mu(\Downarrow a) \text{ for any } a \in \overline{A}.$$

(In particular, the left hand sides are required to exist. We call A *stratified*, if for each $a \in A$ all maximal chains in $\Downarrow a$ have the same ordinal length. For a poset A we say that $a \in A$ *covers* $b \in A$ if we have $a > b$ and if there is no $c \in A$ such that $b < c < a$.

Let A be a partial ordering. A (finite or infinite) sequence (a_i) in A is called *bad* if we have

$$i < j \to a_i \not\leq a_j.$$

The ordering A is called a *well-partial-ordering* (abb. *WPO*) if it contains no infinite bad sequence. We denote by $A^{<\omega}$ the set of all finite sequences in A and put

$$\text{Bad } A = \{(a_0, \ldots, a_n) \in A^{<\omega} \mid i < j \to a_i \not\leq a_j\}.$$

Thus, the empty sequence belongs to Bad A. A mapping $\varphi : \text{Bad } A \to \gamma + 1$ $(\gamma \in \mathrm{Ord})$ is called a *character* of A if we have

$$\varphi(\lambda x) < \varphi(\lambda) \text{ whenever } \lambda x \in \text{Bad } A.$$

Put

$$\begin{aligned}c(A) &= \min\{\gamma \mid \exists \text{ a character } \varphi : \text{Bad } A \to \gamma + 1\}\\ d(A) &= \max\{\gamma \mid \gamma \text{ is isomorphic to a linear extension of } A\}.\end{aligned}$$

(Let us recall that an *extension* means a stronger ordering on the same set.) An extension of a WPO is always WPO (such a statement is not true for WF orderings). Thus, a linear extension of the WPO A is a well-ordering. The existence of a number $d(A)$ was proved in [2].

Now let A be WF, $a \in A$. Define the *height* $h(a)$ of an element $a \in A$ by $h(a) = \overline{\lambda}(\Downarrow a)$. (The reader should be warned that the terminology in the literature differs and some authors use the word "height" in a different meaning.) We call A *almost well-partial ordering* (AWPO) if for any sequence $(a_i)_{i \in \omega}$ in A there exist $i < j$ with

$$a_i \leq a_j \text{ or } h(a_i) \geq h(a_j).$$

For an ordering A and a set $M \subseteq A$ put

$$A_M = \{x \in A \mid (\forall y \in M)\ y \nleq x\}.$$

For a subset $B \subset \mathrm{Ord}$ put

$$M(B) = \sup\{x + 1 \mid x \in B\}.$$

Thus, $M(B)$ is the least ordinal greater than all members of B.

Proposition 2.1. *Let A be WF. Then A is order-perfect iff all the numbers $\lambda(A), \lambda(\Downarrow a)$ $(a \in A)$ exist.*

Proof. If the numbers $\lambda(A)$, $\lambda(\Downarrow a)$ exist, then we have

$$\begin{aligned} \lambda(\Downarrow a) &= M(\{\lambda(\Downarrow b) \mid b < a\}) \\ \lambda(A) &= M(\{\lambda(\Downarrow a) \mid a \in A\}). \end{aligned}$$

We can easily check, however, that analogous recurrent relations hold for μ. □

Remark 2.1. Of course, not all WF orderings have to be order-perfect. Put, e.g., $A_\alpha = \mathrm{Bad}\alpha$ with

$$a \leq b \equiv (\exists c)\ a = bc.$$

Then we have

$$\overline{\lambda}(A_\alpha) = \omega, \mu(A_\alpha) = \alpha.$$

Proposition 2.2. *For each WPO A we have*

$$c(A) + 1 = \mu(\downarrow A) = \lambda(\downarrow A) = d(A) + 1.$$

In particular, $c(A) = d(A)$.

Proof. Once we know that $d(A)$ always exists (see [2] or Section 3 below), the last two equalities are trivial. We present a proof of the first one:
$\leq$: If $\varphi : \downarrow A \to \mu(\downarrow A)$ is strictly monotone then the mapping $\psi : \mathrm{Bad}A \to \mu(\downarrow A)$ given by $\psi(a_1 \ldots a_n) = \varphi(A_{\{a_1 \ldots a_n\}})$ is a character. $\geq$: If

$$\psi : \quad \mathrm{Bad}\, A \to \gamma + 1$$

is a character, then define $\varphi : \downarrow A \to \gamma + 1$ by

$$\varphi(B) = \min\{\psi(T) \mid T \text{ is a bad sequence in } A \backslash B\}.$$

To see that φ is strictly monotone, let $B, C \in \downarrow A$, $\;B \subset C$. There is an element

$$c \in C \backslash B.$$

let T be a bad sequence in $A \backslash C$ with

$$\varphi(C) = \psi(T).$$

The concatenation Tc is a bad sequence, since $C \in \downarrow A$. Thus,

$$\varphi(B) \leq \psi(Tc) < \psi(T) = \varphi(C). \quad \square$$

Theorem 2.1 (Milner and Sauer [3]). *Each AWPO A is order-perfect.* $\square$

Lemma 2.1. *Let I, J, K be well-ordered sets, $K = I \cup J$.*

Then

(a) $c(K) \leq c(I) \oplus c(J)$
(b) If $\min K \in J$, $|I| \geq \omega$ and I, J are closed in K then $c(K) < c(I) \oplus c(J)$.

Proof. (a) is a special case of a theorem proved in [2].
(b) Suppose the contrary. Without loss of generality, $K = \kappa$ is an ordinal. Let λ be the largest limit ordinal with
$\lambda \leq \kappa$.
By our assumptions,

$$I \cap \lambda \neq \emptyset \ \ \& \ \ J \cap \lambda \neq \emptyset. \tag{$*$}$$

There exist $\alpha < \lambda$, $\beta \leq \lambda$ such that β is sum-closed and

$$\alpha + \beta = \lambda.$$

Put

$$\beta' = \{\alpha + i \mid i \in \beta\}.$$

We will show that
One of the sets $I \cap \lambda, J \cap \lambda$ has a maximum. (**)
Indeed, suppose (**) false. Since both I and J are closed in K, the sets
$I \cap \beta', J \cap \beta'$
are cofinal in β'. We conclude that

$$c(I) \oplus c(J) \geq \alpha + 2\beta' > K.$$

Thus, (**) is proved.
Now let, say, $\epsilon = \max(I \cap \lambda)$. Define a mapping $\varphi : \kappa \to \kappa + 1$ by

$$\begin{aligned}
\varphi(\beta) &= \beta \ \text{ for } \ \beta < \epsilon \\
\varphi(\epsilon) &= \lambda \\
\varphi(\beta) &= \beta - 1 \ \text{ for } \ \epsilon < \beta < \epsilon + \omega \\
\varphi(\beta) &= \beta \ \text{ for } \ \epsilon + \omega \leq \beta < \lambda \\
\varphi(\beta) &= \beta + 1 \ \text{ for } \ \lambda \leq \beta.
\end{aligned}$$

Observe that we have

$$\begin{aligned}
\kappa + 1 &= \varphi(I) \cup \varphi(J) \\
c(\varphi(I)) &= c(I) \ \text{ and } \ c(\varphi(J)) = c(J).
\end{aligned}$$

We conclude that

$$c(I) \oplus c(J) \geq c(\varphi(I)) \oplus c(\varphi(J)) \geq c(\varphi(I) \cup \varphi(J)) \geq \kappa + 1. \quad \square$$

3. Lattices. Positive results.

Observation 3.1. Let A be a WF lattice. Then $\overline{A}$ is a complete lattice.

Proof. Since A has finite meets and is well-founded, it has noempty meets. □

Conditions 3.1. We say that a WF ordering A *satisfies CC* (continuity condition), if there exists a chain $(a_k)_{i \in cf\overline{\lambda}(A)}$ such that

$$\sup_i h(a_i) = \overline{\lambda}(A).$$

A WF-ordering A is said to *satisfy HCC* if for each interval $\langle b, a), b, a \in \overline{A}$, such that $\overline{\lambda}(\langle b, a))$ is sum-closed and of cofinality $> \omega$, satisfies CC.

We say that a lattice A *satisfies the condition WHA* (weak Heyting algebra) if for any set $I, |I| < |A|$, and elements $(b_i)_{i \in I}, a$ of A we have

$$(\bigvee_{i \in I} b_i) \wedge a = \bigvee_{i \in I} (b_i \wedge a). \tag{+}$$

In particular, the joins in question are required to exist.

Theorem 3.1. *A modular WF lattice is order-perfect iff is satisfies HCC.*

Proof. will be given in Proof of Theorem 3.1 below. Till the end of the proof, A is always a well founded modular lattice.

Lemma 3.1. *Let $(c_i)_{i<\lambda}$ be a chain in A, $a \in A$, $M = \{i < \lambda / c_i \not\leq a\}$. Then we have*

$$|(c_i)_{i<\lambda}| \leq |(c_i \wedge a)_{i<\lambda}| \oplus |(c_i \vee a)_{i \in M}|. \tag{+}$$

Moreover, we have

$$|(c_i)_{i<\lambda}| < |(c_i \wedge a)_{i<\lambda}| \oplus |(c_i \vee a)_{i \in M}| \tag{++}$$

whenever

$$|M| \geq \omega \tag{+++}$$

Remark 3.1. The condition (+++) for (++) cannot be weakened. Indeed, consider $A = \omega + 2$, $\lambda = \omega + 2$, $c_i = i$ $(i \in \lambda)$, $a = \omega$, $I = \{\omega + 1\}$, $|(c_i)_{i<\lambda}| = \omega + 2$, $|(c_i \wedge a)_{i \in \lambda}| = \omega + 1$, $|(c_i \vee a)_{i<\lambda}| = 1$ and equality occurs in (+).
Proof: Let $b = (b_i)_{i<\kappa}$ be a chain in A. Put

$$I(b) = \{i < \kappa \mid (\forall j < i)\ b_j < b_i\}.$$

Obviously,

$$|b| = c(I(b)).$$

By modularity of A,

$$I((c_i)_{i<\lambda}) = I((c_i \wedge a)_{i<\lambda}) \cup I((c_i \vee a)_{i \in M}).$$

Moreover, obviously, the right-hand sets are closed in the left-hand one and

$$0 \in I((c_i \wedge a)_{i<\lambda}).$$

The statement of our lemma now directly follows from Lemma 2.1. □

Lemma 3.2. *If we have $b < a$ and there is a chain (c_i) of length α in $\Downarrow a$ then there are chains of lengths β, γ in $\Downarrow b, \langle b, a \rangle$, respectively, such that*

$$\beta \oplus \gamma \geq \alpha.$$

Proof. Assume the chain (c_i) in $\Downarrow a$ is maximal hence continuous. We distinguish two possibilities:

(a) $|(c_i \vee b | c_i \not\leq b)| \geq \omega$. Then take the chains $(c_i \wedge b | c_i \not\geq b)$, $(c_i \vee b | c_i \not\leq b)$ and use 3.1(++).

(b) $|(c_i \vee b \mid c_i \not\leq b) < \omega$. Then take the chain $(c_i \wedge b | c_i \not\geq b)$ and the concatenation of b and $(c_i \vee b \mid c_i \not\leq b)$ and use 3.1(+) together with the commutativity of addition of finite ordinals. □

Remark 3.2. Trying out a few examples, one can see that the cases (a), (b) in the proof of Lemma 3.2 are really governed by different principles. A less careful analysis, however, would cause only a discrepancy by 1, which is still sufficient for the proof of Theorem 3.1.

Lemma 3.3. *For $b < a \in A$ we have*

$$\overline{\lambda}(\Downarrow a) \leq \overline{\lambda}(\Downarrow b) \oplus \overline{\lambda}(\langle b, a \rangle). \tag{+}$$

Moreover, if $\lambda(\Downarrow b)$ exists we have

$$\overline{\lambda}(\Downarrow b) + \overline{\lambda}(\langle b, a \rangle) \leq \overline{\lambda}(\Downarrow a). \tag{++}$$

Proof. (++) is trivial, (+) follows form 3.2. □

Theorem 3.2. *If A (a modular WF lattice) is order-perfect then any interval $\langle b, a \rangle$, a, $b \in \overline{A}$, is order-perfect.*

Proof. First let $\overline{\lambda}(\langle b, a \rangle)$ be sum-closed. We have a chain (a_i) in $\Downarrow a$ of length $\lambda(\Downarrow a)$. Thus, there are chains in $\Downarrow b, \langle b, a \rangle$ of lengths β, γ, respectively, $\beta \oplus \gamma \geq \lambda\ (\Downarrow a)$. By Lemma 3.3(++),

$$\lambda(\Downarrow b) + \overline{\lambda}(\langle b, a \rangle) \leq \lambda(\Downarrow a).$$

Since $\beta \leq \lambda(\Downarrow b)$, we have

$$\lambda(\Downarrow b) \oplus \gamma \geq \lambda(\Downarrow a).$$

Since $\overline{\lambda}(\langle b, a \rangle)$ is sum-closed, we have $\overline{\lambda}(\langle b, a \rangle) \leq \gamma$ and hence

$$\overline{\lambda}(\langle b, a \rangle) = \gamma.$$

Now let $\alpha = \overline{\lambda}(\langle b, a \rangle)$ not be sum-closed. We proceed by induction on a. First realize that $b < c < a \rightarrow \overline{\lambda}(\langle b, c \rangle) < \overline{\lambda}(\langle b, a \rangle)$. In fact, $\lambda(\langle b, c \rangle)$ exists by the inductional hypothesis; use Lemma 3.3(++). Now let β be the last summand of α (in the same sense as in the proof of 2.1). Let (a_i) be a chain in $\langle b, a \rangle$ of length $> \alpha - \beta$ (the least number κ s.t. $\kappa + \beta \geq \alpha$). Put $c = a_{\alpha - \beta}$. We have $\overline{\lambda}(\langle c, a \rangle) = \beta$:

$\leq$ is obvious and $\geq$ follows from the fact that $\overline{\lambda}(\langle b, c \rangle) < \overline{\lambda}(\langle b, a \rangle)$ and

$$\overline{\lambda}(\langle b, c \rangle) \oplus \overline{\lambda}(\langle c, a \rangle) \geq \overline{\lambda}(\langle b, a \rangle) \tag{+}$$

(by Lemma 3.3(+)). Now $\lambda(\langle b, c \rangle)$ exists by the inductional hypothesis and $\lambda(\langle c, a \rangle)$ exists by the first part of the proof. By (+), we have

$$\lambda(\langle b, c \rangle) + \lambda(\langle c, a \rangle) \geq \overline{\lambda}(\langle b, a \rangle)$$

(since $\lambda(\langle c, a \rangle) = \beta$). □

Lemma 3.4. *Let $a, b \in A$. Then we have*

$$h(a) \oplus h(b) \geq h(a \vee b).$$

Proof. Let (c_i) be a chain in $\Downarrow a \vee b$ of length α. By Lemma 3.2, there are chains in $\Downarrow a$, $\langle a, a \vee b)$ of lengths β, γ, respectively, such that

$$\beta \oplus \gamma \geq \alpha.$$

As is well known, $\langle a, a \vee b)$ is isomorphic to $\langle a \wedge b, a)$ via "$\wedge b$". Thus, $\gamma \leq h(b)$. □

Remark 3.3. The discrepancy between natural and ordinal sum prevents us to obtain an exact result as in the finite case.

Lemma 3.5. *Let $\overline{\lambda}(\langle a, b))$ be sum-closed and of cofinality ω, $a, b \in \overline{A}$. Then $\langle a, b)$ satisfies CC.*

Proof. Choose $a_i \in \langle a, b)$ with $\overline{\lambda}(\langle a, b_i)) \nearrow \overline{\lambda}(\langle a, b))$ and put $a_i = b_0 \vee \ldots \vee b_i$. We have $a_i < b$ by Lemma 3.4 and the sum-closedness. □

Proof (of Theorem 3.1). Necessity follows from 3.2. In order to prove sufficiency, we introduce the following statement

S : If A is modular and satisfies HCC and $a \in A$ then $\lambda(\Downarrow a)$ exists.

This implies our result by Proposition 2.1, since neither modularity nor HCC is violated by adding a new greatest element.

We shall prove S by induction on $h(a)$.

Case 3.1. $h(a)$ is not sum-closed. Then let $\alpha + \beta = h(a)$, $\alpha, \beta < h(a)$, β the last summand of $h(a)$. Then there exists a $b < a$ with

$$h(b) \geq \alpha.$$

We may assume $h(b) < h(a)$ (otherwise we replace a by b), hence $\lambda(\Downarrow b)$ exists. By Lemma 3.3,

$$\overline{\lambda}(\langle b, a)) = \beta. \qquad (+)$$

The inductional hypothesis applied to $a \in \langle b, a\rangle$ (a modular lattice satisfying HCC) yields

$$\lambda(\langle b, a)) = \beta. \qquad (++)$$

Now by (+),(++), there exist strict chains

$$(c_i)_{i \in \alpha} \text{ in } \Downarrow b$$
$$(d_i)_{i \in \beta} \text{ in } \langle b, a).$$

The desired chain is obtained by concatenation.

Case 3.2. $\lambda = \overline{\lambda}(\Downarrow a)$ is sum-closed. Then either by HCC or by Lemma 3.5 we have a strict chain $(c_i)_{i \in cf\lambda}$ in $\Downarrow a$ with

$$\lambda(\Downarrow c_i) \nearrow \lambda.$$

By Lemma 3.3(+) and the sum-closedness of λ we have

$$\lim_{\substack{j : j \geq i \\ j \in cf\lambda}} \overline{\lambda}(\langle c_i, c_j)) = \lambda$$

and thus

$$\lim_{\substack{j:j\geq i\\ j\in cf\lambda}} \lambda(\langle c_i, c_j)) = \lambda$$

by the inductional hypothesis. Now take an increasing sequence $i_\alpha(\alpha \in \mathrm{cf}\lambda)$ such that

$$\lim_{\alpha\in cf\lambda} \lambda(\langle c_{i_\alpha}, c_{i_{\alpha+1}}) = \lambda.$$

Now if $(c_i^\alpha)_{i\in\lambda(\langle c_{i_\alpha}, e_{i_{\alpha+1}}))}$ is a strict chain, we obtain a chain of length λ in $\Downarrow a$ by concatenation. □

Theorem 3.3. *If $\overline{A}$ satisfies WHA and is WF then A satisfies CC (and hence, by hereditarity, also HCC).*

Proof: Let $\overline{A} = \langle 0,1\rangle, \lambda = \overline{\lambda}(A)$. If there is a sequence $(c_i)_{i\in cf\lambda}$ such that for $\beta < cf\lambda$

$$\bigvee_{i\in\beta} c_i < 1,$$

the statement is proved. So assume the contrary. We will prove that following fact:
T: For any $c < 1$ we have $\sup\{h(b) \mid c \not\leq b\} < \lambda$.
Really, since A is WF, we can write

$$c = \bigvee_{i\in\beta} c_i$$

where $\beta < cf\lambda$ and c_i are irreducible to sums of size $< cf\lambda$. Now for any $i \in \beta$ the set

$$M_i = \{h(b) \mid c_i \not\leq b\}$$

has supremum $< \lambda$: Otherwise there are $a_\alpha \not\geq c_i$, $\alpha \in cf\lambda$, such that $h(a_\alpha) \nearrow \lambda$. Thus, there is a $\gamma < cf\lambda$ such that

$$\bigvee_{\alpha\in\lambda} a_\alpha = 1.$$

Thus, by WHA,

$$\bigvee_{\alpha\in\gamma} (c_i \wedge a_\alpha) = c_i,$$

contradicting the assumed irreducibility of c_i. Now

$$\sup_{i\in\beta} \sup M_i < \lambda.$$

However, T clearly enables us to choose a sequence (c_i) with $(+)$, contradicting the original assumption. □

Remark 3.4. We have shown that each WF WHA is order-perfect. On the other hand, each countable modular WF lattice is order-perfect, since it satisfies HCC by default.

On might look for weakenings of these conditions. It is fairly easy to find an example of a distributive complete well-founded lattice which is not order-perfect: Take, for instance, the set

$$A = \{\varphi : \omega_1\backslash\{0\} \to \omega_1 \mid |\omega_1\backslash\varphi^{-1}(0)| < \omega \;\&\; (\forall\alpha\in\omega_1\backslash\{0\})\varphi(\alpha) < \alpha\}.$$

and put

$$\varphi \leq \psi \text{ iff } (\forall \alpha \in \omega_1 \setminus \{0\}) \varphi(\alpha) \leq \psi(\alpha).$$

This makes A a well-founded distributive lattice. $\overline{A}$ is a well-founded complete distributive lattice. Obviously,

$$\overline{\lambda}(A) = \omega_1$$

while there is no uncountable chain. An observation of this kind has been first made in Zaguia [6].

This example teaches us that the infinite joins (at least of cardinality $< |A|$) indeed have to be "tightened up" in some way. On the other hand, the (finite) distributivity is obviously not needed.

Note that WHA is equivalent to distributivity + 3.1(+) for *chains* (b_i). One might suggest *modularity + 3.1(+)* for chains (b_i). In the next section we shall show, however, that this condition is not sufficient for order-perfectness. In fact, we shall find a complete well-founded sublattice of Vect$(\mathbb{R}^{(\omega)})$ which is not perfect. This example, because of the strong condition on infinite joins is much more interesting than the one mentioned above.

4. Lattices. A negative result

Theorem 4.1. *There exists a complete sublattice P of Vect $(\mathbb{R}^{(\omega)})$ which is WF but not order-perfect.*

(In this paper, $\mathbb{R}^{(A)}$ means the free $\mathbb{R}$-modul over A. Vect (X) stands for the lattice of all submodules of X.)

Proof. Follows from 4.4, 4.1 and 4.6 below. □

Lemma 4.1. *For any $\alpha < \omega_1$ there exists a WPO B_α such that $\lambda(B_\alpha) = \omega$, $c(B_\alpha) \geq \alpha$, B_α is stratified and*

$$|\{x \in B_\alpha | h(x) = n\}| = n + 1. \tag{+}$$

Proof. Let $\varphi : \alpha \to \omega$ be a bijection. Put $B_\alpha = \{(\alpha, k) \mid \varphi(\alpha) \leq k\}$. Now let $(\alpha, k) < (\beta, m)$ iff $\alpha \leq \beta$ &$k < m$. Clearly $\lambda(B_\alpha) = \omega$ and B_α is stratified (since $h(\alpha, k) = k$). For the same reason B_α satisfies (+) (since φ is a bijection). Now B_α is WPO, since it is the intersection of three (linear) well-orderings: The lexicographical one prefering first coordinate, the lexicographical one prefering second coordinate and the ordering $\prec$ given by

$$(\alpha, k) \prec (\beta, m) \text{ iff } k < m \text{ or } (k = m \text{ and } \beta < \alpha). \ \square$$

The Construction 4.1. In the sequel, we will assume that we are given disjoint sets $B_\alpha | \alpha \in \omega_1$ satisfying the conditions of Lemma 4.1. Let

$$\mathbb{R}^{(\omega)} = \bigoplus_{i \in \omega} V_i, \quad \dim \ V_i = 2^{i+1}(i+1)!$$

Let $\{v(\alpha, c) \mid \alpha \in \omega_1, \ \ c \in B_\alpha, \ \ h(c) = i\}$ be vectors of V_i in general position. (Each subset of cardinality $\leq \dim V_i$ is linearly independent.) For a vector $v \in V_i$ define a subspace $\partial v \subseteq V_{i-1}$ as follows:

If there exists a (in the sequel minimal with respect to inclusion) set

$$\{v(\alpha_1, c_1), \ldots, v(\alpha_n, c_n)\}$$

such that $v \in \langle v(\alpha_1, c_1), \ldots, v(\alpha_n, c_n)\rangle$ (the linear hull) and

$$n \leq 2^i (i+1)!$$

then put

$$\partial v = \langle v(\alpha_i, d) \mid i \leq n \text{ and } c_i \text{ covers } d\rangle.$$

Otherwise put

$$\partial v = V_{i-1}.$$

Because of the general position of $v(\alpha, c), \partial$ is defined correctly. Put for a subspace $V \subseteq V_i$

$$\partial V = \langle \partial v \mid v \in V\rangle.$$

Now let

$$P \subseteq \bigoplus_{i \in \omega} \text{Vect } V_i$$

be the set of all subspaces W satisfying

$$\partial(W \cap V_i) \subseteq W.$$

We will show that P is the desired lattice.

Lemma 4.2. *P is a meet-closed subset of* Vect $(\mathbb{R}^{(\omega)})$.

Proof. It suffices to realize that a space W belongs to P iff it satisfies a set of conditions of the form

$$a \in W \to b \in W.$$

Such conditions are preserved by intersection. □

Lemma 4.3. *Let $v \in V_i$, $\langle v(\beta_1, d_1), \ldots, v(\beta_m, d_m)\rangle \ni v$. Then we have*

$$\partial v \subseteq \langle v(\beta_i, d) \mid i \leq m, \quad d_i \text{ covers } d\rangle$$

Proof. First realize that if $m \geq 2^i (i+1)!$ then at least $2^i\ i!$ of the numbers β_i are different (by 4.1(+) and the Dirichlet principle). Thus, the vectors $v(\beta_j, d)$ such that d_j covers d are distinct and hence they generate V_{i-1}. If, on the other hand, $n < 2^i (1+i)!$ then (by the general position) $\{v(\beta_1, d_1), \ldots, v(\beta_m, d_m)\}$ contains the set $\{v(\alpha_1, c_1), \ldots, v(\alpha_n, c_n)\}$ from the definition of ∂v. □

Lemma 4.4. *P is a complete sublattice of* Vect$(\mathbb{R}^{(\omega)})$.

Proof. Now it suffices to show that P is closed under joins. This, however follows from the relation

$$\partial(\bigvee V_i) = \bigvee \partial V_i$$

which is an easy consequence of 4.1. □

Observation 4.1. Let $B \in \downarrow B_\alpha$. Then

$$V_B = \langle \{v(\alpha, c) \mid c \in B\}\rangle \in P.$$

Thus, P contains a chain of arbitrary countable length. On the other hand, it does not contain an uncountable one, since Vect$(\mathbb{R}^{(\omega)})$ does not. □

Lemma 4.5. *Let $V \in P$ and let an $i \leq k$ exist such that $V \not\supseteq V_i$. Then there exist only finitely many different α with*

$$v(\alpha, c) \in V$$

for some $c \in B_\alpha,\ h(c) \geq k$.

Proof. Because of the stratifiedness of B_α, V contains for each such α a vector $v(\alpha, d)$ with $h(d) = i$. If there were infinitely many such ones then $V_i \subseteq V$ by the general position. □

Lemma 4.6. *P is WF.*

Proof. Let $(W_j)_{j \in \omega}$ be a strictly decreasing chain in P. Let, without loss of generality, $W_0 \not\supseteq V_k$. Now let

$$\{\alpha_1, \ldots, \alpha_n\} = \{\alpha \mid (\exists v(\alpha, c))\ \ h(c) \geq k \ \&\ v(\alpha, c) \in W_0\}.$$

Put $B_j^i = \{c \mid v(\alpha_i, c) \in W_j\}, \quad i \leq n, \quad j \in \omega$. By the definition of P clearly $B_j^i \in \downarrow B_{\alpha_i}$. Moreover, we have

$$B_0^i \supseteq B_1^i \supseteq \ldots.$$

Since B_{α_i} is WPO, it holds without loss of generality that

$$B_0^i = B_1^i = \ldots.$$

Since we have $j \neq j'$ with

$$W_j \cap \bigoplus_{i \in k} V_i = W_{j'} \cap \bigoplus_{i \in k} V,$$

we have $W_j = W_{j'}$. A contradiction. □

References

1. G. Birkhoff: Lattice Theory, Providence, Rhode Island 1967
2. D.H.J. de Jough and R. Parikh: Well-partial orderings and hierarchies, Indag. Math. 39(1977) 195-207.
3. E.C. Milner and N. Sauer: On chains and antichains in well founded partially ordered sets, J. London Math. Soc. (2) 24(1981), 15-33.
4. E. C. Milner and M. Pouzet: unpublished.
5. D. Schmidt: The relation between the height of a well-founded partial ordering and the order-types of its chains and anti-chains, J.C.T. (b) 31(1981) 2, 183-189.
6. N. Zaguia: Chaines d'ideaux de sections initiales d'un ensemble ordonee, Pub. Dept. Math. Lyon, nouvelle serie, 1983 - 7/D, pp. 1-97.

The PCF Theorem Revisited

Saharon Shelah*

[1] Institute of Mathematics, The Hebrew University, Jerusalem, Israel
[2] Department of Mathematics, Rutgers University, New Brunswick, NJ, USA

Dedicated to Paul Erdős

Summary. *The* pcf *theorem (of the possible cofinability theory) was proved for reduced products* $\prod_{i<\kappa} \lambda_i/I$*, where* $\kappa < \min_{i<\kappa} \lambda_i$*. Here we prove this theorem under weaker assumptions such as* $\mathrm{wsat}(I) < \min_{i<\kappa} \lambda_i$*, where* $\mathrm{wsat}(I)$ *is the minimal* θ *such that* κ *cannot be divided to* θ *sets* $\notin I$ *(or even slightly weaker condition). We also look at the existence of exact upper bounds relative to* $<_I$ *(*$<_I$ *−eub) as well as cardinalities of reduced products and the cardinals* $T_D(\lambda)$*. Finally we apply this to the problem of the depth of ultraproducts (and reduced products) of Boolean algebras.*

0. Introduction

An aim of the pcf theory is to answer the question, what are the possible cofinalities (pcf) of the partial orders $\prod_{i<\kappa} \lambda_i/I$, where $\mathrm{cf}(\lambda_i) = \lambda_i$, for different ideals I on κ. For a quick introduction to the pcf theory see [Sh400a], and for a detailed exposition, see [Sh-g] and more history. In §1 and §2 we generalize the basic theorem of this theory by weakening the assumption $\kappa < \min_{i<\kappa} \lambda_i$ to the assumption that I extends a fixed ideal I^* with $\mathrm{wsat}(I^*) < \min_{i<\kappa} \lambda_i$, where $\mathrm{wsat}(I^*)$ is the minimal θ such that κ cannot be divided to θ sets $\notin I^*$ (not just that the Boolean algebra $\mathcal{P}(\kappa)/I^*$ has no θ pairwise disjoint non zero elements). So §1, §2 follow closely [Sh-g, Ch. I=Sh345a], [Sh-g, II 3.1], [Sh-g, VIII §1]. It is interesting to note that some of (as presented in courses and see a forthcoming survey of Kojman) those proofs which look to be superseded when by [Sh420, §1] we know that for regular $\theta < \lambda$, $\theta^+ < \lambda \Rightarrow \exists$ stationary $S \in I[\lambda]$, $S \subseteq \{\delta < \lambda : \mathrm{cf}(\delta) = \theta\}$, give rise to proofs here which seem necessary. Note $\mathrm{wsat}(I^*) \leq |\mathrm{Dom}(I^*)|^+$ (and $\mathrm{reg}_*(I^*) \leq |\mathrm{Dom}(I^*)|^+$ so [Sh-g, I §1, §2, II §1, VII 2.1, 2.2, 2.6] are really a special case of the proofs here.

During the sixties the cardinalities of ultraproducts and reduced products were much investigated (see Chang and Keisler [CK]). For this the notion "regular filter" (and (λ, μ)-regular filter) were introduced, as: if $\lambda_i \geq \aleph_0$, D a regular ultrafilter (or filter) on κ then $\prod_{i<\kappa} \lambda_i/D = (\limsup_D \lambda_i)^\kappa$. We reconsider these problems in §3 (again continuing [Sh-g]). We also draw a conclusion on the depth of the reduced product of Boolean algebras partially answering a problem of Monk; and make it clear that the truth of the full expected result is translated to a problem on pcf. On those problems on Boolean algebras see Monk [M]. In this section we include known proofs for completeness (mainly 3.6).

Let us review the paper in more details. In 1.1, 1.2 we give basic definition of cofinality, true cofinality, $\mathrm{pcf}(\bar\lambda)$ and $J_{<\lambda}[\bar\lambda]$ where usually $\bar\lambda = \langle \lambda_i : i < \kappa \rangle$ a sequence of regular cardinals, I^* a fixed ideal on κ such that we consider only

* Partially supported by the Deutsche Forschungsgemeinschaft, grant Ko 490/7-1. Publication no. 506.

ideals extending it (and filter disjoint to it). Let $\mathrm{wsat}(I^*)$ be the first θ such that we cannot partition κ to θ I^*-positive set (so they are pairwise disjoint, not just disjoint modulo I^*). In 1.3, 1.4 we give the basic properties. In lemma 1.5 we phrase the basic property enabling us to do anything: (1.5(*)): $\liminf_{I^*}(\bar\lambda) \geq \theta \geq \mathrm{wsat}(I^*)$, $\prod \bar\lambda/I^*$ is θ^+-directed; we prove that $\prod \bar\lambda/J_{<\lambda}[\bar\lambda]$ is λ-directed. In 1.6, 1.8 we deduce more properties of $\langle J_{<\lambda}[\bar\lambda] : \lambda \in \mathrm{pcf}(\bar\lambda)\rangle$ and in 1.7 deal with $<_{J_{<\lambda}[\bar\lambda]}$-increasing sequence $\langle f_\alpha : \alpha < \lambda\rangle$ with no $<_{J_{<\lambda}[\bar\lambda]}$-bound in $\prod \bar\lambda$. In 1.9 we prove $\mathrm{pcf}(\bar\lambda)$ has a last element and in 1.10, 1.11 deal with the connection between the true cofinality of $\prod\limits_{i<\kappa} \lambda_i/D^*$ and $\prod\limits_{i<\sigma} \mu_i/E$ when $\mu_i =: \mathrm{tcf}(\prod\limits_{i<\kappa} \lambda_i/D_i)$ and D^* is the E-limit of the D_i's.

In 2.1 we define normality of λ for $\bar\lambda$: $J_{\leq\lambda}[\bar\lambda] = J_{<\lambda}[\bar\lambda] + B_\lambda$ and we define semi-normality: $J_{\leq}[\bar\lambda] = J_{<\lambda}[\bar\lambda] + \{B_\alpha : \alpha < \lambda\}$ where $B_\alpha/J_{<\lambda}[\bar\lambda]$ is increasing. We then (in 2.2) characterize semi normality (there is a $<_{J_{<\lambda}[\bar\lambda]}$-increasing $\bar f = \langle f_\alpha : \alpha < \lambda\rangle$ cofinal in $\prod \bar\lambda/D$ for every ultrafilter D (disjoint to I^* of course) such that $\mathrm{tcf}(\prod \bar\lambda/D) = \lambda$) and when semi normality implies normality (if some such $\bar f$ has a $<_{J_{<\lambda}[\bar\lambda]}$ $-$eub).

We then deal with continuity system $\bar a$ and $<_{J_{<\lambda}[\bar\lambda]}$- increasing sequence obeying $\bar a$, in a way adapted to the basic assumption $(*)$ of 1.5.

Here as elsewhere if $\min(\bar\lambda) \geq \theta^+$ our life is easier than when we just assume $\lim sup_{I^*}(\bar\lambda) \geq \theta$, $\prod \bar\lambda/I^*$ is θ^+-directed (where $\theta \geq \mathrm{wsat}(I^*)$ of course). In 2.3 we give the definitions, in 2.4 we quote existence theorem, show existence of obedient sequences (in 2.5), essential uniqueness (in 2.7) and better consequence to 1.7 (in the direction to normality). We define (2.9) generating sequence and draw a conclusion (2.10(1)). Now we get some desirable properties: in 2.8 we prove semi normality, in 2.10(2) we compute $\mathrm{cf}(\prod \bar\lambda/I^*)$ as $\max \mathrm{pcf}(\bar\lambda)$. Next we relook at the whole thing: define several variants of the pcf-th (Definition 2.11). Then (in 2.12) we show that e.g. if $\min(\bar\lambda) > \theta^+$, we get the strongest version (including normality using 2.6, i.e. obedience). Lastly we try to map the implications between the various properties when we do not use the basic assumption 1.5 $(*)$ (in fact there are considerable dependence, see 2.13, 2.14).

In 3.1, 3.3 we present measures of regularity of filters, in 3.2 we present measures of hereditary cofinality of $\prod \bar\lambda/D$: allowing to decrease $\bar\lambda$ and/or increase the filter. In 3.4 - 3.8 we try to estimate reduced products of cardinalities $\prod\limits_{i<\kappa} \lambda_i/D$ and in 3.9 we give a reasonable upper bound by hereditary cofinality $(\leq (\theta^\kappa/D + \mathrm{hcf}_{D,\theta}(\prod\limits_{i<\kappa} \lambda_I))^{<\theta}$ when $\theta \geq \mathrm{reg}_\otimes(D))$.

In 3.10-3.11 we return to existence of eub's and obedience and in 3.12 draw conclusion on "downward closure". On $T_D(f)$, starting with Galvin and Hajnal [GH] see [Sh-g].

In 3.13 - 3.14 we estimate $T_D(\bar\lambda)$ and in 3.15 try to translate it more fully to pcf problem (countable cofinality is somewhat problematic (so we restrict ourselves to $T_D(\bar\lambda) > \mu = \mu^{\aleph_0}$). We also mention $\aleph_1$-complete filters; (3.16, 3.17) and see what can be done without relaying on pcf (3.20)).

Now we deal with depth: define it (3.18, see 3.19), give lower bound (3.22), compute it for ultraproducts of interval Boolean algebras of ordinals (3.24). Lastly we connect the problem "does $\lambda_i < \mathrm{Depth}^+(B_i)$ for $i < \kappa$ implies $\mu < \mathrm{Depth}^+(\prod\limits_{i<\kappa} B_i/D)$" at least when $\mu > 2^\kappa$ and $(\forall \alpha < \mu)[|\alpha|^{\aleph_0} < \mu]$, to a pcf problem (in 3.26). This is continued in [Sh589].

In the last section we phrase a reason why 1.5(∗) works (see 4.1), analyze the case we weaken 1.5(∗) to $\liminf_{I^*}(\bar\lambda) \geq \theta \geq \mathrm{wsat}(I^*)$ proving the pseudo pcf-th (4.3).

1. Basic pcf

Notation 1.0. I, J denote ideals on a set $\mathrm{Dom}(I)$, $\mathrm{Dom}(J)$ resp., called its domain (possibly $\bigcup_{A\in I} A \subset \mathrm{Dom} I$). If not said otherwise the domain is an infinite cardinal denoted by κ and also the ideal is proper i.e. $\mathrm{Dom}(I) \notin I$. Similarly D denotes a filter on a set $\mathrm{Dom} D$; we do not always distinguish strictly between an ideal on κ and the dual filter on κ. Let $\bar\lambda$ denote a sequence of the form $\langle \lambda_i : i < \kappa\rangle$. We say $\bar\lambda$ is regular if every λ_i is regular, $\min\bar\lambda = \min\{\lambda_i : i < \kappa\}$ (of course also in $\bar\lambda$ we can replace κ by another set), and let $\prod\bar\lambda = \prod\limits_{i<\kappa} \lambda_i$; usually we are assuming $\bar\lambda$ is regular. Let I^* denote a fixed ideal on κ. Let $I^+ = \mathcal{P}(\kappa) \setminus I$ (similarly $D^+ = \{A \subseteq \kappa : \kappa \setminus A \notin D\}$), let

$$\liminf_I \bar\lambda = \min\{\mu : \{i < \kappa : \lambda_i \leq \mu\} \in I^+\} \quad \text{and}$$

$$\limsup_I \bar\lambda = \min\{\mu : \{i < \kappa : \lambda_i > \mu\} \in I\} \quad \text{and}$$

$$\mathrm{atom}_I \bar\lambda = \{\mu : \{i : \lambda_i = \mu\} \in I^+\}.$$

For a set A of ordinals with no last element, $J_A^{\mathrm{bd}} = \{B \subseteq A : \sup(B) < \sup(A)\}$, i.e. the ideal of bounded subsets. Generally, if $\mathrm{inv}(X) = \sup\{|y| : \models \varphi[X,y]\}$ then $\mathrm{inv}^+(X) = \sup\{|y|^+ : \models \varphi[X,y]\}$, and any y such that $\models \varphi[X,y]$ is a witness for $|y| \leq \mathrm{inv}(X)$ (and $|y| < \mathrm{inv}^+(X)$), and it exemplifies this. Let $\bar A^*_\theta[\bar\lambda] = \langle A^*_\alpha : \alpha < \theta\rangle = \langle A^*_{\theta,\alpha}[\bar\lambda] : \alpha < \theta\rangle$ be defined by: $A^*_\alpha = \{i < \kappa : \lambda_i > \alpha\}$. Let Ord be the class of ordinals.

Definition 1.1.

(1) For a partial order* P:
  - (a) P is λ-directed if: for every $A \subseteq P$, $|A| < \lambda$ there is $q \in P$ such that $\bigwedge_{p\in A} p \leq q$, and we say: q is an upper bound of A;
  - (b) P has true cofinality λ if there is $\langle p_\alpha : \alpha < \lambda\rangle$ cofinal in P, i.e.: $\bigwedge_{\alpha<\beta} p_\alpha < p_\beta$ and $\forall q \in P[\bigvee_{\alpha<\lambda} q \leq p_\alpha]$ [and one writes $\mathrm{tcf}(P) = \lambda$ for the minimal such λ] (note: if P is linearly ordered it always has a true cofinality but e.g. $(\omega, <) \times (\omega_1, <)$ does not).
  - (c) P is called endless if $\forall p \in P \exists q \in P[q > p]$ (so if P is endless, in clauses (a), (b), (d) above we can replace $\leq$ by $<$).
  - (d) $A \subseteq P$ is a cover if: $\forall p \in P \exists q \in A[p \leq q]$; we also say "$A$ is cofinal in P".
  - (e) $\mathrm{cf}(P) = \min\{|A| : A \subseteq P \text{ is a cover}\}$.
  - (f) We say that, in P, p is a lub (least upper bound) of $A \subseteq P$ if:
    - (α) p is an upper bound of A (see (a))
    - (β) if p' is an upper bound of A then $p \leq p'$.

(2) If D is a filter on S, α_s (for $s \in S$) are ordinals, $f, g \in \prod\limits_{s\in S} \alpha_s$, then: $f/D < g/D$, $f <_D g$ and $f < g \bmod D$ all mean $\{s \in S : f(s) < g(s)\} \in D$. Also if f, g are partial functions from S to ordinals, D a filter on S <u>then</u> $f < g \bmod D$ means

* actually we do not require $p \leq q \leq p \Rightarrow p = q$ so we should say quasi partial order

$\{i \in \mathrm{Dom}(D) : i \notin \mathrm{Dom}(f)$ <u>or</u> $f(i) < g(i)$ (so both are defined)$\}$ belongs to D. We write $X = A \mod D$ if $\mathrm{Dom}(D) \setminus [(X \setminus A) \cup (A \setminus X)]$ belongs to D. Similarly for $\leq$, and we do not distinguish between a filter and the dual ideal in such notions. So if J is an ideal on κ and $f, g \in \prod \bar{\lambda}$, then $f < g \mod J$ iff $\{i < \kappa : \neg f(i) < g(i)\} \in J$. Similarly if we replace the α_s's by partial orders.

(3) For $f, g: S \to$ Ordinals, $f < g$ means $\bigwedge_{s \in S} f(s) < g(s)$; similarly $f \leq g$. So $(\prod \bar{\lambda}, \leq)$ is a partial order, we denote it usually by $\prod \bar{\lambda}$; similarly $\prod f$ or $\prod\limits_{i<\kappa} f(i)$.

(4) If I is an ideal on κ, $F \subseteq {}^{\kappa}\mathrm{Ord}$, we call $g \in {}^{\kappa}\mathrm{Ord}$ an $\leq_I$-eub (exact upper bound) of F if:
   - (α) g is an $\leq_I$-upper bound of F (in ${}^{\kappa}\mathrm{Ord}$)
   - (β) if $h \in {}^{\kappa}\mathrm{Ord}$, $h <_I \mathrm{Max}\{g, 1\}$ <u>then</u> for some $f \in F$, $h < \max\{f, 1\} \mod I$.
   - (γ) if $A \subseteq \kappa$, $A \neq \emptyset \mod I$ and $[f \in F \Rightarrow f \restriction A =_I 0_A$, i.e. $\{i \in A : f(i) \neq 0\} \in I]$ then $g \restriction A =_J 0_A$.

(5) (a) We say the ideal I (on κ) is θ-weakly saturated if κ cannot be divided to θ pairwise disjoint sets from I^+ (which is $\mathcal{P}(\kappa) \setminus I$)
   (b) $\mathrm{wsat}(I) = \min\{\theta : I$ is θ-weakly saturated$\}$

Remark 1.1A.

(1) Concerning 1.1(4), note: $g' = \mathrm{Max}\{g, 1\}$ means $g'(i) = \mathrm{Max}\{g(i), 1\}$ for each $i < \kappa$; if there is $f \in F$, $\{i < \kappa : f(i) = 0\} \in I$ we can replace $\mathrm{Max}\{g, 1\}$, $\mathrm{Max}\{f, 1\}$ by g, f respectively in clause (β) and omit clause (γ).

(2) Considering $\prod\limits_{i<\kappa} f(i), <_I$ formally if $(\exists i) f(i) = 0$ then $\prod\limits_{i<\kappa} f(i) = \emptyset$; but we usually ignore this, particularly when $\{i : f(i) = 0\} \in I$.

Definition 1.2. Below if Γ is "a filter disjoint to I", we write I instead Γ.

(1) For a property Γ of ultrafilters:

$$\mathrm{pcf}_{\Gamma}(\bar{\lambda}) = \mathrm{pcf}(\bar{\lambda}, \Gamma) = \{\mathrm{tcf}(\prod \bar{\lambda}/D) : D \text{ is an ultrafilter on } \kappa \text{ satisfying } \Gamma\}$$

(so $\bar{\lambda}$ is a sequence of ordinals, usually of regular cardinals, note: as D is an ultrafilter, $\prod \bar{\lambda}/D$ is linearly ordered hence has true cofinality).

(1A) More generally, for a property Γ of ideals on κ we let $\mathrm{pcf}_{\Gamma}(\bar{\lambda}) = \{\mathrm{tcf}(\prod \bar{\lambda}/J) : J$ is an ideal on κ satisfying Γ such that $\prod \bar{\lambda}/J$ has true cofinality$\}$. Similarly below.

(2) $J_{<\lambda}[\bar{\lambda}, \Gamma] = \{B \subseteq \kappa$: for no ultrafilter D on κ satisfying Γ to which B belongs, is $\mathrm{tcf}(\prod \bar{\lambda}/D) \geq \lambda\}$.

(3) $J_{\leq\lambda}[\bar{\lambda}, \Gamma] = J_{<\lambda^+}[\bar{\lambda}, \Gamma]$.

(4) $\mathrm{pcf}_{\Gamma}(\bar{\lambda}, I) = \{\mathrm{tcf}(\prod \bar{\lambda}/D) : D$ a filter on κ disjoint to I satisfying $\Gamma\}$.

(5) If $B \in I^+$, $\mathrm{pcf}_I(\bar{\lambda} \restriction B) = \mathrm{pcf}_{I + (\kappa \setminus B)}(\bar{\lambda})$ (so if $B \in I$ it is $\emptyset$), also $J_{<\lambda}(\bar{\lambda} \restriction B, I) \subseteq \mathcal{P}(B)$ is defined similarly.

(6) If $I = I^*$ we may omit it, similarly in (2), (4).
If $\Gamma = \Gamma_{I^*} = \{D : D$ a filter on κ disjoint to $I^*\}$ we may omit it.

Remark. We mostly use $\mathrm{pcf}(\bar{\lambda})$, $J_{<\lambda}[\bar{\lambda}]$.

Claim 1.3.

(0) $(\prod\bar{\lambda}, <_J)$ and $(\prod\bar{\lambda}, \leq_J)$ are endless (even when each λ_i is just a limit ordinal);

(1) $\min(\mathrm{pcf}_I(\bar{\lambda})) \geq \liminf_I(\bar{\lambda})$ for $\bar{\lambda}$ regular;

(2) (i) If $B_1 \subseteq B_2$ are from I^+ <u>then</u> $\mathrm{pcf}_I(\bar{\lambda} \restriction B_1) \subseteq \mathrm{pcf}_I(\bar{\lambda} \restriction B_2)$;

(ii) if $I \subseteq J$ <u>then</u> $\mathrm{pcf}_J(\bar{\lambda}) \subseteq \mathrm{pcf}_I(\bar{\lambda})$; and

(iii) for $B_1, B_2 \subseteq \kappa$ we have $\mathrm{pcf}_I(\bar{\lambda} \restriction (B_1 \cup B_2)) = \mathrm{pcf}_I(\bar{\lambda} \restriction B_1) \bigcup \mathrm{pcf}_I(\bar{\lambda} \restriction B_2)$. Also

(iv) $A \in J_{<\lambda}[\bar{\lambda} \restriction (B_1 \cup B_2)] \Leftrightarrow A \cap B_1 \in J_{<\lambda}[\bar{\lambda} \restriction B_1]$ & $A \cap B_2 \in J_{<\lambda}[\bar{\lambda} \restriction B_2]$

(v) If $A_1, A_2 \in I^+$, $A_1 \cap A_2 = \emptyset$, $A_1 \cup A_2 = \kappa$, and $\mathrm{tcf}(\prod \bar{\lambda} \restriction A_\ell, <_I) = \lambda$ for $\ell = 1, 2$ <u>then</u> $\mathrm{tcf}(\prod \bar{\lambda}, <_I) = \lambda$; and if the sequence $\bar{f} = \langle f_\alpha : \alpha < \lambda \rangle$ witness both assumptions then it witness the conclusion.

(3) (i) if $B_1 \subseteq B_2 \subseteq \kappa$, B_1 finite and $\bar{\lambda}$ regular <u>then</u>

$$\mathrm{pcf}_I(\bar{\lambda} \restriction B_2) \setminus \mathrm{Rang}(\bar{\lambda} \restriction B_1) \subseteq \mathrm{pcf}_I(\bar{\lambda} \restriction (B_2 \setminus B_1)) \subseteq \mathrm{pcf}_I(\bar{\lambda} \restriction B_2)$$

(ii) if in addition $i \in B_1 \Rightarrow \lambda_i < \min(\mathrm{Rang}[\bar{\lambda} \restriction (B_2 \setminus B_1)])$, <u>then</u> $\mathrm{pcf}_I(\bar{\lambda} \restriction B_2) \setminus \mathrm{Rang}(\bar{\lambda} \restriction B_1) = \mathrm{pcf}_I(\bar{\lambda} \restriction (B_2 \setminus B_1))$.

(4) Let $\bar{\lambda}$ be regular (i.e. each λ_i is regular);

(i) If $\theta = \liminf_I \bar{\lambda}$ <u>then</u> $\prod \bar{\lambda}/I$ is θ-directed

(ii) If $\theta = \liminf_I \bar{\lambda}$ is singular <u>then</u> $\prod \bar{\lambda}/I$ is θ^+-directed

(iii) If $\theta = \liminf_I \bar{\lambda}$ is inaccessible (i.e. a limit regular cardinal), the set $\{i < \kappa : \lambda_i = \theta\}$ is in the ideal I and for some club E of θ, $\{i < \kappa \colon \lambda_i \in E\} \in I$ <u>then</u> $\prod \bar{\lambda}/I$ is θ^+-directed. We can weaken the assumption to "I is not weakly normal for $\bar{\lambda}$" (defined in the next sentence). Let "I is not medium normal for $(\theta, \bar{\lambda})$" mean: for some $h \in \prod \bar{\lambda}$, for no $j < \theta$ is $\{i < \kappa \colon \lambda_i \leq \theta \Rightarrow h(i) < j\} = \kappa \bmod I$; and let "$I$ is not weakly normal for $(\theta, \bar{\lambda})$" mean: for some $h \in \prod \bar{\lambda}$, for no $\zeta < \liminf_I(\bar{\lambda}) = \theta$, is $\{i < \kappa \colon \lambda_i \leq \theta \Rightarrow h(i) < \zeta\} \in I^+$.

(iv) If $\{i : \lambda_i = \theta\} = \kappa \bmod I$ and I is medium normal for $\bar{\lambda}$ <u>then</u> $(\prod \bar{\lambda}, <_I)$ has true cofinality θ.

(v) If $\prod \bar{\lambda}/I$ is θ-directed <u>then</u> $\mathrm{cf}(\prod \bar{\lambda}/I) \geq \theta$ and $\min \mathrm{pcf}_I(\prod \bar{\lambda}) \geq \theta$.

(vi) $\mathrm{pcf}_I(\bar{\lambda})$ is non empty set of regular cardinals. [see part (7)]

(5) Assume $\bar{\lambda}$ is regular and: if $\theta' =: \limsup_I(\bar{\lambda})$ is regular <u>then</u> I is not medium normal for $(\theta', \bar{\lambda})$. <u>Then</u> $\mathrm{pcf}_I(\bar{\lambda}) \not\subseteq (\limsup_I(\bar{\lambda}))^+$; in fact for some ideal J extending I, $\prod \bar{\lambda}/J$ is $(\limsup_I(\bar{\lambda}))^+$-directed.

(6) If D is a filter on a set S and for $s \in S$, α_s is a limit ordinal <u>then</u>:

(i) $\mathrm{cf}(\prod_{s \in S} \alpha_s, <_D) = \mathrm{cf}(\prod_{s \in S} \mathrm{cf}(\alpha_s), <_D) = \mathrm{cf}(\prod_{s \in S} (\alpha_s, <)/D)$, and

(ii) $\mathrm{tcf}(\prod_{s \in S} \alpha_s, <_D) = \mathrm{tcf}(\prod_{s \in S} (\mathrm{cf}(\alpha_s), <_D)) = \mathrm{tcf}(\prod_{s \in S} (\alpha_s, <)/D)$.

In particular, if one of them is well defined, <u>then</u> so are the others. This is true even if we replace α_s by linear orders or even partial orders with true cofinality.

(7) If D is an ultrafilter on a set S, λ_s a regular cardinal, <u>then</u> $\theta =: \mathrm{tcf}(\prod_{s \in S} \lambda_s, <_D)$ is well defined and $\theta \in \mathrm{pcf}(\{\lambda_s \colon s \in S\})$.

(8) If D is a filter on a set S, for $s \in S$, λ_s is a regular cardinal, $S^* = \{\lambda_s \colon s \in S\}$ and

$$E =: \{B \colon \quad B \subseteq S^* \text{ and } \{s \colon \lambda_s \in B\} \in D\}$$

and $\lambda_s > |S|$ or at least $\lambda_s > |\{t : \lambda_t = \lambda_s\}|$ for any $s \in S$ <u>then</u>:

(i) E is a filter on S^*, and if D is an ultrafilter on S <u>then</u> E is an ultrafilter on S^*.

(ii) S^* is a set of regular cardinals and if $s \in S \Rightarrow \lambda_s > |S|$ <u>then</u> $(\forall \lambda \in S^*) \lambda > |S^*|$,

(iii) $F = \{f \in \prod_{s \in S} \lambda_s \colon s = t \Rightarrow f(s) = f(t)\}$ is a cover of $\prod_{s \in S} \lambda_s$,

(iv) $\mathrm{cf}(\prod_{s \in S} \lambda_s/D) = \mathrm{cf}(\prod S^*/E)$ and $\mathrm{tcf}(\prod_{s \in S} \lambda_s/D) = \mathrm{tcf}(\prod S^*/E)$.

(9) Assume I is an ideal on κ, $F \subseteq {}^\kappa\mathrm{Ord}$ and $g \in {}^\kappa\mathrm{Ord}$. If g is a $\leq_I$-eub of F <u>then</u> g is a $\leq_I$-lub of F.

(10) $\sup \mathrm{pcf}_I(\bar\lambda) \leq |\prod \bar\lambda / I|$

(11) If I is an ideal on S and $(\prod_{s\in S} \alpha_s, <_I)$ has true cofinality λ as exemplified by $\bar f = \langle f_\alpha : \alpha < \lambda\rangle$ <u>then</u> the function $\langle \alpha_s : s \in S\rangle$ is a $<_I$ $-$eub (hence $<_I$ $-$lub) of $\bar f$.

(12) The inverse of (11) holds: <u>if</u> I is an ideal on S and $f_\alpha \in {}^S\mathrm{Ord}$ for $\alpha < \lambda = \mathrm{cf}(\lambda)$, $\langle f_\alpha : \alpha < \lambda\rangle$ is $<_I$-increasing with $<_I$ $-$eub f <u>then</u> $\mathrm{tcf}(\prod_i f(i), <_I) = \mathrm{tcf}(\prod \mathrm{cf}[f(i)], <_I) = \lambda$.

(13) If $I \subseteq J$ are ideals on κ <u>then</u>
  (a) $\mathrm{wsat}(I) \geq \mathrm{wsat}(J)$
  (b) $\liminf_I(\bar\lambda) \leq \liminf_J(\bar\lambda)$
  (c) if $\lambda = \mathrm{tcf}(\prod_{i<\kappa} \lambda_i, <_I)$ then $\lambda = \mathrm{tcf}(\prod_{i<\kappa} \lambda_i, <_J)$

(14) If f_1, f_2 are $<_I$ $-$lub of F <u>then</u> $f_1 =_I f_2$.

Proof. They are all very easy, e.g.

(0) We shall show $(\prod\bar\lambda, <_J)$ is endless (assuming, of course, that J is a proper ideal on κ). Let $f \in \prod\bar\lambda$, then $g =: f+1$ (defined $(f+1)(\gamma) = f(\gamma)+1$) is in $\prod\bar\lambda$ too as each λ_α being an infinite cardinal is a limit ordinal and $f < g \bmod J$.

(5) Let $\theta' =: \limsup_I(\bar\lambda)$ and define

$$J =: \{A \subseteq \kappa : \text{ for some } \theta < \theta', \{i < \kappa : \lambda_i > \theta \text{ and } i \in A\} \text{ belongs to } I\}.$$

Clearly J is an ideal on κ extending I (and $\kappa \notin J$) and $\limsup_J(\bar\lambda) = \liminf_J(\bar\lambda) = \theta'$.

<u>Case 1</u>: θ' is singular

By part (4) clause (ii), $\prod \bar\lambda/J$ is $(\theta')^+$-directed and we get the desired conclusion.

<u>Case 2</u>: θ' is regular.

Let $h \in \prod\bar\lambda$ witness that "I is not medium normal for $(\theta', \bar\lambda)$" and let

$$J^* = \{A \subseteq \kappa : \text{ for some } j < \theta' \text{ we have } \{i \in A : h(i) < j\} = A \mod I\}.$$

Note that if $A \in J$ then for some $\theta < \theta'$, $A' =: \{i \in A : \theta_i > \theta\} \in I$ hence the choice $j =: \theta$ witness $A \in J^*$. So $J \subseteq J^*$. Also $J^* \subseteq \mathcal{P}(\kappa)$ by its definition. J^* is closed under subsets (trivial) and under union [why? assume A_0, $A_1 \in J^*$, $A = A_0 \cup A_1$; choose j_0, $j_1 < \theta'$ such that $A'_\ell =: \{i \in A_\ell : h(i) < j_\ell\} = A_\ell \mod I$, so $j =: \max\{j_0, j_1\} < \theta$ and $A' = \{i \in A : h(i) < j\} = A \mod I$; so $A \in J^*$]. Also $\kappa \notin J^*$ [why? as h witness that I is not medium normal for $(\theta', \bar\lambda)$]. So together J^* is an ideal on κ extending I. Now J^* is not weakly normal for $(\theta, \bar\lambda)$, as witnessed by h. Lastly $\prod\bar\lambda/J^*$ is $(\theta')^+$-directed (by part (4) clause (iii)), and so $\mathrm{pcf}_J(\bar\lambda)$ is disjoint to $(\theta')^+$.

(9) Let us prove g is a $\leq_I$ $-$lub of F in $({}^\kappa\mathrm{Ord}, \leq_I)$. As we can deal separately with $I + A$, $I + (\kappa \setminus A)$ where $A =: \{i : g(i) = 0\}$, and the later case is trivial we can assume $A = \emptyset$. So assume g is not a $\leq_I$ $-$lub, so there is an upper bound g' of F, but not $g \leq_I g'$. Define $g'' \in {}^\kappa\mathrm{Ord}$:

$$g''(i) = \begin{cases} 0 & \text{if } g(i) \leq g'(i) \\ g'(i) & \text{if } g'(i) < g(i) \end{cases}.$$

Clearly $g'' <_I g$. So, as g in an $\leq_I$ $-$eub of F for I, there is $f \in F$ such that $g'' <_I \max\{f, 1\}$, but $B =: \{i : g'(i) < g(i)\} \neq \emptyset \bmod I$ (as not $g \leq_I g'$) so $g' \restriction B =$

$g'' \restriction B <_I \max\{f, 1\} \restriction B$. But we know that $f \le_I g'$ (as g' is an upper bound of F) hence $f \restriction B \le_I g' \restriction B$, so by the previous sentence necessarily $f \restriction B =_I 0_B$ hence $g' \restriction B =_I 0_B$; as g' is a $\le_I$-upper bound of F we know $[f' \in F \Rightarrow f' \restriction B =_I 0_B]$, hence by ($\gamma$) of Definition 1.1(4) we have $g \restriction B =_I 0_B$, a contradiction to $B \notin I$ (see above). $\blacksquare_{1.3}$

Remark 1.3A. In 1.3 we can also have the straight monotonicity properties of

$$\mathrm{pcf}_I(\prod \bar{\lambda}, \Gamma) \text{ in } \Gamma \text{ and in } I.$$

CLAIM 1.4:

(1) $J_{<\lambda}[\bar{\lambda}]$ is an ideal (of $\mathcal{P}(\kappa)$ i.e. on κ, but the ideal may not be proper).
(2) if $\lambda \le \mu$, then $J_{<\lambda}[\bar{\lambda}] \subseteq J_{<\mu}[\bar{\lambda}]$
(3) if λ is singular, $J_{<\lambda}[\bar{\lambda}] = J_{<\lambda^+}[\bar{\lambda}] = J_{\le\lambda}[\bar{\lambda}]$
(4) if $\lambda \notin \mathrm{pcf}(\bar{\lambda})$, then we have $J_{<\lambda}[\bar{\lambda}] = J_{\le\lambda}[\bar{\lambda}]$.
(5) If $A \subseteq \kappa$, $A \notin J_{<\lambda}[\bar{\lambda}]$, and $f_\alpha \in \prod \bar{\lambda} \restriction A$, $\langle f_\alpha : \alpha < \lambda \rangle$ is $<_{J_{<\lambda}[\bar{\lambda}]}$-increasing cofinal in $(\prod \bar{\lambda} \restriction A)/J_{<\lambda}[\bar{\lambda}]$ then $A \in J_{\le\lambda}[\bar{\lambda}]$. Also this holds when we replace $J_{<\lambda}[\bar{\lambda}]$ by any ideal J on κ, $I^* \subseteq J \subseteq J_{\le\lambda}[\bar{\lambda}]$.
(6) The earlier parts hold for $J_{<\lambda}[\bar{\lambda}, \Gamma]$ too.

Proof. Straight.

LEMMA 1.5: Assume

($*$) $\bar{\lambda}$ is regular and
  (α) $\min \bar{\lambda} > \theta \ge \mathrm{wsat}(I^*)$ (see 1.1(5)(b)) or at least
  (β) $\liminf_{I^*}(\bar{\lambda}) \ge \theta \ge \mathrm{wsat}(I^*)$, and $\prod \bar{\lambda}/I^*$ is θ^+-directed**.

If λ is a cardinal $\ge \theta$, and $\kappa \notin J_{<\lambda}[\bar{\lambda}]$ then $(\prod \bar{\lambda}, <_{J_{<\lambda}[\bar{\lambda}]})$ is λ-directed (remember: $J_{<\lambda}[\bar{\lambda}] = J_{<\lambda}[\bar{\lambda}, I^*]$).

Proof. Note: if $f \in \prod \bar{\lambda}$ then $f < f + 1 \in \prod \bar{\lambda}$, (i.e. $(\prod \bar{\lambda}, <_{J_\lambda[\bar{\lambda}]})$ is endless) where $f + 1$ is defined by $(f+1)(i) = f(i) + 1$. Let $F \subseteq \prod \bar{\lambda}$, $|F| < \lambda$, and we shall prove that for some $g \in \prod \bar{\lambda}$ we have $(\forall f \in F)(f \le g \bmod J_{<\lambda}[\bar{\lambda}])$, this suffices. The proof is by induction on $|F|$. If $|F|$ is finite, this is trivial. Also if $|F| \le \theta$, when (α) of ($*$) holds it is easy: let $g \in \prod \bar{\lambda}$ be $g(i) = \sup\{f(i) : f \in F\} < \lambda_i$; when ($\beta$) of ($*$) holds use second clause of (β). So assume $|F| = \mu$, $\theta < \mu < \lambda$ so let $F = \{f^0_\alpha : \alpha < \mu\}$. By the induction hypothesis we can choose by induction on $\alpha < \mu$, $f^1_\alpha \in \prod \bar{\lambda}$ such that:

(a) $f^0_\alpha \le f^1_\alpha \bmod J_{<\lambda}[\bar{\lambda}]$
(b) for $\beta < \alpha$ we have $f^1_\beta < f^1_\alpha \bmod J_{<\lambda}[\bar{\lambda}]$.

If μ is singular, there is $C \subseteq \mu$ unbounded, $|C| = \mathrm{cf}(\mu) < \mu$, and by the induction hypothesis there is $g \in \prod \bar{\lambda}$ such that for $\alpha \in C$, $f^1_\alpha \le g \bmod J_{<\lambda}[\bar{\lambda}]$. Now g is as required: $f^0_\alpha \le f^1_\alpha \le f^1_{\min(C \setminus \alpha)} \le g \bmod J_{<\lambda}[\bar{\lambda}]$. So without loss of generality μ is regular. Let us define $A^*_\varepsilon =: \{i < \kappa : \lambda_i > |\varepsilon|\}$ for $\varepsilon < \theta$, so $\varepsilon < \zeta < \theta \Rightarrow A^*_\zeta \subseteq A^*_\varepsilon$ and $\varepsilon < \theta \Rightarrow A^*_\varepsilon = \kappa \bmod I^*$. Now we try to define by induction on $\varepsilon < \theta$, g_ε, $\alpha_\varepsilon = \alpha(\varepsilon) < \mu$ and $\langle B^\varepsilon_\alpha : \alpha < \mu \rangle$ such that:

** note if $\mathrm{cf}(\theta) < \theta$ then "θ^+-directed" follows from "θ-directed" which follows from "$\liminf_{I^*}(\bar{\lambda}) \ge \theta$, i.e. first part of clause (β) implies clause (β). Note also that if clause (α) holds then $\prod \bar{\lambda}/I^*$ is θ^+-directed (even $(\prod \bar{\lambda}, <)$ is θ^+-directed), so clause (α) implies clause (β).

(i) $g_\varepsilon \in \prod \bar{\lambda}$
(ii) for $\varepsilon < \zeta$ we have $g_\varepsilon \restriction A^*_\zeta \leq g_\zeta \restriction A^*_\zeta$
(iii) for $\alpha < \mu$ let $B^\varepsilon_\alpha =: \{i < \kappa : f^1_\alpha(i) > g_\varepsilon(i)\}$
(iv) for each $\varepsilon < \theta$, for every $\alpha \in [\alpha_{\varepsilon+1}, \mu)$, $B^\varepsilon_\alpha \neq B^{\varepsilon+1}_\alpha \bmod J_{<\lambda}[\bar{\lambda}]$.

We cannot carry this definition: as letting $\alpha(*) = \sup\{\alpha_\varepsilon : \varepsilon < \theta\}$, then $\alpha(*) < \mu$ since $\mu = \mathrm{cf}(\mu) > \theta$. We know that $B^\varepsilon_{\alpha(*)} \cap A^*_{\varepsilon+1} \neq B^{\varepsilon+1}_{\alpha(*)} \cap A^*_{\varepsilon+1}$ for $\alpha < \theta$ (by (iv) and as $A^*_{\varepsilon+1} = \kappa \bmod I^*$ and $I^* \subseteq J_{<\lambda}[\bar{\lambda}]$) and $B^\varepsilon_{\alpha(*)} \subseteq \kappa$ (by (iii)) and $[\varepsilon < \zeta \Rightarrow B^\zeta_{\alpha(*)} \cap A^*_\zeta \subseteq B^\varepsilon_{\alpha(*)}]$ (by (ii)), together $\langle A^*_{\varepsilon+1} \cap (B^\varepsilon_{\alpha(*)} \setminus B^{\varepsilon+1}_{\alpha(*)}) : \varepsilon < \theta \rangle$ is a sequence of θ pairwise disjoint members of $(I^*)^+$, a contradiction*** to the definition of $\theta = \mathrm{wsat}(I^*)$.
Now for $\varepsilon = 0$ let g_i be f^1_0 and $\alpha_\varepsilon = 0$.
For ε limit let $g_\varepsilon(i) = \bigcup_{\zeta<\varepsilon} g_\zeta(i)$ for $i \in A^*_\varepsilon$ and zero otherwise (note: $g_\varepsilon \in \prod \bar{\lambda}$ as $\varepsilon < \theta$, $\lambda_i > \varepsilon$ for $i \in A^*_\varepsilon$ and $\bar{\lambda}$ is a sequence of regular cardinals) and let $\alpha_\varepsilon = 0$.
For $\varepsilon = \zeta + 1$, suppose that g_ζ hence $\langle B^\zeta_\alpha : \alpha < \mu \rangle$ are defined. If $B^\zeta_\alpha \in J_{<\lambda}[\bar{\lambda}]$ for unboundedly many $\alpha < \mu$ (hence for every $\alpha < \mu$) then g_ζ is an upper bound for F mod $J_{<\lambda}[\bar{\lambda}]$ and the proof is complete. So assume this fails, then there is a minimal $\alpha(\varepsilon) < \mu$ such that $B^\zeta_{\alpha(\varepsilon)} \notin J_{<\lambda}[\bar{\lambda}]$. As $B^\zeta_{\alpha(\varepsilon)} \notin J_{<\lambda}[\bar{\lambda}]$, by Definition 1.2(2) for some ultrafilter D on κ disjoint to $J_{<\lambda}[\bar{\lambda}]$ we have $B^\zeta_{\alpha(\varepsilon)} \in D$ and $\mathrm{cf}(\prod \bar{\lambda}/D) \geq \lambda$. Hence $\{f^1_\alpha/D : \alpha < \mu\}$ has an upper bound h_ε/D where $h_\varepsilon \in \prod \bar{\lambda}$. Let us define $g_\varepsilon \in \prod \bar{\lambda}$:

$$g_\varepsilon(i) = \mathrm{Max}\{g_\zeta(i), h_\varepsilon(i)\}.$$

Now (i), (ii) hold trivially and B^ε_α is defined by (iii). Why does (iv) hold (for ζ) with $\alpha_{\zeta+1} = \alpha_\varepsilon =: \alpha(\varepsilon)$? Suppose $\alpha(\varepsilon) \leq \alpha < \mu$. As $f^1_{\alpha(\varepsilon)} \leq f^1_\alpha \bmod J_{<\lambda}[\bar{\lambda}]$ clearly $B^\zeta_{\alpha(\varepsilon)} \subseteq B^\zeta_\alpha \bmod J_{<\lambda}[\bar{\lambda}]$. Moreover $J_{<\lambda}[\bar{\lambda}]$ is disjoint to D (by its choice) so $B^\zeta_{\alpha(\varepsilon)} \in D$ implies $B^\zeta_\alpha \in D$.
On the other hand B^ε_α is $\{i < \kappa : f^1_\alpha(i) > g_\varepsilon(i)\}$ which is equal to $\{i \in \bar{\lambda} : f^1_\alpha(i) > g_\zeta(i), h_\varepsilon(i)\}$ which does not belong to D (h_ε was chosen such that $f^1_\alpha \leq h_\varepsilon \bmod D$). We can conclude $B^\varepsilon_\alpha \notin D$, whereas $B^\zeta_\alpha \in D$; so they are distinct $\bmod J_{<\lambda}[\bar{\lambda}]$ as required in clause (iv).

Now we have said that we cannot carry the definition for all $\varepsilon < \theta$, so we are stuck at some ε; by the above ε is successor, say $\varepsilon = \zeta + 1$, and g_ζ is as required: an upper bound for F modulo $J_{<\lambda}[\bar{\lambda}]$. $\blacksquare_{1.5}$

LEMMA 1.6: If (*) of 1.5, D is an ultrafilter on κ disjoint to I^* and $\lambda = \mathrm{tcf}(\prod \bar{\lambda}, <_D)$, then for some $B \in D$, $(\prod \bar{\lambda} \restriction B, <_{J_{<\lambda}[\bar{\lambda}]})$ has true cofinality λ. (So $B \in J_{\leq\lambda}[\bar{\lambda}] \setminus J_{<\lambda}[\bar{\lambda}]$ by 1.4(5).)

Proof. By the definition of $J_{<\lambda}[\bar{\lambda}]$ clearly we have $D \cap J_{<\lambda}[\bar{\lambda}] = \emptyset$.
Let $\langle f_\alpha/D : \alpha < \lambda \rangle$ be increasing unbounded in $\prod \bar{\lambda}/D$ (so $f_\alpha \in \prod \bar{\lambda}$). By 1.5 without loss of generality $(\forall \beta < \alpha)(f_\beta < f_\alpha \bmod J_{<\lambda}[\bar{\lambda}])$.
Now 1.6 follows from 1.7 below: its hypothesis clearly holds. If $\bigwedge_{\alpha<\lambda} B_\alpha = \emptyset \bmod D$, (see (A) of 1.7) then (see (D) of 1.7) $J \cap D = \emptyset$ hence (see (D) of 1.7) g/D contradicts the choice of $\langle f_\alpha/D : \alpha < \lambda \rangle$. So for some $\alpha < \lambda$, $B_\alpha \in D$; by (C) of 1.7 and 1.4(5) we get the desired conclusion. $\blacksquare_{1.6}$

*** in fact note that for no $B_\varepsilon \subseteq \kappa$ $(\varepsilon < \theta)$ do we have: $B_\varepsilon \neq B_{\varepsilon+1}$ mod I^* and $\varepsilon < \zeta < \theta \Rightarrow B_\varepsilon \cap A_\zeta \subseteq B_\zeta$ where $A_\zeta = \kappa$ mod I^* (e.g. $A_\zeta = A^*_\zeta$)

LEMMA 1.7: Suppose (*) of 1.5, $\mathrm{cf}(\lambda) > \theta$, $f_\alpha \in \prod \bar{\lambda}$, $f_\alpha < f_\beta \bmod J_{<\lambda}[\bar{\lambda}]$ for $\alpha < \beta < \lambda$, and there is no $g \in \prod \bar{\lambda}$ such that for every $\alpha < \lambda$, $f_\alpha < g \bmod J_{<\lambda}[\bar{\lambda}]$. Then there are B_α (for $\alpha < \lambda$) such that:

(A) $B_\alpha \subseteq \kappa$ and for some $\alpha(*) < \lambda$: $\alpha(*) \leq \alpha < \lambda \Rightarrow B_\alpha \notin J_{<\lambda}[\bar{\lambda}]$
(B) $\alpha < \beta \Rightarrow B_\alpha \subseteq B_\beta \bmod J_{<\lambda}[\bar{\lambda}]$ (i.e. $B_\alpha \setminus B_\beta \in J_{<\lambda}[\bar{\lambda}]$)
(C) for each α, $\langle f_\beta \restriction B_\alpha : \beta < \lambda \rangle$ is cofinal in $(\prod \bar{\lambda} \restriction B_\alpha, <_{J_{<\lambda}[\bar{\lambda}]})$ (better restrict yourselves to $\alpha \geq \alpha(*)$ (see (A)) so that necessarily $B_\alpha \notin J_{<\lambda}[\bar{\lambda}]$);.
(D) for some $g \in \prod \bar{\lambda}$, $\bigwedge_{\alpha<\lambda} f_\alpha \leq g \bmod J$ where[†] $J = J_{<\lambda}[\bar{\lambda}] + \{B_\alpha : \alpha < \lambda\}$;

in fact

(D)$^+$ for some $g \in \prod \bar{\lambda}$ for every $\alpha < \lambda$, we have[†] $f_\alpha \leq g \bmod (J_{<\lambda}[\bar{\lambda}] + B_\alpha)$, in fact $B_\alpha = \{i < \kappa : f_\alpha(i) > g(i)\}$
(E) if $g \leq g' \in \prod \bar{\lambda}$, then for arbitrarily large $\alpha < \lambda$:

$$\{i < \kappa : [g(i) \geq f_\alpha(i) \Leftrightarrow g'(i) \geq f_\alpha(i)]\} = \kappa \bmod J_{<\lambda}[\bar{\lambda}]$$

(hence for every large enough $\alpha < \lambda$ this holds)
(F) if δ is a limit ordinal $< \lambda$, f_δ is a $\leq_{J_{<\lambda}[\bar{\lambda}]}$ $-$lub of $\{f_\alpha : \alpha < \delta\}$ then B_δ is a lub of $\{B_\alpha : \alpha < \delta\}$ in $\mathcal{P}(\kappa)/J_{<\lambda}[\bar{\lambda}]$.

Proof of 1.7. Remember that for $\varepsilon < \theta$, $A^*_\varepsilon = \{i < \kappa : \lambda_i > |\varepsilon|\}$ so $A^*_\varepsilon = \kappa \bmod I^*$ and $\varepsilon < \zeta \Rightarrow A^*_\zeta \subseteq A^*_\varepsilon$. We now define by induction on $\varepsilon < \theta$, g_ε, $\alpha(\varepsilon) < \lambda$, $\langle B^\varepsilon_\alpha : \alpha < \lambda \rangle$ such that:

(i) $g_\varepsilon \in \prod \bar{\lambda}$
(ii) for $\zeta < \varepsilon$, $g_\zeta \restriction A^*_\varepsilon \leq g_\varepsilon \restriction A^*_\varepsilon$
(iii) $B^\varepsilon_\alpha =: \{i \in \kappa : f_\alpha(i) > g_\varepsilon(i)\}$
(iv) if $\alpha(\varepsilon) \leq \alpha < \lambda$ then $B^\varepsilon_\alpha \neq B^{\varepsilon+1}_\alpha \bmod J_{<\lambda}[\bar{\lambda}]$

For $\varepsilon = 0$ let $g_\varepsilon = f_0$, and $\alpha(\varepsilon) = 0$.
For ε limit let $g_\varepsilon(i) = \bigcup_{\zeta<\varepsilon} g_\zeta(i)$ if $i \in A^*_\varepsilon$ and zero otherwise; now

$$[\zeta < \varepsilon \Rightarrow g_\zeta \restriction A^*_\varepsilon \leq g_\varepsilon \restriction A^*_\varepsilon]$$

holds trivially and $g_\varepsilon \in \prod \bar{\lambda}$ as each λ_i is regular and $[i \in A^*_\varepsilon \Leftrightarrow \lambda_i > \varepsilon]$), and let $\alpha(\varepsilon) = 0$.
For $\varepsilon = \zeta + 1$, if $\{\alpha < \lambda : B^\zeta_\alpha \in J_{<\lambda}[\bar{\lambda}]\}$ is unbounded in λ, then g_ζ is a bound for $\langle f_\alpha : \alpha < \lambda \rangle \bmod J_{<\lambda}[\bar{\lambda}]$, contradicting an assumption. Clearly

$$\alpha < \beta < \lambda \quad \Rightarrow B^\zeta_\alpha \subseteq B^\zeta_\beta \bmod J_{<\lambda}[\bar{\lambda}]$$

hence $\{\alpha < \lambda : B^\zeta_\alpha \in J_{<\lambda}[\bar{\lambda}]\}$ is an initial segment of λ. So by the previous sentence there is $\alpha(\varepsilon) < \lambda$ such that for every $\alpha \in [\alpha(\varepsilon), \lambda)$, we have $B^\zeta_\alpha \notin J_{<\lambda}[\bar{\lambda}]$ (of course, we may increase $\alpha(\varepsilon)$ later). If $\langle B^\zeta_\alpha : \alpha < \lambda \rangle$ satisfies the desired conclusion, with $\alpha(\varepsilon)$ for $\alpha(*)$ in (A) and g_ζ for g in (D), (D)$^+$ and (E), we are done. Now among the conditions in the conclusion of 1.7, (A) holds by the definition of B^ζ_α and of $\alpha(\varepsilon)$, (B) holds by B^ζ_α's definition as $\alpha < \beta \Rightarrow f_\alpha < f_\beta \bmod J_{<\lambda}[\bar{\lambda}]$, (D)$^+$ holds with $g = g_\zeta$ by the choice of B^ζ_α hence also clause (D) follows. Lastly if (E) fails, say for g', then it can serve as g_ε. Now condition (F) follows immediately from (iii) (if (F) fails for δ, then there is $B \subseteq B^\zeta_\delta$ such that $\bigwedge_{\alpha<\delta} B^\zeta_\alpha \subseteq B \bmod J_{<\lambda}[\bar{\lambda}]$ and

[†] Of course, if $B_\alpha = \kappa \bmod J_{<\lambda}[\bar{\lambda}]$, this becomes trivial.

$B^\zeta_\delta \setminus B \notin J_{<\lambda}[\bar\lambda]$; now the function $g^* =: (g_\zeta \restriction (\kappa\setminus B)) \cup (f_\delta \restriction B)$ contradicts "f_δ is a $\leq_{J_{<\lambda}[\bar\lambda]}$ –lub of $\{f_\alpha : \alpha < \delta\}$", because: $g^* \in \prod \bar\lambda$ (obvious), $\neg(f_\delta \leq g^* \mod J_{<\lambda}[\bar\lambda])$ [why? as $B^\zeta_\delta \setminus B \notin J_{<\lambda}[\bar\lambda]$ and $g^* \restriction (B^\zeta_\delta \setminus B) = g_\zeta \restriction (B^\zeta_\delta \setminus B) < f_\delta \restriction (B^\zeta_\delta \setminus B)$ by the choice of B^ζ_δ], and for $\alpha < \delta$ we have:

$$f_\alpha \restriction B \leq_{J_{<\lambda}[\bar\lambda]} f_\delta \restriction B = g^* \restriction B \quad \text{and}$$

$$f_\alpha \restriction (\kappa \setminus B) \leq_{J_{<\lambda}[\bar\lambda]} g_\zeta \restriction (\kappa \setminus B) = g^* \restriction (\kappa \setminus B)$$

(the $\leq_{J_{<\lambda}[\bar\lambda]}$ holds as $(\kappa\setminus B) \cap B^\zeta_\alpha \in J_{<\lambda}[\bar\lambda]$ and the definition of B^ζ_α). So only clause (C) (of 1.7) may fail, without loss of generality for $\alpha = \alpha(\varepsilon)$. I.e. $\langle f_\beta \restriction B^\zeta_{\alpha(\varepsilon)} : \beta < \lambda\rangle$ is not cofinal in $(\prod \bar\lambda \restriction B^\zeta_{\alpha(\varepsilon)}, <_{J_{<\lambda}[\bar\lambda]})$. As this sequence of functions is increasing w.r.t. $<_{J_{<\lambda}[\bar\lambda]}$, there is $h_\alpha \in \prod(\bar\lambda \restriction B^\zeta_{\alpha(\varepsilon)})$ such that for no $\beta < \lambda$ do we have $h_\alpha \leq f_\beta \restriction B^j_{\alpha(\varepsilon)} \mod J_{<\lambda}[\bar\lambda]$. Let $h'_\varepsilon = h_\varepsilon \cup 0_{(\kappa \setminus B^\zeta_{\alpha(\varepsilon)})}$ and $g_\varepsilon \in \prod \bar\lambda$ be defined by $g_\varepsilon(i) = \mathrm{Max}\{g_\zeta(i), h'_\varepsilon(i)\}$. Now define B^ε_α by (iii) so (i), (ii), (iii) hold trivially, and we can check (iv).

So we can define g_ε, $\alpha(\varepsilon)$ for $\varepsilon < \theta$, satisfying (i)–(iv). As in the proof of 1.5, this is impossible: because (remembering $\mathrm{cf}(\lambda) = \lambda > \theta$) letting $\alpha(*) =: \bigcup_{\varepsilon<\theta} \alpha(\varepsilon) < \lambda$ we have: $\langle B^\varepsilon_{\alpha(*)} \cap A^*_\zeta : \varepsilon < \zeta\rangle$ is $\subseteq$-decreasing, for each $\zeta < \theta$, and $A^*_\varepsilon = \kappa \mod I^*$ and $B^{\varepsilon+1}_{\alpha(*)} \neq B^\varepsilon_{\alpha(*)} \mod J_{<\lambda}[\bar\lambda]$ so $\langle B^\varepsilon_{\alpha(*)} \cap A^*_{\varepsilon+1} \setminus B^{\varepsilon+1}_{\alpha(*)} : \varepsilon < \theta\rangle$ is a sequence of θ pairwise disjoint members of $(J_{<\lambda}[\bar\lambda])^+$ hence of $(I^*)^+$ which give the contradiction to $(*)$ of 1.5; so the lemma cannot fail. $\blacksquare_{1.7}$

LEMMA 1.8: Suppose $(*)$ of 1.5.

(1) For every $B \in J_{\leq\lambda}[\bar\lambda] \setminus J_{<\lambda}[\bar\lambda]$, we have:

$$(\prod \bar\lambda \restriction B, <_{J_{<\lambda}[\bar\lambda]}) \text{ has true cofinality } \lambda \text{ (hence } \lambda \text{ is regular).}$$

(2) If D is an ultrafilter on κ, disjoint to I^*, <u>then</u> $\mathrm{cf}(\prod \bar\lambda / D)$ is $\min\{\lambda : D \cap J_{\leq\lambda}[\bar\lambda] \neq \emptyset\}$.

(3) (i) For λ limit $J_{<\lambda}[\bar\lambda] = \bigcup_{\mu<\lambda} J_{<\mu}[\bar\lambda]$ hence
(ii) for every λ, $J_{<\lambda}[\bar\lambda] = \bigcup_{\mu<\lambda} J_{\leq\mu}[\bar\lambda]$.

(4) $J_{\leq\lambda}[\bar\lambda] \neq J_{<\lambda}[\bar\lambda]$ <u>iff</u> $J_{\leq\lambda}[\bar\lambda] \setminus J_{<\lambda}[\bar\lambda] \neq \emptyset$ <u>iff</u> $\lambda \in \mathrm{pcf}(\bar\lambda)$.

(5) $J_{\leq\lambda}[\bar\lambda]/J_{<\lambda}[\bar\lambda]$ is λ-directed (i.e. if $B_\gamma \in J_{\leq\lambda}[\bar\lambda]$ for $\gamma < \gamma^*$, $\gamma^* < \lambda$ <u>then</u> for some $B \in J_{\leq\lambda}[\bar\lambda]$ we have $B_\gamma \subseteq B \mod J_{<\lambda}[\bar\lambda]$ for every $\gamma < \gamma^*$.)

Proof. (1) Let

$$J = \{B \subseteq \kappa : B \in J_{<\lambda}[\bar\lambda] \text{ or } B \in J_{\leq\lambda}[\bar\lambda] \setminus J_{<\lambda}[\bar\lambda] \text{ and}$$

$$(\prod \bar\lambda \restriction B, <_{J_{<\lambda}[\bar\lambda]}) \text{ has true cofinality } \lambda\}.$$

By its definition clearly $J \subseteq J_{\leq\lambda}[\bar\lambda]$; it is quite easy to check it is an ideal (use 1.3(2)(v)). Assume $J \neq J_{\leq\lambda}[\bar\lambda]$ and we shall get a contradiction. Choose $B \in J_{\leq\lambda}[\bar\lambda] \setminus J$; as J is an ideal, there is an ultrafilter D on κ such that: $D \cap J = \emptyset$ and $B \in D$. Now if $\mathrm{tcf}(\prod \bar\lambda / D) \geq \lambda^+$, then $B \notin J_{\leq\lambda}[\bar\lambda]$ (by the definition of $J_{\leq\lambda}[\bar\lambda]$); contradiction. On the other hand if $F \subseteq \prod \bar\lambda$, $|F| < \lambda$ then there is $g \in \prod \bar\lambda$ such that $(\forall f \in F)(f < g \mod J_{<\lambda}[\bar\lambda])$ (by 1.5), so $(\forall f \in F)[f < g \mod D]$ (as $J_{<\lambda}[\bar\lambda] \subseteq J$, $D \cap J = \emptyset$), and this implies $\mathrm{cf}(\prod \bar\lambda / D) \geq \lambda$. By the last two sentences we know that $\mathrm{tcf}(\prod \bar\lambda / D)$ is λ. Now by 1.6 for some $C \in D$, $\left(\prod(\bar\lambda \restriction C), <_{J_{<\lambda}[\bar\lambda]}\right)$ has true cofinality λ, of course $C \cap B \subseteq C$ and $C \cap B \in D$ hence $C \cap B \notin J_{<\lambda}[\bar{\lambda}]$. Clearly if $C' \subseteq C$, $C' \notin J_{<\lambda}[\bar{\lambda}]$ then also $(\prod \bar{\lambda} \restriction C', <_{J_{<\lambda}[\bar{\lambda}]})$ has true cofinality λ, hence by the last sentence without loss of generality $C \subseteq B$; hence by 1.4(5) we know that $C \in J_{\leq\lambda}[\bar{\lambda}]$ hence by the definition of J we have $C \in J$. But this contradicts the choice of D as disjoint from J.

We have to conclude that $J = J_{\leq\lambda}[\bar{\lambda}]$ so we have proved 1.8(1).

(2) Let λ be minimal such that $D \cap J_{\leq\lambda}[\bar{\lambda}] \neq \emptyset$ (it exists as by 1.3(10) $J_{<(\prod \bar{\lambda})^+}[\bar{\lambda}] = \mathcal{P}(\kappa)$) and choose $B \in D \cap J_{\leq\lambda}[\bar{\lambda}]$. So $[\mu < \lambda \Rightarrow B \notin J_{\leq\mu}[\bar{\lambda}]]$ (by the choice of λ) hence by 1.8(3)(ii) below, we have $B \notin J_{<\lambda}[\bar{\lambda}]$. It similarly follows that $D \cap J_{<\lambda}[\bar{\lambda}] = \emptyset$. Now $(\prod \bar{\lambda} \restriction B, <_{J_{<\lambda}[\bar{\lambda}]})$ has true cofinality λ by 1.8(1). As we know that $B \in D \cap J_{\leq\lambda}[\bar{\lambda}]$, and $J_{<\lambda}[\bar{\lambda}] \cap D = \emptyset$; clearly we have finished the proof.

(3) (i) Let $J =: \bigcup_{\mu<\lambda} J_{<\mu}[\bar{\lambda}]$. Now J is an ideal by 1.4(1)+(2) and $(\prod \bar{\lambda}, <_J)$ is λ-directed; i.e. if $\alpha^* < \lambda$ and $\{f_\alpha : \alpha < \alpha^*\} \subseteq \prod \bar{\lambda}$, then there exists $f \in \prod \bar{\lambda}$ such that

$$(\forall \alpha < \alpha^*)(f_\alpha < f \bmod J).$$

[Why? if $\alpha^* < \theta^+$ as $(*)$ of 1.5 holds, this is obvious, suppose not; λ is a limit cardinal, hence there is μ^* such that $\alpha^* < \mu^* < \lambda$. Without loss of generality $|\alpha^*|^+ < \mu^*$. By 1.5, there is $f \in \prod \bar{\lambda}$ such that $(\forall \alpha < \alpha^*)(f_\alpha < f \bmod J_{<\mu^*}[\bar{\lambda}])$. Since $J_{<\mu^*}[\bar{\lambda}] \subseteq J$, it is immediate that

$$(\forall \alpha < \alpha^*)(f_\alpha < f \bmod J).]$$

Clearly $\bigcup_{\mu<\lambda} J_{<\mu}[\bar{\lambda}] \subseteq J_{<\lambda}[\bar{\lambda}]$ by 1.4(2). On the other hand, let us suppose that there is $B \in (J_{<\lambda}[\bar{\lambda}] \setminus \bigcup_{\mu<\lambda} J_{<\mu}[\bar{\lambda}])$. Choose an ultrafilter D on κ such that $B \in D$ and $D \cap J = \emptyset$. Since $(\prod \bar{\lambda}, <_J)$ is λ-directed and $D \cap J = \emptyset$, one has $\mathrm{tcf}(\prod \bar{\lambda}/D) \geq \lambda$, but $B \in D \cap J_{<\lambda}[\bar{\lambda}]$, in contradiction to Definition 1.2(2).

(3)(ii) If λ limit — by part (i) and 1.4(2); if λ successor — by 1.4(2) and Definition 1.2(3).

(4) Easy.

(5) Let $\langle f^\gamma_\alpha : \alpha < \lambda \rangle$ be $<_{J_{<\lambda}[\bar{\lambda}]+(\kappa \setminus B_\gamma)}$-increasing and cofinal in $\prod \bar{\lambda}$ (for $\gamma < \gamma^*$). Let us choose by induction on $\alpha < \lambda$ a function $f_\alpha \in \prod \bar{\lambda}$, as a $<_{J_{<\lambda}[\bar{\lambda}]}$-bound to $\{f_\beta : \beta < \alpha\} \cup \{f^\gamma_\alpha : \gamma < \gamma^*\}$, such f_α exists by 1.5 and apply 1.7 to $\langle f_\alpha : \alpha < \lambda \rangle$, getting $\langle B'_\alpha : \alpha < \lambda \rangle$, now B'_α for α large enough is as required. $\blacksquare_{1.8}$

CONCLUSION 1.9: If $(*)$ of 1.5, then $\mathrm{pcf}(\bar{\lambda})$ has a last element.

Proof. This is the minimal λ such that $\kappa \in J_{\leq\lambda}[\bar{\lambda}]$. λ exists, since $\lambda^* =: |\prod \bar{\lambda}| \in \{\lambda : \kappa \in J_{\leq\lambda}[\bar{\lambda}]\} \neq \emptyset$ and by 1.4(2); and $\lambda \in pcf(\bar{\lambda})$ by 1.8(4) and $\lambda = \max pcf(\bar{\lambda})$ by 1.4(7)+1.8(4). $\blacksquare_{1.9}$

CLAIM 1.10: Suppose $(*)$ of 1.5 holds. Assume for $j < \sigma$, D_j is a filter on κ extending $\{\kappa \setminus A : A \in I^*\}$, E a filter on σ and $D^* = \{B \subseteq \kappa : \{j < \sigma : B \in D_j\} \in E\}$ (a filter on κ). Let $\mu_j =: \mathrm{tcf}(\prod \bar{\lambda}, <_{D_j})$ be well defined for $j < \sigma$, and assume further $\mu_j > \sigma + \theta$ (where θ is from $(*)$ of 1.5).
Let

$$\lambda = \mathrm{tcf}(\prod \bar{\lambda}, <_{D^*}), \mu = \mathrm{tcf}(\prod_{j<\sigma} \mu_j, <_E).$$

<u>Then</u> $\lambda = \mu$ (in particular, if one is well defined, than so is the other).

Proof. Wlog $\sigma \geq \theta$ (otherwise we can add $\mu_j =: \mu_0$, $D_j =: D_0$ for $j \in \theta \setminus \sigma$, and replace σ by θ and E by $E' = \{A \subseteq \theta: A \cap \sigma \in E\}$). Let $\langle f^j_\alpha: \alpha < \mu_j \rangle$ be an $<_{D_j}$-increasing cofinal sequence in $(\prod \bar{\lambda}, <_{D_j})$.

Now $\ell = 0, 1$, for each $f \in \prod \bar{\lambda}$, define $G_\ell(f) \in \prod_{j<\sigma} \mu_j$ by $G_\ell(f)(j) = \min\{\alpha < \mu_j$: if $\ell = 1$ then $f \leq f^j_\alpha \bmod D_j$ <u>and</u> if $\ell = 0$ then: not $f^j_\alpha < f \bmod D_j\}$ (it is well defined for $f \in \prod \bar{\lambda}$ by the choice of $\langle f^j_\alpha: \alpha < \mu_j \rangle$).
Note that for f^1, $f^2 \in \prod \bar{\lambda}$ and $\ell < 2$ we have:

$$\begin{aligned} f^1 &\leq f^2 \bmod D^* \Leftrightarrow B(f^1, f^2) =: \{i < \kappa: f^1(i) \leq f^2(i)\} \in D^* \\ &\Leftrightarrow A(f^1, f^2) =: \{j < \sigma: B(f^1, f^2) \in D_j\} \in E \\ &\Leftrightarrow \text{ for some } A \in E, \text{ for every } i \in A \text{ we have } f^1 \leq_{D_i} f^2 \\ &\Rightarrow \text{ for some } A \in E \text{ for every } i \in A \text{ we have} \\ &\qquad G_\ell(f^1)(i) \leq G_\ell(f^2)(i) \\ &\Leftrightarrow G_\ell(f^1) \leq G_\ell(f^2) \bmod E. \end{aligned}$$

So

$\otimes_1$ G_ℓ is a mapping from $(\prod \bar{\lambda}, \leq_{D^*})$ into $(\prod_{j<\sigma} \mu_j, \leq_E)$ preserving order.

Next we prove that

$\otimes_2$ for every $g \in \prod_{j<\sigma} \mu_j$ for some $f \in \prod \bar{\lambda}$, we have $g \leq G_0(f) \bmod E$.

[Why? Note that $\min\{\mu_j: j < \sigma\} \geq \sigma^+ \geq \theta^+$ and $J_{\leq\theta}[\bar{\lambda}] \subseteq J_{\leq\sigma}[\bar{\lambda}]$. By 1.5 we know $(\prod \bar{\lambda}, <_{J_{\leq\sigma}[\bar{\lambda}]})$ is σ^+-directed, hence for some function $f \in \prod \bar{\lambda}$:

$(*)_1$ for $j < \sigma$ we have $f^j_{g(j)} < f \bmod J_{\leq\sigma}[\bar{\lambda}]$.

We here assumed $\sigma < \mu_j$, hence $J_{\leq\sigma}[\bar{\lambda}] \subseteq J_{<\mu_j}[\bar{\lambda}]$ (by 1.4(2)) but $J_{<\mu_j}[\bar{\lambda}]$ is disjoint to D_j by the definition of $J_{<\mu_j}[\bar{\lambda}]$ (by 1.8(2) + 1.3(13)(c)) so together with $(*)_1$:

$(*)_2$ for $j < \sigma$, $f^j_{g(j)} < f \bmod D_j$.

So by the definition of G_0 for every $j < \sigma$ we have $g(j) < G_0(f)(j)$ hence clearly $g < G_0(f)$.]

$\otimes_3$ for $f \in \prod \bar{\lambda}$ we have $G_0(f) \leq G_1(f)$ [Why? read the definitions]

$\otimes_4$ if f_1, $f_2 \in \prod \bar{\lambda}$ and $G_1(f_1) <_E G_0(f_2)$ then $f_1 <_{D^*} f_2$

[Why? as $G_1(f_1) <_E G_0(f_2)$ there is $B \in E$ such that: $j \in B \Rightarrow G_1(f_1)(j) < G_0(f_2)(j)$ so for each $j \in B$ we have $f_1 \leq_{D_j} f^j_{G_1(f_1)(j)}$ (by the definition of $G_1(f_1)$) and $f^j_{G_1(f_1)(j)} <_{D_j} f_2$ (as $G_1(f_1)(j) < G_0(f_2)(j)$ and the definition of $G_0(f_2)(j)$) so together $f_1 <_{D_j} f_2$. So $A(f_1, f_2) = \{i < \kappa : f_1(i) < f_2(i)\}$ satisfies: $A(f_1, f_2) \in D_j$ for every $j \in B$ but B was chosen in E, hence $A(f_1, f_2) \in D^*$ (by the definition of D^*) hence $f_1 <_{D^*} f_2$ as required]

Now first assume $\lambda = \operatorname{tcf}(\prod \bar{\lambda}, <_{D^*})$ is well defined, so there is a sequence $\bar{f} = \langle f_\alpha : \alpha < \lambda \rangle$ of members of $\prod \bar{\lambda}$, $<_{D^*}$-increasing and cofinal. So $\langle G_0(f_\alpha) : \alpha < \lambda \rangle$ is $\leq_E$-increasing in $\prod_{j<\sigma} \mu_j$ (by $\otimes_1$), for every $g \in \prod_{j<\sigma} \mu_j$ for some $f \in \prod \bar{\lambda}$ we have $g \leq_E G_0(f)$ (why? by $\otimes_2$), but by the choice of $\bar{f}$ for some $\beta < \lambda$ we have $f <_{D^*} f_\beta$ hence by $\otimes_1$ we have $g \leq_E G_0(f) \leq_E G_0(f_\beta)$, so $\langle G_0(f_\alpha) : \alpha < \lambda \rangle$ is cofinal in $(\prod_{j<\sigma} \mu_j, <_E)$. Also for every $\alpha < \lambda$, applying the previous sentence to $G(f_\alpha) + 1$ $(\in \prod_{j<\sigma} \mu_j)$ we can find $\beta < \lambda$ such that $G(f_\alpha) + 1 \leq_E G(f_\beta)$, so $G(f_\alpha) <_E G(f_\alpha)$, so for some club C of λ, $\langle G_0(f_\alpha):\alpha\in C\rangle$ is $<_E$-increasing cofinal in $(\prod_{j<\sigma}\mu_j,<_E)$. So if λ is well defined then $\mu=\mathrm{tcf}(\prod_{j<\sigma}\mu_j,<_E)$ is well defined and equal to λ.

Lastly assume that μ is well defined i.e. $\prod_{j<\sigma}\mu_j/E$ has true cofinality μ, let $\bar g=\langle g_\alpha:\alpha<\mu\rangle$ exemplifies it. Choose by induction on $\alpha<\mu$, a function f_α and ordinals β_α, γ_α such that

(i) $f_\alpha\in\prod\bar\lambda$ and $\beta_\alpha<\mu$ and $\gamma_\alpha<\mu$
(ii) $g_{\beta_\alpha}<_E G_0(f_\alpha)\le_E G_1(f_\alpha)<_E g_{\gamma_\alpha}$ (so $\beta_\alpha<\gamma_\alpha$)
(iii) $\alpha_1<\alpha_2<\mu\Rightarrow\gamma_{\alpha_1}<\beta_{\alpha_2}$ (so $\beta_\alpha\ge\alpha$)

In stage α, first choose $\beta_\alpha=\bigcup\{\gamma_{\alpha_1}+1:\alpha_1<\alpha\}$, then choose $f_\alpha\in\prod\bar\lambda$ such that $g_{\beta_\alpha}+1<_E G_0(f_\alpha)$ (possible by $\otimes_2$) then choose γ_α such that $G_1(f_\alpha)<_E g_{\gamma_\alpha}$. Now $G_0(f_\alpha)\le_E G_1(f_\alpha)$ by $\otimes_3$. By $\otimes_4$ we have $\alpha_1<\alpha_2\Rightarrow f_{\alpha_1}<_{D^*}f_{\alpha_2}$. Also if $f\in\prod\bar\lambda$ then $G_1(f)\in\prod_{j<\sigma}\mu_j$ hence by the choice of $\bar g$, for some $\alpha<\mu$ we have $G_1(f)<_E g_\alpha$ but $\alpha\le\beta_\alpha$ so $G_1(f)<_E g_\alpha\le_E G_0(f_\alpha)$ hence by $\otimes_4$, $f<_{D^*}f_\alpha$. Altogether, $\langle f_\alpha:\alpha<\mu\rangle$ exemplifies that $(\prod\bar\lambda,<_{D^*})$ has true cofinality μ, so λ is well defined and equal to μ. $\blacksquare_{1.11}$

CONCLUSION 1.11: If $(*)$ of 1.5 holds, and σ, $\bar\mu=\langle\mu_j{:}j<\sigma\rangle$, $\langle D_j{:}j<\sigma\rangle$ are as in 1.10 and $\sigma+\theta<\min(\bar\mu)$, and J is an ideal on σ and I an ideal on κ such that $I^*\subseteq I\subseteq\{A\subseteq\kappa:$ for some $B\in J$ for every $j\in\sigma\setminus A$ we have $B\notin D_j\}$ (e.g. $I=I^*$) then $\mathrm{pcf}_J(\{\mu_j{:}j<\sigma\})\subseteq\mathrm{pcf}_I(\bar\lambda)$.

Proof. Let E be an ultrafilter on σ disjoint to J then we can define an ultrafilter D^* on κ as in 1.10, so clearly D^* is disjoint to I and we apply 1.10. $\blacksquare_{1.11}$

2. Normality of $\lambda\in\mathrm{pcf}(\bar\lambda)$ for $\bar\lambda$

Having found those ideals $J_{<\lambda}[\bar\lambda]$, we would like to know more. As $J_{<\lambda}[\bar\lambda]$ is increasing continuous in λ, the question is how $J_<[\bar\lambda]$, $J_{<\lambda^+}[\bar\lambda]$ are related.

The simplest relation is $J_{<\lambda^+}[\bar\lambda]=J_{<\lambda}[\bar\lambda]+B$ for some $B\subseteq\kappa$, and then we call λ normal (for $\bar\lambda$) and denote $B=B_\lambda[\bar\lambda]$ though it is unique only modulo $J_{<\lambda}[\bar\lambda]$. We give a sufficient condition for existence of such B, using this in 2.8; giving the necessary definition in 2.3 and needed information in 2.4, 2.5, 2.6; lastly 2.7 is the essential uniqueness of cofinal sequences in appropriate $\prod\bar\lambda/I$.

Definition 2.1.

(1) We say $\lambda\in\mathrm{pcf}(\bar\lambda)$ is normal (for $\bar\lambda$) if for some $B\subseteq\kappa$, $J_{\le\lambda}[\bar\lambda]=J_{<\lambda}[\bar\lambda]+B$.
(2) We say $\lambda\in\mathrm{pcf}(\bar\lambda)$ is semi-normal (for $\bar\lambda$) if there are B_α for $\alpha<\lambda$ such that:
  (i) $\alpha<\beta\Rightarrow B_\alpha\subseteq B_\beta \bmod J_{<\lambda}[\bar\lambda]$
  and
  (ii) $J_{\le\lambda}[\bar\lambda]=J_{<\lambda}[\bar\lambda]+\{B_\alpha{:}\alpha<\lambda\}$.
(3) We say $\bar\lambda$ is normal if every $\lambda\in\mathrm{pcf}(\bar\lambda)$ is normal for $\bar\lambda$. Similarly for semi normal.
(4) In (1), (2), (3) instead $\bar\lambda$ we can say $(\bar\lambda,I)$ or $\prod\bar\lambda/I$ or $(\prod\bar\lambda,<_I)$ if we replace I^* by I (an ideal on $\mathrm{Dom}(\bar\lambda)$).

Fact 2.2. Suppose (*) of 1.5 and $\lambda \in \mathrm{pcf}(\bar\lambda)$. Now:

(1) λ is semi-normal for $\bar\lambda$ iff for some $F = \{f_\alpha : \alpha < \lambda\} \subseteq \prod \bar\lambda$ we have: $[\alpha < \beta \Rightarrow f_\alpha < f_\beta \bmod J_{<\lambda}[\bar\lambda]]$ and for every ultrafilter D over κ disjoint to $J_{<\lambda}[\bar\lambda]$, F is unbounded in $(\prod \bar\lambda, <_D)$ whenever $\mathrm{tcf}(\prod \bar\lambda, <_D) = \lambda$.

(2) In 2.1(2), without loss of generality, we may assume that
either: $B_\alpha = B_0 \bmod J_{<\lambda}[\bar\lambda]$ (so λ is normal)
or: $B_\alpha \neq B_\beta \bmod J_{\leq\lambda}[\bar\lambda]$ for $\alpha < \beta < \lambda$ so λ is not normal.

(3) Assume λ is semi normal for $\bar\lambda$. Then λ is normal for $\bar\lambda$ iff for some F as in part (1) (of 2.2), F has a $<_{J_{<\lambda}[\bar\lambda]}$-exact upper bound $g \in \prod_{i<\kappa}(\lambda_i + 1)$ and then $B =: \{i < \kappa : g(i) = \lambda_i\}$ generates $J_{\leq\lambda}[\bar\lambda]$ over $J_{<\lambda}[\bar\lambda]$.

(4) If λ is semi normal for $\bar\lambda$ then for some $\bar f = \langle f_\alpha : \alpha < \lambda\rangle$, $\bar B = \langle B_\alpha : \alpha < \lambda\rangle$ we have: $\bar B$ is increasing modulo $J_{<\lambda}[\bar\lambda]$, $J_{\leq\lambda}[\bar\lambda] = J_{<\lambda}[\bar\lambda] + \{B_\alpha : \alpha < \lambda\}$, and the sequences $\langle f_\alpha : \alpha < \lambda\rangle$ is $<_{J_{<\lambda}[\bar\lambda]}$-increasing and $\bar f$, $\bar B$ are as in 1.7.

Proof. 1) For the direction $\Rightarrow$, given $\langle B_\alpha : \alpha < \lambda\rangle$ as in Definition 2.1(2), for each $\alpha < \lambda$, by 1.8(1) we have $(\prod \bar\lambda \restriction B_\alpha, <_{J_{<\lambda}[\bar\lambda]})$ has true cofinality λ, and let it be exemplified by $\langle f^\alpha_\beta : \beta < \lambda\rangle$. By 1.5 we can choose by induction on $\gamma < \lambda$ a function $f_\gamma \in \prod \bar\lambda$ such that: $\beta, \gamma \leq \alpha \Rightarrow f^\alpha_\beta \leq_{J_{<\lambda}[\bar\lambda]} f_\gamma$ and $\beta < \gamma \Rightarrow f_\beta <_{J_{<\lambda}[\bar\lambda]} f_\gamma$.

Now $F =: \{f_\alpha : \alpha < \lambda\}$ is as required. [Why? First, obviously $\alpha < \beta \Rightarrow f_\alpha < f_\beta \bmod J_{<\lambda}[\bar\lambda]$. Second, if D is an ultrafilter on κ disjoint to I^* and $(\prod \bar\lambda, <_D)$ has true cofinality λ, then by 1.6 for some $B \in J_{\leq\lambda}[\bar\lambda] \setminus J_{<\lambda}[\bar\lambda]$ we have $B \in D$, so for some $\alpha < \lambda$, $B \subseteq B_\alpha \bmod J_{<\lambda}[\bar\lambda]$ hence $B_\alpha \in D$. As $f^\alpha_\beta \leq_{J_{<\lambda}[\bar\lambda]} f_\beta$ for $\beta \in [\alpha, \lambda)$ clearly F is cofinal in $(\prod \bar\lambda, <_D)$.]

The other direction, $\Leftarrow$ follows from 1.7 applied to $F = \{f_\alpha : \alpha < \lambda\}$. [Why? we get there $\langle B_\alpha : \alpha < \lambda\rangle$, $B_\alpha \in J_{\leq\lambda}[\bar\lambda]$ increasing modulo $J_{<\lambda}[\bar\lambda]$ so $J =: J_{<\lambda}[\bar\lambda] + \{B_\alpha : \alpha < \lambda\} \subseteq J_{\leq\lambda}[\bar\lambda]$.

If equality does not hold then for some ultrafilter D over κ, $D \cap J = \emptyset$ but $D \cap J_{\leq\lambda}[\bar\lambda] \neq \emptyset$ so by clause (D) of 1.7, F is bounded in $\prod \lambda / D$ whereas by 1.8(1),(2), $\mathrm{tcf}(\prod \bar\lambda, <_D) = \lambda$ contradicting the assumption on F.]

2) Because we can replace $\langle B_\alpha : \alpha < \lambda\rangle$ by $\langle B_{\alpha_i} : i < \lambda\rangle$ whenever $\langle \alpha_i : i < \lambda\rangle$ is non decreasing, non eventually constant.

3) If λ is normal for $\bar\lambda$, let $B \subseteq \kappa$ be such that $J_{\leq\lambda}[\bar\lambda] = J_{<\lambda}[\bar\lambda] + B$. By 1.8(1) we know that $(\prod(\bar\lambda \restriction B), <_{J_{<\lambda}[\bar\lambda]})$ has true cofinality λ, so let it be exemplified by $\langle f^0_\alpha : \alpha < \lambda\rangle$. Let $f_\alpha = f^0_\alpha \cup 0_{(\kappa \setminus B)}$ for $\alpha < \lambda$ and let $g \in {}^\kappa\mathrm{Ord}$ be defined by $g(i) = \lambda_i$ if $i \in B$ and $g(i) = 0$ if $i \in \kappa \setminus B$. Now $\langle f_\alpha : \alpha < \lambda\rangle$, g are as required by 1.3(11).

Now suppose $\langle f_\alpha : \alpha < \lambda\rangle$ is as in part (1) of 2.2 and g is a $<_{J_{<\lambda}[\bar\lambda]}$ $-$eub of F, $g \in \prod_{i<\kappa} (\lambda_i + 1)$ and $B = \{i : g(i) = \lambda_i\}$. Let D be an ultrafilter on κ disjoint to $J_{<\lambda}[\bar\lambda]$. If $B \in D$ then for every $f \in \prod \bar\lambda$, let $f' = (f \restriction B) \cup 0_{(\kappa\setminus B)}$, now necessarily $f' < \max\{g, 1\}$ (as $[i \in B \Rightarrow f'(i) < \lambda_i = g(i)]$ and $[i \in \kappa \setminus B \Rightarrow f'(i) = 0 \leq g < 1]$), hence (see Definition 1.2(4)) for some $\alpha < \lambda$ we have $f' < \max\{f_\alpha, 1\} \bmod J_{<\lambda}[\bar\lambda]$ hence for some $\alpha < \lambda$, $f' \leq f_\alpha \bmod J_{<\lambda}[\bar\lambda]$ hence $f \leq f' \leq f_\alpha \bmod D$; also $\alpha < \beta \Rightarrow f_\alpha < f_\beta \bmod D$, hence together $\langle f_\alpha : \alpha < \lambda\rangle$ exemplifies $\mathrm{tcf}(\prod \bar\lambda, <_D) = \lambda$. If $B \notin D$ then $\kappa \setminus B \in D$ so $g' = g \restriction (\kappa \setminus B) \cup 0_B = g \bmod D$ and $\alpha < \lambda \Rightarrow f_\alpha <_D f_{\alpha+1} \leq_D g =_D g'$, so $g' \in \prod \bar\lambda$ exemplifies F is bounded in $(\prod \bar\lambda, <_D)$ so as F is as in 2.2(1), $\mathrm{tcf}(\prod \bar\lambda, <_D) = \lambda$ is impossible. As D is disjoint to $J_{<\lambda}[\bar\lambda]$, necessarily $\mathrm{tcf}(\prod \bar\lambda, <_D) > \lambda$. The last two arguments together give, by 1.8(2) that $J_{\leq\lambda}[\bar\lambda] = J_{<\lambda}[\bar\lambda] + B$ as required in the definition of normality.

4) Should be clear. $\blacksquare_{2.2}$

We shall give some sufficient conditions for normality.

Remark. In the following definitions we slightly deviate from [Sh-g, Ch I =Sh345a]. The ones here are perhaps somewhat artificial but enable us to deal also with case (β) of 1.5(*). I.e. in Definition 2.3 below we concentrate on the first θ elements of an a_α and for "obey" we also have $\bar{A}^* = \langle A_\alpha : \alpha < \theta \rangle$ and we want to cover also the case θ is singular.

Definition 2.3. Let there be given regular λ, $\theta < \mu < \lambda$, μ possibly an ordinal, $S \subseteq \lambda$, $\sup(S) = \lambda$ and for simplicity S is a set of limit ordinals or at least have no two successive members.

(1) We call $\bar{a} = \langle a_\alpha : \alpha < \lambda \rangle$ a continuity condition for (S, μ, θ) (or is an (S, μ, θ)-continuity condition) if: S is an unbounded subset of λ, $a_\alpha \subseteq \alpha$, $\mathrm{otp}(a_\alpha) < \mu$, and $[\beta \in a_\alpha \Rightarrow a_\beta = a_\alpha \cap \beta]$ and, for every club E of λ, for some[‡] $\delta \in S$ we have $\theta = \mathrm{otp}\{\alpha \in a_\delta : \mathrm{otp}(a_\alpha) < \theta$ and for no $\beta \in a_\delta \cap \alpha$ is $(\beta, \alpha) \cap E = \emptyset\}$. We say $\bar{a}$ is continuous in S^* if $\alpha \in S^* \Rightarrow \alpha = \sup(a_\alpha)$.

(2) Assume $f_\alpha \in {}^\kappa\mathrm{Ord}$ for $\alpha < \lambda$ and $\bar{A}^* = \langle A^*_\alpha : \alpha < \theta \rangle$ be a decreasing sequence of subsets of κ such that $\kappa \setminus A^*_\alpha \in I^*$. We say $\bar{f} = \langle f_\alpha : \alpha < \lambda \rangle$ obeys $\bar{a} = \langle a_\alpha : \alpha < \lambda \rangle$ for $\bar{A}^*$ if:
   (i) for $\beta \in a_\alpha$, if $\varepsilon =: \mathrm{otp}(a_\alpha) < \theta$ then we have $f_\beta \restriction A^*_\varepsilon \leq f_\alpha \restriction A^*_\varepsilon$ (note: $\bar{A}^*$ determine θ).

(2A) Let κ, $\bar{\lambda}$, I^* be as usual. We say $\bar{f}$ obeys $\bar{a}$ for $\bar{A}^*$ continuously on S^* if: $\bar{a}$ is continuous in S^* and $\bar{f}$ obeys $\bar{a}$ for $\bar{A}^*$ and in addition $S^* \subseteq S$ and for $\alpha \in S^*$ (a limit ordinal) we have $f_\alpha = f_{a_\alpha}$ from (2B), i.e. for every $i < \kappa$ we have $f_\alpha(i) = \sup\{f_\beta(i) : \beta \in a_\alpha\}$ when $|a_\alpha| < \lambda_i$.

(2B) For given $\bar{\lambda} = \langle \lambda_i : i < \kappa \rangle$, $\bar{f} = \langle f_\alpha : \alpha < \lambda \rangle$ where $f_\alpha \in \prod \bar{\lambda}$ and $a \subseteq \lambda$, and θ let $f_a \in \prod \bar{\lambda}$ be defined by: $f_a(i)$ is 0 if $|a| \geq \lambda_i$ and $\cup\{f_\alpha(i) : \alpha \in a\}$ if $|a| < \lambda_i$.

(3) Let (S, θ) stands for $(S, \theta + 1, \theta)$; (λ, μ, θ) stands for "(S, μ, θ) for some unbounded subset S of λ" and (λ, θ) stands for $(\lambda, \theta + 1, \theta)$.
   If each A^*_α is κ then we omit "for $\bar{A}^*$" (but θ should be fixed or said).

(4) We add to "continuity condition" (in part (1)) the adjective "weak" ["θ-weak"] if "$\beta \in a_\alpha \Rightarrow a_\beta = a_\alpha \cap \beta$" is replaced by "$\alpha \in S \& \beta \in a_\alpha \Rightarrow (\exists \gamma < \alpha)[a_\alpha \cap \beta \subseteq a_\gamma$ & $\gamma < \min(a_\alpha \setminus (\beta + 1))$ & $[|a_\alpha \cap \beta| < \theta \Rightarrow |a_\gamma| < \theta]]$" [but we demand that γ exists only if $\mathrm{otp}(a_\alpha \cap \beta) < \theta$]. (Of course a continuity condition is a weak continuity condition which is a θ-weak continuity condition).

Remark 2.3A. There are some obvious monotonicity implications, we state below only 2.4(3).

Fact 2.4.

(1) Let $\theta_r = \begin{cases} \theta & \mathrm{cf}(\theta) = \theta \\ \theta^+ & \mathrm{cf}(\theta) < \theta \end{cases}$ and assume $\lambda = \mathrm{cf}(\lambda) > \theta_r^+$. <u>Then</u> for some stationary $S \subseteq \{\delta < \lambda : \mathrm{cf}(\delta) = \theta_r\}$, there is a continuity condition $\bar{a}$ for (S, θ_r); moreover, it is continuous in S and $\delta \in S \Rightarrow \mathrm{otp}(a_\delta) = \theta_r$; so for every club E of λ for some $\delta \in S$, $\forall \alpha, \beta[\alpha < \beta$ & $\alpha \in a_\delta$ & $\beta \in a_\delta \rightarrow (\alpha, \beta) \cap E \neq \emptyset\}]$.

(2) Assume $\lambda = \theta^{++}$, then for some stationary $S \subseteq \{\delta < \lambda : \mathrm{cf}(\delta) = \mathrm{cf}(\theta)\}$ there is a continuity condition for $(S, \theta + 1, \theta)$. (In fact continuous in S and $[\delta \in S \Rightarrow a_\delta$ closed in $\delta]$ and $[\alpha \in a_\delta$ and $\delta \in S \Rightarrow a_\alpha = a_\delta \cap \alpha]$.)

(3) If $\bar{a}$ is a (λ, μ, θ_1)-continuity condition and $\theta_1 \geq \theta$ <u>then</u> there is a $(\lambda, \theta + 1, \theta)$-continuity condition.

[‡] Note: if $\mathrm{otp}(a_\delta) = \theta$ and $\delta = \sup(a_\delta)$ (holds if $\delta \in S$, $\mu = \theta + 1$ and $\bar{a}$ continuous in S (see below)) and $\delta \in acc(E)$ <u>then</u> δ is as required.

Proof. 1) By [Sh420, §1].

2) By [Sh351, 4.4(2)] and[§] [Sh-g, III 2.14(2), clause (c), p.135-7].

3) Check. $\blacksquare_{2.4}$

Remark 2.4A. Of course also if $\lambda = \theta^+$ the conclusion of 2.4(2) may well hold. We suspect but do not know that the negation is consistent with ZFC.

Fact 2.5. Suppose (*) of 1.5, $f_\alpha \in \prod \bar{\lambda}$ for $\alpha < \lambda$, $\lambda = \text{cf}(\lambda) > \theta$ (of course $\kappa = \text{dom}(\bar{\lambda})$) and $\bar{A}^* = \bar{A}^*[\bar{\lambda}]$ is as in the proof of 1.5, i.e. $A^*_\alpha = \{i < \kappa : \lambda_i > \alpha\}$). Then

(1) Assume $\bar{a}$ is a θ-weak continuity condition for (S, θ), $\lambda = \sup(S)$, then we can find $\bar{f}' = \langle f'_\alpha : \alpha < \lambda \rangle$ such that:
  (i) $f'_\alpha \in \prod \bar{\lambda}$,
  (ii) for $\alpha < \lambda$ we have $f_\alpha \le f'_\alpha$
  (iii) for $\alpha < \beta < \lambda$ we have $f'_\alpha <_{J_{<\lambda}[\bar{\lambda}]} f'_\beta$
  (iv) $\bar{f}'$ obeys $\bar{a}$ for $\bar{A}^*$

(2) If in addition $\min(\bar{\lambda}) \ge \mu$, $S^* \subseteq S$ are stationary subsets of λ but $\bar{a}$ is a continuity condition for (S, μ, θ) and $\bar{a}$ is continuous on S^* then we can find $\bar{f}' = \langle f'_\alpha : \alpha < \lambda \rangle$ such that
  (i) $f'_\alpha \in \prod \bar{\lambda}$
  (ii) for $\alpha \in \lambda \backslash S^*$ we have $f_\alpha \le f'_\alpha$ and $\alpha = \beta + 1 \in \lambda \backslash S^* \ \&\ \beta \in S^* \Rightarrow f_\beta \le f'_\alpha$
  (iii) for $\alpha < \beta < \lambda$ we have $f'_\alpha <_{J_{<\lambda}[\bar{\lambda}]} f'_\beta$
  (iv) $\bar{f}'$ obeys $\bar{a}$ for $\bar{A}^*$ continuously on S^*; moreover 2.3(2)(i) can be strengthened to $\beta \in a_\alpha \Rightarrow f_\beta < f_\alpha$.

(3) Suppose $\langle f'_\alpha : \alpha < \lambda \rangle$ obeys $\bar{a}$ continuously on S^* and satisfies 2.5(2)(ii) (and 2.5(2)'s assumption holds). If $g_\alpha \in \prod \bar{\lambda}$ and $\langle g_\alpha : \alpha < \lambda \rangle$ obeys $\bar{a}$ continuously on S^* and $[\alpha \in \lambda \setminus S^* \Rightarrow g_\alpha \le f_\alpha]$ then $\bigwedge_\alpha g_\alpha \le f'_\alpha$.

(4) If $\zeta < \theta$, for $\varepsilon < \zeta$ we have $\bar{f}^\varepsilon = \langle f^\varepsilon_\alpha : \alpha < \lambda \rangle$, where $f^\varepsilon_\alpha \in \prod \bar{\lambda}$, then in 2.5(1) (and 2.5(2)) we can find $\bar{f}'$ as there for all $\bar{f}^\varepsilon$ simultaneously. Only in clause (ii) we replace $f_\alpha \le f'_\alpha$ by $f_\alpha \restriction A^*_\zeta \le f'_\alpha \restriction A^*_\zeta$ (and $f_\beta \le f'_\alpha$ by $f_\beta \restriction A^*_\zeta \le f'_\alpha \restriction A^*_\zeta$.

Proof. Easy (using 1.5 of course).

Claim 2.5A: In 2.5 we can replace "(*) from 1.5" by "$\prod \bar{\lambda}/J_{<\lambda}[\bar{\lambda}]$ is λ-directed and $\liminf_{I^*}(\bar{\lambda}) \ge \theta$".

Claim 2.6: Assume (*) of 1.5 and let $\bar{A}^*$ be as there,

(1) in 1.7, if $\langle f_\alpha : \alpha < \lambda \rangle$ obeys some (S, θ)-continuity condition or just a θ-weak one for $\bar{A}^*$ (where $S \subseteq \lambda$ is unbounded) then we can deduce also:
  (G) the sequence $\langle B_\alpha / J_{<\lambda}[\bar{\lambda}] : \alpha < \lambda \rangle$ is eventually constant.

(2) If $\theta^+ < \lambda$ then $J_{\le\lambda}[\bar{\lambda}]/J_{<\lambda}[\bar{\lambda}]$ is λ^+-directed (hence if λ is semi normal for $\bar{\lambda}$ then it is normal to $\bar{\lambda}$).

Proof. 1) Assume not, so for some club E of λ we have

(*) $\alpha < \delta < \lambda \& \delta \in E \Rightarrow B_\alpha \ne B_\delta \bmod J_{<\lambda}[\bar{\lambda}]$.

As $\bar{a}$ is a θ-weak (S, θ)-continuity condition, there is $\delta \in S$ such that $b =: \{\alpha \in a_\delta : \text{otp}(a_\delta \cap \alpha) < \theta$ and for no $\beta \in a_\delta \cap \alpha$ is $(\beta, \alpha) \cap E = \emptyset$ and for some $\gamma < \alpha$, $a_\alpha \cap \beta \subseteq a_\gamma$ and $\gamma < \min(a_\alpha \backslash (\beta + 1))$ and $|a_\gamma| < \theta\}$ has order type θ. Let $\{\alpha_\varepsilon : \varepsilon < \theta\}$ list b (increasing with ε). So for every $\varepsilon < \theta$ there is $\gamma_\varepsilon \in (\alpha_\varepsilon, \alpha_{\varepsilon+1}) \cap E$, and let $\beta_\varepsilon < \alpha_{\varepsilon+1}$ be such that $a_\delta \cap \alpha_\varepsilon \subseteq a_{\beta_\varepsilon}$ and $\text{otp}(a_{\beta_\varepsilon} \cap \alpha_\varepsilon) < \theta$; by shrinking b and

---

[§] the definition of B^α_i in the proof of [Sh-g, III 2.14(2)] should be changed as in [Sh351, 4.4(2)]

renaming wlog $\beta_\varepsilon < \gamma_\varepsilon$ and $\alpha_\varepsilon \in a_{\beta_\varepsilon}$. Let $\xi(\varepsilon) =: \mathrm{otp}(a_{\beta_\varepsilon})$. Lastly let $B^0_\varepsilon =: \{i < \kappa: f_{\alpha_\varepsilon}(i) < f_{\beta_\varepsilon}(i) < f_{\gamma_\varepsilon}(i) < f_{\alpha_{\varepsilon+1}}(i)\}$, clearly it is $= \kappa \bmod I^*$ and let (remember $(*)$ above) $B^*_\varepsilon =: A^*_{\xi(\varepsilon)+1} \cap (B_{\gamma_\varepsilon} \setminus B_{\beta_\varepsilon}) \cap B^0_\varepsilon$, now $B_{\alpha_\varepsilon} \subseteq B_{\beta_\varepsilon} \subseteq B_{\gamma_\varepsilon} \bmod J_{<\lambda}[\bar\lambda]$ by clause (B) of 1.7, and $B_{\gamma_\varepsilon} \neq B_{\beta_\varepsilon} \bmod J_{<\lambda}[\bar\lambda]$ by $(*)$ above hence $B_{\gamma_\varepsilon} \setminus B_{\beta_\varepsilon} \neq \emptyset \bmod J_{<\lambda}[\bar\lambda]$. Now B^0_ε, $A^*_{\xi(\varepsilon)+1} = \kappa \bmod I^*$ by the previous sentence and by 1.5$(*)$ which we are assuming respectively and $I^* \subseteq J_{<\lambda}[\bar\lambda]$ by the later's definition; so we have gotten $B^*_\varepsilon \neq \emptyset \bmod J_{<\lambda}[\bar\lambda]$. But for $\varepsilon < \zeta < \theta$ we have $B^*_\varepsilon \cap B^*_\zeta = \emptyset$, for suppose $i \in B^*_\varepsilon \cap B^*_\zeta$, so $i \in A^*_{\xi(\varepsilon)+1}$ and also $f_{\gamma_\varepsilon}(i) < f_{\alpha_{\varepsilon+1}}(i) \leq f_{\beta_\zeta}(i)$ (as $i \in B^0_\varepsilon$ and as $\alpha_{\varepsilon+1} \in a_{\beta_\zeta}$ & $i \in A^*_{\xi(\zeta)+1}$ respectively); now $i \in B^*_\varepsilon$ hence $i \in B_{\gamma_\varepsilon}$ i.e. (where g is from 1.7 clause (D)$^+$) $f_{\gamma_\varepsilon}(i) > g(i)$ hence (by the above) $f_{\beta_\zeta}(i) > g(i)$ hence $i \in B_{\beta_\zeta}$ hence $i \notin B^*_\zeta$, contradiction. So $\langle B^*_\varepsilon : \varepsilon < \theta\rangle$ is a sequence of θ pairwise disjoint members of $(J_{<\lambda}[\bar\lambda])^+$, contradiction.

2) The proof is similar to the proof of 1.8(5), using 2.6(1) instead 1.7 (and $\bar a$ from 2.4(1) if $\lambda > \theta^+_r$ or 2.4(2) if $\lambda = \theta^{++}$). $\blacksquare_{2.6}$

We note also (but shall not use):

Claim 2.7: Suppose $(*)$ of 1.5 and

(a) $f_\alpha \in \prod\bar\lambda$ for $\alpha < \lambda$, $\lambda \in \mathrm{pcf}(\bar\lambda)$ and $\bar f = \langle f_\alpha : \alpha < \lambda\rangle$ is $<_{J_{<\lambda}[\bar\lambda]}$-increasing
(b) $\bar f$ obeys $\bar a$ continuously on S^*, where $\bar a$ is a continuity condition for (S, θ) and $\lambda = \sup(S)$ (hence $\lambda > \theta$ by the last phrase of 2.3(1))
(c) J is an ideal on κ extending $J_{<\lambda}[\bar\lambda]$, and $\langle f_\alpha/J : \alpha < \lambda\rangle$ is cofinal in $(\prod\bar\lambda, <_J)$ (e.g. $J = J_{<\lambda}[\bar\lambda] + (\kappa \setminus B)$, $B \in J_{\leq\lambda}[\bar\lambda] \setminus J_{<\lambda}[\bar\lambda]$).
(d) $\langle f'_\alpha : \alpha < \lambda\rangle$ satisfies (a), (b) above.
(e) $f_\alpha \leq f'_\alpha$ for $\alpha \in \lambda \setminus S^*$ (alternatively: $\langle f'_\alpha : \alpha < \lambda\rangle$ satisfies (c)).
(f) if $\delta \in S^*$ <u>then</u> J is $\mathrm{cf}(\delta)$-indecomposable (i.e. if $\langle A_\varepsilon : \varepsilon < \mathrm{cf}(\delta)\rangle$ is a $\subseteq$-increasing sequence of members, of J then $\bigcup_{\varepsilon < \mathrm{cf}(\delta)} A_\varepsilon \in J$).
<u>Then</u>:
(A) the set

$$\{\delta < \lambda : \text{ if } \delta \in S^* \text{ and } \mathrm{otp}(a_\delta) = \theta \text{ then } f'_\delta = f_\delta \bmod J\}$$

contains a club of λ.
(B) the set

$$\{\delta < \lambda : \text{if } \alpha \in S \text{ and } \delta = \sup(\delta \cap a_\alpha) \text{ and } \mathrm{otp}(\alpha \cap a_\delta) = \theta \text{ then } f'_{\alpha\cap a_\delta} = f_{\alpha \cap a_\delta} \bmod J\}$$

contains a club of λ.

Proof. We concentrate on proving (A). Suppose $\delta \in S^*$, and $f_\delta \neq f'_\delta \bmod J$. Let

$$A_{1,\delta} = \{i < \kappa: f_\delta(i) < f'_\delta(i)\}$$
$$A_{2,\delta} = \{i < \kappa: f_\delta(i) > f'_\delta(i)\},$$

So $A_{1,\delta} \cup A_{2,\delta} \in J^+$, suppose first $A_{1,\delta} \in J^+$. By Definition 2.3(2A), for every $i \in A_{1,\delta}$ for every large enough $\alpha \in a_\delta$, $f_\delta(i) < f'_\alpha(i)$, say for $\alpha \in a_\delta \setminus \beta_i$. As J is $\mathrm{cf}(\delta)$-indecomposable for some $\beta < \alpha$ we have $\{i < \kappa: \beta_i < \beta\} \in J^+$ so $f_\delta \restriction A_{1,\delta} < f'_\beta \restriction A_{1,\delta}$ (and $\beta < \delta$). Now by clause (c), $E =: \{\delta < \lambda$: for every $\beta < \delta$ we have $f'_\beta < f_\delta \bmod J\}$ is a club of λ, and so we have proved

$$\delta \in E \Rightarrow A_{1,\delta} \in J.$$

If $\bigwedge_{\alpha<\lambda} f_\alpha \leq f'_\alpha$ (first possibility in clause (e) implies it) also $A_{2,\delta} \in J$ hence for no $\delta \in S^* \cap E$ do we have $f_\delta \neq f'_\delta \bmod J$. If the second possibility of clause (e) holds, we can interchange $\bar{f}$, $\bar{f}'$ hence $[\delta \in E \Rightarrow A_{2,\delta} \in J]$ and we are done. $\blacksquare_{2.7}$

We now return to investigating the $J_{<\lambda}[\bar{\lambda}]$, first without using continuity conditions.

LEMMA 2.8: Suppose (*) of 1.5 and $\lambda = \mathrm{cf}(\lambda) \in \mathrm{pcf}(\bar{\lambda})$. Then λ is semi normal for $\bar{\lambda}$.

Proof. We assume λ is not semi normal for $\bar{\lambda}$ and eventually get a contradiction. Note that by our assumption $(\prod \bar{\lambda}, <_I)$ is θ^+-directed hence $\min \mathrm{pcf}_I(\bar{\lambda}) \geq \theta^+$ (by 1.3(4)(v)) hence let us define by induction on $\xi \leq \theta$, $\bar{f}^\xi = \langle f^\xi_\alpha : \alpha < \lambda \rangle$, B_ξ and D_ξ such that:

(I) (i) $f^\xi_\alpha \in \prod \bar{\lambda}$
   (ii) $\alpha < \beta < \lambda \Rightarrow f^\xi_\alpha \leq f^\xi_\beta \bmod J_{<\lambda}[\bar{\lambda}]$
   (iii) $\alpha < \lambda \,\&\, \xi < \theta \Rightarrow f^\xi_\alpha \leq f^\theta_\alpha \bmod J_{<\lambda}[\bar{\lambda}]$
   (iv) for $\zeta < \xi < \theta$ and $\alpha < \lambda$: $f^\zeta_\alpha \restriction A^*_\xi \leq f^\xi_\alpha \restriction A^*_\xi$

(II) (i) D_ξ is an ultrafilter on κ such that: $\mathrm{cf}(\prod \bar{\lambda}/D_\xi) = \lambda$
   (ii) $\langle f^\xi_\alpha / D_\xi : \alpha < \lambda \rangle$ is not cofinal in $\prod \bar{\lambda}/D_\xi$
   (iii) $\langle f^{\xi+1}_\alpha / D_\xi : \alpha < \lambda \rangle$ is increasing and cofinal in $\prod \bar{\lambda}/D_\xi$; moreover
   (iii)$^+$ $B_\xi \in D_\xi$ and $\langle f^{\xi+1}_\alpha : \alpha < \lambda \rangle$ is increasing and cofinal in $\prod \bar{\lambda}/(J_{<\lambda}[\bar{\lambda}] + (\kappa \setminus B_\xi))$
   (iv) $f^{\xi+1}_0 / D_\xi$ is above $\{f^\xi_\alpha / D_\xi : \alpha < \lambda\}$.

For $\xi = 0$. No problem. [Use 1.8(1)+(4)].

For ξ *limit* $< \theta$. Let $g^\xi_\alpha \in \prod \bar{\lambda}$ be defined by $g^\xi_\alpha(i) = \sup\{f^\zeta_\alpha(i) : \zeta < \xi\}$ for $i \in A^*_\xi$ and $f^\xi_\alpha(i) = 0$ else, (remember that $\kappa \setminus A^*_\xi \in I^*$). Then choose by induction on $\alpha < \lambda$, $f^\xi_\alpha \in \prod \bar{\lambda}$ such that $g^\xi_\alpha \leq f^\xi_\alpha$ and $\beta < \alpha \Rightarrow f_\beta < f_\alpha \bmod J_{<\lambda}[\bar{\lambda}]$. This is possible by 1.5 and clearly the requirements (I)(i),(ii),(iv) are satisfied. Use 2.2(1) to find an appropriate D_ξ (i.e. satisfying II(i)+(ii)). Now $\langle f^\xi_\alpha : \alpha < \lambda \rangle$ and D_ξ are as required. (The other clauses are irrelevant.)

For $\xi = \theta$. Choose f^θ_α by induction of α satisfying I(i), (ii), (iii) (possible by 1.5).

For $\xi = \zeta + 1$. Use 1.6 to choose $B_\zeta \in D_\zeta \cap J_{\leq\lambda}[\bar{\lambda}] \setminus J_{<\lambda}[\bar{\lambda}]$. Let $\langle g^\xi_\alpha : \alpha < \lambda \rangle$ be cofinal in $(\prod \bar{\lambda}, <_{D_\xi})$ and even in $(\prod \bar{\lambda}, <_{J_{<[\bar{\lambda}]+(\kappa \setminus B_\xi)}})$ and without loss of generality $\bigwedge_{\alpha<\lambda} f^\zeta_\alpha / D_\zeta < g^\xi_0 / D_\zeta$ and $\bigwedge_{\alpha<\lambda} f^\zeta_\alpha \restriction A^*_\xi \leq g^\xi_\alpha \restriction A^*_\xi$. We get $\langle f^\xi_\alpha : \alpha < \lambda \rangle$ increasing and cofinal $\mathrm{mod}(J_{<\lambda}[\bar{\lambda}] + (\kappa \setminus B_\xi))$ such that $g^\xi_\alpha \leq f^\xi_\alpha$ by 1.5 from $\langle g^\xi_\alpha : \alpha < \lambda \rangle$. Then get D_ξ as in the case "ξ limit".

So we have defined the f^ξ_α's and D_ξ's. Now for each $\xi < \theta$ we apply (II) (iii)$^+$ for $\langle f^{\xi+1}_\alpha : \alpha < \lambda \rangle$, $\langle f^\theta_\alpha : \alpha < \lambda \rangle$. We get a club C_ξ of λ such that:

$$\alpha < \beta \in C_\xi \Rightarrow f^\theta_\alpha \restriction B_\xi < f^{\xi+1}_\beta \restriction B_\xi \bmod J_{<\lambda}[\bar{\lambda}] \tag{$*$}$$

So $C =: \bigcap_{\xi<\theta} C_\xi$ is a club of λ. By 2.2(1) applied to $\langle f^\theta_\alpha : \alpha < \lambda \rangle$ (and the assumption "λ is not semi-normal for $\bar{\lambda}$") there is $g \in \prod \bar{\lambda}$ such that

$$\neg g \leq f^\theta_\alpha \bmod J_{<\lambda}[\bar{\lambda}] \text{ for } \alpha < \lambda \tag{$*$}_1$$

(not used) and by 1.5 wlog

$$f_0^\xi < g \bmod J_{<\lambda}[\bar{\lambda}] \quad \text{for } \xi < \theta \qquad (*)_2$$

For each $\xi < \theta$, by II (iii), (iii)$^+$ for some $\alpha_\xi < \lambda$ we have

$$g \restriction B_\xi < f_{\alpha_\xi}^{\xi+1} \restriction B_\xi \bmod J_{<\lambda}[\bar{\lambda}] \qquad (*)_3$$

Let $\alpha(*) = \sup_{\xi<\theta} \alpha_\xi$, so $\alpha(*) < \lambda$ and so

$$g \restriction B_\xi < f_{\alpha(*)}^{\xi+1} \restriction B_\xi \bmod J_{<\lambda}[\bar{\lambda}] \qquad (*)_4$$

For $\zeta < \theta$, let $B_\zeta^* = \{i \in A_\zeta^*: g(i) < f_{\alpha(*)}^\zeta(i)\}$. By $(*)_4$, $B_{\xi+1}^* \in D_\xi$; by (II)(iv)+$(*)_2$ we know $B_\xi^* \notin D_\xi$, hence $B_\xi^* \neq B_{\xi+1}^* \bmod D_\xi$ hence $B_\xi^* \neq B_{\xi+1}^* \bmod J_{<\lambda}[\bar{\lambda}]$. On the other hand by (I)(iv) for each $\zeta < \theta$ we have $\langle B_\xi^* \cap A_\zeta^*: \xi \leq \zeta\rangle$ is $\subseteq$-increasing and (as $A_\zeta^* = \kappa \bmod J_{<\lambda}[\bar{\lambda}]$ for each $\zeta < \theta$) hence by I(iv) we have $\langle B_\xi^*/I^* : \xi < \theta\rangle$ is $\subseteq$-increasing, and by the previous sentence $B_\xi^* \neq B_{\xi+1}^* \bmod J_{<\lambda}[\bar{\lambda}]$ hence $\langle B_\xi^*/I^*: \xi < \theta\rangle$ is strictly $\subseteq$-increasing. Together clearly $\langle B_{\xi+1}^* \cap A_{\xi+1}^* \setminus B_\xi^*: \xi < \theta\rangle$ is a sequence of θ pairwise disjoint members of $(J_{<\lambda}[\bar{\lambda}])^+$, hence of $(I^*)^+$; contradiction to $\theta \geq \mathrm{wsat}(I^*)$. $\blacksquare_{2.8}$

Definition 2.9.

(1) We say $\langle B_\lambda: \lambda \in \mathfrak{c}\rangle$ is a <u>generating sequence</u> for $\bar{\lambda}$ if:
   (i) $B_\lambda \subseteq \kappa$ and $\mathfrak{c} \subseteq \mathrm{pcf}(\bar{\lambda})$
   (ii) $J_{\leq\lambda}[\bar{\lambda}] = J_{<\lambda}[\bar{\lambda}] + B_\lambda$ for each $\lambda \in \mathfrak{c}$

(2) We call $\bar{B} = \langle B_\lambda: \lambda \in \mathfrak{c}\rangle$ <u>smooth</u> if:

$$i \in B_\lambda \& \lambda_i \in \mathfrak{c} \Rightarrow B_{\lambda_i} \subseteq B_\lambda.$$

(3) We call $\bar{B} = \langle B_\lambda: \lambda \in \mathrm{Rang}(\bar{\lambda})\rangle$ <u>closed</u> if for each λ

$$B_\lambda \supseteq \{i < \kappa: \lambda_i \in \mathrm{pcf}(\bar{\lambda} \restriction B_\lambda)\}$$

Fact 2.10. Assume $(*)$ of 1.5.

(1) Suppose $\mathfrak{c} \subseteq \mathrm{pcf}(\bar{\lambda})$, $\bar{B} = \langle B_\lambda: \lambda \in \mathfrak{c}\rangle$ is a generating sequence for $\bar{\lambda}$, and $B \subseteq \kappa$, $\mathrm{pcf}(\bar{\lambda} \restriction B) \subseteq \mathfrak{c}$ <u>then</u> for some finite $\mathfrak{d} \subseteq \mathfrak{c}$, $B \subseteq \bigcup_{\mu\in\mathfrak{d}} B_\mu \bmod I^*$.

(2) $\mathrm{cf}(\prod\bar{\lambda}/I^*) = \max\mathrm{pcf}(\bar{\lambda})$

Remark 2.10A. For another proof of 2.10(2) see 2.12(2)+ 2.12(4) and for another use of the proof of 2.10(2) see 2.14(1).

Proof. (1) If not, then $I = I^* + \{B \cap \bigcup_{\mu\in\mathfrak{d}} B_\mu: \mathfrak{d} \subseteq \mathfrak{c}, \mathfrak{d} \text{ finite}\}$ is a family of subsets of κ, closed under union, $B \notin I$, hence there is an ultrafilter D on κ disjoint from I to which B belongs. Let $\mu =: \mathrm{cf}(\prod_{i<\kappa}\lambda_i/D)$; necessarily $\mu \in \mathrm{pcf}(\bar{\lambda} \restriction B)$, hence by the last assumption of 2.10(1) we have $\mu \in \mathfrak{c}$. By 1.8(2) we know $B_\mu \in D$ hence $B \cap B_\mu \in D$, contradicting the choice of D.

(2) The case $\theta = \aleph_0$ is trivial (as $\mathrm{wsat}(I^*) \leq \aleph_0$ implies $\mathcal{P}(\kappa)/I^*$ is a Boolean algebra satisfying the $\aleph_0$-c.c. (as here we can subtract) hence this Boolean algebra is finite hence also $\mathrm{pcf}(\bar{\lambda})$ is finite) so we assume $\theta > \aleph_0$. For $B \in (I^*)^+$ let $\lambda(B) = \max\mathrm{pcf}_{I^*\restriction B}(\bar{\lambda} \restriction B)$.

We prove by induction on λ that for every $B \in (I^*)^+$, $\mathrm{cf}(\prod\bar{\lambda}, <_{I^*+(\kappa\setminus B)}) = \lambda(B)$ when $\lambda(B) \leq \lambda$; this will suffice (use $B = \kappa$ and $\lambda = |\prod_{i<\kappa}\lambda_i|^+$). Given B let $\lambda = \lambda(B)$, by notational change wlog $B = \kappa$. By 1.9, $\mathrm{pcf}(\prod\bar{\lambda})$ has a last element, necessarily it is $\lambda =: \lambda(B)$. Let $\langle f_\alpha: \alpha < \lambda\rangle$ be $<_{J_{<\lambda}[\bar{\lambda}]}$ increasing cofinal in

$\prod \bar\lambda / J_{<\lambda}[\bar\lambda]$, it clearly exemplifies $\max \text{pcf}(\bar\lambda) \le \text{cf}(\prod \bar\lambda / I^*)$. Let us prove the other inequality. For $A \in J_{<\lambda}[\bar\lambda] \setminus I^*$ choose $F_A \subseteq \prod \bar\lambda$ which is cofinal in $\prod \bar\lambda / (I^* + (\kappa \setminus A))$, $|F_A| = \lambda(A) < \lambda$ (exists by the induction hypothesis). Let χ be a large enough regular, and we now choose by induction on $\varepsilon < \theta$, N_ε, g_ε such that:

(A) (i) $N_\varepsilon \prec (H(\chi), \in, <^*_\chi)$
(ii) $\|N_\varepsilon\| = \lambda$
(iii) $\langle N_\varepsilon : \xi \le \varepsilon \rangle \in N_{\varepsilon+1}$
(iv) $\langle N_\varepsilon : \varepsilon < \theta \rangle$ is increasing continuous
(v) $\{\varepsilon : \varepsilon \le \lambda + 1\} \subseteq N_0$, $\{\bar\lambda, I^*\} \in N_0$, $\langle f_\alpha : \alpha < \lambda \rangle \in N_0$ and the function $A \mapsto F_A$ belongs to N_0.

(B) (i) $g_\varepsilon \in \prod \bar\lambda$ and $g_\varepsilon \in N_{\varepsilon+1}$
(ii) for no $f \in N_\varepsilon \cap \prod \bar\lambda$ does $g_\varepsilon <_{I^*} f$
(iii) $\zeta < \varepsilon \,\&\, \lambda_i > |\varepsilon| \Rightarrow g_\zeta(i) < g_\varepsilon(i)$.

There is no problem to define N_ε, and if we cannot choose g_ε this means that $N_\varepsilon \cap \prod \bar\lambda$ exemplifies $\text{cf}(\prod \bar\lambda, <) \le \lambda$ as required. So assume $\langle N_\varepsilon, g_\varepsilon : \varepsilon < \theta \rangle$ is defined. For each $\varepsilon < \theta$ for some $\alpha(\varepsilon) < \lambda$, $g_\varepsilon < f_{\alpha(\varepsilon)} \mod J_{<\lambda}[\bar\lambda]$ hence $\alpha(\varepsilon) \le \alpha < \lambda \Rightarrow g_\varepsilon <_{J_{<\lambda}[\bar\lambda]} f_\alpha$. As $\lambda = \text{cf}(\lambda) > \theta$, we can choose $\alpha < \lambda$ such that $\alpha > \bigcup_{\varepsilon<\theta} \alpha(\varepsilon)$. Let $B_\varepsilon = \{i < \kappa : g_\varepsilon(i) \ge f_\alpha(i)\}$; so for each $\xi < \theta$ we have $\langle B_\varepsilon \cap A^*_\xi : \varepsilon \le \xi \rangle$ is increasing with ε, (by clause (B)(iii)), hence as usual as $\theta \ge \text{wsat}(I^*)$ (and $\theta > \aleph_0$) we can find $\varepsilon(*) < \theta$ such that $\bigwedge_n B_{\varepsilon(*)+n} = B_{\varepsilon(*)} \mod I^*$ [why do we not demand $\varepsilon \in (\varepsilon(*), \theta) \Rightarrow B_\varepsilon = B_{\varepsilon(*)} \mod I^*$? as θ may be singular]. Now as $g_{\varepsilon(*)} \in N_{\varepsilon(*)+1}$ and $f_\alpha \in N_0 \prec N_{\varepsilon(*)+1}$ clearly, by its definition, $B_{\varepsilon(*)} \in N_{\varepsilon(*)+1}$ hence $F_{B_{\varepsilon(*)}} \in N_{\varepsilon(*)+1}$. Now:

$$\begin{aligned} g_{\varepsilon(*)+1} \restriction (\kappa \setminus B_{\varepsilon(*)}) =_{I^*} g_{\varepsilon(*)+1} \restriction (\kappa \setminus B_{\varepsilon(*)+1}) &< f_\alpha \restriction (\kappa \setminus B_{\varepsilon(*)+1}) \\ &=_{I^*} f_\alpha \restriction (\kappa \setminus B_{\varepsilon(*)}) \end{aligned}$$

[why first equality and last equality? as $B_{\varepsilon(*)+1} = B_{\varepsilon(*)} \mod I^*$, why the $<$ in the middle? by the definition of $B_{\varepsilon(*)+1}$].

But $g_{\varepsilon(*)+1} \restriction B_{\varepsilon(*)} \in \prod_{i \in B_{\varepsilon(*)}} \lambda_i$, and $B_{\varepsilon(*)} \in J_{<\lambda}[\bar\lambda]$ as $g_\varepsilon < f_{\alpha(\varepsilon)} \le f_\alpha \mod J_{<\lambda}[\bar\lambda]$ so for some $f \in F_{B_{\varepsilon(*)}} \subseteq \prod \bar\lambda$ we have $g_{\varepsilon(*)+1} \restriction B_{\varepsilon(*)} < f \restriction B_{\varepsilon(*)} \mod I^*$. By the last two sentences

$$g_{\varepsilon(*)+1} < \max\{f, f_\alpha\} \mod I^* \tag{$*$}$$

Now $f_\alpha \in N_{\varepsilon(*)+1}$ and $f \in N_{\varepsilon(*)+1}$ (as $f \in F_{B_{\varepsilon(*)}}$, $|F_{B_{\varepsilon(*)}}| \le \lambda$, $\lambda + 1 \subseteq N_{\varepsilon(*)+1}$ the function $B \mapsto F_B$ belongs to $N_0 \prec N_{\varepsilon(*)+1}$ and $B_{\varepsilon(*)} \in N_{\varepsilon(*)+1}$ as $\{g_{\varepsilon(*)}, f_\alpha\} \in N_{\varepsilon(*)+1}$) so together

$$\max\{f, f_\alpha\} \in N_{\varepsilon(*)+1}; \tag{$**$}$$

But $(*)$, $(**)$ together contradict the choice of $g_{\varepsilon(*)+1}$ (i.e. clause (B)(ii)). $\blacksquare_{2.10}$

Definition 2.11.

(1) We say that I^* satisfies the pcf-th for (the regular) $(\bar\lambda, \theta)$ if $\prod \bar\lambda / I^*$ is θ-directed and $(\prod \bar\lambda, <_{J_{<\lambda}[\bar\lambda]})$ is λ-directed for each λ and we can find $\langle B_\lambda : \lambda \in \text{pcf}_{I^*}(\bar\lambda) \rangle$, such that:
$B_\lambda \subseteq \kappa$, $J_{<\lambda}[\bar\lambda, I^*] = I^* + \{B_\mu : \mu \in \lambda \cap \text{pcf}_{I^*}(\bar\lambda)\}$, $B_\lambda \notin J_{<\lambda}[\bar\lambda, I^*]$ and $\prod(\bar\lambda \restriction B_\lambda) / J_{<\lambda}[\bar\lambda, I^*]$ has true cofinality λ (so $B_\lambda \in J_{\le\lambda}[\bar\lambda] \setminus J_{<\lambda}[\bar\lambda]$ and $J_{\le\lambda}[\bar\lambda] = J_{<\lambda}[\bar\lambda] + B_\lambda$).

(1A) We say that I^* satisfies the weak pcf-th for $(\bar\lambda, \theta)$ if

$(\prod\bar\lambda,<_{I^*})$ is θ-directed
each $(\prod\bar\lambda,<_{J_{<\lambda}[\bar\lambda]})$ is λ-directed and
there are $B_{\lambda,\alpha}\subseteq\kappa$ for $\alpha<\lambda\in\mathrm{pcf}_{I^*}(\bar\lambda)$ such that

$$\alpha<\beta<\mu\in\mathrm{pcf}_{I^*}(\bar\lambda)\Rightarrow B_{\mu,\alpha}\subseteq B_{\mu,\beta}\mod J_{<\mu}[\bar\lambda,I^*]$$

$$J_{<\lambda}[\bar\lambda]=I^*+\{B_{\mu,\alpha}:\alpha<\mu<\lambda,\mu\in\mathrm{pcf}_{I^*}(\bar\lambda)\}$$

and

$$(\prod(\bar\lambda\restriction B_{\mu,\alpha}),<_{J_{<\lambda}[\bar\lambda]})\text{ has true cofinality }\lambda$$

(1B) We say that I^* satisfies the weaker pcf-th for $(\bar\lambda,\theta)$ <u>if</u> $(\prod\bar\lambda,<_{I^*})$ is θ-directed and each $(\prod\bar\lambda,<_{J_{<\lambda}[\bar\lambda]})$ is λ-directed and for any ultrafilter D on κ disjoint to $J_{<\theta}[\bar\lambda]$ letting $\lambda=\mathrm{tcf}(\prod\bar\lambda,<_D)$ we have: $\lambda\geq\theta$ and for some $B\in D\cap J_{\leq\lambda}[\bar\lambda]\setminus J_{<\lambda}[\bar\lambda]$, the partial order $(\prod(\bar\lambda\restriction B),<_{J_{<\lambda}[\bar\lambda]})$ has true cofinality λ.

(1C) We say that I^* satisfies the weakest pcf-th for $(\bar\lambda,\theta)$ <u>if</u> $(\prod\bar\lambda,<_{I^*})$ is θ-directed and $(\prod\bar\lambda,<_{J_{<\lambda}[\bar\lambda]})$ is λ-directed for any $\lambda\geq\theta$

(1D) Above we write $\bar\lambda$ instead $(\bar\lambda,\theta)$ when we mean

$$\theta=\sup\{\theta:(\prod\bar\lambda,<_{I^*})\text{ is }\theta^+\text{-directed}\}.$$

(2) We say that I^* satisfies the pcf-th for θ if for any regular $\bar\lambda$ such that $\liminf_{I^*}(\bar\lambda)\geq\theta$, we have: I^* satisfies the pcf-th for $\bar\lambda$. We say that I^* satisfies the pcf-th above μ (above μ^-) if it satisfies the pcf-th for $\bar\lambda$ with $\liminf_{I^*}(\bar\lambda)>\mu$ (with $\{i:\lambda_i\geq\mu\}=\kappa\mod I^*$). Similarly (in both cases) for the weak pcf-th and the weaker pcf-th.

(3) Given I^*, θ let $J^{\mathrm{pcf}}_\theta=\{A\subseteq\kappa:A\in I^*\text{ or }A\notin I^*\text{ and }I^*+(\kappa\setminus A)\text{ satisfies the pcf-theorem for }\theta\}$.
$J^{\mathrm{wsat}}_\theta=:\{A\subseteq\kappa:\mathrm{wsat}(I^*\restriction A)\leq\theta\text{ or }A\in I^*\}$;
similarly J^{wpcf}_θ; we may write $J^x_\theta[I^*]$.

(4) We say that I^* satisfies the pseudo pcf-th for $\bar\lambda$ <u>if</u> for every ideal I on κ extending I^*, for some $A\in I^+$ we have $(\prod(\bar\lambda\restriction A),<_I)$ has a true cofinality.

Claim 2.12:

(1) If $(*)$ of 1.5 then I^* satisfies the weak pcf-th for $(\bar\lambda,\theta^+)$.
(2) If $(*)$ of 1.5 holds, and $\prod\bar\lambda/I^*$ is θ^{++}-directed (i.e. $\theta^+<\min\bar\lambda$) or just there is a continuity condition for (θ^+,θ)) <u>then</u> I^* satisfies the pcf-th for $(\bar\lambda,\theta^+)$.
(3) If I^* satisfy the pcf-th for $(\bar\lambda,\theta)$ <u>then</u> I^* satisfy the weak pcf-th for $(\bar\lambda,\theta)$ which implies that I^* satisfies the weaker pcf-th for $(\bar\lambda,\theta)$, which implies that I^* satisfies the weakest pcf-th for $(\bar\lambda,\theta)$.

Proof. (1) Let appropriate $\bar\lambda$ be given. By 1.5, 1.8 most demands holds, but we are left with seminormality. By 2.8, if $\lambda\in\mathrm{pcf}(\bar\lambda)$, then $\bar\lambda$ is semi normal for λ. This finishing the proof of (1).

(2) Let $\lambda\in\mathrm{pcf}(\bar\lambda)$ and let $\bar f$, $\bar B$ be as in 2.2(4). By 2.4(1)+(2) there is $\bar a$, a (λ,θ)-continuity condition; by 2.5(1) wlog $\bar f$ obeys $\bar a$, by 2.6(1) the relevant B_α/I^* are eventually constant which suffices by 2.2(2).

(3) Should be clear. $\blacksquare_{2.12}$

Claim 2.13: Assume $(\prod\bar\lambda,<_{I^*})$ is given (but possibly $(*)$ of 1.5 fails).

(1) If I^*, $\bar\lambda$ satisfies (the conclusion of) 1.6, <u>then</u> I^*, $\bar\lambda$ satisfy (the conclusions of) 1.8(1), 1.8(2), 1.8(3), 1.8(4), 1.9.

(1A) If I^* satisfies the weaker pcf-th for $\bar{\lambda}$ <u>then</u> they satisfy the conclusions of 1.6 and 1.5.
(2) If I^*, $\bar{\lambda}$ satisfies (the conclusion of) 1.5 <u>then</u> I^*, $\bar{\lambda}$ satisfies (the conclusion of) 1.10.
(2A) If I^* satisfies the weakest pcf-th for $\bar{\lambda}$ <u>then</u> I^*, $\bar{\lambda}$ satisfy the conclusion of 1.5.
(3) If I^*, $\bar{\lambda}$ satisfies 1.5, 1.6 <u>then</u> I^*, $\bar{\lambda}$ satisfies 2.2(1) (for 2.2(2) - no assumptions).
(4) If I^*, $\bar{\lambda}$ satisfies 1.8(1), 1.8(2) <u>then</u> I^*, $\bar{\lambda}$ satisfies 2.2(3) when we interpret "seminormal" by the second phrase of 2.2(1)
(5) If I^*, $\bar{\lambda}$ satisfies 1.8(2) <u>then</u> I^*, $\bar{\lambda}$ satisfies 2.10(1).
(6) If I^* $\bar{\lambda}$ satisfy 1.8(1) + 1.8(3)(i) <u>then</u> I^*, $\bar{\lambda}$ satisfies 1.8(2)
(7) If I^*, $\bar{\lambda}$ satisfies 1.8(1) + 1.8(2) and is semi normal <u>then</u> 2.10(2) holds i.e.

$$\mathrm{cf}(\prod \bar{\lambda}, <_{I^*}) \leq \sup \mathrm{pcf}_{I^*}(\lambda).$$

(8) If I^*, $\bar{\lambda}$ satisfies 1.5+1.6 <u>then</u> they satisfy 2.10(2).

Proof. (1) We prove by parts.

Proof of 1.8(2). Let $\lambda = \mathrm{tcf}(\prod \bar{\lambda}/D)$; by the definition of $J_{<\lambda}[\bar{\lambda}]$, clearly $D \cap J_{<\lambda}[\bar{\lambda}] = \emptyset$. Also by 1.6 for some $B \in D$ we have $\lambda = \mathrm{tcf}(\prod(\bar{\lambda} \restriction B), <_{J_{<\lambda}[\bar{\lambda}]})$, so by the previous sentence $B \notin J_{<\lambda}[\bar{\lambda}]$, and by 1.4(5) we have $B \in J_{\leq\lambda}[\bar{\lambda}]$, together we finish.

Proof of 1.8(1). Repeat the proof of 1.8(1) replacing the use of 1.5 by 1.8(2).

Proof of 1.8(3)(i). Let $J =: \bigcup_{\mu<\lambda} J_{<\mu}[\bar{\lambda}]$, so $J \subseteq J_{<\lambda}[\bar{\lambda}]$ is an ideal because $\langle J_{<\mu}[\bar{\lambda}] : \mu < \lambda \rangle$ is $\subseteq$-increasing (by 1.4(2)), if equality fail choose $B \in J_{<\lambda}[\bar{\lambda}] \setminus J$ and choose D an ultrafilter on κ disjoint to J to which B belongs. Now if $\mu = \mathrm{cf}(\mu) < \lambda$ then $\mu^+ < \lambda$ (as λ is a limit cardinal) and $\mu = \mathrm{cf}(\mu)$ & $\mu^+ < \lambda \Rightarrow D \cap J_{\leq\mu}[\bar{\lambda}] = D \cap J_{<\mu^+}[\bar{\lambda}] = \emptyset$ hence by 1.8(2) we have $\mu \neq \mathrm{cf}(\prod \bar{\lambda}/D)$. Also if $\mu = \mathrm{cf}(\mu) \geq \lambda$ then $D \cap J_{<\mu}[\bar{\lambda}] \subseteq D \cap J_{<\lambda}[\bar{\lambda}] \neq \emptyset$ hence by 1.8(2) we have $\mu \neq \mathrm{cf}(\prod \bar{\lambda}/D)$. Together contradiction by 1.3(7).

Proof of 1.8(3)(ii). Follows.

Proof of 1.8(4). Follows.

Proof of 1.9. As in 1.9.
(1A) Check.
(2) Read the proof of 1.10.
(2A) Check.
(3) The direction $\Rightarrow$ is proved directly as in the proof of 2.2(1) (where the use of 1.8(1) is justified by 2.13(1)).

So let us deal with the direction $\Leftarrow$. So assume $\bar{f} = \langle f_\alpha : \alpha < \lambda \rangle$ is a sequence of members of $\prod \bar{\lambda}$ which is $<_{J_{<\lambda}[\bar{\lambda}]}$-increasing such that for every ultrafilter D on κ disjoint to $J_{<\lambda}[\bar{\lambda}]$ we have: $\lambda = \mathrm{tcf}(\prod \bar{\lambda}, <_D)$ iff $\bar{f}$ is unbounded (equivalently cofinal) in $(\prod \bar{\lambda}, <_D)$. By (the conclusion of) 1.5 wlog $\bar{f}$ is $<_{J_{<\lambda}[\bar{\lambda}]}$-increasing.

By 1.5 there is $g \in \prod \bar{\lambda}$ such that $f_\alpha < g \mod J_{\leq\lambda}[\bar{\lambda}]$ for each $\alpha < \lambda$, and let $B_\alpha =: \{i < \kappa : g(i) \leq f_\alpha(i)\}$. Hence $B_\alpha \in J_{\leq\lambda}[\bar{\lambda}]$ (by the previous sentence) and $\langle B_\alpha / J_{<\lambda}[\bar{\lambda}] : \alpha < \lambda \rangle$ is $\subseteq$-increasing (as $\langle f_\alpha : \alpha < \lambda \rangle$ is $<_{J_{<\lambda}[\bar{\lambda}]}$-increasing). Lastly if $B \in J_{\leq\lambda}[\bar{\lambda}]$, but $B \setminus B_\alpha \notin J_{<\lambda}[\bar{\lambda}]$ for each $\alpha < \lambda$, let D be an ultrafilter on κ disjoint to $J_{<\lambda}[\bar{\lambda}] + \{B_\alpha : \alpha < \lambda\}$ but to which B belongs, so $\mathrm{tcf}(\prod \bar{\lambda}, <_D) = \lambda$ (by 1.8(2) which holds by 2.13(1)) but $\{f_\alpha / D : \alpha < \lambda\}$ is bounded by g/D (as

$f_\alpha/D \leq g/D$ by the definition of B_α), contradiction. So the sequence $\langle B_\alpha : \alpha < \lambda\rangle$ is as required.
4) – 6) Left to the reader.
7) Let for $\lambda \in \mathrm{pcf}(\bar\lambda)$, $\langle B_i^\lambda : i < \lambda\rangle$ be such that $J_{\leq\lambda}[\bar\lambda] = J_{<\lambda}[\bar\lambda] + \{B_i^\lambda : i < \lambda\}$ (exists by seminormality; we use only this equality). Let $\langle f_\alpha^{\lambda,i} : \alpha < \lambda\rangle$ be cofinal in $(\prod(\bar\lambda \restriction B_i^\lambda), <_{J_{,\lambda}[\bar\lambda]})$, it exists by 1.8(1). Let F be the closure of $\{f_\alpha^{\lambda,i} : \alpha < \lambda, i < \lambda, \lambda \in \mathrm{pcf}(\bar\lambda)\}$, under the operation $\max\{g,h\}$. Clearly $|F| \leq \sup \mathrm{pcf}(\bar\lambda)$, so it suffice to prove that F is a cover of $(\prod \bar\lambda, <_{I^*})$. Let $g \in \prod\bar\lambda$, if $(\exists f \in F)(g \leq f)$ we are done, if not

$$I = \{A \cup \{i < \kappa : f(i) > g(i)\} : f \in F, A \in I^*\}$$

is $\aleph_0$-directed, $\kappa \notin I$, so there is an ultrafilter D on κ disjoint to I, (so $f \in F \Rightarrow f <_D g$) and let $\lambda = \mathrm{tcf}(\prod\bar\lambda/D)$, so by 1.8(2) we have $D \cap J_{\leq\lambda}[\bar\lambda] \setminus J_{<\lambda}[\bar\lambda] \neq \emptyset$, hence for some $i < \lambda$, $B_i^\lambda \in D$, and we get contradiction to the choice of the $\{f_\alpha^{\lambda,\alpha} : \alpha < \lambda\}$ $(\subseteq F)$.
8) Repeat the proof of 2.10(2) (only using $J = \{A \subseteq \kappa$: if $A \notin J_{<\lambda}[\bar\lambda]$ then $\mathrm{cf}(\prod\bar\lambda/I^*) \leq \lambda\}$; if $\kappa \notin J$ let D be an ultrafilter on κ disjoint to J, and use 1.6). $\blacksquare_{2.13}$

CLAIM 2.14: If I^* satisfies pseudo pcf-th <u>then</u>

(1) We can find $\langle(J_\zeta, \theta_\zeta) : \zeta < \zeta^*\rangle$, ζ^* a successor ordinal such that $J_0 = I^*$, $J_{\zeta+1} = \{A \subseteq \kappa :$ if $A \notin J_\zeta$ then $\mathrm{tcf}(\prod(\bar\lambda \restriction A), <_{J_\zeta}) = \theta_\zeta\}$ and for no $A \in (J_\zeta)^+$ does $(\prod(\bar\lambda \restriction A), <_{J_\zeta})$ has true cofinality which is $< \theta_\zeta$.
(2) If I^* satisfies the weaker pcf-th for $\bar\lambda$ <u>then</u> I^* satisfies the pseudo pcf-th for $\bar\lambda$.

Proof. 1) Check (we can also present those ideals in other ways).
2) Check. $\blacksquare_{2.14}$

3. Reduced products of cardinals

We characterize here the cardinalities $\prod_{i<\kappa} \lambda_i/D$ and $T_D(\langle\lambda_i : i < \kappa\rangle)$ using pcf's and the amount of regularity of D (in 3.1-3.4). Later we give sufficient conditions for the existence of $<_D$-lub or $<_D$-eub. Remember the old result of Kanamori [Kn] and Ketonen [Kt]: for D an ultrafilter the sequence $\langle\alpha/D : \alpha < \kappa\rangle$ (i.e. the constant functions) has a $<_D$-lub if $\mathrm{reg}(D) < \kappa$; and see [Sh-g, III 3.3] (for filters). Then we turn to depth of ultraproducts of Boolean algebras.

The questions we would like to answer are (restricting ourselves to "$\lambda_i \geq 2^\kappa$" or "$\lambda_i \geq 2^{2^\kappa}$" and D an ultrafilter on κ will be good enough).

QUESTION A: What can be $\mathrm{Car}_D =: \{\prod_{i<\kappa} \lambda_i/D : \lambda_i$ a cardinal for $i < \kappa\}$ i.e. characterize it by properties of D; (or at least $\mathrm{Card}_D \setminus 2^\kappa$) (for D a filter also $T_D(\prod\lambda_i)$ is natural).

QUESTION B: What can be $\mathrm{DEPTH}_D^+ = \{\mathrm{Depth}^+(\prod_{i<\kappa} \lambda_i/D) : \lambda_i$ a regular cardinal$\}$ (at least $\mathrm{DEPTH}_D^+ \setminus 2^\kappa$, see Definition 3.18).

If D is an $\aleph_1$-complete ultrafilter, the answer is clear. For D a regular ultrafilter on κ, $\lambda_i \geq \aleph_0$ the answer to question A is known ([CK]) in fact it was the reason for defining "regularity of filters" (for $\lambda_i < \aleph_0$ see [Sh7], [Sh-a, VI §3 Th 3.12 and pp 357-370] better [Sh-c VI§3] and Koppleberg [Ko].) For D a regular ultrafilter on κ, the answer to the question is essentially completed in 3.22(1), the remaining problem can be answered by *pp* (see [Sh-g]) except the restriction $(\forall\alpha < \lambda)(|\alpha|^{\aleph_0} < \lambda)$, which can be removed if the cov $=$ *pp* problem is completed (see [Sh-g, AG]). So the problem is for the other ultrafilters D, on which we give a reasonable amount on information translating to a pcf problem, sometimes depending on the pcf theorem.

Definition 3.1.

(1) For a filter D let $\text{reg}(D) = \min\{\theta\colon D \text{ is not } \theta\text{-regular}\}$ (see below).
(2) A filter D is θ-regular if there are $A_\varepsilon \in D$ for $\varepsilon < \theta$ such that the intersection of any infinitely many A_ε-s' is empty.
(3) For a filter D let

$$\text{reg}_*(D) = \min\{\theta\colon \text{there are no } A_\varepsilon \in D^+ \text{ for } \varepsilon < \theta \text{ such that no } i < \kappa \text{ belongs to infinitely many } A_\varepsilon\text{'s}\}$$

and

$$\text{reg}_\otimes(D) =: \{\theta : \text{there are no } A_\varepsilon \in D^+ \text{ for } \varepsilon < \theta \text{ such that}: \varepsilon < \zeta \Rightarrow A_\zeta \subseteq A_\varepsilon \bmod D \text{ and no } i < \kappa \text{ belongs to infinitely many } A_\varepsilon\text{'s}\}.$$

(4) $\text{reg}^\sigma(D) = \min\{\theta : D \text{ is not } (\theta,\sigma)\text{-regular}\}$ where "D is (θ,σ)-regular" means that there are $A_\varepsilon \in D$ for $\alpha < \theta$ such that the intersection of any σ of them is empty. Lastly $\text{reg}^\sigma_*(D)$, $\text{reg}^\sigma_\otimes(D)$ are defined similarly using $A_\varepsilon \in D^+$. Of course $\text{reg}(I)$ etc. means $\text{reg}(D)$ where D is the dual filter.

Definition 3.2.

(1) Let

$$\text{htcf}_{D,\mu}(\prod\gamma_i) = \sup\{\text{tcf}(\prod_{i<\kappa}\lambda_i/D)\colon \mu \le \lambda_i = \text{cf}\,\lambda_i \le \gamma_i \text{ for } i < \kappa \text{ and } \text{tcf}(\prod\lambda_i/D) \text{ is well defined}\} \text{ and}$$

$$\text{hcf}_{D,\mu}(\prod_{i<\kappa}\gamma_i) = \sup\{\text{cf}(\prod_{i<\kappa}\lambda_i/D)\colon \mu \le \lambda_i = \text{cf}\,\lambda_i \le \gamma_i\};$$

if $\mu = \aleph_0$ we may omit it.
(2) For E a family of filters on κ let $\text{htcf}_{E,\mu}(\prod_{i<\kappa}\alpha_i)$ be

$$\sup\{\text{tcf}(\prod_{i<\kappa}\lambda_i/D)\colon D \in E \text{ and } \mu \le \lambda_i = \text{cf}\,\lambda_i \le \alpha_i \text{ for } i < \kappa \text{ and } \text{tcf}(\prod_{i<\kappa}\lambda_i/D) \text{ is well defined}\}.$$

Similarly for $\text{hcf}_{E,\mu}$ (using cf instead tcf).
(3) $\text{hcf}^*_{D,\mu}(\prod_{i<\kappa}\alpha_i)$ is $\text{hcf}_{E,\mu}(\prod_{i<\kappa}\alpha_i)$ for $E = \{D'\colon D' \text{ a filter on } \kappa \text{ extending } D\}$. Similarly for $\text{htcf}^*_{D,\mu}$.
(4) When we write I e.g. in $\text{hcf}_{I,\mu}$ we mean $\text{hcf}_{D,\mu}$ where D is the dual filter.

Claim 3.3:

(1) $\text{reg}(D)$ is always regular
(2) If $\theta < \text{reg}_*(D)$ then some filter extending D is θ-regular.

(3) $\mathrm{wsat}(D) \le \mathrm{reg}_*(D)$
(4) $\mathrm{reg}(D) \le \mathrm{reg}_\otimes(D) \le \mathrm{reg}_*(D)$
(5) $\mathrm{reg}_*(D) = \min\{\theta : \text{no ultrafilter } D_1 \text{ on } \kappa \text{ extending } D \text{ is } \theta\text{-regular}\}$
(6) If $D \subseteq E$ are filters on κ then:
 (a) $\mathrm{reg}(D) \le \mathrm{reg}(E)$
 (b) $\mathrm{reg}_*(D) \ge \mathrm{reg}_*(E)$

Proof. Should be clear. E.g (2) let $\langle u_\varepsilon : \varepsilon < \theta\rangle$ list the finite subsets of θ, and let $\{A_\varepsilon : \varepsilon < \theta\} \subseteq D^+$ exemplify "$\theta < \mathrm{reg}_*(D)$". Now let $D^* =: \{A \subseteq \kappa$: for some finite $u \subseteq \theta$, for every $\varepsilon < \theta$ we have: $u \subseteq u_\varepsilon \Rightarrow A_\varepsilon \subseteq A \bmod D\}$, and let $A^*_\varepsilon = \bigcup\{A_\zeta : \varepsilon \in u_\zeta\}$. Now D^* is a filter on κ extending D and for $\varepsilon < \theta$ we have $A^*_\varepsilon \in D$. Finally the intersection of $A^*_{\varepsilon_0} \cap A^*_{\varepsilon_1} \cap \ldots$ for distinct $\varepsilon_n < \theta$ is empty, because for any member j of it we can find $\zeta_n < \theta$ such that $j \in A_{\zeta_n}$ and $\varepsilon_n \in u_{\zeta_n}$. Now <u>if</u> $\{\zeta_n : n < \omega\}$ is infinite then there is no such j by the choice of $\langle A_\varepsilon : \varepsilon < \theta\rangle$, and <u>if</u> $\{\zeta_n : n < \omega\}$ is finite then wlog $\bigwedge_{n,\omega} \zeta_n = \zeta_0$ contradicting "u_{ζ_0} is finite" as $\bigwedge_{n<\omega} \varepsilon_n \in u_{\zeta_n}$. Lastly $\emptyset \notin D^*$ because $A^*_\varepsilon \ne \emptyset \bmod D$. $\blacksquare_{3.3}$

Observation 3.4. $|\prod_{i<\kappa} \lambda_i / I| \ge |\aleph_0^\kappa / I|$ holds when $\bigwedge_{i<\kappa} \lambda_i \ge \aleph_0$.

Observation 3.5.

(1) $|\prod_{i<\kappa} \lambda_i / I| \ge \mathrm{htcf}^*_I(\prod_{i<\kappa} \lambda_i)$.
(2) If I^* satisfies the pcf-th for $\bar\lambda$ or even the weaker pcf-th for $\bar\lambda$ (see Definition 2.11) <u>then</u>: $\mathrm{cf}(\prod \bar\lambda / I^*) = \max \mathrm{pcf}_{I^*}(\bar\lambda)$.
(3) If I^* satisfies the pcf-th for μ for and $\min(\bar\lambda) \ge \mu$ <u>then</u>

$$\mathrm{hcf}_{D,\mu}(\prod \bar\lambda) = \mathrm{hcf}^*_{D,\mu}(\prod \bar\lambda) = \mathrm{htcf}^*_{D,\mu}(\prod \bar\lambda)$$

 whenever D is disjoint to I^*.
(4) $\mathrm{hcf}_{E,\mu}(\prod_{i<\kappa} \lambda_i) = \mathrm{hcf}^*_{E,\mu}(\prod_{i<\kappa} \lambda_i)$.
(5) $\prod_{i<\kappa} \lambda_i / I \ge \mathrm{hcf}_{I,\mu}(\prod_{i<\kappa} \lambda_i) = \mathrm{hcf}^*_{I,\mu}(\prod_{i<\kappa} \lambda_i) \ge \mathrm{htcf}^*_{I,\mu}(\prod_{i<\kappa} \lambda_i)$ and $\mathrm{hcf}_{I,\mu}(\prod_{i<\kappa} \lambda_i) \ge \mathrm{htcf}_{I,\mu}(\prod_{i<\kappa} \lambda_i)$.

Remark 3.5A. In 3.5(3) concerning $\mathrm{htcf}_{D,\mu}$ see 3.10.

Proof. 1) By the definition of htcf^*_I it suffices to show $|\prod_{i<\kappa} \lambda_i / I| \ge \mathrm{tcf}(\prod \lambda'_i / I')$, when I' is an ideal on κ extending I, $\lambda'_i = \mathrm{cf}\, \lambda'_i \le \lambda_i$ for $i < \kappa$ and $\mathrm{tcf}(\prod_{i<\kappa} \lambda'_i / I')$ is well defined. Now $|\prod_{i<\kappa} \lambda_i / I| \ge |\prod_{i<\kappa} \lambda'_i / I| \ge |\prod_{i<\kappa} \lambda'_i / I'| \ge \mathrm{cf}(\prod \lambda'_i / I')$, so we have finished.

2) By 2.13(1A)clearly I^*, $\bar\lambda$ satisfies 1.5, 1.6 hence by 2.13(1), (2) also 1.8(1), (2), (3), (4) and 1.9 and 1.10. Now by 2.13(8) also (the conclusion of) 2.10(2) holds which is what we need.

3) Left to the reader (see Definition 2.11(2) and part (2)).

4), 5) Check. $\blacksquare_{3.5}$

Claim 3.6: If $\lambda = |\prod_{i<\kappa} \lambda_i / I|$ (and $\lambda_i \ge \aleph_0$ and, of course, I an ideal on κ) and $\theta < \mathrm{reg}(I)$ <u>then</u> $\lambda = \lambda^\theta$.

Proof. For each $i < \kappa$, let $\langle \eta^i_\alpha : \alpha < \lambda_i \rangle$ list the finite sequences from λ_i. Let $M_i = (\lambda_i, F_i, G_i)$ where $F_i(\alpha) = \ell g(\eta^i_\alpha)$, $G_i(\alpha, \beta)$ is $\eta^i_\alpha(\beta)$ if $\beta < \ell g(\eta^i_\alpha)$ $(= F_i(\alpha))$, and $F(\alpha, \beta) = 0$ otherwise; let $M = \prod_{i<\kappa} M_i / I$ so $\|M\| = |\prod \lambda_i / I|$ and let $M = (\prod \lambda_i / I, F, G)$. Let $\langle A_i : i < \theta \rangle$ exemplifies I is θ-regular. Now

- $(*)_1$ We can find $f \in {}^\kappa\omega$ and $f_\varepsilon \in \prod_{i<\kappa} f(i)$ for $\varepsilon < \theta$ such that: $\varepsilon < \zeta < \theta \Rightarrow f_\varepsilon <_I f_\zeta$ [just for $i < \kappa$ let $w_i = \{\varepsilon < \theta : i \in A_\varepsilon\}$, it is finite and let $f(i) = |w_i| + 1$ and $f_\varepsilon(i) = |\varepsilon \cap w_i| \le f(i)$, and note $\varepsilon < \zeta \& i \in A_\varepsilon \cap A_\zeta \Rightarrow f_\varepsilon(i) < f_\zeta(i)$].
- $(*)_2$ For every sequence $\bar g = \langle g_\varepsilon : \varepsilon < \theta \rangle$ of members of $\prod_{i<\kappa} \lambda_i$, there is $h \in \prod_{i<\kappa} \lambda_i$ such that $\varepsilon < \theta \Rightarrow M \models F(h/I, f_\varepsilon/I) = g_\varepsilon/I$ [why? let, in the notation of $(*)_1$, $h(i)$ be such that $\eta^i_{h(i)} = \langle g_\varepsilon(i) : \varepsilon \in w_i \rangle$ (in the natural order)].

So in M, every θ-sequence of members is coded using f/I, f_ε/I (for $\varepsilon < \theta$) by at least one member so $\|M\|^\theta = \|M\|$, but $\|M\| = |\prod_{i<\kappa} \lambda_i / I|$ hence we have proved 3.6. $\blacksquare_{3.6}$

Fact 3.7.

(1) For D a filter on κ, $\langle A_1, A_2 \rangle$ a partition of κ and (non zero) cardinals λ_i for $i < \kappa$ we have

$$|\prod_{i<\kappa} \lambda_i / D| = |\prod_{i<\kappa} \lambda_i / (D + A_1)| \times |\prod_{i<\kappa} \lambda_i / (D + A_2)|$$

(note: $|\prod_{i<\kappa} \lambda_i / \mathcal{P}(\kappa)| = 1$).

(2) $D^{[\mu]} =: \{A \subseteq \kappa : |\prod_{i<\kappa} \lambda_i / (D + (\kappa \setminus A))| < \mu\}$ is a filter on κ (μ an infinite cardinal of course) and if $\aleph_0 \le \mu \le \prod_{i<\kappa} \lambda_i / D$ <u>then</u> $D^{[\mu]}$ is a proper filter.

(3) If $\lambda \le |\prod\limits_{i<\kappa} \lambda_i / I|$, ($\lambda_i$ infinite, of course, I an ideal on κ) and $A \in I^+ \Rightarrow |\prod\limits_{i \in A} \lambda_i / I| \ge \lambda$ and $\sigma < \mathrm{reg}_*(I)$ <u>then</u> $|\prod \lambda_i / I| \ge \lambda^\sigma$

Proof. Check (part (3): by the proof of 3.3(2) we can find $A_\varepsilon \in I^+$ for $\varepsilon < \sigma$ such that for finite $u \subseteq \sigma$, $\cap_{\varepsilon \in u} A_\varepsilon \in I^+$ and continue as in the proof of 3.6).

Claim 3.8: If $D \subseteq E$ are filters on κ then

$$|\prod_{i<\kappa} \lambda_i / D| \le |\prod_{i<\kappa} \lambda_i / E| + \sup_{A \in E \setminus D} |\prod_{i<\kappa} \lambda_i / (D + (\kappa \setminus A))| + (2^\kappa / D) + \aleph_0.$$

We can replace $2^\kappa / D$ by $|\mathcal{P}|$ if $\mathcal{P}$ is a maximal subset of E such that $A \ne B \in \mathcal{P} \Rightarrow (A \setminus B) \cup (B \setminus A) \ne \emptyset \bmod D$.

Proof. Think.

Lemma 3.9: $|\prod_{i<\kappa} \lambda_i / D| \le (\theta^\kappa / D + \mathrm{hcf}_{D,\theta}(\prod_{i<\kappa} \lambda_i))^{<\theta}$ (see Definition 3.2(1)) provided that:

$$\theta \ge \mathrm{reg}_\otimes(D) \qquad (*)$$

Remark 3.9A. 1) If $\theta = \theta_1^+$, we can replace θ^κ / D by θ_1^κ / D. In general we can replace θ^κ / D by $\sup\{\prod\limits_{i<\kappa} f(i) / D : f \in \theta^\kappa\}$.

2) If D satisfies the pcf-th above θ (see 2.11(1A), 2.12(2)) then by 3.5(3) we can use htcf* (sometime even htcf, see 3.10). But by 3.7(1) we can ignore the $\lambda_i \le \theta$, and when $i < \kappa \Rightarrow \lambda_i > \theta$ we know that 1.5(*)(α) holds by 3.3(3).

Proof. Let $\lambda = \theta^\kappa/D + \mathrm{hcf}_{D,\theta}(\prod_{i<\kappa} \lambda_i)$. Let for $\zeta < \theta$, $\mu_\zeta =: \lambda^{|\zeta|}$ i.e. $\mu_\zeta =: (\theta^\kappa/D + \mathrm{hcf}_{D,\theta} \prod_{i<\kappa} \lambda_i)^{|\zeta|}$, clearly $\mu_\zeta = \mu_\zeta^{|\zeta|}$. Let $\chi = \beth_8(\sup_{i<\kappa} \lambda_i)^+$ and $N_\zeta \prec (H(\chi), \in, <^*_\chi)$ be such that $\|N_\zeta\| = \mu_\zeta$, $N^{\leq|\zeta|} \subseteq N_\zeta$, $\lambda + 1 \subseteq N_\zeta$ and $\{D, \langle \lambda_i : i < \kappa\rangle\} \in N_\zeta$ and $[\varepsilon < \zeta \Rightarrow N_\varepsilon \prec N_\zeta]$. Let $N = \cup\{N_\zeta : \zeta < \theta\}$. Let $g^* \in \prod_{i<\kappa} \lambda_i$ and we shall find $f \in N$ such that $g^* = f \bmod D$, this will suffice. We shall choose by induction on $\zeta < \theta$, $f^e_\zeta (e < 3)$ and $\bar{A}^\zeta$ such that:

(a) $f^e_\zeta \in \prod_{i<\kappa} (\lambda_i + 1)$
(b) $f^1_\zeta \in N_\zeta$ and $f^2_\zeta \in N_\zeta$.
(c) $\bar{A}^\zeta = \langle A^\zeta_i : i < \kappa\rangle \in N_\zeta$.
(d) $\lambda_i \in A^\zeta_i \subseteq \lambda_i + 1$, $|A^\zeta_i| \leq |\zeta| + 1$, and $\langle A^\zeta_i : \zeta < \theta\rangle$ is increasing continuous (in ζ).
(e) $f^0_\zeta(i) = \min(A^\zeta_i \setminus g^*(i))$; note: it is well defined as $g^*(i) < \lambda_i \in A^\zeta_i$
(f) $f^1_\zeta = f^0_\zeta \bmod D$
(g) $g^* < f^2_\zeta < f^1_\zeta \bmod (D + \{i < \kappa : g^*(i) \neq f^1_\zeta(i)\})$.
(h) $f^2_\zeta(i) \in A^{\zeta+1}_i$

So assume everything is defined for every $\varepsilon < \zeta$. If $\zeta = 0$, let $A^\zeta_i = \{\lambda_i\}$, if ζ limit $A^\zeta_i = \bigcup_{\varepsilon<\zeta} A^\varepsilon_i$, for $\zeta = \varepsilon + 1$, A^ζ_i will be defined in stage ε. So arriving to ζ, $\bar{A}^\zeta$ is well defined and it belongs to N_ζ: for $\zeta = 0$ check, for $\zeta = \varepsilon + 1$, done in stage ε, for ζ limit it belongs to N_ζ as we have $N_\zeta^{\leq|\zeta|} \subseteq N_\zeta$ and: $\xi < \zeta \Rightarrow N_\xi \prec N_\zeta$ and $\bar{A}^\xi \in N_\xi$. Now use clause (e) to define f^0_ζ/D. As $\langle A^\zeta_i : i < \kappa\rangle \in N_\zeta$, $|A^\zeta_i| \leq |\zeta| + 1 < \theta$ and $\theta^\kappa/D \leq \lambda < \lambda + 1 \subseteq N_\zeta$, clearly $|\prod_{i<\kappa} |A^\zeta_i|/D| \leq \lambda$ hence $\{f/D : f \in \prod_{i<\kappa} A^\zeta_i\} \subseteq N_\zeta$ hence $f^0_\zeta/D \in N_\zeta$ hence there is $f^1_\zeta \in N_\zeta$ such that $f^1_\zeta \in f^0_\zeta/D$ i.e. clause (f) holds. As $g^* \leq f^0_\zeta$ clearly $g^* \leq f^1_\zeta \bmod D$, let $y^\zeta_0 =: \{i < \kappa : g^*(i) \geq f^1_\zeta(i)\}$, $y^\zeta_1 =: \{i < \kappa : i \notin y^\zeta_0$ and $\mathrm{cf}(f^1_\zeta(i)) < \theta\}$ and $y^\zeta_2 =: \kappa \setminus y^\zeta_0 \setminus y^\zeta_1$. So $\langle y^\zeta_e : e < 3\rangle$ is a partition of κ and $g^* < f^1_\zeta \bmod (D + y^\zeta_e)$ for $e = 1, 2$.

Let $y^\zeta_4 = \{i < \kappa : \mathrm{cf}(f^1_\zeta(i)) \geq \theta\}$ so $f^1_\zeta \in N_\zeta$, and $\theta \in N_\zeta$ hence $y^\zeta_4 \in N_\zeta$, so $(\prod_{i<\kappa} f^1_\zeta(i), <_{D+y^\zeta_4}) \in N_\zeta$. Clearly $y^\zeta_2 \subseteq y^\zeta_4 \subseteq y^\zeta_0 \cup y^\zeta_2$. Now

$$\mathrm{cf}(\prod_{i<\kappa} f^1_\zeta(i), <_{D+y^\zeta_4}) \leq \mathrm{hcf}_{D+y^\zeta_4,\theta}(\prod_{i<\kappa} \lambda_i) \leq \mathrm{hcf}_{D,\theta}(\prod_{i<\kappa} \lambda_i) \subseteq \lambda < \lambda+1 \subseteq N_\zeta$$

hence there is $F \in N_\zeta$, $|F| \leq \lambda$, $F \subseteq \prod_{i\in y^\zeta_4} f^1_\zeta(i)$ such that:

$$(\forall g)[g \in \prod_{i\in y^\zeta_4} f^1_\zeta(i) \Rightarrow (\exists f \in F)(g < f \mod (D + y^\zeta_4)))].$$

As $\lambda + 1 \subseteq N$ necessarily $F \subseteq N_\zeta$. Apply the property of F to $(g^* \restriction y^\zeta_2) \cup 0_{(\kappa \setminus y^\zeta_2)}$ and get $f^\zeta_4 \in F \subseteq N_\zeta$ such that $g^* < f^\zeta_4 \mod (D + y^\zeta_2)$. Now use similarly $\prod_{i<\kappa} \mathrm{cf}(f^1_\zeta(i))/(D + (\kappa \setminus y^\zeta_4)) \leq |\theta^\kappa/D| \leq \lambda$; by the proof of 3.7(1) there is a function $f^2_\zeta \in N_\zeta \cap \prod_{i<\kappa} f^1_\zeta(i)$ such that $g^* \restriction (y^\zeta_1 + y^\zeta_2) < f^2_\zeta \bmod D$. Let $A^{\zeta+1}_i$ be: $A^\zeta_i \cup \{f^2_\zeta(i)\}$.

It is easy to check clauses (g), (h). So we have carried the definition. Let

$$X_\zeta =: \{i < \kappa : f^0_{\zeta+1}(i) < f^0_\zeta(i)\}.$$

Note that by the choice of f^1_ζ, $f^1_{\zeta+1}$ we know $X_\zeta = y^\zeta_1 \cup y^\zeta_2 \mod D$, if this last set is not D-positive then $g^* \geq f^1_\zeta \bmod D$, hence $g^*/D = f^1_\zeta/D \in N_\zeta$, contradiction, so $y^\zeta_1 \cup y^\zeta_2 \neq \emptyset \bmod D$ hence $X_\zeta \in D^+$. Also $\langle (y^\zeta_1 \cup y^\zeta_2)/D : \zeta < \theta\rangle$ is $\subseteq$-decreasing hence $\langle X_\zeta/D : \zeta < \theta\rangle$ is $\subseteq$-decreasing.

Also if $i \in X_{\zeta_1} \cap X_{\zeta_2}$ and $\zeta_1 < \zeta_2$ then $f^0_{\zeta_2}(i) \leq f^0_{\zeta_1+1}(i) < f^0_{\zeta_1}(i)$ (first inequality: as $A^{\zeta_1+1}_i \subseteq A^{\zeta_2}_i$ and clause (e) above, second inequality by the definition of X_{ζ_1}), hence for each ordinal i the set $\{\zeta < \theta : i \in X_\zeta\}$ is finite. So $\theta < \mathrm{reg}_\otimes(D)$, contradiction to the assumption (*). $\blacksquare_{3.9}$

Note we can conclude

CLAIM 3.9B:

$$\prod_{i<\kappa} \lambda_i/D = \sup\{(\prod_{i<\kappa} f(i))^{<\mathrm{reg}_\otimes(D_1)} + \mathrm{hcf}_{D_1}(\prod_{i<\kappa}\lambda_i)^{<\mathrm{reg}_\otimes(D_1)} : D_1 \text{ is a filter on } \kappa \text{ extending } D \text{ such that } A \in D^+_1 \Rightarrow \prod_{i<\kappa}\lambda_i/(D_1+A) = \prod_{i<\kappa}\lambda_i/D_1 \text{ and } f \in \theta^\kappa, f(i) \leq \lambda_i\}$$

Proof. The inequality $\geq$ should be clear by 3.7(3). For the other direction let μ be the right side cardinality and let $D_0 = \{\kappa \setminus A : A \subseteq \kappa$ and if $A \in D^+$ then $\prod_{i<\kappa} \lambda_i/(D+A) \leq \mu\}$, so we know by 3.7(2) that D_0 is a filter on κ extending D. If $\emptyset \in D_0$ we are done so assume not. Now $\mu \geq 2^\kappa/D$ (by the term $(\prod_i f(i)/D_0)^{<\mathrm{reg}_\otimes(D_1)}$) so by 3.8 we have $\prod_{i<\kappa} \lambda_i/D_0 > \mu$ (use 3.8 with D, D_0 here corresponding to D, E there). Now the same holds for $D_0 + A$ for every $A \in D^+_0$. Also $A \subseteq B \subseteq \kappa$ and $A \in D^+_0 \Rightarrow \prod_{i<\kappa}\lambda_i/(D_0+A) \leq \prod_{i<\kappa}\lambda_i/(D_1+B)$ so for some $B \in D^+_0$, $D_1 =: D_0 + B$ satisfies the requirement inside the definition of μ, so $\mu \geq \mathrm{hcf}_{D_1}(\prod_{i<\kappa}\lambda_1)^{<\mathrm{reg}_\otimes(D_1)}$. By 3.9 (see 3.9A(1)) we get a contradiction. $\blacksquare_{3.9B}$

Next we deal with existence of $<_D$ −eub.

CLAIM 3.10: 1) Assume D a filter on κ, $g^*_\alpha \in {}^\kappa\mathrm{Ord}$ for $\alpha < \delta$, $\bar g^* = \langle g^*_\alpha : \alpha < \delta\rangle$ is $\leq_D$-increasing, and

$$\mathrm{cf}(\delta) \geq \theta \geq \mathrm{reg}_*(D). \tag{*}$$

Then at least one of the following holds:

(A) $\langle g^*_\alpha : \alpha < \delta\rangle$ has a $<_D$-eub $g \in {}^\kappa\mathrm{Ord}$; moreover $\theta \leq \liminf_D \langle \mathrm{cf}[g(i)] : i < \kappa\rangle$

(B) $\mathrm{cf}(\delta) = \mathrm{reg}_*(D)$

(C) for some club C of δ and some $\theta_1 < \theta$ and $\gamma_i < \theta^+_1$ and $w_i \subseteq \mathrm{Ord}$ of order type γ_i for $i < \kappa$, there are $f_\alpha \in \prod_{i<\kappa} w_i$ (for $\alpha \in C$) such that $f_\alpha(i) = \min(w_i \setminus g^*_\alpha(i))$ and $\alpha \in C \mathbin{\&} \beta \in C \mathbin{\&} \alpha < \beta \Rightarrow f_\alpha \leq_D f_\beta \mathbin{\&} \neg f_\alpha =_D f_\beta \mathbin{\&} \neg f_\alpha \leq_D g^*_\beta \mathbin{\&} g^*_\alpha \leq f_\alpha$.

2) In (C) above if for simplicity D is an ultrafilter we can find $w_i \subseteq \mathrm{Ord}$, $\mathrm{otp}(w_i) = \gamma_i$, $\langle \alpha_\xi : \xi < \mathrm{cf}(\delta)\rangle$ increasing continuous with limit δ, and $h_\varepsilon \in \prod_{i<\kappa} w_i$ such that $f_{\alpha_\varepsilon} <_D h_\varepsilon <_D f_{\alpha_{\varepsilon+1}}$, moreover, $\bigwedge_{i<\kappa} \gamma_i < \omega$.

Proof. 1) Let $\sigma = \mathrm{reg}_*(D)$. We try to choose by induction on $\zeta < \sigma$, g_ζ, $f_{\alpha,\zeta}$ (for $\alpha < \delta$), $\bar{A}^\zeta$, α_ζ such that

(a) $\bar{A}^\zeta = \langle A_i^\zeta : i < \kappa \rangle$.
(b) $A_i^\zeta = \{f_{\alpha_\varepsilon,\varepsilon}(i), g_\varepsilon(i) : \varepsilon < \zeta\} \cup \{[\sup_{\alpha<\delta} g_\alpha^*(i)] + 1\}$.
(c) $f_{\alpha,\zeta}(i) = \min(A_i^\zeta \setminus g_\alpha^*(i))$ (and $f_{\alpha,\zeta} \in {}^\kappa\mathrm{Ord}$, of course).
(d) α_ζ is the first α, $\bigcup_{\varepsilon<\zeta} \alpha_\varepsilon < \alpha < \delta$ such that $[\beta \in [\alpha,\delta) \Rightarrow f_{\beta,\zeta} = f_{\alpha,\zeta} \bmod D]$ if there is one.
(e) $g_\zeta \le f_{\alpha_\zeta,\zeta}$ moreover $g_\zeta < \max\{f_{\alpha_\zeta,\zeta}, 1_\kappa\}$ but for no $\alpha < \delta$ do we have $g_\zeta < \max\{g_\alpha^*, 1\} \bmod D$.

Let ζ^* be the first for which they are not defined (so $\zeta^* \le \sigma$). Note

$$\varepsilon < \xi < \zeta^* \& \alpha_\xi \le \alpha < \delta \Rightarrow f_{\alpha_\varepsilon,\varepsilon} =_D f_{\alpha,\varepsilon} \& f_{\alpha,\xi} \le f_{\alpha,\varepsilon} \& f_{\alpha,\xi} \ne_D f_{\alpha,\varepsilon}. \qquad (*)$$

[Why last phrase? applying clause (e) above, second phrase with α, ε here standing for α, ζ there we get $A_0 =: \{i < \kappa : \max\{g_\alpha^*(i), 1\} \le g_\varepsilon(i)\} \in D^+$ and applying clause (e) above first phrase with ε here standing for ζ there we get $A_1 = \{i < \kappa : g_\varepsilon(i) < f_{\alpha,\varepsilon}(i)$ or $g_\varepsilon(i) = 0 = f_{\alpha,\varepsilon}(i)\} \in D$, hence $A_0 \cap A_1 \in D^+$, and $g_\varepsilon(i) > 0$ for $i \in A_0 \cap A_1$ (even for $i \in A_0$). Also by clause (c) above $g_\alpha^*(i) \le g_\varepsilon(i) \Rightarrow f_{\alpha,\xi}(i) \le g_\varepsilon(i)$. Now by the last two sentences $i \in A_0 \cap A_1 \Rightarrow g_\alpha^*(i) \le g_\varepsilon(i) < f_{\alpha,\varepsilon}(i) \Rightarrow f_{\alpha,\xi}(i) \le g_\varepsilon(i) < f_{\alpha,\varepsilon}(i)$, together $f_{\alpha,\xi} \ne_D f_{\alpha,\varepsilon}$ as required]

Case A. $\zeta^* = \sigma$ and $\bigcup_{\zeta<\sigma} \alpha_\zeta < \delta$. Let $\alpha(*) = \bigcup_{\zeta<\sigma} \alpha_\zeta$, for $\zeta < \sigma$ let $y_\zeta = \{i < \kappa : f_{\alpha(*),\zeta}(i) \ne f_{\alpha(*),\zeta+1}(i)\} \ne \emptyset \bmod D$. Now for $i < \kappa$, $\langle f_{\alpha(*),\zeta}(i) : \zeta < \sigma \rangle$ is non increasing so i belongs to finitely many y_ζ's only, so $\langle y_\zeta : \zeta < \sigma \rangle$ contradict $\sigma \ge \mathrm{reg}_*(D)$.

Case B. $\zeta^* = \sigma$ and $\bigcup_{\zeta<\sigma} \alpha_\zeta = \delta$. So possibility (B) of Claim 3.10 holds.

Case C. $\zeta^* < \sigma$.

Still $A_i^{\zeta^*}$ $(i < \kappa)$, f_{α,ζ^*} $(\alpha < \delta)$ are well defined.

Subcase C1. α_{ζ^*} cannot be defined.

Then possibility C of 3.10 holds (use $w_i =: A_i^{\zeta^*}$, $f_\beta = f_{\alpha_{\zeta^*}+\beta,\zeta^*}$).

Subcase C2. α_{ζ^*} can be defined.

Then $f_{\alpha_{\zeta^*},\zeta^*}$ is a $<_D$-eub of $\langle g_\alpha^* : \alpha < \delta \rangle$ as otherwise there is g_{ζ^*} as required in clause (e). Now $f_{\alpha_\zeta^*,\zeta^*}$ is almost as required in possibility (A) of Claim 3.10 only the second phrase is missing. If for no $\theta_1 < \theta$, $\{i < \kappa : \mathrm{cf}[f_{\alpha_{\zeta^*},\zeta^*}(i)] \le \theta_1\} \in D^+$, then possibility (A) holds.

So assume $\theta_1 < \theta$ and $B =: \{i < \kappa : \aleph_0 \le \mathrm{cf}[f_{\alpha_{\zeta^*},\zeta^*}(i)] \le \theta_1\}$ belongs to D^+, we shall try to prove that possibility (C) holds, thus finishing. Now we choose w_i for $i < \kappa$: for $i \in \kappa$ we let $w_i^0 =: \{f_{\alpha_{\zeta^*},\zeta^*}(i), [\sup_{\alpha<\delta} g_\alpha^*(i)] + 1\}$, for $i \in B$ let w_i^1 be an unbounded subset of $f_{\alpha_{\zeta^*},\zeta^*}(i)$ of order type $\mathrm{cf}[f_{\alpha_{\zeta^*},\zeta^*}(i)]$ and for $i \in \kappa \setminus B$ let $w_i^1 = \emptyset$, lastly let $w_i = w_i^0 \cup w_i^1$, so $|w_i| \le \theta_1$ as required in possibility (C). Define $f_\alpha \in {}^\kappa\mathrm{Ord}$ by $f_\alpha(i) = \min(w_i \setminus g_\alpha^*(i))$ (by the choice of w_i^0 it is well defined). So $\langle f_\alpha : \alpha < \delta \rangle$ is $\le_D$-increasing; if for some $\alpha^* < \delta$, for every $\alpha \in [\alpha^*, \delta)$ we have $f_\alpha/D = f_{\alpha^*}/D$, we could define $g_{\zeta^*} \in {}^\kappa\mathrm{Ord}$ by:

$g_{\zeta^*} \restriction B = f_{\alpha^*}$ (which is $< f_{\alpha_{\zeta^*},\zeta^*}$),
$g_{\zeta^*} \restriction (\kappa \setminus B) = 0_{\kappa \setminus B}$.

Now g_{ζ^*} is as required in clause (e) so we get contradiction to the choice of ζ^*. So there is no $\alpha^* < \delta$ as above so for some club C of δ we have $\alpha < \beta \in C \Rightarrow f_\alpha \ne_D f_\beta$, so we have actually proved possibility (C).

2) Easy (for $\bigwedge_i \gamma_i < \omega$, wlog $\theta = \text{reg}_*(D)$ but $\text{reg}_*(D) = \text{reg}(D)$ so $\theta_1 < \text{reg}(D)$). $\blacksquare_{3.10}$

CLAIM 3.11:

(1) In 3.10(1), if $\lambda = \delta = \text{cf}(\lambda)$, $\bar{g}^*$ obeys $\bar{a}$ ($\bar{a}$ as in 2.1), $\bar{a}$ a θ-weak (S,θ)-continuity condition, $S \subseteq \lambda$ unbounded, then clause (C) of 3.10 implies:
(C)$'$ there are $\theta_1 < \text{reg}_*(D)$ and $A_\varepsilon \in D^+$ for $\varepsilon < \theta$ such that the intersection of any θ_1^+ of the sets A_ε is empty (equivalently $i < \kappa \Rightarrow (\exists^{\leq \theta_1} \varepsilon)[i \in A_\varepsilon]$ (reminds (σ, θ_1^+)-regularity of ultrafilters).

(2) We can in 3.10(1) weaken the assumption $(*)$ to $(*)'$ below if in the conclusion we weaken clause (A) to (A)$'$ where
$(*)'$ $\text{cf}(\delta) \geq \theta \geq \text{reg}(D)$
(A)$'$ there is a $\leq_D$-upper bound f of $\{g^*_\alpha : \alpha < \delta\}$ such that
no $f' <_D f$ (of course $f' \in {}^\kappa\text{Ord}$) is a $\leq_D$-upper bound of $\{g^*_\alpha : \alpha < \delta\}$
and $\theta \leq \liminf_D \langle \text{cf}[f(i)] : i < \kappa \rangle$

(3) If $g^*_\alpha \in {}^\kappa\text{Ord}$, $\langle g^*_\alpha : \alpha < \delta \rangle$ is $<_D$-increasing and $f \in {}^\kappa\text{Ord}$ satisfies (A)$'$ above and
$(*)''$ $\text{cf}(\delta) \geq \text{wsat}(D)$ and for some $A \in D$ for every $i < \kappa$, $\text{cf}(f(i)) \geq \text{wsat}(D)$
<u>then</u> for some $B \in D^+$ we have $\prod_{i<\kappa} \text{cf}[f(i)]/(D + B)$ has true cofinality $\text{cf}(\delta)$.

Remark. Compare with 2.6.

Proof. 1) By the choice of $\bar{a} = \langle a_\alpha : \alpha < \lambda \rangle$ as C (in clause (c) of 3.11(1)) is a club of λ, we can find $\beta < \lambda$ such that letting $\langle \alpha_\varepsilon : \varepsilon < \theta \rangle$ list $\{\alpha \in a_\beta : \text{otp}(\alpha \cap a_\beta) < \theta\}$ (or just a subset of it) we have $(\alpha_\varepsilon, \alpha_{\varepsilon+1}) \cap C \neq \emptyset$.

Let $\gamma_\varepsilon \in (\alpha_\varepsilon, \alpha_{\varepsilon+1}) \cap C$, and $\xi_\varepsilon \in (\alpha_\varepsilon, \alpha_{\varepsilon+1})$ be such that $\{\alpha_\zeta : \zeta \leq \varepsilon\} \subseteq a_{\xi_\varepsilon}$, and as we can use $\langle \alpha_{2\varepsilon} : \varepsilon < \theta \rangle$, wlog $\xi_\varepsilon < \gamma_\varepsilon$. For $\zeta < \theta$ let $B_\zeta = \{i < \kappa : f_{\alpha_\zeta}(i) < f_{\beta_\zeta}(i) < f_{\gamma_\zeta}(i) < f_{\alpha_{\zeta+1}}(i)$ and $\sup\{f_{\alpha_\xi}(i)+1 : \xi < \zeta\} < \sup\{f_{\alpha_\xi}(i)+1 : \xi < \zeta+1\}$.
2) In the proof of 3.10 we replace clause (e) by
(e$'$) $g_\zeta \leq f_{\alpha_\zeta, \zeta}$ and for $\alpha < \delta$ we have $f_\alpha \leq g_\zeta \bmod D$
3) By 1.8(1) $\blacksquare_{3.11}$

CLAIM 3.12:

(1) Assume $\lambda = \text{tcf}(\prod \bar{\lambda}/D)$ and $\mu = \text{cf}(\mu) < \lambda$ <u>then</u> there is $\bar{\lambda}' <_D \bar{\lambda}$, $\bar{\lambda}'$ a sequence of regular cardinals and $\mu = \text{tcf}(\prod \bar{\lambda}'/D)$ provided that

$$\mu > \text{reg}_*(D),\ \min(\bar{\lambda}) > \text{reg}_*^{\sigma^+}(D) \text{ whenever } \sigma < \text{reg}_*(D) \qquad (*)$$

(2) Let I^* be the ideal dual to D, and assume $(*)$ above. If $(*)(\alpha)$ of 1.5 holds and μ is semi-normal (for $(\bar{\lambda}, I^*)$) <u>*then*</u> it is normal.

Proof. Part (2) follows from part (1) by 2.2(3). Let us prove (1).

Case 1. $\mu < \liminf_D(\bar{\lambda})$
We let

$$\lambda' = \begin{cases} \mu & \text{if } \mu < \lambda_i \\ 1 & \text{if } \mu \geq \lambda_i \end{cases}$$

and we are done.

Case 2. $\liminf_D(\bar\lambda) \geq \theta \geq \text{reg}_*(D)$, $\mu > \theta$, and $(\forall \sigma < \text{reg}_*(D))[\text{reg}^\sigma_*(D) < \theta]$.

Let $\theta =: \text{reg}_*(D)$. There is an unbounded $S \subseteq \mu$ and an (S,θ)-continuity system $\bar a$ (see 2.4). As $\prod \bar\lambda/D$ has true cofinality λ, $\lambda > \mu$ clearly there are $g^*_\alpha \in \prod \bar\lambda$ for $\alpha < \mu$ such that $\bar g^* = \langle g^*_\alpha : \alpha < \mu\rangle$ obeys $\bar a$ for $\bar A^*[\bar\lambda]$ (exists as $\theta \leq \liminf_D(\bar\lambda)$).

Now if in claim 3.10(1) for $\bar g^*$ possibility (A) holds, we are done. By 3.11(1) we get that for some $\sigma < \text{reg}_*(D)$ we have $\text{reg}^\sigma_*(I) \geq \mu$, contradiction.

Case 3. $\liminf_D(\bar\lambda) \geq \theta \geq \text{reg}_*(D)$, $\mu \geq \theta$, and $(\forall \sigma < \text{reg}_*(D))[\text{reg}^\sigma_*(D) < \theta]$.

Like the proof of [Sh-g, Ch II 1.5B] using the silly square. $\blacksquare_{3.12}$

* * *

We turn to other measures of $\prod \bar\lambda/D$.

Definition 3.13.

(a) $T^0_D(\bar\lambda) = \sup\{|F| : F \subseteq \prod\bar\lambda \text{ and } f_1 \neq f_2 \in F \Rightarrow f_1 \neq_D f_2\}$.

(b) $T^1_D(\bar\lambda) = \min\{|F|$: (i) $F \subseteq \prod\bar\lambda$
(ii) $f_1 \neq f_2 \in F \Rightarrow f_1 \neq_D f_2$
(iii) F maximal under (i)+(ii)$\}$

(c) $T^2_D(\bar\lambda) = \min\{|F| : F \subseteq \prod\bar\lambda$ and for every $f_1 \in \prod\bar\lambda$, for some $f_2 \in F$ we have $\neg f_1 \neq_D f_2\}$.

(d) If $T^0_D(\bar\lambda) = T^1_D(\bar\lambda) = T^2_D(\bar\lambda)$ then let $T_D(\bar\lambda) = T^l_D(\bar\lambda)$ for $l < 3$.

(e) for $f \in {}^\kappa\text{Ord}$ and $\ell < 3$ let $T^l_D(f)$ means $T^l_D(\langle f(\alpha) : \alpha < \kappa\rangle)$.

THEOREM 3.14:

(0) If $D_0 \subseteq D_1$ are filters on κ then $T^\ell_{D_0}(\bar\lambda) \leq T^\ell_{D_1}(\bar\lambda)$ for $\ell = 0, 2$. Also if $\kappa = A_0 \cup A_1$, $A_0 \in D^+$, and $A_1 \in D^+$ then $T^\ell_D(\bar\lambda) = \min\{T^\ell_{D+A_0}(\bar\lambda), T^\ell_{D+A_1}(\bar\lambda)\}$ for $\ell = 0, 2$.

(1) $\text{htcf}_D(\prod\bar\lambda) \leq T^2_D(\bar\lambda) \leq T^1_D(\bar\lambda) \leq T^0_D(\bar\lambda)$

(2) If $T^0_D(\bar\lambda) > |\mathcal{P}(\kappa)/D|$ or just $T^0_D(\lambda) > \mu$, and $\mathcal{P}(\kappa)/D$ satisfies the μ^+-c.c. <u>then</u> $T^0_D(\bar\lambda) = T^1_D(\bar\lambda) = T^2_D(\bar\lambda)$ so the supremum in 3.13(a) is obtained (so e.g. $T^0_D(\bar\lambda) > 2^\kappa$ suffice)

(3) $T^0_D(\bar\lambda)^{<\text{reg} D} = T^0_D(\bar\lambda)$ (each λ_i infinite of course).

(4) $[\text{htcf}_D \prod_{i<\kappa} f(i)] \leq T^2_D(f) \leq [\text{htcf}_D \prod_{i<\kappa} f(i)]^{<\theta} + \text{reg}(D)^\kappa/D$ where $\theta = \text{reg}_*(D)$ in fact $\theta = \text{reg}(D) + \text{wsat}(D)$ suffice

(5) If D is an ultrafilter $|\prod\bar\lambda/D| = T^e_D(\bar\lambda)$ for $e \leq 2$.

(6) In (4), if $\bigwedge_{i<\kappa} f(i) \geq 2^\kappa$ (or just $(\text{reg}(D)+2)^\kappa/D \leq \min\limits_{i<\kappa} f(i)$), <u>then</u>
$$[\text{htcf}_D \prod_{i<\kappa} f(i)]^{<\text{reg } D} \leq T^0_D(f)$$

(7) If the sup in the definition of $T^0_D(\bar\lambda)$ is not obtained then it has cofinality $\geq \text{reg}(D)$ and even is regular.

Proof. (0) Check.

(1) First assume $\mu =: T^2_D(\bar\lambda) < \text{htcf}_D(\prod\bar\lambda)$; then we can find $\mu^* = \text{cf}(\mu^*) \in (\mu, \text{htcf}_D(\prod\bar\lambda)]$ and $\bar\mu = \langle \mu_i : i < \kappa\rangle$, a sequence of regular cardinals, $\bigwedge_{i<\kappa} \mu_i \leq \lambda_i$ such that $\mu^* = \text{tcf}(\prod\bar\mu/D)$ and let $\langle f_\alpha : \alpha < \mu^*\rangle$ exemplify this. Now let F exemplify $\mu = T^2_D(\bar\lambda)$, for each $g \in F$ let

$$g' \in \prod_{i<\kappa} \mu_i \text{ be} : g'(i) = \begin{cases} g(i) & \text{if } g(i) < \mu_i \\ 0 & \text{otherwise.} \end{cases}$$

So there is $\alpha(g) < \mu^*$ such that $g' <_D f_{\alpha(g)}$. Let $\alpha^* = \sup\{\alpha(g): g \in F\}$, now $\alpha^* < \mu^*$ (as $\mu^* = \operatorname{cf} \mu^* > \mu = |F|$). So $g \in F \Rightarrow g \neq_D f_{\alpha^*}$, contradiction. So really $T^2_D(\bar\lambda) \leq \operatorname{htcf}_D(\prod \bar\lambda)$ as required.

If F exemplifies the value of $T^1_D(\bar\lambda)$, it also exemplifies $T^2_D(\bar\lambda) \leq |F|$ hence $T^2_D(\bar\lambda) \leq T^1_D(\bar\lambda)$.

Lastly if F exemplifies the value of $T^1_D(f)$ it also exemplifies $T^0_D(\bar\lambda) \geq |F|$, so $T^1_D(\bar\lambda) \leq T^0_D(\bar\lambda)$.

(2) Let μ be $|\mathcal{P}(\kappa)/D|$ or at least μ is such that the Boolean algebra $\mathcal{P}(\kappa)/D$ satisfies the μ^+-c.c. Assume that the desired conclusion fails so $T^2_D(\bar\lambda) < T^0_D(\bar\lambda)$, so there is $F_0 \subseteq \prod \bar\lambda$, such that $[f_1 \neq f_2 \in F_0 \Rightarrow f_1 \neq_D f_2]$, and $|F_0| > T^2_D(\bar\lambda) + \mu$ (by the definition of $T^0_D(\bar\lambda)$). Also there is $F_2 \subseteq \prod \bar\lambda$ exemplifying the value of $T^2_D(\bar\lambda)$. For every $f \in F_0$ there is $g_f \in F_2$ such that $\neg f \neq_D g_f$ (by the choice of F_2). As $|F_0| > T^2_D(\bar\lambda) + \mu$ for some $g \in F_2$, $F^* =: \{f \in F_0: g_f = g\}$ has cardinality $> T^2_D(f) + \mu$. Now for each $f \in F^*$ let $A_f = \{i < \kappa: f(i) = g(i)\}$, clearly $A_f \in D^+$. Now $f \mapsto A_f/D$ is a function from F^* into $\mathcal{P}(\kappa)/D$, hence, if $\mu = |\mathcal{P}(\kappa)/D|$, it is not one to one (by cardinality consideration) so for some $f' \neq f''$ from F^* (hence form F_0) we have $A_{f'}/D = A_{f''}/D$; but so

$$\{i < \kappa: f'(i) = f''(i)\} \supseteq \{i < \kappa: f'(i) = g(i)\} \cap \{i < \kappa: f''(i) = g(i)\} = A_{f'}/D$$

hence is $\neq \emptyset \bmod D$, so $\neg f' \neq_D f''$, contradiction the choice of F_0. If $\mu \neq |\mathcal{P}(\kappa)/D|$ (as $F^* \subseteq F_0$ by the choice of F_0) we have:

$$f_1 \neq f_2 \in F^* \Rightarrow A_{f_1} \cap A_{f_2} = \emptyset \ \bmod\ D$$

so $\{A_f : f \in F^*\}$ contradicts "the Boolean algebra $\mathcal{P}(\kappa)/D$ satisfies the μ^+-c.c.".

(3) Assume that $\theta < \operatorname{reg}(D)$ and¶ $\mu \leq^+ T^0_D(\bar\lambda)$. As $\mu \leq^+ T^0_D(\bar\lambda)$ we can find $f_\alpha \in \prod \bar\lambda$ for $\alpha < \mu$ such that $[\alpha < \beta \Rightarrow f_\alpha \neq_D f_\beta]$. Also (as $\theta < \operatorname{reg}(D)$) we can find $\{A_\varepsilon: \varepsilon < \theta\} \subseteq D$ such that for every $i < \kappa$ the set $w_i =: \{\varepsilon < \theta: i \in A_\varepsilon\}$ is finite. Now for every function $h: \theta \to \mu$ we define g_h, a function with domain κ:

$$g_h(i) = \{(\varepsilon, f_{h(\varepsilon)}(i)): \varepsilon \in w_i\}$$

So $|\{g_h(i): h \in {}^\theta\mu\}| \leq (\lambda_i)^{|w_i|} = \lambda_i$, and if $h_1 \neq h_2$ are from ${}^\theta\mu$ then for some $\varepsilon < \theta$, $h_1(\varepsilon) \neq h_2(\varepsilon)$ so $B_{h_1,h_2} = \{i: f_{h_1(\varepsilon)}(i) \neq f_{h_2(\varepsilon)}(i)\} \in D$ that is $B_{h_1,h_2} \cap A_\varepsilon \in D$ so

$\otimes_1$ if $i \in B_{h_1,h_2} \cap A_\varepsilon$ then $\varepsilon \in w_i$, so $g_{h_1}(i) \neq g_{h_2}(i)$.

$\otimes_2$ $B_{h_1,h_2} \cap A_\varepsilon \in D$

So $\langle g_h: h \in {}^\theta\mu\rangle$ exemplifies $T^0_D(\bar\lambda) \geq \mu^\theta$. If the supremum in the definition of $T^0_D(\bar\lambda)$ is obtained we are done. If not then $T^0_D(\bar\lambda)$ is a limit cardinal, and by the proof above:

$$[\mu < T^0_D(\bar\lambda) \quad \& \quad \theta < \operatorname{reg}(D) \quad \Rightarrow \quad \mu^\theta < T^0_D(\bar\lambda)].$$

So if $T^0_D(\bar\lambda)$ has cofinality $\geq \operatorname{reg}(D)$ we are done; otherwise let it be $\sum_{\varepsilon<\theta} \mu_\varepsilon$ with $\mu_\varepsilon < T^0_D(\bar\lambda)$ and $\theta < \operatorname{reg} D$. Note that by the previous sentence $T^0_D(\bar\lambda)^\theta = T^0_D(\bar\lambda)^{<\operatorname{reg}(D)} = \prod_{\varepsilon<\theta} \mu_\varepsilon$, and let $\{f^\varepsilon_\alpha: \alpha < \mu_\varepsilon\} \subseteq \prod \bar\lambda$ be such that $[\alpha < \beta \Rightarrow f^\varepsilon_\alpha \neq_D f^\varepsilon_D]$ and repeat the previous proof with $f^\varepsilon_{h(\varepsilon)}$ replacing $f_{h(\varepsilon)}$.

¶ $\leq^+$ means here that the right side is a supremum, right bigger than the left or equal but the supremum is obtained

(4) For the first inequality. assume it fails so $\mu =: T^2_D(f) < \mathrm{htcf}_D(\prod_{i<\kappa} f(i))$ hence for some $g \in \prod_{i<f(i)} (f(i)+1)$, $\mathrm{tcf}(\prod_{i<\kappa} g(i), <_D)$ is λ with $\lambda = \mathrm{cf}(\lambda) > \mu$. Let $\langle f_\alpha : \alpha < \lambda \rangle$ exemplifies this. Let F be as in the definition of $T^2_D(f)$, now for each $h \in F$, there is $\alpha(h) < \lambda$ such that

$$\{i < \kappa : \text{ if } h(i) < g(i) \text{ then } h(i) < f_{\alpha(g)}(i)\} \in D.$$

Let $\alpha^* = \sup\{\alpha(h)+1 : h \in F\}$, now $f_{\alpha^*} \in \prod_{i<\kappa} f(i)$ and $h \in F \Rightarrow h \neq_D f_{\alpha^*}$ contradicting the choice of F.

For the second inequality. Repeat the proof of 3.9 except that here we prove $F =: \bigcup_{\zeta<\theta}(N_\zeta \cap \prod_{i<\kappa} f(i))$ exemplifies $T^2_D(f) \leq \lambda$. So let $g^* \in \prod_{i<\kappa} \lambda_i$, and we should find $f \in N$ such that $(g^* \neq_D f)$; we replace clause (g) in the proof by
(g)′ $g^* < f^2_{\zeta+1} < f^1_\zeta \bmod D$
the construction is for $\zeta < \mathrm{reg}(D)$ and if we are stuck in ζ then $\neg f^1_\zeta \neq_D g^*$ and so we are done.

(5) Straightforward.

(6) Note that all those cardinals are $\geq 2^\kappa$ and $2^\kappa \geq \mathrm{reg}(D)^\kappa/D$. Now write successively inequalities from (2), (4), (1) and (3):

$$T^0_D(f) = T^2_D(f) \leq [\mathrm{htcf}_D \prod_{i<\kappa} f(i)]^{<\mathrm{reg}(D)} \leq [T^0_D(f)]^{<\mathrm{reg}(D)} = T^0_D(f).$$

(7) See proof of part (3). Moreover, let $\mu = \sum_{\varepsilon<\tau} \mu_\varepsilon$, $\tau < T^0_D(\bar\lambda)$, $\mu_\varepsilon < T^0_D(\bar\lambda)$ as exemplified by $\{f_\varepsilon : \varepsilon < \tau\}$, $\{f^\varepsilon_\alpha : \alpha < \mu_\varepsilon\}$ respectively. Let g_α be: if $\sum_{\varepsilon<\zeta} \mu_\varepsilon < \alpha < \sum_{\varepsilon\leq\zeta} \mu_\varepsilon$ then $g_\alpha(i) = (f_\varepsilon(i), f^\varepsilon_\alpha(i))$. So $\{g_\alpha : \alpha < \mu\}$ show: if $T^0_D(\bar\lambda)$ is singular then the supremum is obtained. $\blacksquare_{3.14}$

CLAIM 3.15: Assume D is a filter on κ, $f \in {}^\kappa\mathrm{Ord}$, $\mu^{\aleph_0} = \mu$ and $2^\kappa < \mu$, $T_D(f)$, (see Definition 3.13(d) and Theorem 3.14(2)) and $\mathrm{reg}_*(D) = \mathrm{reg}(D)$. If $\mu < T_D(f)$ <u>then</u> for some sequence $\bar\lambda \leq f$ of regulars, $\mu^+ = \mathrm{tcf}(\prod \bar\lambda/D)$, or at least

(∗) there are $\langle\langle\lambda_{i,n} : n < n_i\rangle : i < \kappa\rangle$, $\lambda_{i,n} = \mathrm{cf}(\lambda_{i,n}) < f(i)$ and a filter D^* on $\bigcup_{i<\kappa}\{i\} \times n_i$ such that: $\mu^+ = \mathrm{tcf}(\prod_{(i,n)} \lambda_{i,n}/D^*)$ and $D = \{A \subseteq \kappa : \bigcup_{i\in A}\{i\} \times n_i \in D^*\}$.

Also the inverse is true.

Remark 3.15A. (1) It is not clear whether the first possibility may fail. We have explained earlier the doubtful role of $\mu^{\aleph_0} = \mu$.

(2) We can replace μ^+ by any regular μ such that $\bigwedge_{\alpha<\mu} |\alpha|^{\aleph_0} < \mu$ and then we use 3.14(4) to get $\mu \leq^+ T_D(f)$.

(3) The assumption $2^\kappa < \mu$ can be omitted.

Proof. The inverse should be clear (as in the proof of 3.6, by 3.14(3)).

Wlog $f(i) > 2^\kappa$ for $i < \kappa$, and trivially $(\mathrm{reg}(D))^\kappa/D \leq 2^\kappa$, so by 3.14(4)

$$T_D(f) \leq [\mathrm{htcf}_D(\prod_{i<\kappa} f(i))]^{<\mathrm{reg}_*(D)}.$$

If $\mu < \mathrm{htcf}_D(\prod_{i<\kappa} f(i))$ we are done (by 3.12(1)), so assume $\mathrm{htcf}_D(\prod_{i<\kappa} f(i)) \leq \mu$, but we have assumed $\mu < T_D(f)$ so by 3.14(4) as $\mathrm{reg}_*(D) = \mathrm{reg}(D)$ we have $\mu^{<\mathrm{reg}(D)} \geq \mu^+$. Let $\chi \leq \mu$ be minimal such that $\bigvee_{\theta<\mathrm{reg}(D)} \chi^\theta \geq \mu$, and let $\theta =: \mathrm{cf}(\chi)$ so, as $\mu > 2^\kappa$ we know $\chi^{\mathrm{cf}\,\chi} = \chi^{<\mathrm{reg}(D)} = \mu^{<\mathrm{reg}(D)} \geq \mu^+$, $\chi > 2^\kappa$, $\bigwedge_{\alpha<\chi} |\alpha|^{<\mathrm{reg}(D)} < \chi$. By the assumption $\mu = \mu^{\aleph_0}$ we know $\theta > \aleph_0$ (of course θ is regular). By [Sh-g, VIII 1.6(2), IX 3.5] and [Sh513, 6.12] there is a strictly increasing sequence $\langle \mu_\varepsilon : \varepsilon < \theta \rangle$ of regular cardinals with limit χ such that $\mu^+ = \mathrm{tcf}(\prod_{\varepsilon<\theta} \mu_\varepsilon / J^{\mathrm{bd}}_\theta)$.

As clearly $\chi \leq \mathrm{htcf}_D(\prod_{i<\kappa} f(i))$, by 2.12(1) there is for for each each $\varepsilon < \theta$, a sequence $\bar\lambda^\varepsilon = \langle \lambda^\varepsilon_i : i < \kappa \rangle$ such that $\lambda^\varepsilon_i = \mathrm{cf}(\lambda^\varepsilon_i) \leq f(i)$, and $\mathrm{tcf}(\prod_{i<\kappa} \lambda^\varepsilon_i / D) = \mu_\varepsilon$, also wlog $\lambda^\varepsilon_i > 2^\kappa$. Let $\langle A_\varepsilon : \varepsilon < \theta \rangle$ exemplify $\theta < \mathrm{reg}(D)$ and $n_i = |\{\varepsilon < \theta : i \in A_\varepsilon\}|$ and $\{\lambda_{i,n} : n < \omega\}$ enumerate $\{\lambda^\varepsilon_i : \varepsilon \text{ satisfies } i \in A_\varepsilon\}$, so we have gotten $(*)$. ■3.15

Conclusion 3.16. Suppose D is an $\aleph_1$-complete filter on κ and $\mathrm{reg}_*(D) = \mathrm{reg}(D)$. If $\lambda_i \geq 2^\kappa$ for $i < \kappa$ and $\sup_{A\in D^+} T_{D+A}(\bar\lambda) > \mu^{\aleph_0}$ <u>then</u> for some $\lambda'_i = \mathrm{cf}(\lambda'_i) \leq \lambda_i$ we have

$$\sup_{A\in D^+} \mathrm{htcf}_{D+A}(\prod_{i<\kappa} \lambda'_i) > \mu.$$

Conclusion 3.17. Let D be an $\aleph_1$-complete filter on κ and $\mathrm{reg}_*(D) = \mathrm{reg}(D)$. If for $i < \kappa$, B_i is a Boolean algebra and $\lambda_i < \mathrm{Depth}^+(B_i)$ (see below) and

$$2^\kappa < \mu^{\aleph_0} < \sup_{A\in D^+} T_{D+A}(\bar\lambda)$$

<u>then</u> $\mu^+ < \mathrm{Depth}^+(\prod_{i<\kappa} B_i/D)$.

Proof. Use 3.25 below and 3.16 above.

Definition 3.18. For a partial order P (e.g. a Boolean algebra) let $\mathrm{Depth}^+(P) = \min\{\lambda : \text{we cannot find } a_\alpha \in P \text{ for } \alpha < \lambda \text{ such that } \alpha < \beta \Rightarrow a_\alpha <_P a_\beta\}$.

Discussion 3.19.

(1) We conjecture that in 3.16 (and 3.17) the assumption "D is $\aleph_1$-complete" can be omitted. See [Sh589].

(2) Note that our results are for $\mu = \mu^{\aleph_0}$ only; to remove this we need first to improve the theorem on $pp = cov$ (i.e. to prove $\mathrm{cf}(\lambda) = \aleph_0 < \lambda \Rightarrow pp(\lambda) = cov(\lambda, \lambda, \aleph_1, 2)$ (or $\sup\{pp(\mu) : \mathrm{cf}\,\mu = \aleph_0 < \mu < \lambda\} = \mathrm{cf}(S_{\leq\aleph_0}(\lambda), \subseteq)$ (see [Sh-g], [Sh430, §1]), which seems to me a very serious open problem (see [Sh-g, Analitic guide, 14]).

(3) In 3.17, if we can find $f_\alpha \in \prod_{i<\kappa} \lambda_i$ for $\alpha < \lambda$: $[\alpha < \beta < \lambda \Rightarrow f_\alpha \leq f_\beta \bmod D]$ and $\neg f_\alpha =_D f_{\alpha+1}$ then $\lambda < \mathrm{Depth}^+(\prod_{i<\kappa} B_i/D)$. But this does not help for λ regular $> 2^\kappa$.

(4) We can approach 3.15 differently, by 3.20–3.23 below.

CLAIM 3.20: If $2^{2^\kappa} \le \mu < T_D(\bar\lambda)$, (or at least $2^{|D|+\kappa} \le \mu < T_D(\bar\lambda)$) and $\mu^{<\theta} = \mu$, then for some θ-complete filter $E \subseteq D$ we have $T_E(\bar\lambda) > \mu$.

Proof. Wlog θ is regular (as $\mu^{<\theta} = \mu$ &, $\mathrm{cf}(\theta) < \theta \Rightarrow \mu^{<\theta^+} = \mu$). Let $\{f_\alpha : \alpha < \mu^+\} \subseteq \prod \bar\lambda$, be such that $[\alpha < \beta \Rightarrow f_\alpha \ne_D f_\beta]$. We choose by induction on ζ, $\alpha_\zeta < \mu^+$ as follows: α_ζ is the minimal ordinal $\alpha < \mu^+$ such that $E_{\zeta,\alpha} \subseteq D$ where $E_{\zeta,\alpha} =$ the θ-complete filter generated by

$$\big\{\{i < \kappa : f_{\alpha_\varepsilon}(i) \ne f_\alpha(i)\} : \varepsilon < \zeta\big\}$$

(note: each generator of $E_{\zeta,\alpha}$ is in D but not necessarily $E_{\zeta,\alpha} \subseteq D$!).
Let α_ζ be well defined if $\zeta < \zeta^*$, clearly $\varepsilon < \zeta \Rightarrow \alpha_\varepsilon < \alpha_\zeta$. Now if $\zeta^* < \mu^+$, then clearly $\alpha^* = \bigcup_{\zeta<\zeta^*} \alpha_\zeta < \mu^+$ and for every $\alpha \in (\alpha^*, \mu^+)$, $E_{\zeta^*,\alpha} \not\subseteq D$, so for every such α there are $A_\alpha \in D^+$ and $a_\alpha \in [\zeta^*]^{<\theta}$ such that $A_\alpha = \bigcup_{\varepsilon\in a_\alpha} \{i < \kappa : f_{\alpha_\varepsilon}(i) = f_\alpha(i)\}$. But for every $A \in D^+$, $a \in [\zeta^*]^{<\theta}$ we have

$$\{\alpha : \alpha \in (\alpha^*, \mu^+), A_\alpha = A, a_\alpha = a\} \subseteq \{\alpha : f_\alpha \restriction A \in \prod_{i<\kappa} \{f_{\alpha_\varepsilon}(i) : \varepsilon \in a_\alpha\}\},$$

hence has cardinality $\le \theta^\kappa \le 2^\kappa < \mu$. Also $|[\zeta^*]^{<\theta}| \le \mu^{<\theta} < \mu^+$, $|D^+| \le 2^\kappa < \mu^\kappa$ so we get easy contradiction.

So $\zeta^* = \mu^+$, but the number of possible E's is $\le 2^{2^\kappa}$, hence for some E we have $|\{\varepsilon < \mu^+ : E_{\varepsilon,\alpha_\varepsilon} = E\}| = \mu^+$. Necessarily $E \subseteq D$ and E is θ-complete, and $\{f_{\alpha_\varepsilon} : \varepsilon < \mu^+$, and $E_{\alpha_\varepsilon} = E\}$ exemplifies $T_E(\bar\lambda) > \mu$, so E is as required. $\blacksquare_{3.20}$

Fact 3.21. 1. In 3.20 we can replace μ^+ by μ^* if $2^{2^\kappa} < \mathrm{cf}(\mu^*) \le \mu^* \le T^0_D(\bar\lambda)$ and $\bigwedge_{\alpha<\mu^*} |\alpha|^{<\theta} < \mu^*$.

Proof. The same proof as 3.20.

CLAIM 3.22:

(1) If $2^\kappa < |\prod \bar\lambda/D|$, D an ultrafilter on κ, $\mu = \mathrm{cf}(\mu) \le |\prod \bar\lambda/D|$, $\bigwedge_{i<\kappa} |i|^{\aleph_0} < \mu$, and D is regular then $\mu < \mathrm{Depth}^+(\prod_{i<\kappa} \lambda_i/D)$

(2) Similarly for D just a filter but $A \in D^+ \Rightarrow \prod \bar\lambda/(D + A) = \prod \bar\lambda/D$.

Proof. 1) Wlog $\lambda =: \lim_D \bar\lambda = \sup(\bar\lambda)$, so $|\prod \bar\lambda/D| = \lambda^\kappa$ (see 3.6, by [CK]). If $\mu \le \lambda$ we are done; otherwise let $\chi = \min\{\chi : \chi^\kappa = \lambda^\kappa\}$, so $\chi^{\mathrm{cf}(\chi)} = \lambda^\kappa$, $\mathrm{cf}(\chi) \le \kappa$ but $\lambda < \mu \le \lambda^\kappa$ hence $\lambda^{\aleph_0} < \mu$ hence $\mathrm{cf}(\chi) > \aleph_0$, also by $\chi's$ minimality $\bigwedge_{i<\chi} |i|^{\mathrm{cf}\,\chi} \le |i|^\kappa < \chi$, and remember $\chi < \mu = \mathrm{cf}\,\mu \le \chi^{\mathrm{cf}\,\chi}$ so by [Sh-g. VIII 1.6(2)] there is $\langle \mu_\varepsilon : \varepsilon < \mathrm{cf}(\chi)\rangle$ strictly increasing sequence of regular cardinals with limit χ, $\prod_{\varepsilon<\mathrm{cf}(\chi)} \mu_\varepsilon / J^{bd}_{\mathrm{cf}\,\chi}$ has true cofinality μ. Let $\chi_\varepsilon = \sup\{\mu_\zeta : \zeta < \varepsilon\} + 2^\kappa$, let $\mathrm{i} : \kappa \to \mathrm{cf}(\chi)$ be $\mathrm{i}(i) = \sup\{\varepsilon + 1 : \lambda_i \ge \chi_\varepsilon\}$. If there is a function $h \in \prod_{i<\kappa} \mathrm{i}(i)$ such that $\bigwedge_{j<\mathrm{cf}(\chi)} \{i < \kappa : h(i) < j\} = \emptyset \bmod D$ then $\prod_{i<\kappa} \mu_{h(i)}/D$ has true cofinality μ as required; if not (D, i) is weakly normal (i.e. there is no such h - see [Sh420]). But for D regular, D is $\mathrm{cf}(\chi)$-regular, some $\langle A_\varepsilon : \varepsilon < \mathrm{cf}(\kappa)\rangle$ exemplifies it and $h(i) = \max\{\varepsilon : \varepsilon < \mathrm{i}(i)$ and $i \in A_\varepsilon\}$ (maximum over a finite set) is as required.

2) Similarly using $\lambda =: \liminf_D(\bar\lambda)$. $\blacksquare_{3.22}$

Discussion 3.23.

1. In 3.20 (or 3.21) we can apply [Sh 410, §6] so $\mu = \mathrm{tcf}(\prod \bigcup_{i<\mu} \mathfrak{a}_i / D^*)$, where $D = \{A \subseteq \kappa : \bigcup_{i \in A} \mathfrak{a}_i \in D^*\}$ and each $\mathfrak{a}_i$ is finite.

   See also in 3.15.

CLAIM 3.24: If D is a filter on κ, B_i is the interval Boolean algebra on the ordinal α_i, and $|\prod_{i<\kappa} \alpha_i/D| > 2^\kappa$ <u>then</u> for regular μ we have: $\mu < \mathrm{Depth}^+(\prod_{i<\kappa} B_i/D)$ iff for some $\mu_i \le \alpha_i$ (for $i < \kappa$) and $A \in D^+$, the true cofinality of $\prod_{i<\kappa} \mu_i/(D+A)$ is well defined and equal to μ.

Proof. The $\Leftarrow$ (i.e. if direction) is clear. For the $\Rightarrow$ direction assume μ is regular $< \mathrm{Depth}^+(\prod_{i<\kappa} B_i/D)$ so there are $f_\alpha \in \prod_{i<\kappa} B_i$ such that $\prod_{i<\kappa} B_i/D \vDash f_\alpha/D < f_\beta/D$ for $\alpha < \beta$.

Wlog $\mu > 2^\kappa$. Let $f_\alpha(i) = \bigcup_{\ell < n(\alpha,i)} [j_{\alpha,i,2\ell}, j_{\alpha,i.2\ell+1})$ where $j_{\alpha,i,\ell} < j_{\alpha,i,\ell+1} < \alpha_i$ for $\ell < 2n(\alpha,i)$. As $\mu = \mathrm{cf}(\mu) > 2^\kappa$ wlog $n_{\alpha,i} = n_i$. By [Sh430, 6.6D] (see more [Sh513, 6.1]) we can find $A \subseteq A^* =: \{(i,\ell) : i < \kappa, \ell < 2n_\alpha\}$ and $\langle \gamma^*_{i,\ell} : i < \kappa, \ell < 2n_i \rangle$ such that $(i,\ell) \in A \Rightarrow \gamma^*_{i,\ell}$ is a limit ordinal and

$(*)$ for every $f \in \prod_{(i,\ell)\in A} \gamma^*_{i,\ell}$ and $\alpha < \mu$ there is $\beta \in (\alpha,\mu)$ such that

$$(i,\ell) \in A^* \setminus A \Rightarrow j_{\alpha,i,\ell} = \gamma^*_{i,\ell}$$
$$(i,\ell) \in A \Rightarrow f(i,\ell) < j_{\beta,i,\ell} < \gamma^*_{i,\ell}$$
$$(i,\ell) \in A \Rightarrow \mathrm{cf}(\gamma^*_{i,\ell}) > 2^\kappa$$

Let $\ell(i) = \max\{\ell < 2n(i) : (i,\ell) \in A\}$ and let $B = \{i : \ell(i) \text{ well defined}\}$. Clearly $B \in D^+$ (otherwise we can find $\alpha < \beta < \mu$ such that $f_\alpha/D = f_\beta/D$, contradiction). For $(i,\ell) \in A$ define $\beta^*_{i,\ell}$ by $\beta^*_{i,\ell} = \sup\{\gamma^*_{j,m} + 1 : (j,m) \in A^*$ and $\gamma^*_{j,m} < \gamma^*_{i,\ell}\}$. Now $\beta^*_{i,\ell} < \gamma^*_{i,\ell}$ as $\mathrm{cf}(\gamma^*_{i,\ell}) > 2^\kappa$. Let

$$Y = \{\alpha < \mu : \text{if } (i,\ell) \in A^* \setminus A \text{ then } j_{\alpha,i,\ell} = \gamma^*_{i,\ell}$$
$$\text{and if } (i,\ell) \in A \text{ then } \beta^*_{i,\ell} < j_{\alpha,\ell,i} < \gamma^*_{\ell,i}\}$$

Let $B_1 = \{i \in B : \ell(i) \text{ is odd}\}$. Clearly $B_1 \subseteq B$ and $B \setminus B_1 = \emptyset \bmod D$ (otherwise as in $(*)_1$, $(*)_2$ below get contradiction) hence $B_1 \in D^+$. Now

$(*)_1$ for $\alpha < \beta$ from Y we have

$$\langle j_{\alpha,i,\ell(i)} : i \in B_1 \rangle \le \langle j_{\beta,i,\ell(i)} : i \in B_1 \rangle \bmod (D \restriction B_1)$$

[Why? as f_α/D was non decreasing in $\prod_{i<\kappa} B_i/D$]

$(*)_2$ for every $\alpha \in Y$ for some β, $\alpha < \beta \in Y$ we have

$$\langle j_{\alpha,i,\ell(i)} : i \in B_1 \rangle < \langle j_{\beta,i,\ell(i)} : i \in B_1 \rangle \bmod (D \restriction B_1)$$

[Why? by $(*)$ above]

Together for some unbounded $Z \subseteq Y$, $\big\langle \langle j_{\alpha,\ell,\ell(i)} : i \in B_1 \rangle/(D \restriction B_1) : \alpha \in Z \big\rangle$ is $<_{D\restriction B_1}$-increasing, so it has a $<_{(D\restriction B_1)}$ $-$eub (as $\mu > 2^\kappa$, see 3.10, and more in [Sh-g, II §1]), say $\langle j^*_i : i \in B_1 \rangle$ hence $\prod_{i\in B_1} j^*_i/(D \restriction B_1)$ has true cofinality μ by 1.3(12) and clearly $j^*_i \le \gamma^*_{i,\ell(i)} \le \alpha_i$, so we have finished. $\blacksquare_{3.24}$

CLAIM 3.25: If D is a filter on κ, B_i a Boolean algebra, $\lambda_i < \mathrm{Depth}^+(B_i)$ <u>then</u>

(a) $\mathrm{Depth}(\prod_{i<\kappa} B_i/D) \geq \sup_{A\in D^+} \mathrm{tcf}(\prod_{i<\kappa}\lambda_i/(D+A))$ (i.e. on the cases tcf is well defined).
(b) $\mathrm{Depth}^+(\prod_{i<\kappa} B_i/D)$ is $\geq \mathrm{Depth}^+(\mathcal{P}(\kappa)/D)$ and is at least

$$\sup\{[\mathrm{tcf}(\prod_{i<\kappa}\lambda'_i/(D+A))]^+ : \lambda'_i < \mathrm{Depth}^+(B_i), A \in D^+\}.$$

Proof. Check.

CLAIM 3.26: Let D be a filter on κ, $\langle\lambda_i : i<\kappa\rangle$ a sequence of cardinals and $2^\kappa < \mu = \mathrm{cf}(\mu)$. Then $(\alpha) \Leftrightarrow (\beta) \Rightarrow (\gamma) \Rightarrow (\delta)$, and if $(\forall\sigma<\mu)(\sigma^{\aleph_0}<\mu)$ and $\mathrm{reg}_*(D) = \mathrm{reg}(D)$ we also have $(\gamma) \Leftrightarrow (\delta)$ where

(α) if B_i is a Boolean algebra, $\lambda_i < \mathrm{Depth}^+(B_i)$ <u>then</u> $\mu < \mathrm{Depth}^+(\prod_{i<\kappa} B_i/D)$
(β) there are $\mu_i = \mathrm{cf}(\mu_i) \leq \lambda_i$ for $i<\kappa$ and $A \in D^+$ such that $\mu = \mathrm{tcf}(\prod \mu_i/(D+A))$
(γ) there are $\langle\langle\lambda_{i,n} : n<n_i\rangle : i<\kappa\rangle$, $\lambda_{i,n} = \mathrm{cf}(\lambda_{i,n}) < \lambda_i$, $A^* \in D^+$ and a filter D^* on $\bigcup_{i<\kappa}\{i\}\times n_i$ such that:

$$\mu = \mathrm{tcf}(\prod_{(i,n)}\lambda_{i,n}/D^*) \text{ and } D{+}A^* = \{A \subseteq \kappa \colon \text{the set } \bigcup_{i\in A}\{i\}\times n_i \text{ belongs to } D^*\}.$$

(δ) for some $A \in D^+$, $\mu \leq T_{D+A}(\langle\lambda_i : i<\kappa\rangle)$

Remark. So the question whether $(\alpha) \Leftrightarrow (\delta)$ assuming $(\forall\sigma<\mu)(\sigma^{\aleph_0}<\mu)$ is equivalent to $(\beta) \leftrightarrow (\gamma)$ which is a "pure" pcf problem.

Proof. Note $(\gamma) \Rightarrow (\delta)$ is easy (as in 3.15, i.e. as in the proof of 3.6, only easier). Now $(\beta) \Rightarrow (\gamma)$ is trivial and $(\beta) \Rightarrow (\alpha)$ by 3.25. Next $(\alpha) \Rightarrow (\beta)$ holds as we can use (α) for $B_i =$: the interval Boolean algebra of the order λ_i and use 3.24. Lastly assume $(\forall\sigma<\mu)(\sigma^{\aleph_0}<\mu)$ and $\mathrm{reg}_*(D) = \mathrm{reg}(D)$, now $(\gamma) \Leftrightarrow (\delta)$ by 3.15. ■3.26

Discussion. We would like to have (letting B_i denote Boolean algebra)

$$\mathrm{Depth}^{(+)}(\prod_{i<\kappa} B_i/D) \geq \prod_{i<\kappa}\mathrm{Depth}^{(+)}(B_i)/D$$

if D is just filter we should use T_D and so by the problem of attainment (serious by Magidor Shelah [MgSh433]), we ask

⊗ for D an ultrafilter on κ, does $\lambda_i < \mathrm{Depth}^+(B_i)$ for $i<\kappa$ implies

$$\prod_{i<\kappa}\lambda_i/D < \mathrm{Depth}^+(\prod_{i<\kappa} B_i/D)$$

at least when $\lambda_i > 2^\kappa$;

⊗′ for D a filter on κ, does $\lambda_i < \mathrm{Depth}^+(B_i)$ for $i<\kappa$ implies, assuming $\lambda_i > 2^\kappa$ for simplicity,

$$T_D(\langle\lambda_i : i<\kappa\rangle) < \mathrm{Depth}^+(\prod_{i<\kappa} B_i/D)$$

As explained in 3.26 this is a pcf problem.

In [Sh589] we deal with this under reasonable assumption (e.g. $\mu = \chi^+$ and $\chi = \chi^{\aleph_0}$). We also deal with a variant, changing the invariant (closing under homomorphisms, see [M]).

4. Remarks on the conditions for the pcf analysis

We consider a generalization whose interest is not so clear.

CLAIM 4.1: Suppose $\bar{\lambda} = \langle \lambda_i : i < \kappa \rangle$ is a sequence of regular cardinals, and θ is a cardinal and I^* is an ideal on κ; and H is a function with domain κ. We consider the following statements:

$(**)_H$ $\liminf_{I^*}(\bar{\lambda}) \geq \theta \geq \mathrm{wsat}(I^*)$ and H is a function from κ to $\mathcal{P}(\theta)$ such that:

(a) for every $\varepsilon < \theta$ we have $\{i < \kappa : \varepsilon \in H(i)\} = \kappa \bmod I^*$

(b) for $i < \kappa$ we have $\mathrm{otp}(H(i)) \leq \lambda_i$ or at least $\{i < \kappa : |H(i)| \geq \lambda_i\} \in I^*$

$(**)^+$ similarly but

$(b)^+$ for $i < \kappa$ we have $\mathrm{otp}(H(i)) < \lambda_i$

(1) In 1.5 we can replace the assumption $(*)$ by $(**)_H$ above.
(2) Also in 1.6, 1.7, 1.8, 1.9, 1.10, 1.11 we can replace 1.5$(*)$ by $(**)_H$.
(3) Suppose in Definition 2.3(2) we say $\bar{f}$ obeys $\bar{a}$ for H (instead of for $\bar{A}^*$) if
(i) for $\beta \in a_\alpha$ such that $\varepsilon =: \mathrm{otp}(a_\alpha) < \theta$ we have

$$\mathrm{otp}(a_\beta), \mathrm{otp}(a_\alpha) \in H(i) \Rightarrow f_\beta(i) \leq f_\alpha(i)$$

and in 2.3(2A), $f_\alpha(i) = \sup\{f_\beta(i) : \beta \in a_\alpha \text{ and } \mathrm{otp}(a_\beta), \mathrm{otp}(a_\alpha) \in H(i)\}$.

Then we can replace 1.5$(*)$ by $(**)_H$ in 2.5, 2.5A, 2,6; and replace 1.5$(*)$ by $(**)_H^+$ in 2.7 (with the natural changes).

Proof. (1) Like the proof of 1.5, but defining the g_ε's by induction on ε we change requirement (ii) to

(ii)$'$ if $\zeta < \varepsilon$, and $\{\zeta, \varepsilon\} \subseteq H(i)$ then $g_\zeta(i) < g_\varepsilon(i)$.

We can not succeed as

$$\langle (B^\varepsilon_{\alpha(*)} \setminus B^{\varepsilon+1}_{\alpha(*)}) \cap \{i < \kappa : \varepsilon, \varepsilon+1 \in H(i)\} : \varepsilon < \theta \rangle$$

is a sequence of θ pairwise disjoint member of $(I^*)^+$.

In the induction, for ε limit let $g_\varepsilon(i) < \cup\{g_\zeta(i) : \zeta \in H(i) \text{ and } \varepsilon \in H(i)\}$ (so this is a union at most $\mathrm{otp}(H(i) \cap \varepsilon)$ but only when $\varepsilon \in H(i)$ hence is $< \mathrm{otp}(H(i)) \leq \lambda_i$).

(2) The proof of 1.6 is the same, in the proof of 1.7 we again replace (ii) by (ii)$'$. Also the proof of the rest is the same.

(3) Left to the reader. $\blacksquare_{4.1}$

We want to see how much weakening $(*)$ of 1.5 to "$\liminf_{I^*}(\bar{\lambda}) \geq \theta \geq \mathrm{wsat}(I^*)$ suffices. If θ singular or $\liminf_{I^*}(\bar{\lambda}) > \theta$ or just $(\prod \bar{\lambda}, <_{I^*})$ is θ^+-directed then case (β) of 1.5 applies. This explains $(*)$ of 4.2 below.

CLAIM 4.2: Suppose $\bar{\lambda} = \langle \lambda_i : i < \kappa \rangle$, $\lambda_i = \mathrm{cf}(\lambda_i)$, I^* an ideal on κ, and

$$\liminf_I(\bar{\lambda}) = \theta \geq \mathrm{wsat}(I^*), \quad \theta \text{ regular} \tag{$*$}$$

<u>Then</u> we can define a sequence $\bar{J} = \langle J_\zeta : \zeta < \zeta(*) \rangle$ and an ordinal $\zeta(*) \leq \theta^+$ such that

(a) $\bar{J}$ is an increasing continuous sequence of ideals on κ.
(b) $J_0 = I^*$, $J_{\zeta+1} =: \{A : A \subseteq \kappa$, and: $A \in J_\zeta$ or we can find $h : A \to \theta$ such that $\lambda_i > h(i)$ and $\varepsilon < \theta \Rightarrow \{i : h(i) < \varepsilon\} \in J_\zeta\}$.
(c) for $\zeta < \zeta(*)$ and $A \in J_{\zeta+1} \setminus J_\zeta$, the pair $(\prod \bar{\lambda}, J_\zeta + (\kappa \setminus A))$ (equivalently $(\prod \bar{\lambda} \restriction A, J_\zeta \restriction A)$) satisfies condition 1.5$(*)$ (case (β)) hence its consequences, (in particular it satisfies the weak pcf-th for θ).
(d) if $\kappa \notin \cup_{\zeta < \zeta(*)} J_\zeta$ then $(\prod \bar{\lambda}, \cup_{\zeta < \zeta(*)} J_\zeta)$ has true cofinality θ.

Proof. Straight. (We define J_ζ for $\zeta \leq \theta^+$ by clause (b) for $\zeta = 0$, ζ successor and as $\bigcup_{\varepsilon<\zeta} J_\varepsilon$ for ζ limit. Clause (c) holds by claim 4.4 below. It should be clear that $J_{\theta^++1} = J_{\theta^+}$, and let $\zeta(*) = \min\{\zeta: J_{\zeta+1} = \bigcup_{\varepsilon<\zeta} J_\varepsilon\}$ so we are left with checking clause (d). If $A \in J^+_{\zeta(*)}$, $h \in \prod_{i\in A} \lambda_i$, choose by induction on $\zeta < \theta$, $\varepsilon(\zeta) < \theta$ increasing with ζ such that $\{i < \kappa: h(i) \in (\varepsilon(\zeta), \varepsilon(\zeta+1)) \in J^+_{\zeta(*)}$. If we succeed we contradict $\theta \geq \mathrm{wsat}(I^*)$ as θ is regular. So for some $\zeta < \theta$, $\varepsilon(\zeta)$ is well defined but not $\varepsilon(\zeta+1)$. As $J_{\zeta(*)} = J_{\zeta(*)+1}$, clearly $\{i < \kappa: h(i) \leq \varepsilon(\zeta)\} = \kappa \mod J_{\zeta(*)}$. So $g_\varepsilon(i) = \begin{cases} \varepsilon & \text{if } \varepsilon < \lambda_i \\ 0 & \text{if } \varepsilon \geq \lambda_i \end{cases}$ exemplifies $\mathrm{tcf}(\prod \bar\lambda / J_{\zeta(*)}) = \theta$. $\blacksquare_{4.2}$

Now:

CONCLUSION 4.3: Under the assumptions of 4.2, I^* satisfies the pseudo pcf-th (see Definition 2.11(4)).

CLAIM 4.4: Under the assumption of 4.2, if J is an ideal on κ extending I^* the following conditions are equivalent

(a) for some $h \in \prod \bar\lambda$, for every $\varepsilon < \theta$ we have $\{i \in A : h(i) < \varepsilon\} \in J$

(b) $(\prod \bar\lambda, <_{J+(\kappa\setminus A)})$ is θ^+-directed.

Proof. <u>$(a) \Rightarrow (b)$</u>

Let $f_\zeta \in \prod \bar\lambda$ for $\zeta < \theta$, we define $f^* \in \prod \bar\lambda$ by

$$f^*(i) = \sup\{f_\zeta(i) + 1 : \zeta < h(i)\}.$$

Now $f^*(i) < \lambda_i$ as $h(i) < \lambda_i = \mathrm{cf}(\lambda_i)$ and $f_\zeta \restriction A <_J f^* \restriction A$ as $\{i \in A : h(i) < \zeta\} \in J$.

<u>$(b) \Rightarrow (a)$</u>

Let f_ζ be the following function with domain κ:

$$f_\zeta(i) = \begin{cases} \zeta & \text{if } \zeta < \lambda_i \\ 0 & \text{if } \zeta \geq \lambda_i \end{cases}$$

As $\liminf_{I^*} \geq \theta$, clearly $\varepsilon < \zeta \Rightarrow f_\varepsilon <_{I^*} f_\zeta$ and of course $f_\zeta \in \prod \bar\lambda$. By our assumption (b) there is $h \in \prod \bar\lambda$ such that $\zeta < \theta \Rightarrow f_\zeta \restriction A < h \restriction A \mod J$. Clearly h is as required. $\blacksquare_{4.4}$

References

[CK] C. C. Chang and H. J. Keisler, *Model Theory*, North Holland Publishing Company (1973).

[GH] F. Galvin and A. Hajnal, *Inequalities for cardinal power*, Annals of Math., **10** (1975) 491–498.

[Kn] A. Kanamori, *Weakly normal filters and irregular ultra-filter*, Trans of A.M.S., **220** (1976) 393–396.

[Ko] S. Koppelberg, *Cardinalities of ultraproducts of finite sets*, The Journal of Symbolic Logic, **45** (1980) 574–584.

[Kt] J. Ketonen, *Some combinatorial properties of ultra-filters,* Fund Math. VII (1980) 225–235.

[M] J. D. Monk, *Cardinal Function on Boolean Algebras,* Lectures in Mathematics, ETH Zürich, Bikhäuser, Verlag, Baser, Boston, Berlin, 1990.

[Sh-b] S. Shelah, *Proper forcing* Springer Lecture Notes, 940 (1982) 496+xxix.

[Sh-g] S. Shelah, *Cardinal Arithmetic,* volume 29 of Oxford Logic Guides, General Editors: Dov M. Gabbai, Angus Macintyre and Dana Scott, Oxford University Press, 1994.

[Sh7] S. Shelah, *On the cardinality of ultraproduct of finite sets*, Journal of Symbolic Logic, **35** (1970) 83–84.

[Sh345] S. Shelah, *Products of regular cardinals and cardinal invariants of Boolean Algebra,* Israel Journal of Mathematics, **70** (1990) 129–187.

[Sh400a] S. Shelah, *Cardinal arithmetic for skeptics*, American Mathematical Society. Bulletin. New Series, **26** (1992) 197–210.

[Sh410] S. Shelah, *More on cardinal arithmetic*, Archive of Math Logic, **32** (1993) 399–428.

[Sh420] S. Shelah, *Advances in Cardinal Arithmetic,* Proceedings of the Conference in Banff, Alberta, April 1991, ed. N. W. Sauer et al., Finite and Infinite Combinatorics, Kluwer Academic Publ., (1993) 355–383.

[Sh430] S. Shelah, *Further cardinal arithmetic*, Israel Journal of Mathematics, **accepted**.

[MgSh433] M. Magidor and S. Shelah, λ_i *inaccessible* $> \kappa$, $\prod_{i<\kappa} \lambda_i / D$ *of order type* μ^+, preprint.

[Sh589] S. Shelah, *PCF theory: Application*, in preparation.

Set Theory: Geometric and Real

Péter Komjáth*

Dept. of Computer Science, Eötvös University, Budapest, Muzeum krt. 6-8, 1088, Hungary

1. Introduction

In this Chapter we consider P. Erdős' research on what can be called as the borderlines of set theory with some of the more classical branches of mathematics as geometry and real analysis. His continuing interest in these topics arose from the world view in which the prime examples of sets are those which are subsets of some Euclidean spaces. 'Abstract' sets of arbitrary cardinality are of course equally existing. Paul only uses his favorite game for inventing new problems; having solved some problems find new ones by adding and/or deleting some structure on the sets currently under research. A good example is the one about set mappings. This topic was initiated by P. Turán who asked if a finite set $f(x)$ is associated to every point x of the real line does necessarily exist an infinite free set, i.e., when $x \notin f(y)$ holds for any two distinct elements. Clearly the underlying structure has nothing to do with the question and eventually a nice theory emerged which culminated in the results of Erdős, G. Fodor, and A. Hajnal. But Paul and his collaborators kept returning to the original setup when the condition is e.g. changed to: let $f(x)$ be nowhere dense, etc. Several nice and hard results have recently been proved. (See Section 8. in this Chapter.)

But this is not Paul's secret yet. He has the gift to ask the right questions and–always–ask the right person.

2. Sierpiński's decomposition

One direction of theorems and problems originated from W. Sierpiński's famous decomposition theorem ([34]). This paradoxical statement says that CH (the continuum hypothesis) holds if and only if $\mathbf{R}^2$ can be decomposed as $A \cup B$ where A, B have countable intersection with every horizontal, resp. vertical line. (See [14] for a lucid description of Paul Erdős' astonishment when first hearing on this result.) This shows that the generalization of Fubini's theorem $\int\int f(x,y)\,dx\,dy = \int\int f(x,y)\,dy\,dx$ may be false if we only assume the existence of the two double integrals and not the measurability of f (take the characteristic function of A). H. Friedman showed that it can also be true (if one adds $\aleph_2$ random reals)(see [21]). Later, C. Freiling ([20]) proved that

* Research supported by Hungarian National OTKA Fund No. 2117.

if there is a counter-example to this statement then necessarily there is one of the Sierpiński type, i.e., then $\mathbf{R}^2 = A \cup B$ such that the intersection of A (or B) with every horizontal (vertical) line is a measure zero set. See [36] which is a very thorough survey paper on this result and several variants. [11] generalizes this as follows. Suppose the lines of $\mathbf{R}^2$ are decomposed into two classes (and CH holds, of course). Then a decomposition as above exists with $A \cap \ell$ countable if ℓ is in the first class, and $B \cap \ell$ is countable if ℓ is in the second class.

3. Linear equations

An old result of R. Rado states that every $\mathbf{R}^n$ in fact every vector space over $\mathbf{Q}$, the rationals, is the union of countably many pieces none containing a 3-element arithmetic progression. This observation initiated several different research areas. In [18] Erdős and Kakutani showed that every vector space of cardinal at most $\aleph_1$ is the union of countably many bases and this is sharp. Various other results hold if one want to cover a linear space with countably many sets omitting the solutions of some linear equations (and not all as in the above case)[27].

One way of proving the above mentioned Erdős-Kakutani result is via a combinatorial lemma of Erdős and Hajnal which also gives that if the continuum hypothesis fails and $\mathbf{R}$ is decomposed into countably many pieces then one of them contains a, $a + x$, b, $b + x$ for some a, b, x, $x \neq 0$. Erdős then asked if the following complementary result holds. If CH holds then $\mathbf{R}^n$ can be written as the union of countably many sets none containing the same distance twice. For $n = 1$ this is a very special case of the quoted Erdős-Kakutani result. For $n = 2$ this was established by R. O. Davies [7] and finally for the general case by K. Kunen [29].

4. Euclidean Ramsey theory

Perhaps the central topic in our discussion is the one that may be called Euclidean Ramsey theory. This deals with decompositions of $\mathbf{R}^n$ where a configuration similar to a given one is excluded. Though in the finite version, i.e., when the number of classes is finite, there are several positive results (given by P.Erdős, R.L.Graham, P.Frankl, V.Rödl, and others), the infinite case abounds in counterexamples.

To begin with, if the excluded configuration is infinite then there is always a decomposition of $\mathbf{R}^n$ into two classes with no copy of the configuration in one class. This nice observation which uses an earlier result of P.Erdős and A.Hajnal is well hidden in [15]. As for the case of of finite configurations J.Ceder already in 1969 proved that $\mathbf{R}^2$ can be colored with countably many colors without a monocolored equilateral triangle ([4]). The tricky proof uses that $\mathbf{C}$, the complex plane is a vector space over the countable field $\mathbf{Q}(\sqrt{-3})$. In this vector space any node of an equilateral triangle is a convex linear combination of the other two. The proof now goes like R.Rado's quoted proof. It "defines" the coloring of a vector from its coordinates over a Hamel-basis.

P.Erdős asked if Ceder's result holds for $\mathbf{R}^n$, as well. Obviously, the above proof won't generalize. The more classic inductive argument, in the spirit of Davies' results, makes possible to show that a similar statement holds for regular tetrahedra in $\mathbf{R}^3$, see [23]. For regular simplices in $\mathbf{R}^n$ this was shown by J.Schmerl [31]. Then he extended his result to equilateral triangles ([32]). Finally, adding some really nice and deep arguments from higher dimensional calculus and logic J.Schmerl was able to show that P.Erdős's conjecture is true, there is a countable decomposition of every $\mathbf{R}^n$ with no isosceles triangles in one part ([33]).

One might think that excluding right triangles is similar to the case of the isosceles triangles. In fact, it was shown in [19] that CH is actually *equivalent* to the existence of a countable decomposition of $\mathbf{R}^2$ not containing the three nodes of of a right triangle in one piece. Interestingly enough the 'only if' part is an easy corollary of an above quoted Erdős-Hajnal result, the 'if' part is rather complicated.

Another variant, also raised by P.Erdős, if one can exclude those triangles with rational, nonzero areas. Clearly, we cannot multicolor triangles with zero area, and the statement is obvious for one given value of area. In unpublished work K. Kunen showed that if CH holds then there is a decomposition of $\mathbf{R}^2$ omitting rational areas as asked. In [28] we extended this to a broad class of configurations namely we showed that there is, under CH, a countable decomposition of $\mathbf{R}^n$ with no different points $\overline{a_1}$,... $\overline{a_t}$, satisfying any $p(\overline{a_1},\ldots,\overline{a_t}) = 0$ where p is a polynomial with rational coefficients such that $p(\overline{a},\ldots,\overline{a}) \neq 0$ holds for every $\overline{a} \in \mathbf{R}^n$. (Again, this is trivial for *one* such polynomial.) This has recently been extended to ZFC proofs in [33] by J. Schmerl.

Paul Erdős asked if the following asymmetric variant of the above quoted result of [15] is true. $\mathbf{R}$ can be decomposed into two pieces, A and B such that A omits 3-element while B omits infinite arithmetic progressions. This was proved (for arbitrary vector spaces) by J. E. Baumgartner [3] using ideas somewhat similar to the proofs of the results of Rado, Ceder, and Schmerl.

5. The Hilbert space

The situation changes radically if we replace $\mathbf{R}^n$ with the Hilbert space ℓ^∞ of infinite real vectors $(x_0, x_1, \ldots)$ with $\sum x_i^2$ finite. An observation due to Erdős, Kakutani, Oxtoby, L.M.Kelly, Nordhaus, and possibly many others is that in this case there are continuum many points with pairwise rational distance.

As it is easy not to find a proof I sketch one. Work in the Hilbert space where an orthonormal basis is $\{b_s\}$ where s can be any finite 0-1 sequence. To every infinite 0-1 sequence z associate $a(z) = \sum \lambda_n b_{z|n}$ where $z|n$ denotes the string of the first n terms of z and $\lambda_n = \sqrt{3} \cdot 2^{-(n+1)}$. If $z \neq z'$ first differ at the $(n+1)$-st position then $(a(z) - a(z'))^2 = \sum\{\lambda_i^2 : i > n\} = 4^{-n}$ so the distance between $a(z)$ and $a(z')$ is 2^{-n}. It is easy to see that every triangle in this construction is isosceles. I don't know if there is a similar (or any) construction of continuum many points such that all three-element subsets form a triangle with nonzero rational area. (See some problems and several remarks on this topic in [12].)

6. Games

Yet another variant of these problems is to decompose the spaces by games. Let V be the κ-dimensional space over the rationals and let two players alternatively choose previously unchosen vectors in a transfinite series of steps. At limit stages the first player chooses and he has to cover an arithmetic progression as long as possible. Erdős and Hajnal proved that the second player can always prevent her opponent from selecting an infinite arithmetic progression. If the second player is allowed to select $\aleph_0$ vectors she can prevent the first player from selecting a 3-element (4-element) AP if $\kappa \leq \aleph_1$ ($\kappa \leq \aleph_2$) but the first player can always cover a 3-element arithmetic progression if $\kappa \geq \aleph_2$. This was generalized by Fred Galvin and Zsiga Nagy who showed that the first player can cover an $(n+2)$-element AP for $\kappa = \aleph_n$ but his opponent can prevent him from selecting longer.

7. Large subsets

Another type of problems asked repeatedly by P. Erdős is the following. Let X be a subset of the n-dimensional Euclidean space with some infinite (usually uncountable) cardinal κ. Given a property P can one always find a subset $Y \subseteq X$ of cardinal κ with property P. Already in [9] (see also [13], [5]) Erdős proves this if P is the property that no distance occurs twice. Notice that this is a property depending on finite sets. In [9] it is also proved that if κ is regular then there is a Y such that all $r \leq n$-dimensional simplices formed by $r+1$ elements of Y are of different area (but this fails for e.g., $\aleph_\omega$).

8. Set mappings

As I already sketched in the introduction the theory of set mappings originated from a question of P. Turán concerning a problem in approximation theory. As usual, a set mapping is any function f on a set X such that $f(x)$ is a subset of X excluding x. A subset $Y \subseteq X$ is *free* if $x \notin f(y)$ for $x, y \in Y$. Turán inquired about free sets if $X = \mathbf{R}$ and $f(x)$ is always finite. If one allows $f(x)$ to be countable then there may be no 2-element free subset if $|X| \leq \aleph_1$, a moment's reflection shows that this is just a reformulation of Sierpiński's statement about the paradoxical decomposition of $\mathbf{R}^2$. P. Erdős proved ([10]) that if $f(x)$ is nowhere dense for $x \in \mathbf{R}$ then there is always an infinite free set. Bagemihl extended this to finding a dense free set [2]. Hechler proved [22] that under CH there is a set mapping f for which an uncountable free set does not exist and $f(x)$ is an ω-sequence converging to x! Uri Abraham observed that this may not occur if MA_{ω_1} holds and the existence of an uncountable free set if $f(x)$ is assumed to be nowhere dense is consistent with and independent of MA_{ω_1} (see [1]).

C. Freiling investigated what happens if a set mapping is defined on a set of size $\aleph_2$. One of his results can be formulated as follows. If $A \subseteq \mathbf{R}$ has cardinal $\leq \aleph_2$ and f is a set mapping on A with no free subsets of size 2 such that $f(x)$ is always of first category then either A is the union of $\aleph_1$ first category sets or every subset of A of cardinal $< |A|$ is of first category. It is easy to see that in either case there is a well order $\prec$ of A in which every proper initial segment is first category. In recent work the author (of this paper) proved that it is consistent that there is a set $A \subseteq \mathbf{R}$ with $|A| = \aleph_3$ with a set mapping as above but with no well order with first category segments [28].

In [16] Erdős and Hajnal proved that if f is a set mapping on $\mathbf{R}$ such that $f(x)$ is always of measure zero and not everywhere dense then there is a free pair but not necessarily a free triplet. They also proved that if $f(x)$ is bounded and of outer measure at most 1 then for every $k < \omega$ there is a free set of size k and asked if an infinite free set exists. This was finally proved in [30] .

References

1. U. Abraham: Free sets for nowhere-dense set mappings, *Israel Journal of Mathematics,* **39** (1981), 167–176.
2. F. Bagemihl: The existence of an everywhere dense independent set , *Michigan Mathematical Journal* **20** (1973), 1–2.
3. J. E.Baumgartner: Partitioning vector spaces,*J. Comb. Theory, Ser. A.* **18** (1975), 231–233.
4. J. Ceder: Finite subsets and countable decompositions of Euclidean spaces, *Rev. Roumaine Math. Pures Appl.* **14** (1969), 1247–1251.
5. Z. Daróczy: Jelentés az 1965. évi Schweitzer Miklós matematikai emlékversenyről, *Matematikai Lapok* **17** (1966), 344–366. (in Hungarian)
6. R. O. Davies: Covering the plane with denumerably many curves, *Bull. London Math. Soc.* **38** (1963), 343–348.

7. R. O. Davies: Partitioning the plane into denumerably many sets without repeated distances, *Proc. Camb. Phil. Soc.* **72** (1972), 179–183.
8. R. O. Davies: Covering the space with denumerably many curves, *Bull. London Math. Soc.* **6** (1974), 189–190.
9. P. Erdős: Some remarks on set theory, *Proc. Amer. Math. Soc.* **1** (1950), 127–141.
10. P. Erdős: Some remarks on set theory, III, *Michigan Mathematical Journal* **2** (1953–54), 51–57.
11. P. Erdős: Some remarks on set theory IV, *Mich. Math.* **2** (1953–54), 169–173.
12. P.Erdős: Hilbert térben levő ponthalmazok néhány geometriai és halmazelméleti tulajdonságáról, *Matematikai Lapok* **19** (1968), 255–258 (Geometrical and set theoretical properties of subsets of Hilbert spaces, in Hungarian).
13. P.Erdős: Set-theoretic, measure-theoretic, combinatorial, and number-theoretic problems concerning point sets in Euclidean space, *Real Anal. Exchange* **4** (1978-79), 113- 138.
14. P.Erdős: My Scottish Book "Problems", in: *The Scottish Book, Mathematics from the Scottish Café* (ed. R.D.Mauldin), Birkhauser, 1981, 35–43.
15. P. Erdős, R. L. Graham, P. Montgomery, B. L. Rothschild, J. Spencer, E. G. Straus: Euclidean Ramsey theorems II, in: *Infinite and Finite Sets,* Keszthely (Hungary), 1973, Coll. Math. Soc. J. Bolyai *10*, 529–557.
16. P. Erdős, A. Hajnal: Some remarks on set theory, VIII, *Michigan Mathematical Journal* **7** (1960), 187–191.
17. P. Erdős, S. Jackson, R. D. Mauldin: On partitions of lines and space, *Fund. Math.* **145** (1994), 101–119.
18. P. Erdős, S. Kakutani: On non-denumerable graphs, *Bull. Amer. Math. Soc.*, **49** (1943), 457–461.
19. P. Erdős, P. Komjáth: Countable decompositions of $\mathbf{R}^2$ and $\mathbf{R}^3$, *Discrete and Computational Geometry* **5** (1990), 325–331.
20. C. Freiling: Axioms of symmetry: Throwing darts at the real number line, *Journal of Symbolic Logic* **51** (1986), 190–200.
21. H. Friedman: A consistent Fubini-Tonelli theorem for nonmeasurable functions, *Illinois Journal of Mathematics,* **24** (1980), 390–395.
22. S. H. Hechler: Directed graphs over topological spaces: some set theoretical aspects, *Israel Journal of Mathematics* **11** (1972), 231–248.
23. P. Komjáth: Tetrahedron free decomposition of $\mathbf{R}^3$, *Bull. London Math. Soc.* **23** (1991), 116–120.
24. P. Komjáth: The master coloring, *Comptes Rendus Mathématiques de l'Academie des sciences, la Société royale du Canada,* **14**(1992), 181–182.
25. P. Komjáth: Set theoretic constructions in Euclidean spaces, in: *New trends in Discrete and Computational Geometry*, (J. Pach, ed.), Springer, *Algorithms and Combinatorics,* **10** (1993), 303–325.
26. P. Komjáth: A decomposition theorem for $\mathbf{R}^n$, *Proc. Amer. Math. Soc.* **120** (1994), 921–927.
27. P. Komjáth: Partitions of vector spaces, *Periodica Math. Hung.* **28** (1994), 187–193.
28. P. Komjáth: A note on set mappings with meager images, *Studia Math. Hung.*, accepted.
29. K. Kunen: Partitioning Euclidean space, *Math. Proc. Camb. Phil. Soc.* **102**, (1987), 379–383.
30. L. Newelski, J. Pawlikowski, W. Seredyński: Infinite free set for small measure set mappings, *Proc. Amer. Math. Soc.* **100** (1987), 335–339.
31. J. H. Schmerl: Partitioning Euclidean space, *Discrete and Computational Geometry*

32. J. H. Schmerl: Triangle-free partitions of Euclidean space, to appear.
33. J. H. Schmerl: Countable partitions of Euclidean space, to appear.
34. W. Sierpiński: Sur un théorème équivalent à l'hypothèse du continu, *Bull. Int. Acad. Sci. Cracovie A* (1919), 1–3.
35. W. Sierpiński: Hypothèse du continu, Warsaw, 1934.
36. J. C. Simms: Sierpiński's theorem, *Simon Stevin* **65** (1991), 69–163.

Paul Erdős: The Master of Collaboration

Jerrold W. Grossman

Department of Mathematical Sciences, Oakland University, Rochester, MI 48309-4401

Over a span of more than 60 years, Paul Erdős has taken the art of collaborative research in mathematics to heights never before achieved. In this brief look at his collaborative efforts, we will explore the breadth of Paul's interests, the company he has kept, and the influence of his collaboration in the mathematical community. Rather than focusing on the mathematical content of his work or the man himself, we will see what conclusions can be drawn by looking mainly at publication lists. Thus our approach will be mostly bibliographical, rather than either mathematical or biographical. The data come mainly from the bibliography in this present volume and records kept by *Mathematical Reviews* (*MR*) [13]. Additional useful sources of information include *The Hypertext Bibliography Project* (a database of articles in theoretical computer science) [11], *Zentralblatt* [16], the *Jahrbuch* [10], various necrological articles too numerous to list, and personal communications. Previous articles on these topics can be found in [3,4,7,8,14].

Paul has certainly become a legend, whose fame (as well as his genius and eccentricity) has spread beyond the circles of research mathematicians. We find a popular videotape about him [2], articles in general circulation magazines [9,15] (as well as in mathematical publications—see [1] for a wonderful example), and graffiti on the Internet (e.g., his quotation that a mathematician is a device for turning coffee into theorems, on a World Wide Web page designed as a sample of the use of the html language [12]). But even within the academic (and corporate research) community, his style and output have created a lot of folklore.

The reader is probably familiar with the concept of *Erdős number*, defined inductively as follows. Paul has Erdős number 0. For each $n \geq 0$, a person not yet assigned an Erdős number who has written a joint mathematical paper with a person having Erdős number n has Erdős number $n + 1$. Anyone who is not assigned an Erdős number by this process is said to have Erdős number ∞. Thus a person's Erdős number is just the distance from that person to Paul Erdős in the *collaboration graph* C (in which two authors are joined by an edge if they have published joint research—of course one need not restrict the field to mathematics). For example, Albert Einstein has Erdős number 2, since he did not collaborate with Paul Erdős, but he did publish joint research with Ernst Straus, who was one of Paul's major collaborators. Purists can argue over how to count papers with more than two authors, but here we will adopt the liberal attitude that each of the $\binom{k}{2}$ pairs of authors in a k-author paper are adjacent in C.

H. Abbott 74; J. Aczél 65; R. Aharoni 88; M. Aigner 87; M. Ajtai 81; L. Alaoglu 44:2; Y. Alavi 87:7; K. Alladi 77:5; N. Alon 85:2; J. Anderson 85; B. Andrásfai 74; N. Ankeny 54; N. Anning 45; B. Aronov 94; J. Ash 74; D. Avis 88:2; L. Babai 80:3; G. Babu 75:3; F. Bagemihl 53:4; A. Balog 90; L. Bankoff 73; A. Barak 84; P. Bateman 50:5; J. Baumgartner 79:2; L. Beasley 87; M. Behzad 91; S. Benkoski 74; M. Berger 88; E. Bertram 94; A. Bialostocki 95; A. Blass 92; M. Bleicher 75:3; A. Boals 87:2; R. Boas, Jr. 48; D. Boes 81; B. Bollobás 62:14; D. Bonar 77; J. Bondy 73; R. Bonnet 74; I. Borosh 78; J. Bosák 71; J. Bovey 75; J. Brenner 87; J. Brillhart 83; B. Brindza 91; T. Brown 85:2; W. Brown 73:6; R. Buck 48; S. Burr 75:24; D. Busolini 77; L. Caccetta 85:4; P. Cameron 90; E. Canfield 83; F. Carroll 77; F. Cater 78; M. Cates 76; P. Catlin 80; J. Chalk 59; G. Chartrand 87:5; C. Chen 76; G. Chen 93:2; H. Chen 92; R. Chen 88; P. Chinn 81; S. Choi 74:3; S. Chowla 50:3; C. Chui 78; F. Chung 79:13; K.-L. Chung 47:4; V. Chvátal 72:3; B. Clark 85; L. Clark 93; J. Clarkson 43; J. Clunie 67; S. Cohen 76; C. Colbourn 85; J. Conway 79; A. Copeland 46; H. Croft 79; E. Csáki 85; I. Csiszár 65; J. Czipszer 62; D. Darling 56:2; R. Darst 81; H. Davenport 36:7; R. Davies 75; D. Daykin 76:2; N. de Bruijn 48:6; D. de Caen 86; J.-M. De Koninck 81; P. Deheuvels 87; J. Dénes 69; J. Deshouillers 76; M. Deza 75:4; H. Diamond 78:3; G. Dirac 63; J. Dixmier 87; Y. Dowker 59; D. Drake 90; U. Dudley 83; R. Duke 77:7; A. Dvoretzky 50:8; E. Ecklund, Jr. 74:2; A. Edrei 85; R. Eggleton 72:7; M. El-Zahar 85; G. Elekes 81; P. Elliott 69:3; R. Entringer 72:3; M. Erné 86; A. Evans 89; V. Faber 81:3; S. Fajtlowicz 77:4; R. Faudree 76:41; L. Fejes Tóth 56; E. Feldheim 36; W. Feller 49; A. Felzenbaum 88; L. Few 64; P. Fishburn 91:4; G. Fodor 56:3; D. Fon Der Flaass 92; J. Fowler 85:2; A. Fraenkel 88; P. Frankl 78:6; A. Freedman 90; G. Freiman 90; G. Freud 74; R. Freud 83:4; E. Fried 72; H. Fried 47; W. Fuchs 56; Z. Füredi 82:9; I. Gál 48:3; J. Galambos 74; T. Gallai 36:8; F. Galvin 75:6; L. Gerencsér 70; J. Gillis 37; L. Gillman 55; J. Gimbel 90:4; A. Ginzburg 61:2; W. Goddard 94; C. Godsil 88; M. Goldberg 88; M. Golomb 55; A. Goodman 66; B. Gordon 64; R. Gould 87:3; R. Graham 72:26; A. Granville 90; D. Grieser 87; K. Grill 87; P. Gruber 89; B. Grünbaum 73:2; G. Grünwald 37:3; D. Gunderson 95; H. Gupta 76; M. Guy 79; R. Guy 70:4; A. Gyárfás 88:9; E. Győri 92:2; K. Győry 80; A. Hajnal 58:54; G. Halász 91; R. Hall 73:13; J. Hammer 89; H. Hanani 62:2; D. Hanson 74; F. Harary 65:2; G. Hardy 78; W. Hare 87; C. Harner 73; S. Hartman 67; E. Härtter 66; E. Harzheim 86; J. Hattingh 93; S. Hechler 75:2; S. Hedetniemi 87; Z. Hedrlín 72; N. Hegyvári 83:3; H. Heilbronn 64; P. Hell 89; R. Hemminger 84; M. Henning 93; M. Henriksen 55; F. Herzog 50:6; M. Herzog 70; D. Hickerson 89:2; D. Higgs 84; A. Hildebrand 87; N. Hindman 76:2; A. Hobbs 77:3; A. Hoffman 80; V. Hoggatt, Jr. 78; D. Holton 84; R. Holzman 94; P. Horák 94; M. Horváth 91; E. Howorka 80; D. Hsu 92; G. Hunt 53; J. Hwang 78:2; K.-H. Indlekofer 87; A. Ingham 64; A. Ivić 80:7; E. Jabotinsky 58:3; S. Jackson 94; M. Jacobson 87:2; V. Jarník 37; G. Jin 93; F. Jones 82:2; I. Joó 87:9; M. Joó 92; M. Kac 39:5; P. Kainen 78; S. Kakutani 43:7; I. Kaplansky 46:2; J. Karamata 56; I. Kátai 69:7; Zhao Ke 38:4; P. Kelly 63:2; J. Kennedy 87; H. Kestelman 63; S. Khare 76; H. Kierstead 91; P. Kiss 88:2; M. Klamkin 73:2; M. Klawe 80; D. Kleitman 68:6; M. Klugerman 94; J. Knappenberger 93; J. Koksma 49:2; P. Komjáth 86:2; J. Komlós 70:3; V. Komornik 90:2; I. Koren 88; A. Kostochka 92; T. Kővári 56; S. Krantz 88; D. Kratsch 91; A. Kroó 89; E. Kubicka 90; G. Kubicki 91; K. Kunen 81; C. Lacampagne 85:3; J. Larson 79:5; R. Laskar 83:3; H. Lefmann 95; J. Lehel 91; J. Lehner 41; B. Lengyel 38; W. LeVeque 63; M. Lewin 95; D. Lick 91; N. Linial 87:2; J. Liu 92; M. Loebl 95; G. Lorentz 58; L. Lovász 73:6; J. Loxton 79; T. Luczak 92:3; A. Macintyre 54; M. Magidor 76; K. Mahler 38:2; H. Maier 87; E. Makai 91:2; M. Makkai 66; P. Malde 87:2; S. Marcus 57; A. Máté 70:3; R. Mauldin 76:4; T. Maxsein 90; M. Mays 88; J. McCanna 92; R. McEliece 71; B. McKay 84; A. Meir 71:5; G. Mills 81; E. Milner 66:15; H. Minc 73; L. Mirsky 52; P. Montgomery 73:3; J. Moon 64:4; S. Moran 87:2; P. Morton 83; L. Moser 64:3; R. Mullin 83; M. Ram Murty 87; V. Kumar Murty 87; M. Nathanson 75:18; J. Nešetřil 83:3; E. Netanyahu 73; J. Neveu 63; D. Newman 74:5; P. Ney 74; J.-L. Nicolas 75:16; I. Niven 45:6; D. Norton 88; P. O'Neil 73; R. Obláth 37; A. Odlyzko 79:3; O. Oellermann 87:4; A. Offord 56; E. Ordman 85:6; J. Oxtoby 55; J. Pach 80:16; P. Pálfy 87:2; Z. Palka 83:2; E. Palmer 83; Z. Papp 80; T. Parsons 88; C. Payan 82; D. Penney 78; K. Phelps 85; A. Pinkus 85; G. Piranian 47:14; R. Pollack 89; H. Pollard 49; C. Pomerance 78:20; L. Pósa 62:4; K. Prachar 61; D. Preiss 76; P. Pudaite 87; N. Pullman 85:2; G. Purdy 71:8; L. Pyber 88:2; R. Rado 50:18; K. Ramachandra 75:2; S. Rao 92; A. Reddy 73:11; T. Reid 95; A. Rényi 50:32; P. Révész 75:9; B. Reznick 87; I. Richards 77; L. Richmond 76:5; G. Rieger 75; H. Riesel 88; R. Robinson 83; V. Rödl 83:8; C. Rogers 53:7; A. Rosa 71; P. Rosenbloom 46; B. Rothschild 73:8; C. Rousseau 76:30; L. Rubel 64:2; A. Rubin 80; M. Rudin 75; I. Ruzsa 73:5; C. Ryavec 72; H. Sachs 63; B. Saffari 79; E. Saias 95; M. Saks 86; P. Salamon 88; A. Sárközy 66:57; N. Sauer 75; J. Schaer 75; R. Schelp 76:34; P. Scherk 58; A. Schinzel 60:2; E. Schmutz 91; J. Schönheim 70:3; L. Schulman 94; S. Schuster 81:2; A. Schwenk 87:4; S. Segal 78; W. Seidel 53; J. Selfridge 67:13; S. Selkow 80; Á. Seress 86; J. Shallit 91; H. N. Shapiro 51:3; H. S. Shapiro 65; A. Sharma 65; S. Shelah 72:3; T. Sheng 75; A. Shields 65; O. Shisha 85; T. Shorey 76; G. Silberman 88; R. Silverman 83; A. Simmons 73; M. Simonovits 66:21; N. Singhi 77:2; J. Šíráň 94; T. Sirao 59; D. Skilton 85:3; B. Smith 81; P. Smith 90; A. Soifer 93:2; V. Sós 66:33; E. Specker 61; J. Spencer 71:22; C. Spiro-Silverman 90; W. Staton 91:2; A. Stein 83; S. Stein 63:2; C. Stewart 76:4; D. Stinson 83; A. Stone 45:3; H. Straight 90; E. Straus 53:20; M. Subbarao 72:2; H. Sun 93; J. Surányi 59:3; H. Swart 93; J. Szabados 78:4; M. Szalay 77:6; M. Szegedy 87; G. Szegő 42:2; L. Székely 87:2; G. Szekeres 34:5; E. Szemerédi 66:29; P. Szüsz 58:2; A. Tarski 43:2; A. Taylor 92; H. Taylor 71:3; S. Taylor 57:7; G. Tenenbaum 81:6; P. Tetali 90; C. Thomassen 89; R. Tijdeman 88; W. Trotter 78:2; P. Turán 34:29; J. Turk 84; W. Tutte 65; Z. Tuza 89:8; S. Ulam 68:3; K. Urbanik 58; J. Vaaler 87:2; E. van Kampen 40; J. van Lint 66:2; A. Varma 86:2; R. Vaughan 74:2; E. Vázsonyi 36:2; P. Vértesi 80:7; K. Vesztergombi 88:2; K. Vijayan 85:3; I. Vincze 58:2; B. Volkmann 66; S. Wagstaff, Jr. 80; G. Weiss 83; D. West 85:2; A. Williamson 87; R. M. Wilson 85; R. J. Wilson 77; P. Winkler 89; A. Wintner 39:3; N. Wormald 86; F. Yao 79; A. Zaks 90; S. Zaks 88; Y. Zalcstein 87:3; S. Zaremba 73; Z. Zhang 93:2; A. Ziv 61

Fig. 1. Coauthors of Paul Erdős.

A common variant is to give a person who has written $p > 0$ papers with Paul the Erdős number $1/p$. András Sárközy, with $\frac{1}{57}$, and András Hajnal, with $\frac{1}{54}$, seem to have the smallest positive Erdős numbers under this definition, followed in order by Faudree, Schelp, Sós, Rényi, Rousseau, Szemerédi, Turán, Graham, Burr, Spencer, Simonovits, Pomerance, Straus, Nathanson,

Rado, Nicolas, Pach, Milner, Bollobás, Piranian, F. Chung, Hall, Selfridge, and Reddy, who all have a value under 0.1.

The 458 people currently known to have Erdős number 1 are listed in Figure 1. The entry "$x\ y{:}\,n$" indicates that x first published with Erdős in year 19y and to date has n joint papers with him (with or without other coauthors); n is omitted when it equals 1. About 60% of the coauthors have just one joint paper. The mean number of papers per coauthor is 3.2, with a standard deviation of 6.1.

The author maintains electronic lists of coauthors of Paul Erdős and coauthors of these coauthors (i.e., all people with Erdős number not exceeding 2) and updates these lists annually. He intends to make them available indefinitely via anonymous ftp [6] and on the World Wide Web [5]. (Difficulties in author identification, among other problems, surely make the data less than 100% accurate, but we believe that the number of errors is not large.)

As was pointed out in [7], the average number of authors per research article in the mathematical sciences has increased dramatically over Paul Erdős's lifetime. (One can speculate whether his existence is part of the reason for this.) Specifically, the fraction of all authored items reviewed in *MR* having two or more authors has increased, as a function of time. While over 90% of all papers 56 years ago (when *MR* began) were the work of just one mathematician, today scarcely more than half of them are solo works. In the same period, the fraction of two-author papers has risen from under 10% to about one third. Also, in 1940 there were virtually no papers with three authors, let alone four or more; now about 10% of all papers in the mathematical sciences have three or more authors, including about 2% with four or more.

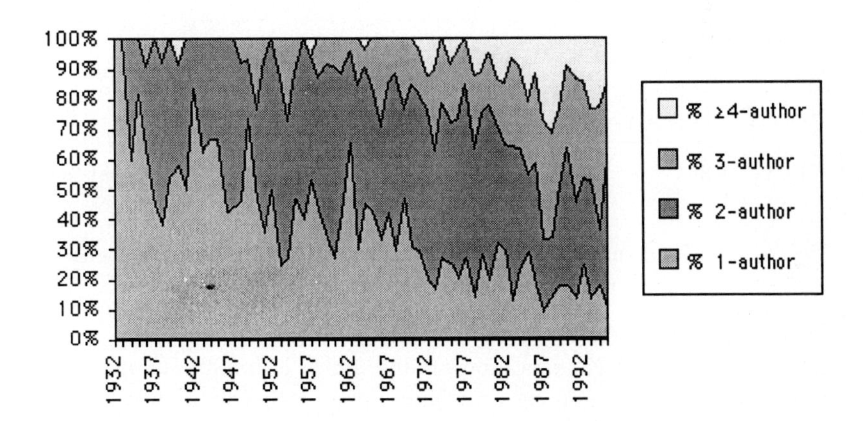

Fig. 2. The number of coauthors on Paul Erdős's papers over the years.

The same trend can be seen in Paul's work, but with an even greater amount of collaboration. The graph in Figure 2 shows the fraction of 1-, 2-, 3-, and ≥4-author items in Paul's publication list, year by year. (Almost all of these are research papers. The rest are books, articles about people, or other writings.) For reference, Figure 3 shows the absolute sizes we are talking about—the number of publications year by year. (The last two figures—for 1994 and 1995—are probably too low, due to incomplete publication data.) Cumulatively (Figure 4), fewer than one third of Paul's 1400 works are solo ventures. In fact, the mean number of authors (including Erdős) is almost exactly two.

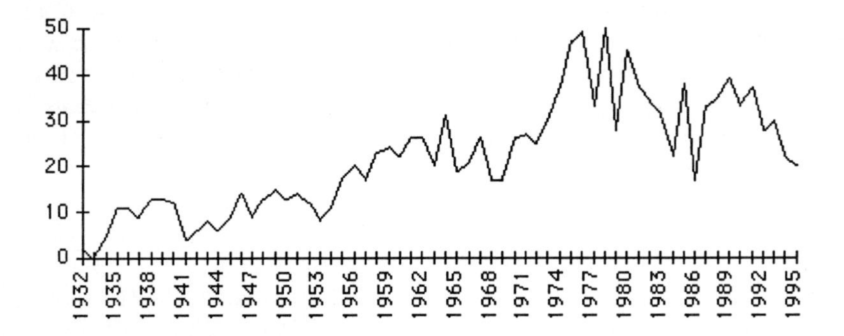

Fig. 3. The number of papers by Paul Erdős over the years.

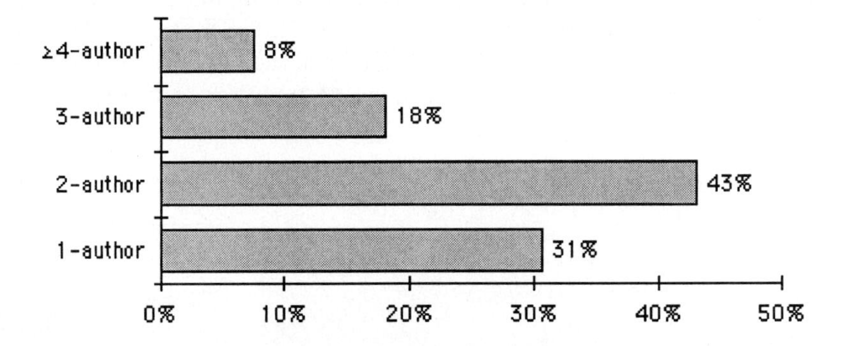

Fig. 4. The fraction of Paul Erdős's papers with different numbers of authors.

Paul's current mode of operation (dating from his departure from a permanent position at Notre Dame around 1954) is unique among mathematicians. Rather than staying at a home institution (research institute or academic department), he is constantly on the move, visiting mathematicians at conferences and research centers around the world. He often spends some summer months in Budapest, where he is a member of the Hungarian Academy of Sciences and where he can work with several of his most prolific collaborators. Some of his favorite haunts are Memphis, Tennessee, the New York City area, and other places too numerous to list. He is a permanent fixture at the annual Southeastern Combinatorics Conference (in Boca Raton, Florida, or Baton Rouge, Louisiana) and other regular meetings in his various fields. For example, in the few months around the time this article is being written, Paul reports having been (or planning to go) at least to Atlanta, Memphis, three cities in Texas, New Jersey, New Haven, Baton Rouge, Colorado, France, Germany, Kalamazoo, and Pennsylvania, in that order.

As he has met and worked with ever-increasing numbers of people on his travels, it is not surprising that Paul has added new coauthors every year since 1936. (Only two—George Szekeres and Pál Turán—published with him before that, in 1934.) Figure 5 shows how the cumulative number of coauthors has increased, while Figure 6 shows the discrete time derivative of this function. Paul usually leaves the actual writing up of the papers to his collaborators—partly, he says, because he does not type.

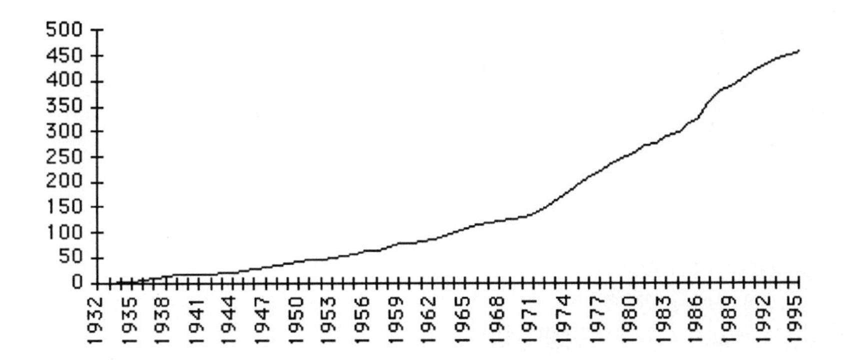

Fig. 5. The cumulative number of Paul Erdős's coauthors as a function of time.

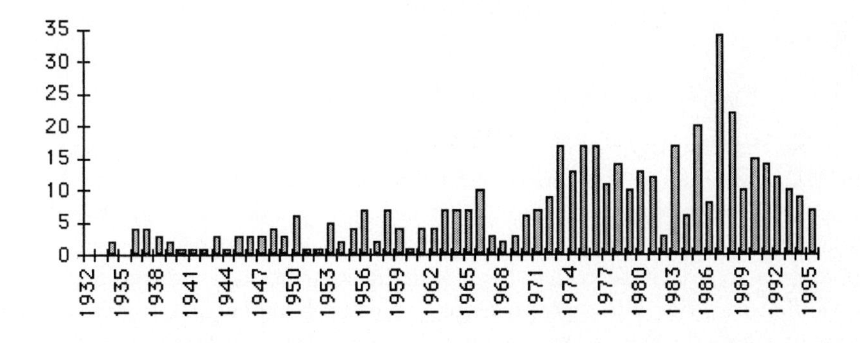

Fig. 6. The number of new coauthors added each year.

As the present collection shows, Paul Erdős's papers span many branches of mathematics and exploit relationships among them. (One fine example of the latter is his application of probability to combinatorics.) *Mathematical Reviews* currently has about 60 broad subject classifications, ranging from "Mathematical Logic and Foundations" through "Information and Communication, Circuits" (plus a section on history and biography), one of which it assigns to every item it records as its primary subject area. This list has varied slightly over time, with some categories being added or discarded as *MR* tries to keep up with current trends. Paul's works, although often spanning two or more subject areas and therefore difficult to pinpoint into one category, have been given primary classification in about 40% of these subject areas or their equivalent predecessors. They include not only the two main areas of number theory and combinatorics, and substantial work in approximation theory, geometry, set theory, and probability theory, but also papers in mathematical logic and foundations, lattices and ordered algebraic structures, linear algebra, group theory, topological groups, polynomials, measure and integration, functions of a complex variable, finite differences and functional equations, sequences and series, Fourier analysis, functional analysis, general and algebraic topology, statistics, numerical analysis, computer science, and information theory.

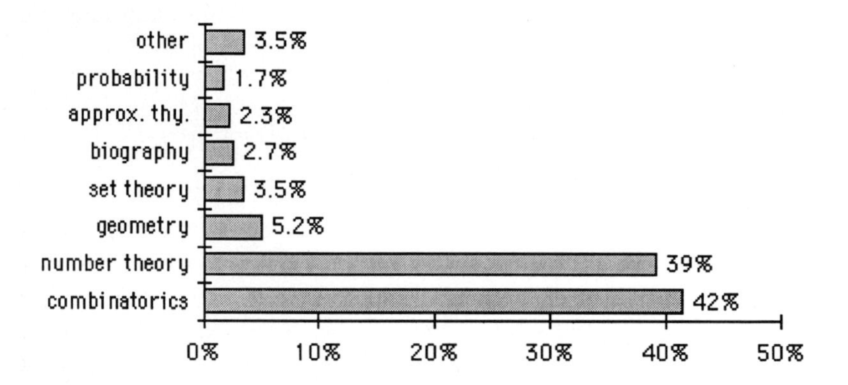

Fig. 7. Paul Erdős's papers since 1979 by broad category.

The chart in Figure 7 shows the fractions of Paul's publications reviewed in *MR* since 1980 in the various categories. Such a tabulation is easy to do, since the *MR* review number includes the category code as the first two digits following the colon. Figure 8 is a less accurate pie chart covering all years. It can be seen that there has been a slight trend toward increased work in combinatorics (including graph theory, of course) in the past 15 years, with a comparable decline in the output in number theory. Indeed, nearly all his early papers were in number theory (61 of the first 64, by one count, covering the period 1932–1939).

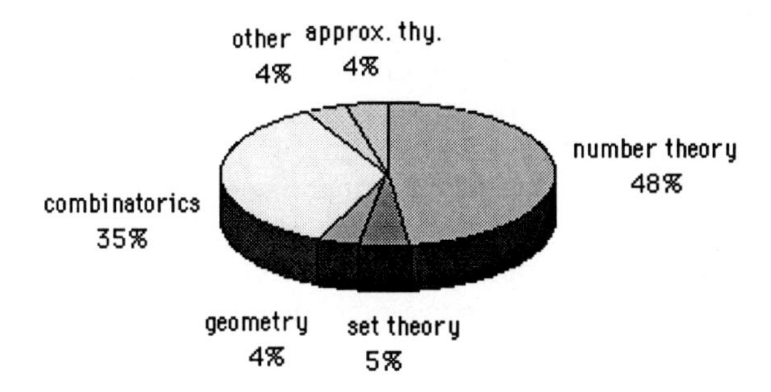

Fig. 8. Paul Erdős's papers by broad category (approximate).

Since Erdős's coauthors work in such varied fields, one would expect the set of people with Erdős number 2, 3, or a little higher to range over essentially all of mathematics. Indeed, this is the case. All Fields Medalists over the past three cycles (1986–1994) are in the Erdős component of the collaboration graph, with Erdős number at most 9. This group includes people working in theoretical physics; for instance, there is the path Edward Witten—Chiara Nappi—Robert Israel—Robert Phelps—David Preiss—Paul Erdős. Thus one can conjecture that many (if not most) physicists are also in the Erdős component, as are, therefore, many (or most) scientists in general. The large number of applications of graph theory and statistics to the social sciences might also lead one to suspect that many researchers in other academic areas are included as well.

It is interesting to explore the publication lists (or at least the coauthor lists) of Erdős's coauthors, to see how much collaboration goes on after Paul has left town. Let E_1 be the subgraph of C induced by people with Erdős number 1. According to data collected through 1995, E_1 contains 458 vertices and 1218 edges; thus an average Erdős coauthor collaborated with over 5 other Erdős coauthors. (The median, as opposed to the mean, of this statistic is only 3, however. Its standard deviation is about 6, and it takes values over 30 in four cases—Ron Graham, Frank Harary, Vojtěch Rödl, and Joel Spencer). There are only 40 isolated vertices in E_1 (less than 9%), and three components with two vertices each. The remaining 412 vertices in E_1 induce a connected subgraph. Paul's style seems to rub off.

Looking at it more broadly, we find that people with Erdős number 1 have a mean of 20 other collaborators (median 15, standard deviation 22), and only six of them collaborated with no one except Erdős. Five of them have over 100 coauthors (Frank Harary, Saharon Shelah, Ron Graham, Noga Alon, and Dan Kleitman).

Another 4546 people have felt Paul's influence second-hand—by doing joint research with one of the honored 458. Three quarters of the people with Erdős number 2 have only one coauthor with Erdős number 1 (i.e., each such person has a unique path of length 2 to Erdős in C). However, their mean number of Erdős number 1 coauthors is 1.5 (standard deviation 1.2), and the count ranges as high as 13 (for Dwight Duffus and Linda Lesniak).

The accompanying bibliography lists about 1400 papers, and it is probably incomplete, especially with regard to recent works. Most of the papers published since 1939 appear in the *Mathematical Reviews* database. Reviews of Paul Erdős's papers appear in every volume that *MR* has published, including a review of a joint paper with Tibor Gallai on page 1 of volume 1, written by George Pólya. It is interesting to note that the second most prolific writer in the *MR* database is Leonard Carlitz, with about 735 items. Carlitz has Erdős number 2 (via seven different coauthors) and has written the *MR* review of several of Paul's papers. In all, nearly 500 different people have reviewed Erdős's papers for *MR*. Paul writes for *MR* as well and to date has over 700 reviews to his credit.

Readers with additions or corrections to any of the information in this article, the accompanying bibliography, or the current coauthor lists are urged to communicate with the author (`grossman@oakland.edu`).

References

1. László Babai, In and out of Hungary: Paul Erdős, his friends and times, *Combinatorics, Paul Erdős is Eighty, Volume 2*, pp. 7–93, J. Bolyai Mathematical Society, Budapest, 1996.
2. George Paul Csicsery, *N is a number, a portrait of Paul Erdős*, 57-min. videotape, George Paul Csicsery, Oakland, CA, 1993.
3. Paul Erdős, On the fundamental problem of mathematics, *Amer. Math. Monthly* **79** (1972), 149–150.
4. Casper Goffman, And what is your Erdős number?, *Amer. Math. Monthly* **76** (1969), 791.
5. Jerrold W. Grossman, Lists of people with Erdős number at most 2, available on the World Wide Web: `http://www.acs.oakland.edu/~grossman/erdoshp.html`, Oakland University, Rochester, MI, 1996 (updated annually).
6. Jerrold W. Grossman, Lists of people with Erdős number at most 2, available via anonymous ftp to `vela.acs.oakland.edu` in directory `pub/math/erdos`, Oakland University, Rochester, MI, 1996 (updated annually).
7. Jerrold W. Grossman and Patrick D. F. Ion, On a portion of the well-known collaboration graph, *Proceedings of the Twenty-sixth Southeastern Conference on Combinatorics, Graph Theory and Computing (Boca Raton, FL, 1995), Congressus Numerantium*, to appear.
8. Frank Harary, The collaboration graph of mathematicians and a conjecture of Erdős, *Journal of Recreational Mathematics* **4** (1971), 212–213.
9. Paul Hoffman, The man who loves only numbers, *The Atlantic Monthly* **260** (Nov., 1987) no. 1, 60–74.
10. *Jahrbuch über die Fortschritte der Mathematik*, Berlin, 1868–1942.
11. David M. Jones, The hypertext bibliography project, World Wide Web: `http://theory.lcs.mit.edu/%7Edmjones/hbp/`, 1996.

12. Otmar Lendl, Otmar's list of HTML tags, World Wide Web:
`http://wwwcip.informatik.uni-erlangen.de`[*continued on next line*]
`/CIP/Manuals/www/html/taglist.html`, 1996.
13. *Mathematical Reviews*, American Mathematical Society, Providence, RI, 1940–.
14. Tom Odda [=Ronald L. Graham], On properties of a well-known graph or what is your Ramsey number?, *Topics in Graph Theory (New York, 1977), Ann. New York Acad. Sci., Volume 328*, pp. 166–172, New York Acad. Sci., New York, 1979, **MR** 81d:05055.
15. John Tierney, Paul Erdős is in town, his brain is open, *Science* (Amer. Assoc. Adv. Science) **5** (Oct., 1984) no. 8, 40–47.
16. *Zentralblatt für Mathematik und Ihre Grenzgebiete*, Springer, Berlin-New York, 1931–.

List of Publications of Paul Erdős

This bibliography was prepared by Jerrold Grossman (Oakland University, Rochester, Michigan), updating previous bibliographies, most recently the list prepared by Dezső Miklós in *Combinatorics, Paul Erdős is Eighty, Volume 1, Bolyai Society Mathematical Studies*, pp. 471–527, János Bolyai Mathematical Society, Budapest, 1993. Thanks are due not only to Dezső Miklós (Mathematical Institute, Hungarian Academy of Sciences) and the János Bolyai Mathematical Society, but also to László Babai (University of Chicago), Ronald Graham (AT&T Research), Patrick Ion (*Mathematical Reviews*), and Jaroslav Nešetřil (Charles University, Prague) for additional advice and data, *Zentralblatt für Mathematik und Ihre Grenzgebiete* for providing the Zbl numbers and making other additions and corrections, Springer-Verlag for the preparation of this list in its final form, and Paul Erdős for several helpful conversations.

Generally, the bibliographic style of *Mathematical Reviews* has been followed, at least approximately. Coauthor names, if any, are given in parentheses near the end of each entry. The *Mathematical Reviews* review number (**MR**) is included with each item for which it exists; otherwise, the *Current Mathematical Publications* control number (**CMP**) is included when it exists. Similarly, the **Zbl** (or **MATH** for a few items) numbers are provided where possible. This list incorporates corrections of some errors and omissions that appeared in the 1993 list (for example, the omission of Turán as a coauthor of the last paper listed for 1935). In some cases items have been shifted to a different year to reflect more accurately the actual publication date. Items have necessarily been renumbered in order to maintain a year-by-year list, alphabetical by title. *Please send further additions or corrections to* `grossman@oakland.edu`.

1932.01 Beweis eines Satzes von Tschebyschef (in German), *Acta Litt. Sci. Szeged* **5** (1932), 194–198; **Zbl.** 4,101.

1932.01 Egy Kürschák-féle elemi számelméleti tétel általánosítása (Generalization of an elementary number-theoretic theorem of Kürschák, in Hungarian), *Mat. Fiz. Lapok* **39** (1932), 17–24; **Zbl.** 7,103.

1934.01 A theorem of Sylvester and Schur, *J. London Math. Soc.* **9** (1934), 282–288; **Zbl.** 10,103.

1934.02 Bizonyos számtani sorok törzsszámairól (On primes in some arithmetic progressions, in Hungarian), *Bölcsészdoktori értekezés*, Sárospatak, 1934.

1934.03 On a problem in the elementary theory of numbers, *Amer. Math. Monthly* **41** (1934), 608–611 (P. Turán); **Zbl.** 10,294.

1934.04 On the density of the abundant numbers, *J. London Math. Soc.* **9** (1934), 278–282; **Zbl.** 10,103.

1934.05 Über die Anzahl der Abelschen Gruppen gegebener Ordnung und über ein verwandtes zahlentheoretisches Problem (in German), *Acta Litt. Sci. Szeged* **7** (1934), 95–102 (G. Szekeres); **Zbl.** 10,294.

1935.01 A combinatorial problem in geometry, *Compositio Math.* **2** (1935), 463–470 (G. Szekeres); **Zbl.** 12,270.

1935.02 Ein zahlentheoretischer Satz (in German), *Mitt. Forsch.-Inst. Math. Mech. Univ. Tomsk* **1** (1935), 101–103 (P. Turán); **Zbl.** 12,12.

1935.03 Note on consecutive abundant numbers, *J. London Math. Soc.* **10** (1935), 128–131; **Zbl.** 12,11.

1935.04 Note on sequences of integers no one of which is divisible by any other, *J. London Math. Soc.* **10** (1935), 126–128; **Zbl.** 12,52.

1935.05 On primitive abundant numbers, *J. London Math. Soc.* **10** (1935), 49–58; **Zbl.** 10,391.

1935.06 On the density of some sequences of numbers, *J. London Math. Soc.* **10** (1935), 120–125; **Zbl.** 12,10.

1935.07 On the difference of consecutive primes, *Quart. J. Math., Oxford Ser.* **6** (1935), 124–128; **Zbl.** 12,11.

1935.08 On the normal number of prime factors of $p-1$ and some related problems concerning Euler's φ-function, *Quart. J. Math., Oxford Ser.* **6** (1935), 205–213; **Zbl.** 12,149.

1935.09 The representation of an integer as the sum of the square of a prime and of a square-free integer, *J. London Math. Soc.* **10** (1935), 243–245; **Zbl.** 13,104.

1935.10 Über die Primzahlen gewisser arithmetischer Reihen (in German), *Math. Z.* **39** (1935), 473–491; **Zbl.** 10,293.

1935.11 Über die Vereinfachung eines Landauschen Satzes (in German), *Mitt. Forsch.-Inst. Math. Mech. Univ. Tomsk* **1** (1935), 144–147 (P. Turán); **Zbl.** 13,6.

1936.01 A generalization of a theorem of Besicovitch, *J. London. Math. Soc.* **11** (1936), 92–98; **Zbl.** 14,11.

1936.02 Note on some additive properties of integers, *Publ. de Congrès International des Math.*, Oslo, 1936, 1–2.

1936.03 On a problem of Chowla and some related problems, *Proc. Cambridge Philos. Soc.* **32** (1936), 530–540; **Zbl.** 15,246.

1936.04 On sequences of positive integers, *Acta Arith.* **2** (1936), 147–151 (H. Davenport); **Zbl.** 15,100.

1936.05 On some sequences of integers, *J. London Math. Soc.* **11** (1936), 261–264 (P. Turán); **Zbl.** 15,152.

1936.06 On the arithmetical density of the sum of two sequences one of which forms a basis for the integers, *Acta Arith.* **1** (1936), 197–200; **Zbl.** 13,150.

1936.07 On the integers which are the totient of a product of three primes, *Quart. J. Math., Oxford Ser.* **7** (1936), 16–19; **Zbl.** 13,246.

1936.08 On the integers which are the totient of a product of two primes, *Quart. J. Math., Oxford Ser.* **7** (1936), 227–229; **Zbl.** 15,5.

1936.09 On the representation of an integer as the sum of k kth powers, *J. London Math. Soc.* **11** (1936), 133–136; **Zbl.** 13,390.

1936.10 Sur le mode de convergence pour l'interpolation de Lagrange (in French), *C. R. Acad. Sci. Paris* **203** (1936), 913–915 (E. Feldheim); **Zbl.** 15,252.

1936.11 Végtelen gráfok Euler vonalairól (On Euler lines of infinite graphs, in Hungarian), *Mat. Fiz. Lapok* **43** (1936), 129–140 (T. Grünwald [=T. Gallai]; E. Weiszfeld [=E. Vázsonyi]); **Zbl.** 15,178.

1937.01 Eine Bemerkung über lineare Kongruenzen (in German), *Acta Arith.* **2** (1937), 214–220 (V. Jarník); **Zbl.** 18,6.

1937.02 Note on the number of prime divisors of integers, *J. London Math. Soc.* **12** (1937), 308–314; **Zbl.** 17,246.

1937.03 Note on the transfinite diameter, *J. London Math. Soc.* **12** (1937), 185–192 (J. Gillis); **Zbl.** 17,115.

1937.04 On interpolation, I. Quadrature and mean convergence in the Lagrange interpolation, *Ann. of Math. (2)* **38** (1937), 142–155 (P. Turán); **Zbl.** 16,106.

1937.05 On the density of some sequences of numbers, II., *J. London Math. Soc.* **12** (1937), 7–11; **Zbl.** 16,12.

1937.06 On the easier Waring problem for powers of primes, I., *Proc. Cambridge Philos. Soc.* **33** (1937), 6–12; **Zbl.** 16,102.

1937.07 On the sum and difference of squares of primes, *J. London Math. Soc.* **12** (1937), 133–136; **Zbl.** 16,201.

1937.08 On the sum and difference of squares of primes, II., *J. London Math. Soc.* **12** (1937), 168–171; **Zbl.** 17,103.

1937.09 Über diophantische Gleichungen der Form $n! = x^p \pm y^p$ und $n! \pm m! = x^p$ (in German), *Acta Litt. Sci. Szeged* **8** (1937), 241–255 (R. Obláth); **Zbl.** 17,4.

1938.01 Note on an elementary problem of interpolation, *Bull. Amer. Math. Soc.* **44** (1938), 515–518 (G. Grünwald); **Zbl.** 19,111.

1938.02 Note on the Euclidean algorithm, *J. London Math. Soc.* **13** (1938), 3–8 (Chao Ko [=Zhao Ke]); **Zbl.** 18,106.

1938.03 On additive properties of squares of primes, I., *Nederl. Akad. Wetensch., Proc.* **41** (1938), 37–41; **Zbl.** 23,9.

1938.04 On definite quadratic forms which are not the sum of two definite or semi-definite forms, *Acta Arith.* **3** (1938), 102–122 (Chao Ko [=Zhao Ke]); **Zbl.** 19,151.

1938.05 On fundamental functions of Lagrangean interpolation, *Bull. Amer. Math. Soc.* **44** (1938), 828–834 (B. A. Lengyel); **Zbl.** 20,12.

1938.06 On interpolation, II. On the distribution of the fundamental points of Lagrange and Hermite interpolation, *Ann. of Math. (2)* **39** (1938), 703–724 (P. Turán); **Zbl.** 19,404.

1938.07 On sequences of integers no one of which divides the product of two others and on some related problems, *Mitt. Forsch.-Inst. Math. Mech. Univ. Tomsk* **2** (1938), 74–82; **Zbl.** 20,5.

1938.08 On the asymptotic density of the sum of two sequences one of which forms a basis for the integers, II., *Trav. Inst. Math. Tbilissi* **3** (1938), 217–223; **Zbl.** 19,104.

1938.09 On the density of some sequences of numbers, III., *J. London Math. Soc.* **13** (1938), 119–127; **Zbl.** 18,293.

1938.10 On the number of integers which can be represented by a binary form, *J. London Math. Soc.* **13** (1938), 134–139 (K. Mahler); **Zbl.** 18,344.

1938.11 Some results on definite quadratic forms, *J. London Math. Soc.* **13** (1938), 217–224 (Chao Ko [=Zhao Ke]); **Zbl.** 19,151.

1938.12 Über die arithmetischen Mittelwerte der Lagrangeschen Interpolationspolynome (in German), *Studia Math.* **7** (1938), 82–95 (G. Grünwald); **Zbl.** 18,118.

1938.13 Über die Reihe $\sum \frac{1}{p}$ (in German), *Mathematica, Zutphen B* **7** (1938), 1–2; **Zbl.** 18,343.

1938.14 Über einen Faber'schen Satz (in German), *Ann. of Math. (2)* **39** (1938), 257–261 (G. Grünwald); **Zbl.** 18,397.

1938.15 Über Euler-Linien unendlicher Graphen (in German), *J. Math. Phys. Mass. Inst. Tech.* **17** (1938), 59–75 (T. Grünwald [=T. Gallai]; E. Vázsonyi); **Zbl.** 19,236.

1939.01 Additive arithmetical functions and statistical independence, *Amer. J. Math.* **61** (1939), 713–721 (A. Wintner); **MR** 1,40c; **Zbl.** 22,9.

1939.02 An extremum-problem concerning trigonometric polynomials, *Acta Litt. Sci. Szeged* **9** (1939), 113–115; **Zbl.** 21,17.

1939.03 Note on products of consecutive integers, *J. London Math. Soc.* **14** (1939), 194–198; **MR** 1,4e; **Zbl.** 21,207.

1939.04 Note on the product of consecutive integers, II., *J. London Math. Soc.* **14** (1939), 245–249; **MR** 1,39d; **Zbl.** 26,388.

1939.05 On a family of symmetric Bernoulli convolutions, *Amer. J. Math.* **61** (1939), 974–976; **MR** 1,52a; **Zbl.** 22,354.

1939.06 On polynomials with only real roots, *Ann. of Math. (2)* **40** (1939), 537–548 (T. Grünwald =[T. Gallai]); **MR** 1,1g; **Zbl.** 21,395.

1939.07 On sums of positive integral kth powers, *Ann. of Math. (2)* **40** (1939), 533–536 (H. Davenport); **MR** 1,5d; **Zbl.** 21,207.

1939.08 On the easier Waring problem for powers of primes, II., *Proc. Cambridge Philos. Soc.* **35** (1939), 149–165; **Zbl.** 21,106.

1939.09 On the Gaussian law of errors in the theory of additive functions, *Proc. Nat. Acad. Sci. U. S. A.* **25** (1939), 206–207 (M. Kac); **Zbl.** 21,207.

1939.10 On the integers of the form x^k+y^k, *J. London Math. Soc.* **14** (1939), 250–254; **MR** 1,42b; **Zbl.** 26,297.

1939.11 On the smoothness of the asymptotic distribution of additive arithmetical functions, *Amer. J. Math.* **61** (1939), 722–725; **MR** 1,41a; **Zbl.** 22,10.

1939.12 Some arithmetical properties of the convergents of a continued fraction, *J. London Math. Soc.* **14** (1939), 12–18 (K. Mahler); **Zbl.** 20,294.

1940.01 Additive functions and almost periodicity (B^2), *Amer. J. Math.* **62** (1940), 635–645 (A. Wintner); **MR** 2,41f; **Zbl.** 24,16.

1940.02 Note on some elementary properties of polynomials, *Bull. Amer. Math. Soc.* **46** (1940), 954–958; **MR** 2,242b; **Zbl.** 24,306.

1940.03 On a conjecture of Steinhaus, *Univ. Nac. Tucumán. Revista A.* **1** (1940), 217–220; **MR** 2,360c; **Zbl.** 25,158.

1940.04 On extremal properties of the derivatives of polynomials, *Ann. of Math. (2)* **41** (1940), 310–313; **MR** 1,323g; **Zbl.** 24,4.

1940.05 On interpolation, III. Interpolatory theory of polynomials, *Ann. of Math. (2)* **41** (1940), 510–553 (P. Turán); **MR** 1,333e; **Zbl.** 24,391.

1940.06 On the distribution of normal point groups, *Proc. Nat. Acad. Sci. U. S. A.* **26** (1940), 294–297; **MR** 1,333f; **Zbl.** 63,Index.

1940.07 On the smoothness properties of a family of Bernoulli convolutions, *Amer. J. Math.* **62** (1940), 180–186; **MR** 1,139e; **Zbl.** 22,354.

1940.08 On the uniformly-dense distribution of certain sequences of points, *Ann. of Math. (2)* **41** (1940), 162–173 (P. Turán); **MR** 1,217c; **Zbl.** 23,22.

1940.09 Ramanujan sums and almost periodic functions, *Studia Math.* **9** (1940), 43–53 (M. Kac; E. R. van Kampen; A. Wintner); **MR** 3,69f; **Zbl.** 63,Index.

1940.10 The difference of consecutive primes, *Duke Math. J.* **6** (1940), 438–441; **MR** 1,292h; **Zbl.** 23,298.

1940.11 The dimension of the rational points in Hilbert space, *Ann. of Math. (2)* **41** (1940), 734–736; **MR** 2,178a; **Zbl.** 25,187.

1940.12 The Gaussian law of errors in the theory of additive number theoretic functions, *Amer. J. Math.* **62** (1940), 738–742 (M. Kac); **MR** 2,42c; **Zbl.** 24,102.

1941.01 On a problem of Sidon in additive number theory and on some related problems, *J. London Math. Soc.* **16** (1941), 212–215 (P. Turán); **MR** 3,270e; **Zbl.** 61,73.

1941.02 On divergence properties of the Lagrange interpolation parabolas, *Ann. of Math. (2)* **42** (1941), 309–315; **MR** 2,283d; **Zbl.** 24,307.

1941.03 On some asymptotic formulas in the theory of the "factorisatio numerorum", *Ann. of Math. (2)* **42** (1941), 989–993; **MR** 3,165b; **Zbl.** 61,79.

1941.04 The distribution of the number of summands in the partitions of a positive integer, *Duke Math. J.* **8** (1941), 335–345 (J. Lehner); **MR** 3,69a; **Zbl.** 25,107.

1942.01 On a problem of I. Schur, *Ann. of Math. (2)* **43** (1942), 451–470 (G. Szegő); **MR** 4,41d; **Zbl.** 60,55.

1942.02 On an elementary proof of some asymptotic formulas in the theory of partitions, *Ann. of Math. (2)* **43** (1942), 437–450; **MR** 4,36a; **Zbl.** 61,79.

1942.03 On the asymptotic density of the sum of two sequences, *Ann. of Math. (2)* **43** (1942), 65–68; **MR** 3,165c.

1942.04 On the law of the iterated logarithm, *Ann. of Math. (2)* **43** (1942), 419–436; **MR** 4,16j; **Zbl.** 63,Index.

1942.05 On the uniform distribution of the roots of certain polynomials, *Ann. of Math. (2)* **43** (1942), 59–64; **MR** 3,236a; **Zbl.** 60,55.

1942.06 Some set-theoretical properties of graphs, *Univ. Nac. Tucumán. Revista A.* **3** (1942), 363–367; **MR** 5,151d; **Zbl.** 63,Index.

1943.01 A note on Farey series, *Quart. J. Math., Oxford Ser.* **14** (1943), 82–85; **MR** 5,236b; **Zbl.** 61,128.

1943.02 Approximation by polynomials, *Duke Math. J.* **10** (1943), 5–11 (J. A. Clarkson); **MR** 4,196e; **Zbl.** 63,Index.

1943.03 Corrections to two of my papers, *Ann. of Math. (2)* **44** (1943), 647–651; **MR** 5,172c and 5,180d; **Zbl.** 61,79 and 63,Index.

1943.04 On families of mutually exclusive sets, *Ann. of Math. (2)* **44** (1943), 315–329 (A. Tarski); **Zbl.** 60,126.

1943.05 On non-denumerable graphs, *Bull. Amer. Math. Soc.* **49** (1943), 457–461 (S. Kakutani); **MR** 4,249f; **Zbl.** 63,Index.

1943.06 On some convergence properties in the interpolation polynomials, *Ann. of Math. (2)* **44** (1943), 330–337; **MR** 4,273e; **Zbl.** 63,Index.

1943.07 On the convergence of trigonometric series, *J. Math. Phys. Mass. Inst. Tech.* **22** (1943), 37–39; **MR** 4,271e; **Zbl.** 60,178.

1943.08 Some remarks on set theory, *Ann. of Math. (2)* **44** (1943), 643–646; **MR** 5,173c; **Zbl.** 60,131.

1944.01 A conjecture in elementary number theory, *Bull. Amer. Math. Soc.* **50** (1944), 881–882 (L. Alaoglu); **MR** 6,117b; **Zbl.** 61,78.

1944.02 Addendum. On a problem of Sidon in additive number theory and on some related problems [*J. London Math. Soc.* **16** (1941), 212–215], *J. London Math. Soc.* **19** (1944), 208; **MR** 7,242f; **Zbl.** 61,73.

1944.03 On highly composite and similar numbers, *Trans. Amer. Math. Soc.* **56** (1944), 448–469 (L. Alaoglu); **MR** 6,117c; **Zbl.** 61,79.

1944.04 On highly composite numbers, *J. London Math. Soc.* **19** (1944), 130–133; **MR** 7,145d; **Zbl.** 61,79.

1944.05 On the maximum of the fundamental functions of the ultraspherical polynomials, *Ann. of Math. (2)* **45** (1944), 335–339; **MR** 5,264e; **Zbl.** 63,Index.

1944.06 Some remarks on connected sets, *Bull. Amer. Math. Soc.* **50** (1944), 442–446; **MR** 6,43a; **Zbl.** 61,401.

1945.01 Integral distances, *Bull. Amer. Math. Soc.* **51** (1945), 598–600 (N. H. Anning); **MR** 7,164a; **Zbl.** 63,Index.

1945.02 Integral distances, *Bull. Amer. Math. Soc.* **51** (1945), 996; **MR** 7,164b; **Zbl.** 63,Index.

1945.03 Note on the converse of Fabry's gap theorem, *Trans. Amer. Math. Soc.* **57** (1945), 102–104; **MR** 6,148f; **Zbl.** 60,203.

1945.04 On a lemma of Littlewood and Offord, *Bull. Amer. Math. Soc.* **51** (1945), 898–902; **MR** 7,309j; **Zbl.** 63,Index.

1945.05 On certain variations of the harmonic series, *Bull. Amer. Math. Soc.* **51** (1945), 433–436 (I. Niven); **MR** 7,11i; **Zbl.** 61,129.

1945.06 On the least primitive root of a prime p, *Bull. Amer. Math. Soc.* **51** (1945), 131–132; **MR** 6,170b; **Zbl.** 61,66.

1945.07 Some remarks on almost periodic transformations, *Bull. Amer. Math. Soc.* **51** (1945), 126–130 (A. H. Stone); **MR** 6,165b; **Zbl.** 63,Index.

1945.08 Some remarks on Euler's φ-function and some related problems, *Bull. Amer. Math. Soc.* **51** (1945), 540–544; **MR** 7,49f; **Zbl.** 61,80.

1945.09 Some remarks on the measurability of certain sets, *Bull. Amer. Math. Soc.* **51** (1945), 728–731; **MR** 7,197f; **Zbl.** 63,Index.

1946.01 Note on normal numbers, *Bull. Amer. Math. Soc.* **52** (1946), 857–860 (A. H. Copeland); **MR** 8,194b; **Zbl.** 63,Index.

1946.02 On certain limit theorems of the theory of probability, *Bull. Amer. Math. Soc.* **52** (1946), 292–302 (M. Kac); **MR** 7,459b; **Zbl.** 63,Index.

1946.03 On sets of distances of n points, *Amer. Math. Monthly* **53** (1946), 248–250; **MR** 7,471c; **Zbl.** 60,348.

1946.04 On some asymptotic formulas in the theory of partitions, *Bull. Amer. Math. Soc.* **52** (1946), 185–188; **MR** 7,273i; **Zbl.** 61,79.

1946.05 On the coefficients of the cyclotomic polynomial, *Bull. Amer. Math. Soc.* **52** (1946), 179–184; **MR** 7,242e; **Zbl.** 61,18.

1946.06 On the distribution function of additive functions, *Ann. of Math. (2)* **47** (1946), 1–20; **MR** 7,416c; **Zbl.** 61,79.

1946.07 On the Hausdorff dimension of some sets in Euclidean space, *Bull. Amer. Math. Soc.* **52** (1946), 107–109; **MR** 7,377a; **Zbl.** 63,Index.

1946.08 On the structure of linear graphs, *Bull. Amer. Math. Soc.* **52** (1946), 1087–1091 (A. H. Stone); **MR** 8,333b; **Zbl.** 63,Index.

1946.09 Sequences of plus and minus, *Scripta Math.* **12** (1946), 73–75 (I. Kaplansky); **MR** 8,126i; **Zbl.** 60,29.

1946.10 Some properties of partial sums of the harmonic series, *Bull. Amer. Math. Soc.* **52** (1946), 248–251 (I. Niven); **MR** 7,413e; **Zbl.** 61,65.

1946.11 Some remarks about additive and multiplicative functions, *Bull. Amer. Math. Soc.* **52** (1946), 527–537; **MR** 7,507g; **Zbl.** 61,79.

1946.12 The asymptotic number of Latin rectangles, *Amer. J. Math.* **68** (1946), 230–236 (I. Kaplansky); **MR** 7,407b; **Zbl.** 60,28.

1946.13 The $\alpha + \beta$ hypothesis and related problems, *Amer. Math. Monthly* **53** (1946), 314–317 (I. Niven); **MR** 7,507f.

1946.14 Toeplitz methods which sum a given sequence, *Bull. Amer. Math. Soc.* **52** (1946), 463–464 (P. C. Rosenbloom); **MR** 8,146i; **Zbl.** 61,121.

1947.01 A note on transforms of unbounded sequences, *Bull. Amer. Math. Soc.* **53** (1947), 787–790 (G. Piranian); **MR** 9,234b; **Zbl.** 31,294.

1947.02 On the connection between gaps in power series and the roots of their partial sums, *Trans. Amer. Math. Soc.* **62** (1947), 53–61 (H. Fried); **MR** 9,84f; **Zbl.** 32,65.

1947.03 On the lower limit of sums of independent random variables, *Ann. of Math. (2)* **48** (1947), 1003–1013 (K.-L. Chung); **MR** 9,292f; **Zbl.** 29,152.

1947.04 On the number of positive sums of independent random variables, *Bull. Amer. Math. Soc.* **53** (1947), 1011–1020 (M. Kac); **MR** 9,292g; **Zbl.** 32,35.

1947.05 Over-convergence on the circle of convergence, *Duke Math. J.* **14** (1947), 647–658 (G. Piranian); **MR** 9,232e; **Zbl.** 30,152.

1947.06 Some asymptotic formulas for multiplicative functions, *Bull. Amer. Math. Soc.* **53** (1947), 536–544; **MR** 9,12d; **Zbl.** 37,311.

1947.07 Some remarks and corrections to one of my papers, *Bull. Amer. Math. Soc.* **53** (1947), 761–763; **MR** 9,12e.

1947.08 Some remarks on polynomials, *Bull. Amer. Math. Soc.* **53** (1947), 1169–1176; **MR** 9,281g; **Zbl.** 32,386.

1947.09 Some remarks on the theory of graphs, *Bull. Amer. Math. Soc.* **53** (1947), 292–294; **MR** 8,479d; **Zbl.** 32,192.

1948.01 On a combinatorial problem, *Nederl. Akad. Wetensch., Proc.* **51** (1948), 1277–1279 = *Indag. Math.* **10** (1948), 421–423 (N. G. de Bruijn); **MR** 10,424a; **Zbl.** 32,244.

1948.02 On a problem in the theory of uniform distribution, I., *Nederl. Akad. Wetensch., Proc.* **51** (1948), 1146–1154 = *Indag. Math.* **10** (1948), 370–478 (P. Turán); **MR** 10,372c; **Zbl.** 31,254.

1948.03 On a problem in the theory of uniform distribution, II., *Nederl. Akad. Wetensch., Proc.* **51** (1948), 1262–1269 = *Indag. Math.* **10** (1948), 406–413 (P. Turán); **MR** 10,372d; **Zbl.** 32,16.

1948.04 On arithmetical properties of Lambert series, *J. Indian Math. Soc. (N.S.)* **12** (1948), 63–66; **MR** 10,594c; **Zbl.** 32,17.

1948.05 On some new questions on the distribution on prime numbers, *Bull. Amer. Math. Soc.* **54** (1948), 371–378 (P. Turán); **MR** 9,498k; **Zbl.** 32,269.

1948.06 On the density of some sequences of integers, *Bull. Amer. Math. Soc.* **54** (1948), 685–692; **MR** 10,105b; **Zbl.** 32,13.

1948.07 On the difference of consecutive primes, *Bull. Amer. Math. Soc.* **54** (1948), 885–889; **MR** 10,235b; **Zbl.** 32,269.

1948.08 On the integers having exactly k prime factors, *Ann. of Math. (2)* **49** (1948), 53–66; **MR** 9,333b; **Zbl.** 30,296.

1948.09 On the representation of $1, 2, \ldots, N$ by differences, *Nederl. Akad. Wetensch., Proc.* **51** (1948), 1155–1158 = *Indag. Math.* **10** (1949), 379–382 (I. S. Gál); **MR** 11,14a; **Zbl.** 32,13.

1948.10 On the roots of a polynomial and its derivative, *Bull. Amer. Math. Soc.* **54** (1948), 184–190 (I. Niven); **MR** 9,582g; **Zbl.** 32,248.

1948.11 Some asymptotic formulas in number theory, *J. Indian Math. Soc. (N.S.)* **12** (1948), 75–78; **MR** 10,594d; **Zbl.** 41,368.

1948.12 Some remarks on Diophantine approximations, *J. Indian Math. Soc. (N.S.)* **12** (1948), 67–74; **MR** 10,513b; **Zbl.** 32,16.

1948.13 The set on which an entire function is small, *Amer. J. Math.* **70** (1948), 400–402 (R. P. Boas, Jr.; R. C. Buck); **MR** 9,577a; **Zbl.** 36,46.

1949.01 A property of power series with positive coefficients, *Bull. Amer. Math. Soc.* **55** (1949), 201–204 (W. Feller; H. Pollard); **MR** 10,367d; **Zbl.** 32,278.

1949.02 On a new method in elementary number theory which leads to an elementary proof of the prime number theorem, *Proc. Nat. Acad. Sci. U. S. A.* **35** (1949), 374–384; **MR** 10,595c; **Zbl.** 34,314.

1949.03 On a Tauberian theorem connected with the new proof of the prime number theorem, *J. Indian Math. Soc. (N.S.)* **13** (1949), 131–144; **MR** 11,420a; **Zbl.** 34,315.

1949.04 On a theorem of Hsu and Robbins, *Ann. Math. Statistics* **20** (1949), 286–291; **MR** 11,40f; **Zbl.** 33,290.

1949.05 On some applications of Brun's method, *Acta Univ. Szeged. Sect. Sci. Math.* **13** (1949), 57–63; **MR** 10,684c; **Zbl.** 34,24.

1949.06 On the coefficients of the cyclotomic polynomial, *Portugaliae Math.* **8** (1949), 63–71; **MR** 12,11f; **Zbl.** 38,10.

1949.07 On the converse of Fermat's theorem, *Amer. Math. Monthly* **56** (1949), 623–624; **MR** 11,331g; **Zbl.** 33,250.

1949.08 On the number of terms of the square of a polynomial, *Nieuw Arch. Wiskunde (2)* **23** (1949), 63–65; **MR** 10,354b; **Zbl.** 32,2.

1949.09 On the strong law of large numbers, *Trans. Amer. Math. Soc.* **67** (1949), 51–56; **MR** 11,375c; **Zbl.** 34,72.

1949.10 On the uniform distribution modulo 1 of lacunary sequences, *Nederl. Akad. Wetensch., Proc.* **52** (1949), 264–273 = *Indag. Math.* **11** (1949), 79–88 (J. F. Koksma); **MR** 11,14b; **Zbl.** 33,165.

1949.11 On the uniform distribution modulo 1 of sequences $(f(n, \theta))$, *Nederl. Akad. Wetensch., Proc.* **52** (1949), 851–854 = *Indag. Math.* **11** (1949), 299–302 (J. F. Koksma); **MR** 11,331f; **Zbl.** 35,321.

1949.12 Problems and results on the differences of consecutive primes, *Publ. Math. Debrecen* **1** (1949), 33–37; **MR** 11,84a; **Zbl.** 33,163.

1949.13 Sequences of points on a circle, *Nederl. Akad. Wetensch., Proc.* **52** (1949), 14–17 = *Indag. Math.* **11** (1949), 46–49 (N. G. de Bruijn); **MR** 11,423i; **Zbl.** 31,348.

1949.14 Supplementary note, *J. Indian Math. Soc. (N.S.)* **13** (1949), 145–147; **MR** 11,420b; **Zbl.** 34,315.

1950.01 A combinatorial theorem, *J. London Math. Soc.* **25** (1950), 249–255 (R. Rado); **MR** 12,322f; **Zbl.** 38,153.

1950.02 Az $\frac{1}{x_1} + \frac{1}{x_2} + \cdots + \frac{1}{x_n} = \frac{a}{b}$ egyenlet egész számú megoldásairól (On a Diophantine equation, in Hungarian), *Mat. Lapok* **1** (1950), 192–210; **MR** 13,280b.

1950.03 Convergence fields of row-finite and row-infinite Toeplitz transformations, *Proc. Amer. Math. Soc.* **1** (1950), 397–401 (G. Piranian); **MR** 12,92a; **Zbl.** 37,327.

1950.04 Double points of paths of Brownian motion in n-space, *Acta Sci. Math. Szeged* **12** (1950), Leopoldo Fejer et Frederico Riesz LXX annos natis dedicatus, Pars B, 75–81 (A. Dvoretzky; S. Kakutani); **MR** 11,671e; **Zbl.** 36,90.

1950.05 On a problem in elementary number theory, *Math. Student* **17** (1949), 32–33, 1950; **MR** 11,642d; **Zbl.** 36,23.

1950.06 On almost primes, *Amer. Math. Monthly* **57** (1950), 404–407; **MR** 12,80i; **Zbl.** 38,181.

1950.07 On integers of the form $2^k + p$ and some related problems, *Summa Brasil. Math.* **2** (1950), 113–123; **MR** 13,437i; **Zbl.** 41,368.

1950.08 On the distribution of roots of polynomials, *Ann. of Math. (2)* **51** (1950), 105–119 (P. Turán); **MR** 11,431b; **Zbl.** 36,15.

1950.09 Remark on my paper "On a theorem of Hsu and Robbins" [*Ann. Math. Statistics* **20** (1949), 286–291], *Ann. Math. Statistics* **21** (1950), 138; **MR** 11,375b; **Zbl.** 35,214.

1950.10 Remarks on the size of $L(1,\chi)$, *Publ. Math. Debrecen* **1** (1950), 165–182 (P. T. Bateman; S. Chowla); **MR** 12,244b; **Zbl.** 36,307.

1950.11 Schlicht gap series whose convergence on the unit circle is uniform but not absolute, *Proc. Internat. Congr. Math.* **1** (1950) (F. Herzog; G. Piranian).

1950.12 Some problems and results on consecutive primes, *Simon Stevin* **27** (1950), 115–125 (A. Rényi); **MR** 11,644d; **Zbl.** 38,182.

1950.13 Some remarks on set theory, *Proc. Amer. Math. Soc.* **1** (1950), 127–141; **MR** 12,14c; **Zbl.** 39,49.

1950.14 Some theorems and remarks on interpolation, *Acta Sci. Math. Szeged* **12** (1950), Leopoldo Fejer et Frederico Riesz LXX annos natis dedicatus, Pars A, 11–17; **MR** 12,164c; **Zbl.** 37,177.

1951.01 A colour problem for infinite graphs and a problem in the theory of relations, *Nederl. Akad. Wetensch. Proc. Ser. A.* **54** = *Indag. Math.* **13** (1951), 369–373 (N. G. de Bruijn); **MR** 13,763g; **Zbl.** 44,382.

1951.02 A theorem on the distribution of the values of L-functions, *J. Indian Math. Soc. (N.S.)* **15** (1951), 11–18 (S. Chowla); **MR** 13,439a; **Zbl.** 43,46.

1951.03 Geometrical extrema suggested by a lemma of Besicovitch, *Amer. Math. Monthly* **58** (1951), 306–314 (P. T. Bateman); **MR** 12,851a; **Zbl.** 43,162.

1951.04 On a conjecture of Klee, *Amer. Math. Monthly* **58** (1951), 98–101; **MR** 12,674h; **Zbl.** 42,275.

1951.05 On a diophantine equation, *J. London Math. Soc.* **26** (1951), 176–178; **MR** 12,804d; **Zbl.** 43,43.

1951.06 On a theorem of Rådström, *Proc. Amer. Math. Soc.* **2** (1951), 205–206; **MR** 12,815b; **Zbl.** 42,311.

1951.07 On sequences of positive integers, *J. Indian Math. Soc. (N.S.)* **15** (1951), 19–24 (H. Davenport); **MR** 13,326c; **Zbl.** 43,49.

1951.08 On some problems of Bellman and a theorem of Romanoff, *J. Chinese Math. Soc. (N.S.)* (1951), 409–421; **MR** 17,238e.

1951.09 On the changes of sign of a certain error function, *Canadian J. Math.* **3** (1951), 375–385 (H. N. Shapiro); **MR** 13,535i; **Zbl.** 44,39.

1951.10 Probability limit theorems assuming only the first moment, I., *Mem. Amer. Math. Soc.* (1951) no. 6, 19 pp. (K.-L. Chung); **MR** 12,722g; **Zbl.** 42,376.

1951.11 Schlicht Taylor series whose convergence on the unit circle is uniform but not absolute, *Pacific J. Math.* **1** (1951), 75–82 (F. Herzog; G. Piranian); **MR** 13,335d; **Zbl.** 43,80.

1951.12 Some linear and some quadratic recursion formulas, I., *Nederl. Akad. Wetensch. Proc. Ser. A.* **54** = *Indag. Math.* **13** (1951), 374–382 (N. G. de Bruijn); **MR** 13,836f; **Zbl.** 44,60.

1951.13 Some problems and results in elementary number theory, *Publ. Math. Debrecen* **2** (1951), 103–109; **MR** 13,627a; **Zbl.** 44,36.

1951.14 Some problems on random walk in space, *Proceedings of the Second Berkeley Symposium on Mathematical Statistics and Probability, 1950*, pp. 353–367, University of California Press, Berkeley and Los Angeles, 1951 (A. Dvoretzky); **MR** 13,852b; **Zbl.** 44,140.

1952.01 A theorem on the Riemann integral, *Nederl. Akad. Wetensch. Proc. Ser. A.* **55** = *Indag. Math.* **14** (1952), 142–144; **MR** 13,830e; **Zbl.** 47,62.

1952.02 Combinatorial theorems on classifications of subsets of a given set, *Proc. London Math. Soc. (3)* **2** (1952), 417–439 (R. Rado); **MR** 16,455d; **Zbl.** 48,282.

1952.03 Egy kongruenciarendszerekről szóló problémáról (On a problem concerning congruence-systems, in Hungarian), *Mat. Lapok* **3** (1952), 122–128; **MR** 17,14d.

1952.04 Note on normal decimals, *Canadian J. Math.* **4** (1952), 58–63 (H. Davenport); **MR** 13,825g; **Zbl.** 46,49.

1952.05 On a Tauberian theorem for Euler summability, *Acad. Serbe Sci. Publ. Inst. Math.* **4** (1952), 51–56; **MR** 14,265g; **Zbl.** 47,301.

1952.06 On the application of the Borel-Cantelli lemma, *Trans. Amer. Math. Soc.* **72** (1952), 179–186 (K.-L. Chung); **MR** 13,567b; **Zbl.** 46,352.

1952.07 On the greatest prime factor of $\prod_{k=1}^{x} f(k)$, *J. London Math. Soc.* **27** (1952), 379–384; **MR** 13,914a; **Zbl.** 46,41.

1952.08 On the sum $\sum_{k=1}^{x} d(f(k))$, *J. London Math. Soc.* **27** (1952), 7–15; **MR** 13,438f; **Zbl.** 46,41.

1952.09 On the uniform but not absolute convergence of power series with gaps, *Ann. Soc. Polon. Math.* **25** (1952), 162–168; **MR** 15,417a; **Zbl.** 48,310.

1952.10 Some linear and some quadratic recursion formulas, II., *Nederl. Akad. Wetensch. Proc. Ser. A.* **55** = *Indag. Math.* **14** (1952), 152–163 (N. G. de Bruijn); **MR** 13,836g; **Zbl.** 47,63.

1952.11 The distribution of quadratic and higher residues, *Publ. Math. Debrecen* **2** (1952), 252–265 (H. Davenport); **MR** 14,1063h; **Zbl.** 50,43.

1952.12 The distribution of values of the divisor function $d(n)$, *Proc. London Math. Soc. (3)* **2** (1952), 257–271 (L. Mirsky); **MR** 14,249e; **Zbl.** 47,46.

1953.01 A problem on ordered sets, *J. London Math. Soc.* **28** (1953), 426–438 (R. Rado); **MR** 15,410b; **Zbl.** 51,40.

1953.02 Arithmetical properties of polynomials, *J. London Math. Soc.* **28** (1953), 416–425; **MR** 15,104f; **Zbl.** 51,277.

1953.03 Changes of sign of sums of random variables, *Pacific. J. Math.* **3** (1953), 673–687 (G. A. Hunt); **MR** 15,444e; **Zbl.** 51,103.

1953.04 On a conjecture of Hammersley, *J. London Math. Soc.* **28** (1953), 232–236; **MR** 14,726f; **Zbl.** 50,270.

1953.05 On a recursion formula and on some Tauberian theorems, *J. Research Nat. Bur. Standards* **50** (1953), 161–164 (N. G. de Bruijn); **MR** 14,973e; **Zbl.** 53,369.

1953.06 On linear independence of sequences in a Banach space, *Pacific J. Math.* **3** (1953), 689–694 (E. G. Straus); **MR** 15,437d; **Zbl.** 53,80.

1953.07 Sur quelques propriétés frontières des fonctions holomorphes définies par certains produits dans le cercle-unité (in French), *Ann. Sci. Ecole Norm. Sup.(3)* **70** (1953), 135–147 (F. Bagemihl; W. Seidel); **MR** 15,412g; **Zbl.** 53,238.

1953.08 The covering of n-dimensional space by spheres, *J. London Math. Soc.* **28** (1953), 287–293 (C. A. Rogers); **MR** 14,1066b; **Zbl.** 50,389.

1954.01 Integral functions with gap power series, *Proc. Edinburgh Math. Soc. (2)* **10** (1954), 62–70 (A. J. Macintyre); **MR** 16,579a; **Zbl.** 58,63.

1954.02 Intersections of prescribed power, type, or measure, *Fund. Math.* **41** (1954), 57–67 (F. Bagemihl); **MR** 16,20f; **Zbl.** 56,50.

1954.03 Multiple points of paths of Brownian motion in the plane, *Bull. Res. Council Israel* **3** (1954), 364–371 (A. Dvoretzky; S. Kakutani); **MR** 16,725b.

1954.04 On a problem of Sidon in additive number theory, *Acta Sci. Math. Szeged* **15** (1954), 255–259; **MR** 16,336c; **Zbl.** 57,39.

1954.05 On Taylor series of functions regular in Gaier regions, *Arch. Math.* **5** (1954), 39–52 (F. Herzog; G. Piranian); **MR** 15,946b; **Zbl.** 55,68.

1954.06 Rearrangements of C_1-summable series, *Acta Math.* **92** (1954), 35–53 (F. Bagemihl); **MR** 16,583c; **Zbl.** 56,282.

1954.07 Sets of divergence of Taylor series and of trigonometric series, *Math. Scand.* **2** (1954), 262–266 (F. Herzog; G. Piranian); **MR** 16,691d; **Zbl.** 57,58.

1954.08 Some remarks on set theory, III., *Michigan Math. J.* **2** (1954), 51–57; **MR** 16,20e; **Zbl.** 56,51.

1954.09 Some results on additive number theory, *Proc. Amer. Math. Soc.* **5** (1954), 847–853; **MR** 16,336b; **Zbl.** 56,270.

1954.10 The insolubility of classes of diophantine equations, *Amer. J. Math.* **76** (1954), 488–496 (N. C. Ankeny); **MR** 15,934a; **Zbl.** 56,35.

1954.11 The number of multinomial coefficients, *Amer. Math. Monthly* **61** (1954), 37–39 (I. Niven); **MR** 15,387e.

1955.01 An isomorphism theorem for real-closed fields, *Ann. of Math. (2)* **61** (1955), 542–554 (L. Gillman; M. Henriksen); **MR** 16,993e; **Zbl.** 65,23.

1955.02 Functions which are symmetric about several points, *Nieuw Arch. Wisk. (3)* **3** (1955), 13–19 (M. Golomb); **MR** 16,931e; **Zbl.** 64,121.

1955.03 On amicable numbers, *Publ. Math. Debrecen* **4** (1955), 108–111; **MR** 16,998h; **Zbl.** 65,27.

1955.04 On consecutive integers, *Nieuw Arch. Wisk. (3)* **3** (1955), 124–128; **MR** 17,461f; **Zbl.** 65,276.

1955.05 On power series diverging everywhere on the circle of convergence, *Michigan Math. J.* **3** (1955), 31–35 (A. Dvoretzky); **MR** 17,138e; **Zbl.** 73,62.

1955.06 On the law of the iterated logarithm, I., *Nederl. Akad. Wetensch. Proc. Ser. A.* **58** = *Indag. Math.* **17** (1955), 65–76 (I. S. Gál); **MR** 16,1016g; **Zbl.** 68,54.

1955.07 On the law of the iterated logarithm, II., *Nederl. Akad. Wetensch. Proc. Ser. A.* **58** = *Indag. Math.* **17** (1955), 77–84 (I. S. Gál); **MR** 16,1016g; **Zbl.** 68,54.

1955.08 On the product of consecutive integers, III., *Nederl. Akad. Wetensch. Proc. Ser. A.* **58** = *Indag. Math.* **17** (1955), 85–90; **MR** 16,797d; **Zbl.** 68,37.

1955.09 On the role of the Lebesgue functions in the theory of the Lagrange interpolation, *Acta Math. Acad. Sci. Hungar.* **6** (1955), 47–66 (P. Turán); **MR** 17,148b; **Zbl.** 64,301.

1955.10 Partitions of the plane into sets having positive measure in every non-null measurable product set, *Trans. Amer. Math. Soc.* **79** (1955), 91–102 (J. C. Oxtoby); **MR** 17,352f; **Zbl.** 66,298.

1955.11 Polynomials whose zeros lie on the unit circle, *Duke Math. J.* **22** (1955), 347–351 (F. Herzog; G. Piranian); **MR** 16,1093c; **Zbl.** 68,58.

1955.12 Some problems on the distribution of prime numbers, *Teoria dei numeri, Math. Congr. Varenna, 1954*, 8 pp., 1955; **Zbl.** 67,275.

1955.13 Some remarks on number theory (in Hebrew), *Riveon Lematematika* **9** (1955), 45–48; **MR** 17,460d.

1955.14 Some remarks on set theory, IV., *Michigan Math. J.* **2** (1953–54), 169–173 (1955); **MR** 16,682a; **Zbl.** 58,45.

1955.15 Some theorems on graphs (in Hebrew), *Riveon Lematematika* **9** (1955), 13–17; **MR** 18,408c.

1955.16 The existence of a distribution function for an error term related to the Euler function, *Canad. J. Math.* **7** (1955), 63–76 (H. N. Shapiro); **MR** 16,448f; **Zbl.** 67,276.

1955.17 Über die Anzahl der Lösungen von $[p-1, q-1] \leq x$. (Aus einem Brief von P. Erdős an K. Prachar.) (in German), *Monatsh. Math.* **59** (1955), 318–319; **MR** 17,461g; **Zbl.** 67,23.

1956.01 A limit theorem for the maximum of normalized sums of independent random variables, *Duke Math. J.* **23** (1956), 143–156 (D. A. Darling); **MR** 17,635c; **Zbl.** 70,138.

1956.02 A partition calculus in set theory, *Bull. Amer. Math. Soc.* **62** (1956), 427–489 (R. Rado); **MR** 18,458a; **Zbl.** 71,51.

1956.03 Megjegyzések a Matematikai Lapok két feladatához (Remarks on two problems of the Matematikai Lapok, in Hungarian, Russian and English summaries), *Mat. Lapok* **7** (1956), 10–17; **MR** 20#4534; **Zbl.** 75,31.

1956.04 Megjegyzések Kőváry Tamás egy dolgozatához (Remarks on a paper of T. Kőváry, in Hungarian), *Mat. Lapok* **7** (1956), 214–217; **MR** 20#4645.

1956.05 Monotonicity of partition functions, *Mathematika* **3** (1956), 1–14 (P. T. Bateman); **MR** 18,195a; **Zbl.** 74,35.

1956.06 On a high-indices theorem in Borel summability, *Acta Math. Acad. Sci. Hungar.* **7** (1956), 265–281; **MR** 19,135g; **Zbl.** 74,46.

1956.07 On a problem of additive number theory, *J. London Math. Soc.* **31** (1956), 67–73 (W. H. J. Fuchs); **MR** 17,586d; **Zbl.** 70,41.

1956.08 On additive arithmetical functions and applications of probability to number theory, *Proceedings of the International Congress of Mathematicians, 1954, Amsterdam, vol. III*, pp. 13–19, Erven P. Noordhoff N. V., Groningen; North-Holland Publishing Co., Amsterdam, 1956; **MR** 19,393d; **Zbl.** 73,267.

1956.09 On perfect and multiply perfect numbers, *Ann. Mat. Pura Appl. (4)* **42** (1956), 253–258; **MR** 18,563b; **Zbl.** 72,275.

1956.10 On pseudoprimes and Carmichael numbers, *Publ. Math. Debrecen* **4** (1956), 201–206; **MR** 18,18e; **Zbl.** 74,271.

1956.11 On some combinatorial problems, *Publ. Math. Debrecen* **4** (1956), 398–405 (A. Rényi); **MR** 18,3c; **Zbl.** 70,11.

1956.12 On the maximum modulus of entire functions, *Acta Math. Acad. Sci. Hungar.* **7** (1956), 305–317 (T. Kővári); **MR** 18,884a; **Zbl.** 72,74.

1956.13 On the number of real roots of a random algebraic equation, *Proc. London Math. Soc. (3)* **6** (1956), 139–160 (A. C. Offord); **MR** 17,500f; **Zbl.** 70,17.

1956.14 On the number of zeros of successive derivatives of analytic functions, *Acta Math. Acad. Sci. Hungar.* **7** (1956), 125–144 (A. Rényi); **MR** 18,201b; **Zbl.** 70,296.

1956.15 Partitions into primes, *Publ. Math. Debrecen* **4** (1956), 198–200 (P. T. Bateman); **MR** 18,15c; **Zbl.** 73,31.

1956.16 Pontok elhelyezése egy tartományban (The distribution of points in a region, in Hungarian), *Magyar Tud. Akad. Mat. Fiz. Oszt. Közl.* **6** (1956), 185–190 (L. Fejes Tóth); **MR** 20#1953; **Zbl.** 75,177.

1956.17 Problems and results in additive number theory, *Colloque sur la Théorie des Nombres, Bruxelles, 1955*, pp. 127–137, George Thone, Liège; Masson and Cie, Paris, 1956; **MR** 18,18a; **Zbl.** 73,31.

1956.18 Some remarks on set theory, V., *Acta Sci. Math. Szeged* **17** (1956), 250–260 (G. Fodor); **MR** 18,711a; **Zbl.** 72,41.

1956.19 Sur la majorabilité C des suites de nombres réels (in French), *Acad. Serbe. Sci. Publ. Inst. Math.* **10** (1956), 37–52 (J. Karamata); **MR** 18,478e; **Zbl.** 75,47.

1956.20 Über arithmetische Eigenschaften der Substitutionswerte eines Polynoms für ganzzahlige Werte des Arguments (in German), *Revue Math. Pures et Appl.* **1** (1956), 189–194.

1957.01 A probabilistic approach to problems of diophantine approximation, *Illinois J. Math.* **1** (1957), 303–315 (A. Rényi); **MR** 19,636d; **Zbl.** 99,39.

1957.02 Einige Bemerkungen zur Arbeit von A. Stöhr: "Gelöste und ungelöste Fragen über Basen der natürlichen Zahlenreihe" (in German), *J. Reine Angew. Math.* **197** (1957), 216–219; **MR** 19,122b; **Zbl.** 77,263.

1957.03 Néhány geometriai problémáról (On some geometrical problems, in Hungarian), *Mat. Lapok* **8** (1957), 86–92; **MR** 20#6056; **Zbl.** 102,370.

1957.04 On a perfect set, *Colloq. Math.* **4** (1957), 195–196 (S. Kakutani); **MR** 19,734e; **Zbl.** 77,271.

1957.05 On the distribution function of additive arithmetical functions and on some related problems, *Rend. Sem. Mat. Fis. Milano* **27** (1957), 45–49; **MR** 20#7004; **Zbl.** 81,42.

1957.06 On the growth of the cyclotomic polynomial in the interval $(0, 1)$, *Proc. Glasgow Math. Assoc.* **3** (1957), 102–104; **MR** 19,1039d; **Zbl.** 81,17.

1957.07 On the irrationality of certain series, *Nederl. Akad. Wetensch. Proc. Ser. A.* **60** = *Indag. Math.* **19** (1957), 212–219; **MR** 19,252e; **Zbl.** 79,74.

1957.08 On the least primitive root of a prime, *Pacific J. Math.* **7** (1957), 861–865 (H. N. Shapiro); **MR** 20#3830; **Zbl.** 79,63.

1957.09 On the number of zeros of successive derivatives of entire functions of finite order, *Acta Math. Acad. Sci. Hungar.* **8** (1957), 223–225 (A. Rényi); **MR** 19,539d; **Zbl.** 78,263.

1957.10 On the set of points of convergence of a lacunary trigonometric series and the equidistribution properties of related sequences, *Proc. London Math. Soc.* **7** (1957), 598–615 (S. J. Taylor); **MR** 19,1050b; **Zbl.** 111,268.

1957.11 Remarks on a theorem of Ramsey, *Bull. Res. Council Israel, Sect. F* **7F** (1957/1958), 21–24; **MR** 21#3347; **Zbl.** 88,157.

1957.12 Some remarks on set theory, VI., *Acta Sci. Math. Szeged* **18** (1957), 243–260 (G. Fodor); **MR** 19,1152a; **Zbl.** 78,42.

1957.13 Some unsolved problems, *Michigan Math. J.* **4** (1957), 291–300; **MR** 20#5157; **Zbl.** 81,1.

1957.14 Sur la décomposition de l'espace Euclidien en ensembles homogènes (in French), *Acta Math. Acad. Sci. Hungar.* **8** (1957), 443–452 (S. Marcus); **MR** 20#1958; **Zbl.** 79,78.

1957.15 Triple points of Brownian paths in 3-space, *Proc. Cambridge Philos. Soc.* **53** (1957), 856–862 (A. Dvoretzky; S. Kakutani; S. J. Taylor); **MR** 20#1364; **Zbl.** 208,441.

1957.16 Über eine Art von Lakunarität (in German), *Colloq. Math.* **5** (1957), 6–7; **MR** 19,1160a; **Zbl.** 81,39.

1957.17 Über eine Fragestellung von Gaier und Meyer-König (in German), *Jber. Deutsch. Math. Verein.* **60** (1957), Abt. 1, 89–92; **MR** 19,1045b; **Zbl.** 78,261.

1958.01 Asymptotic formulas for some arithmetical functions, *Canad. Math. Bull.* **1** (1958), 149–153; **MR** 21#30; **Zbl.** 85,34.

1958.02 Concerning approximation with nodes, *Colloq. Math.* **6** (1958), 25–27; **MR** 21#2142; **Zbl.** 85,52.

1958.03 Előadókörúton Kanadában (in Hungarian), *Magyar Tudomány* **3**, 8–9 (1958), 335–341.

1958.04 Konvex, zárt síkgörbék megközelítéséről (Über die Annäherung geschlossener, konvexer Kurven = Approximation of convex, closed curves, in Hungarian), *Mat. Lapok* **9** (1958), 19–36 (I. Vincze); **MR** 20#6070; **Zbl.** 91,354.

1958.05 Metric properties of polynomials, *J. Analyse Math.* **6** (1958), 125–148 (F. Herzog; G. Piranian); **MR** 21#123; **Zbl.** 88,253.

1958.06 On an elementary problem in number theory, *Canad. Math. Bull.* **1** (1958), 5–8; **MR** 20#1654; **Zbl.** 83,37.

1958.07 On Engel's and Sylvester's series, *Ann. Univ. Sci. Budapest. Eötvös Sect. Math.* **1** (1958), 7–32, (A. Rényi; P. Szüsz); **MR** 21#1288; **Zbl.** 107,270.

1958.08 On sequences of integers generated by a sieving process, I., *Nederl. Akad. Wetensch. Proc. Ser. A.* **61** = *Indag. Math.* **20** (1958), 115–123 (E. Jabotinsky); **MR** 21#2628; **Zbl.** 80,263.

1958.09 On sequences of integers generated by a sieving process, II., *Nederl. Akad. Wetensch. Proc. Ser. A.* **61** = *Indag. Math.* **20** (1958), 124–128 (E. Jabotinsky); **MR** 21#2628; **Zbl.** 80,263.

1958.10 On sets which are measured by multiples of irrational numbers, *Bull. Acad. Polon. Sci. Sér. Sci. Math. Astr. Phys.* **6** (1958), 743–748 (K. Urbanik); **Zbl.** 84,45.

1958.11 On singular radii of power series, *Magyar Tud. Akad. Mat. Kutató Int. Közl.* **3** (1958), 159–169 (A. Rényi); **MR** 21#5011; **Zbl.** 89,49.

1958.12 On the structure of set-mappings, *Acta Math. Acad. Sci. Hungar.* **9** (1958), 111–131 (A. Hajnal); **MR** 20#1630; **Zbl.** 102,284.

1958.13 Points of multiplicity **c** of plane Brownian paths, *Bull. Res. Council Israel, Sect. F* **7F** (1958), 175–180 (A. Dvoretzky; S. Kakutani); **MR** 23#A3594.

1958.14 Problems and results on the theory of interpolation, I., *Acta Math. Acad. Sci. Hungar.* **9** (1958), 381–388; **MR** 21#423; **Zbl.** 83,290.

1958.15 Remarks on the theory of diophantine approximation, *Colloq. Math.* **6** (1958), 119–126 (P. Szüsz; P. Turán); **MR** 21#1290; **Zbl.** 87,43.

1958.16 Solution of two problems of Jankowska, *Bull. Acad. Polon. Sci. Sér. Sci. Math. Astr. Phys.* **6** (1958), 545–547; **MR** 20#7003; **Zbl.** 83,37.

1958.17 Some remarks on a paper of McCarthy, *Canad. Math. Bull.* **1** (1958), 71–75; **MR** 20#3093; **Zbl.** 81,270.

1958.18 Some remarks on Euler's φ function, *Acta Arith.* **4** (1958), 10–19; **MR** 22#1539; **Zbl.** 81,42.

1958.19 Sur certaines séries à valeur irrationnelle (in French), *Enseignement Math. (2)* **4** (1958), 93–100; **MR** 20#5187; **Zbl.** 80,33.

1958.20 The topologization of a sequence space by Toeplitz matrices, *Michigan Math. J.* **5**, 139–148 (G. Piranian); **MR** 21#812; **Zbl.** 84,54.

1959.01 A remark on the iteration of entire functions, *Riveon Lematematika* **13** (1959), 13–16; **MR** 22#111.

1959.02 A theorem on partial well-ordering of sets of vectors, *J. London Math. Soc.* **34** (1959), 222–224 (R. Rado); **MR** 21#2604; **Zbl.** 85,38.

1959.03 About an estimation problem of Zahorski, *Colloq. Math.* **7**(1959/60), 167–170; **MR** 22#3919; **Zbl.** 106,277.

1959.04 Divergence of random power series, *Michigan Math. J.* **6** (1959), 343–347 (A. Dvoretzky); **MR** 22#97; **Zbl.** 95,122.

1959.05 Egy additív számelméleti probléma (On a problem in additive number theory = Über ein Problem aus der additiven Zahlentheorie, in Hungarian, Russian and German summaries), *Mat. Lapok* **10** (1959), 284–290 (J. Surányi); **MR** 23#A3122; **Zbl.** 100,272.

1959.06 Graph theory and probability, *Canad. J. Math.* **11** (1959), 34–38; **MR** 21#876; **Zbl.** 84,396.

1959.07 Megjegyzések egy versenyfeladathoz (Remarks to a problem = Bemerkungen zu einer Aufgabe eines mathematischen Wettbewerbs, in Hungarian, Russian and German summaries), *Mat. Lapok* **10** (1959), 39–48 (J. Surányi); **MR** 26#2388; **Zbl.** 93,257.

1959.08 On a question of additive number theory, *Acta Arith.* **5** (1958), 45–55, 1959 (P. Scherk); **MR** 21#2631; **Zbl.** 83,39.

1959.09 On Cantor's series with convergent $\sum 1/q_n$, *Ann. Univ. Sci. Budapest. Eötvös Sect. Math.* **2** (1959), 93–109 (A. Rényi); **MR** 23#A3710; **Zbl.** 95,265.

1959.10 On maximal paths and circuits of graphs, *Acta Math. Acad. Sci. Hungar.* **10** (1959), 337–356 (unbound insert) (T. Gallai); **MR** 22#5591; **Zbl.** 90,394.

1959.11 On random graphs, I., *Publ. Math. Debrecen* **6** (1959), 290–297 (A. Rényi); **MR** 22#10924; **Zbl.** 92,157.

1959.12 On random interpolation, *J. Austral. Math. Soc.* **1** (1959/1961), 129–133; **MR** 22#3912; **Zbl.** 108,57.

1959.13 On the central limit theorem for samples from a finite population, *Magyar Tud. Akad. Mat. Kutató Int. Közl.* **4** (1959), 49–61, (A. Rényi); **MR** 21#6019; **Zbl.** 86,340.

1959.14 On the distribution of primitive lattice points in the plane, *Canad. Math. Bull.* **2** (1959), 91–96 (J. H. H. Chalk); **MR** 21#4145; **Zbl.** 88,257.

1959.15 On the Lipschitz's condition for Brownian motion, *J. Math. Soc. Japan* **11** (1959), 263–274 (K.-L. Chung; T. Sirao); **MR** 22#12602; **Zbl.** 91,133.

1959.16 On the probability that n and $g(n)$ are relatively prime, *Acta Arith.* **5** (1958), 35–44, 1959 (G. G. Lorentz); **MR** 21#37; **Zbl.** 85,31.

1959.17 On the product $\prod_{k=1}^{n}(1 - z^{a_k})$, *Acad. Serbe Sci. Publ. Inst. Math.* **13** (1959), 29–34 (G. Szekeres); **MR** 23#A3721; **Zbl.** 97,33.

1959.18 On the structure of inner set mappings, *Acta Sci. Math. Szeged* **20** (1959), 81–90 (G. Fodor; A. Hajnal); **MR** 21#3334; **Zbl.** 94,32.

1959.19 Partition relations connected with the chromatic number of graphs, *J. London Math. Soc.* **34** (1959), 63–72 (R. Rado); **MR** 21#652; **Zbl.** 84,197.

1959.20 Remarks on number theory, I. On primitive α-abundant numbers, *Acta Arith.* **5** (1958), 25–33, 1959; **MR** 21#24; **Zbl.** 83,263.

1959.21 Remarks on number theory, II. Some problems on the σ function, *Acta Arith.* **5** (1959), 171–177; **MR** 21#6348; **Zbl.** 92,46.

1959.22 Sequences of linear fractional transformations, *Michigan Math. J.* **6** (1959), 205–209 (G. Piranian); **MR** 22#114; **Zbl.** 87,45.

1959.23 Some examples in ergodic theory, *Proc. London Math. Soc. (3)* **9** (1959), 227–241 (Y. N. Dowker); **MR** 21#1374; **Zbl.** 84,341.

1959.24 Some further statistical properties of the digits in Cantor's series, *Acta Math. Acad. Sci. Hungar.* **10** (1959), 21–29 (unbound insert) (A. Rényi); **MR** 21#6356; **Zbl.** 88,258.

1959.25 Some remarks on prime factors of integers, *Canad. J. Math.* **11** (1959), 161–167; **MR** 21#3387; **Zbl.** 92,43.

1959.26 Some results on diophantine approximation, *Acta Arith.* **5** (1959), 359–369; **MR** 22#12091; **Zbl.** 97,35.

1959.27 Über einige Probleme der additiven Zahlentheorie (in German), *Sammelband zu Ehren des 250. Geburtstages Leonhard Eulers*, pp. 116–119, Akademie–Verlag, Berlin, 1959; **MR** 31#1240; **Zbl.** 105,266.

1960.01 A construction of graphs without triangles having pre-assigned order and chromatic number, *J. London Math. Soc.* **35** (1960), 445–448 (R. Rado); **MR** 25#3853; **Zbl.** 97,164.

1960.02 Additive properties of random sequences of positive integers, *Acta Arith.* **6** (1960), 83–110 (A. Rényi); **MR** 22#10970; **Zbl.** 91,44.

1960.03 Distributions of the values of some arithmetical functions, *Acta Arith.* **6** (1960/1961), 473–485 (A. Schinzel); **MR** 23#A3706; **Zbl.** 104,272.

1960.04 Intersection theorems for systems of sets, *J. London Math. Soc.* **35** (1960), 85–90 (R. Rado); **MR** 22#2554; **Zbl.** 103,279.

1960.05 Megjegyzések a Matematikai Lapok két problémájához (Remarks on two problems, in Hungarian), *Mat. Lapok* **11** (1960), 26–32; **MR** 23#A863; **Zbl.** 100,272.

1960.06 Ob odnom asimptoticheskom neravenstve v teorii tschisel (An asymptotic inequality in the theory of numbers, in Russian), *Vestnik Leningrad. Univ.* **15** (1960) no. 13, 41–49; **MR** 23#A3720; **Zbl.** 104,268.

1960.07 On analytic iteration, *J. Analyse Math.* **8** (1960/1961), 361–376 (E. Jabotinsky); **MR** 23#A3240; **Zbl.** 126,88.

1960.08 On sets of distances of n points in Euclidean space, *Magyar Tud. Akad. Mat. Kutató Int. Közl.* **5** (1960), 165–169; **MR** 25#A4420; **Zbl.** 94,168.

1960.09 On some extremum problems in elementary geometry, *Ann. Univ. Sci. Budapest. Eötvös Sect. Math.* **3–4** (1960/1961), 53–62 (G. Szekeres); **MR** 24#A3560; **Zbl.** 103,155.

1960.10 On the evolution of random graphs, *Magyar Tud. Akad. Mat. Kutató Int. Közl.* **5** (1960), 17–61 (A. Rényi); **MR** 23#A2338; **Zbl.** 103,163.

1960.11 On the maximum number of pairwise orthogonal Latin squares of a given order, *Canad. J. Math.* **12** (1960), 204–208 (S. Chowla; E. G. Straus); **MR** 23#A70; **Zbl.** 93,320.

1960.12 Remarks and corrections to my paper "Some remarks on a paper of McCarthy" [*Canad. Math. Bull.* **1** (1958), 71–75], *Canad. Math. Bull.* **3** (1960), 127–129; **MR** 24#A1238; **Zbl.** 93,51.

1960.13 Remarks on number theory, III. On addition chains, *Acta Arith.* **6** (1960), 77–81; **MR** 22#12085; **Zbl.** 219.10064.

1960.14 Restricted cluster sets, *Math. Nachr.* **22** (1960), 155–158 (G. Piranian); **MR** 23#A1041; **Zbl.** 113,55.

1960.15 Scales of functions, *J. Austral. Math. Soc.* **1** (1960), 396–418 (C. A. Rogers; S. J. Taylor); **Zbl.** 158,50.

1960.16 Some intersection properties of random walk paths, *Acta Math. Acad. Sci. Hungar.* **11** (1960), 231–248 (S. J. Taylor); **MR** 23 #A3595; **Zbl.** 96,333.

1960.17 Some problems concerning the structure of random walk paths, *Acta Math. Acad. Sci. Hungar.* **11** (1960), 137–162 (unbound insert) (S. J. Taylor); **MR** 22#12599; **Zbl.** 91,133.

1960.18 Some remarks on set theory, VII., *Acta Sci. Math. (Szeged)* **21** (1960), 154–163 (A. Hajnal); **MR** 24#A3071; **Zbl.** 102,285.

1960.19 Some remarks on set theory, VIII., *Michigan Math. J.* **7** (1960), 187–191 (A. Hajnal); **MR** 22#6718; **Zbl.** 95,39.

1960.20 Über die kleinste quadratfreie Zahl einer arithmetischen Reihe (in German), *Monatsh. Math.* **64** (1960), 314–316; **MR** 22#9476; **Zbl.** 97,28.

1960.21 Válogatott fejezetek a számelméletböl (Selected chapters from number theory, in Hungarian), *Tankonyvkiado Vallalat*, Budapest, 1960, 250 pp. (J. Surányi); **MR** 28#5022; **Zbl.** 95,29.

1961.01 A problem about prime numbers and the random walk, II., *Illinois J. Math.* **5** (1961), 352–353; **MR** 22#12080; **Zbl.** 98,324.

1961.02 An extremal problem in the theory of interpolation, *Acta Math. Acad. Sci. Hungar.* **12** (1961), 221–234 (P. Turán); **MR** 26#4093; **Zbl.** 98,271.

1961.03 Correction to "On a problem of I. Schur" [*Ann. of Math. (2)* **43** (1942), 451–470], *Ann. of Math. (2)* **74** (1961), 628 (G. Szegő); **MR** 24#1341; **Zbl.** 99,251.

1961.04 Covering space with convex bodies, *Acta Arith.* **7** (1961/1962), 281–285 (C. A. Rogers); **MR** 26#6863; **Zbl.** 213,58.

1961.05 Gráfok előírt fokú pontokkal (Graphs with points of prescribed degrees = Graphen mit Punkten vorgeschriebenen Grades, in Hungarian), *Mat. Lapok* **11** (1961), 264–274 (T. Gallai); **Zbl.** 103,397.

1961.06 Graph theory and probability, II., *Canad. J. Math.* **13** (1961), 346–352; **MR** 22#10925; **Zbl.** 97,391.

1961.07 Intersection theorems for systems of finite sets, *Quart. J. Math. Oxford Ser. (2)* **12** (1961), 313–320 (Chao Ko [=Zhao Ke]; R. Rado); **MR** 25#3839; **Zbl.** 100,19.

1961.08 Nonincrease everywhere of the Brownian motion process, *Proc. 4th Berkeley Sympos. Math. Statist. and Prob., Vol. II*, pp. 103–116,

Univ. California Press, Berkeley, 1961 (A. Dvoretzky; S. Kakutani); **MR** 24#A2448; **Zbl.** 111,150.

1961.09 On a classical problem of probability theory, *Magyar Tud. Akad. Mat. Kutató Int. Közl.* **6** (1961), 215–220 (A. Rényi); **MR** 27#794; **Zbl.** 102,352.

1961.10 On a problem of G. Golomb, *J. Austral. Math. Soc.* **2** (1961/1962), 1–8; **MR** 23#A864; **Zbl.** 100,271.

1961.11 On a property of families of sets, *Acta Math. Acad. Sci. Hungar.* **12** (1961), 87–123 (A. Hajnal); **MR** 27#50; **Zbl.** 201,328.

1961.12 On a theorem in the theory of relations and a solution of a problem of Knaster, *Colloq. Math.* **8** (1961), 19–21 (E. Specker); **MR** 24#A49; **Zbl.** 97,42.

1961.13 On Note 2921, *Math. Gaz.* **45** (1961), 39; **Zbl.** 127,267.

1961.14 On some problems involving inaccessible cardinals, *Essays on the foundations of mathematics*, pp. 50–82, Magnes Press, Hebrew Univ., Jerusalem, 1961 (A. Tarski); **MR** 29#4695; **Zbl.** 212,325.

1961.15 On the evolution of random graphs, *Bull. Inst. Internat. Statist.* **38** (1961) no. 4, 343–347 (A. Rényi); **MR** 26#5564; **Zbl.** 106,120.

1961.16 On the Hausdorff measure of Brownian paths in the plane, *Proc. Cambridge Philos. Soc.* **57** (1961), 209–222 (S. J. Taylor); **MR** 23#A4186; **Zbl.** 101,112.

1961.17 On the minimal number of vertices representing the edges of a graph, *Magyar Tud. Akad. Mat. Kutató Int. Közl.* **6** (1961), 181–203 (T. Gallai); **MR** 26#1878; **Zbl.** 101,410.

1961.18 On the representation of large integers as sums of distinct summands taken from a fixed set, *Acta Arith.* **7** (1961/1962), 345–354; **MR** 26#2387; **Zbl.** 106,38.

1961.19 On the strength of connectedness of a random graph, *Acta Math. Acad. Sci. Hungar.* **12** (1961), 261–267 (A. Rényi); **MR** 24#A54; **Zbl.** 103,163.

1961.20 Problems and results on the theory of interpolation, II., *Acta Math. Acad. Sci. Hungar.* **12** (1961), 235–244; **MR** 26#2779; **Zbl.** 98,41.

1961.21 Sätze und Probleme über p_k/k (in German), *Abh. Math. Sem. Univ. Hamburg* **25** (1961/1962), 251–256 (K. Prachar); **MR** 25#3901; **Zbl.** 107,266.

1961.22 Some unsolved problems, *Magyar Tud. Akad. Mat. Kutató Int. Közl.* **6** (1961), 221–254; **MR** 31#2106; **Zbl.** 100,20.

1961.23 Számelméleti megjegyzések, I. (Remarks on number theory, I., in Hungarian), *Mat. Lapok* **12** (1961), 10–17; **MR** 26#2410; **Zbl.** 154,294.

1961.24 Számelméleti megjegyzések, II. Az Euler-féle φ-függvény néhány tulajdonságáról (Some remarks on number theory, II., in Hungarian), *Mat. Lapok* **12** (1961), 161–169; **MR** 26#2411; **Zbl.** 154,294.

1961.25 Theorem in the additive number theory, *Bull. Res. Council Israel* **10** (1961) (A. Ginzburg; A. Ziv).

1961.26 Útiélmények Moszkva-Peking-Singapore (in Hungarian), *Magyar Tudomány* **8–9** (1968), 193–197.

1961.27 Über einige Probleme der additiven Zahlentheorie, *J. Reine Angew. Math.* **206** (1961), 61–66; **MR** 24#A707; **Zbl.** 114,263.

1962.01 An inequality for the maximum of trigonometric polynomials, *Ann. Polon. Math.* **12** (1962), 151–154; **MR** 25#5330; **Zbl.** 106,277.

1962.02 Applications of probability to combinatorial problems, *Colloq. on Combinatorial Methods in Probability Theory (Aarhus, 1962)*, pp. 90–92, Matemisk Institut, Aarhus Universitet, Aarhus, 1962; **Zbl.** 142,251.

1962.03 Beantwortung einer Frage von E. Teuffel (in German), *Elem. Math.* **17** (1962), 107–108; **Zbl.** 106,33.

1962.04 Gráfelméleti szélsőértékekre vonatkozó problémákról (Extremal problems in graph theory, in Hungarian), *Mat. Lapok* **13** (1962), 143–152 (B. Bollobás); **MR** 26#3036; **Zbl.** 117,412.

1962.05 Néhány elemi geometriai problémáról (On some problems in elementary geometry, Hungarian), *Középisk. Mat. Lapok* **24/5** (1962), 1–9.

1962.06 On a classification of denumerable order types and an application to the partition calculus, *Fund. Math.* **51** (1962/1963), 117–129 (A. Hajnal); **MR** 25#5000; **Zbl.** 111,11.

1962.07 On a problem of A. Zygmund, *Studies in mathematical analysis and related topics*, pp. 110–116, Stanford Univ. Press, Stanford, California, 1962 (A. Rényi); **MR** 26#2586; **Zbl.** 171,316.

1962.08 On a problem of Sierpiński, *Atti Accad. Naz. Lincei Rend. Cl. Sci. Fis. Mat. Nat. (8)* **33** (1962), 122–124; **MR** 27#93; **Zbl.** 111,47.

1962.09 On a theorem of Rademacher-Turán, *Illinois J. Math.* **6** (1962), 122–127; **MR** 25#1111; **Zbl.** 99,394.

1962.10 On circuits and subgraphs of chromatic graphs, *Mathematika* **9** (1962), 170–175; **MR** 26#3035; **Zbl.** 109,165.

1962.11 On C_1-summability of series, *Michigan Math. J.* **9** (1962), 1–14 (H. Hanani); **MR** 25#362; **Zbl.** 111,261.

1962.12 On the integers relatively prime to n and on a number-theoretic function considered by Jacobsthal, *Math. Scand.* **10** (1962), 163–170; **MR** 26#3651; **Zbl.** 202,330.

1962.13 On the maximal number of disjoint circuits of a graph, *Publ. Math. Debrecen* **9** (1962), 3–12 (L. Pósa); **MR** 27#743; **Zbl.** 133,167.

1962.14 On the number of complete subgraphs contained in certain graphs, *Magyar Tud. Akad. Mat. Kutató Int. Közl.* **7** (1962), 459–464; **MR** 27#1937; **Zbl.** 116,12.

1962.15 On the topological product of discrete λ-compact spaces, *General Topology and its Relations to Modern Analysis and Algebra (Proc. Sympos., Prague, 1961)*, pp. 148–151, Academic Press, New York; Publ. House Czech. Acad. Sci., Prague, 1962 (A. Hajnal); **MR** 26#6927; **Zbl.** 114,141.

1962.16 On trigonometric sums with gaps, *Magyar Tud. Akad. Mat. Kutató Int. Közl.* **7** (1962), 37–42; **MR** 26#2797; **Zbl.** 116,47.

1962.17 Remarks on a paper of Pósa, *Magyar Tud. Akad. Mat. Kutató Int. Közl.* **7** (1962), 227–229; **MR** 32#2348; **Zbl.** 114,400.

1962.18 Representations of real numbers as sums and products of Liouville numbers, *Michigan Math. J.* **9** (1962), 59–60; **MR** 24#A3134; **Zbl.** 114,263.

1962.19 Some extremal problems on infinite graphs (in Russian), *Magyar Tud. Akad. Mat. Kutató Int. Közl.* **7** (1962), 441–457 (J. Czipszer; A. Hajnal); **MR** 27#744; **Zbl.** 114,13.

1962.20 Some remarks concerning our paper "On the structure of set mappings" [*Acta Math. Acad. Sci. Hungar.* **9** (1958), 111–131]. Nonexistence of a two-valued σ-measure for the first uncountable inaccessible cardinal, *Acta Math. Acad. Sci. Hungar.* **13** (1962), 223–226 (A. Hajnal); **MR** 25#5001; **Zbl.** 134,16.

1962.21 Some remarks on the functions φ and σ, *Bull. Acad. Polon. Sci. Sér. Sci. Math. Astr. Phys.* **10** (1962), 617–619; **MR** 26#3652; **Zbl.** 106,40.

1962.22 Számelméleti megjegyzések, III. Néhány additív számelméleti problémáról (some remarks on number theory, III., in Hungarian), *Mat. Lapok* **13** (1962), 28–38; **MR** 26#2412; **Zbl.** 123,255.

1962.23 Számelméleti megjegyzések, IV. Extremális problémák a számelméletben, I. (Remarks on number theory, IV. Extremal problems in number theory, I., in Hungarian), *Mat. Lapok* **13** (1962), 228–255; **MR** 33#4020; **Zbl.** 127,22.

1962.24 The construction of certain graphs, *Canad. J. Math.* **14** (1962), 702–707 (C. A. Rogers); **MR** 25#5010; **Zbl.** 194,253.

1962.25 Über ein Extremalproblem in der Graphentheorie (in German), *Arch. Math.* **13** (1962), 222–227; **MR** 25#2974; **Zbl.** 105,175.

1962.26 Verchu niakoy geometritchesky zadatchy (On some geometric problems, in Bulgarian), *Fiz.-Mat. Spis. Bŭlgar. Akad. Nauk.* **5(38)** (1962), 205–212; **MR** 31#2648.

1963.01 A theorem on uniform distribution, *Magyar Tud. Akad. Mat. Kutató Int. Közl.* **8** (1963), 3–11 (H. Davenport); **MR** 29#4750; **Zbl.** 122,59.

1963.02 An elementary inequality between the probabilities of events, *Math. Scand.* **13** (1963), 99–104 (J. Neveu; A. Rényi); **MR** 29#4075; **Zbl.** 129,314.

1963.03 An intersection property of sets with positive measure, *Colloq. Math.* **11** (1963), 75–80 (H. Kestelman; C. A. Rogers); **MR** 28#2182; **Zbl.** 122,299.

1963.04 Asymmetric graphs, *Acta Math. Acad. Sci. Hungar.* **14** (1963), 295–315 (A. Rényi); **MR** 27#6258; **Zbl.** 118,189.

1963.05 Egy gráfelméleti problémáról (On a problem in the theory of graphs, in Hungarian), *Magyar Tud. Akad. Mat. Kutató Int. Közl.* **7** (1962), 623–641, 1963 (A. Rényi); **MR** 33#1246; **Zbl.** 131,210.

1963.06 On a combinatorial problem, *Nordisk Mat. Tidskr.* **11** (1963), 5–10, 40; **MR** 26#6061; **Zbl.** 116,11.

1963.07 On a limit theorem in combinatorial analysis, *Publ. Math. Debrecen* **10** (1963), 10–13 (H. Hanani); **MR** 29#3394; **Zbl.** 122,248.

1963.08 On a problem in graph theory, *Math. Gaz.* **47** (1963), 220–223; **MR** 28#2536; **Zbl.** 117,174.

1963.09 On some properties of Hamel bases, *Colloq. Math.* **10** (1963), 267–269; **MR** 28#4068; **Zbl.** 123,320.

1963.10 On the maximal number of independent circuits in a graph, *Acta Math. Acad. Sci. Hungar.* **14** (1963), 79–94 (G. A. Dirac); **MR** 26#5563; **Zbl.** 122,249.

1963.11 On the structure of linear graphs, *Israel J. Math.* **1** (1963), 156–160; **MR** 28#4533.

1963.12 On two problems of information theory, *Magyar Tud. Akad. Mat. Kutató Int. Közl.* **8** (1963), 229–243 (A. Rényi); **MR** 29#3268; **Zbl.** 119,340.

1963.13 On Weyl's criterion for uniform distribution, *Michigan Math. J.* **10** (1963), 311–314 (H. Davenport; W. J. LeVeque); **MR** 27#3618; **Zbl.** 119,282.

1963.14 Quelques problèmes de théorie des nombres (in French), *Monographies de l'Enseignement Mathématique, No. 6*, pp. 81–135, L'Enseignement Mathématique, Université, Geneva, 1963; **MR** 28#2070; **Zbl.** 117,29.

1963.15 Ramsey és Van der Waerden tételével kapcsolatos kombinatorikai kérdésekről (On combinatorial questions connected with a theorem of Ramsey and van der Waerden, in Hungarian), *Mat. Lapok* **14** (1963), 29–37; **MR** 34#7409; **Zbl.** 115,10.

1963.16 Reguläre Graphen gegebener Taillenweite mit minimaler Knotenzahl (in German), *Wiss. Z. Martin-Luther-Univ. Halle-Wittenberg Math.-Natur. Reihe* **12** (1963), 251–257 (H. Sachs); **MR** 29#2797; **Zbl.** 116,150.

1963.17 Remarks on a problem of Obreanu, *Canad. Math. Bull.* **6** (1963), 267–273 (A. Rényi); **MR** 31#2528; **Zbl.** 121,296.

1963.18 Sums of distinct unit fractions, *Proc. Amer. Math. Soc.* **14** (1963), 126–131 (S. Stein); **MR** 26#71; **Zbl.** 115,265.

1963.19 The Hausdorff measure of the intersection of sets of positive Lebesgue measure, *Mathematika* **10** (1963), 1–9 (S. J. Taylor); **MR** 27#3765; **Zbl.** 141,55.

1963.20 The minimal regular graph containing a given graph, *Amer. Math. Monthly* **70** (1963), 1074–1075 (P. Kelly).

1964.01 A problem concerning the zeros of a certain kind of holomorphic function in the unit disk, *J. Reine Angew. Math.* **214/215** (1964), 340–344 (F. Bagemihl); **MR** 31#3580; **Zbl.** 131,75.

1964.02 A problem in graph theory, *Amer. Math. Monthly* **71** (1964), 1107–1110 (A. Hajnal; J. W. Moon); **MR** 30#577; **Zbl.** 126,394.

1964.03 A problem on tournaments, *Canad. Math. Bull.* **7** (1964), 351–356 (L. Moser); **MR** 29#4046; **Zbl.** 129,347.

1964.04 An interpolation problem associated with the continuum hypothesis, *Michigan Math. J.* **11** (1964), 9–10; **MR** 29#5744; **Zbl.** 121,258.

1964.05 Arithmetical Tauberian theorems, *Acta Arith.* **9** (1964), 341–356 (A. E. Ingham); **MR** 31#1228; **Zbl.** 127,271.

1964.06 Extremal problems in graph theory, *Theory of Graphs and its Applications (Proc. Sympos. Smolenice, 1963)*, pp. 29–36, Publ. House Czech. Acad. Sci., Prague, 1964; **MR** 31#4735; **Zbl.** 161,205.

1964.07 Laconicity and redundancy of Toeplitz matrices, *Math. Z.* **83** (1964), 381–394 (G. Piranian); **MR** 29#1471; **Zbl.** 129,42.

1964.08 On a combinatorial problem, II., *Acta Math. Acad. Sci. Hungar.* **15** (1964), 445–447; **MR** 29#4700; **Zbl.** 201,337.

1964.09 On a combinatorial problem in Latin squares, *Magyar Tud. Akad. Mat. Kutató Int. Közl.* **8** (1963), 407–411, 1964 (A. Ginzburg); **MR** 29#2197; **Zbl.** 125,282.

1964.10 On a problem in elementary number theory and a combinatorial problem, *Math. Comp.* **18** (1964), 644–646; **MR** 30#1087; **Zbl.** 127,22.

1964.11 On an extremal problem in graph theory, *Colloq. Math.* **13** (1964/65), 251–254; **MR** 31#3353; **Zbl.** 137,181.

1964.12 On complete topological subgraphs of certain graphs, *Ann. Univ. Sci. Budapest. Eötvös Sect. Math.* **7** (1964), 143–149 (A. Hajnal); **MR** 30#3460; **Zbl.** 125,405.

1964.13 On extremal problems of graphs and generalized graphs, *Israel J. Math.* **2** (1964), 183–190; **MR** 32#1134; **Zbl.** 129,399.

1964.14 On random matrices, *Magyar Tud. Akad. Mat. Kutató Int. Közl.* **8** (1964), 455–461 (A. Rényi); **MR** 29#4769; **Zbl.** 133,260.

1964.15 On some applications of probability to analysis and number theory, *J. London Math. Soc.* **39** (1964), 692–696; **MR** 30#1997; **Zbl.** 125,86.

1964.16 On some divisibility properties of $\binom{2n}{n}$, *Canad. Math. Bull.* **7** (1964), 513–518; **MR** 30#52; **Zbl.** 125,23.

1964.17 On subgraphs of the complete bipartite graph, *Canad. Math. Bull.* **7** (1964), 35–39 (J. W. Moon); **MR** 28#1612; **Zbl.** 122,419.

1964.18 On the addition of residue classes mod p, *Acta Arith.* **9** (1964), 149–159 (H. Heilbronn); **MR** 29#3463; **Zbl.** 156,48.

1964.19 On the irrationality of certain Ahmes series, *J. Indian Math. Soc. (N.S.)* **27** (1964), 129–133 (E. G. Straus); **MR** 31#124; **Zbl.** 131,49.

1964.20 On the multiplicative representation of integers, *Israel J. Math.* **2** (1964), 251–261; **MR** 31#5847; **Zbl.** 146,53.

1964.21 On the number of triangles contained in certain graphs, *Canad. Math. Bull.* **7** (1964), 53–56; **MR** 28#2537; **Zbl.** 121,403.

1964.22 On the representation of directed graphs as unions of orderings, *Magyar Tud. Akad. Mat. Kutató Int. Közl.* **9** (1964), 125–132 (L. Moser); **MR** 29#5756; **Zbl.** 136,449.

1964.23 On two problems of S. Marcus concerning functions with the Darboux property, *Rev. Roumaine Math. Pures Appl.* **9** (1964), 803–804; **MR** 31#5944; **Zbl.** 128,279.

1964.24 Problems and results on diophantine approximations, *Compositio Math.* **16** (1964), 52–65; **MR** 31#3382; **Zbl.** 131,48.

1964.25 Solution of a problem of Dirac, *Theory of Graphs and its Applications (Proc. Sympos. Smolenice, 1963)*, pp. 167–168, Publ. House Czech. Acad. Sci., Prague, 1964 (T. Gallai); **Zbl.** 161,433.

1964.26 Some applications of probability to graph theory and combinatorial problems, *Theory of Graphs and its Applications (Proc. Sympos. Smolenice, 1963)*, pp. 133–136, Publ. House Czech. Acad. Sci., Prague, 1964; **MR** 30#3459; **Zbl.** 161,433.

1964.27 Some remarks on Ramsey's theorem, *Canad. Math. Bull.* **7** (1964), 619–622; **MR** 30#576; **Zbl.** 129,401.

1964.28 Some remarks on set theory, IX. Combinatorial problems in measure theory and set theory, *Michigan Math. J.* **11** (1964), 107–127 (A. Hajnal); **MR** 30#1940; **Zbl.** 199,23.

1964.29 Tauberian theorems for sum sets, *Acta Arith.* **9** (1964), 177–189 (B. Gordon; L. A. Rubel; E. G. Straus); **MR** 29#3417; **Zbl.** 135,100.

1964.30 The amount of overlapping in partial coverings of space by equal spheres, *Mathematika* **11** (1964), 171–184 (L. Few; C. A. Rogers); **MR** 31#664; **Zbl.** 127,276.

1964.31 The star number of coverings of space with convex bodies, *Acta Arith.* **9** (1964), 41–45 (C. A. Rogers); **MR** 29#5165; **Zbl.** 132,32.

1965.01 A problem on independent r-tuples, *Ann. Univ. Sci. Budapest. Eötvös Sect. Math.* **8** (1965), 93–95; **MR** 41#5223; **Zbl.** 136,213.

1965.02 Extremal problems in number theory, *Proc. Sympos. Pure Math., Vol. VIII*, pp. 181–189, Amer. Math. Soc., Providence, R.I., 1965; **MR** 30#4740; **Zbl.** 144,281.

1965.03 Large and small subspaces of Hilbert space, *Michigan Math. J.* **12** (1965), 169–178 (H. S. Shapiro; A. L. Shields); **MR** 31#2607; **Zbl.** 132,349.

1965.04 On a problem of Sierpiński (Extract from a letter to W. Sierpiński), *Acta Arith.* **11** (1965), 189–192; **MR** 32#5620; **Zbl.** 129,28.

1965.05 On independent circuits contained in a graph, *Canad. J. Math.* **17** (1965), 347–352 (L. Pósa); **MR** 31#86; **Zbl.** 129,399.

1965.06 On sets of consistent arcs in a tournament, *Canad. Math. Bull.* **8** (1965), 269–271 (J. W. Moon); **MR** 32#57; **Zbl.** 137,433.

1965.07 On some extremal problems in graph theory, *Israel J. Math.* **3** (1965), 113–116; **MR** 32#7443; **Zbl.** 134,434.

1965.08 On some problems of a statistical group-theory, I., *Z. Wahrscheinlichkeitstheorie und Verw. Gebiete* **4** (1965), 175–186 (P. Turán); **MR** 32#2465; **Zbl.** 137,256.

1965.09 On the dimension of a graph, *Mathematika* **12** (1965), 118–122 (F. Harary; W. T. Tutte); **MR** 32#5537; **Zbl.** 151,332.

1965.10 On the distribution of divisors of integers in the residue classes (mod d), *Bull. Soc. Math. Grèce (N.S.)* **6 I** (1965), fasc. 1, 27–36; **MR** 34#7474; **Zbl.** 133,299.

1965.11 On the function $g(t) = \limsup_{t\to\infty}(f(x+t) - f(x))$, *Magyar Tud. Akad. Mat. Kutató Int. Közl.* **9** (1965), 603–606 (I. Csiszár); **Zbl.** 133,304.

1965.12 On the mean value of nonnegative multiplicative number-theoretical functions, *Michigan Math. J.* **12** (1965), 321–338 (A. Rényi); **MR** 34#2537; **Zbl.** 131,43.

1965.13 On Tschebycheff quadrature, *Canad. J. Math.* **17** (1965), 652–658 (A. Sharma); **MR** 31#3774; **Zbl.** 156,71.

1965.14 Partition relations for cardinal numbers, *Acta Math. Acad. Sci. Hungar.* **16** (1965), 93–196 (A. Hajnal; R. Rado); **MR** 34#2475; **Zbl.** 158,266.

1965.15 Probabilistic methods in group theory, *J. Analyse Math.* **14** (1965), 127–138 (A. Rényi); **MR** 34#2690; **Zbl.** 247.20045.

1965.16 Remarks on a theorem of Zygmund, *Proc. London Math. Soc. (3)* **14a** (1965), 81–85; **MR** 31#5031; **Zbl.** 148,54.

1965.17 Some recent advances and current problems in number theory, *Lectures on Modern Mathematics, Vol. III*, pp. 196–244, Wiley, New York, 1965; **MR** 31#2191; **Zbl.** 132,284.

1965.18 Some remarks on number theory, *Israel J. Math.* **3** (1965), 6–12; **MR** 32#1181; **Zbl.** 131,39.

1965.19 The non-existence of a Hamel-basis and the general solution of Cauchy's functional equation for non-negative numbers, *Publ. Math. Debrecen* **12** (1965), 259–263 (J. Aczél); **MR** 32#4022; **Zbl.** 151,210.

1966.01 A limit theorem in graph theory, *Studia Sci. Math. Hungar.* **1** (1966), 51–57 (M. Simonovits); **MR** 34#5702; **Zbl.** 178,273.

1966.02 Additive Gruppen mit vorgegebener Hausdorffscher Dimension (in German), *J. Reine Angew. Math.* **221** (1966), 203–208 (B. Volkmann); **MR** 32#4238; **Zbl.** 135,102.

1966.03 An example concerning open everywhere discontinuous functions, *Rev. Roumaine Math. Pures Appl.* **11** (1966), 621–622; **MR** 33#5796; **Zbl.** 163,299.

1966.04 Konstruktion von nichtperiodischen Minimalbasen mit der Dichte $\frac{1}{2}$ für die Menge der nichtnegativen ganzen Zahlen (in German), *J. Reine Angew. Math.* **221** (1966), 44–47 (E. Härtter); **MR** 32#7533; **Zbl.** 135,97.

1966.05 On a problem of B. Jónsson, *Bull. Acad. Polon. Sci. Sér. Sci. Math. Astr. Phys.* **14** (1966), 19–23 (A. Hajnal); **MR** 35#64; **Zbl.** 171,265.

1966.06 On a problem of graph theory, *Studia Sci. Math. Hungar.* **1** (1966), 215–235 (A. Rényi; V. T. Sós); **MR** 36#6310; **Zbl.** 144,233.

1966.07 On chromatic number of graphs and set-systems, *Acta Math. Acad. Sci. Hungar.* **17** (1966), 61–99 (A. Hajnal); **MR** 33#1247; **Zbl.** 151,337.

1966.08 On cliques in graphs, *Israel J. Math.* **4** (1966), 233–234; **MR** 34#5700; **Zbl.** 163,182.

1966.09 On divisibility properties of sequences of integers, *Studia Sci. Math. Hungar.* **1** (1966), 431–435 (A. Sárközy; E. Szemerédi); **MR** 34#4233; **Zbl.** 146,271.

1966.10 On some applications of probability methods to function theory and on some extremal properties of polynomials, *Contemporary Problems in Theory Anal. Functions (Internat. Conf., Erevan, 1965) (Russian)*, pp. 359–362, Izdat. "Nauda", Moscow, 1966; **MR** 34#6034; **Zbl.** 174,366.

1966.11 On some properties of prime factors of integers, *Nagoya Math. J.* **27** (1966), 617–623; **MR** 34#4220; **Zbl.** 151,35.

1966.12 On the complete subgraphs of graphs defined by systems of sets, *Acta Math. Acad. Sci. Hungar.* **17** (1966), 159–229 (A. Hajnal; E. C. Milner); **MR** 36#6298; **Zbl.** 151,337.

1966.13 On the construction of certain graphs, *J. Combinatorial Theory* **1** (1966), 149–153; **MR** 34#5701; **Zbl.** 144,454.

1966.14 On the difference of consecutive terms of sequences defined by divisibility properties, *Acta Arith.* **12** (1966/1967), 175–182; **MR** 34#7488; **Zbl.** 147,26.

1966.15 On the divisibility properties of sequences of integers, I., *Acta Arith.* **11** (1966), 411–418 (A. Sárközy; E. Szemerédi); **MR** 34#5791; **Zbl.** 146,271.

1966.16 On the existence of a factor of degree one of a connected random graph, *Acta Math. Acad. Sci. Hungar.* **17** (1966), 359–368 (A. Rényi); **MR** 34#85; **Zbl.** 203,569.

1966.17 On the number of positive integers $\leq x$ and free of prime factors $> y$, *Simon Stevin* **40** (1966/1967), 73–76 (J. H. van Lint); **MR** 35#2836; **Zbl.** 146,53.

1966.18 On the solvability of the equations $[a_i, a_j] = a_r$ and $(a'_i, a'_j) = a'_r$ in sequences of positive density, *J. Math. Anal. Appl.* **15** (1966), 60–64 (A. Sárközy; E. Szemerédi); **MR** 33#4035; **Zbl.** 151,35.

1966.19 Some remarks on set theory, X., *Studia Sci. Math. Hungar.* **1** (1966), 157–159 (M. Makkai); **MR** 35#70; **Zbl.** 199,23.

1966.20 Számelméleti megjegyzések, V. Extremális problémák a számelméletben, II. (Remarks on number theory, V. Extremal problems in number theory, II., in Hungarian), *Mat. Lapok* **17** (1966), 135–155; **MR** 36#133; **Zbl.** 146,272.

1966.21 The representation of a graph by set intersections, *Canad. J. Math.* **18** (1966), 106–112 (A. W. Goodman; L. Pósa); **MR** 32#4034; **Zbl.** 137,432.

1967.01 A statisztikus csoportelmélet egyes problémáiról (Certain problems of statistical group theory, in Hungarian), *Magyar Tud. Akad. Mat. Fiz. Oszt. Közl.* **17** (1967), 51–57 (P. Turán); **MR** 35#6744; **Zbl.** 146,254.

1967.02 Applications of probabilistic methods to graph theory, *A Seminar on Graph Theory*, pp. 60–64, Holt, Rinehart and Winston, New York, 1967; **MR** 36#68; **Zbl.** 159,541.

1967.03 Asymptotische Untersuchungen über die Anzahl der Teiler von n (in German), *Math. Ann.* **169** (1967), 230–238; **MR** 34#5771; **Zbl.** 149,288.

1967.04 Essential Hausdorff cores of sequences, *J. Indian Math. Soc. (N.S.)* **30** (1966), 93–115, 1967 (G. Piranian); **MR** 36#5560; **Zbl.** 148,289.

1967.05 Extremal problems in graph theory, *A Seminar on Graph Theory*, pp. 54–59, Holt, Rinehart and Winston, New York, 1967; **MR** 36#6311; **Zbl.** 159,541.

1967.06 Gráfok páros körüljárású részgráfjairól (On even subgraphs of graphs, in Hungarian), *Mat. Lapok* **18** (1967), 283–288; **MR** 39#95; **Zbl.** 193,243.

1967.07 Kromatikus gráfokról (On chromatic graphs, in Hungarian), *Mat. Lapok* **18** (1967), 1–4 (A. Hajnal); **MR** 37#2635; **Zbl.** 152,412.

1967.08 O rozreshymost'y niekotorych urevnienych v plotnych posledovatiel'nostyach tzelych tshysel (On the solvability of certain equations in the dense sequences of integers, in Russian), *Dokl. Akad. Nauk SSSR* **176** (1967), 541–544 (Russian) [English translation in *Soviet Math. Dokl.* **8** (1967), 1160–1164] (A. Sárközy; E. Szemerédi); **MR** 37#2716; **Zbl.** 159,60.

1967.09 On a theorem of Behrend, *J. Austral. Math. Soc.* **7** (1967), 9–16 (A. Sárközy; E. Szemerédi); **MR** 35#148; **Zbl.** 146,271.

1967.10 On an extremal problem concerning primitive sequences, *J. London Math. Soc.* **42** (1967), 484–488 (A. Sárközy; E. Szemerédi); **MR** 36#1412; **Zbl.** 166,51.

1967.11 On decomposition of graphs, *Acta Math. Acad. Sci. Hungar.* **18** (1967), 359–377 (A. Hajnal); **MR** 36#6309; **Zbl.** 169,266.

1967.12 On sequences of distances of a sequence, *Colloq. Math.* **17** (1967), 191–193 (S. Hartman); **MR** 36#2584; **Zbl.** 161,47.

1967.13 On some applications of graph theory to geometry, *Canad. J. Math.* **19** (1967), 968–971; **MR** 36#2520; **Zbl.** 161,206.

1967.14 On some problems of a statistical group-theory, II., *Acta Math. Acad. Sci. Hungar.* **18** (1967), 151–164 (P. Turán); **MR** 34#7624; **Zbl.** 189,313.

1967.15 On some problems of a statistical group-theory, III., *Acta Math. Acad. Sci. Hungar.* **18** (1967), 309–320 (P. Turán); **MR** 35#6743; **Zbl.** 235.20003.

1967.16 On the boundedness and unboundedness of polynomials, *J. Analyse Math.* **19** (1967), 135–148; **MR** 36#4216; **Zbl.** 186,379.

1967.17 On the divisibility properties of sequences of integers, II., *Acta Arith.* **14** (1967/1968), 1–12 (A. Sárközy; E. Szemerédi); **MR** 37#2717; **Zbl.** 186,80.

1967.18 On the partial sums of power series, *Proc. Roy. Irish Acad. Sect. A* **65** (1967), 113–123 (J. Clunie); **MR** 36#5314.

1967.19 Partition relations and transitivity domains of binary relations, *J. London Math. Soc.* **42** (1967), 624–633 (R. Rado); **MR** 36#1335; **Zbl.** 204,9.

1967.20 Problems and results on the convergence and divergence properties of the Lagrange interpolation polynomials and some extremal problems, *Mathematica (Cluj)* **10 (33)** (1967), 65–73; **MR** 38#1437; **Zbl.** 159,356.

1967.21 Some problems on the prime factors of consecutive integers, *Illinois J. Math.* **11** (1967), 428–430 (J. L. Selfridge); **MR** 37#5144; **Zbl.** 149,289.

1967.22 Some recent results on extremal problems in graph theory. Results, *Theory of Graphs (Internat. Sympos., Rome, 1966)*, pp. 117–123 (English), pp. 124–130 (French), Gordon and Breach, New York; Dunod, Paris, 1967; **MR** 37#2634; **Zbl.** 187,210.

1967.23 Some remarks on chromatic graphs, *Colloq. Math.* **16** (1967), 253–256; **MR** 35#1504; **Zbl.** 156,223.

1967.24 Some remarks on number theory, II., *Israel J. Math.* **5** (1967), 57–64; **MR** 35#2851; **Zbl.** 147,302.

1967.25 Some remarks on the iterates of the φ and σ functions, *Colloq. Math.* **17** (1967), 195–202; **MR** 36#2573; **Zbl.** 173,39.

1967.26 The minimal regular graph containing a given graph, *A Seminar on Graph Theory*, pp. 65–69, Holt, Rinehart and Winston, New York, 1967 (P. Kelly); **MR** 36#6312; **Zbl.** 159,541.

1968.01 Egy kombinatorikus problémáról (On a combinatorial problem, in Hungarian, English summary), *Mat. Lapok* **19** (1968), 345–348 (A. Hajnal); **MR** 39#5378; **Zbl.** 179,28.

1968.02 Hilbert térben levő ponthalmazok néhány geometriai és halmazelméleti tulajdonságáról (Geometrical and set-theoretical properties of subsets of Hilbert space, in Hungarian, English summary) *Mat. Lapok* **19** (1968), 255–258; **MR** 40#708; **Zbl.** 182,331.

1968.03 On a problem of P. Erdős and S. Stein, *Acta Arith.* **15** (1968), 85–90 (E. Szemerédi); **MR** 38#3218; **Zbl.** 186,79.

1968.04 On chromatic number of infinite graphs, *Theory of Graphs (Proc. Colloq., Tihany, 1966)*, pp. 83–98, Academic Press, New York, 1968 (A. Hajnal); **MR** 41#8294; **Zbl.** 164,248.

1968.05 On coloring graphs to maximize the proportion of multicolored k-edges, *J. Combinatorial Theory* **5** (1968), 164–169 (D. J. Kleitman); **MR** 37#3956; **Zbl.** 167,223.

1968.06 On equations with sets as unknowns, *Proc. Nat. Acad. Sci. U.S.A.* **60** (1968), 1189–1195 (S. Ulam); **MR** 38#3152; **Zbl.** 162,20.

1968.07 On random matrices, II., *Studia Sci. Math. Hungar.* **3** (1968), 459–464 (A. Rényi); **MR** 39#5389; **Zbl.** 174,41.

1968.08 On sets of almost disjoint subsets of a set, *Acta Math. Acad. Sci. Hungar.* **19** (1968), 209–218 (A. Hajnal; E. C. Milner); **MR** 37#84; **Zbl.** 174,18.

1968.09 On some applications of graph theory to number theoretic problems, *Publ. Ramanujan Inst. No.* **1** (1968/1969), 131–136; **MR** 42#4520; **Zbl.** 208,56.

1968.10 On some new inequalities concerning extremal properties of graphs, *Theory of Graphs (Proc. Colloq., Tihany, 1966)*, pp. 77–81, Academic Press, New York, 1968; **MR** 38#1026; **Zbl.** 161,433.

1968.11 On some problems of a statistical group-theory, IV., *Acta Math. Acad. Sci. Hungar.* **19** (1968), 413–435 (P. Turán); **MR** 38#1156; **Zbl.** 235.20004.

1968.12 On the distribution of prime divisors (Short communication), *Aequationes Math.* **1** (1968), 208–209.

1968.13 On the recurrence of a certain chain, *Proc. Amer. Math. Soc.* **19** (1968), 336–338 (D. A. Darling); **MR** 36#6012; **Zbl.** 164,475.

1968.14 On the solvability of certain equations in sequences of positive upper logarithmic density, *J. London Math. Soc.* **43** (1968), 71–78 (A. Sárközy; E. Szemerédi); **MR** 37#183; **Zbl.** 155,88.

1968.15 Problems, *Theory of Graphs (Proc. Colloq., Tihany, 1966)*, pp. 361–362, Academic Press, New York, 1968; **MR** 38#1016 (for entire book); **Zbl.** 155,2 (for entire book).

1968.16 Some remarks on the large sieve of Yu. V. Linnik, *Ann. Univ. Sci. Budapest. Eötvös Sect. Math.* **11** (1968), 3–13 (A. Rényi); **MR** 39#2318; **Zbl.** 207,359.

1968.17 Über eine geometrische Frage von Fejes-Tóth (in German), *Elem. Math.* **23** (1968), 11–14 (E. G. Straus); **MR** 37#2068; **Zbl.** 158,406.

1969.01 A problem on well ordered sets, *Acta Math. Acad. Sci. Hungar.* **20** (1969), 323–329 (A. Hajnal; E. C. Milner); **MR** 41#5222; **Zbl.** 193,308.

1969.02 Intersection theorems for systems of sets, II., *J. London Math. Soc.* **44** (1969), 467–479 (R. Rado); **MR** 39#6757; **Zbl.** 172,296.

1969.03 On a combinatorial problem, III., *Canad. Math. Bull.* **12** (1969), 413–416; **MR** 40#2551; **Zbl.** 199,318.

1969.04 On random entire functions, *Zastos. Mat.* **10** (1969), 47–55 (A. Rényi); **MR** 39#5794; **Zbl.** 256.30025.

1969.05 On some extremal properties of sequences of integers, *Ann. Univ. Sci. Budapest. Eötvös Sect. Math.* **12** (1969), 131–135 (A. Sárközy; E. Szemerédi); **MR** 41#5326; **Zbl.** 188,345.

1969.06 On some statistical properties of the alternating group of degree n, *Enseignement Math. (2)* **15** (1969), 89–99 (J. Dénes; P. Turán); **MR** 40#214; **Zbl.** 186,42.

1969.07 On the distribution of prime divisors, *Aequationes Math.* **2** (1969), 177–183; **MR** 39#5495; **Zbl.** 174,81.

1969.08 On the growth of $d_k(n)$, *Fibonacci Quart.* **7** (1969), 267–274 (I. Kátai); **MR** 40#5559; **Zbl.** 188,81.

1969.09 On the irrationality of certain series, *Math. Student* **36** (1968), 222–226, 1969; **MR** 41#6787; **Zbl.** 198,67.

1969.10 On the number of complete subgraphs and circuits contained in graphs, *Časopis Pěst. Mat.* **94** (1969), 290–296; **MR** 40#5474; **Zbl.** 177,525.

1969.11 On the sum $\sum d_4(n)$, *Acta Sci. Math. (Szeged)* **30** (1969), 313–324 (I. Kátai); **MR** 41#162; **Zbl.** 186,358.

1969.12 On the sum $\sum_{n=1}^{x} d[d(n)]$, *Math. Student* **36** (1968), 227–229, 1969; **MR** 41#6802; **Zbl.** 198,67.

1969.13 Problems and results in chromatic graph theory, *Proof Techniques in Graph Theory (Proc. Second Ann Arbor Graph Theory Conf., Ann Arbor, Mich., 1968)*, pp. 27–35, Academic Press, New York, 1969; **MR** 40#5494; **Zbl.** 194,251.

1969.14 Some applications of graph theory to number theory, *The Many Facets of Graph Theory (Proc. Conf., Western Mich. Univ., Kalamazoo, Mich., 1968)*, pp. 77–82, Springer, Berlin, 1969; **MR** 40#4149; **Zbl.** 187,210.

1969.15 Some matching theorems, *J. Indian Math. Soc. (N. S.)* **32** (1968), 215–219, 1969 (P. D. T. A. Elliott); **MR** 44#103; **Zbl.** 194,254.

1969.16 Über die in Graphen enthaltenen saturierten planaren Graphen (in German), *Math. Nachr.* **40** (1969), 13–17; **MR** 42#5851; **Zbl.** 194,254.

1969.17 Über Folgen ganzer Zahlen (in German), *Number Theory and Analysis (Papers in Honor of Edmund Landau)*, pp. 77–86, Plenum, New York, 1969 (A. Sárközy; E. Szemerédi); **MR** 41#8372; **Zbl.** 208,314.

1970.01 An extremal problem in graph theory, *J. Austral. Math. Soc.* **11** (1970), 42–47 (L. Moser); **MR** 42#5831; **Zbl.** 187,210.

1970.02 An extremal problem on the set of noncoprime divisors of a number, *Israel J. Math.* **8** (1970), 408–412 (M. Herzog; J. Schönheim); **MR** 43#178; **Zbl.** 217,307.

1970.03 Distinct distances between lattice points, *Elem. Math.* **25** (1970), 121–123 (R. K. Guy); **MR** 43#7406; **Zbl.** 222.10053.

1970.04 Extremal problems among subsets of a set, *Proc. Second Chapel Hill Conf. on Combinatorial Mathematics and its Applications (Univ. North Carolina, Chapel Hill, N.C., 1970)*, pp. 146–170, Univ. North Carolina, Chapel Hill, N.C., 1970 (D. J. Kleitman); **MR** 42#1667; **Zbl.** 215,330.

1970.05 Nonaveraging sets, II., *Combinatorial theory and its applications, II (Proc. Colloq., Balatonfüred, 1969)*, pp. 405–411, North-Holland, Amsterdam, 1970 (E. G. Straus); **MR** 47#4804; **Zbl.** 216,15.

1970.06 On a lemma of Hajnal-Folkman, *Combinatorial theory and its applications, I (Proc. Colloq., Balatonfüred, 1969)*, pp. 311–316, North-Holland, Amsterdam, 1970; **MR** 45#6655; **Zbl.** 209,280.

1970.07 On a new law of large numbers, *J. Analyse Math.* **23** (1970), 103–111 (A. Rényi); **MR** 42#6907; **Zbl.** 225.60015.

1970.08 On a problem of Moser, *Combinatorial theory and its applications, I (Proc. Colloq., Balatonfüred, 1969)*, pp. 365–367, North-Holland, Amsterdam, 1970 (J. Komlós); **MR** 45#6636; **Zbl.** 215,330.

1970.09 On divisibility properties of sequences of integers, *Number Theory (Colloq., János Bolyai Math. Soc., Debrecen, 1968)*, pp. 35–49, North Holland, Amsterdam, 1970 (A. Sárközy; E. Szemerédi); **MR** 43#4790; **Zbl.** 212,397.

1970.10 On sets of distances of n points, *Amer. Math. Monthly* **77** (1970), 739–740.

1970.11 On some applications of probability methods to additive number theoretic problems, *Contributions to Ergodic Theory and Probability (Proc. Conf., Ohio State Univ., Columbus, Ohio, 1970), Lecture Notes in Math., 160*, pp. 37–44, Springer, Berlin, 1970 (A. Rényi); **MR** 43#1938; **Zbl.** 209,354.

1970.12 On the distribution of the convergents of almost all real numbers, *J. Number Theory* **2** (1970), 425–441; **MR** 42#5941; **Zbl.** 205,349.

1970.13 On the divisibility properties of sequences of integers, *Proc. London Math. Soc. (3)* **21** (1970), 97–101 (A. Sárközy); **MR** 42#222; **Zbl.** 201,51.

1970.14 On the set of non pairwise coprime divisors of a number, *Combinatorial theory and its applications, I (Proc. Colloq., Balatonfüred, 1969)*, pp. 369–376, North-Holland, Amsterdam, 1970 (J. Schönheim); **MR** 47#1620; **Zbl.** 222.05007.

1970.15 On the sum of two Borel sets, *Proc. Amer. Math. Soc.* **25** (1970), 304–306 (A. H. Stone); **MR** 41#5578; **Zbl.** 192,403.

1970.16 Problems and results in finite and infinite combinatorial analysis, *Ann. New York Acad. Sci.* **175** (1970), 115–124 (A. Hajnal); **MR** 41#8276; **Zbl.** 236.05120.

1970.17 Problems in combinatorial set theory, *Combinatorial Structures and their Applications (Proc. Calgary Internat. Conf., Calgary, Alta. 1969)*, pp. 97–100, Gordon and Breach, New York, 1970; **MR** 41#8241; **Zbl.** 251.04004.

1970.18 Problems of graph theory concerning optimal design, *Combinatorial theory and its applications, I (Proc. Colloq., Balatonfüred, 1969)*, pp. 317–325, North-Holland, Amsterdam, 1970 (L. Gerencsér; A. Máté); **MR** 48#170; **Zbl.** 209,280.

1970.19 Set mappings and polarized partition relations, *Combinatorial theory and its applications, I (Proc. Colloq., Balatonfüred, 1969)*, pp. 327–363, North-Holland, Amsterdam, 1970 (A. Hajnal; E. C. Milner); **MR** 45#8585; **Zbl.** 215,329.

1970.20 Some applications of graph theory to number theory, *Proc. Second Chapel Hill Conf. on Combinatorial Mathematics and its Applications (Univ. North Carolina, Chapel Hill, N.C., 1970)*, pp. 136–145, Univ. North Carolina, Chapel Hill, N.C., 1970; **MR** 42#1748; **Zbl.** 214,306.

1970.21 Some extremal problems in combinatorial number theory, *Mathematical Essays Dedicated to A. J. Macintyre*, pp. 123–133, Ohio Univ. Press, Athens, Ohio, 1970; **MR** 43#1942; **Zbl.** 214,306.

1970.22 Some extremal problems in graph theory, *Combinatorial theory and its applications, I (Proc. Colloq., Balatonfüred, 1969)*, pp. 377–390, North-Holland, Amsterdam, 1970 (M. Simonovits); **MR** 46#84; **Zbl.** 209,280.

1970.23 Some problems in additive number theory, *Amer. Math. Monthly* **77** (1970), 619–621; **MR** 42#3040.

1970.24 Some remarks on Ramsey's and Turán's theorem, *Combinatorial theory and its applications, II (Proc. Colloq., Balatonfüred, 1969)*, pp. 395–404, North-Holland, Amsterdam, 1970 (V. T. Sós); **MR** 45#8560; **Zbl.** 209,280.

1970.25 Some results and problems on certain polarized partitions, *Acta Math. Acad. Sci. Hungar.* **21** (1970), 369–392 (A. Hajnal); **MR** 43#7357; **Zbl.** 214,27.

1971.01 An extremal graph problem, *Acta Math. Acad. Sci. Hungar.* **22** (1971/72), 275–282 (M. Simonovits); **MR** 45#1790; **Zbl.** 234.05118.

1971.02 Child prodigies, *Proceedings of the Washington State University Conference on Number Theory (Pullman, Wash., 1971)*, pp. 1–12, Dept. Math., Washington State Univ., Pullman, Wash., 1971; **MR** 47#4754; **Zbl.** 242.01017.

1971.03 Complete prime subsets of consecutive integers, *Proceedings of the Manitoba Conference on Numerical Mathematics (Univ. Manitoba, Winnipeg, Man., 1971)*, pp. 1–14, Dept. Comput. Sci., Univ. Manitoba, Winnipeg, Man., 1971 (J. L. Selfridge); **MR** 49#2597; **Zbl.** 267.10054.

1971.04 Decompositions of complete graphs into factors with diameter two, *Mat. Časopis Sloven. Akad. Vied* **21** (1971), 14–28 (J. Bosák; A. Rosa); **MR** 48#165; **Zbl.** 213,510.

1971.05 Imbalances in k-colorations, *Networks* **1** (1971/72), 379–385 (J. H. Spencer); **MR** 45#8573; **Zbl.** 248.05114.

1971.06 Non complete sums of multiplicative functions, *Period. Math. Hungar.* **1** (1971) no. 3, 209–212 (I. Kátai); **MR** 44#6625; **Zbl.** 227.10005.

1971.07 On collections of subsets containing no 4-member Boolean algebra, *Proc. Amer. Math. Soc.* **28** (1971), 87–90 (D. J. Kleitman); **MR** 42#5807; **Zbl.** 214,28.

1971.08 On some applications of graph theory, II., *Studies in Pure Mathematics (Presented to Richard Rado)*, pp. 89–99, Academic Press, London, 1971 (A. Meir; V. T. Sós; P. Turán); **MR** 44#3887; **Zbl.** 218.52005.

1971.09 On some extremal problems on r-graphs, *Discrete Math.* **1** (1971/72) no. 1, 1–6; **MR** 45#6656; **Zbl.** 211,270.

1971.10 On some general problems in the theory of partitions, I., *Acta Arith.* **18** (1971), 53–62 (P. Turán); **MR** 44#6636; **Zbl.** 217,322.

1971.11 On some problems of a statistical group theory, V., *Period. Math. Hungar.* **1** (1971) no. 1, 5–13 (P. Turán); **MR** 44#6638; **Zbl.** 223.10005.

1971.12 On some problems of a statistical group theory, VI., *J. Indian Math. Soc. (N.S.)* **34** (1970) no. 3–4, 175–192, 1971 (P. Turán); **MR** 58#21974; **Zbl.** 235.10008.

1971.13 On the application of combinatorial analysis to number theory, geometry and analysis, *Actes du Congrès International des Mathématiciens (Nice, 1970), Tome 3*, pp. 201–210, Gauthier-Villars, Paris, 1971; **MR** 54#7278; **Zbl.** 231.05003.

1971.14 On the sum $\sum_{d|2^n-1} d^{-1}$, *Israel J. Math.* **9** (1971), 43–48; **MR** 42#4508; **Zbl.** 209,343.

1971.15 Ordinary partition relations for ordinal numbers, *Period. Math. Hungar.* **1** (1971) no. 3, 171–185 (A. Hajnal); **MR** 45#8531; **Zbl.** 257.04004.

1971.16 Partition relations for η_α-sets, *J. London Math. Soc. (2)* **3** (1971), 193–204 (E. C. Milner; R. Rado); **MR** 43#60; **Zbl.** 212,22.

1971.17 Polarized partition relations for ordinal numbers, *Studies in Pure Mathematics (Presented to Richard Rado)*, pp. 63–87, Academic Press, London, 1971 (A. Hajnal; E. C. Milner); **MR** 43#3123; **Zbl.** 228.04002.

1971.18 Problems and results in combinatorial analysis, *Combinatorics (Proc. Sympos. Pure Math., Vol. XIX, Univ. California, Los Angeles, Calif., 1968)*, pp. 77–89, Amer. Math. Soc., Providence, R. I., 1971; **MR** 55#5450; **Zbl.** 231.05002.

1971.19 Ramsey bounds for graph products, *Pacific J. Math.* **37** (1971), 45–46 (R. J. McEliece; H. Taylor); **MR** 46#3376; **Zbl.** 207,228 and 213,262.

1971.20 Some extremal problems in geometry, *J. Combinatorial Theory Ser. A* **10** (1971), 246–252 (G. B. Purdy); **MR** 43#1045; **Zbl.** 219.05006.

1971.21 Some number theoretic results, *Pacific J. Math.* **36** (1971), 635–646 (E. G. Straus); **MR** 43#7413; **Zbl.** 216,322.

1971.22 Some probabilistic remarks on Fermat's last theorem, *Rocky Mountain J. Math.* **1** (1971), 613–616 (S. Ulam); **MR** 44#2724; **Zbl.** 228.10035.

1971.23 Some problems in number theory, *Computers in number theory (Proc. Atlas Sympos., Oxford, 1969)*, pp. 405–414, Academic Press, London, 1971; **Zbl.** 217,31.

1971.24 Some problems on the prime factors of consecutive integers, II., *Proceedings of the Washington State University Conference on Number Theory (Pullman, Wash., 1971)*, pp. 13–21, Dept. Math., Washington State Univ., Pullman, Wash., 1971 (J. L. Selfridge); **MR** 47#6625; **Zbl.** 228.10028.

1971.25 Some unsolved problems in graph theory and combinatorial analysis, *Combinatorial Mathematics and its Applications (Proc. Conf., Oxford, 1969)*, pp. 97–109, Academic Press, London, 1971; **MR** 43#3125; **Zbl.** 221.05051.

1971.26 Topics in combinatorial analysis, *Proceedings of the Second Louisiana Conference on Combinatorics, Graph Theory and Computing (Louisiana State Univ., Baton Rouge, 1971)*, pp. 2–20, Louisiana State University, Baton Rouge, LA, 1971; **MR** 47#4750 (for entire conference proceedings); **Zbl.** 289.05001.

1971.27 Turán Pál gráf tételéről (On the graph theorem of Turán, in Hungarian), *Mat. Lapok* **21** (1970), 249–251, 1971; **MR** 46#7090; **Zbl.** 231.05110.

1971.28 Unsolved problems in set theory, *Axiomatic Set Theory (Proc. Sympos. Pure Math., Vol. XIII, Part I, Univ. California, Los Angeles, Calif,. 1967)*, pp. 17–48, Amer. Math. Soc., Providence, R.I., 1971 (A. Hajnal); **MR** 43#6101; **Zbl.** 228.04001.

1972.01 A characterization of finitely monotonic additive functions, *J. London Math. Soc. (2)* **5** (1972), 362–367 (C. Ryavec); **MR** 48#8362; **Zbl.** 238.10002.

1972.02 A note on Hamiltonian circuits, *Discrete Math.* **2** (1972), 111–113 (V. Chvátal); **MR** 45#6654; **Zbl.** 233.05123.

1972.03 A theorem in the partition calculus, *Canad. Math. Bull.* **15** (1972), 501–505 (E. C. Milner); **MR** 48#10819; **Zbl.** 271.04003.

1972.04 Erdős és Hajnal egy problémájáról (On a certain problem of Erdős and Hajnal, in Hungarian, English summary), *Mat. Lapok* **22** (1971), 1–2, 1972 (J. H. Spencer); **MR** 48#112; **Zbl.** 247.05007.

1972.05 Extremal problems in number theory, *Proceedings of the Number Theory Conference (Univ. Colorado, Boulder, Colo., 1972)*, pp. 80–86, Univ. Colorado, Boulder, Colo., 1972; **MR** 52#13713; **Zbl.** 325.10001.

1972.06 On a linear diophantine problem of Frobenius, *Acta Arith.* **21** (1972), 399–408 (R. L. Graham); **MR** 47#127; **Zbl.** 246.10010.

1972.07 On a problem of Grünbaum, *Canad. Math. Bull.* **15** (1972), 23–25; **MR** 47#5709; **Zbl.** 233.05017.

1972.08 On a Ramsey type theorem, Collection of articles dedicated to the memory of Alfréd Rényi, I., *Period. Math. Hungar.* **2** (1972), 295–299 (E. Szemerédi); **MR** 48#3793; **Zbl.** 242.05122.

1972.09 On problems of Moser and Hanson, *Graph theory and applications (Proc. Conf., Western Michigan Univ., Kalamazoo, Mich., 1972; dedicated to the memory of J. W. T. Youngs), Lecture Notes in Math., 303*, pp. 75–79, Springer, Berlin, 1972 (S. Shelah); **MR** 49#2415; **Zbl.** 249.05004.

1972.10 On Ramsey like theorems. Problems and results, *Combinatorics (Proc. Conf. Combinatorial Math., Math. Inst., Oxford, 1972)*, pp. 123–140, Inst. Math. Appl., Southend-on-Sea, 1972 (A. Hajnal); **MR** 49#2405; **Zbl.** 469.05001 (for entire book).

1972.11 On some applications of graph theory, I., *Discrete Math.* **2** (1972) no. 3, 207–228 (A. Meir; V. T. Sós; P. Turán); **MR** 46#5053; **Zbl.** 236.05119.

1972.12 On some applications of graph theory, III., *Canad. Math. Bull.* **15** (1972), 27–32 (A. Meir; V. T. Sós; P. Turán); **MR** 50#4393; **Zbl.** 232.05003.

1972.13 On some problems of a statistical group theory, VII., Collection of articles dedicated to the memory of Alfréd Rényi, I., *Period. Math. Hungar.* **2** (1972), 149–163 (P. Turán); **MR** 56#5470; **Zbl.** 247.20008.

1972.14 On sums of Fibonacci numbers, *Fibonacci Quart.* **10** (1972), 249–254 (R. L. Graham); **MR** 46#5228; **Zbl.** 235.10006.

1972.15 On the distribution of the roots of orthogonal polynomials, *Proceedings of the Conference on the Constructive Theory of Functions (Approximation Theory) (Budapest, 1969)*, pp. 145–150, Akadémiai Kiadó, Budapest, 1972; **MR** 53#13987; **Zbl.** 234.33014.

1972.16 On the fundamental problem of mathematics, *Amer. Math. Monthly* **79** (1972), 149–150; **Zbl.** 231.00006.

1972.17 On the iterates of some arithmetic functions, *The theory of arithmetic functions (Proc. Conf., Western Michigan Univ., Kalamazoo, Mich. 1971), Lecture Notes in Math., 251*, pp. 119–125, Springer, Berlin, 1972 (M. V. Subbarao); **MR** 48#11012; **Zbl.** 228.10033.

1972.18 On the number of unique subgraphs of a graph, *J. Combinatorial Theory Ser. B* **13** (1972), 112–115 (R. Entringer); **MR** 47#6539; **Zbl.** 241.05111.

1972.19 Partition relations for η_α and for $\aleph_\alpha$-saturated models, *Theory of sets and topology (in honour of Felix Hausdorff, 1868–1942)*, pp. 95–108, VEB Deutsch. Verlag Wissensch., Berlin, 1972 (A. Hajnal; E. C. Milner); **MR** 49#7143; **Zbl.** 277.04006.

1972.20 Ramsey's theorem and self-complementary graphs, *Discrete Math.* **3** (1972), 301–304 (V. Chvátal; Z. Hedrlín); **MR** 47#1674; **Zbl.** 244.05114.

1972.21 Separability properties of almost-disjoint families of sets, *Israel J. Math.* **12** (1972), 207–214 (S. Shelah); **MR** 47#8312; **Zbl.** 246.05002.

1972.22 Simple one-point extensions of tournaments, *Mathematika* **19** (1972), 57–62 (A. Hajnal; E. C. Milner); **MR** 52#2947; **Zbl.** 242.05113.

1972.23 Some problems on consecutive prime numbers, *Mathematika* **19** (1972), 91–95; **MR** 47#4949; **Zbl.** 245.10032.

1972.24 Some remarks on simple tournaments, *Algebra Universalis* **2** (1972), 238–245 (E. Fried; A. Hajnal; E. C. Milner); **MR** 46#5161; **Zbl.** 267.05104.

1972.25 Two combinatorial problems in group theory, *Acta Arith.* **21** (1972), 111–116 (R. B. Eggleton); **MR** 46#3643; **Zbl.** 248.20068.

1973.01 A remark on polynomials and the transfinite diameter, *Israel J. Math.* **14** (1973), 23–25 (E. Netanyahu); **MR** 47#7006; **Zbl.** 259.30004.

1973.02 A triangle inequality, *Univ. Beograd. Publ. Elektrotehn. Fak. Ser. Mat. Fiz.* No. 412–460 (1973), 117–118 (M. S. Klamkin); **MR** 48#4208; **Zbl.** 278.50008.

1973.03 Chain conditions on set mappings and free sets, *Acta Sci. Math. (Szeged)* **34** (1973), 69–79 (A. Hajnal; A. Máté); **MR** 47#6492; **Zbl.** 274.04005.

1973.04 Corrigendum: "On some applications of graph theory, I." [*Discrete Math.* **2** (1972) no. 3, 207–228], *Discrete Math.* **4** (1973), 90 (A. Meir; V. T. Sós; P. Turán); **MR** 46#7093; **Zbl.** 245.05130.

1973.05 Crossing number problem, *Amer. Math. Monthly* **80** (1973), 52–58 (R. K. Guy); **MR** 52#2894; **Zbl.** 264.05109.

1973.06 Diagonals of nonnegative matrices, *Linear and Multilinear Algebra* **1** (1973) no. 2, 89–95 (H. Minc); **MR** 48#2164; **Zbl.** 277.15011.

1973.07 Dissection of graphs of planar point sets, *A survey of combinatorial theory (Proc. Internat. Sympos., Colorado State Univ., Fort Collins,*

Colo., 1971), pp. 139–149, North-Holland, Amsterdam, 1973 (L. Lovász; A. Simmons; E. G. Straus); **MR** 51#241; **Zbl.** 258.05112.

1973.08 Euclidean Ramsey theorems, I., *J. Combinatorial Theory Ser. A* **14** (1973), 341–363 (R. L. Graham; P. Montgomery; B. L. Rothschild; J. H. Spencer; E. G. Straus); **MR** 47#4825; **Zbl.** 276.05001.

1973.09 Extremal problems for directed graphs, *J. Combinatorial Theory Ser. B* **15** (1973), 77–93 (W. G. Brown; M. Simonovits); **MR** 52#7952; **Zbl.** 257.05112 and 253.05124.

1973.10 On a combinatorial game, *J. Combinatorial Theory Ser. A* **14** (1973), 298–301 (J. L. Selfridge); **MR** 48#5655; **Zbl.** 293.05004.

1973.11 On a generalization of Ramsey numbers, *Discrete Math.* **4** (1973), 29–35 (P. E. O'Neil); **MR** 46#8849; **Zbl.** 249.05003.

1973.12 On a valence problem in extremal graph theory, *Discrete Math.* **5** (1973), 323–334 (M. Simonovits); **MR** 49#7175; **Zbl.** 268.05121.

1973.13 "On chromatic number of graphs and set-systems" by Erdős and Hajnal [*Acta Math. Acad. Sci. Hungar.* **17** (1966), 61–99], *Cambridge Summer School in Mathematical Logic (Cambridge, England, 1971), Lecture Notes in Math., 337*, pp. 531–538, Springer, Berlin, 1973 (A. Hajnal; B. L. Rothschild); **MR** 52#7949; **Zbl.** 289.04002.

1973.14 On the capacity of graphs, Collection of articles dedicated to the memory of Alfréd Rényi, II., *Period. Math. Hungar.* **3** (1973), 125–133 (J. Komlós); **MR** 49#126; **Zbl.** 245.05112.

1973.15 On the existence of triangulated spheres in 3-graphs, and related problems, *Period. Math. Hungar.* **3** (1973) no. 3–4, 221–228 (W. G. Brown; V. T. Sós); **MR** 48#2003; **Zbl.** 269.05111.

1973.16 On the number of solutions of $f(n) = a$ for additive functions, Collection of articles dedicated to Carl Ludwig Siegel on the occasion of his seventy-fifth birthday, I., *Acta Arith.* **24** (1973), 1–9 (I. Z. Ruzsa; A. Sárközy); **MR** 48#11013; **Zbl.** 261.10007.

1973.17 On the number of solutions of $m = \sum_{i=1}^{k} \chi_i^k$, *Analytic number theory (Proc. Symp. Pure Math., Vol. XXIV, St. Louis Univ., St. Louis, Mo., 1972)*, pp. 83–90, Amer. Math. Soc., Providence, R.I., 1973, 83–90 (E. Szemerédi); **MR** 49#2529; **Zbl.** 264.10016.

1973.18 On the structure of edge graphs, *Bull. London Math. Soc.* **5** (1973), 317–321 (B. Bollobás); **MR** 49#129; **Zbl.** 277.05135.

1973.19 On the values of Euler's φ-function, *Acta Arith.* **22** (1973), 201–206 (R. R. Hall); **MR** 53#13143; **Zbl.** 252.10007.

1973.20 Osculation vertices in arrangements of curves, *Geometriae Dedicata* **1** (1973) no. 3, 322–333 (B. Grünbaum); **MR** 47#5705; **Zbl.** 257.52016.

1973.21 Problems and results on combinatorial number theory, *A survey of combinatorial theory (Proc. Internat. Sympos., Colorado State Univ., Fort Collins, Colo., 1971)*, pp. 117–138, North-Holland, Amsterdam, 1973; **MR** 50#12957; **Zbl.** 263.10001.

1973.22 Ramsey numbers for cycles in graphs, *J. Combinatorial Theory Ser. B* **14** (1973), 46–54 (J. A. Bondy); **MR** 47#6540; **Zbl.** 248.05127.

1973.23 Rational approximation to certain entire functions in $[0, +\infty)$, *Bull. Amer. Math. Soc.* **79** (1973), 992–993 (A. R. Reddy); **MR** 49#3390; **Zbl.** 272.41007.

1973.24 Résultats et problèmes en théorie des nombres (in French), *Séminaire Delange-Pisot-Poitou (14e année: 1972/73), Théorie des nombres, Fasc. 2, Exp. No. 24*, 7 pp., Secrétariat Mathematique, Paris, 1973; **MR** 53#243; **Zbl.** 319.10002.

1973.25 Some extremal problems on r-graphs, *New directions in the theory of graphs (Proc. Third Ann Arbor Conf., Univ. Michigan, Ann Arbor, Mich., 1971)*, pp. 53–63, Academic Press, New York, 1973 (W. G. Brown; V. T. Sós); **MR** 50#4376; **Zbl.** 258.05132.

1973.26 Some extremal properties concerning transitivity in graphs, *Period. Math. Hungar.* **3** (1973) no. 3–4, 275–279 (R. Entringer; C. C. Harner); **MR** 48#169; **Zbl.** 263.05108.

1973.27 *The art of counting: Selected writings*, edited by Joel Spencer and with a dedication by Richard Rado, Mathematicians of Our Time, Vol. 5, xxiii+742 pp., MIT Press, Cambridge, Mass.-London, 1973; **MR** 58#27144; **Zbl.** 287.01028.

1973.28 The asymmetric propeller, *Math. Mag.* **46** (1973), 270–272 (L. Bankoff; M. S. Klamkin); **MR** 48#7099; **Zbl.** 274.50006.

1973.29 Über die Anzahl der Primfaktoren von $\binom{n}{k}$ (in German), *Arch. Math. (Basel)* **24** (1973), 53–56; **MR** 47#8422; **Zbl.** 251.10010.

1973.30 Über die Zahlen der form $\sigma(n) - n$ und $n - \varphi(n)$ (in German), *Elem. Math.* **28** (1973), 83–86; **MR** 49#2502; **Zbl.** 272.10003.

1974.01 A new function associated with the prime factors of $\binom{n}{k}$, *Math. Comp.* **28** (1974), 647–649 (E. F. Ecklund, Jr.; J. L. Selfridge); **MR** 49#2501; **Zbl.** 279.10034.

1974.02 Asymptotic distribution of normalized arithmetical functions, *Proc. Amer. Math. Soc.* **46** (1974), 1–8 (J. Galambos); **MR** 50#9828; **Zbl.** 287.10048 and 275.10028.

1974.03 Bounds for the r-th coefficients of cyclotomic polynomials, *J. London Math. Soc. (2)* **8** (1974), 393–400 (R. C. Vaughan); **MR** 50#9835; **Zbl.** 295.10014.

1974.04 Chebyshev rational approximation to entire functions in $[0, +\infty)$, *Mathematical Structures (papers dedicated to Professor L. Iliev's 60th anniversary)*, pp. 225–234, Sofia, 1974 (A. R. Reddy); **Zbl.** 269.41014.

1974.05 Complete subgraphs of chromatic graphs and hypergraphs, *Utilitas Math.* **6** (1974), 343–347 (B. Bollobás; E. G. Straus); **MR** 52#162; **Zbl.** 333.05116.

1974.06 Correction to: "Osculation vertices in arrangements of curves" [*Geometriae Dedicata* **1** (1973), 322–333], *Geometriae Dedicata* **3** (1974), 130 (B. Grünbaum); **MR** 48#12240.

1974.07 Corrigendum: "A theorem in the partition calculus" [*Canad. Math. Bull.* **15** (1972), 501–505], *Canad. Math. Bull.* **17** (1974), 305 (E. C. Milner); **MR** 50#12734; **Zbl.** 289.04001.

1974.08 Exhausting an area with discs, *Proc. Amer. Math. Soc.* **45** (1974), 305–308 (D. J. Newman); **MR** 50#8307; **Zbl.** 298.52010.

1974.09 Extremal problems among subsets of a set, *Discrete Math.* **8** (1974), 281–294 (D. J. Kleitman); **MR** 48#10821; **Zbl.** 281.04002.

1974.10 Extremal problems on graphs and hypergraphs, *Hypergraph Seminar (Proc. First Working Sem., Ohio State Univ., Columbus, Ohio, 1972; dedicated to Arnold Ross), Lecture Notes in Math., 411*, pp. 75–84, Springer, Berlin, 1974; **MR** 50#12800; **Zbl.** 301.05123.

1974.11 Intersection theorems for systems of sets, III., Collection of articles dedicated to the memory of Hanna Neumann, IX., *J. Austral. Math. Soc.* **18** (1974), 22–40 (E. C. Milner; R. Rado); **MR** 51#169; **Zbl.** 331.04002.

1974.12 On abundant-like numbers, *Canad. Math. Bull.* **17** (1974), 599–602; **MR** 52#8013; **Zbl.** 312.10003.

1974.13 On orthogonal polynomials with regularly distributed zeros, *Proc. London Math. Soc. (3)* **29** (1974), 521–537 (G. Freud); **MR** 54#8134; **Zbl.** 294.33006.

1974.14 On products of integers, Collection of articles dedicated to K. Mahler on the occasion of his seventieth birthday, *J. Number Theory* **6** (1974), 416–421 (S. L. G. Choi); **MR** 51#12740; **Zbl.** 294.10029.

1974.15 On refining partitions, *J. London Math. Soc. (2)* **9** (1974/75), 565–570 (R. K. Guy; J. W. Moon); **MR** 50#12752; **Zbl.** 312.05008.

1974.16 On sets of consistent arcs in a tournament (in Russian), *Teor. Graf. Pokryt. Ukladki Turniry*, pp. 160–162, 1974; **Zbl.** 289.05114.

1974.17 On some general properties of chromatic numbers, *Topics in topology (Proc. Colloq., Keszthely, 1972); Colloq. Math. Soc. János Bolyai, Vol. 8*, pp. 243–255, North-Holland, Amsterdam, 1974 (A. Hajnal; S. Shelah); **MR** 50#9662; **Zbl.** 299.02083.

1974.18 On the connection between chromatic number, maximal clique and minimal degree of a graph, *Discrete Math.* **8** (1974), 205–218 (B. Andrásfai; V. T. Sós); **MR** 49#4831; **Zbl.** 284.05106.

1974.19 On the distribution of numbers of the form $\sigma(n)/n$ and on some related questions, *Pacific J. Math.* **52** (1974), 59–65; **MR** 50#7079; **Zbl.** 291.10040.

1974.20 On the distribution of values of certain divisor functions, *J. Number Theory* **6** (1974), 52–63 (R. R. Hall); **MR** 49#2606; **Zbl.** 274.10043.

1974.21 On the existence of a factor of degree one of a connected random graph (in Russian), *Teor. Graf. Pokryt. Ukladki Turniry*, pp. 12–23, 1974; **Zbl.** 289.05128.

1974.22 On the irrationality of certain series, *Pacific J. Math.* **55** (1974), 85–92 (E. G. Straus); **MR** 51#3067; **Zbl.** 294.10024 and 279.10026.

1974.23 On the number of times an integer occurs as a binomial coefficient, *Amer. Math. Monthly* **81** (1974), 256–261 (H. L. Abbott; D. Hanson); **MR** 49#65; **Zbl.** 276.05005.

1974.24 On weird and pseudoperfect numbers, *Math. Comp.* **28** (1974), 617–623 (S. J. Benkoski); **MR** 50#228; **Zbl.** 279.10005.

1974.25 *Probabilistic methods in combinatorics, Probability and Mathematical Statistics, Vol. 17*, 106 pp., Academic Press [A subsidiary of Harcourt Brace Jovanovich, Publishers], New York-London, 1974 (J. H. Spencer); **MR** 52#2895; **Zbl.** 308.05001. Also translated into Russian, along with a translation of Erdős's paper "Extremal problems among subsets of a set" [*Discrete Math.* **8** (1974), 281–294 (D. J. Kleitman)], Izdat. "Mir", Moscow, 1976, 131 pp.; **MR** 54#2470.

1974.26 Remark on a theorem of Lindström, *J. Combinatorial Theory Ser. A* **17** (1974), 129–130; **MR** 49#7144; **Zbl.** 285.05004.

1974.27 Remarks on some problems in number theory, Papers presented at the Fifth Balkan Mathematical Congress (Belgrade, 1974), *Math. Balkanica* **4** (1974), 197–202; **MR** 55#2715; **Zbl.** 313.10045.

1974.28 Some distribution problems concerning the divisors of integers, *Acta Arith.* **26** (1974/75), 175–188 (R. R. Hall); **MR** 50#7070; **Zbl.** 292.10027 and 272.10021.

1974.29 Some matching theorems (in Russian), *Teor. Graf. Pokryt. Ukladki Turniry*, pp. 7–11, 1974; **Zbl.** 289.05124.

1974.30 Some new applications of probability methods to combinatorial analysis and graph theory, *Proceedings of the Fifth Southeastern Conference on Combinatorics, Graph Theory and Computing (Florida Atlantic Univ., Boca Raton, Fla., 1974)*, pp. 39–51, *Congress. Numer. X*, Utilitas Math., Winnipeg, Man., 1974; **MR** 51#275; **Zbl.** 312.05126.

1974.31 Some problems in graph theory, *Hypergraph Seminar (Proc. First Working Sem., Ohio State Univ., Columbus, Ohio, 1972; dedicated to Arnold Ross), Lecture Notes in Math., 411*, pp. 187–190, Springer, Berlin, 1974; **MR** 52#193; **Zbl.** 297.05133.

1974.32 Some problems on random intervals and annihilating particles, *Ann. Probability* **2** (1974), 828–839 (P. Ney); **MR** 51#9270; **Zbl.** 297.60052.

1974.33 Some remarks on set theory, XI., Collection of articles dedicated to Andrzej Mostowski on the occasion of his sixtieth birthday, III., *Fund. Math.* **81** (1974), 261–265 (A. Hajnal); **MR** 50#114; **Zbl.** 285.04002.

1974.34 The arithmetic function $\sum_{d|n} \log d/d$, Collection of articles dedicated to Stanislaw Golab on his 70th birthday, II., *Demonstratio Math.* **6** (1973), 575–579, 1974 (S. K. Zaremba); **MR** 50#4513; **Zbl.** 287.10005.

1974.35 The chromatic index of an infinite complete hypergraph: a partition theorem, *Hypergraph Seminar (Proc. First Working Sem., Ohio State Univ., Columbus, Ohio, 1972; dedicated to Arnold Ross), Lecture Notes in Math., 411*, pp. 54–60, Springer, Berlin, 1974 (R. Bonnet); **MR** 51#10145; **Zbl.** 311.05113.

1974.36 Unsolved and solved problems in set theory, *Proceedings of the Tarski Symposium (Proc. Sympos. Pure Math., Vol. XXV, Univ. California, Berkeley, Calif., 1971)*, pp. 269–287, Amer. Math. Soc., Providence, R.I., 1974 (A. Hajnal); **MR** 50#9590; **Zbl.** 334.04003.

1974.37 Very slowly varying functions, *Aequationes Math.* **10** (1974), 1–9 (J. M. Ash; L. A. Rubel); **MR** 48#8698; **Zbl.** 273.26001.

1975.01 A non-normal box product, *Infinite and finite sets (Colloq., Keszthely, 1973; dedicated to P. Erdős on his 60th birthday), Vol. II; Colloq. Math. Soc. János Bolyai, Vol. 10*, pp. 629–631, North-Holland, Amsterdam, 1975 (M. E. Rudin); **MR** 52#15338; **Zbl.** 328.54017.

1975.02 A note on rational approximation, *Period. Math. Hungar.* **6** (1975) no. 3, 241–244 (A. R. Reddy); **MR** 52#11418; **Zbl.** 307.41008 and 273.41012.

1975.03 An asymptotic formula in additive number theory, *Acta Arith.* **28** (1975/76) no. 4, 405–412 (G. J. Babu; K. Ramachandra); **MR** 53#7974; **Zbl.** 315.10042 and 278.10047.

1975.04 An extremal problem of graphs with diameter 2, *Math. Mag.* **48** (1975), 281–283 (B. Bollobás); **Zbl.** 353.05045.

1975.05 Anti-Ramsey theorems, *Infinite and finite sets (Colloq., Keszthely, 1973; dedicated to P. Erdős on his 60th birthday), Vol. II; Colloq. Math. Soc. János Bolyai, Vol. 10*, pp. 633–643, North-Holland, Amsterdam, 1975 (M. Simonovits; V. T. Sós); **MR** 52#164; **Zbl.** 316.05111.

1975.06 Conditions for a zero sum modulo n, *Canad. Math. Bull.* **18** (1975), 27–29 (J. D. Bovey; I. Niven); **MR** 52#13714; **Zbl.** 314.10040.

1975.07 Consecutive integers, *Eureka, The Archimedeans' Journal* **38** (1975/76), 3–8.

1975.08 Distribution of rational points on the real line, *J. Austral. Math. Soc.* **20** (1975), 124–128 (T. K. Sheng); **MR** 52#309; **Zbl.** 307.10020.

1975.09 Edge decompositions of the complete graph into copies of a connected subgraph, *Proceedings of the Conference on Algebraic Aspects of Combinatorics (Univ. Toronto, Toronto, Ont., 1975), Congress. Numer. XIII*, pp. 271–278, Utilitas Math., Winnipeg, Man., 1975 (J. Schönheim); **MR** 52#10501; **Zbl.** 323.05130.

1975.10 Ein Nachtrag über befreundete Zahlen (in German), *J. Reine Angew. Math.* **273** (1975), 220 (G. J. Rieger); **MR** 51#412; **Zbl.** 298.10004.

1975.11 Euclidean Ramsey theorems, II., *Infinite and finite sets (Colloq., Keszthely, 1973; dedicated to P. Erdős on his 60th birthday), Vol.*

I; Colloq. Math. Soc. János Bolyai, Vol. 10, pp. 529–557, North-Holland, Amsterdam, 1975 (R. L. Graham; P. Montgomery; B. L. Rothschild; J. H. Spencer; E. G. Straus); **MR** 52#2935; **Zbl.** 313.05002.

1975.12 Euclidean Ramsey theorems, III., *Infinite and finite sets (Colloq., Keszthely, 1973; dedicated to P. Erdős on his 60th birthday), Vol. I; Colloq. Math. Soc. János Bolyai, Vol. 10*, pp. 559–583, North-Holland, Amsterdam, 1975 (R. L. Graham; P. Montgomery; B. L. Rothschild; J. H. Spencer; E. G. Straus); **MR** 52#2936; **Zbl.** 313.05002.

1975.13 Extensions de quelques théorèmes de densité (in French), *Séminaire Delange-Pisot-Poitou (16e année: 1974/75), Théorie des Nombres, Fasc. I, Exp. No. 8*, 2 pp., Secrétariat Mathematique, Paris, 1975; **Zbl.** 316.10037.

1975.14 Extension de quelques théorèmes sur les densités de séries d' éleménts de N à des séries de sous-ensembles finis de N (in French, English summary), *Discrete Math.* **12** (1975) no. 4, 295–308 (M. Deza); **MR** 55#5569; **Zbl.** 308.05004.

1975.15 Factorizing the complete graph into factors with large star number, *J. Combinatorial Theory Ser. B.* **18** (1975), 180–183 (N. Sauer; J. Schaer; J. H. Spencer); **MR** 51#276; **Zbl.** 295.05116 and 279.05122.

1975.16 How abelian is a finite group?, *Linear and Multilinear Algebra* **3** (1975/76) no. 4, 307–312 (E. G. Straus); **MR** 53#10933; **Zbl.** 335.20011.

1975.17 Maximal asymptotic nonbases, *Proc. Amer. Math. Soc.* **48** (1975), 57–60 (M. B. Nathanson); **MR** 50#9831; **Zbl.** 296.10031.

1975.18 Méthodes probabilistes en théorie des nombres (in French), *Séminaire Delange-Pisot-Poitou (15e année: 1973/74), Théorie des nombres, Fasc. 1, Exp. No. 1*, 4 pp., Secrétariat Mathematique, Paris, 1975; **MR** 53#13154; **Zbl.** 329.10042.

1975.19 On complete subgraphs of r-chromatic graphs, *Discrete Math.* **13** (1975), 97–107 (B. Bollobás; E. Szemerédi); **MR** 52#10470; **Zbl.** 306.05121.

1975.20 On graphs of Ramsey type, *Proceedings of the Sixth Southeastern Conference on Combinatorics, Graph Theory and Computing (Florida Atlantic Univ., Boca Raton, Fla., 1975), Congress. Numer. XIV*, p. 643, Utilitas Math., Winnipeg, Man., 1975 (S. A. Burr; L. Lovász); **Zbl.** 338.05115.

1975.21 On maximal almost-disjoint families over singular cardinals, *Infinite and finite sets (Colloq., Keszthely, 1973; dedicated to P. Erdős on his 60th birthday), Vol. I; Colloq. Math. Soc. János Bolyai, Vol. 10*, pp. 597–604, North-Holland, Amsterdam, 1975 (S. H. Hechler); **MR** 51#12530; **Zbl.** 326.02050.

1975.22 On packing squares with equal squares, *J. Combinatorial Theory Ser. A.* **19** (1975), 119–123 (R. L. Graham); **MR** 51#6595; **Zbl.** 324.05018.

1975.23 On partition theorems for finite graphs, *Infinite and finite sets (Colloq., Keszthely, 1973; dedicated to P. Erdős on his 60th birthday), Vol. I; Colloq. Math. Soc. János Bolyai, Vol. 10*, pp. 515–527, North-Holland, Amsterdam, 1975 (R. L. Graham); **MR** 51#10159; **Zbl.** 324.05124.

1975.24 On set-systems having large chromatic number and not containing prescribed subsystems, *Infinite and finite sets (Colloq., Keszthely, 1973; dedicated to P. Erdős on his 60th birthday), Vol. I; Colloq. Math. Soc. János Bolyai, Vol. 10*, pp. 425–513, North-Holland, Amsterdam, 1975 (F. Galvin; A. Hajnal); **MR** 53#2727; **Zbl.** 324.04005.

1975.25 On some problems of elementary and combinatorial geometry, *Ann. Mat. Pura Appl. (4)* **103** (1975), 99–108; **MR** 54#113; **Zbl.** 303.52006.

1975.26 On the magnitude of generalized Ramsey numbers for graphs, *Infinite and finite sets (Colloq., Keszthely, 1973; dedicated to P. Erdős on his 60th birthday), Vol. I; Colloq. Math. Soc. János Bolyai, Vol. 10*, pp. 215–240, North-Holland, Amsterdam, 1975 (S. A. Burr); **MR** 51#7918; **Zbl.** 316.05110.

1975.27 On the prime factors of $\binom{2n}{n}$, Collection of articles dedicated to Derrick Henry Lehmer on the occasion of his seventieth birthday, *Math. Comp.* **29** (1975), 83–92 (R. L. Graham; I. Z. Ruzsa; E. G. Straus); **MR** 51#5523; **Zbl.** 296.10008.

1975.28 On the structure of edge graphs, II., *J. London Math. Soc. (2)* **12** (1975/76) no. 2, 219–224 (B. Bollobás; M. Simonovits); **MR** 58#1805; **Zbl.** 318.05110.

1975.29 Oscillations of bases for the natural numbers, *Proc. Amer. Math. Soc.* **53** (1975), 253–258 (M. B. Nathanson); **MR** 52#5612; **Zbl.** 319.10066.

1975.30 Problems and results in combinatorial number theory, *Journées Arithmétiques de Bordeaux (Conf., Univ. Bordeaux, Bordeaux, 1974), Astérisque, Nos. 24–25*, pp. 295–310, Soc. Math. France, Paris, 1975; **MR** 51#10275; **Zbl.** 305.10050.

1975.31 Problems and results on Diophantine approximations, II., Répartition modulo 1 *(Actes Colloq., Marseille-Luminy, 1974), Lecture Notes in Math., 475*, pp. 89–99, Springer, Berlin, 1975; **MR** 54#265; **Zbl.** 308.10019.

1975.32 Problems and results on finite and infinite combinatorial analysis, *Infinite and finite sets (Colloq., Keszthely, 1973; dedicated to P. Erdős on his 60th birthday), Vol. I; Colloq. Math. Soc. János Bolyai, Vol. 10*, pp. 403–424, North-Holland, Amsterdam, 1975; **MR** 52#10438; **Zbl.** 361.05038.

1975.33 Problems and results on finite and infinite graphs, *Recent advances in graph theory (Proc. Second Czechoslovak Sympos., Prague, 1974)*, pp. 183–192 (loose errata), Academia, Prague, 1975; **MR** 52#10500; **Zbl.** 347.05116.

1975.34 Problems and results on 3-chromatic hypergraphs and some related questions, *Infinite and finite sets (Colloq., Keszthely, 1973; dedicated to P. Erdős on his 60th birthday), Vol. II; Colloq. Math. Soc. János Bolyai, Vol. 10*, pp. 609–627, North-Holland, Amsterdam, 1975 (L. Lovász); **MR** 52#2938; **Zbl.** 315.05117.

1975.35 Ramsey theorems for multiple copies of graphs, *Trans. Amer. Math. Soc.* **209** (1975), 87–99 (S. A. Burr; J. H. Spencer); **MR** 53#13015; **Zbl.** 302.05105 and 273.05111.

1975.36 Rational approximation on the positive real axis, *Proc. London Math. Soc (3)* **31** (1975) no. 4, 439–456 (A. R. Reddy); **MR** 53#821; **Zbl.** 347.41009.

1975.37 Répartition des nombres superabondants (in French, English summary), *Bull. Soc. Math. France* **103** (1975) no. 1, 65–90 (J.-L. Nicolas); **MR** 54#257; **Zbl.** 306.10025.

1975.38 Répartition des nombres superabondants (in French), *Séminaire Delange-Pisot-Poitou (15e année: 1973/74), Théorie des nombres, Fasc. 1, Exp. No. 5*, 18 pp., Secrétariat Mathematique, Paris, 1975 (J.-L. Nicolas); **MR** 53#303; **Zbl.** 321.10036.

1975.39 Some additive and multiplicative problems in number theory, Collection of articles in memory of Juriĭ Vladimirovič Linnik, *Acta Arith.* **27** (1975), 37–50 (S. L. G. Choi; E. Szemerédi); **MR** 51#5540; **Zbl.** 303.10057.

1975.40 Some extremal problems in geometry, III., *Proceedings of the Sixth Southeastern Conference on Combinatorics, Graph Theory and Computing (Florida Atlantic Univ., Boca Raton, Fla., 1975), Congress. Numer. XIV*, pp. 291–308, Utilitas Math., Winnipeg, Man., 1975 (G. B. Purdy); **MR** 52#13650; **Zbl.** 328.05018.

1975.41 Some problems on elementary geometry, *Austral. Math. Soc. Gaz.* **2** (1975), 2–3; **Zbl.** 429.05032.

1975.42 Some recent progress on extremal problems in graph theory, *Proceedings of the Sixth Southeastern Conference on Combinatorics, Graph Theory and Computing (Florida Atlantic Univ., Boca Raton, Fla., 1975), Congress. Numer. XIV*, pp. 3–14, Utilitas Math., Winnipeg, Man., 1975; **MR** 52#13488; **Zbl.** 323.05126.

1975.43 Splitting almost-disjoint collections of sets into subcollections admitting almost-transversals, *Infinite and finite sets (Colloq., Keszthely, 1973; dedicated to P. Erdős on his 60th birthday), Vol. I; Colloq. Math. Soc. János Bolyai, Vol. 10*, pp. 307–322, North-Holland, Amsterdam, 1975 (R. O. Davies); **MR** 51#5318; **Zbl.** 304.04003.

1975.44 Strong embeddings of graphs into colored graphs, *Infinite and finite sets (Colloq., Keszthely, 1973; dedicated to P. Erdős on his 60th birthday), Vol. I; Colloq. Math. Soc. János Bolyai, Vol. 10*, pp. 585–595, North-Holland, Amsterdam, 1975 (A. Hajnal; L. Pósa); **MR** 52#2937; **Zbl.** 312.05123.

1975.45 The number of distinct subsums of $\sum_1^N 1/i$, Collection of articles dedicated to Derrick Henry Lehmer on the occasion of his seventieth birthday, *Math. Comp.* **29** (1975), 29–42 (M. N. Bleicher); **MR** 51#3041; **Zbl.** 298.10012.

1975.46 The product of consecutive integers is never a power, *Illinois J. Math.* **19** (1975), 292–301 (J. L. Selfridge); **MR** 51#12692; **Zbl.** 295.10017.

1975.47 Varga Tamás egy problémájáról (On a problem of T. Varga, in Hungarian), *Mat. Lapok* **24** (1973), 273–282, 1975 (P. Révész); **Zbl.** 373.60035.

1976.01 A note on regular methods of summability and the Banach-Saks property, *Proc. Amer. Math. Soc.* **59** (1976) no. 2, 232–234 (M. Magidor); **MR** 55#3601; **Zbl.** 355.40007.

1976.02 Alternating Hamiltonian cycles, *Israel J. Math.* **23** (1976) no. 2, 126–131 (B. Bollobás); **MR** 54#128; **Zbl.** 325.05114.

1976.03 Asymptotic enumeration of K_n-free graphs (Italian summary), *Colloquio Internazionale sulle Teorie Combinatorie (Rome, 1973), Tomo II, Atti dei Convegni Lincei, No. 17*, pp. 19–27, Accad. Naz. Lincei, Rome, 1976 (D. J. Kleitman; B. L. Rothschild); **MR** 57#2984; **Zbl.** 358.05027.

1976.04 Bemerkungen zu einer Aufgabe in den Elementen [*Elem. Math.* **26** (1971), 43, by G. Jaeschke] (in German), *Arch. Math. (Basel)* **27** (1976) no. 2, 159–163; **MR** 53#7969; **Zbl.** 328.10004.

1976.05 Cliques in random graphs, *Math. Proc. Cambridge Philos. Soc.* **80** (1976) no. 3, 419–427 (B. Bollobás); **MR** 58#16408; **Zbl.** 344.05155.

1976.06 Computation of sequences maximizing least common multiples, *Proceedings of the Fifth Manitoba Conference on Numerical Mathematics (Univ. Manitoba, Winnipeg, Man., 1975), Congress Numer. XVI*, pp. 293–303, Utilitas Math., Winnipeg, Man., 1976 (R. B. Eggleton; J. L. Selfridge); **MR** 54#5106; **Zbl.** 332.10002.

1976.07 Concerning periodicity in the asymptotic behaviour of partition functions, *J. Austral. Math. Soc. Ser. A* **21** (1976) no. 4, 447–456 (L. B. Richmond); **MR** 53#10748; **Zbl.** 326.10042.

1976.08 Decomposition of spheres in Hilbert spaces, *Comment. Math. Univ. Carolinae* **17** (1976) no. 4, 791–795 (D. Preiss); **MR** 58#330; **Zbl.** 346.05106.

1976.09 Denominators of Egyptian fractions, *J. Number Theory* **8** (1976), 157–168 (M. N. Bleicher); **MR** 53#7925; **Zbl.** 328.10010.

1976.10 Denominators of Egyptian fractions, II., *Illinois J. Math.* **20** (1976) no. 4, 598–613 (M. N. Bleicher); **MR** 54#7359; **Zbl.** 336.10007.

1976.11 Distinct values of Euler's φ-function, *Mathematika* **23** (1976) no. 1, 1–3 (R. R. Hall); **MR** 54#2603; **Zbl.** 329.10036.

1976.12 Extremal problems on polynomials, *Approximation theory, II. (Proc. Internat. Sympos., Univ. Texas, Austin, Tex., 1976)*, pp. 347–355, Academic Press, New York, 1976; **MR** 57#16548; **Zbl.** 354.41003.

1976.13 Extremal Ramsey theory for graphs, *Utilitas Math.* **9** (1976), 247–258 (S. A. Burr); **MR** 55#2633; **Zbl.** 333.05119.

1976.14 Families of sets whose pairwise intersections have prescribed cardinals or order types, *Math. Proc. Cambridge Philos. Soc.* **80** (1976) no. 2, 215–221 (E. C. Milner; R. Rado); **MR** 54#2469; **Zbl.** 337.04004.

1976.15 Generalized Ramsey theory for multiple colors, *J. Combinatorial Theory Ser. B* **20** (1976) no. 3, 250–264 (R. J. Faudree; C. C. Rousseau; R. H. Schelp); **MR** 54#5030; **Zbl.** 329.05116.

1976.16 Méthodes probabilistes et combinatoires en théorie des nombres, *Bull. Sci. Math. (2)* **100** (1976) no. 4, 301–320 (J.-L. Nicolas); **MR** 58#21998; **Zbl.** 343.10037.

1976.17 Müntz's theorem and rational approximation, *J. Approximation Theory* **17** (1976) no. 4, 393–394 (A. R. Reddy); **MR** 54#13399; **Zbl.** 352.41013.

1976.18 On a problem of Graham, *Publ. Math. Debrecen* **23** (1976) no. 1–2, 123–127 (E. Szemerédi); **MR** 54#12609; **Zbl.** 349.10046.

1976.19 On a problem of M. D. Hirschhorn [*Amer. Math. Monthly* **80** (1973) no. 6, part 1, 675–677], *Amer. Math. Monthly* **83** (1976), 23–26 (M. Simonovits); **MR** 52#10619b; **Zbl.** 329.10005.

1976.20 On a Ramsey-Turán type problem, *J. Combinatorial Theory Ser. B.* **21** (1976) no. 1–2, 166–168 (B. Bollobás); **MR** 54#12572; **Zbl.** 337.05134.

1976.21 On additive bases, *Acta Arith.* **30** (1976) no. 2, 121–132 (J. M. Deshouillers; A. Sárközy); **MR** 54#5165; **Zbl.** 349.10047.

1976.22 On asymptotic properties of aliquot sequences, *Math. Comput.* **30** (1976) no. 135, 641–645; **MR** 53#7919; **Zbl.** 337.10005.

1976.23 On graphs of Ramsey type, *Ars Combinatoria* **1** (1976) no. 1, 167–190 (S. A. Burr; L. Lovász); **MR** 54#7308; **Zbl.** 333.05120.

1976.24 On multiplicative representations of integers, *J. Austral. Math. Soc. Ser. A* **21** (1976) no. 4, 418–427 (E. Szemerédi); **MR** 54#5167; **Zbl.** 327.10004.

1976.25 On products of factorials, *Bull. Inst. Math. Acad. Sinica* **4** (1976) no. 2, 337–355 (R. L. Graham); **MR** 57#256; **Zbl.** 346.10004.

1976.26 On sparse graphs with dense long paths, *Computers and mathematics with applications*, pp. 365–369, Pergamon, Oxford, 1976 (R. L. Graham; E. Szemerédi); **MR** 57#16131; **Zbl.** 328.05123.

1976.27 On the greatest and least prime factors of $n! + 1$, *J. London Math. Soc. (2)* **13** (1976) no. 3, 513–519 (C. L. Stewart); **MR** 53#13093; **Zbl.** 332.10028.

1976.28 On the greatest prime factor of $2^p - 1$ for a prime p and other expressions, *Acta Arith.* **30** (1976) no. 3, 257–265 (T. N. Shorey); **MR** 54#7402; **Zbl.** 333.10026 and 296.10021.

1976.29 On the number of distinct prime divisors of $\binom{n}{k}$, *Utilitas Math.* **10** (1976), 51–60 (H. Gupta; S. P. Khare); **MR** 55#2729; **Zbl.** 339.10006.

1976.30 On the prime factors of $\binom{n}{k}$, *Fibonacci Quart.* **14** (1976) no. 4, 348–352 (R. L. Graham); **MR** 54#10118; **Zbl.** 354.10010.

1976.31 Partition theorems for subsets of vector spaces, *J. Combinatorial Theory Ser. A* **20** (1976) no. 3, 279–291 (M. L. Cates; N. Hindman; B. L. Rothschild); **MR** 53#10583; **Zbl.** 363.04009.

1976.32 Partitions of the natural numbers into infinitely oscillating bases and nonbases, *Comment. Math. Helv.* **51** (1976) no. 2, 171–182 (M. B. Nathanson); **MR** 54#226; **Zbl.** 329.10045.

1976.33 Prime polynomial sequences, *J. London Math. Soc. (2)* **14** (1976) no. 3, 559–562 (S. D. Cohen; M. B. Nathanson); **MR** 55#290; **Zbl.** 346.10027.

1976.34 Probabilistic methods in group theory, II., *Houston J. Math.* **2** (1976) no. 2, 173–180 (R. R. Hall); **MR** 58#10791; **Zbl.** 336.20041.

1976.35 Problems and results in combinatorial analysis (Italian summary), *Colloquio Internazionale sulle Teorie Combinatorie (Rome, 1973), Tomo II, Atti dei Convegni Lincei, No. 17*, pp. 3–17, Accad. Naz. Lincei, Rome, 1976; **MR** 57#5764; **Zbl.** 361.05037.

1976.36 Problems and results in graph theory and combinatorial analysis, *Proceedings of the Fifth British Combinatorial Conference (Univ. Aberdeen, Aberdeen, 1975), Congress. Numer. XV*, pp. 169–192, Utilitas Math., Winnipeg, Man., 1976; **MR** 53#13006; **Zbl.** 335.05002.

1976.37 Problems and results in rational approximation, *Period. Math. Hungar.* **7** (1976) no. 1, 27–35 (A. R. Reddy); **MR** 54#13379; **Zbl.** 337.41020.

1976.38 Problems and results on consecutive integers, *Publ. Math. Debrecen* **23** (1976) no. 3–4, 271–282; **MR** 56#11931; **Zbl.** 353.10032.

1976.39 Problems and results on number theoretic properties of consecutive integers and related questions, *Proceedings of the Fifth Manitoba Conference on Numerical Mathematics (Univ. Manitoba, Winnipeg, Man., 1975), Congress. Numer. XVI*, pp. 25–44, Utilitas Math., Winnipeg, Man., 1976; **MR** 54#10138; **Zbl.** 337.10001.

1976.40 Proof of a conjecture about the distribution of divisors of integers in residue classes, *Math. Proc. Cambridge Philos. Soc.* **79** (1976), 281–287 (R. R. Hall); **MR** 53#304; **Zbl.** 368.10033.

1976.41 Rational approximation, *Advances in Math.* **21** (1976) no. 1, 78–109 (A. R. Reddy); **MR** 53#13938; **Zbl.** 334.30019.

1976.42 Sets of independent edges of a hypergraph, *Quart. J. Math. Oxford Ser. (2)* **27** (1976) no. 105, 25–32 (B. Bollobás; D. E. Daykin); **MR** 54#159; **Zbl.** 337.05135.

1976.43 Some extremal problems in geometry, IV., *Proceedings of the Seventh Southeastern Conference on Combinatorics, Graph Theory, and Computing (Louisiana State Univ., Baton Rouge, La., 1976), Congress. Numer. XVII*, pp. 307–322, Utilitas Math., Winnipeg, Man., 1976 (G. B. Purdy); **MR** 55#10292; **Zbl.** 345.52007.

1976.44 Some problems and results on the irrationality of the sum of infinite series, *J. Math. Sci.* **10** (1975), 1–7, 1976; **MR** 80k:10029; **Zbl.** 372.10023.

1976.45 Some recent problems and results in graph theory, combinatorics and number theory, *Proceedings of the Seventh Southeastern Conference on Combinatorics, Graph Theory, and Computing (Louisiana State Univ., Baton Rouge, La., 1976), Congress. Numer. XVII*, pp. 3–14, Utilitas Math., Winnipeg, Man., 1976; **MR** 54#10023; **Zbl.** 352.05024.

1976.46 Subgraphs with all colours in a line-coloured graph, *Proceedings of the Fifth British Combinatorial Conference (Univ. Aberdeen, Aberdeen, 1975), Congress. Numer. XV*, pp. 101–112, Utilitas Math., Winnipeg, Man., 1976 (C. C. Chen; D. E. Daykin); **MR** 53#183; **Zbl.** 344.05118.

1976.47 The nonexistence of certain invariant measures, *Proc. Amer. Math. Soc.* **59** (1976) no. 2, 321–322 (R. D. Mauldin); **MR** 54#516; **Zbl.** 361.28013.

1976.48 The powers that be, *Amer. Math. Monthly* **83** (1976) no. 10, 801–805 (R. B. Eggleton; J. L. Selfridge); **MR** 58#5476; **Zbl.** 359.10001.

1977.01 A note of welcome, *J. Graph Theory* **1** (1977), 3.

1977.02 Addendum to: "Rational approximation" [*Adv. Math.* **21** (1976), 78–109], *Adv. Math.* **25** (1977) no. 1, 91–93 (A. R. Reddy); **MR** 56#932; **Zbl.** 374.30030.

1977.03 An asymptotic formula in additive number theory, II., *J. Indian Math. Soc. (N.S.)* **41** (1977) no. 3–4, 281–291 (G. J. Babu; K. Ramachandra); **MR** 81g:10062; **Zbl.** 438.10036.

1977.04 Approximation by rational functions, *J. London Math. Soc. (2)* **15** (1977) no. 2, 319–328 (D. J. Newman; A. R. Reddy); **MR** 55#10918; **Zbl.** 372.41008.

1977.05 Bases for sets of integers, *J. Number Theory* **9** (1977) no. 4, 420–425 (D. J. Newman); **MR** 56#11941; **Zbl.** 359.10045.

1977.06 Characterizing cliques in hypergraphs, *Ars Combinatoria* **4** (1977), 81–118 (B. L. Rothschild; N. M. Singhi); **MR** 58#10607; **Zbl.** 408.04002.

1977.07 Corrigenda: "Families of sets whose pairwise intersections have prescribed cardinals or order types" [*Math. Proc. Cambridge Philos. Soc.* **80** (1976) no. 2, 215–221], *Math. Proc. Cambridge Philos. Soc.* **81** (1977) no. 3, 523 (E. C. Milner; R. Rado); **MR** 55#7784; **Zbl.** 349.04006.

1977.08 Corrigendum: "Rational approximation on the positive real axis" [*Proc. London Math. Soc. (3)* **31** (1975) no. 4, 439–456], *Proc.*

London Math. Soc. (3) **35** (1977) no. 2, 290 (A. R. Reddy); **MR** 57#6969; **Zbl.** 353.41005.

1977.09 Density functions for prime and relatively prime numbers, *Monatsh. Math.* **83** (1977) no. 2, 99–112 (I. Richards); **MR** 55#12669; **Zbl.** 355.10034.

1977.10 Euler's φ-function and its iterates, *Mathematika* **24** (1977) no. 2, 173–177 (R. R. Hall); **MR** 57#12356; **Zbl.** 367.10004.

1977.11 Hamiltonian cycles in regular graphs of moderate degree, *J. Combinatorial Theory Ser. B* **23** (1977) no. 1, 139–142 (A. M. Hobbs); **MR** 58#10594; **Zbl.** 374.05037.

1977.12 Néhány személyes és matematikai emlékem Kalmár Lászlóról (My personal and mathematical reminiscences of László Kalmár, in Hungarian), *Mat. Lapok* **25** (1974) no. 3–4, 253–255, 1977; **MR** 82e:01088; **Zbl.** 376.01005.

1977.13 Nonbases of density zero not contained in maximal nonbases, *J. London Math. Soc. (2)* **15** (1977) no. 3, 403–405 (M. B. Nathanson); **MR** 58#10807; **Zbl.** 357.10029.

1977.14 On a problem in extremal graph theory, *J. Combinatorial Theory Ser. B* **23** (1977) no. 2–3, 251–254 (D. T. Busolini); **MR** 58#27621; **Zbl.** 364.05033.

1977.15 On an additive arithmetic function, *Pacific J. Math.* **71** (1977) no. 2, 275–294 (K. Alladi); **MR** 56#5401; **Zbl.** 359.10038.

1977.16 On composition of polynomials, *Algebra Universalis* **7** (1977) no. 3, 357–360 (S. Fajtlowicz); **MR** 56#330; **Zbl.** 364.08006.

1977.17 On differences and sums of integers, II., *Bull. Soc. Math. Grèce (N. S.)* **18** (1977) no. 2, 204–223 (A. Sárközy); **MR** 82m:10079; **Zbl.** 413.10049.

1977.18 On products of consecutive integers, *Number theory and algebra*, pp. 63–70, Academic Press, New York, 1977 (E. G. Straus); **MR** 57#3051; **Zbl.** 386.10002.

1977.19 On spanned subgraphs of graphs, *Beiträge zur Graphentheorie und deren Anwendungen (Kolloq., Oberhof (DDR), 1977) (Contributions to graph theory and its applications (Internat. Colloq., Oberhof, 1977))*, pp. 80–96, Tech. Hochschule Ilmenau, Ilmenau, 1977 (A. Hajnal); **MR** 82d:05082; **Zbl.** 405.05031.

1977.20 On the chromatic index of almost all graphs, *J. Combinatorial Theory Ser. B* **23** (1977) no. 2–3, 255–257 (R. J. Wilson); **MR** 57#2986; **Zbl.** 378.05032.

1977.21 On the length of the longest head-run, *Topics in information theory (Second Colloq., Keszthely, 1975), Colloq. Math. Soc. János Bolyai, Vol. 16*, pp. 219–228, North-Holland, Amsterdam, 1977 (P. Révész); **MR** 57#17788; **Zbl.** 362.60044.

1977.22 Paul Turán, 1910–1976: his work in graph theory, *J. Graph Theory* **1** (1977) no. 2, 97–101; **MR** 56#61; **Zbl.** 383.05022.

1977.23 Problèmes extrémaux et combinatoires en théorie des nombres (rédigé par Jean-Louis Nicolas), *Séminaire Delange-Pisot-Poitou (17e année: 1975/76), Théorie des nombres, Fasc. 2, Exp. No. 67*, 5 pp., Secrétariat Mathematique, Paris, 1977; **Zbl.** 368.10003.

1977.24 Problems and results in combinatorial analysis, *Proceedings of the Eighth Southeastern Conference on Combinatorics, Graph Theory and Computing (Louisiana State Univ., Baton Rouge, La., 1977), Congress. Numer. XIX*, pp. 3–12, Utilitas Math., Winnipeg, Man., 1977; **MR** 58#27542; **Zbl.** 416.05059.

1977.25 Problems and results in combinatorial analysis, II., *Creation Math.* **10** (1977), 27–30; **Zbl.** 415.05006.

1977.26 Problems and results on combinatorial number theory, II., *J. Indian Math. Soc. (N.S.)* **40** (1976) no. 1–4, 285–298, 1977; **MR** 56#2902; **Zbl.** 434.10003.

1977.27 Problems and results on combinatorial number theory, III., *Number theory day (Proc. Conf., Rockefeller Univ., New York, 1976), Lecture Notes in Math., 626*, pp. 43–72, Springer, Berlin, 1977; **MR** 57#12442; **Zbl.** 368.10002.

1977.28 Problems in number theory and combinatorics, *Proceedings of the Sixth Manitoba Conference on Numerical Mathematics (Univ. Manitoba, Winnipeg, Man., 1976), Congress. Numer. XVIII*, pp. 35–58, Utilitas Math., Winnipeg, Man., 1977; **MR** 80e:10005; **Zbl.** 471.10002.

1977.29 Propriétés probabilistes des diviseurs d'un nombre (in French), *Journées Arithmétiques de Caen (Univ. Caen, Caen, 1976), Astérisque No. 41–42*, pp. 203–214, Soc. Math. France, Paris, 1977 (J.-L. Nicolas); **MR** 56#8510; **Zbl.** 346.10033.

1977.30 Some extremal problems in geometry, V., *Proceedings of the Eighth Southeastern Conference on Combinatorics, Graph Theory and Computing (Louisiana State Univ., Baton Rouge, La., 1977), Congress. Numer. XIX*, pp. 569–578, Utilitas Math., Winnipeg, Man., 1977 (G. B. Purdy); **MR** 57#16104; **Zbl.** 403.52006.

1977.31 Strongly annular functions with small coefficients, and related results, *Proc. Amer. Math. Soc.* **67** (1977) no. 1, 129–132 (D. D. Bonar; F. W. Carroll); **MR** 56#15931; **Zbl.** 377.30023.

1977.32 Systems of finite sets having a common intersection, *Proceedings of the Eighth Southeastern Conference on Combinatorics, Graph Theory and Computing (Louisiana State Univ., Baton Rouge, La., 1977), Congress. Numer. XIX*, pp. 247–252, Utilitas Math., Winnipeg, Man., 1977 (R. Duke); **MR** 58#5248; **Zbl.** 443.05002.

1977.33 Turán Pál matematikai munkássága, I. Statisztikus csoportelmélet és partíció- elmélet (Mathematical works of Paul Turán, I. Statistical group theory and theory of partitio numerorum, in Hungarian), *Mat. Lapok* **25** (1974), 229–238, 1977 (M. Szalay); **Zbl.** 383.10031.

1978.01 A class of Hamiltonian regular graphs, *J. Graph Theory* **2** (1978) no. 2, 129–135 (A. M. Hobbs); **MR** 58#5366; **Zbl.** 416.05055.

1978.02 A class of Ramsey-finite graphs, *Proceedings of the Ninth Southeastern Conference on Combinatorics, Graph Theory, and Computing (Florida Atlantic Univ., Boca Raton, Fla., 1978), Congress. Numer. XXI*, pp. 171–180, Utilitas Math., Winnipeg, Man., 1978 (S. A. Burr; R. J. Faudree; R. H. Schelp); **MR** 80m:05081; **Zbl.** 432.05038.

1978.03 A measure of the nonmonotonicity of the Euler phi function, *Pacific J. Math.* **77** (1978) no. 1, 83–101 (H. G. Diamond); **MR** 80e:10035; **Zbl.** 383.10034 and 352.10027.

1978.04 A note on Ingham's summation method, *J. Number Theory* **10** (1978) no. 1, 95–98 (S. L. Segal); **MR** 58#549; **Zbl.** 367.40005.

1978.05 A property of 70, *Math. Mag.* **51** (1978) no. 4, 238–240; **MR** 80a:10009; **Zbl.** 391.10004.

1978.06 Biased positional games, *Ann. Discrete Math.* **2** (1978), 221–229 (V. Chvátal); **MR** 81a:05017; **Zbl.** 374.90086.

1978.07 Combinatorial problems on subsets and their intersections, *Studies in foundations and combinatorics, Advances in Math. Suppl. Stud., 1*, pp. 259–265, Academic Press, New York-London, 1978, (M. Deza; N. M. Singhi); **MR** 80e:05006; **Zbl.** 434.05001.

1978.08 Combinatorial properties of systems of sets, *J. Combinatorial Theory Ser. A* **24** (1978) no. 3, 308–313 (E. Szemerédi); **MR** 58#10467; **Zbl.** 383.05002.

1978.09 Embedding theorems for graphs establishing negative partition relations, *Period. Math. Hungar.* **9** (1978) no. 3, 205–230 (A. Hajnal); **MR** 80g:04004; **Zbl.** 381.04004.

1978.10 Extremal graphs without large forbidden subgraphs, Advances in graph theory (Cambridge Combinatorial Conf., Trinity Coll., Cambridge, 1977), *Ann. Discrete Math.* **3** (1978), 29–41 (B. Bollobás; M. Simonovits; E. Szemerédi); **MR** 80a:05119; **Zbl.** 375.05034.

1978.11 Intersection properties of systems of finite sets, *Combinatorics (Proc. Fifth Hungarian Colloq., Keszthely, 1976). Vol. I, Colloq. Math. Soc. János Bolyai, Vol. 18*, pp. 251–256, North-Holland, Amsterdam, 1978 (M. Deza; P. Frankl); **MR** 80c:05004; **Zbl.** 384.05001.

1978.12 Intersection properties of systems of finite sets, *Proc. London Math. Soc. (3)* **36** (1978) no. 2, 369–384 (M. Deza; P. Frankl); **MR** 57#16096; **Zbl.** 407.05006.

1978.13 On a class of relatively prime sequences, *J. Number Theory* **10** (1978) no. 4, 451–474 (D. E. Penney; C. Pomerance); **MR** 80a:10059; **Zbl.** 399.10042.

1978.14 On a geometric property of lemniscates, *Aequationes Math.* **17** (1978) no. 2–3, 344–347 (J. S. Hwang); **MR** 80c:30007; **Zbl.** 383.30001.

1978.15 On a geometric property of lemniscates (Short communication), *Aequationes Math.* **17** (1978), 118–119 (J. S. Hwang).

1978.16 On additive partitions of integers, *Discrete Math.* **22** (1978) no. 3, 201–211 (K. Alladi; V. E. Hoggatt, Jr.); **MR** 80a:10025; **Zbl.** 376.10011.

1978.17 On changes of signs in infinite series (Russian summary), *Anal. Math.* **4** (1978) no. 1, 3–12 (I. Borosh; C. K. Chui); **MR** 58#1805; **Zbl.** 391.10038.

1978.18 On cycle–complete graph Ramsey numbers, *J. Graph Theory* **2** (1978) no. 1, 53–64 (R. J. Faudree; C. C. Rousseau; R. H. Schelp); **MR** 58#16412; **Zbl.** 383.05027.

1978.19 On differences and sums of integers, I., *J. Number Theory* **10** (1978) no. 4, 430–450 (A. Sárközy); **MR** 82m:10078; **Zbl.** 404.10029.

1978.20 On finite superuniversal graphs, *Discrete Math.* **24** (1978) no. 3, 235–249 (S. H. Hechler; P. Kainen); **MR** 80c:05109; **Zbl.** 397.05019.

1978.21 On multigraph extremal problems, *Problèmes combinatoires et théorie des graphes (Colloq. Internat. CNRS, Univ. Orsay, Orsay, 1976), Colloq. Internat. CNRS, 260*, pp. 63–66, CNRS, Paris, 1978 (W. G. Brown; M. Simonovits); **Zbl.** 413.05019.

1978.22 On partitions of N into summands coprime to N, *Aequationes Math.* **18** (1978) no. 1–2, 178–186 (L. B. Richmond); **MR** 58#16572; **Zbl.** 401.10057.

1978.23 On partitions of N into summands coprime to N (Short communication), *Aequationes Math.* **17** (1978), 382 (L. B. Richmond).

1978.24 On products of integers, II., *Acta Sci. Math. (Szeged)* **40** (1978) no. 3–4, 243–259 (A. Sárközy); **MR** 80a:10006; **Zbl.** 401.10068.

1978.25 On set systems having paradoxical covering properties, *Acta Math. Acad. Sci. Hungar.* **31** (1978) no. 1–2, 89–124 (A. Hajnal; E. C. Milner); **MR** 58#21628; **Zbl.** 378.04002.

1978.26 On some unconventional problems on the divisors of integers, *J. Austral. Math. Soc. Ser. A* **25** (1978) no. 4, 479–485 (R. R. Hall); **MR** 58#21975; **Zbl.** 393.10047.

1978.27 On the density of λ-box products, *General Topology Appl.* **9** (1978) no. 3, 307–312 (F. S. Cater; F. Galvin); **MR** 80e:54004; **Zbl.** 394.54002.

1978.28 On the integral of the Lebesgue function of interpolation, *Acta Math. Acad. Sci. Hungar.* **32** (1978) no. 1–2, 191–195 (J. Szabados); **MR** 80a:41002; **Zbl.** 391.41003.

1978.29 On the largest prime factors of n and $n+1$, *Aequationes Math.* **17** (1978) no. 2–3, 311–321 (C. Pomerance); **MR** 58#476; **Zbl.** 379.10027.

1978.30 On the largest prime factors of n and $n+1$ (Short communication), *Aequationes Math.* **17** (1978), 115 (C. Pomerance).

1978.31 On the prime factorization of binomial coefficients, *J. Austral. Math. Soc. Ser. A* **26** (1978) no. 3, 257–269 (E. F. Ecklund, Jr.; R. B. Eggleton; J. L. Selfridge); **MR** 80e:10009; **Zbl.** 393.10005.

1978.32 On the Schnirelmann density of k-free integers, *Indian J. Math.* **20** (1978) no. 1, 45–56 (G. E. Hardy; M. V. Subbarao); **MR** 82j:10089; **Zbl.** 417.10047.

1978.33 On total matching numbers and total covering numbers of complementary graphs, *Discrete Math.* **19** (1977) no. 3, 229–233, 1978 (A. Meir); **MR** 57#2985; **Zbl.** 374.05047.

1978.34 Partitions into summands of the form $[m\alpha]$, *Proceedings of the Seventh Manitoba Conference on Numerical Mathematics and Computing (Univ. Manitoba, Winnipeg, Man., 1977), Congress. Numer. XX*, pp. 371–377, Utilitas Math., Winnipeg, Man., 1978 (L. B. Richmond); **MR** 80g:10048; **Zbl.** 483.10014.

1978.35 Problems and results in combinatorial analysis, *Creation Math.* **11** (1978), 17–20; **Zbl.** 439.05009.

1978.36 Problems and results in combinatorial analysis and combinatorial number theory, *Proceedings of the Ninth Southeastern Conference on Combinatorics, Graph Theory, and Computing (Florida Atlantic Univ., Boca Raton, Fla., 1978), Congress. Numer. XXI*, pp. 29–40, Utilitas Math., Winnipeg, Man., 1978; **MR** 80h:05001; **Zbl.** 423.05001.

1978.37 Problems and results in graph theory and combinatorial analysis (French summary), *Problèmes combinatoires et théorie des graphes (Colloq. Internat. CNRS, Univ. Orsay, Orsay, 1976), Colloq. Internat. CNRS, 260*, pp. 127–129, CNRS, Paris, 1978; **MR** 80m:05023; **Zbl.** 412.05033.

1978.38 Ramsey-minimal graphs for multiple copies, *Nederl. Akad. Wetensch. Indag. Math.* **40** (1978) no. 2, 187–195 (S. A. Burr; R. J. Faudree; C. C. Rousseau; R. H. Schelp); **MR** 58#5387; **Zbl.** 382.05043.

1978.39 Rational approximation, II., *Adv. in Math.* **29** (1978) no. 2, 135–156 (D. J. Newman; A. R. Reddy); **MR** 80e:41008; **Zbl.** 386.30020.

1978.40 Set-theoretic, measure-theoretic, combinatorial, and number-theoretic problems concerning point sets in Euclidean space, *Real Anal. Exchange* **4** (78/79) no. 2, 113–138; **MR** 80g:04005; **Zbl.** 418.04002.

1978.41 Sets of natural numbers with no minimal asymptotic bases, *Proc. Amer. Math. Soc.* **70** (1978) no. 2, 100–102 (M. B. Nathanson); **MR** 58#5573; **Zbl.** 389.10038.

1978.42 Some combinatorial problems in the plane, *J. Combin. Theory Ser. A* **25** (1978) no. 2, 205–210 (G. B. Purdy); **MR** 58#21645; **Zbl.** 422.05023.

1978.43 Some extremal problems on families of graphs and related problems, *Combinatorial mathematics (Proc. Internat. Conf. Combinatorial Theory, Australian Nat. Univ., Canberra, 1977), Lecture Notes in Math., 686*, pp. 13–21, Springer, Berlin, 1978; **MR** 80m:05063; **Zbl.** 422.05039.

1978.44 Some more problems on elementary geometry, *Austral. Math. Soc. Gaz.* **5** (1978) no. 2, 52–54; **MR** 80b:52005; **Zbl.** 417.52002.

1978.45 Some new results in probabilistic group theory, *Comment. Math. Helv.* **53** (1978) no. 3, 448–457 (R. R. Hall); **MR** 58#16561; **Zbl.** 385.20045.

1978.46 Some number theoretic problems on binomial coefficients, *Austral. Math. Soc. Gaz.* **5** (1978) no. 3, 97–99 (G. Szekeres); **MR** 80e:10010; **Zbl.** 401.10003.

1978.47 Some solved and unsolved problems in combinatorial number theory, *Math. Slovaca* **28** (1978) no. 4, 407–421 (A. Sárközy); **MR** 80i:10001; **Zbl.** 395.10002.

1978.48 The size Ramsey number, *Period. Math. Hungar.* **9** (1978), 145–161 (R. J. Faudree; C. C. Rousseau; R. H. Schelp); **MR** 80h:05037; **Zbl.** 368.05033 and 331.05122.

1978.49 When the Cartesian product of directed cycles is Hamiltonian, *J. Graph Theory* **2** (1978) no. 2, 137–142 (W. T. Trotter); **MR** 80e:05063; **Zbl.** 406.05048.

1979.01 Additive arithmetic functions bounded by monotone functions on thin sets, *Annales Univ. Sci. Budapest. Eötvös Sect. Math.* **22/23** (1979/80), 97–111 (P. D. T. A. Elliott); **MR** 81m:10098; **Zbl.** 439.10037.

1979.02 Bases and nonbases of squarefree integers, *J. Number Theory* **11** (1979) no. 2, 197–208 (M. B. Nathanson); **MR** 80h:10062; **Zbl.** 409.10042.

1979.03 Colorful partitions of cardinal numbers, *Canad. J. Math.* **31** (1979) no. 3, 524–541 (J. E. Baumgartner; F. Galvin; J. A. Larson); **MR** 81d:04002; **Zbl.** 419.04002.

1979.04 Combinatorial problems in geometry and number theory, *Relations between combinatorics and other parts of mathematics (Proc. Sympos. Pure Math., Ohio State Univ., Columbus, Ohio, 1978), Proc. Sympos. Pure Math., XXXIV*, pp. 149–162, Amer. Math. Soc., Providence, R.I., 1979; **MR** 80i:05001; **Zbl.** 406.05012.

1979.05 Evolution of the n-cube, *Comput. Math. Appl.* **5** (1979) no. 1, 33–39 (J. H. Spencer); **MR** 80g:05054; **Zbl.** 399.05041.

1979.06 Minimal decompositions of two graphs into pairwise isomorphic subgraphs, *Proceedings of the Tenth Southeastern Conference on Combinatorics, Graph Theory and Computing (Florida Atlantic Univ., Boca Raton, Fla., 1979), Congress. Numer. XXIII*, pp. 3–18, Utilitas Math., Winnipeg, Man., 1978 (F. R. K. Chung; R. L. Graham; S. Ulam; F. F. Yao); **MR** 82b:05080; **Zbl.** 434.05046.

1979.07 Old and new problems and results in combinatorial number theory: van der Waerden's theorem and related topics, *Enseign. Math. (2)* **25** (1979) no. 3–4, 325–344 (R. L. Graham); **MR** 81f:10005; **Zbl.** 434.10002.

1979.08 Old and new problems in combinatorial analysis and graph theory, *Second International Conference on Combinatorial Mathematics (New York, 1978), Ann. N. Y. Acad. Sci. 319*, pp. 177–187, New York Acad. Sci., New York, 1979; **MR** 81a:05033; **Zbl.** 481.05002.

1979.09 On the asymptotic behavior of large prime factors of integers, *Pacific J. Math.* **82** (1979) no. 2, 295–315 (K. Alladi); **MR** 81c:10049; **Zbl.** 419.10042.

1979.10 On the asymptotic density of sets of integers, II., *J. London Math. Soc. (2)* **19** (1979) no. 1, 17–20 (B. Saffari; R. C. Vaughan); **MR** 80i:10074; **Zbl.** 409.10043.

1979.11 On the concentration of distribution of additive functions, *Acta Sci. Math. (Szeged)* **41** (1979) no. 3–4, 295–305 (I. Kátai); **MR** 81d:10044; **Zbl.** 431.10031.

1979.12 On the density of odd integers of the form $(p-1)2^{-n}$ and related questions, *J. Number Theory* **11** (1979) no. 2, 257–263 (A. M. Odlyzko); **MR** 80i:10077; **Zbl.** 405.10036.

1979.13 On the distribution of values of angles determined by coplanar points, *J. London Math. Soc. (2)* **19** (1979) no. 1, 137–143 (J. H. Conway; H. T. Croft; M. J. T. Guy); **MR** 80h:51021; **Zbl.** 414.05006.

1979.14 On the growth of some additive functions on small intervals, *Acta Math. Acad. Sci. Hungar.* **33** (1979) no. 3–4, 345–359 (I. Kátai); **MR** 81b:10029; **Zbl.** 417.10039.

1979.15 On the prime factors of $\binom{n}{k}$ and of consecutive integers, *Utilitas Math.* **16** (1979), 197–215 (A. Sárközy); **MR** 81k:10066; **Zbl.** 419.10040.

1979.16 On the product of the point and line covering numbers of a graph, *Second International Conference on Combinatorial Mathematics (New York, 1978), Ann. N. Y. Acad. Sci. 319*, pp. 597–602, New York Acad. Sci., New York, 1979 (F. R. K. Chung; R. L. Graham); **MR** 81h:05082; **Zbl.** 483,05053.

1979.17 Problems and results in graph theory and combinatorial analysis, *Graph theory and related topics (Proc. Conf., Univ. Waterloo, Waterloo, Ont., 1977)*, pp. 153–163, Academic Press, New York-London, 1979; **MR** 81a:05034; **Zbl.** 457.05024.

1979.18 Some old and new problems in various branches of combinatorics, *Proceedings of the Tenth Southeastern Conference on Combinatorics, Graph Theory and Computing (Florida Atlantic Univ., Boca Raton, Fla., 1979), Congress. Numer. XXIII*, pp. 19–37, Utilitas Math., Winnipeg, Man., 1978; **MR** 81f:05001; **Zbl.** 429.05047.

1979.19 Some problems in partitio numerorum, *J. Austral. Math. Soc. Ser. A* **27** (1979) no. 3, 319–331 (J. H. Loxton); **MR** 80h:10021; **Zbl.** 403.10004.

1979.20 Some remarks on subgroups of real numbers, *Colloq. Math.* **42** (1979), 119–120; **MR** 82c:04004; **Zbl.** 433.04002.

1979.21 Some unconventional problems in number theory, *Journées Arithmétiques de Luminy (Colloq. Internat. CNRS, Centre Univ. Luminy, Luminy, 1978), Astérisque 61*, pp. 73–82, Soc. Math. France, Paris, 1979; **MR** 81h:10001; **Zbl.** 399.10001.

1979.22 Some unconventional problems in number theory, *Math. Mag.* **52** (1979) no. 2, 67–70; **MR** 80b:10002; **Zbl.** 407.10001.

1979.23 Some unconventional problems in number theory, Special issue dedicated to George Alexits on the occasion of his 80th birthday, *Acta Math. Acad. Sci. Hungar.* **33** (1979) no. 1–2, 71–80; **MR** 80b:10001; **Zbl.** 393.10046.

1979.24 Strong independence of graphcopy functions, *Graph theory and related topics (Proc. Conf., Univ. Waterloo, Waterloo, Ont., 1977)*, pp. 165–172, Academic Press, New York-London, 1979 (L. Lovász; J. H. Spencer); **MR** 81b:05060; **Zbl.** 462.05057.

1979.25 Systems of distinct representatives and minimal bases in additive number theory, *Number theory, Carbondale 1979 (Proc. Southern Illinois Conf., Southern Illinois Univ., Carbondale, Ill., 1979), Lecture Notes Math., 751*, pp. 89–107, Springer, Berlin, 1979 (M. B. Nathanson); **MR** 81k:10089; **Zbl.** 414.10053.

1979.26 The propinquity of divisors, *Bull. London Math. Soc.* **11** (1979) no. 3, 304–307 (R. R. Hall); **MR** 81m:10102; **Zbl.** 421.10027.

1979.27 The tails of infinitely divisible laws and a problem in number theory, *J. Number Theory* **11** (1979) no. 4, 542–551 (P. D. T. A. Elliott); **MR** 81b:10034; **Zbl.** 409.10039.

1979.28 Transversals and multitransversals, *J. London Math. Society (2)* **20** (1979), 387–395 (F. Galvin; R. Rado); **MR** 81c:04001; **Zbl.** 429.04005.

1980.01 A problem on complements and disjoint edges in hypergraphs, Proceedings of the Eleventh Southeastern Conference on Combinatorics, Graph Theory and Computing (Florida Atlantic Univ., Boca Raton, Fla., 1980), Vol. I., *Congr. Numer.* **28** (1980), 369–375 (R. Duke); **MR** 82g:05065; **Zbl.** 453.05050.

1980.02 A $\sigma(n)$, $\varphi(n)$, $d(n)$ és $\nu(n)$ függvények néhány új tulajdonságáról (On some new properties of functions $\sigma(n)$, $\varphi(n)$, $d(n)$ and $\nu(n)$, in Hungarian, English summary), *Mat. Lapok* **28** (1980) no. 1–3, 125–131 (K. Győry; Z. Papp); **MR** 82a:10004; **Zbl.** 453.10004.

1980.03 A survey of problems in combinatorial number theory, Combinatorial mathematics, optimal designs and their applications (Proc. Sympos. Combin. Math. and Optimal Design, Colorado State Univ., Fort Collins, Colo., 1978), *Ann. Discrete Mathematics* **6** (1980), 89–115; **MR** 81m:10001; **Zbl.** 448.10002.

1980.04 An extremal problem in generalized Ramsey theory, *Ars Combin.* **10** (1980), 193–203 (S. A. Burr; R. J. Faudree; C. C. Rousseau; R. H. Schelp); **MR** 82b:05096; **Zbl.** 458.05045.

1980.05 An extremal problem in graph theory, *Ars Combin.* **9** (1980), 249–251 (E. Howorka); **MR** 81j:05074; **Zbl.** 501.05043.

1980.06 Binomiális együtthatók prímfaktorairól (On prime factors of binomial coefficients, in Hungarian, English summary), *Mat. Lapok* **28** (1977/80) no. 4, 287–296; **MR** 82k:10010; **Zbl.** 473.10004.

1980.07 Choosability in graphs, *Proceedings of the West Coast Conference on Combinatorics, Graph Theory and Computing (Humboldt State Univ., Arcata, Calif., 1979), Congress. Numer. XXVI*, pp. 125–157, Utilitas Math., Winnipeg, Man., 1980 (A. L. Rubin; H. Taylor); **MR** 82f:05038; **Zbl.** 469.05032.

1980.08 Estimates for sums involving the largest prime factor of an integer and certain related additive functions, *Studia Sci. Math. Hungar.* **15** (1980) no. 1–3, 183–199 (A. Ivić); **MR** 84a:10046; **Zbl.** 499.10051 and 455.10031.

1980.09 Generalized Ramsey numbers involving subdivision graphs, and related problems in graph theory, Combinatorics 79 (Proc. Colloq., Univ. Montreal, Montreal, Que., 1979), Part II, *Ann. Discrete Math.* **9** (1980), 37–42 (S. A. Burr); **MR** 82c:05070; **Zbl.** 456.05047.

1980.10 Hadwiger's conjecture is true for almost every graph, *European J. Comb.* **1** (1980) no. 3, 195–199 (B. Bollobás; P. A. Catlin); **MR** 82b:05107; **Zbl.** 457.05041.

1980.11 How many pairs of products of consecutive integers have the same prime factors? (Research problem), *Amer. Math. Monthly* **87** (1980), 391–392; **Zbl.** 427.10004.

1980.12 Lagrange's theorem with $N^{1/3}$ squares, *Proc. Amer. Math. Soc.* **79** (1980) no. 2, 203–205 (S. L. G. Choi; M. B. Nathanson); **MR** 81k:10077; **Zbl.** 436.10024.

1980.13 Matching the natural numbers up to n with distinct multiples in another interval, *Nederl. Akad. Wetensch. Indag. Math.* **42** (1980) no. 2, 147–161 (C. Pomerance); **MR** 81i:10053; **Zbl.** 426.10048.

1980.14 Maximum degree in graphs of diameter 2, *Networks* **10** (1980) no. 1, 87–90 (S. Fajtlowicz; A. J. Hoffman); **MR** 81b:05061; **Zbl.** 427.05042.

1980.15 Megjegyzések az American Mathematical Monthly egy problémájáról (Remarks on a problem of the American Mathematical Monthly, in Hungarian, English summary), *Mat. Lapok* **28** (1980) no. 1–3, 121–124 (E. Szemerédi); **MR** 82c:10066; **Zbl.** 476.10045.

1980.16 Minimal asymptotic bases for the natural numbers, *J. Number Theory* **12** (1980) no. 2, 154–159 (M. B. Nathanson); **MR** 81i:10069; **Zbl.** 426.10057.

1980.17 Multiplicative functions whose values are uniformly distributed in $(0, \infty)$, *Proceedings of the Queen's Number Theory Conference, 1979 (Kingston, Ont., 1979), Queen's Papers in Pure and Appl. Math., 54*, pp. 329–378, Queen's Univ., Kingston, Ont., 1980 (H. G. Diamond); **MR** 83e:10063; **Zbl.** 449.10033.

1980.18 Néhány elemi geometriai problémáról (On some problems in elementary geometry, in Hungarian), *Köz. Mat. Lapok* **61** (1980), 49–54.

1980.19 Noen mindre kjente problemer i kombinatorisk tallteori (Nine little known problems in combinatorial number theory, in Norwegian, English summary), *Normat Nordisk. Matematisk Tidskrift* **28** (1980) no. 4, 155–164, 180; **MR** 81k:10001; **Zbl.** 455.10002.

1980.20 Old and new problems and results in combinatorial number theory, *Monographies de L'Enseignement Mathématique [Monographs of L'Enseignement Mathématique], 28*, 128 pp., Université de Genève, L'Enseignement Mathématique, Geneva, 1980 (R. L. Graham); **MR** 82j:10001; **Zbl.** 434.10001.

1980.21 On a problem of L. Fejes Tóth, *Discrete Math.* **30** (1980) no. 2, 103–109 (J. Pach); **MR** 81e:51008; **Zbl.** 444.52008.

1980.22 On bases with an exact order, *Acta Arith.* **37** (1980), 201–207 (R. L. Graham); **MR** 82e:10093; **Zbl.** 443.10036.

1980.23 On some extremal properties of sequences of integers, II., *Publ. Math. Debrecen* **27** (1980) no. 1–2, 117–125 (A. Sárközy; E. Szemerédi); **MR** 82b:10077; **Zbl.** 461.10047.

1980.24 On the almost everywhere divergence of Lagrange interpolation polynomials for arbitrary systems of nodes, *Acta Math. Acad. Sci. Hungar.* **36** (1980) no. 1–2, 71–89 (P. Vértesi); **MR** 82e:41006a; **Zbl.** 463.41002.

1980.25 On the chromatic number of geometric graphs, *Ars Combin.* **9** (1980), 229–246 (M. Simonovits); **MR** 82c:05048; **Zbl.** 466.05031.

1980.26 On the maximal value of additive functions in short intervals and on some related questions, *Acta Math. Acad. Sci. Hungar.* **3** (1980) no. 1–2, 257–278 (I. Kátai); **MR** 83a:10090a; **Zbl.** 456.10024.

1980.27 On the Möbius function, *J. Reine Angew. Math.* **315** (1980), 121–126 (R. R. Hall); **MR** 82c:10065; **Zbl.** 419.10006.

1980.28 On the number of prime factors of integers, *Acta Sci. Math. (Szeged)* **42** (1980) no. 3–4, 237–246 (A. Sárközy); **MR** 82c:10053; **Zbl.** 448.10040.

1980.29 On the small sieve, I. Sifting by primes, *J. Number Theory* **12** (1980) no. 3, 385–394 (I. Z. Ruzsa); **MR** 81k:10075; **Zbl.** 435.10028.

1980.30 Problems and results in number theory and graph theory, *Proceedings of the Ninth Manitoba Conference on Numerical Mathematics and Computing (Univ. Manitoba, Winnipeg, Man., 1979), Congress. Numer., XXVII*, pp. 3–21, Utilitas Math., Winnipeg, Man., 1980; **MR** 81m:10006; **Zbl.** 441.10001.

1980.31 Problems and results on polynomials and interpolation, *Aspects of contemporary complex analysis (Proc. NATO Adv. Study Inst., Univ. Durham, Durham, 1979)*, pp. 383–391, Academic Press, London-New York, 1980; **MR** 82g:30013; **Zbl.** 494.30002.

1980.32 Problems and results on Ramsey-Turán type theorems (preliminary report), *Proceedings of the West Coast Conference on Combinatorics, Graph Theory and Computing (Humboldt State Univ., Arcata, Calif., 1979), Congress. Numer. XXVI*, pp. 17–23, Utilitas Math., Winnipeg, Man., 1980 (V. T. Sós); **MR** 82a:05055; **Zbl.** 463.05050.

1980.33 Proof of a conjecture of Offord, *Proc. Roy. Soc. Edinburgh Sect. A* **86** (1980) no. 1–2, 103–106; **MR** 82d:05009; **Zbl.** 432.10003.

1980.34 Ramsey-minimal graphs for the pair star, connected graph, *Studia Sci. Math. Hungar.* **15** (1980) no. 1–3, 265–273 (S. A. Burr; R. J. Faudree; C. C. Rousseau; R. H. Schelp); **MR** 84b:05072; **Zbl.** 502.05042 and 438.05046.

1980.35 Random graph isomorphism, *SIAM J. Comput.* **9** (1980) no. 3, 628–635 (L. Babai; S. M. Selkow); **MR** 83c:68071; **Zbl.** 454.05038.

1980.36 Remarks on the differences between consecutive primes, *Elem. Math.* **35** (1980) no. 5, 115–118 (E. G. Straus); **MR** 84a:10052; **Zbl.** 435.10025.

1980.37 Residually complete graphs, Combinatorial mathematics, optimal designs and their applications (Proc. Sympos. Combin. Math. and Optimal Design, Colorado State Univ., Fort Collins, Colo., 1978), *Ann. Discrete Math.* **6** (1980), 117–123 (F. Harary; M. Klawe); **MR** 82c:05058; **Zbl.** 451.05040.

1980.38 Rotations of the circle, *Measure theory, Oberwolfach 1979 (Proc. Conf., Oberwolfach, 1979), Lecture Notes in Math., 794*, pp. 53–56, Springer, Berlin, 1980 (R. D. Mauldin); **MR** 81k:28019; **Zbl.** 472.28009.

1980.39 Some applications of Ramsey's theorem to additive number theory, *Europ. J. Combin.* **1** (1980) no. 1, 43–46; **MR** 82a:10067; **Zbl.** 442.10037.

1980.40 Some asymptotic formulas on generalized divisor functions, IV., *Studia Sci. Math. Hungar.* **15** (1980) no. 4, 467–479 (A. Sárközy); **MR** 84m:10038c; **Zbl.** 512.10037.

1980.41 Some combinational problems in geometry, *Geometry and differential geometry (Proc. Conf., Univ. Haifa, Haifa, 1979), Lecture Notes in Math., 792*, pp. 46–53, Springer, Berlin, 1980; **MR** 82d:51002; **Zbl.** 428.05008.

1980.42 Some notes on Turán's mathematical work, *J. Approx. Theory* **29** (1980) no. 1, 2–5; **MR** 82f:01091; **Zbl.** 444.01024.

1980.43 Some personal reminiscences of the mathematical work of Paul Turán, *Acta Arith.* **37** (1980), 4–8; **MR** 82f:01090a; **Zbl.** 438.10001 and 368.10004.

1980.44 Sur la fonction "nombre de facteurs premiers de n" (On the function "number of prime factors of n", in French), *Séminaire Delange-Pisot-Poitou, 20e année: 1978/1979. Théorie des nombres, Fasc. 2, Exp. No. 32*, 19 pp., Secrétariat Math., Paris, 1980 (J.-L. Nicolas); **MR** 83a:10074a; **Zbl.** 422.10035.

1980.45 The fractional parts of the Bernoulli numbers, *Illinois J. Math.* **24** (1980) no. 1, 104–112 (S. S. Wagstaff, Jr.); **MR** 81c:10064; **Zbl.** 449.10010 and 405.10011.

1980.46 Values of the divisor function on short intervals, *J. Number Theory* **12** (1980) no. 2, 176–187 (R. R. Hall); **MR** 81j:10064; **Zbl.** 435.10027.

1981.01 A correction to the paper "On the maximal value of additive functions in short intervals and on some related questions" [*Acta Math. Acad. Sci. Hungar.* **3** (1980) no. 1–2, 257–278], *Acta Math. Acad. Sci. Hungar.* **37** (1981) no. 4, 499 (I. Kátai); **MR** 83a:10090b; **Zbl.** 456.10025.

1981.02 Completeness properties of perturbed sequences, *J. Number Theory* **13** (1981) no. 4, 446–455 (S. A. Burr); **MR** 83f:10055; **Zbl.** 469.10037.

1981.03 Correction of some misprints in our paper: "On almost everywhere divergence of Lagrange interpolatory polynomials for arbitrary system of nodes" [*Acta Math. Acad. Sci. Hungar.* **36** (1980) no. 1–2, 71–89], *Acta Math. Acad. Sci. Hungar.* **38** (1981) no. 1–4, 263 (P. Vértesi); **MR** 82e:41006b; **Zbl.** 463.41003 and 486.41003.

1981.04 Existence of complementary graphs with specified independence numbers, *The theory and applications of graphs (Kalamazoo, Mich., 1980)*, pp. 343–349, Wiley, New York, 1981 (S. Schuster); **MR** 82m:05080; **Zbl.** 479.05054.

1981.05 Fat, symmetric, irrational Cantor sets, *Amer. Math. Monthly* **88** (1981) no. 5, 340–341 (D. Boes; R. Darst); **MR** 83i:26003; **Zbl.** 468.26003.

1981.06 Finite abelian group cohesion, *Israel J. Math.* **39** (1981) no. 3, 177–185 (B. Smith); **MR** 82m:20057; **Zbl.** 464.20034.

1981.07 Grandes valeurs d'une fonction liée au produit d'entiers consécutifs, *Publ. Math. Orsay* **81.01** (1981), 30–34 (J.-L. Nicolas); **Zbl.** 446.10033.

1981.08 Lagrange's theorem and thin subsequences of squares, *Contributions to probability*, pp. 3–9, Academic Press, New York-London, 1981 (M. B. Nathanson); **MR** 82k:10062; **Zbl.** 539.10038.

1981.09 Many old and on some new problems of mine in number theory, Proceedings of the Tenth Manitoba Conference on Numerical Mathematics and Computing, Vol. I (Winnipeg, Man., 1980), *Congr. Numer.* **30** (1981), 3–27; **MR** 82m:10003; **Zbl.** 515.10002.

1981.10 Minimal decompositions of graphs into mutually isomorphic subgraphs, *Combinatorica* **1** (1981), 13–24 (F. R. K. Chung; R. L. Graham); **MR** 82j:05071; **Zbl.** 491.05049.

1981.11 On sharp elementary prime number estimates, *Enseign. Math. (2)* **26** (1980) no. 3–4, 313–321, 1981 (H. G. Diamond); **MR** 83i:10055; **Zbl.** 453.10007.

1981.12 On some partition properties of families of sets, *Studia Sci. Math. Hungar.* **13** (1978) no. 1–2, 151–155, 1981 (G. Elekes; A. Hajnal); **MR** 84h:03110; **Zbl.** 474.04002.

1981.13 On some problems in number theory, Séminaire Delange-Pisot-Poitou, Paris, 1979–80, *Prog. Mathematics* **12** (1981), 71–75; **Zbl.** 448.10003.

1981.14 On the almost everywhere divergence of Lagrange interpolation, *Approximation and function spaces (Gdańsk, 1979)*, pp. 270–278,

North-Holland, Amsterdam-New York, 1981 (P. Vértesi); **MR** 84d:41002; **Zbl.** 491.41001.

1981.15 On the bandwidths of a graph and its complement, *The theory and application of graphs (Kalamazoo, Mich., 1980)*, pp. 243–253, Wiley, New York, 1981 (P. Z. Chinn; F. R. K. Chung; R. L. Graham); **MR** 83j:05043; **Zbl.** 467.05038.

1981.16 On the combinatorial problems which I would most like to see solved, *Combinatorica* **1** (1981) no. 1, 25–42; **MR** 82k:05001; **Zbl.** 486.05001.

1981.17 On the conjecture of Hajós, *Combinatorica* **1** (1981) no. 2, 141–143 (S. Fajtlowicz); **MR** 83d:05042; **Zbl.** 504.05052.

1981.18 On the Lebesgue function of interpolation, *Functional analysis and approximation (Oberwolfach, 1980), Internat. Ser. Numer. Math., 60*, pp. 299–309, Birkhäuser, Basel-Boston, Mass., 1981 (P. Vértesi); **MR** 83k:41002; **Zbl.** 499.41004.

1981.19 On Turán's theorem for sparse graphs, *Combinatorica* **1** (1981) no. 4, 313–317 (M. Ajtai; J. Komlós; E. Szemerédi); **MR** 83d:05052; **Zbl.** 491.05038.

1981.20 Problems and results in graph theory, *The theory and application of graphs (Kalamazoo, Mich., 1980)*, pp. 331–341, Wiley, New York, 1981; **MR** 83c:05112; **Zbl.** 463.05036.

1981.21 Problems and results in number theory, *Recent progress in analytic number theory, Vol. 1 (Durham, 1979)*, pp. 1–13, Academic Press, London-New York, 1981; **MR** 84j:10001; **Zbl.** 459.10002.

1981.22 Problems and results on finite and infinite combinatorial analysis, II., *Enseign. Math. (2)* **27** (1981) no. 1–2, 163–176; **MR** 83c:04006a; **Zbl.** 459.05003.

1981.23 Ramsey-minimal graphs for matchings, *The theory and application of graphs (Kalamazoo, Mich., 1980)*, pp. 159–168, Wiley, New York, 1981 (S. A. Burr; R. J. Faudree; C. C. Rousseau; R. H. Schelp); **MR** 83c:05092; **Zbl.** 469.05048.

1981.24 Ramsey-minimal graphs for star-forests, *Discrete Math.* **33** (1981) no. 3, 227–237 (S. A. Burr; R. J. Faudree; C. C. Rousseau; R. H. Schelp); **MR** 83e:05084; **Zbl.** 456.05046.

1981.25 Reciprocals of certain large additive functions, *Canad. Math. Bull.* **24** (1981) no. 2, 225–231 (J.-M. De Koninck; A. Ivić); **MR** 82k:10053; **Zbl.** 463.10032.

1981.26 Sets of natural numbers of positive density and cylindric set algebras of dimension 2, *Algebra Universalis* **12** (1981) no. 1, 81–92 (V. Faber; J. A. Larson); **MR** 82h:03069; **Zbl.** 473.03057.

1981.27 Solved and unsolved problems in combinatorics and combinatorial number theory, Proceedings of the Twelfth Southeastern Conference on Combinatorics, Graph Theory and Computing, Vol. I (Baton Rouge, La., 1981), *Congr. Numer.* **32** (1981), 49–62; **MR** 84i:10011; **Zbl.** 486.05002.

1981.28 Some additive properties of sets of real numbers, *Fund. Math.* **113** (1981) no. 3, 187–199 (K. Kunen; R. D. Mauldin); **MR** 85f:04003; **Zbl.** 482.28001.

1981.29 Some applications of graph theory and combinatorial methods to number theory and geometry, *Algebraic methods in graph theory, Vol. I, II (Szeged, 1978), Colloq. Math. Soc. János Bolyai, 25*, pp. 137–148, North-Holland, Amsterdam-New York, 1981; **MR** 83g:05001; **Zbl.** 472.10001.

1981.30 Some bounds for the Ramsey-Paris-Harrington numbers, *J. Comb. Theory Ser. A* **30** (1981) no. 1, 53–70 (G. Mills); **MR** 84c:03099; **Zbl.** 471.05045.

1981.31 Some extremal problems on divisibility properties of sequences of integers, *Fibonacci Quart.* **19** (1981) no. 3, 208–213; **MR** 82i:10077; **Zbl.** 469.10036.

1981.32 Some new problems and results in graph theory and other branches of combinatorial mathematics, *Combinatorics and graph theory (Calcutta, 1980), Lecture Notes in Math., 885*, pp. 9–17, Springer, Berlin-New York, 1981; **MR** 83k:05038; **Zbl.** 477.05049.

1981.33 Some problems and results on additive and multiplicative number theory, *Analytic number theory (Proc. Conf., Temple Univ., Phila., 1980), Lecture Notes in Math., 899*, pp. 171–182, Springer, Berlin-New York, 1981; **MR** 84c:10048; **Zbl.** 472.10002.

1981.34 Sur la fonction: nombre de facteurs premiers de N (in French), *Enseign. Math (2)* **27** (1981) no. 1–2, 3–27 (J.-L. Nicolas); **MR** 83a:10074b; **Zbl.** 466.10037.

1981.35 Sur la structure de la suite des diviseurs d'un entier (in French), *Ann. Inst. Fourier (Grenoble)* **31** (1981) no. 1, ix, 17–37 (G. Tenenbaum); **MR** 82h:10061; **Zbl.** 456.10022 and 437.10020.

1981.36 Sur l'irrationalité d'une certaine série (in French), *C. R. Acad. Sci. Paris, Sér. I Math.* **292** (1981) no. 17, 765–768; **MR** 82g:10052; **Zbl.** 466.10028.

1981.37 The arithmetic mean of the divisors of an integer, *Analytic number theory (Proc. Conf., Temple Univ., Phila., 1980), Lecture Notes in Math., 899*, pp. 197–220, Springer, Berlin-New York, 1981 (P. T. Bateman; C. Pomerance; E. G. Straus); **MR** 84b:10066; **Zbl.** 478.10027.

1982.01 Another property of 239 and some related questions, Proceedings of the Eleventh Manitoba Conference on Numerical Mathematics and Computing (Winnipeg, Man., 1981), *Congr. Numer.* **34** (1982), 243–257 (R. K. Guy; J. L. Selfridge); **MR** 84f:10023; **Zbl.** 536.10007.

1982.02 Compactness results in extremal graph theory, *Combinatorica* **2** (1982) no. 3, 275–288 (M. Simonovits); **MR** 84g:05083; **Zbl.** 508.05043.

1982.03 Disjoint cliques and disjoint maximal independent sets of vertices in graphs, *Discrete Math.* **42** (1982) no. 1, 57–61 (A. M. Hobbs; C. Payan); **MR** 84g:05084; **Zbl.** 493.05049.

1982.04 Families of finite sets in which no set is covered by the union of two others, *J. Combin. Theory Ser. A* **33** (1982) no. 2, 158–166 (P. Frankl; Z. Füredi); **MR** 84e:05002; **Zbl.** 489.05003.

1982.05 Grandes valeurs d'une fonction liée au produit d'entiers consécutifs (Large values of a function related to the product of consecutive integers, in French, English summary), *Ann. Fac. Sci. Toulouse Math. (5)* **3** (1981) no. 3–4, 173–199, 1982 (J.-L. Nicolas); **MR** 83i:10056; **Zbl.** 488.10045.

1982.06 Graph with certain families of spanning trees, *J. Combin. Theory Ser. B* **32** (1982) no. 2, 162–170 (R. J. Faudree; C. C. Rousseau; R. H. Schelp); **MR** 84c:05071; **Zbl.** 478.05076 and 465.05058.

1982.07 Minimal decompositions of hypergraphs into mutually isomorphic subhypergraphs, *J. Combin. Theory Ser. A* **32** (1982) no. 2, 241–251 (F. R. K. Chung; R. L. Graham); **MR** 83j:05057; **Zbl.** 493.05048.

1982.08 Miscellaneous problems in number theory, Proceedings of the Eleventh Manitoba Conference on Numerical Mathematics and Computing (Winnipeg, Man., 1981), *Congr. Numer.* **34** (1982), 25–45; **MR** 84f:10002; **Zbl.** 563.10002.

1982.09 On a problem in combinatorial geometry, *Discrete Math.* **40** (1982) no. 1, 45–52 (G. B. Purdy; E. G. Straus); **MR** 84g:52015; **Zbl.** 501.52009.

1982.10 On a problem of R. R. Hall (in Hungarian, English summary) *Mat. Lapok* **30** (1978/82) no. 1–3, 23–31 (A. Sárközy); **MR** 86e:11090; **Zbl.** 542.10040.

1982.11 On almost bipartite large chromatic graphs, *Annals of Discrete Math.* **12** (1982), *Theory and practice of combinatorics, North-Holland Math. Stud., 60*, pp. 117–123, North-Holland, Amsterdam, New York, 1982 (A. Hajnal; E. Szemerédi); **MR** 86j:05060; **Zbl.** 501.05033.

1982.12 On graphs which contain all sparse graphs, *Annals of Discrete Math.* **12** (1982), *Theory and practice of combinatorics, North-Holland Math. Stud., 60*, pp. 21–26, North-Holland, Amsterdam-New York, 1982 (L. Babai; F. R. K. Chung; R. L. Graham; J. H. Spencer); **MR** 86m:05057; **Zbl.** 495.05035.

1982.13 On pairwise balanced block designs with the sizes of blocks as uniform as possible, *Annals of Discrete Math.* **15** (1982), *Algebraic and geometric combinatorics, North-Holland Math. Stud., 65*, pp. 129–134, North-Holland, Amsterdam-New York, 1982 (J. A. Larson); **MR** 85m:05012; **Zbl.** 499.05014.

1982.14 On prime factors of binomial coefficients, II. (in Hungarian), *Mat. Lapok* **30** (1978/82) no. 4, 307–316; **MR** 85f:11012; **Zbl.** 541.10002.

1982.15 On Ramsey-Turán type theorems for hypergraphs, *Combinatorica* **2** (1982) no. 3, 289–295 (V. T. Sós); **MR** 85d:05185; **Zbl.** 511.05049.

1982.16 On some unusual nonconventional problems in additive number theory (in Hungarian), *Mat. Lapok* **30** (1978/82) no. 1–3, 9–14; **MR** 85c:11093; **Zbl.** 542.10011.

1982.17 On sums involving reciprocals of certain arithmetical functions, *Publ. Inst. Math. (Beograd) (N.S.)* **32(46)** (1982), 49–56 (A. Ivić); **MR** 85g:11083; **Zbl.** 506.10035.

1982.18 On the approximation of convex, closed plane curves by multifocal ellipses, Essays in statistical science, *J. Appl. Probab. Spec. Vol.* **19** *A* (1982) or *Z. Phys. C.* **10** (1981) no. 2, 89–96 (I. Vincze); **MR** 83a:52006; **Zbl.** 483.51010.

1982.19 On the average ratio of the smallest and largest prime divisor of n, *Nederl. Akad. Wetensch. Indag. Math.* **44** (1982) no. 2, 127–132 (J. H. van Lint); **MR** 83m:10075; **Zbl.** 489.10041.

1982.20 On the covering of the vertices of a graph by cliques, *J. Math. Res. Exposition* **2** (1982) no. 1, 93–96; **MR** 84b:05058; **Zbl.** 485.05052.

1982.21 On Turán-Ramsey type theorems, II., *Studia Sci. Math. Hungar.* **14** (1979) no. 1–3, 27–36, 1982 (V. T. Sós); **MR** 84j:05081; **Zbl.** 487.05054.

1982.22 Personal reminiscences and remarks on the mathematical work of Tibor Gallai, *Combinatorica* **2** (1982) no. 3, 207–212; **MR** 84e:01073; **Zbl.** 505.01012.

1982.23 Problems and results on block designs and set systems, Proceedings of the thirteenth Southeastern conference on combinatorics, graph theory and computing (Boca Raton, Fla., 1982), *Congr. Numer.* **35** (1982), 3–16; **MR** 84m:05014; **Zbl.** 517.05018.

1982.24 Problems and results on finite and infinite combinatorial analysis, II., *Logic and algorithmic (Zürich, 1980), Monograph. Enseign. Math.*, pp. 131–144, Univ. Genève, Geneva, 1982; **MR** 83c:04006b; **Zbl.** 477.05012.

1982.25 Projective $(2n, n, \lambda, 1)$-designs, *J. Statistical Planning and Inference* **7** (1982/83) no. 2, 181–191 (V. Faber; F. Jones); **MR** 85f:05017; **Zbl.** 496.05009.

1982.26 Ramsey-minimal graphs for forests, *Discrete Math.* **38** (1982) no. 1, 23–32 (S. A. Burr; R. J. Faudree; C. C. Rousseau; R. H. Schelp); **MR** 84h:05091; **Zbl.** 489.05039.

1982.27 Ramsey numbers for brooms, Proceedings of the thirteenth Southeastern conference on combinatorics, graph theory and computing (Boca Raton, Fla., 1982), *Congr. Numer.* **35** (1982), 283–293 (R. J. Faudree; C. C. Rousseau; R. H. Schelp); **MR** 84m:05056; **Zbl.** 513.05038.

1982.28 Ramsey numbers for the pair sparse graph–path or cycle. *Trans. Amer. Math. Soc.* **269** (1982) no. 2, 501–512 (S. A. Burr; R. J. Faudree; C. C. Rousseau; R. H. Schelp); **MR** 83c:05093; **Zbl.** 543.05047.

1982.29 Representation of group elements as short products, *Annals of Discrete Math.* **12** (1982), *Theory and practice of combinatorics, North-*

Holland Math. Stud., 60, pp. 27–30, North-Holland, Amsterdam, New York, 1982 (L. Babai); **MR** 87a:20027.

1982.30 Some asymptotic formulas on generalized divisor functions, II., *J. Number Theory* **15** (1982) no. 1, 115–136 (A. Sárközy); **MR** 84m:10038a; **Zbl.** 488.10043.

1982.31 Some asymptotic formulas on generalized divisor functions, III., *Acta Arith.* **41** (1982) no. 4, 395–411 (A. Sárközy); **MR** 84m:10038b; **Zbl.** 492.10037.

1982.32 Some new problems and results in number theory, *Number theory (Mysore, 1981), Lecture Notes in Math., 938*, pp. 50–74, Springer, Berlin-New York, 1982; **MR** 84g:10002; **Zbl.** 484.10001.

1982.33 Some of my favourite problems which recently have been solved, *Proceedings of the International Mathematical Conference, Singapore 1981 (Singapore, 1981), North-Holland Math. Stud., 74*, pp. 59–79, North-Holland, Amsterdam-New York, 1982; **MR** 84f:10003.

1982.34 Some problems on additive number theory, *Annals of Discrete Math.* **12** (1982), *Theory and practice of combinatorics, North-Holland Math. Stud., 60*, pp. 113–116, North-Holland, Amsterdam, New York, 1982; **MR** 86k:11055; **Zbl.** 491.10044.

1982.35 Subgraphs in which each pair of edges lies in a short common cycle, Proceedings of the thirteenth Southeastern conference on combinatorics, graph theory and computing (Boca Raton, Fla., 1982), *Congr. Numer.* **35** (1982), 253–260 (R. Duke); **MR** 85d:05183; **Zbl.** 515.05048.

1983.01 An analogue of Grimm's problem of finding distinct prime factors of consecutive integers, *Utilitas Math.* **24** (1983), 45–65 (C. Pomerance); **MR** 85b:11072; **Zbl.** 525.10023.

1983.02 Arithmetical properties of permutations of integers, *Acta Math. Hungar.* **41** (1983) no. 1–2, 169–176 (R. Freud; N. Hegyvári); **MR** 85d:11002; **Zbl.** 518.10063.

1983.03 Combinatorial problems in geometry, *Math. Chronicle* **12** (1983), 35–54; **MR** 84i:52014; **Zbl.** 537.51017.

1983.04 Dot product rearrangements, *Internat. J. Math. Math. Sci.* **6** (1983) no. 3, 409–418 (G. Weiss); **MR** 85d:40001; **Zbl.** 539.40001.

1983.05 Finite linear spaces and projective planes, *Discrete Math.* **47** (1983) no. 1, 49–62 (R. C. Mullin; V. T. Sós; D. Stinson); **MR** 84k:05026; **Zbl.** 521.51005.

1983.06 Generalizations of a Ramsey-theoretic result of Chvátal, *J. Graph Theory* **7** (1983) no. 1, 39–51 (S. A. Burr); **MR** 84d:05125; **Zbl.** 513.05040.

1983.07 Intersection properties of families containing sets of nearly the same size, *Ars Combin.* **15** (1983), 247–259 (R. Silverman; A. Stein); **MR** 84j:05032; **Zbl.** 521.05004.

1983.08 Local connectivity of a random graph, *J. Graph Theory* **7** (1983) no. 4, 411–417 (E. M. Palmer; R. W. Robinson); **MR** 85d:05210; **Zbl.** 529.05053.

1983.09 More results on Ramsey-Turán type problems, *Combinatorica* **3** (1983) no. 1, 69–81 (A. Hajnal; V. T. Sós; E. Szemerédi); **MR** 85b:05129; **Zbl.** 526.05031.

1983.10 On a generalization of Turán's graph-theorem, *Studies in pure mathematics, To the memory of Paul Turán*, pp. 181–185, Birkhäuser, Basel-Boston, Mass., 1983 (V. T. Sós); **MR** 86m:05058; **Zbl.** 518.05044.

1983.11 On a problem of Oppenheim concerning "factorisatio numerorum", *J. Number Theory* **17** (1983) no. 1, 1–28 (E. R. Canfield; C. Pomerance); **MR** 85j:11012; **Zbl.** 513.10043.

1983.12 On a quasi-Ramsey problem, *J. Graph Theory* **7** (1983) no. 1, 137–147 (J. Pach); **MR** 84d:05127; **Zbl.** 511.05047.

1983.13 On almost divisibility properties of sequences of integers, I., *Acta Math. Hungar.* **41** (1983) no. 3–4, 309–324 (A. Sárközy); **MR** 85b:11075; **Zbl.** 523.10023.

1983.14 On maximum chordal subgraph, Proceedings of the fourteenth Southeastern conference on combinatorics, graph theory and computing (Boca Raton, Fla., 1983), *Congr. Numer.* **39** (1983), 367–373 (R. Laskar); **MR** 85i:05132; **Zbl.** 534.05037.

1983.15 On some of my conjectures in number theory and combinatorics, Proceedings of the fourteenth Southeastern conference on combinatorics, graph theory and computing (Boca Raton, Fla., 1983), *Congr. Numer.* **39** (1983), 3–19; **MR** 85j:11004; **Zbl.** 539.05001.

1983.16 On some problems of J. Dénes and P. Turán, *Studies in pure mathematics, To the memory of Paul Turán*, pp. 187–212, Birkhäuser, Basel-Boston, Mass., 1983 (M. Szalay); **MR** 87g:11131; **Zbl.** 523.10029.

1983.17 On some problems related to partitions of edges of a graph, *Graphs and other combinatorial topics (Prague, 1982), Teubner-Texte Math., 59*, pp. 54–63, Teubner, Leipzig, 1983 (J. Nešetřil; V. Rödl); **CMP** 737 014; **Zbl.** 522.05070.

1983.18 On sums and products of integers, *Studies in pure mathematics, To the memory of Paul Turán*, pp. 213–218, Birkhäuser, Basel-Boston, Mass., 1983 (E. Szemerédi); **MR** 86m:11011; **Zbl.** 526.10011.

1983.19 On sums of Rudin-Shapiro coefficients, II., *Pacific J. Math.* **107** (1983) no. 1, 39–69 (J. Brillhart; P. Morton); **MR** 85i:11080; **Zbl.** 505.10029 and 469.10034.

1983.20 On the decomposition of graphs into complete bipartite subgraphs, *Studies in pure mathematics, To the memory of Paul Turán*, pp. 95–101, Birkhäuser, Basel-Boston, Mass., 1983 (F. R. K. Chung; J. H. Spencer); **MR** 87k:05097; **Zbl.** 531.05042.

1983.21 On unavoidable graphs, *Combinatorica* **3** (1983) no. 2, 167–176 (F. R. K. Chung); **MR** 85a:05047; **Zbl.** 527.05042.

1983.22 Polychromatic Euclidean Ramsey theorems, *J. Geom.* **20** (1983) no. 1, 28–35 (B. L. Rothschild; E. G. Straus); **MR** 85d:05073; **Zbl.** 541.05010.

1983.23 Preface. Personal reminiscences, *Studies in pure mathematics, To the memory of Paul Turán*, pp. 11–12, Birkhäuser, Basel-Boston, Mass., 1983; **MR** 87f:01042a; **Zbl.** 515.01012.

1983.24 Problems and results on polynomials and interpolation, *Functions, series, operators, Vol. I, II (Budapest, 1980), Colloq. Math. Soc. János Bolyai, 35*, pp. 485–495, North-Holland, Amsterdam-New York, 1983; **MR** 85i:30071; **Zbl.** 556.41002.

1983.25 Projective $(2n, n, \lambda, 1)$-designs, *J. Statist. Plann. Inference* **7** (82/83) no. 2, 181–191 (V. Faber; F. Jones); **MR** 85f:05017; **Zbl.** 496.05009.

1983.26 Some asymptotic formulas on generalized divisor functions, I., *Studies in pure mathematics, To the memory of Paul Turán*, pp. 165–179, Birkhäuser, Basel-Boston, Mass., 1983 (A. Sárközy); **MR** 87b:11092; **Zbl.** 517.10048.

1983.27 Some remarks and problems in number theory related to the work of Euler, *Math. Mag.* **56** (1983) no. 5, 292–298 (U. Dudley); **MR** 86a:01018; **Zbl.** 526.01014.

1983.28 Supersaturated graphs and hypergraphs, *Combinatorica* **3** (1983) no. 2, 181–192 (M. Simonovits); **MR** 85e:05125; **Zbl.** 529.05027.

1983.29 Sur les diviseurs consécutifs d'un entier (The consecutive divisors of an integer, in French, English summary), *Bull. Soc. Math. France* **111** (1983) no. 2, 125–145 (G. Tenenbaum); **MR** 86a:11037; **Zbl.** 526.10036.

1983.30 The greatest angle among n points in the $d-$dimensional Euclidean space, *Annals of Discrete Math.* **17** (1983), *Combinatorial mathematics (Marseille-Luminy, 1981), North-Holland Math. Stud., 75*, pp. 275–283, North-Holland, Amsterdam-New York, 1983 (Z. Füredi); **MR** 87g:52018; **Zbl.** 534.52007.

1983.31 Trees in random graphs, *Discrete Math.* **46** (1983) no. 2, 145–150 (Z. Palka); **MR** 84i:05103; **Zbl.** 535.05049.

1984.01 Addendum to: "Trees in random graphs" [*Discrete Math.* **46** (1983) no. 2, 145–150], *Discrete Math.* **48** (1984) no. 2–3, 331 (Z. Palka); **MR** 85b:05151; **Zbl.** 546.05052.

1984.02 *Combinatorial set theory: partition relations for cardinals, Studies in Logic and the Foundations of Mathematics, 106*, North-Holland Publishing Co., Amsterdam-New York, 1984, 347 pp., ISBN: 0-444-86157-2 (A. Hajnal; A. Máté; R. Rado); **MR** 87g:04002; **Zbl.** 573.03019.

1984.03 Cross-cuts in the power set of an infinite set, *Order* **1** (1984) no. 2, 139–145 (J. E. Baumgartner; D. Higgs); **MR** 85m:04002; **Zbl.** 559.04009.

1984.04 Cube-supersaturated graphs and related problems, *Progress in graph theory (Waterloo, Ont., 1982)*, pp. 203–218, Academic Press, Toronto, ON, 1984 (M. Simonovits); **MR** 86b:05041; **Zbl.** 565.05042.

1984.05 Enumeration of intersecting families, *Discrete Math.* **48** (1984) no. 1, 61–65 (N. Hindman); **MR** 85h:05007; **Zbl.** 529.05044.

1984.06 Extremal problems in number theory, combinatorics and geometry, *Proceedings of the International Congress of Mathematicians, Vol. 1, 2 (Warsaw, 1983)*, pp. 51–70, PWN, Warsaw, 1984; **MR** 87a:11001; **Zbl.** 563.10003.

1984.07 Inverse extremal digraph problems, *Finite and infinite sets, Vol. I, II (Eger, 1981), Colloq. Math. Soc. János Bolyai, 37*, pp. 119–156, North-Holland, Amsterdam-New York, 1984 (W. G. Brown; M. Simonovits); **MR** 87f:05087; **Zbl.** 569.05023.

1984.08 Minimal decomposition of all graphs with equinumerous vertices and edges into mutually isomorphic subgraphs, *Finite and infinite sets, Vol. I, II (Eger, 1981), Colloq. Math. Soc. János Bolyai, 37*, pp. 171–179, North-Holland, Amsterdam-New York, 1984 (F. R. K. Chung; R. L. Graham); **MR** 87b:05071; **Zbl.** 565.05040.

1984.09 More results on subgraphs with many short cycles, Proceedings of the fifteenth Southeastern conference on combinatorics, graph theory and computing (Baton Rouge, La., 1984), *Congr. Numer.* **43** (1984), 295–300 (R. Duke; V. Rödl); **MR** 86f:05079; **Zbl.** 559.05038.

1984.10 On disjoint sets of differences, *J. Number Theory* **18** (1984) no. 1, 99–109 (R. Freud); **MR** 85g:11018; **Zbl.** 544.10060.

1984.11 On some problems in graph theory, combinatorial analysis and combinatorial number theory, *Graph theory and combinatorics (Cambridge, 1983)*, pp. 1–17, Academic Press, London-New York, 1984; **MR** 86e:05001; **Zbl.** 546.05002.

1984.12 On the chessmaster problem, *Progress in graph theory (Waterloo, Ont., 1982)*, pp. 532–536, Academic Press, Toronto, ON, 1984 (R. L. Hemminger; D. A. Holton; B. D. McKay); **MR** 85k:05005 (for entire book); **Zbl.** 546.00007 (for entire book).

1984.13 On the favourite points of a random walk, *Mathematical structure—computational mathematics—mathematical modelling, 2*, pp. 152–157, Bulgar. Acad. Sci., Sofia, 1984 (P. Révész); **MR** 86k:60126; **Zbl.** 593.60072.

1984.14 On the maximal number of strongly independent vertices in a random acyclic directed graph, *SIAM J. Algebraic Discrete Methods* **5** (1984) no. 4, 508–514 (A. B. Barak); **MR** 86g:05083; **Zbl.** 558.05026.

1984.15 On the statistical theory of partitions, *Topics in classical number theory, Vol. I, II (Budapest, 1981), Colloq. Math. Soc. János Bolyai, 34*, pp. 397–450, North-Holland, Amsterdam-New York, 1984 (M. Szalay); **MR** 86f:11075; **Zbl.** 548.10010.

1984.16 On two unconventional number theoretic functions and on some related problems *Calcutta Mathematical Society, Diamond-cum-platinum jubilee commemoration volume (1908–1983), Part I*, pp. 113–121, Calcutta Math. Soc., Calcutta, 1984; **MR** 87k:11007; **Zbl.** 593.10036.

1984.17 Products of integers in short intervals, *Acta Arith.* **44** (1984) no. 2, 147–174 (J. Turk); **MR** 86d:11073; **Zbl.** 547.10036 and 497.10033.

1984.18 Research problems, *Period. Math. Hungar.* **15** (1984), 101–103; **Zbl.** 537.05015.

1984.19 Selectivity of hypergraphs, *Finite and infinite sets, Vol. I, II (Eger, 1981), Colloq. Math. Soc. János Bolyai, 37*, pp. 265–284, North-Holland, Amsterdam-New York, 1984 (J. Nešetřil; V. Rödl); **MR** 87d:05123; **Zbl.** 569.05041.

1984.20 Size Ramsey numbers involving matchings, *Finite and infinite sets, Vol. I, II (Eger, 1981), Colloq. Math. Soc. János Bolyai, 37*, pp. 247–264, North-Holland, Amsterdam-New York, 1984 (R. J. Faudree); **MR** 87i:05145; **Zbl.** 563.05043.

1984.21 Some new and old problems on chromatic graphs, *Combinatorics and applications (Calcutta, 1982)*, pp. 118–126, Indian Statist. Inst., Calcutta, 1984; **CMP** 852 032; **Zbl.** 716.05023.

1984.22 Some old and new problems in combinatorial geometry, *Annals of Discrete Math.* **20** (1984), *Convexity and graph theory (Jerusalem, 1981), North-Holland Math. Stud., 87*, pp. 129–136, North-Holland, Amsterdam-New York, 1984; **MR** 87b:52018; **Zbl.** 562.51008.

1984.23 Some results in combinatorial number theory, *Topics in classical number theory, Vol. I, II (Budapest, 1981), Colloq. Math. Soc. János Bolyai, 34*, pp. 389–396, North-Holland, Amsterdam, 1984 (R. Freud; N. Hegyvári); **CMP** 781 148; **Zbl.** 562.10029.

1984.24 Tree–multipartite graph Ramsey numbers, *Graph theory and combinatorics (Cambridge, 1983)*, pp. 155–160, Academic Press, London-New York, 1984 (R. J. Faudree; C. C. Rousseau; R. H. Schelp); **MR** 86e:05063; **Zbl.** 547.05045.

1985.01 A conjecture on dominating cycles, Proceedings of the sixteenth Southeastern international conference on combinatorics, graph theory and computing (Boca Raton, Fla., 1985), *Congr. Numer.* **47** (1985), 189–197 (B. N. Clark; C. J. Colbourn); **MR** 87k:05104; **Zbl.** 622.05042.

1985.02 A note on the interval number of a graph, *Discrete Math.* **55** (1985) no. 2, 129–133 (D. B. West); **MR** 86k:05062; **Zbl.** 576.05019.

1985.03 A note on the size of a chordal subgraph, Proceedings of the sixteenth Southeastern international conference on combinatorics, graph theory and computing (Boca Raton, Fla., 1985), *Congr. Numer.* **48** (1985), 81–86 (R. Laskar); **MR** 87k:05098; **Zbl.** 647.05034.

1985.04 A property of random graphs, *Ars Combin.* **19** (1985) A, 287–294 (L. Caccetta; K. Vijayan); **MR** 87b:05069; **Zbl.** 572.05036.

1985.05 A Ramsey-type property in additive number theory, *Glasgow Math. J.* **27** (1985), 5–10 (S. A. Burr); **MR** 87b:11014; **Zbl.** 578.10055.

1985.06 Algorithmic solution of extremal digraph problems, *Trans. Amer. Math. Soc.* **292** (1985) no. 2, 421–449 (W. G. Brown; M. Simonovits); **MR** 87a:05083; **Zbl.** 607.05040.

1985.07 An application of graph theory to additive number theory, *European J. Combin.* **6** (1985) no. 3, 201–203 (N. Alon); **MR** 87d:11015; **Zbl.** 581.10029.

1985.08 Chromatic number of finite and infinite graphs and hypergraphs (French summary), Special volume on ordered sets and their applications (L'Arbresle, 1982), *Discrete Math.* **53** (1985), 281–285 (A. Hajnal); **MR** 86k:05050; **Zbl.** 566.05029.

1985.09 Colouring the real line, *J. Combin. Theory Ser. B* **39** (1985) no. 1, 86–100 (R. B. Eggleton; D. K. Skilton); **MR** 87b:05057; **Zbl.** 564.05029 and 549.05029.

1985.10 E. Straus (1921–1983), Number theory (Winnipeg, Man., 1983), *Rocky Mountain J. Math.* **15** (1985) no. 2, 331–341; **MR** 87f:01043; **Zbl.** 581.01021.

1985.11 Entire functions bounded outside a finite area, *Acta Math. Hungar.* **45** (1985) no. 3–4, 367–376 (A. Edrei); **MR** 86f:30027; **Zbl.** 578.30018.

1985.12 Extremal problems for pairwise balanced designs, Proceedings of the sixteenth Southeastern international conference on combinatorics, graph theory and computing (Boca Raton, Fla., 1985), *Congr. Numer.* **48** (1985), 55–66 (R. Duke; J. C. Fowler; K. T. Phelps); **MR** 87g:05058; **Zbl.** 628.05005.

1985.13 Extremal subgraphs for two graphs, *J. Combin. Theory Ser. B* **38** (1985) no. 3, 248–260 (F. R. K. Chung; J. H. Spencer); **MR** 87b:05073; **Zbl.** 554.05037.

1985.14 Families of finite sets in which no set is covered by the union of r others, *Israel J. Math.* **51** (1985) no. 1–2, 79–89 (P. Frankl; Z. Füredi); **MR** 87a:05008; **Zbl.** 587.05021.

1985.15 Multipartite graph–sparse graph Ramsey numbers, *Combinatorica* **5** (1985) no. 4, 311–318 (R. J. Faudree; C. C. Rousseau; R. H. Schelp); **MR** 88e:05081; **Zbl.** 593.05050.

1985.16 On locally repeated values of certain arithmetic functions, I., *J. Number Theory* **21** (1985) no. 3, 319–332 (A. Sárközy; C. Pomerance); **MR** 87b:11007; **Zbl.** 574.10012.

1985.17 On some of my problems in number theory I would most like to see solved, *Number theory (Ootacamund, 1984), Lecture Notes in Math., 1122*, pp. 74–84, Springer, Berlin, 1985; **CMP** 797 781; **Zbl.** 559.10001.

1985.18 On the existence of two nonneighboring subgraphs in a graph, *Combinatorica* **5** (1985) no. 4, 295–300 (M. El-Zahar); **MR** 87g:05120; **Zbl.** 596.05027.

1985.19 On the length of the longest excursion, *Z. Wahrsch. Verw. Gebiete* **68** (1985) no. 3, 365–382 (E. Csáki; P. Révész); **MR** 86f:60086; **Zbl.** 547.60074 and 537.60062.

1985.20 On the normal number of prime factors of $\varphi(n)$, Number theory (Winnipeg, Man., 1983), *Rocky Mountain J. Math.* **15** (1985) no. 2, 343–352 (C. Pomerance); **MR** 87e:11112; **Zbl.** 617.10037.

1985.21 On the Schnirelmann and asymptotic densities of sets of nonmultiples, Proceedings of the sixteenth Southeastern international conference on combinatorics, graph theory and computing (Boca Raton, Fla., 1985), *Congr. Numer.* **48** (1985), 67–79, (C. B. Lacampagne; C. Pomerance; J. L. Selfridge); **MR** 87j:11013; **Zbl.** 627.10035.

1985.22 On 2-designs, *J. Combin. Theory Ser. A* **38** (1985) no. 2, 131–142 (J. C. Fowler; V. T. Sós; R. M. Wilson); **MR** 86k:05026; **Zbl.** 575.05008.

1985.23 Problems and results in combinatorial geometry, *Discrete geometry and convexity (New York, 1982), Ann. New York Acad. Sci., 440*, pp. 1–11, New York Acad. Sci., New York, 1985; **MR** 87g:52001; **Zbl.** 568.51011.

1985.24 Problems and results on additive properties of general sequences, I., *Pacific J. Math.* **118** (1985) no. 2, 347–357 (A. Sárközy); **MR** 86j:11015; **Zbl.** 569.10032.

1985.25 Problems and results on additive properties of general sequences, IV., *Number theory (Ootacamund, 1984), Lecture Notes in Math., 1122*, pp. 85–104, Springer, Berlin-New York, 1985 (A. Sárközy; V. T. Sós); **MR** 88i:11011a; **Zbl.** 588.10056.

1985.26 Problems and results on chromatic numbers in finite and infinite graphs, *Graph theory with applications to algorithms and computer science (Kalamazoo, Mich., 1984)*, pp. 201–213, Wiley-Intersci. Publ., Wiley, New York, 1985; **MR** 87f:05068; **Zbl.** 573.05021.

1985.27 Problems and results on consecutive integers and prime factors of binomial coefficients, Number theory (Winnipeg, Man., 1983), *Rocky Mountain J. Math.* **15** (1985) no. 2, 353–363; **MR** 87g:11006; **Zbl.** 581.10019.

1985.28 Quantitative forms of a theorem of Hilbert, *J. Combin. Theory Ser. A* **38** (1985) no. 2, 210–216 (T. C. Brown; F. R. K. Chung; R. L. Graham); **MR** 86f:05013; **Zbl.** 577.05007.

1985.29 Ramsey minimal graphs for star forests, *Creation Math.* **18** (1985), 13–14 (S. A. Burr; R. J. Faudree; C. C. Rousseau; R. H. Schelp); **Zbl.** 574.05037.

1985.30 Remarks on stars and independent sets, *Aspects of topology, London Math. Soc. Lecture Note Ser., 93*, pp. 307–313, Cambridge Univ. Press, Cambridge-New York, 1985 (J. Pach); **MR** 86k:05061; **Zbl.** 587.05032.

1985.31 Some applications of probability methods to number theory, *Mathematical statistics and applications, Vol. B (Bad Tatzmannsdorf,*

1983), pp. 1–18, Reidel, Dordrecht-Boston, Mass.-London, 1985; **MR** 88b:11052; **Zbl.** 593.10039.

1985.32 Some of my old and new problems in elementary number theory and geometry, Proceedings of the Sundance conference on combinatorics and related topics (Sundance, Utah, 1985), *Congr. Numer.* **50** (1985), 97–106; **CMP** 833 541; **Zbl.** 601.10001.

1985.33 Some problems and results in number theory, *Number theory and combinatorics, Japan 1984 (Tokyo, Okayama and Kyoto, 1984)*, pp. 65–87, World Sci. Publishing, Singapore, 1985; **MR** 87g:11003; **Zbl.** 603.10001.

1985.34 Some solved and unsolved problems of mine in number theory, *Topics in analytic number theory (Austin, Tex., 1982)*, pp. 59–75, Univ. Texas Press, Austin, TX, 1985; **CMP** 804 242; **Zbl.** 596.10001.

1985.35 The closed linear span of $\{x^k - c_k\}_1^\infty$, *J. Approx. Theory* **43** (1985) no. 1, 75–80 (J. M. Anderson; A. Pinkus; O. Shisha); **MR** 86m:41005; **Zbl.** 576.41022.

1985.36 The difference between the clique numbers of a graph, *Ars Combin.* **19** (1985) A, 97–106 (L. Caccetta; E. T. Ordman; N. J. Pullman); **MR** 86g:05048; **Zbl.** 573.05034.

1985.37 The Ramsey number for the pair complete bipartite graph–graph of limited degree, *Graph theory with applications to algorithms and computer science (Kalamazoo, Mich., 1984)*, pp. 163–174, Wiley-Intersci. Publ., Wiley, New York, 1985 (S. A. Burr; R. J. Faudree; C. C. Rousseau; R. H. Schelp); **MR** 87f:05117; **Zbl.** 571.05035.

1985.38 Ulam, the man and the mathematician, *J. Graph Theory* **9** (1985) no. 4, 445–449; **MR** 88e:01053; **Zbl.** 668.01033. Also appears in *Creation Math.* **19** (1986), 13–16; **Zbl.** 599.01015.

1985.39 Welcoming address, *Annals of Discrete Math.* **28** (1985), *Random Graphs '83 (Poznań, 1983), North-Holland Math. Stud., 118*, pp. 1–5, North-Holland, Amsterdam-New York, 1985; **MR** 87i:05174; **Zbl.** 592.05001.

1986.01 An extremum problem concerning algebraic polynomials, *Acta Math. Hungar.* **47** (1986) no. 1–2, 137–143 (A. K. Varma); **MR** 87f:26017; **Zbl.** 624.26011.

1986.02 Clique numbers of graphs, *Discrete Math.* **59** (1986) no. 3, 235–242 (M. Erné); **MR** 87i:05183; **Zbl.** 586.05024.

1986.03 Coloring graphs with locally few colors, *Discrete Math.* **59** (1986) no. 1–2, 21–34 (Z. Füredi; A. Hajnal; P. Komjáth; V. Rödl; Á. Seress); **MR** 87f:05069; **Zbl.** 591.05030.

1986.04 Congruent subsets of finite sets of natural numbers, *J. Reine Angew. Math.* **367** (1986), 207–214 (E. Harzheim); **MR** 87h:11012; **Zbl.** 575.10041.

1986.05 Erratum: "Colouring the real line" [*J. Combin. Theory Ser. B* **39** (1985) no. 1, 86–100], *J. Combin Theory Ser. B* **41** (1986) no. 1, 139 (R. B. Eggleton; D. K. Skilton); **MR** 87j:05077; **Zbl.** 591.05031.

1986.06 Extremal clique coverings of complementary graphs, *Combinatorica* **6** (1986) no. 4, 309–314 (D. de Caen; N. J. Pullman; N. C. Wormald); **MR** 88d:05090; **Zbl.** 616.05042.

1986.07 Independence of solution sets in additive number theory, *Probability, statistical mechanics, and number theory, Adv. Math. Suppl. Stud., 9*, pp. 97–105, Academic Press, Orlando, Fla., 1986 (M. B. Nathanson); **MR** 88e:11011; **Zbl.** 608.10050.

1986.08 Maximum induced trees in graphs, *J. Combin. Theory Ser. B* **41** (1986) no. 1, 61–79 (M. Saks; V. T. Sós); **MR** 87k:05062; **Zbl.** 603.05023.

1986.09 On some metric and combinatorial geometric problems, *Discrete Math.* **60** (1986), 147–153; **MR** 88f:52011; **Zbl.** 595.52013.

1986.10 On sums involving reciprocals of the largest prime factor of an integer, *Glas. Mat. Ser. III* **21(41)** (1986) no. 2, 283–300 (A. Ivić; C. Pomerance); **MR** 89a:11090; **Zbl.** 615.10055.

1986.11 On the number of false witnesses for a composite number, *Math. Comp.* **46** (1986) no. 173, 259–279 (C. Pomerance); **MR** 87i:11183; **Zbl.** 586.10003.

1986.12 Problems and results on additive properties of general sequences, II., *Acta Math. Hungar.* **48** (1986) no. 1–2, 201–211 (A. Sárközy); **MR** 88c:11016; **Zbl.** 621.10041.

1986.13 Problems and results on additive properties of general sequences, V., *Monatsh. Math.* **102** (1986) no. 3, 183–197 (A. Sárközy; V. T. Sós); **MR** 88i:11011b; **Zbl.** 597.10055.

1986.14 Problems and results on intersections of set systems of structural type, *Utilitas Math.* **29** (1986), 61–70 (V. T. Sós); **MR** 87f:05004; **Zbl.** 638.05030.

1986.15 Some problems on number theory, *Analytic and elementary number theory (Marseille, 1983), Publ. Math. Orsay, 86-1*, pp. 53–67, Univ. Paris XI, Orsay, 1986; **MR** 87i:11006; **Zbl.** 584.10002.

1986.16 Some problems on number theory, Proceedings of the seventeenth Southeastern international conference on combinatorics, graph theory, and computing (Boca Raton, Fla., 1986), *Congr. Numer.* **54** (1986), 225–244; **MR** 88g:11001; **Zbl.** 615.10001.

1986.17 The asymptotic number of graphs not containing a fixed subgraph and a problem for hypergraphs having no exponent, *Graphs Combin.* **2** (1986) no. 2, 113–121 (P. Frankl; V. Rödl); **MR** 89b:05102; **Zbl.** 593.05038.

1987.01 $a \pmod p \leq b \pmod p$ for all primes p implies $a = b$, *Amer. Math. Monthly* **94** (1987) no. 2, 169–170 (P. P. Pálfy; M. Szegedy); **MR** 87k:11003; **Zbl.** 616.10003.

1987.02 A Ramsey problem of Harary on graphs with prescribed size, *Discrete Math.* **67** (1987) no. 3, 227–233 (R. J. Faudree; C. C. Rousseau; R. H. Schelp); **MR** 88k:05139; **Zbl.** 624.05047.

1987.03 Bounds on threshold dimension and disjoint threshold coverings, *SIAM J. Algebraic Discrete Methods* **8** (1987) no. 2, 151–154 (E. T. Ordman; Y. Zalcstein); **MR** 88d:05092; **Zbl.** 626.05045.

1987.04 Extremal problems on permutations under cyclic equivalence, *Discrete Math.* **64** (1987) no. 1, 1–11 (N. Linial; S. Moran); **MR** 88e:05002; **Zbl.** 656.05002.

1987.05 Generation of alternating groups by pairs of conjugates, *Period. Math. Hungar.* **18** (1987) no. 4, 259–269 (L. B. Beasley; J. L. Brenner; M. Szalay; A. G. Williamson); **MR** 88j:20006; **Zbl.** 617.20045. Also appears, in two parts, in *Creation Math. Sci.* **21** (1988), 20–23, and **22** (1989), 16–18; **Zbl.** 651.20002 and 672.20003.

1987.06 Goodness of trees for generalized books, *Graphs Combin.* **3** (1987) no. 1, 1–6 (S. A. Burr; R. J. Faudree; C. C. Rousseau; R. H. Schelp; R. J. Gould; M. S. Jacobson); **MR** 89b:05122; **Zbl.** 612.05046.

1987.07 Highly irregular graphs, *J. Graph Theory* **11** (1987) no. 2, 235–249 (Y. Alavi; G. Chartrand; F. R. K. Chung; R. L. Graham; O. R. Oellermann); **MR** 88m:05062; **Zbl.** 665.05043.

1987.08 k-connectivity in random graphs, *European J. Combin.* **8** (1987) no. 3, 281–286 (J. W. Kennedy); **MR** 88m:05071; **Zbl.** 674.05056.

1987.09 Many heads in a short block, *Mathematical statistics and probability theory, Vol. A (Bad Tatzmannsdorf, 1986)*, pp. 53–67, Reidel, Dordrecht-Boston, Mass.-London, 1987 (P. Deheuvels; K. Grill; P. Révész); **MR** 89b:60073; **Zbl.** 632.60027.

1987.10 Multiplicative functions and small divisors, *Analytic number theory and Diophantine problems (Stillwater, OK, 1984), Progr. Math., 70*, pp. 1–13, Birkhäuser Boston, Boston, MA, 1987 (K. Alladi; J. D. Vaaler); **MR** 90h:11089; **Zbl.** 626.10004.

1987.11 Multiplikative Funktionen auf kurzen Intervallen (Multiplicative functions in short intervals, in German), *J. Reine Angew. Math.* **381** (1987), 148–160 (K.-H. Indlekofer); **MR** 89a:11091; **Zbl.** 618.10041.

1987.12 My joint work with Richard Rado, *Surveys in combinatorics 1987 (New Cross, 1987), London Math. Soc. Lecture Note Ser., 123*, pp. 53–80, Cambridge Univ. Press, Cambridge-New York, 1987; **MR** 88k:01032; **Zbl.** 623.01010.

1987.13 On divisibility properties of integers of the form $a + a'$, *Acta Math. Hungar.* **50** (1987) no. 1–2, 117–122 (A. Sárközy); **MR** 88i:11065; **Zbl.** 625.10038.

1987.14 On locally repeated values of certain arithmetic functions, II., *Acta Math. Hungar.* **49** (1987) no. 1–2, 251–259 (C. Pomerance; A. Sárközy); **MR** 88c:11008; **Zbl.** 609.10034.

1987.15 On locally repeated values of certain arithmetic functions, III., *Proc. Amer. Math. Soc.* **101** (1987) no. 1, 1–7 (C. Pomerance; A. Sárközy); **MR** 88k:11006; **Zbl.** 631.10029.

1987.16 On the distribution of the number of prime factors of sums $a + b$, *Trans. Amer. Math. Soc.* **302** (1987) no. 1, 269–280, (H. Maier; A. Sárközy); **MR** 88d:11090; **Zbl.** 617.10038.

1987.17 On the enumeration of finite groups, *J. Number Theory* **25** (1987) no. 3, 360–378 (M. Ram Murty; V. Kumar Murty); **MR** 88d:11087; **Zbl.** 612.10038.

1987.18 On the equality of the Grundy and ochromatic numbers of a graph, *J. Graph Theory* **11** (1987) no. 2, 157–159 (W. R. Hare; S. T. Hedetniemi; R. Laskar); **MR** 88g:05057; **Zbl.** 708.05021.

1987.19 On the number of false witnesses for a composite number, *Number theory (New York, 1984–1985), Lecture Notes in Math., 1240*, pp. 97–100, Springer, Berlin-New York, 1987 (C. Pomerance); **MR** 89a:11010.

1987.20 On the order of directly indecomposable groups (in Hungarian, English and Russian summaries), *Mat. Lapok* **33** (1982/86) no. 4, 289–298, 1987 (P. P. Pálfy); **MR** 89j:11093; **Zbl.** 649.20024.

1987.21 On the representing number of intersecting families, *Arch. Math. (Basel)* **49** (1987) no. 2, 114–118 (M. Aigner; D. Grieser); **MR** 88g:05004; **Zbl.** 629.05006.

1987.22 On the residues of products of prime numbers, *Period. Math. Hungar.* **18** (1987) no. 3, 229–239 (A. M. Odlyzko; A. Sárközy); **MR** 88j:11057; **Zbl.** 625.10035.

1987.23 On unavoidable hypergraphs, *J. Graph Theory* **11** (1987) no. 2, 251–263 (F. R. K. Chung); **MR** 88c:05084; **Zbl.** 725.05062.

1987.24 Problems and results on additive properties of general sequences, III., *Studia Sci. Math. Hungar.* **22** (1987) no. 1–4, 53–63 (A. Sárközy; V. T. Sós); **MR** 89b:11015; **Zbl.** 669.10078.

1987.25 Problems and results on minimal bases in additive number theory, *Number theory (New York, 1984–1985), Lecture Notes in Math., 1240*, pp. 87–96, Springer, Berlin-New York, 1987 (M. B. Nathanson); **MR** 88j:11006; **Zbl.** 622.10041.

1987.26 Problems and results on random walks, *Mathematical statistics and probability theory, Vol. B (Bad Tatzmannsdorf, 1986)*, pp. 59–65, Reidel, Dordrecht-Boston, MA-London, 1987 (P. Révész); **MR** 89h:60112; **Zbl.** 629.60081.

1987.27 Some combinatorial and metric problems in geometry, *Intuitive geometry (Siófok, 1985), Colloq. Math. Soc. János Bolyai, 48*, pp. 167–177, North-Holland, Amsterdam-New York, 1987; **MR** 89i:52012; **Zbl.** 625.52008.

1987.28 Some problems on finite and infinite graphs, *Logic and combinatorics (Arcata, Calif., 1985), Contemp. Math. 65*, pp. 223–228, Amer. Math. Soc., Providence, R. I., 1987; **MR** 88g:04003; **Zbl.** 638.04005.

1987.29 Some remarks on infinite series, *Studia Sci. Math. Hungar.* **22** (1987) no. 1–4, 395–400 (I. Joó; L. A. Székely); **MR** 89g:40001; **Zbl.** 648.40001.

1987.30 Sur le nombre d'invariants fondamentaux des formes binaires (On the number of fundamental invariants of binary forms, in French, English summary) *C. R. Acad. Sci. Paris Sér. I Math.* **305** (1987) no. 8, 319–322 (J. Dixmier; J.-L. Nicolas); **MR** 89a:11040; **Zbl.** 642.10021.

1987.31 The ascending subgraph decomposition problem, Eighteenth Southeastern International Conference on Combinatorics, Graph Theory, and Computing (Boca Raton, Fla., 1987), *Congr. Numer.* **58** (1987), 7–14 (Y. Alavi; A. J. Boals; G. Chartrand; O. R. Oellermann); **MR** 89d:05136; **Zbl.** 641.05046.

1987.32 The asymptotic behavior of a family of sequences, *Pacific J. Math.* **126** (1987) no. 2, 227–241 (A. Hildebrand; A. M. Odlyzko; P. Pudaite; B. Reznick); **MR** 89c:11024; **Zbl.** 558.10010.

1987.33 The vertex independence sequence of a graph is not constrained, Eighteenth Southeastern International Conference on Combinatorics, Graph Theory, and Computing (Boca Raton, Fla., 1987), *Congr. Numer.* **58** (1987), 15–23 (Y. Alavi; P. J. Malde; A. J. Schwenk); **MR** 89e:05181; **Zbl.** 679.05061.

1988.01 A lower bound for the counting function of Lucas pseudoprimes, *Math. Comp.* **51** (1988) no. 183, 315–323 (P. Kiss; A. Sárközy); **MR** 89e:11011; **Zbl.** 658.10003.

1988.02 A tribute to Torrence Parsons, *J. Graph Theory* **12** (1988) no. 2, v–vi; **MR** 89i:01073; **Zbl.** 639.01021.

1988.03 An extremal problem for complete bipartite graphs, *Studia Sci. Math. Hungar.* **23** (1988) no. 3–4, 319–326 (R. J. Faudree; C. C. Rousseau; R. H. Schelp); **MR** 90f:05076; **Zbl.** 674.05038.

1988.04 Clique partitions and clique coverings, Proceedings of the First Japan Conference on Graph Theory and Applications (Hakone, 1986), *Discrete Math.* **72** (1988) no. 1–3, 93–101 (R. J. Faudree; E. T. Ordman); **MR** 89m:05090; **Zbl.** 679.05040.

1988.05 Cutting a graph into two dissimilar halves, *J. Graph Theory* **12** (1988) no. 1, 121–131 (M. K. Goldberg; J. Pach; J. H. Spencer); **MR** 89j:05043; **Zbl.** 655.05059.

1988.06 Cycles in graphs without proper subgraphs of minimum degree 3, Eleventh British Combinatorial Conference (London, 1987), *Ars Combin.* **25** (1988) B, 195–201 (R. J. Faudree; A. Gyárfás; R. H. Schelp); **MR** 89e:05126; **Zbl.** 657.05048.

1988.07 Extremal problems for degree sequences, *Combinatorics (Eger, 1987), Colloq. Math. Soc. János Bolyai, 52*, pp. 183–193, North-Holland, Amsterdam, 1988 (S. A. Burr; R. J. Faudree; A. Gyárfás; R. H. Schelp); **MR** 94h:05042; **Zbl.** 685.05025.

1988.08 Extremal theory and bipartite graph–tree Ramsey numbers, Proceedings of the First Japan Conference on Graph Theory and Applications (Hakone, 1986), *Discrete Math.* **72** (1988) no. 1–3, 103–112

(R. J. Faudree; C. C. Rousseau; R. H. Schelp); **MR** 90e:05043; **Zbl.** 659.05066.

1988.09 Graphs with unavoidable subgraphs with large degrees, *J. Graph Theory* **12** (1988) no. 1, 17–27 (L. Caccetta; K. Vijayan); **MR** 89d:05105; **Zbl.** 659.05075.

1988.10 Has every Latin square of order n a partial Latin transversal of size $n-1$?, *Amer. Math. Monthly* **95** (1988) no. 5, 428–430 (D. Hickerson; D. A. Norton; S. Stein); **Zbl.** 655.05018.

1988.11 How to define an irregular graph, *College Math. J.* **19** (1988) no. 1, 36–42 (G. Chartrand; O. R. Oellermann); **CMP** 931 654.

1988.12 How to make a graph bipartite, *J. Combin. Theory Ser. B* **45** (1988) no. 1, 86–98 (R. J. Faudree; J. Pach; J. H. Spencer); **MR** 89f:05134; **Zbl.** 729.05025.

1988.13 Intersection graphs for families of balls in $\mathbf{R}^n$, *European J. Combin.* **9** (1988) no. 5, 501–505 (C. D. Godsil; S. G. Krantz; T. Parsons); **MR** 89i:05225; **Zbl.** 659.05079.

1988.14 Isomorphic subgraphs in a graph, *Combinatorics (Eger, 1987), Colloq. Math. Soc. János Bolyai, 52*, pp. 553–556, North-Holland, Amsterdam, 1988 (J. Pach; L. Pyber); **CMP** 1 221 596; **Zbl.** 703.05044.

1988.15 k-path irregular graphs, Nineteenth Southeastern Conference on Combinatorics, Graph Theory, and Computing (Baton Rouge, LA, 1988), *Congr. Numer.* **65** (1988), 201–210 (Y. Alavi; A. J. Boals; G. Chartrand; O. R. Oellermann); **MR** 90c:05120; **Zbl.** 669.05046.

1988.16 Minimal asymptotic bases with prescribed densities, *Illinois J. Math.* **32** (1988) no. 3, 562–574 (M. B. Nathanson); **MR** 89j:11010; **Zbl.** 651.10035.

1988.17 Minimum-diameter cyclic arrangements in mapping data-flow graphs onto VLSI arrays, *Math. Systems Theory* **21** (1988) no. 2, 85–98 (I. Koren; S. Moran; G. M. Silberman; S. Zaks); **MR** 89k:68077; **Zbl.** 659.68078.

1988.18 Nearly disjoint covering systems, Eleventh British Combinatorial Conference (London, 1987), *Ars Combin.* **25** (1988) B, 231–246 (M. A. Berger; A. Felzenbaum; A. S. Fraenkel); **MR** 89i:11004; **Zbl.** 675.10003.

1988.19 On admissible constellations of consecutive primes, *BIT* **28** (1988) no. 3, 391–396 (H. Riesel); **MR** 90b:11145; **Zbl.** 655.10004.

1988.20 On nilpotent but not abelian groups and abelian but not cyclic groups, *J. Number Theory* **28** (1988) no. 3, 363–368 (M. E. Mays); **MR** 89d:11080; **Zbl.** 635.10040.

1988.21 On the area of the circles covered by a random walk, *J. Multivariate Anal.* **27** (1988) no. 1, 169–180 (P. Révész); **MR** 89m:60158; **Zbl.** 655.60055. Reprinted in *Multivariate statistics and probability, Essays in memory of Paruchuri R. Krishnaiah*, pp. 169–180, Academic Press, Inc., Boston, MA, 1989; **MR** 90m:62004 (for entire book); **Zbl.** 704.60072.

1988.22 On the irrationality of certain series: problems and results, *New advances in transcendence theory (Durham, 1986)*, pp. 102–109, Cambridge Univ. Press, Cambridge-New York, 1988; **MR** 89k:11057; **Zbl.** 656.10026.

1988.23 On the mean distance between points of a graph, 250th Anniversary Conference on Graph Theory (Fort Wayne, IN 1986), *Congr. Numer.* **64** (1988), 121–124 (J. Pach; J. H. Spencer); **MR** 90i:05087; **Zbl.** 677.05053.

1988.24 Optima of dual integer linear programs, *Combinatorica* **8** (1988) no. 1, 13–20 (R. Aharoni; N. Linial); **MR** 89f:90093; **Zbl.** 648.90054.

1988.25 Partitions of bases into disjoint unions of bases, *J. Number Theory* **29** (1988) no. 1, 1–9 (M. B. Nathanson); **MR** 89f:11025; **Zbl.** 645.10045.

1988.26 Prime factors of binomial coefficients and related problems, *Acta Arith.* **49** (1988) no. 5, 507–523 (C. B. Lacampagne; J. L. Selfridge); **MR** 90f:11009; **Zbl.** 669.10011.

1988.27 Problems and results in combinatorial analysis and graph theory, Proceedings of the First Japan Conference on Graph Theory and Applications (Hakone, 1986), *Discrete Math.* **72** (1988) no. 1–3, 81–92; **MR** 89k:05048; **Zbl.** 661.05037.

1988.28 Random walks on $\mathbf{Z}_2^n$, *J. Multivariate Anal.* **25** (1988) no. 1, 111–118 (R. W. Chen); **MR** 89d:60124; **Zbl.** 651.60072.

1988.29 Recollections on Kurt Gödel, *Jahrb. Kurt-Gödel-Ges.* **1988**, 94–95; **MR** 90j:01060; **Zbl.** 673.01014.

1988.30 Repeated distances in space, *Graphs Combin.* **4** (1988) no. 3, 207–217 (D. Avis; J. Pach); **MR** 90b:05068; **Zbl.** 656.05039.

1988.31 Some Diophantine equations with many solutions, *Compositio Math.* **66** (1988) no. 1, 37–56 (C. L. Stewart; R. Tijdeman); **MR** 89j:11027; **Zbl.** 639.10014.

1988.32 Some old and new problems in combinatorial geometry, *Applications of discrete mathematics (Clemson, SC, 1986)*, pp. 32–37, SIAM, Philadelphia, PA, 1988; **MR** 90a:52017; **Zbl.** 663.05013.

1988.33 Sumsets containing infinite arithmetic progressions, *J. Number Theory* **28** (1988) no. 2, 159–166 (M. B. Nathanson; A. Sárközy); **MR** 89e:11006; **Zbl.** 633.10047.

1988.34 The book–tree Ramsey numbers, *Scientia, A: Mathematics* **1** (1988), 111–117 (R. J. Faudree; C. C. Rousseau; R. H. Schelp); **Zbl.** 695.05048.

1988.35 The chromatic number of the graph of large distances, *Combinatorics (Eger, 1987), Colloq. Math. Soc. János Bolyai, 52*, pp. 547–551, North-Holland, Amsterdam, 1988 (L. Lovász; K. Vesztergombi); **CMP** 1 221 595; **Zbl.** 683.05020.

1988.36 The solution to a problem of Grünbaum, *Canad. Math. Bull.* **31** (1988) no. 2, 129–138 (P. Salamon); **MR** 89f:52022; **Zbl.** 606.05005.

1989.01 A note on the distribution function of additive arithmetical functions in short intervals, *Canad. Math. Bull.* **32** (1989) no. 4, 441–445 (G. J. Babu); **MR** 90i:11106; **Zbl.** 681.10034 and 638.10049.

1989.02 A problem of Leo Moser about repeated distances on the sphere, *Amer. Math. Monthly* **96** (1989) no. 7, 569–575 (D. Hickerson; J. Pach); **MR** 90h:52008; **Zbl.** 737.05006.

1989.03 Additive bases with many representations, *Acta Arith.* **52** (1989) no. 4, 399–406 (M. B. Nathanson); **MR** 91e:11015; **Zbl.** 692.10045.

1989.04 An extremal result for paths, *Graph theory and its applications: East and West (Jinan, 1986), Ann. New York Acad. Sci., 576*, pp. 155–162, New York Acad. Sci., New York, 1989 (R. J. Faudree; R. H. Schelp; M. Simonovits); **MR** 92k:05068; **Zbl.** 792.05076.

1989.05 Bandwidth versus bandsize, *Graph theory in memory of G. A. Dirac (Sandbjerg, 1985), Ann. Discrete Math., 41*, pp. 117–129, North-Holland, Amsterdam-New York, 1989 (P. Hell; P. M. Winkler); **MR** 90k:05084; **Zbl.** 684.05043.

1989.06 Disjoint edges in geometric graphs, *Discrete Comput. Geom.* **4** (1989) no. 4, 287–290 (N. Alon); **MR** 90c:05112; **Zbl.** 692.05037.

1989.07 Domination in colored complete graphs, *J. Graph Theory* **13** (1989) no. 6, 713–718 (R. J. Faudree; A. Gyárfás; R. H. Schelp); **MR** 90i:05049; **Zbl.** 708.05057.

1989.08 Grandes valeurs de fonctions liées aux divisieurs premiers consécutifs d'un entier (Large values of functions associated with the consecutive prime divisors of an integer, in French, English summary), *Théorie des nombres (Quebec, PQ, 1987)*, pp. 169–200, de Gruyter, Berlin-New York, 1989 (J.-L. Nicolas); **MR** 90i:11098; **Zbl.** 683.10035.

1989.09 *Lattice points, Pitman Monographs and Surveys in Pure and Applied Mathematics* **39**, Longman Scientific & Technical, Harlow; copublished in the United States with John Wiley & Sons, Inc., New York, 1989, viii+184 pp., ISBN 0-582-01478-6 and 0-470-21154-7 (P. M. Gruber; J. Hammer); **MR** 90g:11081; **Zbl.** 683.10025.

1989.10 Maximal anti-Ramsey graphs and the strong chromatic number, *J. Graph Theory* **13** (1989) no. 3, 263–282 (S. A. Burr; R. L. Graham; V. T. Sós); **MR** 90c:05146; **Zbl.** 682.05046.

1989.11 Monochromatic sumsets, *J. Combin. Theory Ser. A* **50** (1989) no. 1, 162–163 (J. H. Spencer); **MR** 89j:05008; **Zbl.** 666.10036.

1989.12 Multipartite graph–tree Ramsey numbers, *Graph theory and its applications: East and West (Jinan, 1986), Ann. New York Acad. Sci., 576*, pp. 146–154, New York Acad. Sci., New York, 1989 (R. J. Faudree; C. C. Rousseau; R. H. Schelp); **MR** 92d:05107; **Zbl.** 792.05103.

1989.13 Multiplicative functions and small divisors, II., *J. Number Theory* **31** (1989) no. 2, 183–190 (K. Alladi; J. D. Vaaler); **MR** 90h:11090; **Zbl.** 664.10025.

1989.14 On a conjecture of Roth and some related problems, I., *Irregularities of partitions (Fertőd, 1986), Algorithms Combin.: Study Res. Texts, 8*, pp. 47–59, Springer, Berlin-New York, 1989 (A. Sárközy; V. T. Sós); **MR** 90d:11017; **Zbl.** 689.10061.

1989.15 On certain saturation problems, *Acta Math. Hungar.* **53** (1989) no. 1–2, 197–203 (P. Vértesi); **MR** 90c:41049; **Zbl.** 682.41031.

1989.16 On convergent interpolatory polynomials, *J. Approx. Theory* **58** (1989) no. 2, 232–241 (A. Kroó; J. Szabados); **MR** 90k:41003; **Zbl.** 692.41004.

1989.17 On functions connected with prime divisors of an integer, *Number theory and applications (Banff, AB, 1988), NATO Adv. Sci. Inst. Ser. C: Math. Phys. Sci., 265*, pp. 381–391, Kluwer Acad. Publ., Dordrecht, 1989 (J.-L. Nicolas); **MR** 92i:11096; **Zbl.** 687.10031.

1989.18 On graphs with adjacent vertices of large degree, *J. Combin. Math. Combin. Comput.* **5** (1989), 217–222 (L. Caccetta; K. Vijayan); **MR** 90f:05075; **Zbl.** 674.05040.

1989.19 On some aspects of my work with Gabriel Dirac, *Graph theory in memory of G. A. Dirac (Sandbjerg, 1985), Ann. Discrete Math., 41*, pp. 111–116, North-Holland, Amsterdam-New York, 1989; **MR** 90b:01072; **Zbl.** 664.01008.

1989.20 On some problems and results in elementary number theory (Chinese summary), *Sichuan Daxue Xuebao* **26** (1989), Special Issue, 1–6; **MR** 91f:11001; **Zbl.** 709.11001.

1989.21 On the difference between consecutive Ramsey numbers, *Utilitas Math.* **35** (1989), 115–118 (S. A. Burr; R. J. Faudree; R. H. Schelp); **MR** 90c:05147; **Zbl.** 678.05039.

1989.22 On the graph of large distances, *Discrete Comput. Geom.* **4** (1989) no. 6, 541–549 (L. Lovász; K. Vesztergombi); **MR** 90i:05048; **Zbl.** 694.05031.

1989.23 On the iterates of the enumerating function of finite abelian groups, *Bull. Acad. Serbe Sci. Arts Cl. Sci. Math. Natur.* No. 17 (1989), 13–22 (A. Ivić); **MR** 92e:11098; **Zbl.** 695.10040.

1989.24 On the number of distinct induced subgraphs of a graph, Graph theory and combinatorics (Cambridge, 1988), *Discrete Math.* **75** (1989) no. 1–3, 145–154 (A. Hajnal); **MR** 90g:05099; **Zbl.** 668.-05037.

1989.25 On the number of partitions of n without a given subsum, I., Graph theory and combinatorics (Cambridge, 1988), *Discrete Math.* **75** (1989) no. 1–3, 155–166 (J.-L. Nicolas; A. Sárközy); **MR** 90e:11151; **Zbl.** 673.05007.

1989.26 Partitions into parts which are unequal and large, *Number theory (Ulm, 1987), Lecture Notes in Math., 1380*, pp. 19–30, Springer, New York-Berlin, 1989 (J.-L. Nicolas; M. Szalay); **MR** 90g:11140; **Zbl.** 679.10013.

1989.27 Problems and results on extremal problems in number theory, geometry, and combinatorics, Proceedings of the 7th Fischland Colloquium, I (Wustrow, 1988), *Rostock. Math. Kolloq.*, No. 38 (1989), 6–14; **MR** 91d:05088; **Zbl.** 718.11001.

1989.28 Radius, diameter, and minimum degree, *J. Combin. Theory Ser. B* **47** (1989) no. 1, 73–79 (J. Pach; R. Pollack; Zs. Tuza); **MR** 90f:05077; **Zbl.** 686.05029.

1989.29 Ramanujan and I, *Number theory, Madras 1987, Lecture Notes in Math., 1395*, pp. 1–20, Springer, Berlin-New York, 1989; **MR** 91a:11003; **Zbl.** 685.10002.

1989.30 Ramsey-type theorems, Combinatorics and complexity (Chicago, IL, 1987), *Discrete Appl. Math.* **25** (1989) no. 1–2, 37–52 (A. Hajnal); **MR** 90m:05091; **Zbl.** 715.05052.

1989.31 Representations of graphs and orthogonal Latin square graphs, *J. Graph Theory* **13** (1989) no. 5, 593–595 (A. B. Evans); **MR** 90g:05049; **Zbl.** 691.05053.

1989.32 Some complete bipartite graph–tree Ramsey numbers, *Graph theory in memory of G. A. Dirac (Sandbjerg, 1985), Ann. Discrete Math., 41*, pp. 79–89, North-Holland, Amsterdam-New York, 1989 (S. A. Burr; R. J. Faudree; C. C. Rousseau; R. H. Schelp); **MR** 90d:05166; **Zbl.** 672.05063.

1989.33 Some old and new problems on additive and combinatorial number theory, *Combinatorial Mathematics: Proceedings of the Third International Conference (New York, 1985), Ann. New York Acad. Sci., 555*, pp. 181–186, New York Acad. Sci., New York, 1989; **MR** 90i:11016; **Zbl.** 709.11002.

1989.34 Some personal and mathematical reminiscences of Kurt Mahler, *Austral. Math. Soc. Gaz.* **16** (1989) no. 1, 1–2; **MR** 90c:01068; **Zbl.** 673.01015.

1989.35 Some problems and results on combinatorial number theory, *Graph theory and its applications: East and West (Jinan, 1986), Ann. New York Acad. Sci., 576*, pp. 132–145, New York Acad. Sci., New York, 1989; **MR** 92g:11011; **Zbl.** 790.11015.

1989.36 Sur les densités de certaines suites d'entiers (On the densities of certain sequences of integers, in French), *Proc. London Math. Soc. (3)* **59** (1989) no. 3, 417–438 (G. Tenenbaum); **MR** 90h:11087; **Zbl.** 694.10040.

1989.37 Sur les fonctions arithmétiques liées aux diviseurs consécutifs (On arithmetical functions associated with consecutive divisors, in French), *J. Number Theory* **31** (1989) no. 3, 285–311 (G. Tenenbaum); **MR** 90i:11103; **Zbl.** 676.10030.

1989.38 The size of chordal, interval and threshold subgraphs, *Combinatorica* **9** (1989) no. 3, 245–253 (A. Gyárfás; E. T. Ordman; Y. Zalcstein); **MR** 90i:05047; **Zbl.** 738.05051.

1989.39 Tight bounds on the chromatic sum of a connected graph, *J. Graph Theory* **13** (1989) no. 3, 353–357 (C. Thomassen; Y. Alavi; P. J. Malde; A. J. Schwenk); **MR** 90h:05054; **Zbl.** 677.05028.

1990.01 A new law of iterated logarithm, *Acta Math. Hungar.* **55** (1990) no. 1–2, 125–131 (P. Révész); **MR** 91m:60052; **Zbl.** 711.60025.

1990.02 Bounds on the number of pairs of unjoined points in a partial plane, *Coding theory and design theory, Part I, IMA Vol. Math. Appl., 20*, pp. 102–112, Springer, New York-Berlin, 1990 (D. A. Drake); **MR** 91b:51010; **Zbl.** 732.51007.

1990.03 Characterization of the unique expansions $1 = \sum_{i=1}^{\infty} q^{-n_i}$ and related problems (French summary), *Bull. Soc. Math. France* **118** (1990) no. 3, 377–390 (I. Joó; V. Komornik); **MR** 91j:11006; **Zbl.** 721.11005.

1990.04 Chromatic number versus cochromatic number in graphs with bounded clique number, *European J. Combin.* **11** (1990) no. 3, 235–240 (J. G. Gimbel; H. J. Straight); **MR** 91k:05042; **Zbl.** 721.05020.

1990.05 *Collected papers of Paul Turán* (editor and contributor of personal reminiscences), Akademiai Kiado (Publishing House of the Hungarian Academy of Sciences), Budapest, 1990, xxxviii+2665 pp. (3 volumes), ISBN: 963-05-4298-6; **MR** 91i:01145; **Zbl.** 703.01019.

1990.06 Colouring prime distance graphs, *Graphs Combin.* **6** (1990) no. 1, 17–32 (R. B. Eggleton; D. K. Skilton); **MR** 91f:05052; **Zbl.** 698.05033.

1990.07 Countable decompositions of $\mathbf{R}^2$ and $\mathbf{R}^3$, *Discrete Comput. Geom.* **5** (1990) no. 4, 325–331 (P. Komjáth); **MR** 91b:04002; **Zbl.** 723.52005.

1990.08 Graphs that require many colors to achieve their chromatic sum, Proceedings of the Twentieth Southeastern Conference on Combinatorics, Graph Theory, and Computing (Boca Raton, FL, 1989), *Congr. Numer.* **71** (1990), 17–28 (E. Kubicka; A. J. Schwenk); **MR** 91b:05098; **Zbl.** 704.05020.

1990.09 Introduction to *How does one cut a triangle?* by Alexander Soifer, Center for Excellence in Mathematical Education, Colorado Springs, CO, 1990; **MR** 91f:51002 (for entire book); **Zbl.** 691.52001 (for entire book).

1990.10 Monochromatic coverings in colored complete graphs, Proceedings of the Twentieth Southeastern Conference on Combinatorics, Graph Theory, and Computing (Boca Raton, FL, 1989), *Congr. Numer.* **71** (1990), 29–38 (R. J. Faudree; R. J. Gould; A. Gyárfás; C. C. Rousseau; R. H. Schelp); **MR** 91a:05043; **Zbl.** 704.05039.

1990.11 On a conjecture of Roth and some related problems, II., *Number theory (Banff, AB, 1988)*, pp. 125–138, de Gruyter, Berlin, 1990 (A. Sárközy); **MR** 93e:11017; **Zbl.** 699.10068.

1990.12 On a problem of Straus, *Disorder in physical systems, Oxford Sci. Publ.*, pp. 55–66, Oxford Univ. Press, New York, 1990 (A. Sárközy); **MR** 91i:11012; **Zbl.** 725.11012.

1990.13 On a theorem of Besicovitch: values of arithmetic functions that divide their arguments, *Indian J. Math.* **32** (1990) no. 3, 279–287 (C. Pomerance); **MR** 92a:11091; **Zbl.** 723.11046.

1990.14 On arithmetic functions involving consecutive divisors, *Analytic number theory (Allerton Park, IL, 1989), Progr. Math., 85*, pp. 77–90, Birkhäuser Boston, Boston, MA, 1990 (A. Balog; G. Tenenbaum); **MR** 92b:11069; **Zbl.** 718.11041.

1990.15 On Pisier type problems and results (combinatorial applications to number theory), *Mathematics of Ramsey theory, Algorithms Combin., 5*, pp. 214–231, Springer, Berlin, 1990 (J. Nešetřil; V. Rödl); **CMP** 1 083 603; **Zbl.** 727.11009.

1990.16 On some problems of the statistical theory of partitions, *Number theory, Vol. I (Budapest, 1987), Colloq. Math. Soc. János Bolyai, 51*, pp. 93–110, North-Holland, Amsterdam, 1990 (M. Szalay); **MR** 91g:11113; **Zbl.** 707.11071.

1990.17 On the greatest prime factor of $\prod_{k=1}^{x} f(k)$, *Acta Arith.* **55** (1990) no. 2, 191–200 (A. Schinzel); **MR** 91h:11100; **Zbl.** 715.11050.

1990.18 On the normal behavior of the iterates of some arithmetic functions, *Analytic number theory (Allerton Park, IL, 1989), Progr. Math., 85*, pp. 165–204, Birkhäuser Boston, Boston, MA, 1990 (A. Granville; C. Pomerance; C. A. Spiro-Silverman); **MR** 92a:11113; **Zbl.** 721.11034.

1990.19 On the number of partitions of n without a given subsum, II., *Analytic number theory (Allerton Park, IL, 1989), Progr. Math., 85*, pp. 205–234, Birkhäuser Boston, Boston, MA, 1990 (J.-L. Nicolas; A. Sárközy); **MR** 92c:11108; **Zbl.** 727.11038.

1990.20 On the number of sets of integers with various properties, *Number theory (Banff, AB, 1988)*, pp. 61–79, de Gruyter, Berlin, 1990 (P. J. Cameron); **MR** 92g:11010; **Zbl.** 695.10048.

1990.21 On two additive problems, *J. Number Theory* **34** (1990) no. 1, 1–12 (G. Freiman); **MR** 91c:11011; **Zbl.** 697.10047.

1990.22 Primzahlpotenzen in rekurrenten Folgen (Prime powers in recurrent sequences, in German, English summary), *Analysis* **10** (1990) no. 1, 71–83 (T. Maxsein; P. R. Smith); **MR** 91i:11015; **Zbl.** 709.11015.

1990.23 Problems and results on graphs and hypergraphs: similarities and differences, *Mathematics of Ramsey theory, Algorithms Combin., 5*, pp. 12–28, Springer, Berlin, 1990; **CMP** 1 083 590; **Zbl.** 725.05051.

1990.24 Quasi-progressions and descending waves, *J. Combin. Theory Ser. A.* **53** (1990) no. 1, 81–95 (T. C. Brown; A. R. Freedman); **MR** 91a:11016; **Zbl.** 699.10069.

1990.25 Rainbow Hamiltonian paths and canonically colored subgraphs in infinite complete graphs, *Math. Pannon.* **1** (1990) no. 1, 5–13 (Zs. Tuza); **MR** 93e:05062; **Zbl.** 724.05048.

1990.26 Reducible sums and splittable sets, *J. Number Theory* **36** (1990) no. 1, 89–94 (A. Zaks); **MR** 91k:11025; **Zbl.** 715.11014.

1990.27 Representations of integers as the sum of k terms, *Random Structures Algorithms* **1** (1990) no. 3, 245–261 (P. Tetali); **MR** 92c:11012; **Zbl.** 725.11007.

1990.28 Some applications of probability methods to number theory. Successes and limitations, *Sequences (Naples/Positano, 1988)*, pp. 182–194, Springer, New York, 1990; **MR** 91d:11084; **Zbl.** 697.10002.

1990.29 Some of my favourite unsolved problems, *A tribute to Paul Erdős*, pp. 467–478, Cambridge University Press, Cambridge, 1990; **MR** 92f:11003; **Zbl.** 709.11003.

1990.30 Some of my old and new combinatorial problems, *Paths, flows, and VLSI-layout (Bonn, 1988), Algorithms Combin., 9*, pp. 35–45, Springer, Berlin, 1990; **MR** 91j:05001; **Zbl.** 734.05002.

1990.31 Subgraphs of minimal degree k, *Discrete Math.* **85** (1990) no. 1, 53–58 (R. J. Faudree; C. C. Rousseau; R. H. Schelp); **MR** 91i:05065; **Zbl.** 714.05033.

1990.32 The distribution of values of a certain class of arithmetic functions at consecutive integers, *Number theory, Vol.I (Budapest, 1987), Colloq. Math. Soc. János Bolyai, 51*, pp. 45–91, North-Holland, Amsterdam, 1990 (A. Ivić); **MR** 91f:11068; **Zbl.** 704.11032.

1990.33 Uses of and limitations of computers in number theory, *Computers in mathematics (Stanford, CA, 1986), Lecture Notes in Pure and Appl. Math., 125*, pp. 241–260, Dekker, New York, 1990; **MR** 91k:11112; **Zbl.** 708.11002.

1990.34 Variations on the theme of repeated distances, *Combinatorica* **10** (1990) no. 3, 261–269 (J. Pach); **MR** 92b:52037; **Zbl.** 722.52009.

1991.01 A note on the largest H-free subgraph in a random graph, *Graph theory, combinatorics and applications, Vol. 1 (Kalamazoo, MI, 1988), Wiley-Intersci. Publ.*, pp. 435–437, Wiley, New York, 1991 (J. G. Gimbel); **MR** 93d:05137; **Zbl.** 787.05004 (for entire conference proceedings).

1991.02 Absorbing common subgraphs, *Graph theory, combinatorics, algorithms, and applications (San Francisco, CA, 1989)*, pp. 96–105, SIAM, Philadelphia, PA, 1991 (G. Chartrand; G. Kubicki); **MR** 92i:05169; **Zbl.** 751.05070.

1991.03 Carmichael's lambda function, *Acta Arith.* **58** (1991) no. 4, 363–385 (C. Pomerance; E. Schmutz); **MR** 92g:11093; **Zbl.** 734.11047.

1991.04 Degree sequences in triangle-free graphs, *Discrete Math.* **92** (1991) no. 1–3, 85–88 (S. Fajtlowicz; W. Staton); **MR** 92m:05080; **Zbl.** 752.05028.

1991.05 Distances in convex polygons, *Geombinatorics* **1** (1991) no. 3, 4; **CMP** 1 208 435.

1991.06 Distinct distances determined by subsets of a point set in space, *Comput. Geom.* **1** (1991) no. 1, 1–11 (D. Avis; J. Pach); **MR** 92m:52038; **Zbl.** 732.52004.

1991.07 Double vertex graphs, *J. Combin. Inform. System Sci.* **16** (1991) no. 1, 37–50 (Y. Alavi; M. Behzad; D. R. Lick); **MR** 93d:05126; **Zbl.** 764.05077.

1991.08 Edge conditions for the existence of minimal degree subgraphs, *Graph theory, combinatorics and applications, Vol. 1 (Kalamazoo, MI, 1988), Wiley-Intersci. Publ.*, pp. 419–434, Wiley, New York, 1991 (R. J. Faudree; C. C. Rousseau; R. H. Schelp); **MR** 93c:05062; **Zbl.** 787.05004 (for entire conference proceedings).

1991.09 Existence of complementary graphs having specified edge domination numbers, *J. Combin. Inform. System Sci.* **16** (1991) no. 1, 7–10 (S. Schuster); **MR** 93d:05068; **Zbl.** 767.05089.

1991.10 Extremal non-Ramsey graphs, *Graph theory, combinatorics, algorithms, and applications (San Francisco, 1989)*, pp. 42–66, SIAM, Philadelphia, PA, 1991 (S. A. Burr); **MR** 93b:05118; **Zbl.** 745.05043.

1991.11 Extremal problems for cycle-connected graphs, Proceedings of the Twenty-second Southeastern Conference on Combinatorics, Graph Theory, and Computing (Baton Rouge, LA, 1991), *Congr. Numer.* **83** (1991), 147–151 (R. Duke; V. Rödl); **MR** 93a:05073; **Zbl.** 772.05052.

1991.12 Further results on maximal anti-Ramsey graphs, *Graph theory, combinatorics and applications, Vol. 1 (Kalamazoo, MI, 1988), Wiley-Intersci. Publ.*, pp. 193–206, Wiley, New York, 1991 (S. A. Burr; V. T. Sós; P. Frankl; R. L. Graham); **MR** 93i:05093; **Zbl.** 787.05004 (for entire conference proceedings).

1991.13 Gaps in difference sets, and the graph of nearly equal distances, *Applied geometry and discrete mathematics, DIMACS Ser. Discrete Math. Theoret. Comput. Sci., 4*, pp. 265–273, Amer. Math. Soc., Providence, RI, 1991 (E. Makai; J. Pach; J. H. Spencer); **MR** 92i:52021; **Zbl.** 741.52010.

1991.14 Graphs realizing the same degree sequences and their respective clique numbers, *Graph theory, combinatorics and applications, Vol. 1 (Kalamazoo, MI, 1988), Wiley-Intersci. Publ.*, pp. 439–449, Wiley, New York, 1991 (M. S. Jacobson; J. Lehel); **MR** 93d:05149; **Zbl.** 787.05004 (for entire conference proceedings).

1991.15 Introduction to *Geometric etudes in combinatorial mathematics* by Vladimir Boltyanskij and Alexander Soifer, Center for Excellence in Mathematical Education, Colorado Springs, CO, 1991; **MR** 92f:-52031 (for entire book); **Zbl.** 727.52001 (for entire book).

1991.16 Local constraints ensuring small representing sets, *J. Combin. Theory Ser. A* **58** (1991) no. 1, 78–84 (A. Hajnal; Zs. Tuza); **MR** 92k:05128; **Zbl.** 728.05059.

1991.17 Lopsided Lovász local lemma and Latin transversals, ARIDAM III (New Brunswick, NJ, 1988), *Discrete Appl. Math.* **30** (1991) no. 2–3, 151–154 (J. H. Spencer); **MR** 92c:05160; **Zbl.** 717.05017.

1991.18 Matchings from a set below to a set above, Directions in infinite graph theory and combinatorics (Cambridge, 1989), *Discrete Math.*

95 (1991) no. 1–3, 169–182 (J. A. Larson); **MR** 93a:04001; **Zbl.** 761.04003.

1991.19 Midpoints of diagonals of convex n-gons, *SIAM J. Discrete Math.* **4** (1991) no. 3, 329–341 (P. C. Fishburn; Z. Füredi); **MR** 92f:52032; **Zbl.** 737.52006.

1991.20 New bounds on the length of finite Pierce and Engel series, *Sém. Théor. Nombres Bordeaux (2)* **3** (1991) no. 1, 43–53 (J. O. Shallit); **MR** 92f:11016; **Zbl.** 727.11003.

1991.21 Odd cycles in graphs of given minimum degree, *Graph theory, combinatorics and applications, Vol. 1 (Kalamazoo, MI, 1988), Wiley-Intersci. Publ.*, pp. 407–418, Wiley, New York, 1991 (R. J. Faudree; A. Gyárfás; R. H. Schelp); **MR** 93d:05085; **Zbl.** 787.05004 (for entire conference proceedings).

1991.22 On prime divisors of Mersenne numbers, *Acta Arith.* **57** (1991) no. 3, 267–281 (P. Kiss; C. Pomerance); **MR** 92d:11104; **Zbl.** 733.11003.

1991.23 On some diophantine problems involving powers and factorials, *J. Austral. Math. Soc. Ser. A* **51** (1991) no. 1, 1–7 (B. Brindza); **MR** 92i:11036; **Zbl.** 746.11021.

1991.24 On some of my favourite problems in graph theory and block designs, Graphs, designs and combinatorial geometries (Catania, 1989), *Matematiche (Catania)* **45** (1990) no. 1, 61–73, 1991; **MR** 93h:05052; **Zbl.** 737.05001.

1991.25 On sums of a Sidon-sequence, *J. Number Theory* **38** (1991) no. 2, 196–205 (R. Freud); **MR** 92g:11028; **Zbl.** 731.11008.

1991.26 On the arithmetic means of Lagrange interpolation, *Approximation theory (Kecskemét, 1990), Colloq. Math. Soc. János Bolyai, 58*, pp. 263–274, North-Holland, Amsterdam, 1991 (G. Halász); **MR** 94f:41003; **Zbl.** 767.41004.

1991.27 On the expansion $1 = \sum q^{-n_i}$, *Period. Math. Hungar.* **23** (1991) no. 1, 27–30 (I. Joó); **MR** 92i:11030; **Zbl.** 747.11006.

1991.28 On the uniqueness of the expansions $1 = \sum q^{-n_i}$, *Acta Math. Hungar.* **58** (1991) no. 3–4, 333–342 (M. Horváth; I. Joó); **MR** 93e:11012; **Zbl.** 747.11005.

1991.29 Point distances determined by n points in the plane, *Geombinatorics* **1** (1991) no. 2, 3–4; **CMP** 1 208 430.

1991.30 Problems and results in combinatorial analysis and combinatorial number theory, *Graph theory, combinatorics and applications, Vol. 1 (Kalamazoo, MI, 1988), Wiley-Intersci. Publ.*, pp. 397–406, Wiley, New York, 1991; **MR** 93g:05136; **Zbl.** 787.05004 (for entire conference proceedings).

1991.31 Problems and results on polynomials and interpolation, *Approximation theory (Kecskemét, 1990), Colloq. Math. Soc. János Bolyai, 58*, pp. 253–261, North-Holland, Amsterdam, 1991; **MR** 94g:41002; **Zbl.** 768.41007.

1991.32 Saturated r-uniform hypergraphs, *Discrete Math.* **98** (1991) no. 2, 95–104 (Z. Füredi; Zs. Tuza); **MR** 92k:05095; **Zbl.** 766.05060.

1991.33 Some extremal results in cochromatic and dichromatic theory, *J. Graph Theory* **15** (1991) no. 6, 579–585 (J. G. Gimbel; D. Kratsch); **MR** 92i:05118; **Zbl.** 743.05047.

1991.34 Some Ramsey-type theorems, *Discrete Math.* **87** (1991) no. 3, 261–269 (F. Galvin); **MR** 92b:05078; **Zbl.** 759.05095.

1991.35 Sommes de sous-ensembles (Sums of subsets, in French, English summary), *Sém. Théor. Nombres Bordeaux (2)* **3** (1991) no. 1, 55–72 (J.-L. Nicolas; A. Sárközy); **MR** 92k:11025; **Zbl.** 742.11008.

1991.36 The dimension of random ordered sets, *Random Structures Algorithms* **2** (1991) no. 3, 253–275 (H. A. Kierstead; W. T. Trotter); **MR** 92g:06006; **Zbl.** 741.06001.

1991.37 Three problems on the random walk in $\mathbf{Z}^d$, *Studia Sci. Math. Hungar.* **26** (1991) no. 2–3, 309–320 (P. Révész); **MR** 93k:60171; **Zbl.** 774.60036.

1991.38 Vertex coverings by monochromatic cycles and trees, *J. Combin. Theory Ser. B* **51** (1991) no. 1, 90–95 (A. Gyárfás; L. Pyber); **MR** 92g:05142; **Zbl.** 766.05062.

1992.01 Appendix to *The probabilistic method* by Noga Alon and Joel H. Spencer, *Wiley-Interscience Series in Discrete Mathematics and Optimization*, John Wiley & Sons, Inc., New York, 1992; **MR** 93h:60002 (for entire book); **Zbl.** 767.05001 (for entire book).

1992.02 Arithmetic progressions in subset sums, *Discrete Math.* **102** (1992) no. 3, 249–264 (A. Sárközy); **MR** 93g:11015; **Zbl.** 758.11007.

1992.03 Bounds for arrays of dots with distinct slopes on lengths, *Combinatorica* **12** (1992) no. 1, 39–44 (R. L. Graham; I. Z. Ruzsa; H. Taylor); **MR** 93k:05182; **Zbl.** 774.05020.

1992.04 Corrigendum: "On the expansion $1 = \sum q^{-n_i}$" [*Period. Math. Hungar.* **23** (1991) no. 1, 27–30], *Period. Math. Hungar.* **25** (1992) no. 1, 113 (I. Joó); **MR** 93k:11017; **Zbl.** 761.11002.

1992.05 Covering the cliques of a graph with vertices, Topological, algebraical and combinatorial structures, Frolík's memorial volume, *Discrete Math.* **108** (1992) no. 1–3, 279–289 (T. Gallai; Zs. Tuza); **MR** 93h:05124; **Zbl.** 766.05063.

1992.06 Cycle-connected graphs, Topological, algebraical and combinatorial structures, Frolík's memorial volume, *Discrete Math.* **108** (1992) no. 1–3, 261–278 (R. Duke; V. Rödl); **MR** 94a:05106; **Zbl.** 776.05057.

1992.07 Diameters of point sets, *Geombinatorics* **1** (1992) no. 4, 4; **CMP** 1 208 439.

1992.08 Distances determined by points in the plane, *Geombinatorics* **2** (1992) no. 1, 7; **CMP** 1 208 444.

1992.09 Distances determined by points in the plane, II., *Geombinatorics* **2** (1992) no. 2, 24; **CMP** 1 208 447.

1992.10 Distributed loop network with minimum transmission delay, *Theoret. Comput. Sci.* **100** (1992) no. 1, 223–241 (D. F. Hsu); **MR** 93d:68006; **Zbl.** 780.68005.

1992.11 Diverse homogeneous sets, *J. Combin Theory Ser. A* **59** (1992) no. 2, 312–317 (A. Blass; A. Taylor); **MR** 92j:05178; **Zbl.** 757.05010.

1992.12 Extremal problems involving vertices and edges on odd cycles, Special volume to mark the centennial of Julius Petersen's "Die Theorie der regulären Graphs", Part II, *Discrete Math.* **101** (1992) no. 1–3, 23–31 (R. J. Faudree; C. C. Rousseau); **MR** 93g:05074; **Zbl.** 767.05056.

1992.13 How many edges should be deleted to make a triangle-free graph bipartite?, *Sets, graphs and numbers (Budapest, 1991), Colloq. Math. Soc. János Bolyai, 60*, pp. 239–263, North-Holland, Amsterdam, 1992 (E. Győri; M. Simonovits); **MR** 94b:05104; **Zbl.** 785.05052.

1992.14 In memory of Tibor Gallai, *Combinatorica* **12** (1992) no. 4, 373–374; **MR** 93m:01054b; **Zbl.** 760.01009.

1992.15 Obituary of my friend and coauthor Tibor Gallai, *Geombinatorics* **2** (1992) no. 1, 5–6; **CMP** 1 208 443.

1992.16 On a problem of Tamás Varga, *Bull. Soc. Math. France* **120** (1992) no. 4, 507–521 (M. Joó; I. Joó); **MR** 93m:11076; **Zbl.** 787.11002.

1992.17 On prime-additive numbers, *Studia Sci. Math. Hungar.* **27** (1992) no. 1–2, 207–212 (N. Hegyvári); **MR** 94a:11156; **Zbl.** 791.11053 and 688.10043.

1992.18 On some of my favourite problems in various branches of combinatorics, *Fourth Czechoslovakian Symposium on Combinatorics, Graphs and Complexity (Prachatice, 1990), Ann. Discrete Math., 51*, pp. 69–79, North-Holland, Amsterdam, 1992; **MR** 93k:05001; **Zbl.** 760.05025.

1992.19 On some unsolved problems in elementary geometry (in Hungarian), *Mat. Lapok (N.S.)* **2** (1992) no. 2, 1–10; **MR** 95b:52029.

1992.20 On the minimum size of graphs with a given bandwidth, *Bull. Inst. Combin. Appl.* **6** (1992), 22–32 (Y. Alavi; J. Q. Liu; J. McCanna); **MR** 93g:05071; **Zbl.** 829.05055.

1992.21 On the number of expansions $1 = \sum q^{-n_i}$, *Ann. Univ. Sci. Budapest Eötvös Sect. Math.* **35** (1992), 129–132 (I. Joó); **MR** 94a:11012; **Zbl.** 805.11011.

1992.22 On the number of pairs of partitions of n without common subsums, *Colloq. Math.* **63** (1992) no. 1, 61–83 (J.-L. Nicolas; A. Sárközy); **MR** 93c:11087; **Zbl.** 799.11044.

1992.23 On totally supercompact graphs, Combinatorial mathematics and applications (Calcutta, 1988), *Sankhyā Ser. A* **54** (1992), Special Issue, 155–167 (M. Simonovits; V. T. Sós; S. B. Rao); **MR** 94g:05071.

1992.24 Size Ramsey functions, *Sets, graphs and numbers (Budapest, 1991), Colloq. Math. Soc. János Bolyai, 60*, pp. 219–238, North-Holland, Amsterdam, 1992 (R. J. Faudree); **MR** 94e:05185; **Zbl.** 794.05084.

1992.25 Small transversals in uniform hypergraphs, Siberian Advances in Mathematics, *Siberian Adv. Math.* **2** (1992) no. 1, 82–88 (D. Fon Der Flaass; A. V. Kostochka; Zs. Tuza); **MR** 93b:05076.

1992.26 Subgraphs of large minimal degree, *Random graphs, Vol. 2 (Poznań, 1989), Wiley-Intersci Publ.*, pp. 59–66, Wiley, New York, 1992 (T. Luczak; J. H. Spencer); **MR** 94b:05169; **Zbl.** 817.05056.

1992.27 The distribution of quotients of small and large additive functions, II., *Proceedings of the Amalfi Conference on Analytic Number Theory (Maiori, 1989)*, pp. 83–93, Univ. Salerno, Salerno, 1992 (A. Ivić); **MR** 94i:11074; **Zbl.** 791.11051.

1992.28 Tournaments that share several common moments with their complements, *Bull. Inst. Combin. Appl.* **4** (1992), 65–89 (H. Chen; A. J. Schwenk); **MR** 92j:05086; **Zbl.** 829.05033.

1993.01 Asymptotic bounds for irredundant Ramsey numbers, *Quaestiones Math.* **16** (1993) no. 3, 319–331 (J. H. Hattingh); **MR** 94j:05086; **Zbl.** 794.05086.

1993.02 Clique coverings of the edges of a random graph, *Combinatorica* **13** (1993) no. 1, 1–5 (B. Bollobás; J. H. Spencer; D. B. West); **MR** 94g:05076; **Zbl.** 782.05072.

1993.03 Clique partitions of chordal graphs, *Combin. Probab. Comput.* **2** (1993) no. 4, 409–415 (E. T. Ordman; Y. Zalcstein); **MR** 95g:05080; **Zbl.** 793.05081.

1993.04 Errata: "Distances determined by points in the plane, II." [*Geombinatorics* **2** (1992) no. 2, 24], *Geombinatorics* **2** (1993) no. 3, 765; **CMP** 1 208 453.

1993.05 Estimates of the least prime factor of a binomial coefficient, *Math. Comp.* **61** (1993) no. 203, 215–224 (C. B. Lacampagne; J. L. Selfridge); **MR** 93k:11013; **Zbl.** 781.11008.

1993.06 Extremal problems for the Bondy-Chvátal closure of a graph, *Graphs, matrices and designs, Lecture Notes in Pure and Appl. Math., 139*, pp. 73–83, Dekker, New York, 1993 (L. H. Clark; R. Entringer; H. C. Sun; L. A. Székely); **MR** 94a:05105; **Zbl.** 797.05056.

1993.07 Forcing two sums simultaneously, *A tribute to Emil Grosswald: number theory and related analysis, Contemp. Math., 143*, pp. 321–328, Amer. Math. Soc., Providence, RI, 1993 (D. J. Newman; J. Knappenberger); **MR** 94d:41021; **Zbl.** 808.41006.

1993.08 Monochromatic infinite paths, *Discrete Math.* **113** (1993) no. 1–3, 59–70 (F. Galvin); **MR** 94c:05045; **Zbl.** 787.05071.

1993.09 Nearly equal distances in the plane, *Combinat. Probab. Comput* **2** (1993) no. 4, 401–408 (E. Makai; J. Pach); **MR** 95i:52018; **Zbl.** 798.52017.

1993.10 On elements of sumsets with many prime factors, *J. Number Theory* **44** (1993) no. 1, 93–104 (C. Pomerance; A. Sárközy; C. L. Stewart); **MR** 94b:11011; **Zbl.** 780.11040.

1993.11 On graphical partitions, *Combinatorica* **13** (1993) no. 1, 57–63 (L. B. Richmond); **MR** 94g:11088; **Zbl.** 790.05008.

1993.12 On sets of coprime integers in intervals, *Hardy-Ramanujan J.* **16** (1993), 1–20 (A. Sárközy); **MR** 94e:11102; **Zbl.** 776.11011.

1993.13 On the number of expansions $1 = \sum q^{-n_i}$, II., *Ann. Univ. Sci. Budapest. Eötvös Sect. Math.* **36** (1993), 229–233 (I. Joó); **MR** 95c:11012; **Zbl.** 805.11012.

1993.14 Ordinal partition behavior of finite powers of cardinals, *Finite and infinite combinatorics in sets and logic (Banff, AB, 1991), NATO Adv. Sci. Inst. Ser. C Math. Phys. Sci., 411*, pp. 97–115, Kluwer Acad. Publ., Dordrecht, 1993 (A. Hajnal; J. A. Larson); **MR** 95b:03051; **Zbl.** 780.00039 (for entire conference proceedings).

1993.15 Rainbow subgraphs in edge-colorings of complete graphs, *Quo vadis graph theory?, Ann. Discrete Math., 55*, pp. 81–88, North-Holland, Amsterdam, 1993 (Zs. Tuza); **MR** 94b:05078; **Zbl.** 791.05037.

1993.16 Ramsey problems in additive number theory, *Acta Arith.* **64** (1993) no. 4, 341–355 (B. Bollobás; G. P. Jin); **MR** 94g:11009; **Zbl.** 789.11007.

1993.17 Ramsey problems involving degrees in edge-colored complete graphs of vertices belonging to monochromatic subgraphs, *European J. Combin.* **14** (1993) no. 3, 183–189 (G. Chen; C. C. Rousseau; R. H. Schelp); **MR** 94a:05148; **Zbl.** 785.05067.

1993.18 Ramsey size linear graphs, *Combin. Probab. Comput.* **2** (1993) no. 4, 389–399 (R. J. Faudree; C. C. Rousseau; R. H. Schelp); **MR** 95c:05087; **Zbl.** 794.05085.

1993.19 Some of my favorite solved and unsolved problems in graph theory, *Quaestiones Math.* **16** (1993) no. 3, 333–350; **MR** 94i:05045; **Zbl.** 794.05054.

1993.20 Some of my favourite problems in various branches of combinatorics, Combinatorics 92 (Catania, 1992), *Matematiche (Catania)* **47** (1992) no. 2, 231–240, 1993; **MR** 95c:05042; **Zbl.** 797.05001.

1993.21 Some of my forgotten problems in number theory, *Hardy-Ramanujan J.* **15** (1992), 34–50, 1993; **MR** 94b:11001; **Zbl.** 779.11001.

1993.22 Some problems and results in cochromatic theory, *Quo vadis graph theory?, Ann. Discrete Math., 55*, pp. 261–264, North-Holland, Amsterdam, 1993 (J. G. Gimbel); **MR** 94d:05054; **Zbl.** 791.05038.

1993.23 Some solved and unsolved problems in combinatorial number theory, II., *Colloq. Math.* **65** (1993) no. 2, 201–211 (A. Sárközy); **MR** 94j:11012; **MATH** 950.18325.

1993.24 The grid revisited, Graph theory and combinatorics (Marseille-Luminy, 1990), *Discrete Math.* **111** (1993) no. 1–3, 189–196 (Z. Füredi; J. Pach; I. Z. Ruzsa); **MR** 94g:52022; **Zbl.** 794.52004.

1993.25 The size Ramsey number of a complete bipartite graph, *Discrete Math.* **113** (1993) no. 1–3, 259–262 (C. C. Rousseau); **MR** 93k:05117; **Zbl.** 778.05059.

1993.26 The smallest order of a graph with domination number equal to two and with every vertex contained in a K_n, *Ars Combin.* **35** (1993) A, 217–223 (M. A. Henning; H. C. Swart); **MR** 95m:05100.

1993.27 Triangles in convex polygons, *Geombinatorics* **2** (1993) no. 4, 72–74 (A. Soifer); **CMP** 1 214 695.

1993.28 Turán-Ramsey theorems and simple asymptotically extremal structures, *Combinatorica* **13** (1993) no. 1, 31–56 (A. Hajnal; M. Simonovits; V. T. Sós; E. Szemerédi); **MR** 94d:05088; **Zbl.** 774.05050.

1993.29 Upper bound of $\sum 1/(a_i \log a_i)$ for primitive sequences, *Proc. Amer. Math. Soc.* **117** (1993) no. 4, 891–895 (Z. X. Zhang); **MR** 93e:11018; **Zbl.** 776.11013.

1993.30 Upper bound of $\sum 1/(a_i \log a_i)$ for quasi-primitive sequences, *Comput. Math. Appl.* **26** (1993) no. 3, 1–5 (Z. X. Zhang); **MR** 94f:11013; **Zbl.** 781.11011.

1994.01 A local density condition for triangles, Graph theory and applications (Hakone, 1990), *Discrete Math.* **127** (1994) no. 1–3, 153–161 (R. J. Faudree; C. C. Rousseau; R. H. Schelp); **MR** 95b:05109; **Zbl.** 796.05050.

1994.02 A postscript on distances in convex n-gons, *Discrete Comput. Geom.* **11** (1994) no. 1, 111–117 (P. C. Fishburn); **MR** 94j:52035; **Zbl.** 815.52006.

1994.03 Changes of leadership in a random graph process, Proceedings of the Fifth International Seminar on Random Graphs and Probabilistic Methods in Combinatorics and Computer Science (Poznań, 1991), *Random Structures Algorithms* **5** (1994) no. 1, 243–252 (T. Luczak); **MR** 95c:05110; **Zbl.** 792.60009.

1994.04 Clique partitions of split graphs, *Combinatorics, graph theory, algorithms and applications (Beijing, 1993)*, pp. 21–30, World Sci. Publishing, River Edge, NJ, 1994 (G. Chen; E. T. Ordman); **MR** 96a:05114.

1994.05 Crossing families, *Combinatorica* **14** (1994) no. 2, 127–134 (B. Aronov; W. Goddard; D. J. Kleitman; M. Klugerman; J. Pach; L. J. Schulman); **MR** 95e:52025; **Zbl.** 804.52010.

1994.06 Distinct distances between points in the plane, *Geombinatorics* **3** (1994) no. 4, 115–116; **CMP** 1 268 719.

1994.07 Extremal problems and generalized degrees, Graph theory and applications (Hakone, 1990), *Discrete Math.* **127** (1994) no. 1–3, 139–152 (R. J. Faudree; C. C. Rousseau); **MR** 95d:05069; **Zbl.** 796.05049.

1994.08 Independent transversals in sparse partite hypergraphs, *Combin. Probab. Comput.* **3** (1994) no. 3, 293–296 (A. Gyárfás; T. Luczak); **MR** 96b:05125; **Zbl.** 811.05068.

1994.09 Local and global average degree in graphs and multigraphs, *J. Graph Theory* **18** (1994) no. 7, 647–661 (E. Bertram; P. Horák; J. Šíráň; Zs. Tuza); **MR** 96a:05139; **Zbl.** 812.05032.

1994.10 On additive properties of general sequences, Trends in discrete mathematics, *Discrete Math.* **136** (1994) no. 1–3, 75–99 (A. Sárközy; V. T. Sós); **MR** 96d:11014; **Zbl.** 818.11009.

1994.11 On an interpolation theoretical extremal problem, *Studia Sci. Math. Hungar.* **29** (1994) no. 1–2, 55–60 (J. Szabados; A. K. Varma; P. Vértesi); **MR** 95f:41002; **Zbl.** 817.41006.

1994.12 On isolated, respectively consecutive large values of arithmetic functions, *Acta Arith.* **66** (1994) no. 3, 269–295 (A. Sárközy); **MR** 95c:11111; **Zbl.** 802.11035.

1994.13 On maximal triangle-free graphs, *J. Graph Theory* **18** (1994) no. 6, 585–594 (R. Holzman); **MR** 95g:05057; **Zbl.** 807.05040.

1994.14 On partitions of lines and space, *Fund. Math.* **145** (1994) no. 2, 101–119 (S. Jackson; R. D. Mauldin); **MR** 95k:04003; **Zbl.** 809.04004.

1994.15 On prime factors of subset sums, *J. London Math. Soc. (2)* **49** (1994) no. 2, 209–218 (A. Sárközy; C. L. Stewart); **MR** 95d:11128; **MATH** 940.34987.

1994.16 On sum sets of Sidon sets, I., *J. Number Theory* **47** (1994) no. 3, 329–347 (A. Sárközy; V. T. Sós); **MR** 95e:11030; **Zbl.** 811.11014.

1994.17 On the densities of sets of multiples, *J. Reine Angew. Math.* **454** (1994), 119–141 (R. R. Hall; G. Tenenbaum); **MR** 95k:11115; **Zbl.** 814.11043.

1994.18 On the number of q-expansions, *Ann. Univ. Sci. Budapest. Eötvös Sect. Math.* **37** (1994), 109–118 (I. Joó; V. Komornik); **MR** 96d:11011; **Zbl.** 824.11005.

1994.19 Problems and results in discrete mathematics, Trends in discrete mathematics, *Discrete Math.* **136** (1994) no. 1–3, 53–73; **MR** 96a:52025; **Zbl.** 818.52014.

1994.20 Problems and results on set systems and hypergraphs, *Extremal problems for finite sets (Visegrád, 1991), Bolyai Soc. Math. Stud., 3*, pp. 217–227, János Bolyai Math. Soc., Budapest, 1994; **MR** 95k:05131; **Zbl.** 820.05057.

1994.21 Similar configurations and pseudo grids, *Intuitive geometry (Szeged, 1991), Colloq. Math. Soc. János Bolyai, 63*, pp. 85–104, North-Holland, Amsterdam-New York, 1994 (G. Elekes); **Zbl.** 822.52004.

1994.22 Some problems in number theory, combinatorics and combinatorial geometry, *Math. Pannon.* **5** (1994) no. 2, 261–269; **MR** 95j:11018; **Zbl.** 815.11002.

1994.23 Turán-Ramsey theorems and K_p-independence numbers, *Combin. Probab. Comput.* **3** (1994) no. 3, 297–325 (A. Hajnal; M. Simonovits; V. T. Sós; E. Szemerédi); **MR** 96b:05078; **Zbl.** 812.05031.

1995.01 A problem in covering progressions, *Studia Sci. Math. Hungar.* **30** (1995) no. 1–2, 149–154 (J. H. Spencer); **CMP** 1 341 573.

1995.02 Coverings of r-graphs by complete r-partite subgraphs, Proceedings of the Sixth International Seminar on Random Graphs and Probabilistic Methods in Combinatorics and Computer Science, "Random Graphs '93" (Poznań, 1993), *Random Structures Algorithms* **6** (1995) no. 2–3, 319–322 (V. Rödl); **CMP** 1 370 966; **Zbl.** 818.05037.

1995.03 d-complete sequences of integers, *Math. Comp.*, to appear (M. Lewin); **CMP** 1 333 312.

1995.04 Degree sequence and independence in K_4-free graphs, *Discrete Math.* **141** (1995) no. 1–3, 285–290 (R. J. Faudree; T. J. Reid; R. H. Schelp; W. Staton); **MR** 96b:05079; **MATH** 950.32184.

1995.05 Discrepancy of trees, *Studia Sci. Math. Hungar.* **30** (1995) no. 1–2, 47–57 (Z. Füredi; M. Loebl; V. T. Sós); **CMP** 1 341 566.

1995.06 Equal distance sums in the plane, *Normat* **43** (1995) no. 4, 150–161 (I. Beck; N. Bejlegaard; P. C. Fishburn); **CMP** 1 368 376.

1995.07 Extremal graphs for intersecting triangles, *J. Combin. Theory Ser. B* **64** (1995) no. 1, 89–100 (Z. Füredi; R. J. Gould; D. S. Gunderson); **CMP** 1 341 566; **Zbl.** 822.05036.

1995.08 Extremal problems in combinatorial geometry, *Handbook of Combinatorics, Vol. 1, 2*, pp. 809–874, Elsevier, Amsterdam, 1995 (G. B. Purdy); **CMP** 1 373 673.

1995.09 Independence of solution sets and minimal asymptotic bases, *Acta Arith.* **69** (1995) no. 3, 243–258 (M. B. Nathanson); **CMP** 1 316 478; **Zbl.** 828.11006.

1995.10 Intervertex distances in convex polygons, ARIDAM VI and VII (New Brunswick, NJ, 1991/1992), *Discrete Appl. Math.* **60** (1995) no. 1–3, 149–158 (P. C. Fishburn); **CMP** 1 339 082; **Zbl.** 831.52009.

1995.11 Monochromatic and zero-sum sets of nondecreasing diameter, *Discrete Math.* **137** (1995) no. 1–3, 19–34 (A. Bialostocki; H. Lefmann); **CMP** 1 312 441; **Zbl.** 822.05046.

1995.12 Multiplicities of interpoint distances in finite planar sets, ARIDAM VI and VII (New Brunswick, NJ, 1991/1992), *Discrete Appl. Math.* **60** (1995) no. 1–3, 141–147 (P. C. Fishburn); **CMP** 1 339 081; **Zbl.** 831.52008.

1995.13 On practical partitions, *Collect. Math.* **46** (1995) no. 1–2, 57–76 (J.-L. Nicolas); **CMP** 1 366 129.

1995.14 On product representations of powers, I., *European J. Combin.* **16** (1995) no. 6, 567–588 (A. Sárközy; V. T. Sós); **CMP** 1 356 848.

1995.15 On sum sets of Sidon sets, II., *Israel J. Math.* **90** (1995) no. 1–3, 221–233 (A. Sárközy; V. T. Sós); **CMP** 1 336 324; **MATH** 950.38376.

1995.16 On the book size of graphs with large minimum degree, *Studia Sci. Math. Hungar.* **30** (1995) no. 1–2, 25–46 (R. J. Faudree; E. Győri); **CMP** 1 341 565.

1995.17 On the integral of the Lebesgue function of interpolation, II., *Acta Math. Hungar.* **68** (1995) no. 1–2, 1–6 (J. Szabados; P. Vértesi); **MR** 96b:41003; **MATH** 950.42385.

1995.18 On the 120th anniversary of the birth of Schur, *Geombinatorics* **5** (1995) no. 1, 4–5; **CMP** 1 337 152.

1995.19 On the product representation of powers, I., *European J. Combin.* **16** (1995) no. 6, 567–588 (A. Sárközy; V. T. Sós); **CMP** 1 356 848; **MATH** 960.06338.

1995.20 On the size of a random maximal graph, Proceedings of the Sixth International Seminar on Random Graphs and Probabilistic Methods in Combinatorics and Computer Science, "Random Graphs '93" (Poznań, 1993), *Random Structures Algorithms* **6** (1995) no. 2–3, 309–318 (S. Suen; P. M. Winkler); **CMP** 1 370 965; **Zbl.** 820.05054.

1995.21 Some of my favourite problems in number theory, combinatorics, and geometry, Combinatorics Week (Portuguese) (São Paulo, 1994), *Resenhas* **2** (1995) no. 2, 165–186; **CMP** 1 370 501.

1995.22 Some old and new problems in approximation theory: research problems 95-1, *Constr. Approx.* **11** (1995) no. 3, 419–421; **CMP** 1 350 678.

1995.23 Squares in a square, *Geombinatorics* **4** (1995) no. 4, 110–114 (A. Soifer); **CMP** 1 330 337.

1995.24 Sur le graphe divisoriel (The divisor graph, in French), *Acta Arith.* **73** (1995) no. 2, 189–198 (É. Saias); **CMP** 1 358 194.

1995.25 Two combinatorial problems in the plane, *Discrete Comput. Geom.* **13** (1995) no. 3–4, 441–443 (G. B. Purdy); **MR** 96a:52026; **Zbl.** 826.52009.

1995.26 Vertex covering with monochromatic paths, Festschrift for Hans Vogler on the occasion of his 60th birthday, *Math. Pannon.* **6** (1995) no. 1, 7–10 (A. Gyárfás); **MR** 96c:05127; **Zbl.** 828.05040.

1996.01 On a class of aperiodic sum-free sets, *Math. Proc. Cambridge Philos. Soc.* **120** (1996) no. 1, 1–5 (N. J. Calkin); **CMP** 1 373 341.

1996.02 Proof of a conjecture of Bollobás on nested cycles, *J. Combin. Theory Ser. B* **66** (1996) no. 1, 38–43 (G. Chen; W. Staton); **CMP** 1 368 515; **MATH** 960.07461.

1996.03 Sure monochromatic subset sums, *Acta Arith.* **74** (1996) no. 3, 269–272 (N. Alon); **CMP** 1 373 713; **MATH** 960.06960.

NOTE: The total number of items in this list is 1414.

Postscript

It is fitting that Paul Erdős himself should have the last word. As is amply illustrated in this collection, Paul's profound influence on so many mathematicians and fields of mathematics through his prolific research and incessant traveling is destined to leave a legacy that may never be equalled. Here then are some very special lines written by Paul for the Postscript of these volumes:

"Determine or estimate as well as you can the number of solutions in positive integers of

$$\frac{1}{x_1} + \frac{1}{x_2} + \cdots + \frac{1}{x_h} = 1, \quad x_1 < x_2 < \cdots x_k \ .$$

Can the squares be decomposed as the finite union of sum-free sets (or Sidon) sequences? Apologies if this is trivial or trivially false. Let $x_1, \ldots, x_n$ be n points in the plane with at most t on a line (t fixed, n large). $f(n;t)$ is defined to be size of the largest subset with no three on a line. We can only prove

$$c_2\sqrt{\frac{n}{t}} < f(n_1 t) < \frac{c_1 n}{t} \ .$$

The lower bound is easy by the greedy algorithm. Füredi showed

$$f(n;t) > g(n)\sqrt{\frac{n}{t}}$$

where $g(n) \to \infty$ slowly. What is the truth here?

(Erdős-Turán.) Let $p_{n+1} - p_n = d_n$. Prove that for infinitely many n,

$$d_n > d_{n+1} > d_{n+2} \ \text{ or } \ d_n < d_{n+1} < d_{n+2} \ .$$

This surely holds since if not, then there is an n for which

$$d_n \geq d_{n+1}, \ d_{n+1} \leq d_{n+2}, \ d_{n+2} \geq d_{n+3}, \ldots, \text{etc} \ .$$

I offer 100 dollars for a proof and 25000 dollars for a counterexample. This is of course a joke since the conjecture surely holds.

If I live I hope to have some more conjectures and even proofs. Will there be a celebration for my 90th birthday or only a meeting for my memory. May my theorems and problems live forever.

My mother was very glad to read the many eulogies written for my 50th birthday. I am only sorry that my mother and father are not reading this volume (if you believe in survival after death then you can believe that perhaps they are reading it). Let me add another problem: In 1934 Turán and I proved (Amer Math Monthly 1934): Let $a_1 < a_2 < \dots < a_n$ be any set of n integers. Then the number of distinct prime factors of $\prod_{1 \le i < j \le n} (a_i + a_j)$ is greater than $c \log n$. It does not have to be greater than $\frac{cn}{\log n}$ (trivially). Try to improve both the upper and lower bounds. I offer 500 dollars for this.

I offer \$100 for a proof and \$25000 for a counterexample. This of course is a joke since the conjecture surely holds.

If I live I hope to have some more conjectures and even proofs. Will there be a celebration for my 90th birthday or only a meeting for my memory. May my theorems and problems live forever.

My mother was very glad to read the many eulogies written for my 50th birthday. I am only sorry that my mother and father are not reading this volume (if you believe in survival after death then you can believe that perhaps they are reading it). Let me add another problem: In 1934 Turán and I proved (Amer Math Monthly 1934): Let $a_1 < a_2 < \ldots < a_n$ be any set of n integers. Then the number of distinct prime factors of $\prod_{1 \le i < j \le n} (a_i + a_j)$ is greater than $c \log n$. It does not have to be greater than $\frac{cn}{\log n}$ (trivially). Try to improve both the upper and lower bounds. I offer 500 dollars for this."

The preceding lines are some of the last lines written by Paul Erdős. They not only represent a fitting conclusion to these volumes but they capture Paul's style and his vision of life as a scholar, a style which influenced us all from around the world, a world which will not be the same without him.

Volume 9:
B. Korte, L. Lovász,
H. J. Prömel,
A. Schrijver (Eds.)
Paths, Flows and VLSI-Layout
1990. ISBN 3-540-52685-4

Volume 10:
J. Pach (Ed.)
New Trends in Discrete and Computational Geometry
1993. ISBN 3-540-55713-X

Volume 11:
K. Voss
Discrete Images, Objects, and Functions in Z^n
1993. ISBN 3-540-55943-4

Volume 12:
M. Padberg
Linear Optimization and Extensions
1995. ISBN 3-540-58734-9

Volume 13:
R. L. Graham,
J. Nesětřil (Eds.)
The Mathematics of Paul Erdös I
1996. ISBN 3-540-61032-4

Volume 14:
R. L. Graham,
J. Nesětřil (Eds.)
The Mathematics of Paul Erdös II
1996. ISBN 3-540-61031-6

Springer-Verlag, P. O. Box 31 13 40, D-10643 Berlin, Germany.